HANDBOOK OF THERMODYNAMIC TABLES AND CHARTS

Kuzman Ražnjević

HEMISPHERE PUBLISHING CORPORATION
Washington   London

McGRAW-HILL BOOK COMPANY

New York   St. Louis   San Francisco   Auckland   Düsseldorf   Johannesburg   London   Mexico
Montreal   New Delhi   Panama   Paris   São Paulo   Singapore   Sydney   Tokyo   Toronto

Originally published as Termodinamičke Tablice i dijagrami by Kuzman Ražnjević. Translated by Dr. Marijan Bošković and Professor Rickard Podhorsky.

HANDBOOK OF THERMODYNAMIC TABLES AND CHARTS

2 3 4 5 6 7 8 9 0   H A H A   7 8 4 3 2 1 0 9 8 7

The editor of this book was Harold B. Crawford; the production supervisor was Rebekah McKinney. The printer and binder was Halliday Lithograph.

Library of Congress Cataloging in Publication Data

Ražnjević, Kuzman.
    Handbook of thermodynamic tables and charts.

    Translation of Termodinamičke tablice.
    1. Thermodynamics—Tables, calculations, etc.
I. Title.
QC311.3.R3913 1975   536'.7'0212   75–23056
ISBN 0-07-051270-1

CONTENTS

IMPORTANT NOTE: Users of Liquids, Vapors, and Gases tables are alerted to the fact that the symbol for pressure, *at*, refers to the **technical atmosphere** and not to the **standard atmosphere**, which is designated by *atm*.

The relationship of conversion factors between the two atmospheres is clearly explained in Table 119 on page 337.

VAPORS

GASES

UNITS AND MEASURES

CHARTS

INDEX

PREFACE

This volume provides a comprehensive treatment of basic data for engineering thermodynamics as well as practical and theoretical thermodynamics. Several features will be of immense value to readers. For instance, all numerical values are cited in both technical and SI (International System) units, making conversion from one system to the other unnecessary, and facilitating extensive use of the book and of technical literature in general. In addition, the contents are arranged according to aggregate states: solids, liquids, vapors, and gases. And at the end of the book is a table converting units of measurement (including British-American units) from one system to another; thus the information can be used by specialists in all scientific fields.

The collection and arrangement of the material makes this handbook unique in its field. The simple, clear presentation of the tables will help in rapid solution of a given problem. It is intended for use by all professionals and students in thermodynamics, heat transfer, gas dynamics, heating, refrigeration and air conditioning, turbine design, energetics, combustion systems, chemical process design, aeronautical engineering, chemical engineering, civil engineering, and mechanical engineering. It will also be of use to chemists, physicists, and all others in these scientific and engineering fields. I am convinced that many specialists will find this book invaluable and will use it as a standard handbook.

The editor and publishers have endeavored to remove from this edition as many substances and products as possible that are of local origin to the original source. Nevertheless, some items, such as gabbro stone or specific types of glass, have been left to widen the scope of the Handbook's coverage. The editor will be happy to respond to specific queries regarding such items.

I am indebted to all whose suggestions and advice contributed to this work. I am especially thankful to the translators, Dr. Marijan Bošković and Professor Rickard Podhorsky.

Kuzman Ražnjević

IMPORTANT NOTE: Users of Liquids, Vapors, and Gases tables are alerted to the fact that the symbol for pressure, *at*, refers to the **technical atmosphere** and not to the **standard atmosphere**, which is designated by *atm*.

The relationship of conversion factors between the two atmospheres is clearly explained in Table 119 on page 337.

Table 1-1 Chemical Elements

Name	Latin name	Symbol	Relative atomic mass	Valency	Atomic number	Name	Latin name	Symbol	Relative atomic mass	Valency	Atomic number
Actinium	Actinium	Ac	227	—	89	Mercury	Hydrargyrum	Hg	200.61	1,2	80
Aluminum	Aluminium	Al	26.97	3	13	Molybdenum	Molybdaenum	Mo	95.95	3,6,4	42
Americum	Americium	Am	(243)	—	95	Neodymium	Neodymum	Nd	144.27	3	60
Antimony	Stibium	Sb	121.76	3,5	51	Neon	Neonum	Ne	20.183	0	10
Argon	Argonum	Ar	39.944	0	18	Neptunium	Neptunium	Np	237	—	93
Arsenic	Arsenum	As	74.91	3,5	33	Nickel	Niccolum	Ni	58.69	2,3	28
Astatine	Astatium	At	(210)	—	85	Niobium	Niobium	Nb	92.91	3,5	41
Barium	Barium	Ba	137.36	2	56	Nitrogen	Nitrogenium	N	14.008	3,5	7
Berkelium	Berkelium	Bk	(249)	—	97	Osmium	Osmium	Os	190.2	2,3,4,8	76
Beryllium	Beryllium	Be	9.013	2	4	Oxygen	Oxygenium	O	16.0000	2	8
Bismuth	Bismuthum	Bi	209.00	3,5	83	Palladium	Palladium	Pd	106.7	2,4	46
Boron	Borum	B	10.82	3	5	Phosphorus	Phosphorus	P	30.974	3,5	15
Bromine	Bromum	Br	79.916	1,3,5,7	35	Platinum	Platinum	Pt	195.23	2,4	78
Cadmium	Cadmium	Cd	112.41	2	48	Plutonium	Plutonium	Pu	(244)	—	94
Calcium	Calcium	Ca	40.08	2	20	Polonium	Polonium	Po	210	—	84
Californium	Californium	Cf	(249)	—	98	Potassium	Kalium	K	39.100	1	19
Carbon	Carboneum	C	12.010	4	6	Praseodymium	Praseodymum	Pr	140.92	3	59
Cerium	Cerium	Ce	140.13	3,4	58	Promethium	Promethium	Pm	(145)	—	61
Cesium	Caesium	Cs	132.91	2	55	Protoactinium	Protactinium	Pa	231	—	91
Chlorine	Chlorum	Cl	35.457	1,3,5,7	17	Radium	Radium	Ra	226.05	2	88
Chromium	Chromium	Cr	52.01	2,3,6	24	Radon	Radonum	Rn	222	0	86
Cobalt	Cobaltum	Co	58.94	2,3	27	Rhenium	Rhenium	Re	186.31	—	75
Copper	Cuprum	Cu	63.54	1,2	29	Rhodium	Rhodium	Rh	102.91	3	45
Curium	Curium	Cm	(248)	—	96	Rubidium	Rubidium	Rb	85.48	1	37
Dysprosium	Dysprosium	Dy	162.46	3	66	Ruthenium	Ruthenium	Ru	101.7	3,4,6,8	44
Einsteinium	Einsteinium	Es	(254)	—	99	Samarium	Samarium	Sm	150.43	3	62
Erbium	Erbium	Er	167.2	3	68	Scandium	Scandium	Sc	44.96	3	21
Europium	Europium	Eu	152.0	2,3	63	Selenium	Selenum	Se	78.96	2,4,6	34
Fermium	Fermium	Fm	(255)	—	100	Silicon	Silicium	Si	28.09	4	14
Fluorine	Fluorum	F	19.00	1	9	Silver	Argentum	Ag	107.880	1	47
Francium	Francium	Fr	(223)	—	87	Sodium	Natrium	Na	22.997	1	11
Gadolinium	Gadolinium	Gd	156.9	3	64	Strontium	Strontium	Sr	87.63	2	38
Gallium	Gallium	Ga	69.72	2,3	31	Sulfur	Sulphur	S	32.066	2,4,6	16
Germanium	Germanium	Ge	72.60	4	32	Tantalum	Tantalum	Ta	180.88	5	73
Gold	Aurum	Au	197.2	1,3	79	Technetium	Technetium	Tc	(98.91)	—	43
Hafnium	Hafnium	Hf	178.6	4	72	Tellurium	Tellurium	Te	127.61	2,4,6	52
Helium	Helium	He	4.003	0	2	Terbium	Terbium	Tb	159.2	3	65
Holmium	Holmium	Ho	164.94	3	67	Thallium	Thallium	Tl	204.39	1,3	81
Hydrogen	Hydrogenium	H	1.0080	1	1	Thorium	Thorium	Th	232.12	4	90
Indium	Indium	In	114.76	3	49	Thulium	Thulium	Tm	169.4	3	69
Iodine	Jodum	I	126.91	1,3,5,7	53	Tin	Stannum	Sn	118.70	2,4	50
Iridium	Iridium	Ir	193.1	3,4	77	Titanium	Titanium	Ti	47.90	3,4	22
Iron	Ferrum	Fe	55.85	2,3	26	Tungsten	Wolframum	W	183.92	6	74
Krypton	Krypton	Kr	83.80	0	36	Uranium	Uranium	U	238.07	4,6	92
Lanthanum	Lanthanum	La	138.92	3	57	Vanadium	Vanadium	V	50.95	3,5	23
Lead	Plumbum	Pb	207.21	2,4	82	Xenon	Xenonum	Xe	131.3	0	54
Lithium	Lithium	Li	6.940	1	3	Ytterbium	Ytterbium	Yb	173.04	3	70
Lutecium	Lutetium	Lu	175.0	3	71	Yttrium	Yttrium	Y	88.92	3	39
Magnesium	Magnesium	Mg	24.32	2	12	Zinc	Zincum	Zn	65.38	2	30
Manganese	Manganum	Mn	54.93	2,3,4,6,7	25	Zirconium	Zirconium	Zr	91.22	4	40
Mendelevium	Mendelevium	Mv	(257)	—	101						

SOLIDS

Table 2-1 Thermal Properties of the Solid Elements

Element	Symbol	Density at 20°C ρ (kg/m³)	Coefficient of linear expansion at 20°C α × 10³ (1/K)	Melting point t (°C)	Melting point T (°K)	Heat of fusion (kcal/kg)	Heat of fusion (kJ/kg)	Boiling point t (°C)	Boiling point T (°K)	Heat of vaporization (kcal/kg)	Heat of vaporization r (kJ/kg)
Aluminum	Al	2700	0.0237	658	931.15	85	355.878	2270	2543.15	2800	11723.040
Antimony	Sb	6690	0.0110	630.5	903.65	40	167.472	1640	1931.15	300	1256.040
Arsenic	As	5720	0.0050	830	1103.15	—	—	625	898.15	400	1674.720
Barium	Ba	3760	—	704	977.15	—	—	1700	1973.15	320	1339.776
Beryllium	Be	1850	0.0130	1278	1551.15	341	1427.699	3000	3273.15	5930	24827.724
Bismuth	Bi	9800	0.0135	271	544.15	13	54.428	1500	1773.15	200	837.360
Boron	B	2340	0.008	2500	2773.15	—	—	—	—	—	—
Cadmium	Cd	8640	0.030	320.9	594.05	13	54.428	767	1040.15	240	1004.832
Calcium	Ca	1540	0.025	851	1124.15	78.5	328.664	1400	1673.15	1000	4186.800
Carbon	C	—	—	3540	3813.15	—	—	4000	4273.15	1200	5024.160
Cerium	Ce	6800	0.010	815	1088.15	—	—	1400	1673.15	—	—
Cesium	Cs	1870	0.097	28	301.15	3.8	15.910	670	943.15	120	502.416
Chromium	Cr	7100	0.008	1800	2073.15	70	293.076	2400	2673.15	1470	6154.596
Cobalt	Co	8800	0.0123	1490	1763.15	67	280.516	3200	3473.15	1550	6489.540
Copper	Cu	8930	0.0166	1083	1356.15	50	209.340	2330	2603.15	1110	4647.348
Gallium	Ga	5900	0.018	29.78	302.93	19.1	79.968	2300	2573.15	—	—
Gold	Au	19290	0.0142	1063	1336.15	16	66.989	2700	2973.15	420	1758.456
Iodine	J	4930	0.093	113.5	386.65	—	—	185	458.15	80	334.944
Iridium	Ir	22500	0.0065	2454	2727.15	—	—	>4800	>5073.15	930	3893.724
Iron	Fe	7860	0.0123	1530	1803.15	65	272.142	2500	2773.15	1520	6363.936
Lead	Pb	11340	0.029	327.3	600.45	5.7	23.865	1730	2003.15	220	921.096
Lithium	Li	534	0.056	180	453.15	33	138.164	1400	1673.15	5100	21352.680
Magnesium	Mg	1740	0.026	650	923.15	50	209.340	1110	1383.15	1350	5652.180
Manganese	Mn	7300	0.023	1250	1523.15	60	251.208	2100	2373.15	1000	4186.800
Mercury	Hg	—	—	-38.83	234.32	—	—	356.95	630.10	—	—
Molybdenum	Mo	10200	0.005	2600	2873.15	70	293.076	3560	3833.15	1700	7117.560
Nickel	Ni	8900	0.013	1455	1728.15	70	293.076	3000	3273.15	1480	6196.464
Osmium	Os	22480	0.0061	2500	2773.15	—	—	—	—	—	—
Palladium	Pd	12000	0.0118	1555	1828.15	36	150.725	—	—	950	3977.460
Phosphorus, white	P	1820	0.125	44.1	317.25	5.2	21.771	280	553.15	400	1674.720
Platinum	Pt	21450	0.009	1773	2046.15	27	113.044	3800	4073.15	600	2512.080
Potassium	K	862	0.083	63	336.15	13	54.428	760	1033.15	490	2051.532
Rhenium	Re	20500	0.009	3150	3423.15	—	—	—	—	—	—
Rhodium	Rh	12400	0.009	1966	2239.15	—	—	—	—	—	—
Rubidium	Rb	1520	0.090	38.5	311.65	6.1	25.539	713	986.15	200	837.360
Selenium	Se	4400	0.037	220	493.15	16.4	68.664	688	961.15	260	1088.568
Silicon	Si	2330	0.0024	1410	1683.15	25	104.670	2350	2623.15	3360	14067.648
Silver	Ag	10500	0.0189	960.5	1233.65	27	113.044	1950	2223.15	520	2177.136
Sodium	Na	971	0.072	97.7	370.85	—	—	880	1153.15	1000	4186.800
Strontium	Sr	—	—	757	1030.15	—	—	1370	1643.15	—	—
Sulfur (monoclinic)	S	1960	0.080	119	392.15	11	46.055	—	—	—	—
Sulfur (rhombic)	S	2060	0.074	112.8	385.95	9.4	39.356	444.60	717.75	70	293.076
Tantalum	Ta	16600	0.0065	3000	3273.15	—	—	—	—	620	2595.816
Tin	Sn	7280	0.027	231.9	505.05	14	58.615	2300	2573.15	—	—
Titanium	Ti	4530	0.0108	1800	2073.15	—	—	—	—	—	—
Tungsten	W	19300	0.0043	3380	3653.15	60	251.208	5000	5273.15	1150	4814.820
Vanadium	V	6000	0.0085	1720	1993.15	—	—	—	—	—	—
Zinc	Zn	7130	0.029	419.4	692.55	26.8	112.206	907	1180.15	430	1800.324
Zirconium	Zr	6530	—	1900	2173.15	—	—	—	—	—	—

1 kcal = 4.1868 kJ

Table 3-1 Thermal Properties of Solid Inorganic Compounds

Substance	Chemical formula	Density at 20°C ρ kg/m³	Coefficient of cubical expansion at 20°C β 1/K	Melting point t °C	T °K	Heat of fusion kcal/kg	kJ/kg	Boiling point t °C	T °K	Heat of vaporization r kcal/kg	kJ/kg
Aluminum oxide	Al_2O_3	4000	0.000005	2050	2323.15	—	—	2980	3253.15	1130	4731.084
Barium chloride	$BaCl_2$	3900	0.000060	955	1228.15	27.8	116.393	1560	1833.15	290	1214.172
Barium oxide	BaO	5700	—	1923	2196.15	—	—	1880	2153.15	590	2470.212
Barium sulphate	$BaSO_4$	4500	0.000075	1580	1853.15	—	—	—	—	—	—
Calcium carbide	CaC_2	2210	—	2300	2573.15	—	—	—	—	—	—
Calcium chloride	$CaCl_2$	2150	0.000067	772	1045.15	—	—	<1600	<1873.15	—	—
Calcium oxide	CaO	3300	0.000060	2572	2845.15	—	—	2850	3123.15	—	—
Chromic oxide	Cr_2O_3	5200	—	2275	2548.15	—	—	—	—	—	—
Cristobalite (silicon dioxide)	SiO_2	2300	—	1710	1983.15	—	—	2590	2863.15	—	—
Ferrous oxide	FeO	5900	—	1370	1643.15	—	—	—	—	—	—
Hematite (ferric oxide)	FeO_3	5200	0.000008	1560	1833.15	—	—	—	—	—	—
Lead monoxide	PbO	9300	0.000055	880	1153.15	10	41.868	1480	1753.15	230	962.964
Magnesium oxide	MgO	3600	0.000040	2800	3073.15	—	—	—	—	—	—
Magnetite (ferroso-ferric oxide)	Fe_3O_4	5100	0.0000096	1550	1823.15	—	—	—	—	—	—
Potassium chloride	KCl	1989	0.000115	770	1043.15	—	—	1413	1686.15	600	2512.080
Potassium hydroxide	KOH	2000	0.000188	360	633.15	—	—	1320	1593.15	550	2320.740
Potassium nitrate	KNO_3	2100	0.000190	337	610.15	—	—	—	—	—	—
Potassium oxide	K_2O	2300	—	—	—	—	—	—	—	390	1632.852
Potassium sulphate	K_2SO_4	2660	0.000126	1067	1340.15	—	—	—	—	—	—
Quartz (silicon dioxide)	SiO_2	2650	—	1470	1743.15	—	—	—	—	—	—
Silver bromide	$AgBr$	6470	—	430	703.15	12	50.242	1330	1603.15	230	962.964
Silver chloride	$AgCl$	5560	—	455	728.15	22	92.110	1554	1827.15	300	1256.040
Sodium chloride	$NaCl$	2164	0.000106	802	1075.15	124	519.163	1440	1713.15	680	2847.024
Sodium hydroxide	$NaOH$	2130	0.000084	328	601.15	40	167.472	1390	1663.15	790	3307.572
Sodium nitrate	$NaNO_3$	2260	0.000110	310	583.15	45	188.406	—	—	—	—
Sodium sulphate	Na_2SO_4	2700	0.000220	884	1157.15	62	259.582	—	—	—	—
Tantalum carbide	TaC	14500	—	3800	4073.15	—	—	—	—	—	—
Zinc oxide	ZnO	5700	0.000023	2000	2273.15	—	—	—	—	—	—

Table 4-1 Thermal Properties of Solid Organic Compounds

Substance	Chemical formula	Density at 20°C ρ kg/m³	Coefficient of cubical expansion at 20°C β 1/K	Melting point t °C	T °K	Heat of fusion kcal/kg	kJ/kg	Boiling point t °C	T °K	Heat of vaporization r kcal/kg	kJ/kg
Anthracene	$C_{14}H_{10}$	1250	0.000210	216	489.15	40	167.472	340	613.15	—	—
Benzoic acid	$C_7H_6O_2$	1340	0.000520	122.4	395.55	33	138.164	250	523.15	168	703.382
Camphor	$C_{10}H_{16}O$	990	0.000485	179	452.15	10	41.868	209	482.15	—	—
Diphenyl	$C_{12}H_{10}$	1040	—	68.5	341.65	29	121.417	255	528.15	74	309.823
Diphenylamine	$C_{12}H_{11}N$	1160	—	54	327.15	24	100.483	302	375.15	—	—
Naphthalene	$C_{10}H_8$	1145	0.000283	80.1	353.25	36	150.725	217.9	491.05	75	314.010
Paraffin	—	—	—	54	327.15	35	146.538	300	573.15	—	—
Phenol	C_6H_6O	1070	—	41	314.15	29	121.417	182	455.15	122	510.790
Phthalic acid	$C_8H_6O_4$	1590	—	194	467.15	75	314.010	—	—	—	—
Picric acid	$C_6H_3O_7N_3$	1767	—	122	395.15	20.4	85.411	—	—	—	—
Resorcinol	$C_6H_6O_2$	1283	—	109	382.15	45	188.406	277	550.15	—	—
Stearic acid	$C_{18}H_{36}O_2$	941	—	70	343.15	48	200.966	380	653.15	56	234.461
Sugar, cane	—	—	—	160	433.15	13.4	56.103	—	—	—	—
Trinitrotoluene	$C_7H_5O_6N_3$	1750	—	81	354.15	21	87.923	—	—	—	—

1 kcal = 4.1868 kJ

Table 5-1 Linear Thermal Expansion Coefficient α of Solids Between 0°C (273.15 K) and t°C (TK)

Substance	Chemical composition in order of percentage of constituents	Temperature t °C	Temperature T °K	Linear expansion coefficient $10^3 \times \alpha$ 1/K
Aluminum	Al	− 253	20.15	−0.0147
		− 190	83.15	−0.0181
		− 100	173.15	−0.0220
		− 50	223.15	−0.0224
		+ 50	323.15	+0.0234
		100	373.15	0.0238
		200	473.15	0.0245
		300	573.15	0.0255
		400	673.15	0.0265
		500	773.15	0.0274
		600	873.15	0.0283
Aluminum alloy	Al, Mg, Mn	100	373.15	0.023
Aluminum alloy, Y-alloy	Al, Cu, Ni	100	373.15	0.023
Amber		50	323.15	0.05400
		75	348.15	0.05600
American alloy	G, Al, Cu	100	373.15	0.025
Anticorodal	Al, Mg, Si	100	373.15	0.022
Antimony ∥	Sb	50	323.15	0.0160
		100	373.15	0.0175
		200	473.15	0.0190
		300	573.15	0.0195
		400	673.15	0.0195
Antimony ⊥	Sb	50	323.15	0.0080
		100	373.15	0.0080
		200	473.15	0.0080
		300	573.15	0.0083
		400	673.15	0.0081
Bakelite		100	373.15	0.021 … 0.036
Basalt		100	373.15	0.009
Beryllium		100	373.15	0.0123
Bismuth	Bi	100	373.15	0.0124
Brass	62Cu, 38Zn	− 253	20.15	−0.01403
		− 190	83.15	−0.01637
Brass	62Cu, 38Zn	+ 100	373.15	+0.0184
		200	473.15	0.01925
		300	573.15	0.02010
		400	673.15	0.02298
Brick		100	373.15	0.0036 … 0.0058
Bronze	85Cu, 9Mn, 6Sn	− 190	83.15	−0.01495
		+ 100	373.15	+0.01750
		200	473.15	0.01790
		300	573.15	0.01833
		400	673.15	0.01878
		500	773.15	0.01922
Bronze red (brass)		100	373.15	0.019
Cadmium	Cd	100	373.15	0.0308
Cement, Portland (concrete)		100	373.15	0.0140
China		− 190	82.15	−0.00168
		+ 100	373.15	+0.00300
		200	473.15	0.00330
		300	573.15	0.00343
		400	673.15	0.00353
		500	773.15	0.00364
		600	873.15	0.00373
		700	973.15	0.00376
		800	1073.15	0.00388
		900	1173.15	0.00440
		1000	1273.15	0.00431
Chromium	Cr	− 50	223.15	0.00940
		+ 100	373.15	0.00800
		200	473.15	0.00875
		300	573.15	0.00733
		400	673.15	0.00788
		500	773.15	0.00700
Clinker		100	373.15	0.0028 … 0.0048
Cobalt	Co	100	373.15	0.0123

Table 5-2 Linear Thermal Expansion Coefficient α of Solids Between 0°C (273.15 K) and t°C (TK) (Continued)

Substance	Chemical composition in order of percentage of constituents	Temperature t °C	Temperature T °K	Linear expansion coefficient $10^3 \times \alpha$ 1/K
Concrete, heaped concrete Blast furnace slag concrete		100	373.15	0.0058 ... 0.0066
Constantan	60Cu, 40Ni	−190	83.15	−0.01189
		+100	373.15	+0.01520
		200	473.15	0.01560
		300	573.15	0.01603
		400	673.15	0.01643
		500	773.15	0.01682
Copper	Cu	−253	20.15	−0.01174
		−190	83.15	−0.01395
		−100	173.15	−0.01550
		+100	373.15	+0.01650
		200	473.15	0.01690
		300	573.15	0.01717
		400	673.15	0.01768
		500	773.15	0.01808
		600	873.15	0.01848
Duralumin	95Al, 4Cu+Mg, Mn, Si, Fe	100	373.15	0.02350
		200	473.15	0.02450
		300	573.15	0.02600
		400	673.15	0.02675
		500	773.15	0.02730
Ebonite		100	373.15	0.0170 ... 0.0280
Electron and Magnevin alloys		100	373.15	0.025
German alloy	G Al, Zn, Cu	100	373.15	0.024
Glass, Jena 16 III		−253	20.15	−0.00478
		−190	83.15	−0.00595
		−80	193.15	−0.00713
		+100	373.15	+0.00810
		200	473.15	0.00835
		300	573.15	0.00867
		400	673.15	0.00898
		500	773.15	0.00926

Substance	Chemical composition in order of percentage of constituents	Temperature t °C	Temperature T °K	Linear expansion coefficient $10^3 \times \alpha$ 1/K
Glass — Jena 59 III		−190	83.15	−0.00432
		+100	373.15	+0.00590
		200	473.15	0.00600
		300	573.15	0.00610
		400	673.15	0.00618
		500	773.15	0.00624
— Jena 1565 III		100	373.15	0.00345
		200	473.15	0.00360
		300	573.15	0.00373
		400	673.15	0.00390
		500	773.15	0.00404
— Jena 2594 III		−253	20.15	−0.00368
		−190	83.15	−0.00447
		−80	192.15	−0.00525
		+100	373.15	+0.00630
		200	473.15	0.00640
		300	573.15	0.00657
		400	673.15	0.00673
		500	773.15	0.00686
Glass, quartz		−259	23.15	+0.00304
		−190	83.15	+0.00158
		−100	173.15	−0.00150
		−50	223.15	−0.00260
		+50	323.15	+0.00044
		100	373.15	0.00051
		200	473.15	0.00585
		300	573.15	0.00627
		400	673.15	0.00635
		500	773.15	0.00612
		600	873.15	0.00600
		700	973.15	0.00571
		800	1073.15	0.00563
		900	1173.15	0.00556
		1000	1273.15	0.00540
Gold	Au	−253	20.15	−0.011739
		−190	83.15	−0.013053
		+100	373.15	+0.0142

Table 5-3 Linear Thermal Expansion Coefficient α of Solids Between 0°C (273.15 K) and t°C (TK) (*Continued*)

Substance	Chemical composition in order of percentage of constituents	t (°C)	T (°K)	$10^3 \times \alpha$ (1/K)
Gold	Au	200	473.15	0.014600
		300	573.15	0.014800
		400	673.15	0.015025
		500	773.15	0.015240
		600	873.15	0.015583
		700	973.15	0.015929
		800	1073.15	0.016250
		900	1173.15	0.016556
Granite		100	373.15	0.0080 ... 00.118
Hydronalium alloy	G Al-Mg	100	373.15	0.020
Indium	Tn	100	373.15	0.044
Invar (Invar steel)	36Ni	100	373.15	0.0015
Iridium	Ir	− 190	83.15	−0.00563
		− 100	173.15	−0.00600
		+ 100	373.15	+0.00650
		1000	1273.15	0.00790
		1500	1773.15	0.00847
		1700	1973.15	0.00871
Iron, cast	Fe	− 190	83.15	−0.008368
		100	373.15	0.0104
		200	473.15	0.011050
		300	573.15	0.011633
		400	673.15	0.012250
		500	773.15	0.012880
		600	873.15	0.013483
		700	973.15	0.014100
		800	1073.15	0.014700
Iron, cast, gray	Fe	100	373.15	0.0104
Iron, pure	Fe	100	373.15	0.012
Lautal alloy	Al, Cu	100	373.15	0.023
Lead	Pb	− 253	20.15	−0.02455
		− 190	83.15	−0.02674
		+ 100	373.15	+0.02900
		200	473.15	0.02965
		300	573.15	0.03110
Limestone		100	373.15	0.0070
Lithium	Li	100	373.15	0.060
Magnesium	Mg	− 190	83.15	−0.02111
		+ 100	373.15	+0.02600
		200	473.15	0.02705
		300	573.15	0.02787
		400	673.15	0.02883
		500	773.15	0.02966
Mangal alloy	Al, Mn	100	373.15	0.022
Manganine	86Cu, 12Mn, 2Ni	100	373.15	0.01750
		200	473.15	0.01867
		300	573.15	0.01888
		400	673.15	0.01940
		500	773.15	0.01983
		600	873.15	0.02043
		700	973.15	0.02100
		800	1073.15	
Marble		100	373.15	0.002 ... 0.02
Mica		100	373.15	0.0135
		200	473.15	0.01350
		300	573.15	0.01383
		400	673.15	0.01400
		500	773.15	0.01380
Molybdenum	Mo	− 190	83.15	−0.004158
		+ 100	373.15	0.005200
		200	473.15	0.005350
		300	573.15	0.005467
		400	673.15	0.005600
Monel metal		100	373.15	0.014
Mortar, cement		100	373.15	0.0085 ... 0.0135
Mortar, lime		100	373.15	0.0073 ... 0.0089

Table 5-4 Linear Thermal Expansion Coefficient α of Solids Between 0°C (273.15 K) and t°C (TK) (Continued)

Substance	Chemical composition in order of percentage of constituents	t (°C)	T (°K)	$10^3 \times \alpha$ (1/K)
Nickel	Ni	− 190	83.15	−0.00995
		+ 100	373.15	+0.01300
		200	473.15	0.01375
		300	573.15	0.01430
		400	673.15	0.01488
		500	773.15	0.01520
		600	873.15	0.01545
		700	973.15	0.01579
		800	1073.15	0.01611
		900	1173.15	0.01644
		1000	1273.15	0.01180
Nickel silver		100	373.15	0.0180
Niobium		100	373.15	0.007
Osmium	Os	100	373.15	0.0067
Palladium	Pd	− 190	83.15	−0.01016
		− 100	173.15	−0.01070
		− 50	223.15	−0.01120
		+ 100	373.15	+0.01190
		200	473.15	0.01210
		300	573.15	0.01233
		400	673.15	0.01255
		500	773.15	0.01276
		600	873.15	0.01288
		700	973.15	0.01320
		800	1073.15	0.01333
		900	1173.15	0.01363
		1000	1273.15	0.01386
Plaster		100	373.15	0.025
Platinum	Pt	− 190	83.15	−0.00795
		− 100	173.15	−0.00840
		− 50	223.15	−0.00860
		+ 100	373.15	+0.00900
		200	473.15	0.00915
		300	573.15	0.00927
		400	673.15	0.00940
Platinum	Pt	500	773.15	0.00954
		600	873.15	0.00967
		700	973.15	0.00980
		800	1073.15	0.00992
		900	1173.15	0.01006
		1000	1273.15	0.01019
Platinum-iridium alloy	90Pt, 10Ir	− 190	83.15	−0.00774
		+ 100	373.15	+0.00900
		200	473.15	0.00910
		300	573.15	0.00923
		400	673.15	0.00938
		500	773.15	0.00950
		600	873.15	0.00963
		700	973.15	0.00976
		800	1073.15	0.00989
		900	1173.15	0.01001
		1000	1273.15	0.01015
		1100	1373.15	0.01032
		1200	1473.15	0.01042
		1300	1573.15	0.01055
		1400	1673.15	0.01068
		1500	1773.15	0.01080
	80Pt, 20Ir	− 190	83.15	−0.00753
		+ 100	373.15	+0.00830
		200	473.15	0.00850
		300	573.15	0.00863
		400	673.15	0.00878
		500	773.15	0.00890
		600	873.15	0.00905
		700	973.15	0.00919
		800	1073.15	0.00934
		900	1173.15	0.00948
		1000	1273.15	0.00962
		1100	1373.15	0.00975
		1200	1473.15	0.00990
		1300	1573.15	0.01004
		1400	1673.15	0.01019
		1500	1773.15	0.01033

Table 5-5 Linear Thermal Expansion Coefficient α of Solids Between 0°C (273.15 K) and t°C (TK) (Continued)

Substance	Chemical composition in order of percentage of constituents	Temperature t (°C)	Temperature T (°K)	Linear expansion coefficient $10^3 \times \alpha$ (1/K)
Platinum-rhodium alloy	80Pt, 20Rh	300	573.15	0.00927
		600	873.15	0.00975
		900	1173.15	0.01023
		1000	1273.15	0.01045
		1500	1773.15	0.01121
Poly(vinylchloride) (PVC)	(C_2H_3Cl)	100	373.15	0.0781
Potassium	K	50	323.15	8.3
Rhodium	Rh	−190	83.15	−0.00684
		−100	173.15	−0.00780
		−50	223.15	−0.00780
		+50	323.15	+0.00900
Sandlime brick		100	373.15	0.0078
Sandstone		100	373.15	00.5...01.2
Silicon	Si	50	323.15	0.0078
Silumin	Al, Si	100	373.15	0.022
	G Al	100	373.15	0.019
Silver	Ag	−253	20.15	−0.01478
		−190	83.15	−0.01695
		+100	373.15	+0.01950
		200	473.15	0.02000
		300	573.15	0.02027
		400	673.15	0.02058
		500	773.15	0.02086
		600	873.15	0.02117
		700	973.15	0.02164
		800	1073.15	0.02206
Sinter corundum		200	473.15	0.00650
		300	573.15	0.00667
		400	673.15	0.00668
		500	773.15	0.00720
		600	873.15	0.00742
		700	973.15	0.00757

Substance	Chemical composition in order of percentage of constituents	Temperature t (°C)	Temperature T (°K)	Linear expansion coefficient $10^3 \times \alpha$ (1/K)
Sinter corundum		800	1073.15	0.00781
		900	1173.15	0.00794
		1000	1273.15	0.00815
		1100	1373.15	0.00832
		1200	1473.15	0.00846
		1300	1573.15	0.00858
		1400	1673.15	0.00868
		1500	1773.15	0.00877
Sinter magnesium		200	473.15	0.01225
		300	573.15	0.01200
		400	673.15	0.01225
		500	773.15	0.01260
		600	873.15	0.01292
		700	973.15	0.01329
		800	1073.15	0.01350
		900	1173.15	0.01352
		1000	1273.15	0.01390
		1100	1373.15	0.01409
		1200	1473.15	0.01429
		1300	1573.15	0.01450
		1400	1673.15	0.01434
		1500	1773.15	0.01507
		1800	2073.15	0.01597
Sodium	Na	50	323.15	0.0720
Steel, carbon	0.1C	100	373.15	0.012
Steel, chrome	0.6C	100	373.15	0.0117
Steel, hard		100	373.15	0.010 ... 0.014
		−190	83.15	−0.00863
		100	373.15	+0.01170
		200	473.15	0.01225
		300	573.15	0.01277
		400	673.15	0.01328
		500	773.15	0.01382
		600	873.15	0.91433
		700	973.15	0.91486

Table 5-6 Linear Thermal Expansion Coefficient α of Solids Between 0°C (273.15 K) and t°C (TK) (Continued)

Substance	Chemical composition in order of percentage of constituents	Temperature t °C	Temperature T °K	Linear expansion coefficient $10^3 \times \alpha$ 1/K
Steel, nickel	77Fe, 23Ni	300	573.15	0.00933
		400	673.15	0.01000
		500	773.15	0.01050
		600	873.15	0.01042
		700	973.15	0.01114
		800	1073.15	0.01156
		900	1173.15	0.01167
		1000	1273.15	0.01185
	64Fe, 36Ni	100	373.15	0.00150
		200	473.15	0.00375
		300	573.15	0.00533
		400	673.15	0.00775
		500	773.15	0.00940
		600	873.15	0.01083
		700	973.15	0.01214
		800	1073.15	0.01313
		900	1173.15	0.01394
Steel, nickel chrome	18Cr, 8Ni	100	373.15	0.0115
		100	373.15	0.016
Steel, nickel chrome/molybdenum	Cr, Ni, Mo	100	373.15	0.011
Steel, soft		− 190	83.15	−0.00879
		+ 100	373.15	+0.01200
		200	473.15	0.01255
		300	573.15	0.01307
		400	673.15	0.01360
		500	773.15	0.01412
		600	873.15	0.01465
		700	973.15	0.01519
Steel, wrought		− 253	20.15	−0.007115
		− 190	83.15	−0.008842
		+ 100	373.15	+0.012200
		200	473.15	0.012650
		300	573.15	0.013100
		400	673.15	0.013575
		500	773.15	0.014040
		600	873.15	0.014517
		700	973.15	0.014986

Substance	Chemical composition in order of percentage of constituents	Temperature t °C	Temperature T °K	Linear expansion coefficient $10^3 \times \alpha$ 1/K
Tantalum	Ta	100	373.15	0.0065
Tin	Sn	− 190	83.15	−0.02232
		+ 100	373.15	+0.02670
Titanium	Ti	100	373.15	0.01080
Tungsten	W	− 190	83.15	−0.003842
		+ 100	373.15	+0.004500
		200	473.15	0.004500
		300	573.15	0.004667
		400	673.15	0.004750
		500	773.15	0.004500
		600	873.15	0.004500
		700	973.15	0.004500
		800	1073.15	0.004500
		900	1173.15	0.004500
		1000	1273.15	0.004600
		1100	1373.15	0.004636
		1200	1473.15	0.004708
		1300	1573.15	0.004785
		1400	1673.15	0.004871
		1500	1773.15	0.004967
		1700	1973.15	0.004765
		2000	2273.15	0.005150
		3000	3273.15	0.006500
Vanadium	V	100	373.15	0.00850
Vinidur plastic		100	373.15	0.0800
Widia		20	373.15	0.0053
Wood, fir ‖		100	373.15	0.0076
Wood, fir ⊥		100	373.15	0.0544
Wood, oak (with the grain) ‖		100	373.15	0.0030
Wood, oak (cross grain) ⊥		100	373.15	0.0580
Zinc	Zn	− 190	83.15	−0.00974
		+ 100	373.15	0.01650

Table 6-1 Melting Points of Alloys

Alloy	Composition in %	Melting point	
		t	T
		°C	°K
Aluminum solder, hard	—	> 540	> 813.15
Aluminum solder, soft	—	250 ... 500	523.15 ... 773.15
American alloy	G Al, Cu	544 ... 640	817.15 ... 913.15
Anticorrodal	Al, Mg, Li	630 ... 650	903.15 ... 923.15
Brass	—	900	1173.15
Bronze	—	≈ 900	≈1173.15
Copper solder	—	1160 ... 1230	1433.15 ... 1503.15
Delta metal	56Cu, 41Zn, 1Fe, Mn, Pb	≈ 950	≈1223.15
Duralumin	Al, Cu, Mg	520 ... 650	793.15 ... 923.15
Electron and Magnevin alloys	—	625 ... 650	898.15 ... 923.15
German alloy	G Al, Zn, Cu	530 ... 630	830.15 ... 903.15
Hard solder	—	820 ... 915	1093.15 ... 1188.15
Hydronalium	Al, Mg	520 ... 630	793.15 ... 903.15
Invar (Invar steel)	—	1425	1698.15
Iron, cast (gray)	—	≈1200	≈1473.15
Iron, cast (white)	—	1130	1403.15
Iron, wrought	—	1300 ... 1500	1573.15 ... 1773.15
Lautal	Al, Cu	650	923.15
Mangal	Al, Mn	650	923.15
Monel metal	—	1315 ... 1350	1588.15 ... 1623.15
Nickel silver	—	950 ... 1180	1223.15 ... 1453.15
Rose's metal	50Bi, 25Pb 25Sn	94	367.15
Silumin	Al, Si	570	843.15
Silver solder	—	720 ... 855	993.15 ... 1128.15
Steel, hard	—	1460	1733.15
Steel, high speed tool	—	1350	1623.15
Steel, soft	—	1520	1793.15
Tin-base solder[1]	25 ... 90Sn, 75 ... 10Pb	181 ... 271	454.15 ... 544.15
Wood's metal	50Bi, 12.5Cd, 25Pb, 12.5Sn	70	343.15

[1] All tin-base solder begin to soften at 181°C. The temperature at which a tin melts completely, lies between 190°C and 275°C, depending upon its composition.

Table 7-1 Melting Points of Miscellaneous Solids

Substance	Melting point	
	t	T
	°C	°K
Asbestos	1150	1423.15
Bauxite	1820	2093.15
Blast furnace slag	1300 ... 1430	1573.15 ... 1703.15
Blubber	44	317.15
Borax	878	1151.15
Butter	31	304.15
China (porcelain)	1550	1823.15
Chromite	≈2180	≈2453.15
Clay, pure	2050	2323.15
Enamel	~ 960	~1233.15
Glass	1200	1473.15
Glass, lead	1100	1373.15
Glauber's salt	884	1157.15
Ice (H$_2$O)	0	273.15
Mineral pitch	50 ... 140	323.15 ... 413.15
Naphthalene	80	353.15
Rubber	≈ 125	≈ 398.15
Shellac	≈ 150	≈ 423.15
Silicon carbide	2537	2810.15
Silicon oxide	1470	1743.15
Stearine	50	323.15
Tallow	40 ... 50	313.15 ... 323.15
Wax	64	337.15
Wax, bees'	60 ... 65	333.15 ... 338.15
Wax, mineral	7 ... 195	280.15 ... 468.15

Table 8-1 Melting Points of Salts for Salt Baths

Substance	Melting point	
	t	T
	°C	°K
Aluminium chloride	190	463.15
Barium chloride	955	1228.15
Barium fluoride	1300	1573.15
Calcium carbonate	891	1164.15
Calcium chloride	772	1045.15
Calcium fluoride	1370	1643.15
Calcium nitrate	337	610.15
Cupric chloride	630	903.15
Ferric chloride	302	575.15
Lead chloride	500	773.15
Lithium carbonate	733	1006.15
Lithium chloride	606	879.15
Lithium fluoride	842	1115.15
Magnesium chloride	718	991.15
Magnesium fluoride	1260	1533.15
Potassium chloride	770	1043.15
Potassium fluoride	846	1119.15
Silver chloride	450	723.15
Sodium carbonate	850	1123.15
Sodium chloride	802	1075.15
Sodium nitrate	310	583.15
Sodium fluoride	995	1268.15
Strontium fluoride	900	1173.15
Zinc chloride	315	588.15

Table 9-1 Specific Heats c and c_m of Solid Elements

Element	Symbol	Temperature		Specific heat		Mean specific heat[1]	
		t	T	c		c_m	
		°C	°K	kcal/kg K	kJ/kg K	kcal/kg K	kJ/kg K
Aluminum	Al	− 200	73.15	0.075	0.314	0.164	0.687
		− 100	173.15	0.175	0.733	0.194	0.812
		0	273.15	0.210	0.879	−	−
		20	293.15	0.214	0.896	0.212	0.888
		100	373.15	0.224	0.938	0.217	0.909
		200	473.15	0.235	0.984	0.223	0.934
		300	573.15	0.241	1.009	0.228	0.955
		400	673.15	0.249	1.043	0.232	0.971
		500	773.15	0.260	1.089	0.237	0.992
Antimony	Sb	− 100	173.15	0.046	0.193	0.048	0.201
		0	273.15	0.0492	0.206	−	−
		20	293.15	0.0496	0.208	0.0494	0.207
		100	373.15	0.0507	0.212	0.0500	0.209
		300	573.15	0.054	0.226	0.0515	0.216
Arsenic	As	− 100	173.15	0.0690	0.289	0.0739	0.309
		0	273.15	0.0778	0.326	−	−
		20	293.15	0.0787	0.330	0.0782	0.327
		100	373.15	0.0810	0.339	0.0796	0.333
Beryllium	Be	− 100	173.15	0.200	0.837	0.305	1.277
		− 50	223.15	0.320	1.340	0.358	1.499
		0	273.15	0.396	1.658	−	−
		20	293.15	0.418	1.750	0.407	1.704
		100	373.15	0.480	2.010	0.442	1.851
		200	473.15	0.535	2.240	0.475	1.989
Bismuth	Bi	− 200	73.15	0.024	0.100	0.0270	0.113
		− 100	173.15	0.0278	0.116	0.0286	0.120
		0	273.15	0.0293	0.123	−	−
		20	293.15	0.0295	0.124	0.0294	0.123
		100	373.15	0.0303	0.127	0.0298	0.125
		200	473.15	0.0320	0.134	0.0304	0.127
Boron	B	− 50	223.15	0.19	0.795	0.21	0.879
		0	273.15	0.23	0.963	−	−
		20	293.15	0.25	1.047	0.24	1.005
		100	373.15	0.29	1.214	0.29	1.089
Cadmium	Cd	− 200	73.15	0.042	0.176	0.0512	0.214
		− 100	173.15	0.0520	0.218	0.0535	0.224
		− 50	223.15	0.0536	0.224	0.0542	0.227
		0	273.15	0.0548	0.229	−	−
		20	293.15	0.0552	0.231	0.0550	0.230
		100	373.15	0.0568	0.238	0.0558	0.234
		200	473.15	0.0586	0.245	0.0568	0.238
Calcium	Ca	− 200	73.15	0.096	0.402	0.136	0.569
		− 100	173.15	0.141	0.590	0.148	0.620
		0	273.15	0.153	0.641	−	−
		20	293.15	0.155	0.649	0.154	0.645
		100	373.15	0.160	0.670	0.157	0.657
		200	473.15	0.170	0.712	0.160	0.670
Carbon, amorphous	C	20	293.15	0.20	0.837	−	−
Carbon, diamond	C	20	293.15	0.120	0.502	−	−
Carbon, graphite		− 200	73.15	0.020	0.084	0.080	0.335
		− 100	173.15	0.080	0.335	0.115	0.481
		− 50	223.15	0.115	0.481	0.134	0.561
		0	273.15	0.153	0.641	−	−
		20	293.15	0.169	0.708	0.161	0.674
		100	373.15	0.223	0.934	0.190	0.795
		200	473.15	0.281	1.176	0.222	0.929
		300	573.15	0.337	1.411	0.252	1.055
		500	773.15	0.390	1.633	0.300	1.256
		1000	1273.15	0.410	1.717	0.340	1.423

[1] Mean specific heat c_m between 0°C (273.15 K) and t°C (TK).　　　　1 kcal = 4.1868 kJ

Table 9-2 Specific Heats c and c_m of Solid Elements (*Continued*)

Element	Symbol	Temperature t °C	Temperature T °K	Specific heat c kcal/kg K	Specific heat c kJ/kg K	Mean specific heat[1] c_m kcal/kg K	Mean specific heat[1] c_m kJ/kg K
Cesium	Cs	20	293.15	0.055	0.230	—	—
Chromium	Cr	— 200	73.15	0.034	0.142	0.071	0.297
		— 100	173.15	0.076	0.318	0.090	0.377
		0	273.15	0.102	0.427	—	—
		20	293.15	0.105	0.440	0.104	0.435
		100	373.15	0.113	0.473	0.108	0.452
		200	473.15	0.119	0.498	—	—
		300	573.15	0.125	0.523	0.116	0.486
		600	873.15	—	—	0.125	0.523
		1000	1273.15	—	—	0.135	0.565
Cobalt	Co	— 200	73.15	0.034	0.142	0.072	0.301
		— 100	173.15	0.075	0.314	0.084	0.352
		0	273.15	0.091	0.381	—	—
		20	293.15	0.093	0.389	0.092	0.385
		100	373.15	0.101	0.423	0.096	0.402
		300	573.15	0.117	0.490	0.105	0.440
		600	873.15	0.140	0.586	0.117	0.490
		900	1173.15	0.164	0.687	0.129	0.540
		1200	1473.15	0.145	0.607	—	—
Copper	Cu	— 200	73.15	0.040	0.167	0.078	0.327
		— 100	173.15	0.082	0.343	0.087	0.364
		0	273.15	0.0906	0.379	—	—
		20	293.15	0.0915	0.383	0.0910	0.381
		100	373.15	0.0947	0.396	0.0926	0.388
		200	473.15	0.0969	0.406	0.0949	0.397
		300	573.15	0.0994	0.416	0.0958	0.401
		400	673.15	0.1020	0.427	0.0970	0.406
		500	773.15	0.1049	0.439	0.0974	0.408
		800	1073.15	0.1120	0.469	0.102	0.427
Gallium	Ga	— 100	173.15	0.082	0.343	0.086	0.360
		0	273.15	0.089	0.373	—	—
		20	293.15	0.090	0.377	0.089	0.373
Gold	Au	— 200	73.15	0.021	0.088	0.0280	0.117
		— 100	173.15	0.0292	0.122	0.0301	0.126
		0	273.15	0.0307	0.129	—	—
		20	293.15	0.0309	0.129	0.0308	0.129
		100	373.15	0.0314	0.131	0.0311	0.130
		300	573.15	0.0322	0.135	0.0316	0.132
		600	873.1	0.0340	0.142	0.0324	0.136
		1000	1273.15	0.0376	0.157	0.0336	0.141
Iridium	Ir	— 200	73.15	0.016	0.067	0.026	0.109
		— 100	173.15	0.027	0.113	0.030	0.126
		0	273.15	0.031	0.130	—	—
		20	293.15	0.032	0.134	0.0315	0.132
		100	373.15	0.033	0.138	0.032	0.134
		1400	1673.15	—	—	0.040	0.167
Iodine	J	— 100	173.15	0.047	0.197	0.049	0.205
		0	273.15	0.051	0.214	—	—
		20	293.15	0.052	0.218	0.052	0.218
		100	373.15	0.056	0.234	0.054	0.226
Iron	Fe	— 200	73.15	0.032	0.134	0.080	0.335
		— 150	123.15	0.067	0.281	0.090	0.377
		— 100	173.15	0.085	0.356	0.096	0.402
		— 50	223.15	0.097	0.406	0.101	0.423
		0	273.15	0.105	0.440	—	—
		20	293.15	0.108	0.452	0.106	0.444
		100	373.15	0.116	0.486	0.111	0.465
		200	473.15	0.127	0.532	0.116	0.486
		300	573.15	0.139	0.582	0.122	0.511
		400	673.15	0.150	0.628	0.127	0.532
		500	773.15	0.162	0.678	0.133	0.557
		600	873.15	0.180	0.754	0.139	0.582
		700	973.15	—	—	0.150	0.628
		800	1073.15	—	—	0.160	0.670
		1000	1273.15	—	—	0.168	0.703
		1400	1673.15	—	—	0.165	0.691

[1] Mean specific heat c_m between 0°C (273.15 K) and t°C (TK).   1 kcal = 4.1868 kJ

Table 9-3   Specific Heats c and c_m of Solid Elements (*Continued*)

Element	Symbol	Temperature		Specific heat		Mean specific heat[1]	
		t	T	c		c_m	
		°C	°K	kcal/kg K	kJ/kg K	kcal/kg K	kJ/kg K
Lead	Pb	− 200	73.15	0.0260	0.109	0.0287	0.120
		− 100	173.15	0.0288	0.121	0.0297	0.124
		0	273.15	0.0306	0.128	−	−
		20	293.15	0.0309	0.129	0.0307	0.129
		100	373.15	0.0320	0.134	0.0313	0.131
		200	473.15	0.0330	0.138	0.0320	0.134
		300	573.15	0.0338	0.142	0.0325	0.136
Lithium	Li	− 200	73.15	0.30	1.256	0.62	2.596
		− 100	173.15	0.65	2.721	0.73	3.056
		0	273.15	0.79	3.308	−	−
		20	293.15	0.81	3.391	0.80	3.349
		100	373.15	0.90	3.768	0.85	3.559
Magnesium	Mg	− 200	73.15	0.13	0.544	0.20	0.837
		− 100	173.15	0.21	0.879	0.225	0.942
		0	273.15	0.239	1.001	−	−
		20	293.15	0.243	1.017	0.241	1.009
		100	373.15	0.255	1.068	0.247	1.034
		200	473.15	0.268	1.122	−	−
		300	573.15	0.276	1.156	0.260	1.089
		500	773.15	0.300	1.256	0.270	1.130
Manganese	Mn	− 100	173.15	0.095	0.398	0.105	0.440
		0	273.15	0.113	0.473	−	−
		20	293.15	0.116	0.486	0.115	0.481
		100	373.15	0.123	0.515	0.119	0.498
		300	573.15	0.140	0.586	0.127	0.532
Mercury	Hg	− 200	73.15	0.0273	0.114	−	−
		− 100	173.15	0.0322	0.135	−	−
Molybdenum	Mo	− 200	73.15	0.020	0.084	0.047	0.197
		− 100	173.15	0.050	0.209	0.055	0.230
		0	273.15	0.059	0.247	−	−
		20	293.15	0.060	0.251	0.060	0.251
		100	373.15	0.062	0.260	0.061	0.255
		400	673.15	0.066	0.276	0.063	0.264
		1000	1273.15	0.074	0.310	0.067	0.281
Nickel	Ni	− 200	73.15	0.036	0.151	0.083	0.348
		− 100	173.15	0.087	0.364	0.099	0.414
		− 50	223.15	0.100	0.419	0.103	0.431
		0	273.15	0.1055	0.422	−	−
		20	293.15	0.1065	0.446	0.106	0.444
		100	373.15	0.1115	0.467	0.108	0.452
		200	473.15	0.123	0.515	0.112	0.469
		300	573.15	0.136	0.569	0.118	0.494
		350	623.15	0.15	0.628	0.121	0.507
		400	673.15	0.13	0.544	0.123	0.515
		700	973.15	0.13	0.544	0.125	0.523
		1000	1273.15	−	−	0.13	0.544
Niobium	Nb	0	273.15	0.0643	0.269	−	−
		20	293.15	0.0645	0.270	0.0644	0.270
		100	373.15	0.0651	0.273	0.0647	0.271
		1000	1273.15	0.0740	0.310	0.0690	0.289
Osmium	Os	0	273.15	0.0309	0.129	−	−
		20	293.15	0.0310	0.130	0.0310	0.130
		100	373.15	0.0314	0.131	0.0312	0.131
		1000	1273.15	0.036	0.151	0.033	0.138
Palladium	Pd	− 100	173.15	0.050	0.209	0.055	0.230
		0	273.15	0.058	0.243	−	−
		20	293.15	0.059	0.247	0.058	0.243
		100	373.15	0.060	0.251	0.059	0.247
		500	773.15	0.065	0.272	0.062	0.260
		1000	1273.15	0.074	0.310	0.066	0.276

[1] Mean specific heat c_m between 0°C (273.15 K) and t°C (TK).   1 kcal = 4.1868 kJ

Table 9-4 Specific Heats c and c_m of Solid Elements (Continued)

Element	Symbol	Temperature		Specific heat		Mean specific heat[1]	
		t	T	c		c_m	
		°C	°K	kcal/kg K	kJ/kg K	kcal/kg K	kJ/kg K
Platinum	Pt	− 200	73.15	0.018	0.075	0.025	0.105
		− 100	173.15	0.028	0.117	0.0306	0.128
		0	273.15	0.0317	0.133	—	—
		20	293.15	0.0318	0.133	0.0318	0.133
		100	373.15	0.0324	0.136	0.0321	0.134
		200	473.15	0.0325	0.136	—	—
		300	573.15	0.0326	0.136	0.0328	0.137
		500	773.15	0.0349	0.146	0.0333	0.139
		1000	1273.15	0.0350	0.147	—	—
		1200	1473.15	0.0393	0.165	0.0355	0.149
Phosphorus	P	20	293.15	0.18	0.754	—	—
Potassium	K	− 200	73.15	0.140	0.586	0.160	0.670
		− 100	173.15	0.160	0.670	0.170	0.712
		0	273.15	0.175	0.733	—	—
		20	293.15	0.177	0.741	0.176	0.737
		50	323.15	0.181	0.758	0.178	0.745
Rhenium	Re	0	273.15	0.0326	0.136	—	—
		20	293.15	0.0327	0.137	0.0326	0.136
		100	373.15	0.0332	0.139	0.0329	0.138
		1000	1273.15	0.039	0.163	0.036	0.151
Rhodium	Rh	0	273.15	0.0589	0.247	—	—
		20	293.15	0.0592	0.248	0.0591	0.247
		100	373.15	0.0603	0.252	0.0596	0.250
		500	773.15	0.069	0.289	0.064	0.268
		1200	1473.15	0.081	0.339	0.070	0.293
Rubidium	Rb	20	293.15	0.083	0.348	—	—
Selenium	Se	20	293.15	0.08	0.335	—	—
Silicon	Si	− 200	73.15	0.040	0.167	0.110	0.461
		− 100	173.15	0.115	0.481	0.142	0.595
		0	273.15	0.162	0.678	—	—
		20	293.15	0.168	0.703	0.165	0.691
		100	373.15	0.189	0.791	0.177	0.741
		400	673.15	0.21	0.879	0.196	0.821
		900	1173.15	0.23	0.963	0.21	0.879
Silver	Ag	− 200	73.15	0.0375	0.157	0.0505	0.211
		− 100	173.15	0.0516	0.216	0.0539	0.226
		0	273.15	0.0556	0.233	—	—
		20	293.15	0.0559	0.234	0.0557	0.233
		100	373.15	0.0568	0.238	0.0562	0.235
		200	473.15	0.0595	0.249	—	—
		300	573.15	0.0589	0.247	0.0572	0.239
		500	773.15	0.0630	0.264	—	—
		700	973.15	0.0642	0.269	0.0597	0.250
Sodium	Na	− 200	73.15	0.21	0.879	0.26	1.089
		− 100	173.15	0.26	1.089	0.272	1.139
		0	273.15	0.284	1.189	—	—
		20	293.15	0.288	1.206	0.282	1.181
		50	323.15	0.294	1.231	0.289	1.210
Sulfur (rhombic)	S	− 100	173.15	0.140	0.586	0.155	0.649
		0	273.15	0.167	0.699	—	—
		20	293.15	0.172	0.720	0.169	0.708
		80	353.15	0.184	0.770	0.175	0.733

[1] Mean specific heat c_m between 0°C (273.15 K) and t°C (T K).   1 kcal = 4.1868 kJ

Table 10-1 Specific Heat c of Solid Organic Compounds

Organic compound	Chemical formula	Temperature t °C	Temperature T °K	Specific heat c kcal/kg K	Specific heat c kJ/kg K
Anthracene	$C_{14}H_{10}$	20	293.15	0.275	1.151
Benzoic acid	$C_7H_6O_2$	20	293.15	0.283	1.185
Benzophenone	$C_{13}H_{10}O$	20	293.15	0.380	1.591
Bromobenzene	C_6H_5Br	20	293.15	0.231	0.967
Bromonaphthalene	$C_{10}H_7Br$	20	293.15	0.250	1.047
Camphor	$C_{10}H_{16}O$	20	293.15	0.410	1.717
Diphenyl	$C_{12}H_{10}$	20	293.15	0.300	1.256
Diphenylamine	$C_{12}H_{11}N$	20	293.15	0.320	1.340
Diphenylmethane	$C_{13}H_{12}$	20	293.15	0.327	1.369
Lactose	$C_{12}H_{22}O_{11}$	20	293.15	0.290	1.214
Naphthalene	$C_{10}H_8$	20	293.15	0.310	1.298
Oxalic acid	$C_2H_2O_4$	20	293.15	0.275	1.151
Palmitic acid	$C_{16}H_{32}O_2$	20	293.15	0.500	2.093
Phenol	C_6H_6O	20	293.15	0.390	1.633
Phthalic acid	$C_8H_6O_4$	20	293.15	0.230	0.963
Picric acid	$C_6H_3O_7N_3$	20	293.15	0.250	1.047
Salicylic acid	$C_7H_6O_3$	20	293.15	0.280	1.172
Stearic acid	$C_{18}H_{36}O_2$	20	293.15	0.400	1.675
Succinic acid	$C_4H_4O_6$	20	293.15	0.310	1.298
Sugar, cane	$C_{12}H_{22}O_{11}$	20	293.15	0.300	1.256
Trinitrotoluene	$C_7H_5O_6N_3$	20	293.15	0.300	1.256
Urea	CH_4ON_2	20	293.15	0.370	1.549

1 kcal = 4.1868 kJ

Table 9-5 Specific Heats c and c_m of Solid Elements (Continued)

Element	Symbol	Temperature t °C	Temperature T °K	Specific heat c kcal/kg K	Specific heat c kJ/kg K	Mean specific heat [1] c_m kcal/kg K	Mean specific heat [1] c_m kJ/kg K
Tantalum	Ta	−200	73.15	0.0200	0.084	0.0290	0.121
		−100	173.15	0.0300	0.126	0.0316	0.132
		0	273.15	0.0328	0.137	−	0.138
		20	293.15	0.0330	0.138	0.0329	0.138
		100	373.15	0.0336	0.141	0.0332	0.139
		400	673.15	0.0350	0.147	0.0340	0.142
		1000	1273.15	0.0380	0.159	0.0350	0.147
Tellurium	Te	20	293.15	0.047	0.197	−	−
Thallium	Tl	−100	173.15	0.0296	0.124	0.0304	0.127
		0	273.15	0.0312	0.131	−	−
		20	293.15	0.0316	0.132	0.0314	0.131
		100	373.15	0.0331	0.139	0.0322	0.135
		200	473.15	0.0350	0.147	0.0330	0.138
Thorium	Th	20	293.15	0.030	0.126	−	−
Tin	Sn	−100	173.15	0.0500	0.209	0.0520	0.218
		0	273.15	0.0538	0.225	−	−
		20	293.15	0.0541	0.227	0.0539	0.226
		100	373.15	0.0560	0.234	0.0549	0.230
		200	473.15	0.0580	0.243	0.0560	0.234
Titanium	Ti	20	293.15	0.146	0.611	0.1462	0.612
		100	373.15	−	−	0.1503	0.629
		200	473.15	−	−	0.1563	0.654
		300	573.15	−	−	−	−
Tungsten	W	−200	73.15	0.0160	0.067	0.026	0.109
		−100	173.15	0.0260	0.109	0.030	0.126
		−50	223.15	0.0300	0.126	0.031	0.130
		0	273.15	0.0320	0.134	−	−
		20	293.15	0.0321	0.134	0.0320	0.134
		100	373.15	0.0325	0.136	0.0323	0.135
		500	773.15	0.0343	0.144	0.0332	0.139
		1000	1273.15	0.0367	0.154	0.0343	0.144
		1500	1773.15	0.0390	0.163	0.0355	0.149
Uranium	U	20	293.15	0.027	0.113	−	−
Vanadium	V	20	293.15	0.120	0.502	−	−
Zinc	Zn	−200	73.15	0.058	0.243	0.082	0.343
		−100	173.15	0.085	0.356	0.088	0.368
		0	273.15	0.091	0.381	−	−
		20	293.15	0.092	0.385	0.091	0.381
		100	373.15	0.095	0.398	0.093	0.389
		200	473.15	0.099	0.414	0.095	0.398
		300	573.15	0.1003	0.420	−	−
		400	673.15	0.1100	0.461	0.100	0.419
Zirconium	Zr	20	293.15	0.065	0.272	−	−

[1] Mean specific heat c_m between 0°C (273.15 K) and t°C (T K).

1 kcal = 4.1868 kJ

Table 11-1 Specific Heats c and c_m of Alloys

Alloy	Composition in %	Temperature		Specific heat		Mean specific heat[1]	
		t	T	c		c_m	
		°C	°K	kcal/kg K	kJ/kg K	kcal/kg K	kJ/kg K
Brass	40Zn	− 100	173.15	0.080	0.335	0.085	0.356
		0	273.15	0.090	0.377	—	—
		20	293.15	0.091	0.381	0.0905	0.379
		100	373.15	0.093	0.389	0.0915	0.383
		200	473.15	0.099	0.414	0.094	0.394
		400	673.15	0.114	0.477	0.099	0.414
Bronze aluminum	12Al	20	293.15	0.100	0.419	—	—
Bronze phosphor	12Sn, 1P	20	293.15	0.086	0.360	—	—
Bronze, red	9Zn, 6Sn, 1Pb	20	293.15	0.090	0.377	—	—
Bronze, tin (Bell metal)	20Sn	20	293.15	0.084	0.352	—	—
Chromium nickel steel		20	293.15	0.114	0.477	—	—
		500	773.15	0.145	0.607	—	—
	18Cr, 8 . . . 36Ni	20	293.15	0.12	0.502	—	—
Constantan	60Cu, 40Ni	20	293.15	0.098	0.410	—	—
Copper - tin	3.9Sn	25	298.15	0.0879	0.368	—	—
	7.9Sn	25	298.15	0.0867	0.363	—	—
	13Sn	25	298.15	0.0857	0.359	—	—
Duralumin	93.2Al, 3.9Cu, 1.3Mn, 0.7Mg, 0.5Si	20	293.15	0.218	0.913	—	—
Gold - copper		20	293.15	0.052	0.218	—	—
Iron, cast	93Fe, 4C, 1P, 1Si, 1Mn	0	273.15	0.127	0.532	—	—
		20	293.15	0.129	0.540	0.128	0.536
		100	373.15	0.133	0.557	0.130	0.544
		300	573.15	0.148	0.620	0.137	0.574
		500	773.15	0.167	0.699	0.145	0.607
		1000	1273.15	—	—	0.174	0.729
Iron, cast (gray)		20	293.15	0.129	0.540	—	—
Iron transformer scheet	95Fe, 4Si, 1Mn	0	273.15	0.108	0.452	—	—
		20	293.15	0.109	0.456	0.108	0.452
		100	373.15	0.114	0.477	0.111	0.465
		300	573.15	0.132	0.553	0.120	0.502
		500	773.15	0.164	0.687	0.130	0.544
		1000	1273.15	—	—	0.168	0.703
Manganese steel	80Fe, 19Mn, 1C	0	273.15	0.117	0.490	—	—
		20	293.15	0.120	0.502	0.119	0.498
		100	373.15	0.127	0.532	0.122	0.511
		300	573.15	0.141	0.590	0.129	0.540
		500	773.15	0.152	0.636	0.135	0.565
Manganine	12Mn, 4Ni	20	293.15	0.097	0.406	—	—
Monel metal	68Ni, 29Cu, 2Fe, 1Mn	− 100	173.15	0.084	0.352	0.093	0.389
		0	273.15	0.100	0.419	—	—
		20	293.15	0.101	0.423	0.100	0.419
		100	373.15	0.105	0.440	0.102	0.427
		1200	1473.15	—	—	0.126	0.528
Nickel silver	15Ni, 22Zn	20	293.15	0.094	0.394	—	—
Nickel steel	67Fe, 31Ni, 1Mn, 1C	20	293.15	0.121	0.507	—	—
Rose's metal	48.9Bi, 27.5Pb, 23.6Sn	20	293.15	0.040	0.167	—	—
Steel	98.5Fe, 1.3C, 0.1Si, 0.1Mn	0	273.15	0.111	0.465	—	—
		20	293.15	0.114	0.477	0.113	0.473
		100	373.15	0.124	0.519	0.118	0.494
		400	673.15	0.15	0.628	0.137	0.574
		800	1073.15	—	—	0.157	0.657
		1200	1473.15	—	—	0.165	0.691
	0.1 . . . 0.6C	20	293.15	0.11	0.461	—	—
	13Cr	20	293.15	0.11	0.461	—	—
Steel, V2A (chromium-nickel)	73Fe, 20Cr, 7Ni	0	273.15	0.111	0.465	—	—
		20	293.15	0.114	0.477	0.113	0.473
		100	373.15	0.121	0.507	0.116	0.486
		400	673.15	0.140	0.586	0.130	0.544
Steel, wrought		0	273.15	0.111	0.465	—	—
		400	673.15	0.15	0.628	—	—
Tin-based solder	64Pb, 36Sn	20	293.15	0.040	0.167	—	—
Wood's metal	52.4Bi, 25.9Pb, 14.7Zn, 7.0Cd	20	293.15	0.350	1.465	—	—

[1] Mean specific heat between 0°C (273.15 K) and t°C (TK).

1 kcal = 4.1868 kJ

Table 12-1 Specific Heats c and c_m of Solid Inorganic Compounds

Inorganic compound	Chemical formula	Temperature		Specific heat		Mean specific heat[1]	
		t	T	c		c_m	
		°C	°K	kcal/kg K	kJ/kg K	kcal/kg K	kJ/kg K
Aluminum oxide	Al_2O_3	0	273.15	.0.16	0.670	—	—
		20	293.15	0.18	0.754	0.17	0.712
		100	373.15	0.22	0.921	0.19	0.795
		500	773.15	0.28	1.172	0.24	1.005
		1000	1273.15	0.31	1.298	0.26	1.089
Ammonium chloride	NH_4Cl	0	273.15	0.36	1.507	—	—
		20	293.15	0.38	1.591	0.37	1.549
		100	373.15	0.44	1.842	0.40	1.675
		150	423.15	0.47	1.968	0.42	1.758
Barium chloride	$BaCl_2$	20	293.15	0.088	0.368	—	—
Calcium carbonate	$CaCO_3$	20	293.15	0.193	0.808	—	—
Calcium chloride	$CaCl_2$	20	293.15	0.15	0.628	—	—
Calcium oxide	CaO	0	273.15	0.180	0.754	—	—
		20	293.15	0.184	0.770	0.182	0.762
		100	373.15	0.193	0.808	0.187	0.783
		400	673.15	0.210	0.879	0.200	0.837
		1200	1473.15	0.230	0.963	0.215	0.900
Calcium sulfate (gypsum)	$CaSO_4\,2H_2O$	20	293.15	0.26	1.089	—	—
Cupric oxide	CuO	— 100	173.15	0.090	0.377	0.110	0.461
		0	273.15	0.125	0.523	—	—
		20	293.15	0.129	0.540	0.127	0.532
		100	373.15	0.139	0.582	0.132	0.553
		500	773.15	0.164	0.687	0.148	0.620
Cupric sulfate	$CuSO_4$	20	293.15	0.150	0.628	—	—
Cuprous oxide	Cu_2O	0	273.15	0.100	0.419	—	—
		20	293.15	0.105	0.440	0.103	0.431
		100	373.15	0.118	0.494	0.111	0.465
		500	773.15	0.136	0.569	0.125	0.523
Ferrous oxide (III)	Fe_2O_3	20	293.15	0.159	0.666	—	—
Ferrous oxide (II, III)	Fe_3O_4	20	293.15	0.153	0.641	—	—
Ice	H_2O	— 80	193.15	0.350	1.465	0.429	1.796
		— 60	213.15	0.396	1.658	0.447	1.871
		— 40	233.15	0.434	1.817	0.463	1.938
		— 20	253.15	0.465	1.947	0.477	1.997
		0	273.15	0.487	2.039	—	—
Lead monoxide	PbO	— 100	173.15	0.039	0.163	0.045	0.188
		0	273.15	0.050	0.209	—	—
		20	293.15	0.051	0.214	0.050	0.209
		100	373.15	0.053	0.222	0.052	0.218
Lead oxide, red (minium)	Pb_3O_4	20	293	0.022	0.921	—	—
Magnesium oxide	MgO	0	273.15	0.226	0.946	—	—
		20	293.15	0.230	0.963	0.228	0.955
		100	373.15	0.242	1.013	0.234	0.980
		500	773.15	0.283	1.185	0.258	1.080
		1000	1273.15	0.308	1.290	0.280	1.172
		1500	1773.15	0.321	1.344	0.293	1.227
Phosphorus pentoxide	P_2O_5	20	293.15	0.17	0.712	—	—
Potassium carbonate	K_2CO_3	20	293.15	0.21	0.879	—	—

[1] Mean specific heat between 0°C (273.15 K) and t°C (TK).   1 kcal = 4.1868 kJ

Table 12-2 Specific Heats c and c_m of Solid Inorganic Compounds (Continued)

Inorganic compound	Chemical formula	Temperature		Specific heat		Mean specific heat[1]	
		t	T	c		c_m	
		°C	°K	kcal/kg K	kJ/kg K	kcal/kg K	kJ/kg K
Potassium chloride	KCl	− 100	173.15	0.151	0.632	0.157	0.657
		0	273.15	0.162	0.678	−	−
		20	293.15	0.163	0.682	0.163	0.682
		100	373.15	0.167	0.699	0.165	0.691
		300	573.15	0.176	0.737	0.169	0.708
		600	873.15	0.191	0.800	0.176	0.737
Potassium nitrate	KNO₃	− 100	173.15	0.180	0.754	0.200	0.837
		0	273.15	0.219	0.917	−	−
		20	293.15	0.225	0.242	0.222	0.929
		100	373.15	0.249	1.043	0.234	0.980
		200	473.15	0.280	1.172	0.320	1.340
Potassium sulfate	K₂SO₄	20	293.15	0.18	0.754	−	−
Silicon carbide (carborundum)	SiC	0	273.15	0.147	0.615	−	−
		20	293.15	0.162	0.678	0.153	0.641
		100	373.15	0.198	0.829	0.176	0.737
		500	773.15	0.270	1.130	0.220	0.921
		1000	1273.15	0.310	1.298	0.260	1.089
Silicon dioxide — α-quartz	SiO₂	− 200	73.15	0.040	0.167	0.112	0.469
		− 100	173.15	0.016	0.067	0.145	0.607
		0	273.15	0.170	0.712	−	−
		20	293.15	0.178	0.745	0.174	0.729
		100	373.15	0.204	0.854	0.187	0.783
		300	573.15	0.252	1.055	0.216	0.904
		500	773.15	0.289	1.210	0.237	0.992
— β-quartz		700	973.15	0.273	1.143	0.254	1.063
— quartz glass		− 200	73.15	0.043	0.180	0.112	0.469
		− 100	173.15	0.116	0.486	0.143	0.599
		0	273.15	0.167	0.699	−	−
		20	293.15	0.174	0.729	0.170	0.712
		100	373.15	0.199	0.833	0.183	0.766
		300	573.15	0.244	1.022	0.210	0.879
		500	773.15	0.266	1.114	0.228	0.955
		1000	1273.15	0.284	1.189	0.251	1.051
Silver bromide	AgBr	20	293.15	0.068	0.285	−	−
Silver chloride	AgCl	20	293.15	0.086	0.360	−	−
Silver nitrate	AgNO₃	20	293.15	0.140	0.586	−	−
Sodium carbonate	Na₂CO₃	− 100	173.15	0.200	0.837	0.220	0.921
		0	273.15	0.243	1.017	−	−
		20	293.15	0.249	1.043	0.246	1.030
		100	373.15	0.270	1.130	0.260	1.089
Sodium chloride	NaCl	− 100	173.15	0.185	0.775	0.195	0.816
		0	273.15	0.203	0.850	−	−
		20	293.15	0.206	0.862	0.204	0.854
		100	373.15	0.212	0.888	0.207	0.867
		300	573.15	0.227	0.950	0.216	0.904
		700	973.15	0.255	1.068	0.230	0.963
Sodium nitrate	NaNO₃	− 100	173.15	0.200	0.837	0.222	0.929
		0	273.15	0.249	1.043	−	−
		20	293.15	0.259	1.084	0.254	1.063
		100	373.15	0.310	1.298	0.275	1.151
		200	473.15	0.380	1.591	0.310	1.298
Sodium sulfate	Na₂SO₄	20	293.15	0.22	0.921	−	−
Sodium tetraborate (borax)	Na₂B₄O₇	20	293.15	0.22	0.921	−	−
Thorium dioxide	ThO₂	20	293.15	0.054	0.226	0.054	0.226
		500	773.15	0.067	0.281	0.062	0.260
		1000	1273.15	0.073	0.306	0.066	0.276
Zinc sulfate	ZnSO₄	20	293.15	0.17	0.712	−	−

[1] Mean specific heat between 0°C (273.15 K) and t°C (TK).   1 kcal = 4.1868 kJ

Table 13-1 Specific Heat c of Miscellaneous Solid Substances

Substance	Temperature t °C	Temperature T °K	Specific heat c kcal/kg K	Specific heat c kJ/kg K
Asbestos	20	293.15	0.19	0.795
Ashes	20	293.15	0.19	0.795
Asphalt	20	293.15	0.22	0.921
Bakelite	20	293.15	0.38	1.591
Basalt	20	293.15	0.19	0.795
	100	373.15	0.23	0.963
Boiler incrustation (sulfate)	300	573.15	0.20	0.837
Brick	20	293.15	0.20	0.837
Brick masonry	20	293.15	0.25	1.047
Calcite	20	293.15	0.19	0.795
Cardboard, dry	20	293.15	0.32	1.340
Cellulose	20	293.15	0.370	1.549
Cement, Portland	20	293.15	0.186	0.779
	100	373.15	0.205	0.858
Cereals	20	293.15	0.50	2.093
Chamotte	20	293.15	0.20	0.837
	500	773.15	0.27	1.130
	1000	1273.15	0.27	1.130
China, porcelain	20	293.15	0.19	0.795
	100	373.15	0.21	0.879
	500	773.15	0.26	1.089
	1000	1273.15	0.31	1.298
Clay	20	293.15	0.21	0.879
Coal				
– anthracite	100	373.15	0.260*	1.089*
– briquettes	20	293.15	0.360	1.507
– brown(lignite), 60% water	20	293.15	0.750	3.140
– brown(lignite), 47.6% water	100	373.15	0.618*	2.587*
– brown(lignite), 20% water	20	293.15	0.500	2.093
– brown(lignite), 12.1% water	100	373.15	0.360*	1.507*
– brown(lignite), 3.4% water	100	373.15	0.297*	1.243*
– brown(lignite), no water	100	373.15	0.306*	1.281*
– channel, gas	100	373.15	0.267*	1.118*
– charcoal	0	273.15	0.24	1.005
	100	373.15	0.24	1.005
	400	673.15	0.37	1.549
	1200	1473.15	0.48	2.010
– gas flame	100	373.15	0.312*	1.306*
– non-baking	100	373.15	0.280*	1.172*
– pit	100	373.15	0.285*	1.193*
Coal dust	30	303.15	0.310	1.298
Coke	0	273.15	0.20	0.837
	20	293.15	0.20	0.837
	100	373.15	0.22	0.921
	1000	1273.15	0.35	1.465
– blast furnace	100	373.15	0.206*	0.862*
– foundry	100	373.15	0.204*	0.854*
– gas	100	373.15	0.201*	0.842*
– low temperature	100	373.15	0.264*	1.105*

*Average specific heat between 0°C (273.15 K) and temperature t°C (TK).

Table 13-2 Specific Heat c of Miscellaneous Solid Substances (Continued)

Substance	Temperature t °C	Temperature T °K	Specific heat c kcal/kg K	Specific heat c kJ/kg K
Colophony	20	293.15	0.29	1.214
Concrete	20	293.15	0.21	0.879
Concrete, cellular	20	293.15	0.19	0.795
Cork	20	293.15	0.45	1.884
— impregnated	20	293.15	0.33	1.382
Cotton	20	293.15	0.31	1.298
Crown	20	293.15	0.159	0.666
Dextrine	20	293.15	0.31	1.298
Dolomite	20	293.15	0.21	0.879
Dry ice (solid CO_2)	20	293.15	0.33	1.382
Ebonite	20	293.15	0.34	1.424
Flax	32	305.15	0.320	1.340
Flint	20	293.15	0.115	0.481
Gelatine	20	293.15	0.51	2.135
Glass	20	293.15	0.20	0.837
— common Thüringia, glass tubes	− 50	223.15	0.158	0.662
	20	293.15	0.184	0.770
	100	373.15	0.214	0.896
— Jena 16 III	20	293.15	0.186	0.779
— Jena 59 III	20	293.15	0.189	0.791
— mirror	20	293.15	0.183	0.766
Glass wool	20	293.15	0.20	0.837
Granite	20	293.15	0.18	0.754
Graphite	20	293.15	0.17	0.712
Gypsum	20	293.15	0.26	1.089
Ice	− 20	253.15	0.510*	2.135*
Iporka	20	293.15	0.33	1.382
Kapok	–	–	0.324	1.357
Kieselguhr	20	293.15	0.20	0.837
Leather	20	293.15	0.36	1.507
Light construction boards	20	293.15	0.40	1.675
Limestone	20	293.15	0.20	0.837
Marble	20	293.15	0.19	0.795
Mica	20	293.15	0.20	0.837
Oil, frozen	20	293.15	0.35	1.465
Paper	20	293.15	0.32	1.340
Paraffin	− 20	253.15	0.377	1.578
	0	273.15	0.535	2.240
	20	293.15	0.694	2.906

*Average specific heat between 0°C (273.15 K) and temperature t°C (TK).

Table 13-3 Specific Heat c of Miscellaneous Solid Substances (Continued)

Substance	Temperature t °C	Temperature T °K	Specific heat c kcal/kg K	Specific heat c kJ/kg K
Peat	20	293.15	0.45	1.884
Plastering	20	293.15	0.20	0.837
Pumice stone	20	293.15	0.24	1.005
Pyrex	20	293.15	0.185	0.775
Quartz	20	293.15	0.18	0.754
	20	293.15	0.174	0.729
Rubber	20	293.15	0.34	1.424
Salt, kitchen	20	293.15	0.21	0.879
Salt, rock	20	293.15	0.22	0.921
Sand (moist)	20	293.15	0.17	0.712
Sandstone	20	293.15	0.17	0.712
Sealing wax	20	293.15	0.25	1.047
Silica, brick	20	293.15	0.22	0.921
	25	298.15	0.19*	0.795*
Silk	20	293.15	0.30	1.256
Slag	20	293.15	0.20	0.837
Slag, blast furnace	20	293.15	0.20	0.837
	500	773.15	0.25	1.047
	1000	1273.15	0.28	1.172
Slag wool	20	293.15	0.18	0.754

Substance	Temperature t °C	Temperature T °K	Specific heat c kcal/kg K	Specific heat c kJ/kg K
Slate	20	293.15	0.18	0.754
Snow	−40	233.15	0.431*	1.805*
Soil	20	293.15	0.44	1.842
Stone	20	293.15	0.21	0.879
Stoneware	20	293.15	0.19	0.795
Styropore (polystyrene foam)	20	293.15	0.33	1.382
Sugar	20	293.15	0.30	1.256
Sulfur (rhombic)	20	293.15	0.17	0.712
Tuff	100	373.15	0.331*	1.386*
Wax, yellow	20	293.15	0.70	2.931
Wood	0	273.15	0.33	1.382
	20	293.15	0.60	2.512
	100	373.15	0.65	2.721
Fir ⊥ fibers	20	293.15	0.65	2.721
Oak ⊥ fibers	20	293.15	0.57	2.386
Pine ⊥ fibers	20	293.15	0.65	2.721
Spruce ⊥ fibers	34	307.15	0.288*	1.206*
Wood resin, coniferous wood	20	293.15	0.44	1.842
Wool	20	293.15	0.45	1.884
	100	373.15	0.40*	1.675*

*Average specific heat between 0°C (273.15 K) and temperature t°C (TK).

Table 14-1 Specific Heat c of Some Foods

Food	Water content,[1] in %	Solids content, in %	Specific heat c Before freezing kcal/kg K	Before freezing kJ/kg K	After freezing kcal/kg K	After freezing kJ/kg K	Heat of solidification or fusion kcal/kg	kJ/kg
Apples	83	17	0.92	3.852	0.42	1.758	67	280.516
Asparagus	94	6	0.93	3.894	0.47	1.968	75	314.010
Bacon	—	—	0.55	2.303	0.31	1.298	17	71.176
Bananas	75	25	0.80	3.349	—	—	60	251.208
Beans, green	89	11	0.92	3.852	0.47	1.968	71	297.263
Beer	89 . . . 91	—	0.90	3.768	—	—	72	301.450
Berries	84 . . . 88	16 . . . 12	0.91	3.810	0.40 . . . 0.50	1.675 . . . 2.093	67 . . . 70	280.516 . . . 293.076
Butter	14 . . . 15	86 . . . 85	0.60 . . . 0.64	2.512 . . . 2.680	0.30	1.256	35 + 12[2]	146.538 + 50.242[2]
Carrots	83	17	0.87	3.643	0.45	1.884	66	276.329
Caviar	50 . . . 60	50 . . . 40	0.70	2.931	0.31	1.298	40 . . . 50	167.472 . . . 209.340
Celery	88 . . . 95	12 . . . 5	0.94	3.936	0.47	1.968	70 . . . 76	293.076 . . . 318.197
Cheese, cream (cottage)	80	20	0.70	2.931	0.45	1.884	64	267.955
Cheese, fat	35 . . . 50	65 . . . 50	0.45 . . . 0.60	1.884 . . . 2.512	0.30	1.256	26 . . . 37	108.857 . . . 154.912
Cheese, skim	53	47	0.68	2.847	0.40	1.675	42	175.846
Cherries	82	18	0.87	3.643	0.44	1.842	66	276.329
Chocolate	1.6	98.4	0.76	3.182	—	—	20 . . . 30	83.736 . . . 125.604
Cocoa powder	0.5	99.5	0.50	2.093	—	—	—	—
Cream	59	41	0.85	3.559	0.36	1.507	47	196.780
Dough	—	—	0.45	1.884	—	—	—	—
Eel	62	38	0.70	2.931	0.39	1.633	50	209.340
Eggs	70	30	0.76	3.182	0.40	1.675	56	234.461
Fat, lard	0.7	99.3	0.60	2.512	0.40	1.675	29 . . . 35	121.417 . . . 146.538
Fat, vegetable	—	—	0.47 . . . 0.50	1.968 . . . 2.093	0.35	1.465	—	—
Fish, dried	—	—	0.54	2.261	0.34	1.424	36	150.725
Fish, fresh, fat	60	40	0.68	2.847	0.38	1.591	50	209.340
Fish, fresh, lean	73	27	0.82	3.433	0.43	1.800	61	255.395
Fish, smoked	—	—	0.76	3.182	—	—	—	—
Flour	12 . . . 13.5	88 . . . 86.5	0.43 . . . 0.45	1.800 . . . 1.884	—	—	—	—
Game	74	26	0.80	3.349	0.40	1.675	59	247.021
Gooseberries	90	10	0.92	3.852	0.46	1.926	72	301.450
Grape	81	19	0.88	3.684	0.45	1.884	63	263.768
Honey	19	81	0.35	1.465	0.26	1.089	14	58.615
Ice cream	60 . . . 65	40 . . . 35	0.78	3.266	0.45	1.884	52	217.714
Ice (water)	100	—	1.00	4.187	0.50	2.093	80	334.944
Kale	91	9	0.93	3.894	0.48	2.010	73	305.636
Leek	91	9	0.93	3.894	0.48	2.010	73	305.636
Lemons	83 . . . 89	17 . . . 11	0.92	3.852	0.46	1.926	66 . . . 71	276.329 . . . 297.263
Lobsters, crabs	77	23	0.81	3.391	0.43	1.800	62	259.582
Margarine	17 . . . 18	83 . . . 82	0.65 . . . 0.70	2.721 . . . 2.931	0.35	1.465	15 + 15	62.802 + 62.802
Meat, beef, fat	51	49	0.608	2.546	0.355	1.486	41	171.659
Meat, beef, lean	72	28	0.776	3.249	0.42	1.758	56	234.461
Meat, mutton, fat	50	50	0.60	2.512	0.35	1.465	40	167.472
Meat, mutton, lean	67	33	0.73	3.056	0.41	1.717	53	221.900
Meat, pork, fat	39 . . . 46	61 . . . 54	0.51	2.135	0.32	1.340	31 . . . 36.6	129.791 . . . 153.237
Meat, veal	63	37	0.704	2.948	0.40	1.675	50	209.340
Milk	88	12	0.94	3.936	0.60	2.512	70	293.076
Oil	—	—	0.40	1.675	0.35	1.465	—	—
Onion	80 . . . 89	20 . . . 11	0.91	3.810	0.46	1.926	64 . . . 71	267.955 . . . 297.263
Oranges	84	16	0.92	3.852	0.44	1.842	68	284.702
Oysters	80	20	0.84	3.517	0.44	1.842	63	263.768
Peaches	87	13	0.92	3.852	0.41	1.717	70	293.076
Pears	83	17	0.92	3.852	0.42	1.758	67	280.516
Peas, green	75	25	0.80	3.349	0.42	1.758	60	251.208
Potatoes	74	26	0.80	3.349	0.42	1.758	58	242.834
Poultry	74	26	0.70 . . . 0.76	2.931 . . . 3.182	0.40	1.675	59	247.021
Strawberries	90	10	0.92	3.852	0.47	1.968	71.6	299.775
Sugar	0.1	99.9	—	—	0.30	1.256	—	—
Tomatoes	94	6	0.93	3.894	0.49	2.052	75	314.010
Walnuts	7.2	94.8	0.25	1.047	0.22	0.921	9	37.681
Watermelons	89	11	0.92	3.852	0.46	1.926	71	297.263
Wine	—	—	0.90	3.768	—	—	—	—

[1] Water content of foods varies considerably with fat content; at the same time this causes differences in specific heat and solidification heat values.

[2] Solidification heat of fat + solidification heat of water.

1 kcal = 4.1868 kJ

Table 15-1 Thermal Conductivities λ of Metals

Substance	Chemical formula	Temperature		Density	Thermal conductivity	
		t	T	ρ	λ	
		°C	°K	kg/m³	kcal/hm K	W/m K
Aluminum, 99.75%	Al	— 190	83.15		220	255.860
		0	273.15	2700	197	229.111
		200	473.15		197	229.111
		300	573.15		191	222.133
		800	1073.15		108	125.604
— 99%		— 100	173.15	—	180	209.340
		0	273.15		180	209.340
		100	373.15		178	207.014
		300	573.15		191	222.133
Antimony, very pure	Sb	— 190	83.15		18	20.934
		— 100	173.15		16.5	19.190
		0	273.15		15.2	17.678
		100	373.15	6690	14	16.282
		300	573.15		13.6	15.817
		500	773.15		16	18.608
Beryllium, 99.5%	Be	— 250	23.15		81	94.203
		— 100	173.15		108	125.604
		0	273.15	1850	138	160.494
		100	373.15		164	190.732
		200	473.15		185	215.155
Bismuth	Bi	— 190	83.15		22	25.586
		— 100	173.15		10.4	12.095
		0	273.15	9800	7.2	8.374
		100	373.15		6.2	7.211
		200	473.15		6.2	7.211
Cadmium, pure	Cd	— 190	83.15		90	104.670
		— 100	173.15		83	96.529
		0	273.15	8620	80	93.040
		100	373.15		79	91.877
		200	473.15		78.5	91.296
		300	573.15		75.5	87.807
Cobalt, 97.1%	Co	20	293.15	≈ 8900	60	69.780
Copper, pure 99.9 to 98%	Cu	— 180	93.15		399	464.037
		— 100	173.15		350	407.050
		0	273.15	8930	332	386.116
		100	373.15		326	379.138
		200	473.15		321	373.323
		400	673.15		313	364.019
		600	873.15		304	353.552
— commercial		20	293.15	8300	320	372.160
— electrolytic, pure		— 180	93.15		420	488.460
		0	273.15	8900	340	395.420
		100	373.15		337	391.931
		300	573.15		328	381.464
		800	1073.15		316	367.508
Gold 99.999%	Au	— 190	83.15		282	327.966
		0	273.15	19290	267	310.521
		100	373.15		267	310.521
		300	573.15		262	304.706
— 99.98%		0	273.15		253	294.239
		100	373.15		253	294.239
Iridium, pure	Ir	0	273.15	22420	51	59.313
		100	373.15		49	56.987
Iron (Armc) 99.92%	Fe	20	293.15	7850	63	73.169
		100	373.15		58	67.454
		200	473.15		53	61.639
		400	673.15		42	48.846
		600	873.15		33	38.379
		800	1073.15		25	29.075

1 kcal/h = 1.163 W

Table 15-2 Thermal Conductivities λ of Metals (*Continued*)

Substance	Chemical formula	Temperature		Density	Thermal conductivity	
		t	T	ρ	λ	
		°C	°K	kg/m³	kcal/hm K	W/m K
Iron (Armc)	Fe					
— cast, 1% Ni		20	293.15	7280	43	50.009
		100	373.15		42.5	49.428
		300	573.15		40	46.520
		500	773.15		32	37.216
— cast, 3% C		20	293.15	7280	48 . . . 55	55.824 . . . 63.965
— steel 99.2%		0	273.15	7800	39	45.357
Fe, 0.2% C		100	373.15		39	45.357
		300	573.15		37	43.031
		500	773.15		32	37.216
		800	1073.15		26	30.238
— wrought, pure		0	273.15	7800	51	59.313
		100	373.15		49	56.987
		200	473.15		45	52.335
		400	673.15		38	44.194
		600	873.15		32	37.216
		800	1073.15		25	29.075
Lead, pure	Pb	— 250	23.15		42	48.846
		— 200	73.15		35	40.705
		— 100	173.15		31.7	36.867
		0	273.15		30.2	35.123
		20	293.15	11340	29.9	34.774
		100	373.15		28.7	33.378
		300	573.15		25.6	29.773
		500	773.15		14.4	16.747
Lithium, pure	Li	0	273.15	530	61	70.943
		100	373.15		61	70.943
Magnesium, pure	Mg	— 190	83.15		160	186.080
		0	273.15	1740	148	172.124
		200	473.15		140	162.820
— 99.6%		0	273.15	≈ 1740	124	144.212
		100	373.15		120	139.560
		300	573.15		113	131.419
		500	773.15		113	131.419
Manganese	Mn	0	273.15	7300	43.2	50.242
Mercury	Hg	— 190	83.15		42	48.846
		— 100	173.15		31	36.053
		— 50	223.15		24	27.912
		0	273.15	13595	7 . . . 9	8.141 . . . 10.467
Molybdenum 99.84%	Mo	— 180	93.15		150	174.450
		— 100	173.15		119	138.397
		0	273.15	10200	118	137.234
		100	373.15		118	137.234
		1000	1273.15		85	98.855
Nickel 99.94%	Ni	— 180	93.15		95	110.485
		0	273.15	8800	80	93.040
		100	373.15		71	82.573
		200	473.15		63	73.269
		300	573.15		55	63.965
		400	673.15		51	59.313
		500	773.15		53	61.639
— 99.2%		0	273.15		58	67.454
		100	373.15		54	62.802
		200	473.15	—	50	58.150
		400	673.15		45	52.335
		600	873.15		49	56.987
		800	1073.15		54	62.802

1 kcal/h = 1.163 W

Table 15-3 Thermal Conductivities λ of Metals (*Continued*)

Substance	Chemical formula	Temperature		Density	Thermal conductivity	
		t	T	ρ	λ	
		°C	°K	kg/m³	kcal/hm K	W/m K
Nickel — 97 to 99%	Ni	— 100	173.15		48	55.824
		0	273.15		50	58.150
		100	373.15		49	56.987
		200	473.15	—	47	54.661
		400	673.15		42	48.846
		600	873.15		46	53.498
		800	1073.15		50	58.150
Palladium, pure	Pd	— 190	83.15		66	76.758
		0	273.15	—	59	68.617
		100	373.15		63	73.269
Platinum, pure	Pt	— 190	83.15		67	77.921
		0	273.15	21400	60.2	70.013
		100	373.15		61.4	71.408
		300	573.15		65	75.595
		500	773.15		68	79.084
		800	1073.15		74	86.062
		1000	1273.15		77	89.551
Potassium, pure	K	0	273.15	860	117	136.071
		100	373.15		102	118.626
Rhodium, pure	Rh	— 190	83.15		183	212.829
		0	273.15	12500	76	88.388
		100	373.15		69	80.247
Silver > 99.98%	Ag	— 190	83.15		366	425.658
		0	273.15	10500	360	418.680
		100	373.15		358	416.354
		300	573.15		350	407.050
— 99.9%		— 100	173.15		361	419.843
		0	273.15	10500	353	410.539
		100	373.15		337	391.931
		300	573.15		311	361.693
		500	773.15		312	362.856
Sodium, pure	Na	— 100	173.15		133	154.679
		0	273.15	970	86	100.018
		50	323.15		80	93.040
		100	373.15		72	83.736
Tantalum	Ta	0	273.15	16650	47	54.661
		100	373.15		46.5	54.080
		1000	1273.15		55	63.965
		1400	1673.15		62	72.106
		1800	2073.15		71	82.573
Thallium, pure	Tl	— 190	83.15		54	62.802
		0	273.15	11840	44	51.172
		100	373.15		36	41.868
Tin, pure	Sn	— 150	123.15		68	79.084
		— 100	173.15		64	74.432
		0	273.15	7300	56.8	66.058
		100	373.15		51	59.313
		200	473.15		49	56.987
Wolfram	W	— 190	83.15		187	217.481
		0	273.15	19300	143	166.309
		100	373.15		130	151.190
		500	773.15		103	119.789
		1000	1273.15		85	98.855
		1500	1773.15		98	113.974
		2000	2273.15		117	136.071
		2400	2673.15		126	146.538
Zinc, pure	Zn	— 100	173.15		99	115.137
		0	273.15	7130	97	112.811
		100	373.15		94.5	109.904
		200	473.15		91	105.833
		300	573.15		87	101.181

1 kcal/h = 1.163 W

Table 16-1 Thermal Conductivities λ of Alloys

Alloy	Composition in %	Temperature		Density	Thermal conductivity	
		t	T	ρ	λ	
		°C	°K	kg/m³	kcal/hm K	W/m K
Aluminum alloys	96Al, 1.8Cu, 0.9Fe, 0.9Cr, 0.4Si	20	293.15	—	90	104.670
Aluminum bronze	95Cu, 5Al	20	293.15	7800	71	82.573
Aluminum magnesium	92Al, 8Mg	−180	93.15		65	75.595
		−100	173.15		73	84.899
		0	273.15		88	102.344
		20	293.15	≈2600	91	105.833
		100	373.15		106	123.278
		200	473.15		127	147.701
Alusil	80Al, 20Si	−180	93.15		105	122.115
		−100	173.15		122	141.886
		0	273.15		136	158.168
		20	293.15	≈2650	138	160.494
		100	373.15		145	168.635
		200	473.15		150	174.450
Bismuth-antimony	80Bi, 20Sb	0	273.15	—	5.68	6.606
		100	373.15		7.41	8.618
	50Bi, 50Sb	0	273.15	—	7.16	8.327
		100	373.15		8.06	9.374
	30Bi, 70Sb	0	273.15	—	8.30	9.653
		100	373.15		10.01	11.660
Brass	90Cu, 10Zn	−100	173.15	≈8600	76	88.388
		0	273.15		88	102.344
		100	373.15		101	117.463
		200	473.15		115	133.745
		300	573.15		128	148.864
		400	673.15		143	166.309
		500	773.15		155	180.265
		600	873.15		168	195.384
	70Cu, 30Zn	0	273.15	≈8600	91	105.833
		100	373.15		94	109.322
		200	473.15		95	110.485
		300	573.15		98	113.974
		400	673.15		100	116.300
		500	773.15		103	119.789
		600	873.15		104	120.952
	66Cu, 33Zn	0	273.15	≈8600	86	100.018
		100	373.15		92	106.996
		200	473.15		97	112.811
		300	573.15		104	120.952
		400	673.15		110	127.930
		500	773.15		116	134.908
		600	873.15		130	151.190
	60Cu, 40Zn	0	273.15	≈8600	91	105.833
		100	373.15		103	119.789
		200	473.15		118	137.234
		300	573.15		131	152.353
		400	673.15		145	168.635
		500	773.15		160	186.080
		600	873.15		172	200.036
	61.5Cu, 38.5Zn	20	293.15		68	79.084
		100	373.15	—	76	88.388
Bronze	90Cu, 10Sn	20	293.15	8766	36	41.868
	75Cu, 25Sn	20	293.15	≈8900	22	25.586
	88Cu, 10Sn, 2Zn	20	293.15	≈8800	41	47.683
	84Cu, 6Sn, 9Zn, 1Pb	20	293.15	—	50	58.150
Bronze (red bass)	86Cu, 7Zn, 6.4Sn	20	293.15	≈8600	52	60.476
		100	373.15		61	70.943

1 kcal/h = 1.163 W

Table 16-2 Thermal Conductivities λ of Alloys (Continued)

Alloy	Composition in %	Temperature		Density	Thermal conductivity	
		t	T	ρ	λ	
		°C	°K	kg/m³	kcal/hm K	W/m K
Chrome-nickel steel	0.8Cr, 3.5Ni, 0.4C	20	293.15	8100 . . . 8700	30	34.890
		100	373.15		31	36.053
		200	473.15		32	37.216
		400	673.15		32	37.216
		600	873.15		27	31.401
	Cr . . . Ni	20	293.15	7900	12	13.956
		200	473.15		15	17.445
		500	773.15		18	20.934
	17 . . . 19Cr, 8Ni, 0.1 . . . 0.2C	20	293.15	8100 . . . 9000	12.5	14.538
		100	373.15		13.5	15.701
		200	473.15		14.5	16.864
		300	573.15		16	18.608
		500	773.15		18	20.934
	10Cr, 34Ni	20	293.15		10.5	12.212
		100	373.15		11.5	13.375
		200	473.15	—	13	15.119
		300	573.15		14	16.282
		500	773.15		16.5	19.190
	15Cr, 27Ni, 3W, 0.5C	20	293.15		9.7	11.281
		100	373.15		11	12.793
		200	473.15	—	12	13.956
		300	573.15		13	15.119
		500	773.15		16	18.608
	15Cr, 13Ni, 2W, 0.5C	20	293.15		10	11.630
		100	373.15		10	11.630
		200	473.15		10	11.630
		300	573.15		10.5	12.212
		500	773.15		11	12.793
		800	1073.15		14	16.282
Chrome steel	0.8Cr, 0.2C	100	373.15	≈7850	34	39.542
		200	473.15		32	37.216
		400	673.15		27	31.401
		600	873.15		23	26.749
	5Cr, 0.5Mn, 0.1C	20	293.15	8100 . . . 9000	32	37.216
		100	373.15		31.5	31.635
		200	473.15		31	31.053
		500	773.15		29	33.727
	15Cr, 0.1C	20	293.15	8100 . . . 9000	22	25.586
		500	773.15		22	25.586
	14Cr, 0.3C	20	293.15	8100 . . . 9000	21	24.423
		100	373.15		21.5	25.005
		200	473.15		22	25.586
		300	573.15		22	25.586
		500	773.15		22	25.586
	16Cr, 0.9C	100	373.15	8100 . . . 9000	20.5	23.842
		200	473.15		20	23.260
		300	573.15		20	23.260
		500	773.15		20	23.260
		800	1073.15		20	23.260
	26Cr, 0.1C	20	293.15	8100 . . . 9000	17	19.771
		100	373.15		18	20.934
		200	473.15		19	22.097
		300	573.15		19.7	22.911
		500	773.15		21	24.423
Cobalt steel	5 . . . 10Co	20	293.15	≈7800	35	40.705
Constantin	60Cu, 40Ni	−100	173.15		18	20.934
		0	273.15		19.1	22.213
		20	293.15	8800	19.5	22.679
		100	373.15		22	25.586
Copper alloys	92Al, 8Cu	−180	93.15		77	89.551
		−100	173.15		94	109.322
		0	273.15		110	127.930
		20	293.15	≈2800	113	131.419
		100	373.15		123	143.049
		200	473.15		131	152.353

1 kcal/h = 1.163 W

Table 16-3 Thermal Conductivities λ of Alloys (*Continued*)

Alloy	Composition in %	Temperature t (°C)	T (°K)	Density ρ (kg/m³)	Thermal conductivity λ (kcal/hm K)	λ (W/m K)
Copper manganese	70Cu, 30Mn	20	293.15	≈7800	11	12.793
Copper-nickel	90Cu, 10Ni	20	293.15	≈8800	50	58.150
		100	373.15		65	75.595
	80Cu, 20Ni	20	293.15	≈8500	29	33.727
		100	373.15		35	40.705
	40Cu, 60Ni	20	293.15	≈8400	19	22.097
		100	373.15		22	25.586
	18Cu, 82Ni	20	293.15		22	25.586
		100	393.15		22	25.586
Duralumin	94 . . . 96Al, 3 . . . 5Cu, 0.5Mg	−180	93.15		78	90.714
		−100	173.15		108	125.604
		0	273.15		137	159.331
		20	293.15	≈2800	142	165.146
		100	373.15		156	181.428
		200	473.15		167	194.221
Electron alloy	93Mg, 4Zn, 0.5Cu	20	293.15	1800	100	116.300
German alloy	88Al, 10Zn, 2Cu	0	273.15	2900	123	143.049
		20	293.15		125	145.375
		100	373.15		133	154.679
Gold-copper alloy	88Au, 12Cu	0	273.15	—	48	55.824
		100	373.15		58	67.454
	27Au, 73Cu	0	273.15	—	78	90.714
		100	373.15		98	113.974
Invar	35Ni, 65Fe	20	293.15	8130	9.5	11.049
Lautal	95Al, 4.5 . . . 5.5Cu, 0.3Si	20	293.15	—	120	139.560
Magnesium-aluminum	92Mg, 8Al	−180	93.15		36	41.868
		−100	173.15		43	50.009
		0	273.15	≈1800	52	60.476
		20	293.15		53	61.639
		100	373.15		60	69.780
		200	473.15		68	79.084
	2.5Al	20	293.15	—	73.6	85.597
	4.2Al	20	293.15	—	59.4	69.082
	6.2Al	20	293.15	—	47.8	55.591
	10.3Al	20	293.15	—	37.4	43.496
Magnesium-aluminum-silicone	88Mg, 10Al, 2Si	−180	93.15		26	30.238
		−100	173.15		35	40.705
		0	273.15	≈1850	48	55.824
		20	293.15		50	58.150
		100	373.15		59	68.617
		200	473.15		65	75.595
Magnesium-copper	92Mg, 8Cu	−180	93.15		76	88.388
		−100	173.15		92	106.996
		0	273.15	≈2400	107	124.441
		20	293.15		108	125.604
		100	373.15		112	130.256
		200	473.15		114	132.582
	93.7Mg, 6.3Cu	20	293.15		113	131.419
Manganese-nickel steel	12Mn, 3Ni, 0.75C	20	293.15		12	13.956
		100	373.15		12.7	14.770
		200	473.15	—	14	16.282
		300	573.15		15	17.445
		500	773.15		17	19.771
Manganese steel	1.6Mn, 0.5C	20	293.15	≈7850	35	40.705
		100	373.15		35	40.705
		300	573.15		32	37.216
		500	773.15		30	34.890
	2Mn	20	293.15	≈7850	28	32.564
	5Mn	20	293.15	≈7850	16	18.608

1 kcal/h = 1.163 W

Table 16-4 Thermal Conductivities λ of Alloys (Continued)

Alloy	Composition in %	Temperature		Density	Thermal conductivity	
		t	T	ρ	λ	
		°C	°K	kg/m³	kcal/hm K	W/m K
Manganine	84Cu, 4Ni, 12Mn	−100	173.15	8400	14	16.282
		0	273.15		18	20.934
		20	293.15		18.8	21.864
		100	373.15		22.7	26.400
Monel	29Cu, 67Ni, 2Fe	20	293.15	8710	19	22.097
		100	373.15		21	24.423
		200	473.15		23.7	27.563
		300	573.15		26	30.238
		400	673.15		29	33.727
New silver	62Cu, 15Ni, 22Zn	−150	123.15	8433	15.2	17.678
		−100	173.15		16.5	19.170
		+ 20	293.15		21.5	25.005
		100	373.15		27	31.401
		200	473.15		34	39.542
		300	573.15		39	45.357
		400	673.15		42	48.846
Nickel alloy	70Ni, 28Cu, 2Fe	20	293.15	≈8200	30.6	34.890
Nickel chrome	90Ni, 10Cr	0	273.15	≈8220	14.7	17.096
		20	293.15		15	17.445
		100	373.15		16.3	18.957
		200	473.15		18	20.934
		300	573.15		19.6	22.795
		400	673.15		21.2	24.656
	80Ni, 20Cr	0	273.15	≈8200	10.5	12.212
		20	293.15		10.8	12.560
		100	373.15		11.9	13.840
		200	473.15		13.4	15.584
		300	573.15		14.8	17.212
		400	673.15		16.3	18.957
		600	873.15		19.4	22.562
Nickel-chrome steel	61Ni, 15Cr, 20Fe, 4Mn	20	293.15	≈8190	10.0	11.630
		100	373.15		10.2	11.863
		200	473.15		10.5	12.212
		300	573.15		10.7	12.444
		400	673.15		10.9	12.677
		600	873.15		11.3	13.142
		800	1073.15		12.0	13.956
	61Ni, 16Cr, 23Fe	0	273.15	≈8190	10.2	11.863
		20	293.15		10.4	12.095
		100	373.15		11.4	13.258
		200	473.15		12.6	14.654
		300	573.15		13.8	16.049
		400	673.15		15.0	17.445
	70Ni, 18Cr, 12Fe	20	293.15	−	9.9	11.514
	62Ni, 12Cr, 26Fe	20	293.15	≈8100	11.6	13.491
Nickel silver	−	0	273.15	−	25.2	29.308
		100	373.15		32.0	37.216
Nickel steel	5Ni	20	293.15	8130	30.0	34.890
	10Ni	20	293.15		24	27.912
	15Ni	20	293.15		19	22.097
	20Ni	20	293.15		16	18.608
	25Ni	20	293.15		13	15.119
	30Ni	20	293.15		10.5	12.212
	35Ni	20	293.15		9.5	11.049
	40Ni	20	293.15		9.5	11.049
	50Ni	20	293.15		12.5	14.538
	60Ni	20	293.15		16.5	19.190
	70Ni	20	293.15		22	25.586
	80Ni	20	293.15		28	32.564
	30Ni, 1Mn, 0.25C	20	293.15	8190	10.4	12.095
		100	373.15		11.7	13.607
	36Ni, 0.8Mn	20	293.15	−	10.4	12.095

1 kcal/h = 1.163 W

Table 16-5 Thermal Conductivities λ of Alloys (*Continued*)

Alloy	Composition in %	Temperature		Density	Thermal conductivity	
		t	T	ρ	λ	
		°C	°K	kg/m³	kcal/hm K	W/m K
Nickel steel	1.4Ni, 0.5Cr, 0.3C	20	293.15	≈7850	39	45.357
		100	373.15		38	44.194
		300	573.15		35	40.705
		500	773.15		32	37.216
Phosphor bronze	92.8Cu, 5Sn, 2Zn, 0.15P	20	293.15	≈8766	68	79.084
	91.7Cu, 8Sn, 0.3P	20	293.15	8800	39	45.357
		100	373.15		45	52.335
		200	473.15		53	61.639
	87.8Cu, 10Sn, 2Zn, 0.2P	20	293.15		36	41.868
	87.2Cu, 12.4Sn, 0.4P	20	293.15	8700	31	36.053
Piston alloy, cast	91.5Al, 4.6Cu, 1.8Ni, 1.5Mg	0	273.15	≈2800	123	143.049
		20	293.15		124	144.212
		100	373.15		130	151.190
		200	473.15		136	158.168
	84Al, 12Si, 1.2Cu, 1Ni	0	273.15	≈2800	116	134.908
		20	293.15		116	134.908
		100	373.15		118	137.234
		200	473.15		120	139.560
Platinum-iridium	90Pt, 10Ir	0	273.15	—	26.6	30.936
		100	373.15		27.0	31.401
Platinum-rhodium	90Pt, 10Rh	0	273.15	—	26	30.238
		100	373.15		26.3	30.587
Rose's metal	50Bi, 25Pb, 25Sn	20	293.15	—	14	16.282
Silumin	86 . . . 89Al, 11 . . . 14Si	0	273.15	2600	137	159.331
		20	293.15		139	161.657
		100	373.15		147	170.961
Steel	0.1C	0	273.15	7850	51	59.313
		100	373.15		45	52.335
		200	473.15		45	52.335
		300	573.15		40	46.520
		400	673.15		38	44.194
		600	873.15		32	37.216
		900	1173.15		29	33.727
	0.2C	20	293.15	7850	43	50.009
	0.6C	20	293.15	7850	40	46.520
— Bessemer	0.52C, 0.34Si	20	293.15	7850	34.6	40.240
Tungsten steel	1W, 0.6Cr, 0.3C	20	293.15	7900	34	39.542
		100	373.15		33	38.379
		300	573.15		31	36.053
		500	773.15		29	33.727
V 1 A steel	—	20	293.15	—	18	20.934
V 2 A steel	—	20	293.15	7860	13	15.119
Wood's metal	48Bi, 26Pb, 13Sn, 13Cd	20	293.15	—	11	12.793

1 kcal/h = 1.163 W

Table 17-1 Thermal Conductivities λ of Building Materials

Building material	Temperature		Density	Thermal conductivity	
	t	T	ρ	λ	
	°C	°K	kg/m³	kcal/hm K	W/m K
Air brick masonry	20	293.15	800	0.30 ... 0.45	0.349 ... 0.523
	20	293.15	1600	0.45 ... 0.65	0.523 ... 0.756
Asbestos slate (Salonite), with high asbestos content	20	293.15	1800 ... 1900	0.15 ... 0.30	0.174 ... 0.349
— 10 ... 50% asbestos, dry	20	293.15	1800	0.55 ... 0.45	0.640 ... 0.523
	20	293.15	2100	0.60	0.698
— 30% asbestos, 10% moisture	20	293.15	2200	0.68	0.791
Asphalt	0	273.15		0.52	0.605
	20	293.15	2120	0.60	0.698
	30	303.15		0.64	0.744
Bark	20	293.15	400	0.047	0.055
	20	293.15	600	0.064	0.074
Basalt	0	273.15		1.42	1.651
	20	293.15	≈2900	1.44	1.675
	100	373.15		1.52	1.768
Bituminous coal	20	293.15	1100	0.15	0.174
Brick	20	293.15	800	0.24	0.279
	20	293.15	1000	0.28	0.326
	20	293.15	1200	0.33	0.384
	20	293.15	1400	0.38	0.442
	20	293.15	1600	0.45	0.523
	20	293.15	1800	0.63	0.733
	20	293.15	2000	1.06	1.233
— dried	20	293.15	1600 ... 1800	0.33 ... 0.45	0.384 ... 0.523
	100	373.15	1400 ... 2000	0.38	0.442
	200	473.15		0.47	0.547
	600	873.15		0.83	0.965
	1000	1273.15		1.11	1.291
— porous	20	293.15	600 ... 800	0.10 ... 0.15	0.116 ... 0.174
— porous, normal moisture	20	293.15	—	0.20 ... 0.30	0.233 ... 0.349
Brick masonry, massive, inside	20	293.15	1600 ... 1800	0.60	0.698
— outside	20	293.15		0.75	0.872
Brick masonry, porous, outside	20	293.15	800	0.34	0.395
	20	293.15	1200	0.48	0.558
Carborundum	20	293.15	—	0.182	0.212
Cement, hard	20	293.15	—	0.90	1.047
Cement, powdered	20	293.15	—	0.06	0.070
	100	373.15	—	0.41	0.477
Clay, 44.7% vol. moisture	23	296.15	1495	1.44	1.675
— dried	25	298.15	1500 ... 1600	0.80	0.930
Clay (48.7% vol. moisture)	23	296.15	1545	1.08	1.256
Clinker brick	20	293.15	1800	0.82	0.954
	20	293.15	2000	0.97	1.128
Cob wall	20	293.15	1700	0.85	0.989
Concrete — air dried	20	293.15	500	0.16	0.186
	20	293.15	1000	0.31	0.361
	20	293.15	1500	0.51	0.593
	20	293.15	2000	0.77	0.896
	20	293.15	2250	0.95	1.105

1 kcal/h = 1.163 W

Table 17-2 Thermal Conductivities λ of Building Materials (*Continued*)

Building material	Temperature		Density	Thermal conductivity	
	t	T	ρ	λ	
	°C	°K	kg/m³	kcal/hm K	W/m K
Concrete					
— completely dry	20	293.15	500	0.11	0.128
	20	293.15	1000	0.20	0.233
	20	293.15	1500	0.35	0.407
	20	293.15	2000	0.57	0.663
	20	293.15	2250	0.72	0.837
— gas and foam, according to composition	20	293.15	600	0.15 ... 0.30	0.174 ... 0.349
	20	293.15	800	0.20 ... 0.45	0.233 ... 0.523
	20	293.15	1000	0.30 ... 0.60	0.349 ... 0.698
	20	293.15	1200	0.40 ... 0.80	0.465 ... 0.930
	20	293.15	1400	0.50 ... 1.00	0.582 ... 1.163
— gravel	20	293.15	1800	0.83	0.965
	20	293.15	2000	1.00	1.163
	20	293.15	2200	1.30	1.512
— gravel, reinforced	20	293.15	1600 ... 1800	0.80	0.930
	20	293.15	1800 ... 2200	1.1 ... 1.3	1.279 ... 1.512
— masonry, light concrete (slag blocks, cell concrete, Aerocrete, porous concrete and others)	20	293.15	800	0.27	0.314
	20	293.15	1000	0.36	0.419
	20	293.15	1200	0.46	0.535
	20	293.15	1600	0.70	0.814
— pumice stone, boards	20	293.15	800	0.32	0.372
	20	293.15	1000	0.44	0.512
	20	293.15	1200	0.54	0.628
— pumice stone, inside	20	293.15	—	0.30	0.349
— pumice stone, outside	20	293.15	—	0.40	0.465
— pumice stone, rammed	20	293.15	800	0.32	0.372
	20	293.15	1000	0.43	0.500
	20	293.15	1200	0.54	0.628
— pumice stone, masonry, cell concrete, porous concrete, and others	20	293.15	800	0.40	0.465
	20	293.15	1000	0.48	0.558
	20	293.15	1200	0.56	0.651
	20	293.15	1400	0.64	0.744
	20	293.15	1600	0.70	0.814
— slag, inside	20	293.15	—	0.50	0.582
— slag, outside	20	293.15	—	0.60	0.698
— steel reinforced	20	293.15	—	1.30	1.512
— with 10% moisture	20	293.15	500	0.22	0.256
	20	293.15	1000	0.42	0.488
	20	293.15	1500	0.68	0.791
	20	293.15	2000	0.98	1.140
	20	293.15	2250	1.15	1.337
Earth, clayey, or clay	20	293.15	1500	1.30	1.512
— with 28% moisture	20	293.15	2000	2.20	2.559
Earth, sandy oil	20	293.15	1500	0.90	1.047
— with 8% moisture	20	293.15	2000	1.50	1.745
Feldspar	20	293.15	—	2.1	2.442
Fire clay, stones	100	373.15	1800 ... 2200	0.40 ... 1.00	0.465 ... 1.163
	500	773.15		0.60 ... 1.20	0.698 ... 1.396
	1000	1273.15		1.50	1.745
Glass, window	20	293.15	2400 ... 3200	0.50 ... 0.90	0.582 ... 1.047

1 kcal/h = 1.163 W

Table 17-3 Thermal Conductivities λ of Building Materials (*Continued*)

Building material	Temperature		Density	Thermal conductivity	
	t	T	ρ	λ	
	°C	°K	kg/m³	kcal/hm K	W/m K
Granite	20	293.15	2600 . . . 2900	2.50 . . . 3.50	2.908 . . . 4.071
Gravel, as filling material	20	293.15	1500 . . . 1800	0.80	0.930
Gypsum (plaster)	20	293.15	800	0.34	0.395
	20	293.15	1000	0.44	0.512
	20	293.15	1200	0.57	0.663
Light building sheets of mineralized wood fiber (Heraklite, Tekton, and others)	20	293.15	200	0.053	0.062
	20	293.15	400	0.071	0.083
	20	293.15	600	0.110	0.128
Lime	20	293.15	—	0.106	0.123
Limestone, amorphous	20	293.15	2550	1.05	1.221
Limestone, calcium carbonate	0	273.15		1.95	2.268
	20	293.15	2650	1.9	2.210
	100	373.15		1.65	1.919
Linoleum	20	293.15	535	0.07	0.081
	20	293.15	1180	0.16	0.186
Magnezite	1000	1273.15	—	1.42	1.651
Marble	20	293.15	2500 . . . 2800	1.80 . . . 3.00	2.093 . . . 3.489
Metallurgical brick	20	293.15	1600	0.56	0.651
	20	293.15	1800	0.65	0.756
	20	293.15	2000	0.74	0.861
	20	293.15	2200	0.87	1.012
Mortar (plaster)	20	293.15	—	0.80	0.930
	20	293.15	1600	0.57	0.663
	20	293.15	1800	0.74	0.861
	20	293.15	2000	0.92	1.070
	20	293.15	2200	1.2	1.396
— with bricks	20	293.15	1600 . . . 1800	0.60 . . . 0.80	0.698 . . . 0.930
— with light concrete blocks	20	293.15	1600 . . . 1800	0.80 . . . 1.00	0.930 . . . 1.163
Natural stone, dense	20	293.15	—	2.50	2.908
Natural stone, porous	20	293.15	—	1.50	1.745
Onyx	20	293.15	—	2.0	2.326
Paving, flagstone	20	293.15	—	0.90	1.047
Paving plates	20	293.15	—	0.08 . . . 0.12	0.093 . . . 0.140
Plaster	20	293.15	1600	0.54	0.628
	20	293.15	1800	0.70	0.814
	20	293.15	2000	0.87	1.012
	20	293.15	2200	1.1	1.279
— inside	20	293.15	1600 . . . 1800	0.60 . . . 0.80	0.698 . . . 0.930
— outside	20	293.15		0.80 . . . 1.00	0.930 . . . 1.163
— supported, inside	20	293.15		0.60	0.698
— supported, outside	20	293.15		0.75	0.872
— unsupported, inside	20	293.15		0.40	0.465
— unsupported, outside	20	293.15		0.60	0.698
Plaster blocks	20	293.15	800	0.27	0.314
— for inner walls	20	293.15	—	0.25	0.291
— light	20	293.15	—	0.15	0.174
— for roofing	20	293.15	—	0.30	0.349
Porphyry, cross grain	20	293.15	2600	1.2	1.396
— parallel to grain	20	293.15		2.0	2.326
Pumice gravel or sand, as filling material	20	293.15	600	0.28	0.326

1 kcal/h = 1.163 W

Table 17-4 Thermal Conductivities λ of Building Materials (Continued)

Building material	Temperature		Density	Thermal conductivity	
	t	T	ρ	λ	
	°C	°K	kg/m³	kcal/hm K	W/m K
Pumice stone	20	293.15	600	0.16 ... 0.27	0.186 ... 0.314
	20	293.15	800	0.23 ... 0.35	0.267 ... 0.407
	20	293.15	1000	0.30 ... 0.40	0.349 ... 0.465
	20	293.15	1400	0.50 ... 0.57	0.582 ... 0.663
Quarry stone masonry	20	293.15	—	1.3 ... 2.1	1.512 ... 2.442
Quartzite	20	293.15	2800	5.2	6.048
Rabitz wall, concrete	20	293.15	—	0.50	0.582
Rabitz wall, gypsum	20	293.15	—	0.25	0.291
Sand	0	273.15	1800 ... 2000	1.50	1.745
— average value, mean value	20	293.15	1500 ... 1800	0.80	0.930
— dry, heaped	20	293.15	—	0.5	0.582
— grown soil	20	293.15	—	2.0	2.326
— moist	20	293.15	1500	0.28	0.326
	20	293.15	1640	0.97	1.128
Sand, bone dry, normally impure	20	293.15		0.28	0.326
— 10% moisture, normally impure	20	293.15		0.83	0.965
— 20% moisture, normally impure	20	293.15		1.14	1.326
— saturated with moisture, normally impure	20	293.15	—	1.62	1.884
Sandlime	20	293.15	400	0.09	0.105
	20	293.15	600	0.125	0.145
	20	293.15	800	0.17	0.198
	20	293.15	1000	0.235	0.273
	20	293.15	1200	0.29	0.337
	20	293.15	1600	0.85	0.989
	20	293.15	1800	1.00	1.163
	20	293.15	2000	1.20	1.396
Sandlime brick	20	293.15	1600	0.70	0.814
— inside	20	293.15		0.80	0.930
— outside	20	293.15		0.90	1.047
Sandstone	20	293.15	2200 ... 2500	1.10 ... 1.80	1.279 ... 2.093
— dry	20	293.15		1.44	1.675
— moist	20	293.15	2250	1.11	1.291
Sawdust	0	273.15	215	0.060	0.070
	30	303.15		0.062	0.072
Sawdust, air-dried	20	293.15	190 ... 215	0.05 ... 0.06	0.058 ... 0.070
Sawdust, as filling material	20	293.15	190 ... 215	0.10	0.116
Sea sand, bone dry	20	293.15	1600	0.27	0.314
— 10% moisture	20	293.15		1.07	1.244
— 20% moisture	20	293.15		1.51	1.756
— saturated with moisture	20	293.15		2.10	2.442
Shell lime	20	293.15	2680	2.10	2.442
— 10% vol. moisture	20	293.15	2680	0.827	0.962
— 20% vol. moisture	20	293.15		1.080	1.256
— 30% vol. moisture	20	293.15		1.260	1.465
Silica brick	100	373.15	1800 ... 2200	0.95	1.105
	500	773.15		0.90 ... 1.10	1.047 ... 1.279
	1000	1273.15		0.95 ... 1.20	1.105 ... 1.396
Slag	20	293.15	—	0.16	0.186
— blast furnace slag	20	293.15	800	0.21 ... 0.33	0.244 ... 0.384
	20	293.15	1000	0.27 ... 0.38	0.314 ... 0.442
	20	293.15	1200	0.32 ... 0.43	0.372 ... 0.500
	20	293.15	1400	0.40 ... 0.49	0.465 ... 0.570

1 kcal/h = 1.163 W

Table 17-5 Thermal Conductivities λ of Building Materials (Continued)

Building material	Temperature		Density	Thermal conductivity	
	t	T	ρ	λ	
	°C	°K	kg/m³	kcal/hm K	W/m K
Slag					
— blast furnace slag, as filling material	20	293.15	1100 . . . 1300	0.50 . . . 0.70	0.582 . . . 0.814
— boiler slag	20	293.15	700 . . . 750	0.28	0.326
— boiler slag, as filling material	20	293.15	700 . . . 750	0.28	0.326
— slag concrete (blocks), masonry	20	293.15	300 . . . 400	0.19	0.221
Slate	20	293.15	–	1.20	1.396
— cross grain	20 100	293.15 373.15	2700	1.30 . . . 1.70 1.70	1.512 . . . 1.977 1.977
— parallel to grain	20	293.15	2700	2.00 . . . 2.90	2.326 . . . 3.373
Stone, gabbro	20	293.15	–	2.20	2.559
Stoneware, ceramics	20	293.15	2200 . . . 2500	0.90 . . . 1.35	1.047 . . . 1.570
Tar paper, roofing — hard	20 20	293.15 293.15	1000 . . . 1200 790	0.12 . . . 0.30 0.13	0.140 . . . 0.349 0.151
Wood — ash — ash ⊥ cross grain — ash ‖ with grain	25 25 25	298.15 298.15 298.15	740	0.14 0.15 0.26	0.163 0.174 0.302
— beech, oak ⊥ cross grain — beech, oak ‖ with grain	20 20	293.15 293.15	700 . . . 900	0.18 . . . 0.23 0.30 . . . 0.32	0.209 . . . 0.267 0.349 . . . 0.372
— birch, cross-grain	25	298.15	680	0.115	0.134
— boxwood	17	290.15	900	0.128	0.149
— dry, inside — dry, outside	20 20	293.15 293.15	– –	0.12 0.18	0.140 0.209
— hardwood	20	293.15	1200 . . . 1400	0.29	0.337
— larch, cross-grain	25	298.15	620	0.12	0.140
— light, balsa, cross-grain	20 25	293.15 298.15	200 . . . 300 100 . . . 200	0.07 . . . 0.09 0.040 . . . 0.057	0.081 . . . 0.105 0.047 . . . 0.066
— mahogany — mahogany — mahogany ⊥ — mahogany ‖	17 25 25 25	290.15 298.15 298.15 298.15	550 700	0.183 0.13 0.13 0.27	0.213 0.151 0.151 0.314
— maple ⊥ — maple ‖	30 30	303.15 303.15	710	0.136 0.360	0.158 0.419
— oak	17	290.15	650	0.209	0.243
— pine, fir, spruce — pine, fir, spruce, with the grain	20 20	293.15 293.15	400 . . . 600	0.11 . . . 0.16 0.24	0.128 . . . 0.186 0.279
— plywood	0 20	273.15 293.15	588	0.094 0.098	0.109 0.114
— teak — teak ⊥	25 25	298.15 298.15	720	0.12 0.14	0.140 0.163
— veneer	0	273.15	600	0.13	0.151
— walnut	70	343.15	700	0.23	0.267
— wood fibre sheets (Celotex, Kapok and others)	20 20 20	293.15 293.15 293.15	200 337 346	0.040 0.064 0.056	0.047 0.074 0.065
Wood cement	20	293.15	–	0.15	0.174
Wood felt	20	293.15	330	0.045	0.052
Wood shavings	30	303.15	140	0.050	0.058

1 kcal/h = 1.163 W

Table 18-1 Thermal Conductivities λ of Miscellaneous Solids

Material	Temperature		Density	Thermal conductivity	
	t	T	ρ	λ	
	°C	°K	kg/m³	kcal/hm K	W/m K
Aluminum oxide, pulverized	20	293.15	—	0.58	0.675
Aluminum oxide, smelted	20	293.15	—	2.85	3.314
Amber	20	293.15	1050 . . . 1100	0.11	0.128
Aniline resin	20	293.15	1210	0.23	0.267
Bakelite	0	273.15		0.193	0.244
	20	293.15	1270	0.200	0.233
	100	373.15		0.230	0.267
Bauxite	600	873.15	—	0.48	0.558
Boiler incrustation:					
— rich in gypsum	20	293.15	2000 . . . 2700	0.60 . . . 2.00	0.698 . . . 2.326
— rich in lime	20	293.15	1000 . . . 2500	0.13 . . . 2.00	0.151 . . . 2.326
— rich in silicates	20	293.15	300 . . . 1200	0.07 . . . 0.20	0.081 . . . 0.233
Carbamide resin	20	293.15	1500	0.30	0.349
Carbon filament	1500	1773.15	—	7.30	8.490
Cardboard	20	293.15	—	0.12 . . . 0.30	0.140 . . . 0.349
Cellon	20	293.15	1300 . . . 1400	0.18 . . . 0.22	0.209 . . . 0.256
Celluloid	20	293.15	1400	0.190	0.221
Chalk	50	323.15	2000	0.80	0.930
Coal:					
— amorphous	20	293.15	—	1.70	1.977
— anthracite	30	303.15	1370	0.205	0.238
— channel	30	303.15	1270	0.168 . . . 0.181	0.195 . . . 0.211
— charcoal	20	293.15	185 . . . 215	0.035 . . . 0.056	0.041 . . . 0.065
— gas	30	303.15	1260	0.187	0.217
— gas, long flame	30	303.15	1280	0.200	0.233
— lignite (brown) 47.6% H_2O	30	303.15	960	0.283	0.329
— lignite, dry, 12.1% H_2O	30	303.15	920	0.142	0.165
— lignite, dry, 3.4% H_2O	30	303.15	965	0.133	0.155
— non-baking	30	303.15	1280	0.182	0.212
— pit	20	293.15	1200 . . . 1350	0.21 . . . 0.23	0.244 . . . 0.267
Coal dust	20	293.15	730	0.10	0.116
Coke	20	293.15	1350	0.14	0.163
Coke	100	373.15	1400	2.5 . . . 3.0	2.908 . . . 3.489
— blast furnace	30	303.15	925	0.834	0.970
— foundry	30	303.15	950	1.04	1.210
— gas	30	303.15	930	0.620	0.721
— low temperature	30	303.15	680	0.130	0.151
Coke dust	20	293.15	1000	0.13	0.151
Compression molding compound, with inorganic filling material	20	293.15	1700 . . . 1900	0.50 . . . 0.80	0.582 . . . 0.930
Compression molding compound, with organic filling material	20	293.15	1310 . . . 1460	0.23 . . . 0.32	0.267 . . . 0.372
Earth, dry	20	293.15	—	0.115	0.134
Ebonite	20	293.15	1200	0.135 . . . 0.15	0.157 . . . 0.174
Fat	20	293.15	—	0.15	0.174
Feathers	20	293.15	109	0.065	0.076
Fiber (plastic)	20	293.15	—	0.20 . . . 0.30	0.233 . . . 0.349

1 kcal/h = 1.163 W

Table 18-2 Thermal Conductivities λ of Miscellaneous Solids (*Continued*)

Material	Temperature		Density	Thermal conductivity	
	t	T	ρ	λ	
	°C	°K	kg/m³	kcal/hm K	W/m K
Glass:					
— common, window	20	293.15	2400 . . . 3200	0.50 . . . 0.90	0.582 . . . 1.047
— crown	20	293.15	2300 . . . 2700	0.90	1.047
— Jena 16 III	20	293.15	2590	0.83	0.965
— lead	20	293.15	2600 . . . 4200	0.66 . . . 0.77	0.768 . . . 0.896
— mirror	20	293.15	2550	0.69	0.802
— plexi	20	293.15	—	0.16	0.186
— quartz	20	293.15	2600 . . . 4700	0.66	0.768
Graphite, strong (electr.)	20	293.15	—	10 . . . 150	11.630 . . . 174.450
Graphite powder	20	293.15	700	1.02	1.186
Guttapercha	20	293.15	—	0.172	0.200
Horn, artificial	20	293.15	1300 . . . 1400	0.14	0.163
Ice	— 100	173.15	928	3.00	3.489
	— 50	223.15	924	2.39	2.780
	— 20	253.15	920	2.10	2.442
	0	273.15	917	1.9	2.210
Igelite	20	293.15	1390	0.13	0.151
Ivory	20	293.15	1800 . . . 1900	0.40 . . . 0.50	0.465 . . . 0.582
Kitchen salt, crystals	20	293.15	—	6.00	6.978
Lampblack	40	313.15	165	0.06 . . . 0.10	0.070 . . . 0.116
Lampblack, coal dust	20	293.15	100	0.025	0.029
	20	293.15	200	0.026	0.030
	20	293.15	400	0.035	0.041
	20	293.15	600	0.050	0.058
	20	293.15	800	0.067	0.078
	20	293.15	1000	0.086	0.010
	20	293.15	1200	0.106	0.123
Lava (volcanic)	20	293.15	—	0.73	0.849
Leather	20	293.15	850 . . . 1000	0.12 . . . 0.15	0.140 . . . 0.174
— excised	20	293.15		0.18	0.209
— fresh	20	293.15		0.18 . . . 0.36	0.209 . . . 0.419
Mica	20	293.15	2600 . . . 3200	0.40 . . . 0.50	0.465 . . . 0.582
	20	293.15	2900	0.45	0.523
Micanite	20	293.15	2480	0.205	0.238
— generally	20	293.15	—	0.18 . . . 0.35	0.209 . . . 0.407
Mipolan	20	293.15	1340	0.18	0.209
Paper	20	293.15	700	0.12	0.140
— hard	20	293.15	790	0.13	0.151
	20	293.15	1000	0.13	0.151
	20	293.15	1300	0.179	0.208
Paraffin	20	293.15	860 . . . 930	0.21 . . . 0.25	0.244 . . . 0.291
Phenolic resin	20	293.15	1320	0.225	0.262
Plexiglass	20	293.15	1180	0.168	0.195
Polystyrene	20	293.15	1050	0.135	0.157
Porcelain, Berlin	20	293.15	2290	0.90 . . . 1.10	1.047 . . . 1.279
— generally	20	293.15	2200 . . . 2500	0.70 . . . 1.60	0.814 . . . 1.861

1 kcal/h = 1.163 W

Table 18-3 Thermal Conductivities λ of Miscellaneous Solids (*Continued*)

Material	Temperature		Density	Thermal conductivity	
	t	T	ρ	λ	
	°C	°K	kg/m³	kcal/hm K	W/m K
Press span, pressboard	20	293.15	1350	0.21 ... 0.24	0.244 ... 0.279
Pumice stone powder	20	293.15	600	0.11	0.128
	20	293.15	800	0.13	0.151
	20	293.15	1000	0.155	0.180
	20	293.15	1200	0.185	0.215
	20	293.15	1400	0.22	0.256
Rubber, Buna	20	293.15	1150	0.20	0.233
	20	293.15	1250	0.40	0.465
— crepe	20	293.15	50	0.034	0.040
	20	293.15	100	0.032	0.037
	20	293.15	200	0.038	0.044
	20	293.15	300	0.05	0.058
	20	293.15	400	0.06	0.070
	20	293.15	500	0.08	0.093
— hard (Ebonite)	20	293.15	1150	0.14	0.163
— hard, normal	— 200	73.15		0.117	0.136
	— 100	173.15		0.129	0.150
	0	273.15	1200	0.135	0.157
	100	373.15		0.138	0.160
— natural	20	293.15	1050	0.14	0.163
	20	293.15	1150	0.24	0.279
— Perbunan	20	293.15	1250	0.25	0.291
	20	293.15	1350	0.38	0.442
— spongy	20	293.15	224	0.047	0.055
— Thiokol	20	293.15	1650	0.25	0.291
— vulcanized, soft: 40% caoutchouc	20	293.15	1100	0.20	0.233
80% caoutchouc	20	293.15		0.13	0.151
100% caoutchouc	20	293.15		0.11	0.128
Rubber powder	20	293.15	134	0.043	0.050
Shellac	20	293.15	—	0.21	0.244
Snow, frost	0	273.15	150	0.10	0.116
	0	273.15	300	0.20	0.233
	0	273.15	500	0.40	0.465
	0	273.15	800	1.10	1.279
Soapstone	20	293.15	2850	2.80	3.256
Stoneware, majolica, semi-porcelain	20	293.15	2100 ... 2400	0.90 ... 1.40	1.047 ... 1.628
Sugar	0	273.15	1600	0.50	0.582
Sulfur, rhombic	20	293.15	—	0.23	0.267
Textile, hard	20	293.15	1310 ... 1330	0.28 ... 0.30	0.326 ... 0.349
Tuff stone	50	323.15	1550 ... 2270	0.54 ... 1.44	0.628 ... 1.675
Vaseline	20	293.15	—	0.15	0.174
Vinidur	20	293.15	1350	0.13	0.151
Vulcano fibre	20	293.15	1100 ... 1450	0.28 ... 0.30	0.326 ... 0.349
Wax, bees'	20	293.15	—	0.033	0.038
Xylolite	20	293.15	715	0.12	0.140

1 kcal/h = 1.163 W

Table 19-1 Thermal Conductivities λ of Some Insulating Materials

Material	Temperature t °C	Temperature T °K	Density ρ kg/m³	Thermal conductivity λ kcal/hm K	Thermal conductivity λ W/m K
Alfol (aluminum foil)	0	273.15		0.026	0.030
	20	293.15	3.6	0.028	0.033
	300	573.15		0.048	0.056
— (aluminum foil)	20	293.15	3.6	0.040	0.047
Aluminum wool	20	293.15	40	0.08	0.093
Asbestos	0	273.15	383	0.096	0.112
	50	323.15		0.099	0.115
	100	373.15		0.102	0.119
— fibrous (asbestos wool)	—200	73.15		0.072	0.084
	—150	123.15		0.101	0.117
	—100	173.15		0.118	0.137
	— 50	223.15		0.128	0.149
	0	273.15	470	0.132	0.154
	20	293.15		0.134	0.156
	100	373.15		0.140	0.163
— fibrous	0	273.15	580	0.172	0.200
	20	293.15		0.174	0.202
	100	373.15		0.182	0.212
	200	473.15		0.190	0.221
— fibrous	—200	73.15		0.134	0.156
	—100	173.15		0.190	0.221
	0	273.15	700	0.200	0.233
	20	293.15		0.202	0.235
	100	373.15		0.210	0.244
Asbestos cotton	25	298.15	140	0.043	0.050
— wool	20	293.15	50	0.050	0.058
	20	293.15	100	0.050	0.058
	20	293.15	300	0.080	0.093
	20	293.15	500	0.138	0.160
	20	293.15	600	0.172	0.200
Asbestos felt (soft, flexible)	20	293.15	420	0.073	0.085
— paper	20	293.15	500	0.06	0.070
	20	293.15	1000	0.13	0.151
Asbestos plates	20	293.15	2000	0.60	0.698
Boiler slag	20	293.15	750	0.28	0.326
Cellular plastics	20	293.15	15	0.030	0.035
Coal slag	0	273.15	700	0.13	0.151
	20	293.15		0.14	0.163
— blast furnace slag	50	323.15	360	0.095	0.110
Copper file dust	20	293.15	3600	0.35	0.41
Cork, boards	20	293.15	150	0.036	0.042
	20	293.15	200	0.041	0.048
	20	293.15	300	0.051	0.059
— expanded (boards)	0	273.15	120	0.0306	0.036
	20	293.15		0.0324	0.038
	50	323.15		0.035	0.041
— expanded, granules, 3 mm	—200	73.15	45	0.008	0.009
	—100	173.15		0.018	0.021
	0	273.15		0.029	0.034
	100	373.15		0.040	0.047
— granulated	20	293.15	50	0.026	0.030
	20	293.15	100	0.030	0.035
	20	293.15	150	0.035	0.041
	20	293.15	200	0.039	0.045
	20	293.15	250	0.044	0.051
	20	293.15	300	0.048	0.056
	20	293.15	350	0.052	0.060

1 kcal/h = 1.163 W

Table 19-2 Thermal Conductivities λ of Some Insulating Materials (Continued)

Material	Temperature t °C	Temperature T °K	Density ρ kg/m³	Thermal conductivity λ kcal/hm K	Thermal conductivity λ W/m K
Cork					
— impregnated (boards)	0	273.15	155	0.035	0.041
	20	293.15		0.037	0.043
	50	323.15		0.039	0.045
— normal, granules, 1 . . . 3 mm	0	273.15	150	0.035	0.041
	100	373.15		0.046	0.053
	200	473.15		0.057	0.066
— roughly granulated, 5 mm	0	273.15	85	0.041	0.048
	100	373.15		0.054	0.063
Cotton	−200	73.15	81	0.028	0.033
	−100	173.15		0.038	0.044
	0	273.15		0.048	0.056
	20	293.15		0.050	0.058
	100	373.15		0.058	0.067
— knitted	20	293.15	330	0.060	0.070
— surgical wool	20	293.15	10	0.035	0.041
	20	293.15	40	0.031	0.036
— woven	20	293.15	245	0.066	0.077
Diatomaceous earth (Diatomite)	200	473.15	466	0.108	0.126
	200	473.15	605	0.147	0.171
	200	473.15	790	0.159	0.185
Felt, hair	20	293.15	270	0.03 . . . 0.07	0.035 . . . 0.081
— rag	20	293.15	200	0.035	0.041
	20	293.15	600	0.075 . . . 0.080	0.087 . . . 0.093
Flax	20	293.15	160	0.047	0.055
Flax	20	293.15	19	0.044	0.051
	20	293.15	25	0.038	0.044
	20	293.15	50	0.034	0.040
	20	293.15	100	0.035	0.041
	20	293.15	150	0.037	0.043
	20	293.15	200	0.040	0.047
Flax, dried, fibers parallel with warm air stream	32	305.15	80	0.0662	0.077
	32	305.15	154	0.1032	0.120
— dried, fibers perpendicular to warm air stream	32	305.15	80	0.0294	0.034
	32	305.15	154	0.0324	0.038
Furnace dust	20	293.15	300	0.040	0.047
	20	293.15	400	0.047	0.055
	20	293.15	500	0.053	0.062
	20	293.15	600	0.060	0.070
	20	293.15	700	0.066	0.077
Glass fiber	0	273.15	220	0.03	0.035
	50	323.15		0.037	0.043
	100	373.15		0.043	0.050
	200	473.15		0.057	0.066
Glass fiber mat	20	293.15	100	0.033	0.038
	20	293.15	200	0.041	0.048
Glass, perpendicular to the warm air stream	0	273.15	186	0.030	0.035
	50	323.15		0.038	0.044
	100	373.15		0.047	0.055
	200	473.15		0.068	0.079
	300	573.15		0.092	0.107
Glass wool	20	293.15	50	0.032	0.037
	20	293.15	100	0.031	0.036
	20	293.15	200	0.034	0.040
	100	373.15	200	0.045	0.052
	300	573.15	200	0.090	0.105
	20	293.15	300	0.037	0.043
	20	293.15	400	0.047	0.055

1 kcal/h = 1.163 W

Table 19-3 Thermal Conductivities λ of Some Insulating Materials (*Continued*)

Material	Temperature		Density	Thermal conductivity	
	t	T	ρ	λ	
	°C	°K	kg/m³	kcal/hm K	W/m K
Gravel, as filling material	20	293.15	1500 . . . 1800	0.80	0.930
Hair	20	293.15	90	0.036	0.042
— animal	20	293.15	176	0.032	0.037
— horse	20	293.15	172	0.045	0.052
	60	333.15		0.045	0.052
Hemp, dry	32	305.15	43	0.065	0.076
— manilla	20	293.15	45	0.042	0.049
Igelite	20	293.15	1390	0.13	0.151
Iporka	0	273.15	15	0.027	0.031
	50	323.15		0.037	0.043
	100	373.15		0.047	0.055
Jute	20	293.15	10	0.052	0.060
	32	305.15	15	0.0457	0.053
	20	293.15	25	0.038	0.044
	20	293.15	50	0.031	0.036
	20	293.15	100	0.032	0.037
	20	293.15	200	0.035	0.041
	20	293.15	300	0.040	0.047
Kapok	20	293.15	5	0.031	0.036
	20	293.15	25	0.030	0.035
	20	293.15	50	0.033	0.038
	20	293.15	100	0.037	0.043
	20	293.15	150	0.041	0.048
Kieselguhr, powdered	−200	73.15	50	0.011	0.013
	0	273.15		0.030	0.035
	100	373.15		0.042	0.049
	0	273.15	200	0.036	0.042
	100	373.15		0.044	0.051
	300	573.15		0.060	0.070
	500	773.15		0.076	0.088
	700	973.15		0.093	0.108
	0	273.15	250	0.047	0.055
	100	373.15		0.055	0.064
	300	573.15		0.071	0.083
	0	273.15	350	0.056	0.065
	100	373.15		0.064	0.074
	300	573.15		0.082	0.095
	500	773.15		0.100	0.116
— burnt	20	293.15	200	0.052	0.060
	20	293.15	300	0.059	0.069
	20	293.15	400	0.070	0.081
	20	293.15	500	0.084	0.098
	20	293.15	600	0.101	0.117
	20	293.15	700	0.120	0.140
	20	293.15	800	0.140	0.163
	20	293.15	900	0.160	0.186
	20	293.15	1000	0.182	0.212
	20	293.15	1200	0.23	0.267
	20	293.15	1400	0.29	0.337
	20	293.15	1600	0.37	0.430
	20	293.15	1800	0.47	0.547
	20	293.15	2000	0.90	0.698
— burnt in forms	0	273.15	350 . . . 700	0.062 . . . 0.113	0.072 . . . 0.131
	50	323.15		0.07 . . . 0.12	0.081 . . . 0.140
	100	373.15		0.076 . . . 0.127	0.088 . . . 0.148
	200	473.15		0.09 . . . 0.141	0.105 . . . 0.164
	300	573.15		0.104 . . . 0.15	0.121 . . . 0.174
	500	773.15		0.12 . . . 0.17	0.140 . . . 0.198

1 kcal/h = 1.163 W

Table 19-4 Thermal Conductivities λ of Some Insulating Materials (*Continued*)

Material	Temperature		Density	Thermal conductivity	
	t	T	ρ	λ	
	°C	°K	kg/m³	kcal/hm K	W/m K
Kieselguhr					
— and magnesia mass	100	373.15	200	0.047	0.055
	100	373.15	300	0.054	0.063
	100	373.15	400	0.063	0.073
	100	373.15	500	0.075	0.087
	100	373.15	600	0.091	0.106
	100	373.15	800	0.135	0.157
	100	373.15	1000	0.190	0.221
Kieselguhr mass	50	323.15	450 . . . 840	0.062 . . . 0.146	0.072 . . . 0.170
	100	373.15		0.064 . . . 0.148	0.074 . . . 0.172
	200	473.15		0.069 . . . 0.151	0.080 . . . 0.176
	300	573.15		0.074 . . . 0.155	0.086 . . . 0.180
Light building blocks in masonry, slag blocks, porous concrete blocks etc.	20	293.15	200	0.053	0.062
	20	293.15	300	0.062	0.072
	20	293.15	400	0.071	0.083
	20	293.15	500	0.090	0.105
	20	293.15	600	0.110	0.128
Light building blocks, mineralized wood wool like Heraklite etc.	20	293.15	600	0.35	0.407
	20	293.15	800	0.41	0.477
	20	293.15	1000	0.49	0.570
	20	293.15	1200	0.57	0.663
	20	293.15	1400	0.67	0.779
Linen	25	298.15	265	0.057	0.066
	25	298.15	590	0.060	0.070
Magnesia, compressed	20	293.15	800	0.052	0.605
— powdered	0	273.15	200	0.063	0.073
Magnesium	−200	73.15		0.018	0.021
	−100	173.15		0.025	0.029
	0	273.15	130	0.033	0.038
	20	293.15		0.035	0.041
	100	373.15		0.042	0.049
	20	293.15	250	0.047	0.055
	20	293.15	500	0.089	0.104
Magnesium carbonate	100	373.15	—	0.083	0.097
Magnesium slag	50	323.15	270	0.063	0.073
	100	373.15		0.066	0.077
	200	473.15		0.072	0.084
Peat, lumps	20	293.15	120	0.04	0.047
— pressed, dry	20	293.15	—	0.03 . . . 0.06	0.035 . . . 0.070
Peat dust	0	273.15	190	0.040	0.047
	20	293.15		0.041	0.048
	50	323.15		0.045	0.052
Peat moss	20	293.15	160	0.036	0.042
Peat plates	0	273.15	210	0.043	0.050
	20	293.15		0.045	0.052
	50	323.15		0.048	0.056
	20	293.15	200 . . . 400	0.04 . . . 0.08	0.047 . . . 0.093
Pine and juniper bark	20	293.15	342	0.069	0.080
Plywood	0	273.15	590	0.094	0.109
	20	293.15		0.098	0.114
Poresta	0	273.15	24	0.027	0.031
	10	283.15		0.029	0.034
	20	293.15		0.030	0.035

1 kcal/h = 1.163 W

Table 19-5 Thermal Conductivities λ of Some Insulating Materials (*Continued*)

Material	Temperature		Density	Thermal conductivity	
	t	T	ρ	λ	
	°C	°K	kg/m³	kcal/hm K	W/m K
Pumice gravel, as filling material	20	293.15	600	0.28	0.326
Pumice stone, natural	0	273.15	300 . . . 600	0.075 . . . 0.15	0.087 . . . 0.174
	20	293.15		0.079 . . . 0.16	0.092 . . . 0.186
	50	323.15		0.085 . . . 0.17	0.099 . . . 0.198
Sawdust	20	293.15	200	0.050	0.058
Sea grass	20	293.15	80	0.030	0.035
Silk, artificial	20	293.15	170	0.042	0.049
	20	293.15	300	0.036	0.042
	20	293.15	464	0.044	0.051
— fibrous	−200	73.15	58	0.011	0.013
	−150	123.15		0.014	0.016
	−100	173.15		0.019	0.022
	− 50	223.15		0.024	0.028
	0	273.15		0.029	0.034
	50	323.15		0.035	0.041
	100	373.15		0.041	0.048
— spun fiber	−200	73.15	100	0.021	0.024
	−100	173.15		0.032	0.037
	0	273.15		0.043	0.050
	100	373.15		0.052	0.060
— woven	30	303.15	—	0.04	0.047
Sisal hemp	20	293.15	109	0.033	0.038
Slag wool	20	293.15	200	0.04	0.047
Slag, wool, stone wool (Silan)	20	293.15	100	0.029	0.034
	20	293.15	200	0.034	0.040
	20	293.15	300	0.040	0.047
	20	293.15	400	0.047	0.055
	20	293.15	500	0.050	0.058
Steel wool	20	293.15	104	0.05	0.058
Straw	0	273.15	140	0.039	0.045
	20	293.15		0.043	0.050
Sugar cane	20	293.15	25	0.039	0.045
	20	293.15	50	0.034	0.040
	20	293.15	100	0.036	0.042
	20	293.15	150	0.040	0.047
	20	293.15	200	0.046	0.053
	20	293.15	250	0.052	0.060
Tree bark	20	293.15	337	0.064	0.074
Wood felt	20	293.15	330	0.045	0.052
	20	293.15	350	0.056	0.065
Wood fibre plates (Celotex, Kapak and others)	20	293.15	200	0.040	0.047
	20	293.15	300	0.044	0.051
	20	293.15	400	0.047	0.055
	20	293.15	500	0.055	0.064
	20	293.15	600	0.064	0.074
Wood shavings	20	293.15	150	0.050	0.058
	20	293.15	200	0.051	0.059
	20	293.15	250	0.053	0.062
	20	293.15	300	0.056	0.065
Wood shavings (stuffing material)	20	293.15	100 . . . 140	0.08	0.093

1 kcal/h = 1.163 W

Table 19-6 Thermal Conductivities λ of Some Insulating Materials (*Continued*)

Material	Temperature		Density	Thermal conductivity	
	t	T	ρ	λ	
	°C	°K	kg/m³	kcal/hm K	W/m K
Wool	20	293.15	50	0.033	0.038
	20	293.15	100	0.031	0.036
	20	293.15	150	0.031	0.036
	20	293.15	200	0.033	0.038
	20	293.15	250	0.035	0.041
	20	293.15	300	0.037	0.043
	20	293.15	350	0.040	0.047
	20	293.15	400	0.043	0.050
— slag	−200	73.15		0.009	0.010
	−100	173.15		0.017	0.020
	0	273.15	95	0.027	0.031
	20	293.15		0.029	0.034
	−200	73.15		0.010	0.012
	−100	173.15		0.018	0.021
	0	273.15	120	0.028	0.033
	20	293.15		0.030	0.035

1 kcal/h = 1.163 W

Table 20-1 Thermal Conductivities λ of Fire Bricks and Ceramic Bricks

Material	Basic constituents in %	Temperature		Density	Thermal conductivity	
		t	T	ρ	λ	
		°C	°K	kg/m³	kcal/hm K	W/m K
Carbon brick	89 C	0	273.15	1200	0.6	0.698
		200	473.15		0.75	0.872
		400	673.15		0.9	1.047
		600	873.15		1.1	1.279
		800	1073.15		1.25	1.454
		1000	1273.15		1.4	1.628
Carborundum	50 SiC	0	273.15	2200	5.0	5.815
		200	473.15		4.5	5.234
		400	673.15		4.0	4.652
		600	873.15		3.7	4.303
		800	1073.15		3.4	3.954
		1000	1273.15		3.1	3.605
		1200	1473.15		2.8	3.256
	75 SiC	0	273.15	2300	14.0	16.282
		200	473.15		11.2	13.026
		400	673.15		9.5	11.049
		600	873.15		8.5	9.886
		800	1073.15		7.6	8.839
		1000	1273.15		6.8	7.908
		1200	1473.15		6.2	7.211
	100 SiC	0	273.15	—	62	72.106
		200	473.15		42	48.846
		400	673.15		32	37.216
		600	873.15		25	29.075
		800	1073.15		20.5	23.412
		1000	1273.15		17	19.771
		1200	1473.15		14.5	16.864
Chamotte	50 . . . 75 SiO₂, 20 . . . 50 Al₂O₃	0	273.15	800	0.18	0.209
		200	473.15		0.21	0.244
		400	673.15		0.24	0.279
		600	873.15		0.27	0.314
		800	1073.15		0.30	0.349
		1000	1273.15		0.33	0.384
		1200	1473.15		0.36	0.419
		0	273.15	1000	0.25	0.291
		200	473.15		0.28	0.326
		400	673.15		0.31	0.361
		600	873.15		0.35	0.407
		800	1073.15		0.38	0.442
		1000	1273.15		0.42	0.488
		1200	1473.15		0.45	0.523
		0	273.15	1200	0.33	0.384
		200	473.15		0.35	0.407
		400	673.15		0.38	0.442
		600	873.15		0.41	0.477
		800	1073.15		0.45	0.523
		1000	1273.15		0.48	0.558
		1200	1473.15		0.52	0.605
		0	273.15	2000	0.92	1.070
		200	473.15		0.97	1.128
		400	673.15		1.03	1.198
		600	873.15		1.08	1.256
		800	1073.15		1.15	1.337
		1000	1273.15		1.21	1.401
		1200	1473.15		1.28	1.489

Table 20-2 Thermal Conductivities λ of Fire Bricks and Ceramic Bricks (*Continued*)

Material	Basic constituents in %	Temperature		Density	Thermal conductivity	
		t	T	ρ	λ	
		°C	°K	kg/m³	kcal/hm K	W/m K
Chamotte		0	273.15	2200	1.34	1.558
		200	473.15		1.42	1.651
		400	673.15		1.50	1.745
		600	873.15		1.58	1.838
		800	1073.15		1.67	1.942
		1000	1273.15		1.76	2.047
		1200	1473.15		1.85	2.152
Chromite	40 . . . 45 Cr_2O_3	0	273.15	2750	1.1	1.279
		200	473.15		1.2	1.396
		400	673.15		1.3	1.512
		600	873.15		1.4	1.628
		800	1073.15		1.4	1.628
		1000	1273.15		1.45	1.686
		1200	1473.15		1.45	1.686
Corundum	80 Al_2O_3	0	273.15	2700	2	2.326
		200	473.15		1.87	2.175
		400	673.15		1.8	2.093
		600	863.15		1.75	2.035
		800	1073.15		1.72	2.000
		1000	1273.15		1.70	1.977
		1200	1473.15		1.68	1.954
	100 Al_2O_3	0	273.15	3750	18	20.934
		200	473.15		9.2	10.700
		400	673.15		6.8	7.908
		600	873.15		5.6	6.513
		800	1073.15		4.9	5.699
		1000	1273.15		4.4	5.117
		1200	1473.15		4.1	4.768
Magnesite	50 MgO	0	273.15	2000	2.3	2.675
		200	473.15		2.2	2.559
		400	673.15		2.0	2.326
		600	873.15		1.8	2.093
		800	1072.15		1.6	1.861
		1000	1273.15		1.45	1.686
		1200	1473.15		1.3	1.512
	75 MgO	0	273.15	2600	4.3	5.001
		200	473.15		3.9	4.536
		400	673.15		3.5	4.071
		600	873.15		3.1	3.605
		800	1073.15		2.7	3.140
		1000	1273.15		2.4	2.791
		1200	1473.15		2.1	2.442
	100 MgO	0	273.15	3500	37	43.031
		200	473.15		22	25.586
		400	673.15		14	16.282
		600	873.15		10	11.630
		800	1073.15		7.5	8.723
		1000	1273.15		5.5	6.397
		1200	1473.15		4.5	5.234
Porcelain	55 Al_2O_3, 45 SiO_2	0	273.15	2350	0.8	0.930
		200	473.15		1.2	1.396
		400	673.15		1.5	1.745
		600	873.15		1.75	2.035
		800	1073.15		1.95	2.268
		1000	1273.15		2.1	2.442

Table 20-3 Thermal Conductivities λ of Fire Bricks and Ceramic Bricks (*Continued*)

Material	Basic constituents in %	Temperature		Density	Thermal conductivity	
		t	T	ρ	λ	
		°C	°K	kg/m³	kcal/hm K	W/m K
Silica	95 SiO_2	0	273.15	1900	0.92	1.070
		200	473.15		1.0	1.163
		400	673.15		1.1	1.279
		600	873.15		1.2	1.396
		800	1073.15		1.32	1.535
		1000	1273.15		1.45	1.686
		1200	1473.15		1.6	1.861
Silimanite	60 . . . 80 Al_2O_3	0	273.15	1000	0.15	0.174
		200	473.15		0.18	0.209
		400	673.15		0.20	0.233
		600	873.15		0.23	0.267
		800	1073.15		0.26	0.302
		1000	1273.15		0.29	0.337
		1200	1473.15		0.32	0.372
		0	273.15	1500	0.40	0.465
		200	473.15		0.42	0.488
		400	673.15		0.45	0.523
		600	873.15		0.48	0.558
		800	1073.15		0.51	0.593
		1000	1273.15		0.54	0.628
		1200	1473.15		0.57	0.663
		0	273.15	2000	0.95	1.105
		200	473.15		0.95	1.105
		400	673.15		0.95	1.105
		600	873.15		0.95	1.105
		800	1073.15		0.95	1.105
		1000	1273.15		0.95	1.105
		1200	1473.15		0.95	1.105
		0	273.15	2500	1.85	2.152
		200	473.15		1.76	2.047
		400	673.15		1.68	1.954
		600	873.15		1.62	1.884
		800	1073.15		1.57	1.826
		1000	1273.15		1.53	1.779
		1200	1473.15		1.50	1.745
Zircon bricks	62 ZrO_2	0	273.15	3600	2.1	2.442
		200	473.15		2.0	2.326
		400	673.15		1.9	2.210
		600	873.15		1.8	2.093
		800	1073.15		1.7	1.977
		1000	1273.15		1.65	1.919
		1200	1473.15		1.65	1.919

Table 21-1 Thermal Conductivity λ of Burnt Kieselguhr (Diatomaceous Earth)

Temperature t (°C)	Temperature T (°K)	Density ρ (kg/m³)	λ (kcal/hm K)	λ (W/m K)
0	273.15	200	0.050	0.058
50	323.15		0.056	0.065
100	373.15		0.062	0.072
200	473.15		0.078	0.091
300	573.15		0.097	0.113
400	673.15		0.116	0.135
500	773.15		0.134	0.156
600	873.15		0.152	0.177
700	973.15		0.171	0.199
0	273.15	300	0.057	0.066
50	323.15		0.063	0.073
100	373.15		0.069	0.080
200	473.15		0.082	0.095
300	573.15		0.096	0.112
400	673.15		0.110	0.128
500	773.15		0.124	0.144
600	873.15		0.137	0.159
700	973.15		0.150	0.174
0	273.15	400	0.068	0.079
50	323.15		0.074	0.086
100	373.15		0.081	0.094
200	473.15		0.094	0.109
300	573.15		0.107	0.124
400	673.15		0.119	0.138
500	773.15		0.130	0.151
600	873.15		0.139	0.162
700	973.15		0.147	0.171
0	273.15	500	0.082	0.095
50	323.15		0.090	0.105
100	373.15		0.098	0.114
200	473.15		0.111	0.129
300	573.15		0.123	0.143
400	673.15	500	0.134	0.156
500	773.15		0.144	0.167
600	873.15		0.153	0.178
700	973.15		0.160	0.186
0	273.15	600	0.098	0.114
50	323.15		0.107	0.124
100	373.15		0.116	0.135
200	474.15		0.130	0.151
300	573.15		0.143	0.166
400	673.15		0.155	0.180
500	773.15		0.164	0.191
600	873.15		0.172	0.200
700	973.15		0.177	0.206
0	273.15	700	0.116	0.135
50	323.15		0.126	0.147
100	373.15		0.137	0.159
200	473.15		0.152	0.177
300	573.15		0.166	0.193
400	673.15		0.178	0.207
500	773.15		0.186	0.216
600	873.15		0.192	0.223
700	973.15		0.197	0.229
0	273.15	800	0.136	0.158
50	323.15		0.148	0.172
100	373.15		0.159	0.185
200	473.15		0.176	0.205
300	573.15		0.190	0.221
400	673.15		0.201	0.234
500	773.15		0.209	0.243
600	873.15		0.216	0.251
700	973.15		0.221	0.257

Table 22-1 Thermal Conductivity λ of Lampblack

Temperature t (°C)	Temperature T (°K)	Density ρ (kg/m³)	λ (kcal/hm K)	λ (W/m K)
0	273.15	100	0.024	0.028
100	373.15		0.030	0.035
200	473.15		0.036	0.042
400	673.15		0.050	0.058
600	873.15		0.069	0.080
800	1073.15		0.093	0.108
1000	1273.15		0.120	0.140
0	273.15	200	0.025	0.029
100	373.15		0.032	0.037
200	473.15		0.039	0.045
400	673.15		0.055	0.064
600	873.15		0.075	0.087
800	1073.15		0.100	0.116
1000	1273.15		0.129	0.150
0	273.15	400	0.034	0.040
100	373.15		0.042	0.049
200	473.15		0.050	0.058
400	673.15		0.069	0.080
600	873.15		0.096	0.112
800	1073.15		0.125	0.145
1000	1273.15		0.157	0.183
0	273.15	600	0.049	0.057
100	373.15		0.058	0.067
200	473.15		0.068	0.079
400	673.15		0.092	0.107
600	873.15	600	0.121	0.141
800	1073.15		0.155	0.180
1000	1273.15		0.190	0.221
0	273.15	800	0.065	0.076
100	373.15		0.076	0.088
200	473.15		0.087	0.101
400	673.15		0.113	0.131
600	873.15		0.145	0.169
800	1073.15		0.185	0.215
1000	1273.15		0.230	0.267
0	273.15	1000	0.084	0.098
100	373.15		0.096	0.112
200	473.15		0.108	0.126
400	673.15		0.138	0.160
600	873.15		0.175	0.204
800	1073.15		0.220	0.256
1000	1273.15		0.270	0.340
0	273.15	1200	0.103	0.120
100	373.15		0.117	0.136
200	473.15		0.130	0.151
400	673.15		0.163	0.190
600	873.15		0.205	0.238
800	1073.15		0.255	0.297
1000	1273.15		0.31	0.361

1 kcal/h = 1.163 W

Table 23-1 Emissivities ϵ_n of Metal Surfaces

Metal	Chemical formula end composition, in %	State of surface	Temperature t °C	Temperature T °K	Emissivity coefficient ϵ_n [1]
Aluminum	Al	aluminium surfaced roofing	43	316.15	0.216
		calorized copper, heated to 600°C	200 600	473.15 873.15	0.180 0.190
		calorized steel, heated to 600°C	200 600	473.15 873.15	0.520 0.570
		oxidized to 600°C	200 600	473.15 873.15	0.110 0.190
		polished plate	23 225 575	296.15 498.15 848.15	0.040 0.039 0.057
		rolled, polished	170	443.15	0.039
		rough plate	25	298.15	0.070
Bismuth	Bi	bright	80	353.15	0.340
Brass	Cu-Zn	burnished	—	—	0.42
		hard rolled	22	295.15	0.06
		oxidized	338	611.15	0.22
		oxidized at 600°C	200 600	473.15 873.15	0.61 0.59
		polished	19 300	292.15 573.15	0.05 0.032
		rolled plate, rubbed with emery	22	295.15	0.20
		tarnished	56 338	329.15 611.15	0.202 0.221
		tube	—	—	0.208
Chromium	Cr	polished	150	423.15	0.058
Copper	Cu	black oxidized	20	293.15	0.780
		eletrolytic, carefully polished	80	353.15	0.018
		lightly tarnished	20	293.15	0.037
		molten	1075 1275	1348.15 1548.15	0.160 0.130
		oxidized	130	403.15	0.760
		oxidized by heating to 600°C	200 600	473.15 873.15	0.570 0.550
		polished	20 115	293.15 388.15	0.030 0.023
		scraped	20	293.15	0.070
		rolled	—	—	0.640
		tube	—	—	0.360
Cuprous oxide		—	800 1100	1073.15 1373.15	0.660 0.540

[1] ϵ_n (normal emissivity), emissivity coefficient for radiation normal to surface. Mean normal total emissivity (ϵ) can be calculated from: $\epsilon = 1.2 \, \epsilon_n$ for bright metal surfaces; $\epsilon = 0.95 \, \epsilon_n$ for other bodies with smooth surface; $\epsilon = 0.98 \, \epsilon_n$ for other rough surfaces.

Table 23-2 Emissivities ϵ_n of Metal Surfaces (*Continued*)

Metal	Chemical formula end composition, in %	State of surface	Temperature t °C	Temperature T °K	Emissivity coefficient ϵ_n [1]
Gold	Au	highly polished	225 625	498.15 898.15	0.018 0.035
		not polished	20	293.15	0.47
		polished	20 130 400	293.15 403.15 673.15	0.025 0.018 0.022
Iron	Fe	completely rusty	20	293.15	0.85
		electrolytic, highly polished	175 225	448.15 498.15	0.052 0.064
		etched, bright	150	423.15	0.128
		fire proof, oxidized	80	353.15	0.613
		ground, bright	20	293.15	0.24
		molten, dull oxidized	20 360	293.15 633.15	0.94 0.94
		oxide	500 1200	773.15 1473.15	0.85 0.89
		oxidized, smooth	125 525	398.15 798.15	0.78 0.82
		polished	425 1020	698.15 1293.15	0.144 0.377
		rolled	20	293.15	0.77
		rubbed with emery	20	293.15	0.242
		rusty, red	20	293.15	0.61
		smooth	900 1040	1173.15 1313.15	0.55 0.60
		unwrought	925 1115	1198.15 1388.15	0.87 0.95
Iron castings		casting skin	100	373.15	0.80
		molten	1330	1603.15	0.28
		oxidized at 600°C	200 600	473.15 873.15	0.64 0.78
		polished	200	473.15	0.21
		rough, strongly oxidized	40 250	313.15 523.15	0.95 0.95
		turned	22 830 990	295.15 1103.15 1263.15	0.435 0.60 0.70
Lead	Pb	gray oxidized	20	293.15	0.28
		oxidized at 200°C	200	473.15	0.63
		polished	130 230	403.15 503.15	0.056 0.074

[1] ϵ_n (normal emissivity), emissivity coefficient for radiation normal to surface. Mean normal total emissivity (ϵ) can be calculated from: $\epsilon = 1.2\,\epsilon_n$ for bright metal surfaces; $\epsilon = 0.95\,\epsilon_n$ for other bodies with smooth surface; $\epsilon = 0.98\,\epsilon_n$ for other rough surfaces.

Table 23-3 Emissivities ϵ_n of Metal Surfaces (Continued)

Metal	Chemical formula end composition, in %	State of surface	Temperature t °C	Temperature T °K	Emissivity coefficient ϵ_n [1]
Manganine		rolled, smooth	118	391.15	0.048
Mercury	Hg	—	20 100	273.15 373.15	0.09 0.12
Molybdenum	Mo	filament	725 2600	998.15 2873.15	0.096 0.292
Nickel	Ni	dull	20 100	293.15 373.15	0.111 0.041
		oxidized	100	373.15	0.41
		oxidized at 600°C	200 600	473.15 873.15	0.370 0.48
		(plated on iron sheet), polished	24	297.15	0.056
		plated on polished iron, polished	23	296.15	0.045
		plated on pickled iron, not polished	20	293.15	0.110
		polished	100 230 375	373.15 503.15 648.15	0.045 0.070 0.087
		wire	185 1000	458.15 1273.15	0.096 0.186
Nickel alloy (Chromenickel)	18% Cr, 8% Ni	after 24 hr heating at 525°C	225 525	498.15 798.15	0.62 0.73
	20% Ni, 25% Cr	brown, weathered	215 525	488.15 798.15	0.90 0.97
	Cr-Ni, 18 . . . 32% Ni, 55 . . . 68% Cr, 20% Zn	gray oxidized	20	293.15	0.262
	18% Cr, 8% Ni	lighly silvery, rough, brown after heating	215 490	488.15 763.15	0.44 0.36
	Ni-Cr	oxidized at 600°C	200 600	473.15 873.15	0.41 0.46
	60% Ni, 12% Cr	smooth, black, firm oxide coat from weathering	270 560	543.15 833.15	0.89 0.82
Nickel-chrome	Cr-Ni	—	52 1035	325.15 1308.15	0.64 0.76
Nickel oxide		—	650 1255	923.15 1528.15	0.59 0.86
Platinum	Pt	filament	25 1230	298.15 1503.15	0.036 0.192
		polished	225 625	498.15 898.15	0.054 0.104
		strip	925 1116	1198.15 1388.15	0.12 0.17
		wire	225 1375	498.15 1648.15	0.073 0.182

[1] ϵ_n (normal emissivity), emissivity coefficient for radiation normal to surface. Mean normal total emissivity (ϵ) can be calculated from: ϵ 1.2 ϵ_n for bright metal surfaces; $\epsilon = 0.95 \epsilon_n$ for other bodies with smooth surface; $\epsilon = 0.98 \epsilon_n$ for other rough surfaces.

Table 23-4 Emissivities ϵ_n of Metal Surfaces (*Continued*)

Metal	Chemical formula end composition, in %	State of surface	Temperature t °C	Temperature T °K	Emissivity coefficient ϵ_n[1]
Silumin, cast		polished	150	423.15	0.186
Silver	Ag	polished	20 38 370 630	293.15 311.15 643.15 903.15	0.025 0.0221 0.0312 0.0320
		polished, clean	225 625	498.15 898.15	0.0198 0.0342
Steel		dense shiny oxide layer	25	298.15	0.82
		ground sheet	940 1100	1 213.15 1373.15	0.520 0.610
		mild, molten	1600 1800	1873.15 2073.15	0.28 0.28
		oxidized at 600°C	200 600	473.15 873.15	0.79 0.79
		oxidized, rough	40 370	313.15 643.15	0.94 0.97
		pipe	0 200	273.15 473.15	0.745 0.800
		rolled sheet	20	293.15	0.057
		thick rough oxide layer	25	298.15	0.80
— tool		galvanized sheet	—	—	0.262
Steel, chrome		(sheet), oxidized	—	—	0.870
Steel casting		polished	770 1040	1043.15 1313.15	0.52 0.56
Tantalum	Ta	filament	1325 2525	1598.15 2798.15	0.193 0.31
Tin	Sn	bright	20	293.15	0.070
		tinned steel sheet	24	297.15	0.056 . . . 0.086
Tungsten	W	—	230 2230	503.15 2503.15	0.053 0.31
		filament	3300	3573.15	0.39
		filament, used	25 3300	298.15 3573.15	0.032 0.035
White metal		tinned, bright	— —	— —	0.056 0.086
Zinc	Zn	galvanized sheet iron, bright galvanized sheet iron, gray oxidized	28 24	301.15 297.15	0.228 0.276
		oxidized by heating at 400°C	400	673.15	0.11
		polished	230 325	503.15 598.15	0.045 0.053
		tarnished	20 50 280	293.15 323.15 553.15	0.25 0.21 0.21

[1] ϵ_n (normal emissivity), emissivity coefficient for radiation normal to surface. Mean normal total emissivity (ϵ) can be calculated from: $\epsilon = 1.2\,\epsilon_n$ for bright metal surfaces; $\epsilon = 0.95\,\epsilon_n$ for other bodies with smooth surface; $\epsilon = 0.98\,\epsilon_n$ for other rough surfaces.

Table 24-1 Emissivities ϵ_n of Nonmetal Surfaces

Substance	State of surface	Temperature t °C	Temperature T °K	Emissivity coefficient ϵ_n [1]
Asbestos board	–	24	297.15	0.96
– paper	–	40 / 370	313.15 / 643.15	0.93 / 0.95
– slate	–	20	293.15	0.96
Brick	red, rough	20	293.15	0.93
Carbon filament	–	1040 / 1405	1313.15 / 1678.15	0.53 / 0.53
Chamotte	–	1200	1473.15	0.60
Chamotte stone	glazed	1000 / 1220	1273.15 / 1493.15	0.75 / 0.60
Clay	burnt	70	343.15	0.91
Coal, pure	ground	125 / 625	398.15 / 898.15	0.81 / 0.79
Corundum powder on paper	rough	80	353.15	0.855
Cotton	–	–	–	0.78
Dinas brick	glazed, rough	1100	1373.15	0.85
	unglazed, rough	1000	1273.15	0.80
Enamel white	fused on iron	20	293.15	0.90
Glass	smooth	20 / 90	293.15 / 363.15	0.90 / 0.94
Granite	ground	–	–	0.427
Gypsum	–	0 / 200	273.15 / 473.15	0.90 / 0.90
Ice	smooth, water	0	273.15	0.966
	rough	0	273.15	0.985
Lampblack	–	0 / 370	273.15 / 643.15	0.945 / 0.945
Linoleum	–	20	293.15	0.885
Magnezite stones	–	1390	1663.15	0.39
Marble	light gray, polished	22	295.15	0.93
	ground smooth	–	–	0.545

Table 24-2 Emissivities ϵ_n of Nonmetal Surfaces (Continued)

Substance	State of surface	Temperature t °C	Temperature T °K	Emissivity coefficient ϵ_n [1]
Masonry	plastered	0 / 200	273.15 / 473.15	0.93 / 0.93
Paper	–	20 / 95	293.15 / 368.15	0.80 / 0.92
Plaster, lime	white, rough	20 / 200	293.15 / 473.15	0.93 / 0.93
Porcelain	glazed	20	293.15	0.93
Quartz	fused, rough	20	293.15	0.93
Refractory materials	high-emissive	500 / 600 / 1000	773.15 / 873.15 / 1273.15	0.80 / 0.85 / 0.90
	low-emissive	500 / 600 / 1000	773.15 / 873.15 / 1273.15	0.65 / 0.70 / 0.75
Roofing cardboard	–	20	293.15	0.93
Rubber, soft	gray	24	297.15	0.86
– hard	black, rough	24	297.15	0.95
Sandstone	ground smooth	–	–	0.576
Silicate stone	rough	1000 / 1220	1273.15 / 1493.15	0.80 / 0.66
Silimanite stone	–	1390	1663.15	0.29
Silk cloth	–	20	293.15	0.77
Slate	ground	60 / 200	333.15 / 473.15	0.665 / 0.665
Wood	planed	20 / 70	293.15 / 343.15	0.90 / 0.925
– beech	planed	70	343.15	0.935
– oak	planed	21	294.15	0.885
Wool cloth	–	20	293.15	0.75

[1] ϵ_n (normal emissivity), emissivity coefficient for radiation normal to surface. Mean normal total emissivity (ϵ) can be calculated from: $\epsilon = 1.2\,\epsilon_n$ for bright metal surfaces; $\epsilon = 0.95\,\epsilon_n$ for other smooth surfaces; $\epsilon = 0.98\,\epsilon_n$ for rough surfaces.

Table 25-1   Emissivities of ϵ_n of Paints and Coatings

| Substance | State of surface | Temperature | | Emissivity coefficient |
| | | t | T | $\epsilon_n{}^1$ |
		°C	°K	
Aluminum bronze	—	100	373.15	0.20 . . . 0.40
Aluminum enamel	rough	20	293.15	0.39
Aluminum paint, after heating to 325°C	—	150 . . . 315	423.15 . . . 588.15	0.35
Aluminum paints, different	rough, smooth	100	373.15	0.27 . . . 0.67
Bakelite enamel	—	80	353.15	0.935
Enamel, white	—	40 95	313.15 368.15	0.80 0.95
— alcohol	black, bright	25	298.15	0.82
— black	bright	25	298.15	0.876
	dull	40 95	313.15 368.15	0.96 0.98
— for irradiators (heating bodies)	—	100	373.15	0.925
Oil	thick layer	—	—	0.82
Oil coating	smooth	—	—	0.78
Oil paint	—	0 200	273.15 473.15	0.885 0.885
Read lead primer	—	20 100	293.15 373.15	0.93 0.93
Shellack, black	bright	21	294.15	0.82
	dull	75 145	348.15 418.15	0.91 0.91
Water	—	—	—	0.8
Water glass-bound lampblack	—	20 100	293.15 373.15	0.96 0.96
White enamel, melted	white, rough	20	293.15	0.90

[1] ϵ_n (normal emissivity), emissivity coefficient for radiation normal to surface. Mean normal total emissivity (ϵ) can be calculated from: $\epsilon = 1.2 \, \epsilon_n$ for bright metal surfaces; $\epsilon = 0.95 \, \epsilon_n$ for other smooth surfaces; $\epsilon = 0.98 \, \epsilon_n$ for rough surfaces.

Table 26-1 Heating Values of Solid Fuels[1]

Fuel	Raw, air-dried fuel									Pure combustible matter (without moisture and ash)							
	Composition, in %							Heating value H_i		Composition, in %						Heating value H_i	
	Carbon	Hydrogen	Sulphur	Oxygen	Nitrogen	Ash	Moisture			Carbon	Hydrogen	Sulphur	Oxygen	Nitrogen	Volatile matter		
	C	H	S	O	N	a	m	kcal/kg	kJ/kg	C	H	S	O	N	v	kcal/kg	kJ/kg
Anthracite	85.6	1.8	0.7	2.0	0.9	8	1	7450	31192	94.0	2.0	0.8	2.2	1.0	5	8200	34332
Brown coal																	
– bright	58.4	4.0	2.4	14.4	0.8	12	8	5550	23237	73.0	5.0	3.0	18.0	1.0	55	7000	29308
– lignite	49.6	3.7	0.4	18.7	0.6	7	20	4700	19678	68.0	5.1	0.5	25.6	0.7	65	6600	27633
– ordinary	52.4	3.9	0.8	17.2	0.7	10	15	4950	20725	70.0	5.2	1.0	22.9	0.9	60	6800	28470
Brown coal coke (low-temperature)	68.9	1.7	0.8	3.0	0.6	20	5	5820	24367	91.9	2.3	1.0	4.0	0.8	–	7800	32657
Charcoal	79.0	3.1	–	11.9	–	1	5	6830	28596	84.0	3.3	–	12.7	–	–	7300	30564
Gas coke	86.0	0.5	0.9	0.9	0.6	9	2	7110	29768	96.6	0.7	1.0	1.0	0.7	–	8000	33494
Peat	40.3	3.8	–	22.1	0.8	8	25	3470	14528	60.0	5.8	–	33.0	1.2	70	5400	22609
Pit coal																	
– coking, bituminous	82.0	4.1	0.7	4.2	1.0	6	2	7720	32322	89.0	4.5	0.8	4.6	1.1	25	8400	35169
– dry	75.2	4.6	0.9	8.8	0.5	8	2	6920	28973	83.5	5.1	1.0	9.8	0.6	40	7700	32238
– forge, bituminous	77.4	4.7	0.7	5.3	0.9	8	3	7470	31275	87.0	5.3	0.8	5.9	1.0	30	8300	34750
– gas	74.8	4.8	0.7	6.6	1.1	10	2	7030	29433	85.0	5.5	0.8	7.5	1.2	35	8000	33494
– non-bituminous	83.8	2.7	0.7	2.9	0.9	7	2	7720	32322	92.0	3.0	0.8	3.2	2.0	15	8500	35588
Smelting coke	87.3	0.5	0.9	0.8	0.5	8	2	7190	30103	97.0	0.5	1.0	0.9	0.6	–	8000	33494
Wood	39.3	4.7	–	34.1	0.4	1.5	20	3410	14277	50.0	6.0	–	43.5	0.5	75	4500	18841

[1] Mean values of composition of some solid fuels.

1 kcal = 4.1868 kJ

LIQUIDS

Table 27-1 Thermal Properties of Liquids

Liquid	Chemical formula	Relative molecular mass M (kg/kmol)	Density at 20°C ρ (kg/m³)	Coefficient of cubical expansion at 20°C β (1/K)	Melting point t (°C)	Melting point T (°K)	Heat of fusion (kcal/kg)	Heat of fusion (kJ/kg)	Boiling point t (°C)	Boiling point T (°K)	Heat of vaporization (kcal/kg)	Heat of vaporization r (kJ/kg)
Acetaldehyde	C₂H₄O	44.05	783	—	−123.5	149.65	17.6	73.688	17.4	290.55	137	573.592
Acetic acid	C₂H₄O₂	60.05	1049	0.00107	16.7	289.15	46.4	194.268	118	391.15	97	406.120
Acetone	C₃H₆O	58.08	791	0.00143	−94.3	178.85	23	96.296	56.1	329.25	125	523.350
n–amyl alcohol	C₅H₁₂O	88.14	810	0.00088	−78	195.15	26.7	111.788	138	411.15	—	—
Amyl benzoate	C₁₂H₁₆O₂	176.24	1010	0.00085	—	—	—	—	260	533.15	—	—
Amyl bromide	C₅H₁₁Br	151.05	1223	—	−95	178.15	22.8	95.459	126	399.15	48	200.966
Amyl chloride	C₅H₁₁Cl	106.59	883	—	—	—	—	—	106	379.15	74	309.823
Aniline	C₆H₇N	93.12	1022	0.00085	−6.2	266.95	27.1	113.462	184	457.15	107	447.988
Arsenic trichloride	AsCl₃	181.28	2170	0.00102	−16	257.15	—	—	130.3	403.45	46	192.593
Benzene	C₆H₆	78.11	879	0.00106	5.5	278.65	30.4	127.279	80.1	353.25	94.5	395.653
Bromine	Br₂	159.83	3120	0.00113	−7.3	265.85	16.2	67.826	58.8	331.95	43	180.032
Bromobenzene	C₆H₅Br	157.02	1495	0.00092	−30.6	242.55	—	—	156	429.15	—	—
Bromoform	CHBr₃	252.77	2890	0.00091	7.9	281.05	—	—	150	423.15	—	—
n–butyl alcohol	C₄H₁₀O	74.12	810	—	−90	183.15	29.9	125.185	117.7	390.85	141	590.339
Carbon disulphide	CS₂	76.13	1263	0.00119	−112	161.15	17.7	74.106	46.3	319.45	89	372.625
Chloral	C₂HOCl₃	147.40	1512	0.00093	−57.5	215.65	—	—	98	371.15	54	226.087
Chlorobenzene	C₆H₅Cl	112.56	1106	0.00098	−45.2	227.95	15.9	66.570	132	405.15	77.6	324.896
Chloroform	CHCl₃	119.39	1489	0.00128	−63.5	209.65	19	79.549	61.20	334.35	59	247.021
m–chlorotoluene	C₇H₇Cl	126.58	1072	—	−47.8	225.35	—	—	162.2	435.35	—	—
o–chlorotoluene	C₇H₇Cl	126.58	1081	0.00089	−36.5	236.65	18.2	76.200	159	432.15	72.6	303.962
Cyclohexane	C₆H₁₂	84.15	778	0.00120	6.4	279.55	7.4	30.982	80.8	353.95	86	360.065
cis–decalin	C₁₀H₁₈	138.24	900	—	−51	222.15	—	—	193	466.15	—	—
trans–decalin	C₁₀H₁₈	138.24	870	—	−36	237.15	—	—	185	458.15	—	—
cis–dichloroethylene	C₂H₂Cl₂	96.95	1265	—	−50	223.15	—	—	48.4	321.55	73	305.636
trans–dichloroethylene	C₂H₂Cl₂	96.95	1283	—	−80	193.15	—	—	60	333.15	74	309.823
Dichlorotetrafluoroethane	C₂F₄Cl₂	170.93	—	—	—	—	—	—	3.5	276.65	30.5	127.697
Diethylamine	C₄H₁₁N	73.13	711	—	−39	234.15	—	—	56	329.15	91	380.999
Diethylene glycol	C₄H₁₀O₃	106.12	1120	—	−10.5	262.65	—	—	245	518.15	150	628.020
Dimethylamine	C₂H₇N	45.08	—	—	−93	180.15	—	—	7.0	280.15	140	586.152
Ethyl acetate	C₄H₈O₂	88.10	900	0.00138	−83.6	189.55	28.4	118.905	77.1	350.25	88	368.438
Ethyl alcohol	C₂H₆O	46.07	789.5	0.00110	−114.5	158.65	25	104.670	78.3	351.45	210	841.547
Ethylamine	C₂H₇N	45.08	—	—	−81	192.15	—	—	16.5	289.65	145	607.086
Ethylbenzene	C₈H₁₀	106.16	868	0.00096	−94	179.15	20.6	86.248	135.4	408.55	81	339.131
Ethyl benzoate	C₉H₁₀O₂	150.17	1047	0.00090	−34.6	238.55	—	—	213.2	486.35	64	267.955
Ethyl bromide	C₂H₅Br	108.98	1450	0.00142	−119	154.15	12.8	53.591	38.4	311.55	60	251.208
Ethyl chloride	C₂H₅Cl	64.52	—	—	−138.7	134.45	—	—	12.2	285.35	92.5	387.279
Ethyl ether	C₄H₁₀O	74.12	714	0.00162	−116.3	156.85	24	100.483	34.48	307.63	86	360.065
Ethyl iodide	C₂H₅J	155.98	1934	0.00117	−111	162.15	—	—	72.5	345.65	45.8	191.755
Ethyl mustard oil (Ethyl isothiocyanate)	C₄H₅NS	99.15	1018	—	−102.5	170.65	—	—	152.0	425.15	—	—
Ethylene glycol	C₂H₆O₂	62.07	1115	—	−12.3	260.85	45	188.406	197	470.15	194	812.239
Formic acid	CH₂O₂	46.03	1220	0.00102	8.4	281.55	66	276.329	100.7	373.85	118	494.042
Glycerol	C₃H₈O₃	92.09	1260	0.00050	18	291.15	47.9	200.548	290	563.15	—	—

Table 27-2 Thermal Properties of Liquids (*Continued*)

Liquid	Chemical formula	Relative molecular mass M (kg/kmol)	Density at 20°C ρ (kg/m³)	Coefficient of cubical expansion at 20°C β (1/K)	Melting point t (°C)	Melting point T (°K)	Heat of fusion (kcal/kg)	Heat of fusion (kJ/kg)	Boiling point t (°C)	Boiling point T (°K)	Heat of vaporization (kcal/kg)	Heat of vaporization r (kJ/kg)
n–heptane	C_7H_{16}	100.19	684	0.00124	−90.6	182.55	33.8	141.514	98.4	371.55	76	318.197
n–heptyl alcohol	$C_7H_{16}O$	116.19	823	—	−34.3	238.85	—	—	176	449.15	105	439.614
n–hexane	C_6H_{14}	86.17	660	0.00135	−95.3	117.85	35	146.538	68.73	341.88	79	330.757
n–hexyl alcohol	$C_6H_{14}O$	102.17	820	—	−50	223.15	36	150.724	157	430.15	—	—
n–hexylene	C_6H_{12}	84.15	683	—	−98.5	174.65	—	—	64	337.15	92.8	388.535
Isoamyl acetate	$C_7H_{14}O_2$	130.18	873	0.00114	—	—	—	—	141	414.15	—	—
Isoamyl alcohol	$C_5H_{12}O$	88.14	810	0.00093	−117	156.15	—	—	131	404.15	—	—
Isobutyl alcohol	$C_4H_{10}O$	74.12	804	0.00094	−108	—	—	108	108	381.15	—	577.778
Isopentane	C_5H_{12}	72.14	621	0.00154	−160.0	113.15	24.4	102.158	28.0	301.15	81	339.131
Isopropyl	C_3H_8O	60.09	786	—	−89	184.15	21.3	89.179	82.3	355.45	160	669.888
Lactic acid	$C_3H_6O_3$	90.08	1240	—	18	291.15	—	—	—	—	—	—
Mercury	Hg	200.61	13545.7	0.000181	−38.83	234.32	2.8	11.723	356.95	630.10	72	301.450
Methyl acetate	$C_3H_6O_2$	74.08	934	—	−98.1	175.05	—	—	57.1	330.25	98	410.306
Methyl alcohol	CH_4O	32.04	792	0.00119	−98	175.15	24	100.483	64.51	337.66	263	1101.128
Methyl benzoate	$C_8H_8O_2$	136.14	1100	0.00090	−12.5	260.65	—	—	199.6	472.75	—	—
Methyl bromide	CH_3Br	94.95	—	—	−93	180.15	—	—	4.0	277.15	62	259.582
Methyl formate	$C_2H_4O_2$	60.05	975	0.00124	−99.8	173.15	30	125.604	31.8	304.95	115	481.482
Methyl iodide	CH_3J	141.95	2279	—	−66.3	206.85	—	—	42.5	315.65	47	196.780
Methylene chloride	CH_2Cl_2	84.94	1336	—	−96.5	176.65	—	—	40	313.15	79	330.757
Methylene iodide	CH_2J_2	267.87	3325	0.00081	6.0	279.15	—	—	180	453.15	—	—
Nitric acid	HNO_3	63.015	1512	0.00124	−41	232.15	9.5	39.775	86	359.15	115	481.482
Nitrobenzene	$C_6H_5O_2N$	123.11	1203	0.00083	5.7	278.85	23.5	98.390	211	484.15	95	397.746
Nitroglycerol	$C_3H_5O_9N_3$	227.09	1600	—	13.2	286.67	23	96.296	—	—	—	—
n–nonyl alcohol	$C_9H_{20}O$	144.25	828	—	−5	268.15	—	—	213.5	486.65	—	—
n–octane	C_8H_{18}	114.22	720	0.00114	−57	216.15	43	180.032	125.7	398.85	71	297.263
n–octyl alcohol	$C_8H_{18}O$	130.22	827	—	−16.5	256.65	—	—	195	468.15	98	410.306
Oleic acid	$C_{18}H_{34}O_2$	298.45	890	—	9	282.15	—	—	370	643.15	57	238.648
n–pentane	C_5H_{12}	72.14	626	0.00160	−129.7	143.45	27.7	115.974	36.1	309.25	85	355.878
Phosphorus trichloride	PCl_3	137.39	1578	—	−91	182.15	—	—	76.0	349.15	51.4	215.202
Propionic acid	$C_3H_6O_2$	74.08	992	0.00109	−21.5	251.65	24.3	101.739	141.4	414.55	100	418.680
n–propyl alcohol	C_3H_8O	60.09	804	0.00098	−126	147.15	20.7	86.667	97.2	370.35	163	682.448
n–propyl chloride	C_3H_7Cl	78.54	892	—	−123	150.15	—	—	46	319.15	84	351.691
Pyridine	C_5H_5N	79.10	983	0.00112	−42.0	231.15	25.0	104.670	115.4	388.55	102	427.054
Quinoline	C_9H_7N	129.15	1093	—	−19.5	253.65	20.0	83.736	242	515.15	—	—
Sulphuric acid	H_2SO_4	98.08	1834	0.00057	10.5	283.65	26.0	108.857	—	—	—	—
Tetrachloromethane	CCl_4	153.84	1595	0.00122	−22.8	250.35	3.75	15.701	76.7	349.85	46	192.593
Tetralin	$C_{10}H_{12}$	132.19	975	—	−35	238.15	—	—	207	480.15	—	—
Toluene	C_7H_8	92.13	866	0.00108	−95	178.15	17.2	72.013	110.7	383.85	85	355.878
Trichloroethylene	C_2HCl_3	131.40	1464	0.00119	−86.4	186.75	—	—	86.8	359.95	57	238.648
Turpentine oil	$C_{10}H_{16}$	136.22	855	0.00097	−10	263.15	—	—	160	433.15	70	293.076
Vinyl bromide	C_2H_3Br	106.96	—	—	−138	135.15	—	—	16	289.15	—	—
Water	H_2O	18.0156	998.2	0.00018	0.00	273.15	79.4	332.432	100.00	373.15	539.1	2257.104
m–xylene	C_8H_{10}	106.16	864	0.00099	−47.9	225.25	25.8	108.019	139.2	412.35	82	343.318
o–xylene	C_8H_{10}	106.16	880	0.00097	−25.3	247.85	29.3	122.673	144	417.15	83	347.504
p–xylene	C_8H_{10}	106.16	861	0.00102	13.3	286.45	38.1	159.517	138.4	411.55	81	339.431

Table 28-1 Critical Constants of Gases

Substance	Chemical formula	Critical temperature		Critical pressure		Critical density
		t_k	T_k	p_k		ρ_k
		°C	°K	kp/cm²	bar	kg/m³
Acetaldehyde	C_2H_4O	188	461.15	—	—	351
Acetic acid	$C_2H_4O_2$	321.6	594.75	57.1	55.99597	351
Acetone	C_3H_6O	236	509.15	60	58.83990	252
n—amyl alcohol	$C_5H_{12}O$	348	621.15	—	—	—
Amyl bromide	$C_5H_{11}Br$	307	580.15	—	—	—
Amyl chloride	$C_5H_{11}Cl$	279	552.15	—	—	—
Aniline	C_6H_7N	425.7	698.85	52.4	51.38685	—
Arsenic trichloride	$AsCl_3$	356	629.15	—	—	—
Benzene	C_6H_6	288.6	561.75	48	47.07192	305
Bromine	Br_2	310	583.15	102	100.02783	1180
Bromobenzene	C_6H_5Br	397	670.15	44.6	43.73766	485
n—butyl alcohol	$C_4H_{10}O$	287	560.15	48.4	47.46419	—
Carbon disulphide	CS_2	277	550.15	75	73.54988	441
Chlorobenzene	C_6H_5Cl	360	633.15	44.6	43.73766	365
Chloroform	$CHCl_3$	260	533.15	54.9	53.83851	496
Cyclohexane	C_6H_{12}	281	554.15	40.6	39.81500	273
cis—dichloroethylene	$C_2H_2Cl_2$	243	516.15	54	52.95591	—
trans—dichloroethylene	$C_2H_2Cl_2$	74	347.15	—	—	—
Diethylamine	$C_4H_{11}N$	223	496.15	38	37.26527	243
Dimethylamine	C_2H_7N	164	437.15	54	52.95591	—
Ethylacetate	$C_4H_8O_2$	250	523.15	38	37.26527	308
Ethylalcohol	C_2H_6O	243	516.15	63	61.78190	280
Ethylamine	C_2H_7N	183.4	456.55	56	54.91724	248
Ethylbenzene	C_8H_{10}	346.4	619.55	38.1	37.36334	—
Ethylbromide	C_2H_5Br	233	506.15	61.5	60.31090	507
Ethylchloride	C_2H_5Cl	185	458.15	53	51.97525	330
Ethylether	$C_4H_{10}O$	194	467.15	36.3	35.59814	265
Ethyliodide	C_2H_5J	281	554.15	—	—	—
n—heptane	C_7H_{16}	266.8	539.95	26.9	26.37989	234
n—heptyl alcohol	$C_7H_{16}O$	365.3	638.45	—	—	—
n—hexane	C_6H_{14}	234.8	507.95	29.8	29.22382	234
n—hexene	C_6H_{12}	244	517.15	—	—	—
Isoamyl acetate	$C_7H_{14}O_2$	326	599.15	—	—	—
Isoamyl alcohol	$C_5H_{12}O$	309	582.15	—	—	—
Isobutyl alcohol	$C_4H_{10}O$	272	545.15	48.3	47.36612	—
Mercury	Hg	1460	1733.15	1042	1021.85293	5000
Methyl acetate	$C_3H_6O_2$	234	507.15	47	46.09126	—
Methyl alcohol	CH_4O	240	513.15	99	97.08584	358
Methyl bromide	CH_3Br	194	467.15	—	—	—
Methyl formate	$C_2H_4O_2$	214	487.15	59	57.85924	349
Methyl iodide	CH_3J	255	528.15	—	—	—
Methylene chloride	CH_2Cl_2	245	518.15	101.4	99.43943	—
n—octane	C_8H_{18}	296.2	569.35	24.7	24.22243	233
n—octyl alcohol	$C_8H_{18}O$	385.5	658.65	—	—	—
i—pentane	C_5H_{12}	188	461.15	33.0	32.36195	234
n—pentane	C_5H_{12}	197	470.15	33.0	32.36195	232
Phosphorus trichloride	PCl_3	286	559.15	—	—	—
Propionic acid	$C_3H_6O_2$	339	612.15	52.9	51.87718	—
i—propyl alcohol	C_3H_8O	240	513.15	53	51.97512	—
n—propyl alcohol	C_3H_8O	264	537.15	50	49.03325	273
n—propyl chloride	C_3H_7Cl	221	494.15	49	48.05259	—
Pyridine	C_5H_5N	344	617.15	60	58.83990	—
Terpentine oil	$C_{10}H_{16}$	376	649.15	—	—	—
Tetrachloromethane (Carbon tetrachloride)	CCl_4	283	556.15	45	44.12993	558
Toluene	C_7H_8	320.6	593.75	41.6	40.79566	—
Water	H_2O	374.15	647.30	225.65	221.28706	315
m—xylene	C_8H_{10}	346	619.15	35.8	35.10781	—
o—xylene	C_8H_{10}	359	632.15	36.9	36.18654	—
p—xylene	C_8H_{10}	345	618.15	35.0	34.32328	—

1 at = 1 kp/cm² = 98065.5 N/m² = 9.800665 N/cm² = 0.980665 bar

Table 29-1 Coefficients β of Cubical Thermal Expansion of Liquids Between 0°C (273.15 K) and t°C (TK) at the Normal Pressure of 760 mm Hg

Liquid	Chemical formula	Temperature		Coefficient of cubical expansion
		t	T	β
		°C	°K	1/K
Acetone	C_3H_6O	10	283.15	0.00135
		20	293.15	0.001375
		30	303.15	0.001433
		40	313.15	0.001463
		50	323.15	0.001500
Alcohol (absolute)		20	293.15	0.00115
Benzene	C_6H_6	10	283.15	0.00120
		20	293.15	0.00120
		30	303.15	0.00123
		40	313.15	0.00125
		50	323.15	0.00127
		60	333.15	0.00129
		70	343.15	0.00131
		80	353.15	0.00133
Glycerol	$C_3H_8O_3$	10	283.15	0.00050
		20	293.15	0.00050
		30	303.15	0.00050
		40	313.15	0.00051
		50	323.15	0.00052
Lubricating oil		20	293.15	0.000740
Mercury	Hg	10	283.15	0.0001819
		20	293.15	0.0001820
		30	303.15	0.0001820
		40	313.15	0.0001821
		50	323.15	0.0001822
		60	333.15	0.0001822
		70	343.15	0.0001823
		80	353.15	0.0001824
		90	363.15	0.0001825
		100	373.15	0.0001826
		150	423.15	0.0001832
		200	473.15	0.0001840
		250	523.15	0.0001851
		300	573.15	0.0001863
Mercury, at 20 at pressure	Hg	200	473.15	0.0001840
		250	523.15	0.0001851
		300	573.15	0.0001863
		350	623.15	0.0001878
		400	673.15	0.0001896
		450	723.15	0.0001916
		500	773.15	0.0001939
Olive oil		18	291.15	0.00072
Paraffin oil		10	283.15	0.00075
		20	293.15	0.00075
Petrol		20	293.15	0.00120
Petroleum		18	291.15	0.00092 . . . 0.0010
Toluene	C_7H_8	20	293.15	0.001075
		30	303.15	0.001073
		40	313.15	0.001100
		50	323.15	0.001112
		60	333.15	0.001133
		70	343.15	0.001150
		80	353.15	0.001171
		90	363.15	0.001189
		100	373.15	0.001205
Transformer oil		20	293.15	0.000690

NOTE:

Volume coefficient of cubical thermal expansion of solids $\beta = 3\alpha$

Volume coefficient of cubical thermal expansion of gases depends upon temperature

Volume coefficient of cubical thermal expansion of ideal gases $\beta = 1/273.16$

Table 30-1 Specific Heat c of Liquids

Liquid	Chemical formula	Temperature		Specific heat	
		t	T	c	
		°C	°K	kcal/kg K	kJ/kg K
Acetic acid	$C_2H_4O_2$	20	293.15	0.485	2.031
Acetone	C_3H_6O	− 50	223.15	0.485	2.031
		0	273.15	0.506	2.119
		20	293.15	0.516	2.160
		50	323.15	0.537	2.248
Airplane motor oil		20	293.15	0.439	1.838
		40	313.15	0.459	1.922
		60	333.15	0.479	2.005
		80	353.15	0.499	2.089
		100	373.15	0.520	2.177
		120	393.15	0.542	2.269
		140	413.15	0.564	2.361
Ammonia	NH_3	− 20	253.15	0.109	0.456
		0	273.15	0.110	0.461
		20	293.15	0.113	0.473
Aniline	C_6H_7N	0	273.15	0.482	2.018
		20	293.15	0.493	2.064
		50	323.15	0.512	2.144
		100	373.15	0.560	2.345
		150	423.15	0.700	2.931
Arsenic trichloride	As_2Cl_3	20	293.15	0.170	0.712
Beer		20	293.15	0.90	3.768
Benzene	C_6H_6	−100	173.15	0.230	0.963
		10	283.15	0.340	1.424
		20	293.15	0.415	1.738
		40	313.15	0.423	1.771
		50	323.15	0.430	1.800
		60	333.15	0.456	1.909
		65	338.15	0.482	2.018
Bromine	Br_2	20	293.15	0.11	0.461
Bromobenzene	C_6H_5Br	20	293.15	0.231	0.967
		40	313.15	0.233	0.976
		60	333.15	0.238	0.996
		80	353.15	0.245	1.026
Bromoform	$CHBr_3$	20	293.15	0.128	0.536
n−butyl alcohol	$C_4H_{10}O$	− 78	195.15	0.442	1.851
		21	294.15	0.565	2.366
		30	303.15	0.582	2.437
		114	387.15	0.689	2.885
Carbon dioxide	CO_2	20	293.15	0.87	3.643
Carbon disulphide	CS_2	−100	173.15	0.194	0.812
		0	273.15	0.238	0.996
		20	293.15	0.243	1.017
Castor oil		20	293.15	0.46	1.926
Chlorobenzene	C_6H_5Cl	20	293.15	0.310	1.298
		40	313.15	0.315	1.319
		60	333.15	0.326	1.365
		80	353.15	0.341	1.428
Chloroform	$CHCl_3$	20	293.15	0.231	0.967
		30	303.15	0.234	0.980
		40	313.15	0.238	0.996
		50	323.15	0.243	1.017

1 kcal = 4.1868 kJ

Table 30-2 Specific Heat _c_ of Liquids (_Continued_)

Liquid	Chemical formula	Temperature		Specific heat	
		_t_	_T_	_c_	
		°C	°K	kcal/kg K	kJ/kg K
_m_—chlorotoluene	C_7H_7Cl	20	293.15	0.29	1.214
_o_—cresol	C_7H_8O	20	293.15	0.50	2.093
Crude oil		20	293.15	0.21	0.879
Dichlorodifluoromethane (Freon 12)	CF_2Cl_2	—150	123.15	0.2003	0.8386
		—140	133.15	0.2009	0.8411
		—130	143.15	0.2011	0.8420
		—120	153.15	0.2014	0.8432
		—110	163.15	0.2018	0.8449
		—100	173.15	0.2024	0.8474
		— 90	183.15	0.2031	0.8503
		— 80	193.15	0.2040	0.8541
		— 70	203.15	0.2051	0.8587
		— 60	213.15	0.2065	0.8646
		— 50	223.15	0.2080	0.8709
		— 40	233.15	0.2099	0.8788
		— 30	243.15	0.2121	0.8880
		— 20	253.15	0.2147	0.8989
		— 10	263.15	0.2178	0.9119
		0	273.15	0.2213	0.9265
		10	283.15	0.2255	0.9441
		20	293.15	0.2306	0.9655
		30	303.15	0.2368	0.9914
		40	313.15	0.2437	1.0203
		50	323.15	0.2538	1.0626
		60	333.15	0.2661	1.1141
		70	343.15	0.2828	1.1840
		80	353.15	0.3067	1.2841
		90	363.15	0.3453	1.4457
		100	373.15	0.4237	1.7739
		110	383.15	0.8737	3.6580
Diphenylmethane	$C_{13}H_{12}$	37.5	310.65	0.390	1.633
		49.4	322.55	0.393	1.645
Diphenyl (mixture by weight 26.5% diphenyl, $C_{12}H_{10}$ 73.5% diphenyl ether, $C_{12}H_{10}O$)		20	293.15	0.370	1.549
		30	303.15	0.383	1.604
		40	313.15	0.395	1.654
		50	323.15	0.406	1.700
		60	333.15	0.416	1.742
		70	343.15	0.425	1.779
		80	353.15	0.433	1.813
		90	363.15	0.440	1.842
		100	373.15	0.447	1.871
		110	383.15	0.453	1.897
		120	393.15	0.459	1.922
		130	403.15	0.464	1.943
		140	413.15	0.469	1.964
		150	423.15	0.474	1.985
		160	433.15	0.478	2.001
		170	443.15	0.482	2.018
		180	453.15	0.486	2.035
		190	463.15	0.490	2.052
		200	473.15	0.494	2.068
		210	483.15	0.498	2.085
		220	493.15	0.502	2.102
		230	503.15	0.506	2.119
		240	513.15	0.510	2.135
		250	523.15	0.515	2.156
		260	533.15	0.520	2.177
		270	543.15	0.525	2.198
		280	553.15	0.530	2.219

1 kcal = 4.1868 kJ

Table 30-3 Specific Heat c of Liquids (*Continued*)

Liquid	Chemical formula	Temperature		Specific heat	
		t	T	c	
		°C	°K	kcal/kg K	kJ/kg K
Diphenyl		290	563.15	0.535	2.240
		300	573.15	0.541	2.265
		310	583.15	0.547	2.290
		320	593.15	0.553	2.315
		330	603.15	0.560	2.345
		340	613.15	0.567	2.374
		350	623.15	0.574	2.403
		360	633.15	0.581	2.433
		370	643.15	0.589	2.466
		380	653.15	0.597	2.500
		390	663.15	0.606	2.537
		400	673.15	0.615	2.575
Dowtherm (mixture by weight 26.5% diphenyl, $C_{12}H_{10}$ 73.5% diphenyl ether, $C_{12}H_{10}O$)		20	293.15	0.37	1.549
Ethylacetate	$C_4H_8O_2$	20	293.15	0.48	2.010
Ethylalcohol	C_2H_6O	−100	173.15	0.45	1.884
		− 50	223.15	0.48	2.010
		0	273.15	0.55	2.303
		20	293.15	0.59	2.470
		40	313.15	0.65	2.721
		50	323.15	0.67	2.805
		80	353.15	0.712	2.981
		120	393.15	0.909	3.806
		160	433.15	1.114	4.664
Ethylbenzene	C_8H_{10}	− 50	223.15	0.360	1.507
		20	293.15	0.413	1.729
		50	323.15	0.450	1.884
Ethylbenzoate	$C_9H_{10}O_2$	20	293.15	0.385	1.612
Ethylbromide	C_2H_5Br	20	293.15	0.21	0.879
Ethylene glycol	$C_2H_6O_2$	20	293.15	0.569	2.382
		40	313.15	0.591	2.474
		60	333.15	0.612	2.562
		80	353.15	0.633	2.650
		100	373.15	0.655	2.742
Ethylether	$C_4H_{10}O$	−100	173.15	0.483	2.022
		0	273.15	0.542	2.269
		20	293.15	0.556	2.328
Ethyliodide	C_2H_5J	− 30	243.15	0.1567	0.656
		0	273.15	0.1616	0.677
		20	293.15	0.1648	0.690
		30	303.15	0.1666	0.698
		60	333.15	0.1715	0.718
Formic acid	CH_2O_2	20	293.15	0.52	3.177
Glue mass		20	293.15	1.00	4.187
Glycerol	$C_3H_8O_3$	0	273.15	0.54	2.261
		20	293.15	0.58	2.428
n−heptane	C_7H_{16}	− 50	223.15	0.491	2.056
		0	273.15	0.522	2.186
		20	293.15	0.530	2.219
n−hexane	C_6H_{14}	20	293.15	0.45	1.884
HT−oil C		20	293.15	0.35	1.465
Hydrochloric acid, 17%	HCl	20	293.15	0.74	3.098

1 kcal = 4.1868 kJ

Table 30-4   Specific Heat c of Liquids (*Continued*)

Liquid	Chemical formula	Temperature		Specific heat	
		t	T	c	
		°C	°K	kcal/kg K	kJ/kg K
Isoamil alcohol	$C_5H_{12}O$	20	293.15	0.56	2.345
Isobutyl alcohol	$C_4H_{10}O$	10	283.15	0.502	2.102
		20	293.15	0.550	2.303
		40	313.15	0.648	2.713
		85	358.15	0.841	3.521
Lubricating oil		20	293.15	0.442	1.851
		40	313.15	0.462	1.934
		60	333.15	0.482	2.018
		80	353.15	0.502	2.102
		100	373.15	0.522	2.186
		120	393.15	0.542	2.269
Magnesium chloride, 20%	$MgCl_2$	− 20	253.15	0.714	2.989
		0	273.15	0.725	3.035
		20	293.15	0.736	3.081
Mercury	Hg	0	273.15	0.0335	0.140
		20	293.15	0.0333	0.139
		100	373.15	0.0328	0.137
		200	473.15	0.0325	0.136
Methyl acetate	$C_3H_6O_2$	20	293.15	0.51	2.135
Methyl alcohol	CH_4O	− 50	223.15	0.55	2.303
		0	273.15	0.58	2.428
		20	293.15	0.59	2.470
		50	323.15	0.61	2.554
Methyl benzoate	$C_8H_8O_2$	20	293.15	0.37	1.549
Methyl chloride	CH_3Cl	− 20	253.15	0.36	1.507
		0	273.15	0.375	1.570
		20	293.15	0.38	1.591
Methylene chloride	CH_2Cl_2	20	293.15	0.29	1.214
Milk		20	293.15	0.94	2.936
Mineral oil		20	293.15	0.45	1.884
Naphthalene	$C_{10}H_3$	90	363.15	0.424	1.775
		120	393.15	0.447	1.871
		190	463.15	0.500	2.093
Nitric acid (100%)	HNO_3	20	293.15	0.41	1.717
Nitrobenzene	$C_6H_5O_2N$	20	293.15	0.36	1.507
n−octane	C_8H_{18}	20	293.15	0.52	2.177
Oleic acid	$C_{18}H_{34}O_2$	20	293.15	0.49	2.052
Olive oil		20	293.15	0.39	1.633
Oxygen	O_2	20	293.15	0.35	1.465
Paraffin oil		20	293.15	0.51	2.135
n−pentane	C_5H_{12}	−100	173.15	0.47	1.968
		0	273.15	0.51	2.135
		20	293.15	0.52	2.177
Petrol (gasoline)		20	293.15	0.500	2.093
Petroleum		20	293.15	0.51	2.135
Phosphorus trichloride	PCl_3	20	293.15	0.20	0.837
Pit coal tar		20	293.15	0.50	2.093

1 kcal = 4.1868 kJ

Table 30-5 Specific Heat *c* of Liquids (*Continued*)

Liquid	Chemical formula	Temperature		Specific heat	
		t	*T*	*c*	
		°C	°K	kcal/kg K	kJ/kg K
Propionic acid	$C_3H_6O_2$	20	293.15	0.52	2.177
n—propyl alcohol	C_3H_8O	20	293.15	0.58	2.428
Pyridine	C_5H_5N	20	293.15	0.41	1.717
Quinoline	C_9H_7N	20	293.15	0.31	1.298
Rubber mass		20	293.15	0.82	3.433
Sodium hydroxide — 20% salt		— 10	263.15	0.73	3.056
— 30% salt		— 10	263.15	0.64	2.680
Sulfur dioxide	SO_2	— 20	253.15	0.304	1.273
		0	273.15	0.324	1.357
		20	293.15	0.332	1.390
Sulfuric acid, 100%	H_2SO_4	20	293.15	0.33	1.382
Tetrachloromethane	CCl_4	20	293.15	0.202	0.846
Tetralin	$C_{10}H_{12}$	20	293.15	0.40	1.675
Toluene	C_7H_8	— 50	223.15	0.36	1.507
		0	273.15	0.39	1.633
		20	293.15	0.40	1.675
		50	323.15	0.43	1.800
		100	373.15	0.47	1.968
Transformer oil		20	293.15	0.452	1.892
		40	313.15	0.476	1.993
		60	333.15	0.500	2.093
		80	353.15	0.525	2.198
		100	373.15	0.548	2.294
Trichloroethylene	C_2HCl_3	20	293.15	0.227	0.950
Trimethylethylene	C_5H_{10}	—129.3	143.85	0.448	1.876
		— 71.8	201.35	0.459	1.921
		— 40.7	232.45	0.472	1.976
		2.2	275.35	0.498	2.085
		20.7	293.85	0.512	2.144
Turpentine oil	$C_{10}H_{16}$	0	273.15	0.41	1.717
		20	293.15	0.43	1.800
		50	323.15	0.46	1.926
		100	373.15	0.50	2.093
Water	H_2O	0	273.15	1.008	4.220
		20	293.15	0.999	4.183
		40	313.15	0.998	4.178
		60	333.15	1.001	4.191
		80	353.15	1.003	4.199
		100	373.15	1.007	4.216
		120	393.15	1.011	4.233
		140	413.15	1.017	4.258
		150	423.15	1.020	4.271
		160	433.15	1.023	4.283
		180	453.15	1.050	4.396
		200	473.15	1.075	4.501
		220	493.15	1.10	4.605
		240	513.15	1.13	4.731
		250	523.15	1.16	4.857
		260	533.15	1.19	4.982
		280	553.15	1.25	5.234
		300	573.15	1.36	5.694
Xylene	C_8H_{10}	20	293.15	0.41	1.717

1 kcal = 4.1868 kJ

Table 31-1 Specific Heat c of Water (H_2O) at Pressure $p = 1$, at $= 0.980665$ N/m²

Temperature		Specific heat		Temperature		Specific heat	
t	T	c		t	T	c	
°C	°K	kcal/kg K	kJ/kg K	°C	°K	kcal/kg K	kJ/kg K
0	273.15	1.0045	4.2056				
1	274.15	1.0041	4.2040	26	299.15	0.9978	4.1776
2	275.15	1.0036	4.2019	27	300.15	0.9977	4.1772
3	276.15	1.0033	4.2006	28	301.15	0.9976	4.1768
4	277.15	1.0029	4.1989	29	302.15	0.9975	4.1763
5	278.15	1.0025	4.1973	30	303.15	0.9975	4.1763
6	279.15	1.0022	4.1960	31	304.15	0.9975	4.1763
7	280.15	1.0018	4.1943	32	305.15	0.9974	4.1759
8	281.15	1.0015	4.1931	33	306.15	0.9974	4.1759
9	282.15	1.0012	4.1918	34	307.15	0.9974	4.1759
10	283.15	1.0009	4.1906	35	308.15	0.9974	4.1759
11	284.15	1.0006	4.1893	36	309.15	0.9975	4.1763
12	285.15	1.0003	4.1881	37	310.15	0.9975	4.1763
13	286.15	1.0001	4.1872	38	311.15	0.9976	4.1768
14	287.15	0.9998	4.1860	39	312.15	0.9976	4.1768
15	288.15	0.9996	4.1851	40	313.15	0.9977	4.1772
16	289.15	0.9993	4.1839	41	314.15	0.9978	4.1776
17	290.15	0.9991	4.1830	42	315.15	0.9979	4.1780
18	291.15	0.9989	4.1822	43	316.15	0.9980	4.1784
19	292.15	0.9987	4.1814	44	317.15	0.9981	4.1788
20	293.15	0.9986	4.1809	45	318.15	0.9983	4.1797
21	294.15	0.9984	4.1801	46	319.15	0.9984	4.1801
22	295.15	0.9982	4.1793	47	320.15	0.9986	4.1809
23	296.15	0.9981	4.1788	48	321.15	0.9988	4.1818
24	297.15	0.9980	4.1784	49	322.15	0.9990	4.1826
25	298.15	0.9979	4.1780	50	323.15	0.9992	4.1835

1 at = 1 kp/cm² = 98066.5 N/m² = 9.80665 N/cm² = 0.980665 bar   1 kcal = 4.1868 kJ

Table 32-1 Specific Heat c of Water (H_2O) at Higher Pressures

Pressure p =		50 at = 49.033250 bar		100 at = 98.066500 bar		150 at = 147.099750 bar		200 at = 196.133000 bar		250 at = 245.166250 bar		300 at = 294.199500 bar	
Temperature		Specific heat											
t	T	c		c		c		c		c		c	
°C	°K	kcal/kg K	kJ/kg K	kcal/kg K	kJ/kg K	kcal/kg K	kJ/kg K	kcal/kg K	kJ/kg K	kcal/kg K	kJ/kg K	kcal/kg K	kJ/kg K
0	273.15	1.004	4.204	1.002	4.195	1.000	4.187	0.998	4.178	0.996	4.170	0.994	4.162
20	293.15	0.996	4.170	0.994	4.162	0.992	4.153	0.989	4.141	0.987	4.132	0.984	4.120
40	313.15	0.994	4.162	0.992	4.153	0.989	4.141	0.986	4.128	0.984	4.120	0.981	4.107
60	333.15	0.995	4.166	0.992	4.153	0.989	4.141	0.986	4.128	0.983	4.116	0.980	4.103
80	353.15	0.999	4.183	0.995	4.166	0.992	4.153	0.989	4.141	0.985	4.124	0.982	4.111
100	373.15	1.004	4.204	1.000	4.187	0.997	4.174	0.993	4.157	0.989	4.141	0.986	4.128
120	393.15	1.011	4.233	1.007	4.216	1.003	4.199	0.999	4.183	0.995	4.166	0.991	4.149
140	413.15	1.019	4.266	1.015	4.250	1.009	4.224	1.006	4.212	1.002	4.195	0.997	4.174
160	433.15	1.033	4.325	1.028	4.304	1.023	4.283	1.018	4.262	1.013	4.241	1.008	4.220
180	453.15	1.050	4.396	1.044	4.371	1.038	4.346	1.032	4.321	1.027	4.300	1.021	4.275
200	473.15	1.071	4.484	1.064	4.455	1.057	4.425	1.050	4.396	1.043	4.367	1.037	4.342
220	493.15	1.097	4.593	1.088	4.555	1.080	4.522	1.072	4.488	1.064	4.455	1.056	4.421
240	513.15	1.132	4.739	1.121	4.693	1.110	4.647	1.100	4.605	1.090	4.564	1.081	4.526
260	533.15	1.181	4.945	1.166	4.882	1.152	4.823	1.139	4.769	1.127	4.719	1.114	4.664
280	553.15	—	—	1.231	5.154	1.212	5.074	1.194	4.999	1.177	4.928	1.161	4.861
300	573.15	—	—	1.352	5.661	1.300	5.443	1.266	5.300	1.242	5.200	1.223	5.120
310	583.15	—	—	—	—	1.372	5.744	1.318	5.518	1.283	5.372	1.257	5.263
320	593.15	—	—	—	—	1.480	6.196	1.391	5.824	1.355	5.673	1.298	5.434
330	603.15	—	—	—	—	1.653	6.921	1.501	6.284	1.409	5.899	1.352	5.661
340	613.15	—	—	—	—	1.939	8.118	1.675	7.013	1.529	6.402	1.425	5.966
350	623.15	—	—	—	—	—	—	1.963	8.219	1.693	7.088	1.536	6.431

1 at = 1 kp/cm² = 98066.5 N/m² = 9.80665 N/cm² = 0.980665 bar   1 kcal = 4.1868 kJ

Table 33-1 Density ρ of Miscellaneous Liquids

Liquid	Chemical formula	Temperature		Density
		t °C	T °K	ρ kg/m³
Acetaldehyde	C_2H_4O	20	293.15	783
Acetic acid	$C_2H_4O_2$	0	273.15	1070
		10	283.15	1059
		20	293.15	1049
		30	303.15	1039
		40	313.15	1028
		50	323.15	1018
		60	333.15	1006
		70	343.15	995
		80	353.15	984
		90	363.15	972
		100	373.15	960
		110	383.15	948
		120	393.15	936
		130	403.15	924
		140	413.15	909
		150	423.15	896
		160	433.15	883
		170	443.15	869
		180	453.15	856
		190	463.15	841
		200	473.15	827
		210	483.15	811
		220	493.15	794
		230	503.15	776
		240	513.15	757
		250	523.15	736
		260	533.15	714
		270	543.15	690
		280	553.15	663
		290	563.15	633
		300	573.15	595
Acetone	C_3H_6O	20	293.15	791
Airplane motor oil		20	293.15	893
		40	313.15	881
		60	333.15	868
		80	353.15	856
		100	373.15	844
		120	393.15	832
		140	413.15	819

Liquid	Chemical formula	Temperature		Density
		t °C	T °K	ρ kg/m³
Ammonia	NH_3	−20	253.15	665
		0	273.15	639
		20	293.15	610
		100	373.15	452
n–amyl alcohol	$C_5H_{12}O$	20	293.15	810
Amyl benzoate	$C_{12}H_{16}O_2$	20	293.15	1010
Amyl bromide	$C_5H_{11}Br$	20	293.15	1223
Amyl chloride	$C_5H_{11}Cl$	20	293.15	883
Amyl iodide	$C_5H_{11}J$	20	293.15	1524
Aniline	C_6H_7N	20	293.15	1022
Anisole	C_7H_8O	20	293.15	994
Benzene	C_6H_6	0	273.15	900
		20	293.15	879
		40	313.15	858
		60	333.15	836
		70	343.15	825
		80	353.15	815
		90	363.15	804
		100	373.15	793
		110	383.15	781
		120	393.15	769
		130	403.15	757
		140	413.15	744
		150	423.15	731
		160	433.15	719
		170	443.15	704
		180	453.15	691
		190	463.15	676
		200	473.15	661
		210	483.15	643
		220	493.15	626
		230	503.15	607
		240	513.15	585
		250	523.15	561
		260	533.15	533
		270	543.15	498
		280	553.15	451

Table 33-2 Density ρ of Miscellaneous Liquids (Continued)

Liquid	Chemical formula	Temperature t (°C)	Temperature T (°K)	Density ρ (kg/m³)
Benzonitrile	C_7H_6N	15	288.15	1010
Bromine	Br_2	20	293.15	3120
Bromobenzene	C_6H_5Br	0	273.15	1522
		20	293.15	1495
		40	313.15	1468
		60	333.15	1441
		80	353.15	1414
		100	373.15	1386
		120	393.15	1358
		140	413.15	1329
		150	423.15	1315
		160	433.15	1299
		170	443.15	1285
		180	453.15	1270
		190	463.15	1253
		200	473.15	1239
		210	483.15	1221
		220	493.15	1204
		230	503.15	1188
		240	513.15	1169
		250	523.15	1151
		260	533.15	1131
		270	543.15	1110
Bromoform	$CHBr_3$	20	293.15	2890
i—butane	C_4H_{10}	0	273.15	582
		8	281.15	573
		16	289.15	563
		24	297.15	553
		32	305.15	543
		40	313.15	533
		48	321.15	522
		56	329.15	511
n—butane	C_4H_{10}	0	273.15	599
		8	281.15	591
		16	289.15	582
		24	297.15	573
		32	305.15	564
		40	313.15	554
		48	321.15	544
		56	329.15	533

Liquid	Chemical formula	Temperature t (°C)	Temperature T (°K)	Density ρ (kg/m³)
n—butyl alcohol	$C_4H_{10}O$	20	293.15	810
n—butyl bromide	C_4H_9Br	20	293.15	1275
n—butyl chloride	C_4H_9Cl	20	293.15	886
n—butyl iodide	C_4H_9J	20	293.15	1615
Carbon dioxide	CO_2	0	273.15	925
		20	293.15	771
		30	303.15	595
Carbon disulfide	CS_2	20	293.15	1263
Castor oil		20	293.15	960
Chlorine	Cl_2	−100	173.15	1717
		−90	183.15	1694
		−80	193.15	1673
		−70	203.15	1646
		−60	213.15	1622
		−50	223.15	1598
		−40	233.15	1574
		−30	243.15	1550
		−20	253.15	1524
		−10	263.15	1496
		0	273.15	1468
		10	283.15	1438
		20	293.15	1408
		30	303.15	1377
		40	313.15	1344
		50	323.15	1310
		60	333.15	1275
		70	343.15	1240
		80	353.15	1199
		90	363.15	1156
		100	373.15	1109
		110	383.15	1059
		120	393.15	998
		130	403.15	920
		140	413.15	750
Chlorobenzene	C_6H_5Cl	0	273.15	1128
		20	293.15	1106
		40	313.15	1085
		60	333.15	1064

Table 33-3 Density ρ of Miscellaneous Liquids (Continued)

Liquid	Chemical formula	t °C	T °K	ρ kg/m³
Chlorobenzene	C_6H_5Cl	80	353.15	1042
		100	373.15	1019
		120	393.15	996
		130	403.15	984
		140	413.15	972
		150	423.15	960
		160	433.15	948
		170	443.15	935
		180	453.15	922
		190	463.15	909
		200	473.15	896
		210	483.15	882
		220	493.15	867
		230	503.15	852
		240	513.15	836
		250	523.15	820
		260	533.15	802
		270	543.15	783
Chloroform	$CHCl_3$	20	293.15	1489
m—cresol	C_7H_8O	20	293.15	1034
o—cresol	C_7H_8O	41	314.15	1027
p—cresol	C_7H_8O	41	314.15	1018
Cyclohexane	C_6H_{12}	7	280.15	791
		16	289.15	782
		20	293.15	779
		25	298.15	774
		40	313.15	759
		51	324.15	744
		79	352.15	722
Cyclohexanol	$C_6H_{10}O$	20	293.15	962
Cylinder oil		0	273.15	890
Dichlorodifluoromethane (Freon 12)	CF_2Cl_2	−150	123.15	1826
		−140	133.15	1798
		−130	143.15	1770
		−120	153.15	1743
		−110	163.15	1715
		−100	173.15	1688

Liquid	Chemical formula	t °C	T °K	ρ kg/m³
Dichlorodifluoromethane	CF_2Cl_2	−90	183.15	1660
		−80	193.15	1632
		−70	203.15	1604
		−60	213.15	1575
		−50	223.15	1546
		−40	233.15	1517
		−30	243.15	1487
		−20	253.15	1456
		−10	263.15	1425
		0	273.15	1394
		10	283.15	1362
		20	293.15	1329
		30	303.15	1293
		40	313.15	1255
		50	323.15	1213
		60	333.15	1167
		70	343.15	1119
		80	353.15	1064
		90	363.15	999
		100	373.15	913
		110	383.15	742
Diethylene glycol	$C_4H_{10}O_3$	20	293.15	1120
Dimethylamine	C_2H_7N	−30	243.15	709
		−20	253.15	699
		−10	263.15	690
		0	273.15	680
		10	283.15	671
		20	293.15	662
		30	303.15	649
		40	313.15	641
Dimethylaniline	$C_8H_{11}N$	20	293.15	956
Diphyl (mixture by weight 26.5% diphenyl, $C_{12}H_{10}$ 73.5% diphenyl ether, $C_{12}H_{10}O$)		20	293.15	1062
		30	303.15	1054
		40	313.15	1046
		50	323.15	1037
		60	333.15	1029
		70	343.15	1021
		80	353.15	1013
		90	363.15	1004

Table 33-4 Density ρ of Miscellaneous Liquids (Continued)

Liquid	Chemical formula	t (°C)	T (°K)	ρ (kg/m³)
Diphenyl		100	373.15	996
		110	383.15	987
		120	393.15	979
		130	403.15	970
		140	413.15	962
		150	423.15	953
		160	433.15	945
		170	443.15	936
		180	453.15	927
		190	463.15	918
		200	473.15	909
		210	483.15	901
		220	493.15	892
		230	503.15	882
		240	513.15	873
		250	523.15	864
		260	533.15	855
		270	543.15	846
		280	553.15	836
		290	563.15	827
		300	573.15	818
		310	583.15	808
		320	593.15	798
		330	603.15	787
		340	613.15	779
		350	623.15	769
		360	633.15	759
		370	643.15	748
		380	653.15	738
		390	663.15	727
		400	673.15	717
Ethyl acetate	$C_4H_8O_2$	0	273.15	924
		20	293.15	901
		40	313.15	876
		60	333.15	851
		70	343.15	838
		80	353.15	825
		90	363.15	811
		100	373.15	797
		110	383.15	783
		120	393.15	768
Ethyl acetate	$C_4H_8O_2$	130	403.15	753
		140	413.15	738
		150	423.15	721
		160	433.15	703
		170	443.15	685
		180	453.15	665
		190	463.15	644
		200	473.15	621
		210	483.15	594
		220	493.15	565
		230	503.15	528
		240	513.15	478
Ethyl alcohol (ethanol)	C_2H_6O	0	273.15	806
		10	283.15	798
		20	293.15	789
		30	303.15	781
		40	313.15	772
		50	323.15	763
		60	333.15	754
		70	343.15	745
		80	353.15	735
		90	363.15	725
		100	373.15	716
		110	383.15	706
		120	393.15	693
		130	403.15	679
		140	413.15	663
		150	423.15	649
		160	433.15	633
		170	443.15	617
		180	453.15	598
		190	463.15	578
		200	473.15	557
		210	483.15	529
		220	493.15	496
		230	503.15	455
		240	513.15	383
Ethylamine	C_2H_7N	0	273.15	708
		10	283.15	695
		20	293.15	683

Table 33-5 Density ρ of Miscellaneous Liquids (Continued)

Liquid	Chemical formula	Temperature t °C	Temperature T °K	Density ρ kg/m³
Ethylamine	C_2H_7N	30	303.15	671
		40	313.15	658
		50	323.15	646
		60	333.15	633
		70	343.15	620
		80	353.15	607
Ethylbenzene	C_8H_{10}	20	293.15	868
Ethyl benzoate	$C_9H_{10}O_2$	20	293.15	1047
Ethyl bromide	C_2H_5Br	20	293.15	1450
Ethyl chloride	C_2H_5Cl	−20	253.15	953
		−10	263.15	933
		0	273.15	919
		10	283.15	907
		20	293.15	892
		30	303.15	878
		40	313.15	862
		50	323.15	846
		60	333.15	829
		70	343.15	813
		80	353.15	796
Ethyl ether	$C_4H_{10}O$	0	273.15	736
		10	283.15	725
		20	293.15	714
		30	303.15	702
		40	313.15	689
		50	323.15	676
		60	333.15	666
		70	343.15	653
		80	353.15	640
		90	363.15	625
		100	373.15	611
		110	383.15	594
		120	393.15	576
		130	403.15	558
		140	413.15	539
		150	423.15	518
		160	433.15	495
		170	443.15	466
		180	453.15	427
		190	463.15	366

Liquid	Chemical formula	Temperature t °C	Temperature T °K	Density ρ kg/m³
Ethyl formate	$C_3H_6O_2$	0	273.15	948
		20	293.15	923
		40	313.15	896
		50	323.15	883
		60	333.15	869
		70	343.15	855
		80	353.15	841
		90	363.15	826
		100	373.15	811
		110	383.15	796
		120	393.15	780
		130	403.15	763
		140	413.15	745
		150	423.15	726
		160	433.15	706
		170	443.15	684
		180	453.15	661
		190	463.15	636
		200	473.15	607
		210	483.15	572
		220	493.15	529
		230	503.15	464
Ethyl iodide	C_2H_5J	20	293.15	1934
Ethylene chloride	$C_2H_4Cl_2$	20	293.15	1257
Ethylene glycol	$C_2H_6O_2$	20	293.15	1113
		40	313.15	1099
		60	333.15	1085
		80	353.15	1070
		100	373.15	1056
Fluorobenzene	C_6H_5F	0	273.15	1047
		20	293.15	1023
		40	313.15	999
		60	333.15	974
		80	353.15	950
		90	363.15	937
		100	373.15	923
		110	383.15	910
		120	393.15	896
		130	403.15	881
		140	413.15	867
		150	423.15	852

Table 33-6 Density ρ of Miscellaneous Liquids (Continued)

Liquid	Chemical formula	Temperature t °C	T °K	Density ρ kg/m³
Fluorobenzene	C_6H_5F	160	433.15	836
		170	443.15	820
		180	453.15	804
		190	463.15	786
		200	473.15	767
		210	483.15	748
		220	493.15	727
		230	503.15	704
		240	513.15	679
		250	523.15	650
		260	533.15	616
		270	543.15	574
		280	553.15	513
Formamide	CH_3ON	20	293.15	1135
Formic acid	CH_2O_2	20	293.15	1220
Glycerol	$C_3H_8O_3$	20	293.15	1260
n—heptane	C_7H_{16}	0	273.15	700
		10	283.15	692
		20	293.15	684
		30	303.15	675
		40	313.15	667
		50	323.15	658
		60	333.15	649
		70	343.15	640
		80	353.15	631
		90	363.15	622
		100	373.15	612
		110	383.15	603
		120	393.15	593
		130	403.15	582
		140	413.15	571
		150	423.15	560
		160	433.15	548
		170	443.15	536
		180	453.15	523
		190	463.15	510
		200	473.15	495
		210	483.15	479
		220	493.15	462

Liquid	Chemical formula	Temperature t °C	T °K	Density ρ kg/m³
n—heptane	C_7H_{16}	230	503.15	441
		240	513.15	418
		250	523.15	388
		260	533.15	346
n—heptyl alcohol	$C_7H_{16}O$	20	293.15	824
n—hexane	C_6H_{14}	0	273.15	677
		10	283.15	668
		20	293.15	660
		30	303.15	651
		40	313.15	641
		50	323.15	632
		60	333.15	622
		70	343.15	612
		80	353.15	602
		90	363.15	592
		100	373.15	581
		110	383.15	570
		120	393.15	559
		130	403.15	547
		140	413.15	534
		150	423.15	521
		160	433.15	506
		170	443.15	491
		180	453.15	475
		190	463.15	457
		200	473.15	437
		210	483.15	412
		220	493.15	381
		230	503.15	333
n—hexyl alcohol	$C_6H_{14}O$	15	288.15	822
Iodobenzene	C_6H_5J	0	273.15	1861
		20	293.15	1831
		40	313.15	1799
		60	333.15	1770
		80	353.15	1739
		100	373.15	1708
		120	393.15	1677
		140	413.15	1645
		160	433.15	1613

Table 33-7 Density ρ of Miscellaneous Liquids (Continued)

Liquid	Chemical formula	Temperature t °C	Temperature T °K	Density ρ kg/m³
Iodobenzene	C_6H_5J	180	453.15	1580
		190	463.15	1563
		200	473.15	1547
		210	483.15	1532
		220	493.15	1512
		230	503.15	1494
		240	513.15	1476
		250	523.15	1458
		260	533.15	1438
		270	543.15	1417
Isoamyl acetate	$C_7H_{14}O_2$	20	293.15	873
Isoamyl alcohol	$C_5H_{12}O$	20	293.15	810
Isobutyl alcohol	$C_4H_{10}O$	20	293.15	804
Isobutyl bromide	C_4H_9Br	20	293.15	1264
Isobutyl chloride	C_4H_9Cl	20	293.15	884
Isobutyl iodide	C_4H_9J	20	293.15	1603
Isopentane	C_5H_{12}	0	273.15	639
		10	283.15	630
		20	293.15	620
		30	303.15	609
		40	313.15	599
		50	323.15	588
		60	333.15	577
		70	343.15	566
		80	353.15	554
		90	363.15	541
		100	373.15	528
		110	383.15	514
		120	393.15	499
		130	403.15	483
		140	413.15	464
		150	423.15	445
		160	433.15	421
		170	443.15	391
		180	453.15	350
Isoprene	C_5H_8	20	293.15	681

Liquid	Chemical formula	Temperature t °C	Temperature T °K	Density ρ kg/m³
Isopropyl alcohol	C_3H_8O	20	293.15	786
Isopropyl bromide	C_3H_7Br	20	293.15	1310
Isopropyl chloride	C_3H_7Cl	20	293.15	859
Isopropyl iodide	C_3H_7J	15	283.15	1714
Lactic acid	$C_3H_6O_3$	20	293.15	1240
Lubricating oil		20	293.15	871
		40	313.15	858
		60	333.15	845
		80	353.15	832
		100	373.15	820
		120	393.15	807
Magnesium chloride, 20%	$MgCl_2$	—20	253.15	1184
		0	273.15	1184
		20	293.15	1184
Mercury	Hg	0	273.15	13595
		20	293.15	13546
Methyl acetate	$C_3H_6O_2$	0	273.15	959
		20	293.15	934
		40	313.15	908
		50	323.15	894
		60	333.15	880
		70	343.15	866
		80	353.15	852
		90	363.15	837
		100	373.15	822
		110	383.15	806
		120	393.15	789
		130	403.15	772
		140	413.15	753
		150	423.15	734
		160	433.15	713
		170	443.15	691
		180	453.15	667
		190	463.15	641
		200	473.15	610
		210	483.15	574

Table 33-8 Density ρ of Miscellaneous Liquids (Continued)

Liquid	Chemical formula	Temperature t °C	Temperature T °K	Density ρ kg/m³
Methyl acetate	$C_3H_6O_2$	220	493.15	528
		230	503.15	453
Methyl alcohol (methanol)	CH_4O	0	273.15	810
		10	283.15	801
		20	293.15	792
		30	303.15	783
		40	313.15	774
		50	323.15	765
		60	333.15	756
		70	343.15	746
		80	353.15	736
		90	363.15	725
		100	373.15	714
		110	383.15	702
		120	393.15	690
		130	403.15	677
		140	413.15	664
		150	423.15	650
		160	433.15	634
		170	443.15	616
		180	453.15	598
		190	463.15	577
		200	473.15	553
		210	483.15	526
		220	493.15	490
		230	503.15	441
Methyl aniline	C_7H_9N	20	293.15	986
Methyl benzoate	$C_8H_8O_2$	20	293.15	1100
Methyl bromide	CH_3Br	−50	223.15	1859
		−40	233.15	1839
		−30	243.15	1802
		−20	253.15	1783
		−10	263.15	1757
		0	273.15	1704
		10	283.15	1706
		20	293.15	1678
		30	303.15	1650
		40	313.15	1621
		50	323.15	1592

Liquid	Chemical formula	Temperature t °C	Temperature T °K	Density ρ kg/m³
Methyl chloride	CH_3Cl	−40	233.15	1025
		−30	243.15	1008
		−20	253.15	997
		−10	263.15	972
		0	273.15	960
		10	283.15	940
		20	293.15	921
		30	303.15	894
		40	313.15	881
		50	323.15	859
		60	333.15	837
Methyl formate	$C_2H_4O_2$	0	273.15	1003
		10	283.15	989
		30	303.15	960
		50	323.15	929
		60	333.15	913
		70	343.15	897
		80	353.15	880
		90	363.15	863
		100	373.15	845
		110	383.15	826
		120	393.15	807
		130	403.15	786
		140	413.15	764
		150	423.15	740
		160	433.15	714
		170	443.15	684
		180	453.15	652
		190	463.15	615
		200	473.15	566
		210	483.15	486
Methyl iodide	CH_3J	20	293.15	2279
Methylamine	CH_5N	−50	223.15	743
		−40	233.15	733
		−30	243.15	722
		−20	253.15	710
		−10	263.15	698
		0	273.15	687
		10	283.15	675
		20	293.15	660

Table 33-9 Density ρ of Miscellaneous Liquids (*Continued*)

Liquid	Chemical formula	Temperature t °C	Temperature T °K	Density ρ kg/m³
Methylamine	CH_5N	30	303.15	647
		40	313.15	633
		50	323.15	618
		60	333.15	603
		70	343.15	587
		80	353.15	571
Methylene chloride	CH_2Cl_2	20	293.15	1336
Nitrobenzene	$C_6H_5NO_2$	20	293.15	1203
Nitromethane	CH_3NO_2	25	293.15	1131
n–nonane	C_9H_{20}	20	293.15	717
Nonyl alcohol	$C_9H_{20}O$	20	293.15	828
n–octane	C_8H_{18}	0	273.15	718
		20	293.15	702
		40	313.15	686
		50	323.15	678
		60	333.15	669
		70	343.15	661
		80	353.15	653
		90	363.15	644
		100	373.15	635
		110	383.15	626
		120	393.15	617
		130	403.15	607
		140	413.15	597
		150	423.15	588
		160	433.15	577
		170	443.15	567
		180	453.15	556
		190	463.15	544
		200	473.15	532
		210	483.15	519
		220	493.15	505
		230	503.15	490
		240	513.15	473
		250	523.15	455
		260	533.15	436
		270	543.15	412
		280	553.15	382
		290	563.15	337

Liquid	Chemical formula	Temperature t °C	Temperature T °K	Density ρ kg/m³
n–octyl alcohol	$C_8H_{18}O$	20	293.15	829
Olive oil		20	293.15	914
Paraffin		0	273.15	880
Paraffin oil		20	293.15	810
n–pentane	C_5H_{12}	0	273.15	645
		10	283.15	636
		20	293.15	626
		30	303.15	617
		40	313.15	606
		50	323.15	596
		60	333.15	585
		70	343.15	574
		80	353.15	562
		90	363.15	550
		100	373.15	538
		110	383.15	525
		120	393.15	511
		130	403.15	496
		140	413.15	479
		150	423.15	460
		160	433.15	439
		170	443.15	416
		180	453.15	387
		190	463.15	345
Petrol, heavy		20	293.15	750
Petrol, light		20	293.15	700
Petroleum		20	293.15	760 … 860
Phenethyl alcohol	$C_8H_{10}O$	13	286.15	1019
Phenol	C_6H_6O	25	298.15	1071
		45	318.15	1054
Phenyl cyanide	C_7H_5N	15	288.15	978
Phenyl isothiocyanate	C_6H_5NS	15	288.15	1138
Propionic acid	$C_3H_6O_2$	20	293.15	993
		190	463.15	800
		200	473.15	786

Table 33-10 Density ρ of Miscellaneous Liquids (Continued)

Liquid	Chemical formula	Temperature t (°C)	Temperature T (°K)	Density ρ (kg/m³)
Propionic anhydride	$C_6H_{10}O_3$	20	293.15	1012
Propyl acetate	$C_5H_{10}O_2$	0	273.15	910
		20	293.15	888
		40	313.15	866
		60	333.15	844
		80	353.15	820
		90	363.15	808
		100	373.15	796
		110	383.15	783
		120	393.15	770
		130	403.15	757
		140	413.15	744
		150	423.15	730
		160	433.15	715
		170	443.15	684
		180	453.15	670
		190	463.15	667
		200	473.15	649
		210	483.15	630
		220	493.15	609
		230	503.15	586
		240	513.15	559
		250	523.15	529
		260	533.15	491
		270	543.15	433
n–propyl alcohol	C_3H_8O	0	273.15	819
		20	293.15	804
		40	313.15	788
		60	333.15	770
		80	353.15	752
		90	363.15	743
		100	373.15	733
		110	383.15	722
		120	393.15	711
		130	403.15	670
		140	413.15	688
		150	423.15	674
		160	433.15	660
		170	443.15	645
		180	453.15	629
		190	463.15	611

Liquid	Chemical formula	Temperature t (°C)	Temperature T (°K)	Density ρ (kg/m³)
n–propyl alcohol	C_3H_8O	200	473.15	592
		210	483.15	572
		220	493.15	549
		230	503.15	523
		240	513.15	492
		250	523.15	453
		260	533.15	391
n–propyl bromide	C_3H_7Br	20	293.15	1353
n–propyl chloride	C_3H_7Cl	20	293.15	890
Propyl formate	$C_4H_8O_2$	0	273.15	929
		20	293.15	906
		40	313.15	883
		60	333.15	859
		70	343.15	847
		80	353.15	834
		90	363.15	821
		100	373.15	808
		110	383.15	795
		120	393.15	781
		130	403.15	767
		140	413.15	752
		150	423.15	737
		160	433.15	721
		170	443.15	705
		180	453.15	687
		190	463.15	669
		200	473.15	649
		210	483.15	626
		220	493.15	602
		230	503.15	576
		240	513.15	544
		250	523.15	503
		260	533.15	440
n–propyl iodide	C_3H_7J	20	293.15	1743
Pyridine	C_5H_5N	20	293.15	983
Sodium	Na	20	293.15	970
Stannic chloride	$SnCl_4$	0	273.15	2279
		20	293.15	2226

Table 33-11 Density ρ of Miscellaneous Liquids (Continued)

Liquid	Chemical formula	t °C	T °K	ρ kg/m³
Stannic chloride	$SnCl_4$	40	313.15	2175
		60	333.15	2123
		80	353.15	2072
		100	373.15	2019
		110	383.15	1992
		120	393.15	1964
		130	403.15	1936
		140	413.15	1907
		150	423.15	1877
		160	433.15	1848
		170	443.15	1818
		180	453.15	1787
		190	463.15	1756
		200	473.15	1722
		210	483.15	1687
		220	493.15	1649
		230	503.15	1609
		240	513.15	1567
		250	523.15	1522
		260	533.15	1475
		270	543.15	1422
		280	553.15	1363
Stearic acid	$C_{18}H_{36}O_2$	69	342.15	847
Sulfur dioxide	SO_2	−20	253.15	1485
		0	273.15	1435
		10	283.15	1409
		20	293.15	1383
Sulfuric acid	H_2SO_4	20	293.15	1834
Tar, brown coal		20	293.15	900
− gas		20	293.15	1000
− road tar		20	293.15	1220
Tetrachloromethane	CCl_4	0	273.15	1633
		20	293.15	1594
		40	313.15	1556
		60	333.15	1517
		70	343.15	1496
		80	353.15	1477
		90	363.15	1455
		100	373.15	1434
		110	383.15	1412
Tetrachloromethane	CCl_4	120	393.15	1390
		130	403.15	1368
		140	413.15	1345
		150	423.15	1322
		160	433.15	1298
		170	443.15	1273
		180	453.15	1247
		190	463.15	1219
		200	473.15	1189
		210	483.15	1157
		220	493.15	1123
		230	503.15	1086
		240	513.15	1044
		250	523.15	998
		260	533.15	941
		270	543.15	867
		280	553.15	763
Thiophene	C_4H_4S	20	293.15	1064
Toluene	C_7H_8	20	293.15	866
		190	463.15	687
		200	473.15	672
n−toluidine	C_7H_9N	20	293.15	989
o−toluidine	C_7H_9N	20	293.15	999
p−toluidine	C_7H_9N	20	293.15	1046
		50	323.15	962
Transformer oil		20	293.15	866
		40	313.15	852
		60	333.15	842
		80	353.15	830
		100	373.15	818
Trichloroethylene	C_2HCl_3	20	293.15	1464
Trichlorotrifluoroethane	$C_2F_3Cl_3$	20	293.15	1576
Turpentine oil	$C_{10}H_{16}$	20	293.15	855
Water	H_2O	0	273.15	1000
		20	293.15	998
		40	313.15	992
		60	333.15	983

Table 33-12 Density ρ of Miscellaneous Liquids (Continued)

Liquid	Chemical formula	Temperature		Density
		t °C	T °K	ρ kg/m³
Water	H_2O	80	353.15	972
		100	373.15	958
		120	393.15	944
		140	413.15	926
		160	433.15	908
		180	453.15	887
		200	473.15	863
		220	493.15	837
		240	513.15	809
		260	533.15	779
		280	553.15	750
		300	573.15	700

Liquid	Chemical formula	Temperature		Density
		t °C	T °K	ρ kg/m³
m—xylene	C_8H_{10}	20	293.15	864
		190	463.15	690
		200	473.15	678
o—xylene	C_8H_{10}	20	293.15	880
		190	463.15	716
		200	473.15	705
p—xylene	C_8H_{10}	20	293.15	861
		190	463.15	620
		200	473.15	612

Table 34-1 Viscosities η of Miscellaneous Liquids

Liquid	Chemical formula	Temperature t °C	Temperature T °K	Viscosity $\eta \times 10^4$ kps/m²	Viscosity $\eta \times 10^4$ Ns/m²
Acetaldehyde	C_2H_4O	0	273.15	0.285	2.797
		10	283.15	0.261	2.557
		20	293.15	0.226	2.220
Acetone	C_3H_6O	−80	193.15	1.516	14.87
		−70	203.15	1.244	12.20
		−60	213.15	1.003	9.84
		−50	223.15	0.823	8.07
		−40	233.15	0.694	6.81
		−30	243.15	0.586	5.75
		−20	253.15	0.517	5.07
		0	273.15	0.407	3.99
		10	283.15	0.369	3.62
		15	288.15	0.344	3.37
		20	293.15	0.338	3.31
		25	298.15	0.322	3.16
		30	303.15	0.301	2.95
		50	323.15	0.261	2.56
Airplane motor oil		20	293.15	811.92	7962.215
		40	313.15	208.08	2040.568
		60	333.15	72.62	712.159
		80	353.15	32.13	315.088
		100	373.15	16.93	166.027
		120	393.15	10.10	99.047
		140	413.15	6.63	65.018
Ammonia	NH_3	−20	253.15	0.260	2.55
		0	273.15	0.245	2.40
		20	293.15	0.224	2.20
n−amyl alcohol	$C_5H_{12}O$	15	288.15	4.742	46.5
		30	303.15	3.049	29.9
Aniline	C_6H_7N	−6	267.15	14.072	138
		0	273.15	10.401	102
		5	278.15	8.219	80.6
		10	283.15	6.628	65.0
		15	288.15	5.415	53.1
		20	293.15	4.517	44.3
		25	298.15	3.783	37.1
		30	303.15	3.294	32.3
		35	308.15	2.845	27.9
		40	313.15	2.417	23.7

Liquid	Chemical formula	Temperature t °C	Temperature T °K	Viscosity $\eta \times 10^4$ kps/m²	Viscosity $\eta \times 10^4$ Ns/m²
Aniline	C_6H_7N	45	318.15	2.203	21.6
		50	323.15	1.886	18.5
		60	333.15	1.591	15.6
		70	343.15	1.295	12.7
		80	353.15	1.111	10.9
		90	363.15	0.953	9.35
		100	373.15	0.841	8.25
		110	383.15	0.743	7.29
		120	393.15	0.668	6.55
		125	398.15	0.502	4.92
Anisole	C_7H_8O	0	273.15	1.815	17.8
		10	283.15	1.540	15.1
		20	293.15	1.346	13.2
		30	303.15	1.234	12.1
		40	313.15	1.142	11.2
		50	323.15	1.060	10.4
		60	333.15	0.989	9.7
Benzene	C_6H_6	0	273.15	0.928	9.1
		10	283.15	0.775	7.6
		20	293.15	0.663	6.5
		30	303.15	0.571	5.6
		40	313.15	0.502	4.92
		50	323.15	0.445	4.36
		60	333.15	0.398	3.90
		70	343.15	0.357	3.50
		80	353.15	0.322	3.16
		90	363.15	0.292	2.86
		100	373.15	0.266	2.61
		110	383.15	0.244	2.39
		120	393.15	0.223	2.19
		130	403.15	0.205	2.01
		140	413.15	0.189	1.89
		150	423.15	0.173	1.70
		160	433.15	0.159	1.56
		170	443.15	0.147	1.44
		180	453.15	0.135	1.32
		190	463.15	0.123	1.21
Benzonitrile	C_7H_5N	0	273.15	1.978	19.4
		20	293.15	1.305	12.8
		25	298.15	1.264	12.4

1 kp = 9.80665 N

Note: Listed values of η are multiplied by 10 000. To obtain viscosity values of η in (kps/m²) and (Ns/m²), multiply by 10^{-4}, i.e., divide by 10 000.

Table 34-2 Viscosities η of Miscellaneous Liquids (Continued)

Liquid	Chemical formula	Temperature t °C	Temperature T °K	Viscosity η × 10⁴ kps/m²	Viscosity η × 10⁴ Ns/m²
Benzonitrile	C₇H₅N	40	313.15	1.020	10.0
		50	323.15	0.893	8.76
		55	328.15	0.842	8.26
		70	343.15	0.679	6.66
Bromine	Br₂	0	273.15	1.265	12.41
		7	280.15	1.162	11.40
		19	292.15	1.020	10.00
		27	300.15	0.943	9.25
		32	305.15	0.906	8.88
Bromobenzene	C₆H₅Br	0	273.15	1.550	15.2
		10	283.15	1.336	13.1
		20	293.15	1.152	11.3
		30	303.15	1.020	10.0
		40	313.15	0.908	8.9
		50	323.15	0.806	7.9
		60	333.15	0.734	7.2
		70	343.15	0.673	6.6
		80	353.15	0.612	6.0
		90	363.15	0.561	5.5
		100	373.15	0.530	5.2
		110	383.15	0.489	4.80
		130	403.15	0.428	4.20
		150	423.15	0.373	3.66
Bromoform	CHBr₃	15	288.15	2.194	21.52
		30	303.15	1.775	17.41
n—butyl alcohol	C₄H₁₀O	−50	223.15	35.384	347
		−40	233.15	22.842	224
		−30	243.15	14.888	146
		−20	253.15	10.503	103
		−10	263.15	7.546	74
		0	273.15	5.292	51.9
		10	283.15	3.946	38.7
		20	293.15	3.008	29.5
		30	303.15	2.325	22.8
		40	313.15	1.815	17.8
		50	323.15	1.438	14.1
		60	333.15	1.162	11.4
		70	343.15	0.948	9.3
		80	353.15	0.775	7.6
		90	363.15	0.642	6.3
		100	373.15	0.551	5.4
		110	383.15	0.469	4.6

Liquid	Chemical formula	Temperature t °C	Temperature T °K	Viscosity η × 10⁴ kps/m²	Viscosity η × 10⁴ Ns/m²
Carbon disulfide	CS₂	0	273.15	0.442	4.33
		10	283.15	0.404	3.96
		20	293.15	0.373	3.66
		30	303.15	0.348	3.41
		40	313.15	0.325	3.19
Castor oil		10	283.15	2488.092	24400
		20	293.15	1006.454	9870
		30	303.15	463.968	4550
		40	313.15	237.592	2330
		50	323.15	131.543	1290
		60	333.15	78.518	770
		70	343.15	49.966	490
		80	353.15	32.631	320
Chlorine	Cl	−74	199.15	0.724	7.10
		−60	213.15	0.622	6.10
		−53	220.15	0.580	5.69
		−45	228.15	0.540	5.30
		−34	239.15	0.499	4.89
Chlorobenzene	C₆H₅Cl	0	273.15	1.081	10.6
		10	283.15	0.925	9.07
		20	293.15	0.815	7.99
		30	303.15	0.719	7.05
		40	313.15	0.643	6.31
		50	323.15	0.578	5.67
		60	333.15	0.525	5.15
		70	343.15	0.480	4.71
		80	353.15	0.439	4.31
		90	363.15	0.405	3.97
		100	373.15	0.374	3.67
		110	383.15	0.346	3.39
		120	393.15	0.319	3.13
		130	403.15	0.299	2.93
		140	413.15	0.279	2.74
		150	423.15	0.261	2.56
		160	433.15	0.244	2.39
		170	443.15	0.227	2.23
		180	453.15	0.213	2.09
		190	463.15	0.200	1.96
		200	473.15	0.189	1.85
		210	483.15	0.176	1.73
		220	493.15	0.166	1.63
		230	503.15	0.156	1.53
		240	513.15	0.147	1.44

1 kp = 9.80665 N

Note: Listed values of η are multiplied by 10 000. To obtain viscosity values of η in (kps/m²) and (Ns/m²), multiply by 10⁻⁴, i.e., divide by 10 000.

Table 34-3 Viscosities η of Miscellaneous Liquids (Continued)

Liquid	Chemical formula	t °C	T °K	Viscosity kps/m²	Viscosity $\eta \times 10^4$ Ns/m²
Chloroform	$CHCl_3$	0	273.15	0.714	7.0
		10	283.15	0.642	6.3
		15	288.15	0.608	5.96
		20	293.15	0.591	5.80
		25	298.15	0.553	5.42
		30	303.15	0.524	5.14
		40	313.15	0.476	4.67
		50	323.15	0.434	4.26
		60	333.15	0.398	3.90
m—cresol	C_7H_8O	0	273.15	96.872	950
		10	283.15	44.765	439
		20	293.15	21.210	208
		30	303.15	10.197	100
		40	313.15	6.302	61.8
		50	323.15	4.466	43.8
		60	333.15	3.436	33.7
		70	343.15	2.549	25
		80	353.15	2.141	21
		90	363.15	1.835	18
		100	373.15	1.632	16
o—cresol	C_7H_8O	20	293.15	9.993	98
		30	303.15	6.220	61
		35	308.15	4.844	47.5
		40	313.15	4.232	41.5
		50	323.15	3.304	32.4
		60	333.15	2.264	22.2
p—cresol	C_5H_8O	20	293.15	20.598	202
		30	303.15	10.503	103
		35	308.15	8.219	80.6
		40	313.15	6.628	65.0
		45	318.15	5.863	57.5
		50	323.15	5.048	49.5
		55	328.15	4.375	42.9
		60	333.15	3.834	37.6
		65	338.15	3.396	33.3
		70	343.15	2.753	27.0
Cyclohexane	C_6H_{12}	15	288.15	1.077	10.56
		20	293.15	0.989	9.7
		30	303.15	0.836	8.2
		40	313.15	0.724	7.1
		50	323.15	0.622	6.1
		60	333.15	0.551	5.4
Cyclohexanol	$C_6H_{12}O$	15	288.15	98.912	970
		20	293.15	69.340	680
		25	298.15	49.966	490
		30	303.15	36.710	360
		35	308.15	27.532	270
		40	313.15	20.394	200
		50	323.15	12.237	120
		60	333.15	7.954	78
		70	343.15	5.099	50
		80	353.15	3.569	35
		90	363.15	2.249	25
Dichlorodifluoromethane (Freon 12)	CF_2Cl_2	−150	123.15	0.9932	9.7400
		−140	133.15	0.9227	9.0486
		−130	143.15	0.8568	8.4023
		−120	153.15	0.7947	7.7933
		−110	163.15	0.7363	7.2206
		−100	173.15	0.6815	6.6832
		−90	183.15	0.6299	6.1772
		−80	193.15	0.5815	5.7026
		−70	203.15	0.5363	5.2593
		−60	213.15	0.4937	4.8415
		−50	223.15	0.4537	4.4493
		−40	233.15	0.4180	4.0992
		−30	243.15	0.3812	3.7383
		−20	253.15	0.3482	3.4147
		−10	263.15	0.3174	3.1126
		0	273.15	0.2884	2.8282
		10	283.15	0.2613	2.5625
		20	293.15	0.2359	2.3134
		30	303.15	0.2120	2.0790
		40	313.15	0.1896	1.8593
		50	323.15	0.1685	1.6524
		60	333.15	0.1485	1.4563
		70	343.15	0.1296	1.2709
		80	353.15	0.1114	1.0925
		90	363.15	0.09357	0.9176
		100	373.15	0.07535	0.7389
		110	383.15	0.05216	0.5115
Dimethyl aniline	$C_8H_{11}N$	10	283.15	1.723	16.9
		20	293.15	1.438	14.1
		30	303.15	1.224	12.0
		40	313.15	1.060	10.4
		50	323.15	0.928	9.1

Note: Listed values of η are multiplied by 10 000. To obtain viscosity values of η in (kps/m²) and (Ns/m²), multiply by 10^{-4}, i.e., divide by 10 000.

1 kp = 9.80665 N

Table 34-4 Viscosities η of Miscellaneous Liquids (Continued)

Liquid	Chemical formula	Temperature t °C	Temperature T °K	Viscosity $\eta \times 10^4$ kps/m²	Viscosity $\eta \times 10^4$ Ns/m²
Dimethyl aniline	$C_8H_{11}N$	60	333.15	0.816	8.0
		70	343.15	0.724	7.1
		80	353.15	0.653	6.4
		90	363.15	0.591	5.8
		98	371.15	0.551	5.4
Diphenyl (mixture by weight: 26.5% diphenyl, $C_{12}H_{10}$ 73.5% diphenyl ether, $C_{12}H_{10}O$)		20	293.15	4.38	42.95
		30	303.15	3.31	32.46
		40	313.15	2.57	25.20
		50	323.15	2.09	20.50
		60	333.15	1.76	17.26
		70	343.15	1.52	14.91
		80	353.15	1.31	12.85
		90	363.15	1.15	11.28
		100	373.15	1.01	9.90
		110	383.15	0.898	8.806
		120	393.15	0.798	7.826
		130	403.15	0.715	7.012
		140	413.15	0.647	6.345
		150	423.15	0.591	5.796
		160	433.15	0.541	5.305
		170	443.15	0.499	4.894
		180	453.15	0.461	4.521
		190	463.15	0.428	4.197
		200	473.15	0.399	3.913
		210	483.15	0.371	3.638
		220	493.15	0.347	3.403
		230	503.15	0.324	3.177
		240	513.15	0.304	2.981
		250	523.15	0.286	2.805
		260	533.15	0.270	2.648
		270	543.15	0.254	2.491
		280	553.15	0.238	2.334
		290	563.15	0.223	2.187
		300	573.15	0.210	2.059
		310	583.15	0.199	1.951
		320	593.15	0.189	1.853
		330	603.15	0.180	1.765
		340	613.15	0.172	1.687
		350	623.15	0.164	1.608
		360	633.15	0.159	1.559
		370	643.15	0.155	1.520

Liquid	Chemical formula	Temperature t °C	Temperature T °K	Viscosity $\eta \times 10^4$ kps/m²	Viscosity $\eta \times 10^4$ Ns/m²
Diphyl		380	653.15	0.151	1.481
		390	663.15	0.147	1.442
		400	673.15	0.143	1.402
Ethyl acetate	$C_4H_8O_2$	0	273.15	0.589	5.78
		10	283.15	0.517	5.07
		20	293.15	0.458	4.49
		30	303.15	0.409	4.01
		40	313.15	0.367	3.60
		50	323.15	0.332	3.26
		60	333.15	0.303	2.97
		70	343.15	0.275	2.70
		80	353.15	0.253	2.48
		90	363.15	0.232	2.28
		100	373.15	0.214	2.10
		110	383.15	0.197	1.93
		120	393.15	0.182	1.78
		130	403.15	0.168	1.65
		140	413.15	0.155	1.52
		150	423.15	0.143	1.40
		160	433.15	0.132	1.29
		170	443.15	0.121	1.19
		180	453.15	0.111	1.09
Ethyl alcohol (ethanol)	C_2H_6O	−100	173.15	47.926	470
		−90	183.15	28.858	283
		−80	193.15	18.457	181
		−70	203.15	12.644	124
		−60	213.15	8.871	87
		−50	223.15	6.526	64
		−40	233.15	4.884	47.9
		−30	243.15	3.722	36.5
		−20	253.15	2.886	28.3
		−10	263.15	2.274	22.3
		0	273.15	1.815	17.8
		10	283.15	1.499	14.7
		20	293.15	1.224	12.0
		30	303.15	1.011	9.91
		40	313.15	0.841	8.25
		50	323.15	0.715	7.01
		60	333.15	0.603	5.91
		70	343.15	0.513	5.03

Note: Listed values of η are multiplied by 10 000. To obtain viscosity values of η in (kps/m²) and (Ns/m²), multiply by 10^{-4}, i.e., divide by 10 000.

1 kp = 9.80665 N

Table 34-5 Viscosities η of Miscellaneous Liquids (Continued)

Liquid	Chemical formula	Temperature t °C	Temperature T °K	Viscosity $\eta \times 10^4$ kps/m²	Viscosity $\eta \times 10^4$ Ns/m²
Ethyl alcohol (ethanol)	C_2H_6O	80	353.15	0.444	4.35
		90	363.15	0.383	3.76
		100	373.15	0.331	3.25
		110	383.15	0.289	2.83
		120	393.15	0.253	2.48
		130	403.15	0.221	2.17
		140	413.15	0.195	1.91
		150	423.15	0.169	1.66
Ethylbenzene	C_8H_{10}	0	273.15	0.891	8.74
		10	283.15	0.775	7.60
		20	293.15	0.679	6.66
		30	303.15	0.602	5.90
		40	313.15	0.537	5.27
		50	323.15	0.484	4.75
		60	333.15	0.441	4.32
		70	343.15	0.402	3.94
		80	353.15	0.367	3.60
		90	363.15	0.338	3.31
		100	373.15	0.311	3.05
		110	383.15	0.288	2.82
		120	393.15	0.267	2.62
Ethyl benzoate	$C_9H_{10}O_2$	10	283.15	2.937	28.8
		15	288.15	2.600	25.5
		20	293.15	2.284	22.4
		25	298.15	2.019	19.8
Ethyl bromide	C_2H_5Br	−120	153.15	5.710	56.0
		−110	163.15	3.936	38.6
		−100	173.15	2.947	28.9
		−90	183.15	2.294	22.5
		−80	193.15	1.846	18.1
		0	273.15	0.494	4.94
		15	288.15	0.426	4.18
		19	292.15	0.405	3.97
		30	303.15	0.355	3.48
		46	319.15	0.310	3.037
		78	351.15	0.238	2.336
		100	373.15	0.202	1.980
		130	403.15	0.164	1.613
		160	433.15	0.128	1.253
Ethylene chloride	$C_2H_4Cl_2$	0	273.15	1.155	11.33
		15	288.15	0.904	8.87
		20	293.15	0.816	8.00
		25	298.15	0.800	7.85
		30	303.15	0.744	7.30
		50	323.15	0.596	5.84
Ethyl chloride	C_2H_5Cl	−20	253.15	0.400	3.92
		−10	263.15	0.361	3.54
		0	273.15	0.326	3.20
		10	283.15	0.297	2.91
		20	293.15	0.271	2.66
		30	303.15	0.249	2.44
		40	313.15	0.228	2.24
Ethyl ether	$C_4H_{10}O$	−120	153.15	4.334	42.5
		−110	163.15	2.590	25.4
		−100	173.15	1.744	17.1
		−90	183.15	1.264	12.4
		−80	193.15	0.989	9.7
		−70	203.15	0.806	7.9
		−60	213.15	0.663	6.5
		−50	223.15	0.561	5.5
		−40	233.15	0.479	4.70
		−30	243.15	0.418	4.10
		−20	253.15	0.371	3.64
		−10	263.15	0.334	3.28
Ethylene glycol	$C_2H_6O_2$	0	273.15	58.113	569.9
		20	293.15	20.292	199
		30	303.15	13.460	132
		40	313.15	9.310	91.3
		60	333.15	5.048	49.5
		80	353.15	3.080	30.2

1 kp = 9.80665 N

Note: Listed values of η are multiplied by 10 000. To obtain viscosity values of η in (kps/m²) and (Ns/m²), multiply by 10^{-4}, i.e., divide by 10 000.

Table 34-6 Viscosities η of Miscellaneous Liquids (Continued)

Liquid	Chemical formula	Temperature t °C	Temperature T °K	Viscosity $\eta \times 10^4$ kps/m²	Viscosity $\eta \times 10^4$ Ns/m²
Ethylene glycol	$C_2H_6O_2$	100	373.15	2.029	19.9
		120	393.15	1.428	14.0
		130	403.15	1.224	12.0
		140	413.15	1.060	10.4
Fluorobenzene	C_6H_5F	0	273.15	0.760	7.45
		10	283.15	0.659	6.46
		20	293.15	0.610	5.98
		30	303.15	0.542	5.32
		40	313.15	0.487	4.78
		50	323.15	0.436	4.28
		60	333.15	0.397	3.89
		70	343.15	0.364	3.57
		80	353.15	0.335	3.29
		90	363.15	0.306	3.00
		100	373.15	0.280	2.75
		110	383.15	0.255	2.50
		120	393.15	0.236	2.31
		130	403.15	0.218	2.14
		140	413.15	0.202	1.98
		150	423.15	0.186	1.82
		160	433.15	0.171	1.68
		170	443.15	0.159	1.56
		180	453.15	0.147	1.44
Formamide	CH_3ON	0	273.15	7.444	73
		10	283.15	5.099	50
		20	293.15	3.824	37.5
		30	303.15	2.998	29.4
		40	313.15	2.478	24.3
		50	323.15	2.080	20.4
		60	333.15	1.744	17.1
		70	343.15	1.448	14.2
		80	353.15	1.193	11.7
		90	363.15	0.999	9.8
		100	373.15	0.846	8.3
		110	383.15	0.734	7.2
		120	393.15	0.642	6.3
Formic acid	CH_2O_2	10	283.15	2.294	22.5
		20	293.15	1.815	17.8
		30	303.15	1.489	14.6
		40	313.15	1.244	12.2
		50	323.15	1.050	10.3

Liquid	Chemical formula	Temperature t °C	Temperature T °K	Viscosity $\eta \times 10^4$ kps/m²	Viscosity $\eta \times 10^4$ Ns/m²
Formic acid	CH_2O_2	60	333.15	0.908	8.9
		70	343.15	0.795	7.8
		80	353.15	0.693	6.8
		90	363.15	0.622	6.1
		100	373.15	0.551	5.4
Glycerol	$C_3H_8O_3$	0	273.15	12338.491	121000
		5	278.15	7188.956	70500
		10	283.15	4027.855	39500
		15	288.15	2396.319	23500
		20	293.15	1509.171	14800
		25	298.15	448.366	4397
		50	323.15	183.548	1800
n–heptane	C_7H_{16}	0	273.15	0.527	5.17
		10	283.15	0.467	4.58
		20	293.15	0.417	4.09
		30	303.15	0.374	3.67
		40	313.15	0.339	3.32
		50	323.15	0.307	3.01
		60	333.15	0.280	2.75
		70	343.15	0.257	2.52
		80	353.15	0.236	2.31
		90	363.15	0.217	2.13
n–hexane	C_6H_{14}	0	273.15	0.405	3.97
		10	283.15	0.362	3.55
		20	293.15	0.326	3.20
		30	303.15	0.296	2.90
		40	313.15	0.269	2.64
		50	323.15	0.246	2.41
		60	333.15	0.225	2.21
Iodobenzene	C_6H_5J	10	283.15	2.009	19.7
		20	293.15	1.519	14.9
		30	303.15	1.479	14.5
		40	313.15	1.290	12.65
		50	323.15	1.142	11.2
		60	333.15	1.015	9.95
		80	353.15	0.831	8.15
		100	373.15	0.704	6.90
		120	393.15	0.597	5.85
		140	413.15	0.520	5.10
Isoamyl alcohol	$C_5H_{12}O$	0	273.15	9.004	88.3

Note: Listed values of η are multiplied by 10 000. To obtain viscosity values of η in (kps/m²) and (Ns/m²), multiply by 10^{-4}, i.e., divide by 10 000.

1 kp = 9.80665 N

Table 34-7 Viscosities η of Miscellaneous Liquids (Continued)

Liquid	Chemical formula	t °C	T °K	η × 10⁴ kps/m²	η × 10⁴ Ns/m²
Isoamyl alcohol	$C_5H_{12}O$	10	283.15	6.322	62.0
		80	353.15	0.904	8.87
Isoamyl alcohol (optically active)	$C_5H_{12}O$	0	273.15	11.319	111
		10	283.15	7.546	74
		20	293.15	5.201	51
		30	303.15	3.661	35.9
		40	313.15	2.661	26.1
		50	323.15	1.987	19.4
		60	333.15	1.499	14.7
		70	343.15	1.173	11.5
		80	353.15	0.928	9.1
		90	363.15	0.755	7.4
		100	373.15	0.622	6.1
		110	383.15	0.520	5.1
		120	393.15	0.438	4.3
Isoamyl alcohol (optically inactive)	$C_5H_{12}O$	0	273.15	8.770	86
		10	283.15	6.220	61
		20	293.15	4.446	43.6
		30	303.15	3.263	32.0
		40	313.15	2.458	24.1
		50	323.15	1.886	18.5
		60	333.15	1.479	14.5
		70	343.15	1.173	11.5
		80	353.15	0.948	9.3
		90	363.15	0.775	7.6
		100	373.15	0.642	6.3
		110	383.15	0.540	5.3
		120	393.15	0.459	4.5
		130	403.15	0.398	3.9
Isobutyl alcohol	$C_4H_{10}O$	−40	233.15	52.311	513
		−30	243.15	30.489	299
		−20	253.15	18.763	184
		−10	263.15	12.542	123
		0	273.15	8.464	83
		10	283.15	5.761	56.5
		20	293.15	4.028	39.5
		30	303.15	2.906	28.5
		40	313.15	2.162	21.4
		50	323.15	1.642	16.1
		60	333.15	1.264	12.4
Isobutyl alcohol	$C_4H_{10}O$	70	343.15	0.989	9.7
		80	353.15	0.795	7.8
		90	363.15	0.642	6.3
		100	373.15	0.530	5.2
Isopentane	C_5H_{12}	0	273.15	0.277	2.72
		10	283.15	0.251	2.46
		20	293.15	0.227	2.23
		30	303.15	0.206	2.02
Isoprene	C_5H_8	0	273.15	0.265	2.60
		10	283.15	0.241	2.36
		20	293.15	0.220	2.16
		30	303.15	0.202	1.98
Isopropyl alcohol	C_3H_8O	−60	213.15	67.403	661
		−50	223.15	38.341	376
		−40	233.15	23.657	232
		−30	243.15	15.194	149
		−20	253.15	10.299	101
		−10	263.15	6.934	68
		0	273.15	4.691	46.0
		10	283.15	3.324	32.6
		20	293.15	2.437	23.9
		30	303.15	1.795	17.6
		40	313.15	1.356	13.3
		60	333.15	0.816	8.0
		80	353.15	0.530	5.2
Isopropyl bromide	C_3H_7Br	0	273.15	0.617	6.05
		10	283.15	0.549	5.38
		20	293.15	0.492	4.82
		30	303.15	0.444	4.35
		40	313.15	0.402	3.94
		50	323.15	0.366	3.59
Isopropyl chloride	C_3H_7Cl	0	273.15	0.410	4.02
		10	283.15	0.365	3.58
		20	293.15	0.328	3.22
		30	303.15	0.298	2.92
Isopropyl iodide	C_3H_7J	15	288.15	0.746	7.32
		30	303.15	0.632	6.20
Lactic acid	$C_3H_6O_3$	25	298.15	41.125	403.3

1 kp = 9.80665 N

Note: Listed values of η are multiplied by 10 000. To obtain viscosity values of η in (kps/m²) and (Ns/m²), multiply by 10^{-4}, i.e., divide by 10 000.

Table 34-8 Viscosities η of Miscellaneous Liquids (Continued)

Liquid	Chemical formula	Temperature t °C	Temperature T °K	Viscosity $\eta \times 10^4$ kps/m²	Viscosity $\eta \times 10^4$ Ns/m²
Lubricating oil		20	293.15	13.31	130.527
		40	313.15	6.94	68.058
		60	333.15	4.26	41.776
		80	353.15	2.89	28.341
		100	373.15	2.04	20.006
		120	393.15	1.57	15.396
Magnesium chloride, 20%	$MgCl_2$	−20	253.15	13.21	129.546
		0	273.15	5.60	54.917
		20	293.15	2.91	28.537
Mercury	Hg	−20	253.15	1.892	18.55
		−10	263.15	1.799	17.64
		0	273.15	1.718	16.85
		10	283.15	1.647	16.15
		20	293.15	1.585	15.54
		30	303.15	1.529	14.99
		40	313.15	1.479	14.50
		50	323.15	1.435	14.07
		60	333.15	1.394	13.67
		70	343.15	1.357	13.31
		80	353.15	1.324	12.98
		90	363.15	1.293	12.68
		100	373.15	1.264	12.40
		110	383.15	1.238	12.14
		120	393.15	1.214	11.91
		130	403.15	1.192	11.69
		140	413.15	1.172	11.49
		150	423.15	1.152	11.30
		160	433.15	1.134	11.12
		170	443.15	1.118	10.96
		180	453.15	1.101	10.80
		190	463.15	1.087	10.66
		200	473.15	1.073	10.52
		210	483.15	1.059	10.39
		220	493.15	1.047	10.27
		230	503.15	1.036	10.16
		240	513.15	1.025	10.05
		250	523.15	1.015	9.95
		260	533.15	1.004	9.85
		270	543.15	0.994	9.75
		280	553.15	0.986	9.67
		290	563.15	0.977	9.58

Liquid	Chemical formula	Temperature t °C	Temperature T °K	Viscosity $\eta \times 10^4$ kps/m²	Viscosity $\eta \times 10^4$ Ns/m²
Mercury	Hg	300	573.15	0.969	9.50
		310	583.15	0.961	9.42
		320	593.15	0.953	9.35
		330	603.15	0.946	9.28
		340	613.15	0.939	9.21
Methyl acetate	$C_3H_6O_2$	20	293.15	0.389	3.81
		30	303.15	0.351	3.44
		40	313.15	0.318	3.12
		50	323.15	0.290	2.84
		60	333.15	0.263	2.58
		70	343.15	0.242	2.37
		80	353.15	0.221	2.17
		90	363.15	0.202	1.98
		100	373.15	0.186	1.82
		110	383.15	0.169	1.66
		120	393.15	0.157	1.54
		130	403.15	0.145	1.42
		140	413.15	0.133	1.30
Methyl alcohol (methanol)	CH_4O	−100	173.15	16.315	160
		−90	183.15	8.973	88
		−80	193.15	5.812	57
		−70	203.15	4.099	40.2
		−60	213.15	3.039	29.8
		−50	223.15	2.305	22.6
		−40	233.15	1.784	17.5
		−30	243.15	1.417	13.9
		−20	253.15	1.183	11.6
		−10	263.15	0.989	9.70
		0	273.15	0.833	8.17
		20	293.15	0.596	5.84
		30	303.15	0.520	5.10
		40	313.15	0.459	4.50
		50	323.15	0.404	3.96
		60	333.15	0.358	3.51
		70	343.15	0.317	3.11
Methylamine	CH_3NH_2	0	273.15	0.241	2.36
Methyl aniline	C_7H_9N	25	298.15	2.039	20.0
		30	303.15	1.581	15.5
		50	323.15	1.509	14.8

Note: Listed values of η are multiplied by 10 000. To obtain viscosity values of η in (kps/m²) and (Ns/m²), multiply by 10^{-4}, i.e., divide by 10 000.

1 kp = 9.80665 N

Table 34-9 Viscosities η of Miscellaneous Liquids (Continued)

Liquid	Chemical formula	Temperature t °C	Temperature T °K	Viscosity $\eta \times 10^4$ kps/m²	Viscosity $\eta \times 10^4$ Ns/m²
Methyl benzoate	$C_8H_8O_2$	20	293.15	2.100	20.50
Methyl bromide	CH_3Br	15	288.15	0.111	1.09
		30	303.15	0.094	0.92
Methyl chloride	CH_3Cl	10	283.15	0.206	2.023
		20	293.15	0.187	1.834
		30	303.15	0.169	1.661
		40	313.15	0.155	1.521
		50	323.15	0.143	1.400
		60	333.15	0.131	1.289
		70	343.15	0.121	1.183
		80	353.15	0.111	1.084
		90	363.15	0.101	0.987
		100	373.15	0.091	0.896
		110	383.15	0.082	0.807
		120	393.15	0.073	0.720
		130	403.15	0.065	0.634
Methyl formate	$C_2H_4O_2$	0	273.15	0.438	4.3
		10	283.15	0.387	3.8
		15	288.15	0.367	3.6
		20	293.15	0.352	3.45
		25	298.15	0.334	3.28
		30	303.15	0.321	3.15
Methyl iodide	CH_3J	0	273.15	0.614	6.025
		20	293.15	0.500	4.900
		40	313.15	0.432	4.240
Methylene chloride	CH_2Cl_2	−20	253.15	0.693	6.8
		−10	263.15	0.614	6.02
		0	273.15	0.548	5.37
		10	283.15	0.490	4.81
		20	293.15	0.444	4.35
		30	303.15	0.404	3.96
		40	313.15	0.370	3.63
Naphthalene	$C_{10}H_8$	80	353.15	0.989	9.7
		90	363.15	0.877	8.6
		100	373.15	0.795	7.8
		110	383.15	0.724	7.1
		120	393.15	0.663	6.5
		130	403.15	0.602	5.9
		140	413.15	0.551	5.4
		150	423.15	0.500	4.9
Nitric acid	HNO_3	0	273.15	2.320	22.75
		10	283.15	1.091	10.7
		20	293.15	0.931	9.13
		40	313.15	0.712	6.98
Nitrobenzene	$C_6H_5O_2N$	0	273.15	3.131	30.7
		10	283.15	2.559	25.1
		20	293.15	2.050	20.1
		30	303.15	1.713	16.8
		40	313.15	1.468	14.4
		50	323.15	1.275	12.5
		60	333.15	1.111	10.9
		70	343.15	0.989	9.7
		80	353.15	0.887	8.7
		90	363.15	0.795	7.8
		100	373.15	0.714	7.0
Nitromethane	CH_3O_2N	0	273.15	0.861	8.44
		10	283.15	0.757	7.42
		20	293.15	0.670	6.57
		30	303.15	0.607	5.95
		40	313.15	0.538	5.28
		50	323.15	0.487	4.78
		60	333.15	0.442	4.33
		70	343.15	0.400	3.92
		80	353.15	0.364	3.57
		85	358.15	0.350	3.43
n−nonane	C_9H_{20}	0	273.15	0.988	9.69
		10	283.15	0.841	8.25
		20	293.15	0.725	7.11
		30	303.15	0.632	6.20
		40	313.15	0.559	5.48
		60	333.15	0.447	4.38
		80	353.15	0.367	3.60
		100	373.15	0.305	2.99
n−octane	C_8H_{18}	0	273.15	0.714	7.0
		10	283.15	0.622	6.1
		20	293.15	0.551	5.4
		30	303.15	0.488	4.79
		40	313.15	0.436	4.28
		50	323.15	0.394	3.86
		60	333.15	0.357	3.50
		70	343.15	0.324	3.18

Note: Listed values of η are multiplied by 10 000. To obtain viscosity values of η in (kps/m²) and (Ns/m²), multiply by 10^{-4}, i.e., divide by 10 000.

1 kp = 9.80665 N

Table 34-10 Viscosities η of Miscellaneous Liquids (*Continued*)

Liquid	Chemical formula	Temperature t (°C)	T (°K)	Viscosity $\eta \times 10^4$ kps/m²	Ns/m²
n–octane	C$_8$H$_{18}$	80	353.15	0.297	2.91
		90	363.15	0.271	2.66
		100	373.15	0.250	2.45
		120	393.15	0.212	2.08
Olive oil		20	293.15	82.393	808
		30	303.15	56.798	557
		40	313.15	37.933	372
		50	323.15	25.799	253
		60	333.15	19.578	192
		70	343.15	14.786	145
		80	353.15	11.829	116
Paraffin oil		18	291.15	103.806	1018
n–pentane	C$_5$H$_{12}$	0	273.15	0.289	2.83
		10	283.15	0.259	2.54
		20	293.15	0.234	2.29
		30	303.15	0.212	2.08
Petrol (gasoline)		18	291.15	0.540	5.3
Phenethyl alcohol (phenyl ethanol)	C$_8$H$_{10}$O	0	273.15	45.071	442
		10	283.15	24.269	238
		20	293.15	14.582	143
		30	303.15	9.157	89.8
		40	313.15	6.230	61.1
		60	333.15	3.324	32.6
		80	353.15	2.050	20.1
		100	373.15	0.137	1.34
Phenol	C$_6$H$_6$O	20	293.15	11.829	116
		30	303.15	7.138	70
		40	313.15	4.864	47.7
		50	323.15	3.498	34.3
		60	333.15	2.610	25.6
		70	343.15	2.039	20.0
		80	353.15	1.621	15.9
		90	363.15	1.315	12.9
		100	373.15	1.071	10.5
		110	383.15	0.897	8.8
		120	393.15	0.795	7.8
		130	403.15	0.734	7.2
		140	413.15	0.704	6.9

Liquid	Chemical formula	Temperature t (°C)	T (°K)	Viscosity $\eta \times 10^4$ kps/m²	Ns/m²
Phenyl cyanide	C$_7$H$_5$N	0	273.15	1.999	19.6
		10	283.15	1.652	16.2
		20	293.15	1.356	13.3
		30	303.15	1.152	11.3
		40	313.15	1.003	9.84
		50	323.15	0.881	8.64
		60	333.15	0.782	7.67
		80	353.15	0.635	6.23
		100	373.15	0.525	5.15
Phenyl isothiocyanate	C$_7$H$_5$NS	0	273.15	2.335	22.9
		10	283.15	1.897	18.6
		20	293.15	1.591	15.6
		30	303.15	1.346	13.2
		40	313.15	1.162	11.4
		60	333.15	0.896	8.79
		80	353.15	0.727	7.13
		100	373.15	0.607	5.95
Phenyl propyl ketone		0	273.15	4.150	40.7
		10	283.15	3.090	30.3
		20	293.15	2.407	23.6
		30	303.15	1.927	18.9
		40	313.15	1.591	15.6
		60	333.15	1.152	11.3
		80	353.15	0.887	8.7
		100	373.15	0.704	6.9
Propionic acid	C$_3$H$_6$O$_2$	0	273.15	1.550	15.2
		10	283.15	1.315	12.9
		20	293.15	1.122	11.0
		30	303.15	0.979	9.6
		40	313.15	0.857	8.4
		50	323.15	0.765	7.5
		60	333.15	0.683	6.7
		70	343.15	0.614	6.02
		80	353.15	0.556	5.45
		90	363.15	0.505	4.95
		100	373.15	0.461	4.52
		110	383.15	0.422	4.14
		120	393.15	0.387	3.80
		130	403.15	0.357	3.50
		140	413.15	0.328	3.22

1 kp = 9.80665 N

Note: Listed values of η are multiplied by 10 000. To obtain viscosity values of η in (kps/m²) and (Ns/m²), multiply by 10^{-4}, i.e., divide by 10 000.

Table 34-11 Viscosities η of Miscellaneous Liquids (Continued)

Liquid	Chemical formula	Temperature t °C	Temperature T °K	Viscosity $\eta \times 10^4$ kps/m²	Viscosity $\eta \times 10^4$ Ns/m²
Propionic anhydride	$C_6H_{10}O_3$	0	273.15	1.642	16.1
		10	283.15	1.356	13.3
		20	293.15	1.142	11.2
		30	303.15	0.979	9.6
		40	313.15	0.846	8.3
		50	323.15	0.744	7.3
		60	333.15	0.663	6.5
		70	343.15	0.591	5.8
		80	353.15	0.530	5.2
		90	363.15	0.481	4.72
		100	373.15	0.438	4.30
		110	383.15	0.407	3.99
		120	393.15	0.367	3.60
		130	403.15	0.338	3.31
		140	413.15	0.312	3.06
		150	423.15	0.290	2.84
		160	433.15	0.269	2.64
Propyl acetate	$C_5H_{10}O_2$	0	273.15	0.785	7.7
		10	283.15	0.683	6.7
		20	293.15	0.591	5.8
		30	303.15	0.520	5.1
		40	313.15	0.469	4.6
		50	323.15	0.418	4.1
		60	333.15	0.375	3.68
		70	343.15	0.341	3.34
		80	353.15	0.310	3.04
		100	373.15	0.255	2.50
n-propyl alcohol	C_3H_8O	− 70	203.15	55.676	546
		− 60	213.15	32.223	316
		− 50	223.15	20.598	202
		− 40	233.15	13.766	135
		− 30	243.15	9.687	95
		− 20	253.15	7.036	69
		− 10	263.15	5.201	51
		0	273.15	3.926	38.5
		10	283.15	2.947	28.9
		20	293.15	2.243	22.0
		30	303.15	1.754	17.2
		40	313.15	1.407	13.8
		60	333.15	0.938	9.2
		80	353.15	0.642	6.3
		90	363.15	0.540	5.3

Liquid	Chemical formula	Temperature t °C	Temperature T °K	Viscosity $\eta \times 10^4$ kps/m²	Viscosity $\eta \times 10^4$ Ns/m²
n-propyl bromide	C_3H_7Br	0	273.15	0.658	6.45
		10	283.15	0.586	5.75
		20	293.15	0.527	5.17
		30	303.15	0.476	4.67
		40	313.15	0.433	4.25
		50	323.15	0.396	3.88
		60	333.15	0.363	3.56
		70	343.15	0.334	3.28
n-propyl chloride	C_3H_7Cl	0	273.15	0.445	4.36
		10	283.15	0.398	3.90
		20	293.15	0.359	3.52
		30	303.15	0.325	3.19
		40	313.15	0.297	2.91
Propylene glycol dinitrate (1.2-dinitro-1.2-propanediol)	$C_3H_6O_6N_2$	10	283.15	5.670	55.6
		20	293.15	4.150	40.7
		30	303.15	3.202	31.4
		40	313.15	2.508	24.6
		50	323.15	2.070	20.3
		60	333.15	1.744	17.1
n-propyl iodide	C_3H_7J	15	288.15	0.853	8.37
		30	303.15	0.683	6.70
Pyridine	C_5H_5N	0	273.15	1.356	13.3
		10	283.15	1.142	11.2
		20	293.15	0.993	9.74
		30	303.15	0.851	8.35
		40	313.15	0.794	7.35
		50	323.15	0.664	6.51
		60	333.15	0.591	5.80
		70	343.15	0.538	5.28
		80	353.15	0.492	4.82
		90	363.15	0.452	4.43
Stannic chloride	$SnCl_4$	− 10	263.15	1.397	13.7
		0	273.15	1.224	12.0
		10	283.15	1.091	10.7
		20	293.15	0.969	9.5
		30	303.15	0.867	8.5
		40	313.15	0.775	7.6
		50	323.15	0.693	6.8
Stearic acid	$C_{18}H_{36}O_2$	70	343.15	11.829	116
		80	353.15	8.107	79.5

1 kp = 9.80665 N

Note: Listed values of η are multiplied by 10 000. To obtain viscosity values of η in (kps/m²) and (Ns/m²), multiply by 10^{-4}, i.e., divide by 10 000.

Table 34-12 Viscosities η of Miscellaneous Liquids (Continued)

Liquid	Chemical formula	Temperature t °C	Temperature T °K	Viscosity $\eta \times 10^4$ kps/m²	Viscosity $\eta \times 10^4$ Ns/m²
Stearic acid	$C_{18}H_{36}O_2$	100	373.15	5.221	51.2
		120	393.15	3.436	33.7
		140	413.15	2.427	23.8
		160	433.15	1.815	17.8
		170	453.15	1.397	13.7
		200	473.15	1.111	10.9
Sulfur dioxide	SO_2	−20	253.15	0.485	4.76
		0	273.15	0.380	3.73
		20	293.15	0.316	3.10
Sulfuric acid	H_2SO_4	0	273.15	49.354	484
		10	283.15	35.894	352
		20	293.15	25.901	254
		30	303.15	16.009	157
		40	313.15	11.727	115
		50	323.15	8.994	88.2
		60	333.15	7.362	72.2
		70	343.15	6.210	60.9
		80	353.15	5.292	51.9
Tetrachloroethane	$C_2H_2Cl_4$	0	273.15	2.712	26.6
		10	283.15	2.192	21.5
		20	293.15	1.784	17.5
		30	303.15	1.509	14.8
		40	313.15	1.305	12.8
		50	323.15	1.152	11.3
		60	333.15	0.989	9.7
		70	343.15	0.867	8.5
		80	353.15	0.765	7.5
Tetrachloroethylene	C_2Cl_4	0	273.15	1.162	11.4
		10	283.15	1.020	10.0
		20	293.15	0.897	8.8
		30	303.15	0.816	8.0
		40	313.15	0.734	7.2
		50	323.15	0.673	6.6
		60	333.15	0.612	6.0
		70	343.15	0.571	5.6
		80	353.15	0.520	5.1
		90	363.15	0.484	4.75
		100	373.15	0.450	4.41
		110	383.15	0.419	4.11
		120	393.15	0.391	3.83

Liquid	Chemical formula	Temperature t °C	Temperature T °K	Viscosity $\eta \times 10^4$ kps/m²	Viscosity $\eta \times 10^4$ Ns/m²
Tetrachloromethane	CCl_4	−10	263.15	1.713	16.80
		0	273.15	1.355	13.29
		10	283.15	1.152	11.30
		20	293.15	0.988	9.69
		30	303.15	0.860	8.43
		40	313.15	0.754	7.39
		50	323.15	0.664	6.51
		60	333.15	0.597	5.85
		70	343.15	0.534	5.24
		80	353.15	0.477	4.68
		90	363.15	0.434	4.26
		100	373.15	0.392	3.84
		110	383.15	0.359	3.52
		120	393.15	0.329	3.23
		130	403.15	0.305	2.99
		140	413.15	0.281	2.76
		150	423.15	0.260	2.55
		160	433.15	0.239	2.34
		170	443.15	0.221	2.17
		180	453.15	0.205	2.01
Thiophene	C_4H_4S	0	273.15	0.887	8.7
		10	283.15	0.765	7.5
		20	293.15	0.673	6.6
		30	303.15	0.591	5.8
		40	313.15	0.530	5.2
		50	323.15	0.477	4.68
		60	333.15	0.432	4.24
		70	343.15	0.394	3.86
		80	353.15	0.357	3.50
Toluene	C_7H_8	0	273.15	0.783	7.68
		10	283.15	0.680	6.67
		20	293.15	0.598	5.86
		30	303.15	0.532	5.22
		40	313.15	0.475	4.66
		50	323.15	0.428	4.20
		60	333.15	0.389	3.81
		70	343.15	0.355	3.48
		80	353.15	0.325	3.19
		90	363.15	0.300	2.94
		100	373.15	0.276	2.71
		110	383.15	0.254	2.49

Note: Listed values of η are multiplied by 10 000. To obtain viscosity values of η in (kps/m²) and (Ns/m²), multiply by 10^{-4}, i.e., divide by 10 000.

1 kp = 9.80665 N

Table 34-13 Viscosities η of Miscellaneous Liquids (Continued)

Liquid	Chemical formula	Temperature t °C	Temperature T °K	Viscosity kps/m³	Viscosity $\eta \times 10^4$ Ns/m²
Toluene	C_7H_6	120	393.15	0.236	2.31
		130	403.15	0.218	2.14
		140	413.15	0.203	1.99
		160	433.15	0.175	1.72
		180	453.15	0.153	1.50
m—toluidine	C_7H_9N	0	273.15	8.831	86.6
		10	283.15	5.619	55.1
		20	293.15	3.885	38.1
		30	303.15	2.845	27.9
		40	313.15	2.182	21.4
		60	333.15	1.428	14.0
		80	353.15	1.020	10.0
		100	373.15	0.785	7.7
o—toluidine	C_7H_9N	0	273.15	10.401	102
		10	283.15	6.557	64.3
		20	293.15	4.477	43.9
		30	303.15	3.263	32.0
		40	313.15	2.488	24.4
		50	323.15	1.958	19.2
		60	333.15	1.611	15.8
		70	343.15	1.326	13.0
		80	353.15	1.132	11.1
		90	363.15	0.969	9.5
		100	373.15	0.846	8.3
p—toluidine	C_7H_9N	50	323.15	1.835	18.0
		60	333.15	1.479	14.5
		70	343.15	1.224	12.0
		80	353.15	1.020	10.0
		90	363.15	0.867	8.5
		100	373.15	0.765	7.5
		110	383.15	0.663	6.5
		120	393.15	0.573	5.62
		130	403.15	0.542	5.32
		140	413.15	0.501	4.91
Transformer oil		20	293.15	32.22	315.970
		40	313.15	14.50	142.196
		60	333.15	7.46	73.158
		80	353.15	4.40	43.149
		100	373.15	3.16	30.989
Trichloroethylene	C_2HCl_3	− 10	263.15	0.806	7.9

Liquid	Chemical formula	Temperature t °C	Temperature T °K	Viscosity kps/m³	Viscosity $\eta \times 10^4$ Ns/m²
Trichloroethylene	C_2HCl_3	0	273.15	0.724	7.1
		10	283.15	0.653	6.4
		20	293.15	0.591	5.8
		30	303.15	0.540	5.3
		40	313.15	0.489	4.8
		50	323.15	0.459	4.5
		60	333.15	0.418	4.1
		70	343.15	0.387	3.8
Trichloromonofluoroethane	$C_2H_2Cl_3F$	20	293.15	1.091	10.7
		30	303.15	0.948	9.3
		40	313.15	0.836	8.2
		50	323.15	0.744	7.3
		60	333.15	0.663	6.5
Trichlorotrifluoroethane	$C_2F_3Cl_3$	0	273.15	0.943	9.25
		10	283.15	0.821	8.05
		20	293.15	0.725	7.11
		30	303.15	0.639	6.27
		40	313.15	0.570	5.59
Triethylcarbinol	$C_7H_{16}O$	0	273.15	32.937	323
		10	283.15	13.970	137
		20	293.15	6.883	67.5
		30	303.15	3.824	37.5
		40	313.15	2.366	23.2
		60	333.15	1.152	11.3
		80	353.15	0.704	6.9
		100	373.15	0.479	4.7
Triisoamylamine	$C_{15}H_{33}N$	0	273.15	5.078	49.8
		10	283.15	3.651	35.8
		20	293.15	2.743	26.9
		30	303.15	2.141	21.0
		40	313.15	1.734	17.0
		60	333.15	1.193	11.7
		80	353.15	0.887	8.7
		100	373.15	0.683	6.7
Tri-n-amylamine	$C_{15}H_{33}N$	0	273.15	4.813	47.2
		10	283.15	3.498	34.3
		20	293.15	2.672	26.2
		30	303.15	2.101	20.6
		40	313.15	1.703	16.7

Note: Listed values of η are multiplied by 10 000. To obtain viscosity values of η in (kps/m²) and (Ns/m²), multiply by 10^{-4}, i.e., divide by 10 000.

1 kp = 9.80665 N

Table 34-14 Viscosities η of Miscellaneous Liquids (Continued)

Liquid	Chemical formula	Temperature t °C	Temperature T °K	Viscosity $\eta \times 10^4$ kps/m²	Viscosity $\eta \times 10^4$ Ns/m²
Tri-n-amylamine	$C_{15}H_{33}N$	60	333.15	1.193	11.7
		80	353.15	0.887	8.7
		100	373.15	0.693	6.8
Tri-n-butylamine	$C_{12}H_{27}N$	0	273.15	2.284	22.4
		10	283.15	1.774	17.4
		20	293.15	1.438	14.1
		30	303.15	1.193	11.7
		40	313.15	1.010	9.9
		60	333.15	0.755	7.4
		80	353.15	0.581	5.7
		100	373.15	0.469	4.6
Turpentine oil		0	273.15	2.292	22.48
		10	283.15	1.818	17.83
		20	293.15	1.516	14.87
		30	303.15	1.297	12.72
		40	313.15	1.092	10.71
		50	323.15	0.944	9.26
		60	333.15	0.837	8.21
		70	343.15	0.742	7.28
		80	353.15	0.684	6.71
Water	H_2O	0	273.15	1.8240	17.887
		5	278.15	1.5454	15.155
		10	283.15	1.3318	13.061
		15	288.15	1.1631	11.406
		20	293.15	1.0244	10.046
		25	298.15	0.9117	8.941
		30	303.15	0.8177	8.019
		35	308.15	0.7245	7.205
		40	313.15	0.6662	6.533
		45	318.15	0.6075	5.958
		50	323.15	0.5605	5.497
		55	328.15	0.5172	5.072
		60	333.15	0.4794	4.701
		65	338.15	0.4445	4.359
		70	343.15	0.4142	4.062
		75	348.15	0.3869	3.794
		80	353.15	0.3626	3.556
		85	358.15	0.3407	3.341
		90	363.15	0.3208	3.146
		95	368.15	0.3040	2.981
		100	373.15	0.2877	2.821

Liquid	Chemical formula	Temperature t °C	Temperature T °K	Viscosity $\eta \times 10^4$ kps/m²	Viscosity $\eta \times 10^4$ Ns/m²
m−xylene	C_8H_{10}	0	273.15	0.816	8.0
		10	283.15	0.714	7.0
		20	293.15	0.622	6.1
		30	303.15	0.561	5.5
		40	313.15	0.500	4.9
		50	323.15	0.452	4.43
		60	333.15	0.411	4.03
		70	343.15	0.376	3.69
		80	353.15	0.346	3.39
		90	363.15	0.319	3.13
		100	373.15	0.295	2.89
		110	383.15	0.274	2.69
		120	393.15	0.255	2.50
		130	403.15	0.238	2.33
o−xylene	C_8H_{10}	0	273.15	1.122	11.0
		10	283.15	0.948	9.3
		20	293.15	0.826	8.1
		30	303.15	0.724	7.1
		40	313.15	0.632	6.2
		50	323.15	0.571	5.6
		60	333.15	0.510	5.0
		70	343.15	0.462	4.53
		80	353.15	0.419	4.11
		90	363.15	0.383	3.76
		100	373.15	0.353	3.46
		110	383.15	0.324	3.18
		120	393.15	0.300	2.94
		130	403.15	0.278	2.73
		140	413.15	0.259	2.54
p−xylene	C_8H_{10}	10	283.15	0.755	7.4
		20	293.15	0.653	6.4
		30	303.15	0.581	5.7
		40	313.15	0.520	5.1
		50	323.15	0.465	4.56
		60	333.15	0.422	4.14
		70	343.15	0.384	3.77
		80	353.15	0.352	3.45
		90	363.15	0.323	3.17
		100	373.15	0.298	2.92
		110	383.15	0.275	2.70
		120	393.15	0.256	2.51
		130	403.15	0.238	2.33

1 kp = 9.80665 N

Note: Listed values of η are multiplied by 10 000. To obtain viscosity values of η in (kps/m²) and (Ns/m²), multiply by 10^{-4}, i.e., divide by 10 000.

Table 35-1 Thermal Conductivities λ of Liquids

Liquid	Chemical formula	t °C	T °K	λ kcal/hm K	λ W/m K
Acetic acid	$C_2H_4O_2$	0	273.15	0.152	0.177
		12	285.15	0.170	0.198
		20	293.15	0.166	0.193
		25	298.15	0.155	0.180
		75	348.15	0.139	0.162
Acetone	C_3H_6O	0	273.15	0.158	0.184
		20	293.15	0.155	0.180
		100	373.15	0.143	0.166
Airplane motor oil		20	293.15	0.125	0.145
		40	313.15	0.123	0.143
		60	333.15	0.121	0.141
		80	353.15	0.120	0.140
		100	373.15	0.118	0.137
		120	393.15	0.117	0.136
		140	413.15	0.115	0.134
Ammonia	NH_3	−20	253.15	0.503	0.585
		0	273.15	0.464	0.540
		20	293.15	0.425	0.494
		100	373.15	0.269	0.313
Amyl acetate	$C_7H_{14}O_2$	12	285.15	0.109	0.127
		33	306.15	0.0899	0.105
Amyl alcohol	$C_5H_{12}O$	0	273.15	0.143	0.166
		20	293.15	0.141	0.164
		50	323.15	0.138	0.160
		100	373.15	0.132	0.154
Amyl benzoate	$C_{12}H_{16}O_2$	33	306.15	0.0913	0.106
Amyl bromide	$C_5H_{11}Br$	12	285.15	0.0853	0.0992
		32	305.15	0.0728	0.0847
Amyl chloride	$C_5H_{11}Cl$	12	285.15	0.102	0.119
Amyl iodide	$C_5H_{11}J$	12	285.15	0.0731	0.0850
Aniline	C_6H_7N	0	273.15	0.148	0.172
		20	293.15	0.148	0.172
		50	323.15	0.148	0.172
		100	373.15	0.144	0.167
		150	423.15	0.136	0.158

Liquid	Chemical formula	t °C	T °K	λ kcal/hm K	λ W/m K
Benzene	C_6H_6	20	293.15	0.132	0.154
		80	353.15	0.130	0.151
Bromobenzene	C_6H_5Br	0	273.15	0.112	0.130
		12	285.15	0.0954	0.111
		20	293.15	0.0960	0.112
		100	373.15	0.104	0.121
n–butyl alcohol	$C_4H_{10}O$	0	273.15	0.146	0.170
		20	293.15	0.144	0.167
		100	373.15	0.137	0.159
Butyric acid (methyl ester)	$C_5H_{10}O_2$	12	285.15	0.122	0.142
Carbon dioxide	CO_2	20	293.15	0.075	0.087
		30	303.15	0.061	0.071
Carbon disulfide	CS_2	0	273.15	0.140	0.162
		30	303.15	0.138	0.160
		75	348.15	0.130	0.151
Castor oil		0	273.15	0.157	0.183
		20	293.15	0.156	0.181
		50	323.15	0.153	0.178
		100	373.15	0.149	0.173
		150	423.15	0.145	0.169
Chlorobenzene	C_6H_5Cl	0	273.15	0.130	0.151
		12	285.15	0.109	0.127
		75	348.15	0.119	0.138
Chloroform	$CHCl_3$	12	285.15	0.104	0.121
		20	293.15	0.111	0.129
		30	303.15	0.119	0.138
Cylinder oil		0	273.15	0.133	0.155
		50	323.15	0.130	0.151
		100	373.15	0.128	0.149
		200	473.15	0.122	0.142
Dichlorodifluoromethane (Freon 12)	CF_2Cl_2	−150	123.15	0.1229	0.1429
		−140	133.15	0.1205	0.1401
		−130	143.15	0.1181	0.1374
		−120	153.15	0.1156	0.1344
		−110	163.15	0.1131	0.1315

1 kcal/h = 1.163 W

Table 35-2 Thermal Conductivities λ of Liquids (Continued)

Liquid	Chemical formula	Temperature t °C	T °K	Thermal conductivity kcal/hm K	λ W/m K
Dichlorodifluoromethane	CF_2Cl_2	−100	173.15	0.1106	0.1286
		−90	183.15	0.1080	0.1256
		−80	193.15	0.1054	0.1226
		−70	203.15	0.1028	0.1196
		−60	213.15	0.1002	0.1165
		−50	223.15	0.0975	0.1134
		−40	233.15	0.0947	0.1101
		−30	243.15	0.0919	0.1069
		−20	253.15	0.0891	0.1036
		−10	263.15	0.0862	0.1003
		0	273.15	0.0833	0.0969
		10	283.15	0.0802	0.0933
		20	293.15	0.0771	0.0897
		30	303.15	0.0738	0.0858
		40	313.15	0.0705	0.0820
		50	323.15	0.0670	0.0779
		60	333.15	0.0633	0.0736
		70	343.15	0.0594	0.0691
		80	353.15	0.0552	0.0642
		90	363.15	0.0504	0.0586
		100	373.15	0.0448	0.0521
		110	383.15	0.0361	0.0420
Diethyl ether	$C_4H_{10}O$	0	273.15	0.121	0.141
		20	293.15	0.119	0.138
		50	323.15	0.117	0.136
		100	373.15	0.114	0.133
Diethylene glycol	$C_4H_{10}O_3$	0	273.15	0.174	0.202
		20	293.15	0.176	0.205
		100	373.15	0.184	0.214
Diphyl (mixture by weight: 26.5% diphenyl, $C_{12}H_{10}$ 73.5% phenyl ether, $C_{12}H_{10}O$)		20	293.15	0.119	0.138
		30	303.15	0.118	0.137
		40	313.15	0.117	0.136
		50	323.15	0.115	0.134
		60	333.15	0.114	0.133
		70	343.15	0.113	0.131
		80	353.15	0.112	0.130
		90	363.15	0.110	0.128
		100	373.15	0.109	0.127
		110	383.15	0.108	0.126

Liquid	Chemical formula	Temperature t °C	T °K	Thermal conductivity kcal/hm K	λ W/m K
Diphyl		120	393.15	0.107	0.124
		130	403.15	0.105	0.122
		140	413.15	0.104	0.121
		150	423.15	0.103	0.120
		160	433.15	0.102	0.119
		170	443.15	0.100	0.116
		180	453.15	0.099	0.115
		190	463.15	0.098	0.114
		200	473.15	0.096	0.112
		210	483.15	0.095	0.110
		220	493.15	0.094	0.109
		230	503.15	0.093	0.108
		240	513.15	0.091	0.106
		250	523.15	0.090	0.105
		260	533.15	0.089	0.104
		270	543.15	0.088	0.102
		280	553.15	0.086	0.100
		290	563.15	0.085	0.099
		300	573.15	0.084	0.098
		310	583.15	0.083	0.097
		320	593.15	0.081	0.094
		330	603.15	0.080	0.093
		340	613.15	0.079	0.092
		350	623.15	0.077	0.090
		360	633.15	0.076	0.088
		370	643.15	0.075	0.087
		380	653.15	0.074	0.086
		390	663.15	0.072	0.084
		400	673.15	0.071	0.083
Dodecane	$C_{12}H_{26}$	0	273.15	0.130	0.151
		75	348.15	0.122	0.142
		100	373.15	0.122	0.142
Ethyl acetate	$C_4H_8O_2$	12	285.15	0.125	0.145
		34	307.15	0.106	0.123
Ethyl alcohol	C_2H_6O	0	273.15	0.159	0.185
		20	293.15	0.1565	0.182
		30	303.15	0.155	0.180
		50	323.15	0.153	0.178
		75	348.15	0.150	0.174

1 kcal/h = 1.163 W

Table 35-3 Thermal Conductivities λ of Liquids (Continued)

Liquid	Chemical formula	Temperature t °C	Temperature T °K	Thermal conductivity λ kcal/hm K	Thermal conductivity λ W/m K
Ethyl benzoate	$C_9H_{10}O_2$	32	305.15	0.104	0.121
Ethyl bromide	C_2H_5Br	0	273.15	0.106	0.123
		20	293.15	0.104	0.121
		50	323.15	0.101	0.117
		100	373.15	0.096	0.112
Ethyl butyrate	$C_6H_{12}O_2$	12	285.15	0.115	0.134
Ethyl formate	$C_3H_6O_2$	12	285.15	0.136	0.158
Ethylene glycol	$C_2H_6O_2$	0	273.15	0.219	0.255
		20	293.15	0.222	0.258
		50	323.15	0.225	0.262
		100	373.15	0.231	0.269
Ethyl iodide	C_2H_5J	0	273.15	0.096	0.112
		20	293.15	0.096	0.112
		50	323.15	0.095	0.110
		100	373.15	0.093	0.108
Ethyl sulfide	$H_4H_{10}S$	12	285.15	0.119	0.138
Ethyl valerate	$C_7H_{14}O_2$	12	285.15	0.112	0.130
Formic acid	CH_2O_2	0	273.15	0.224	0.261
		20	293.15	0.221	0.257
		75	348.15	0.212	0.247
Glycerol	$C_3H_8O_3$	0	273.15	0.243	0.283
		20	293.15	0.245	0.285
		50	323.15	0.247	0.287
		100	373.15	0.250	0.291
n-heptane	C_7H_{16}	0	273.15	0.121	0.141
		20	293.15	0.120	0.140
		60	333.15	0.118	0.137
n-heptyl alcohol	$C_7H_{16}O$	0	273.15	0.143	0.166
		20	293.15	0.141	0.164
		30	303.15	0.140	0.163
		50	323.15	0.138	0.160
		75	348.15	0.135	0.157
		100	373.15	0.133	0.156

Liquid	Chemical formula	Temperature t °C	Temperature T °K	Thermal conductivity λ kcal/hm K	Thermal conductivity λ W/m K
n—hexane	C_6H_{14}	0	273.15	0.119	0.138
		20	293.15	0.118	0.137
		50	323.15	0.117	0.136
		100	373.15	0.116	0.135
n— hexyl alcohol	$C_6H_{14}O$	20	293.15	0.140	0.163
		30	303.15	0.139	0.162
		50	323.15	0.137	0.159
		75	348.15	0.135	0.157
		100	373.15	0.132	0.154
Isoamyl alcohol	$C_5H_{12}O$	0	273.15	0.130	0.151
		20	293.15	0.129	0.150
		50	323.15	0.128	0.149
		100	373.15	0.126	0.147
Isobutyl alcohol	$C_4H_{10}O$	0	273.15	0.133	0.154
		12	285.15	0.122	0.142
Isobutyl bromide	C_4H_9Br	12	285.15	0.1001	0.116
Isobutyl chloride	C_4H_9Cl	12	285.15	0.1001	0.116
Isobutyl iodide	C_4H_9J	12	285.15	0.0749	0.0871
Isopropyl alcohol	C_3H_8O	0	273.15	0.135	0.157
		20	293.15	0.134	0.156
		100	373.15	0.131	0.152
Kerosene		30	303.15	0.129	0.150
		75	348.15	0.121	0.141
Lubricating oil		0	273.15	0.124	0.144
		20	293.15	0.124	0.144
		40	313.15	0.123	0.143
		60	333.15	0.122	0.142
		80	353.15	0.121	0.141
		100	373.15	0.120	0.140
		120	393.15	0.119	0.138
Magnesium chloride, 20%	$MgCl_2$	— 20	253.15	0.337	0.392
		0	273.15	0.389	0.452
Mercury	Hg	0	273.15	9.000	10.467
		20	293.15	8.000	9.304

1 kcal/h = 1.163 W

Table 35-4 Thermal Conductivities λ of Liquids (Continued)

Liquid	Chemical formula	Temperature t °C	Temperature T °K	Thermal conductivity λ kcal/hm K	Thermal conductivity λ W/m K
Methyl acetate	$C_3H_6O_2$	12	285.15	0.139	0.161
		29	302.15	0.117	0.136
Methyl alcohol	CH_4O	0	273.15	0.184	0.214
		20	293.15	0.182	0.212
		30	303.15	0.182	0.212
		75	348.15	0.158	0.184
		100	373.15	0.175	0.204
Methyl chloride	CH_3Cl	−20	253.15	0.168	0.195
		−10	263.15	0.161	0.187
		0	273.15	0.154	0.179
		20	293.15	0.140	0.163
Methylene chloride	CH_2Cl_2	−20	253.15	0.139	0.162
		−10	263.15	0.137	0.159
		0	273.15	0.136	0.158
		20	293.15	0.133	0.155
Methyl valerate	$C_6H_{12}O_2$	12	285.15	0.115	0.134
Nitrobenzene	$C_6H_5NO_2$	0	273.15	0.132	0.154
		12	285.15	0.138	0.160
		20	293.15	0.130	0.151
		125	398.15	0.117	0.136
Nonyl alcohol	$C_9H_{20}O$	0	273.15	0.145	0.169
		20	293.15	0.145	0.169
		100	373.15	0.138	0.160
n−octane	C_8H_{18}	0	273.15	0.128	0.149
		20	293.15	0.126	0.147
		100	373.15	0.118	0.137
n−octyl alcohol	$C_8H_{18}O$	0	273.15	0.148	0.172
		20	293.15	0.144	0.167
		100	373.15	0.136	0.158
Olive oil		0	273.15	0.146	0.170
		20	293.15	0.145	0.169
		50	323.15	0.143	0.166
		100	373.15	0.141	0.164
		200	473.15	0.135	0.157
Paraffin oil		0	273.15	0.108	0.126
		20	293.15	0.107	0.124
		50	323.15	0.105	0.122
		100	373.15	0.102	0.119
Paraffine		0	273.15	0.108	0.126
		50	323.15	0.108	0.126
		100	373.15	0.108	0.126
		200	473.15	0.107	0.124
n−pentane	C_5H_{12}	−200	73.15	0.146	0.170
		−150	123.15	0.140	0.163
		−100	173.15	0.133	0.155
		−50	223.15	0.127	0.148
		0	273.15	0.120	0.140
		20	293.15	0.117	0.136
		100	373.15	0.107	0.124
Petrol		0	273.15	0.125	0.145
		20	293.15	0.113	0.131
		50	323.15	0.095	0.110
Petrolether		0	273.15	0.115	0.134
		30	303.15	0.112	0.131
		50	323.15	0.111	0.129
		75	348.15	0.109	0.127
		100	373.15	0.107	0.124
Petroleum		0	273.15	0.134	0.156
		20	293.15	0.130	0.151
		50	323.15	0.125	0.145
		100	373.15	0.115	0.134
Propionic acid	$C_3H_6O_2$	12	285.15	0.140	0.163
Propyl acetate	$C_5H_{10}O_2$	12	285.15	0.118	0.137
n−propyl alcohol	C_3H_8O	0	273.15	0.150	0.174
		20	293.15	0.148	0.172
		100	373.15	0.139	0.162
Propyl bromide	C_3H_7Br	12	285.15	0.093	0.108
Propyl chloride	C_3H_7Cl	12	285.15	0.102	0.118
Propyl formate	$C_4H_8O_2$	12	285.15	0.129	0.150

1 kcal/h = 1.163 W

Table 35-5 Thermal Conductivities λ of Liquids (Continued)

Liquid	Chemical formula	Temperature t (°C)	Temperature T (°K)	λ (kcal/hm K)	λ (W/m K)
Propyl iodide	C₃H₇J	12	285.15	0.079	0.092
Sodium	Na	100	373.15	72.9	84.783
		210	483.15	68.4	79.549
Sulfur dioxide	SO₂	− 20	253.15	0.192	0.223
		− 10	263.15	0.187	0.217
		0	273.15	0.182	0.212
		20	293.15	0.171	0.199
Sulfuric acid	H₂SO₄	20	293.15	0.270	0.314
Tar		20	293.15	0.120	0.140
Tetrachloromethane	CCl₄	0	273.15	0.094	0.109
		50	323.15	0.092	0.107
		100	373.15	0.090	0.105
Tetralin	C₁₀H₂₂	0	273.15	0.130	0.151
		75	348.15	0.122	0.142
		100	373.15	0.119	0.138
Toluene	C₇H₈	0	273.15	0.130	0.151
		20	293.15	0.130	0.151
		50	323.15	0.126	0.147
		100	373.15	0.118	0.137
Transformer oil		20	293.15	0.107	0.124
		40	313.15	0.106	0.123
		60	333.15	0.105	0.122
		80	353.15	0.103	0.120
		100	373.15	0.102	0.119

Liquid	Chemical formula	Temperature t (°C)	Temperature T (°K)	λ (kcal/hm K)	λ (W/m K)
Trichloroethylene	C₂HCl₃	50	323.15	0.119	0.138
Turpentine		15	288.15	0.110	0.128
Valeric acid	C₅H₁₀O₂	12	285.15	0.119	0.138
Water	H₂O	0	273.15	0.477	0.555
		20	293.15	0.514	0.598
		40	313.15	0.539	0.627
		60	333.15	0.560	0.651
		80	353.15	0.575	0.669
		100	373.15	0.586	0.682
		120	393.15	0.589	0.685
		140	413.15	0.588	0.684
		150	423.15	0.587	0.683
		160	433.15	0.585	0.680
		180	453.15	0.579	0.673
		200	473.15	0.572	0.665
		220	493.15	0.561	0.652
		240	513.15	0.545	0.634
		250	523.15	0.537	0.624
		260	533.15	0.527	0.613
		280	553.15	0.506	0.588
		300	573.15	0.485	0.564
m−xylene	C₈H₁₀	20	293.15	0.125	0.145
		125	398.15	0.097	0.113

1 kcal/h = 1.163 W

Table 36-1 Thermodynamic Properties of Water at the Saturation Pressure

Temperature		Density	Coefficient of volumetric thermal expansion	Specific heat		Thermal conductivity		Thermal diffusivity	Absolute viscosity		Kinematic viscosity	Prandtl number
t	T	ρ	$\beta \times 10^4$	c_p		λ		$\alpha \times 10^6$	$\eta \times 10^6$		$\nu \times 10^6$	
°C	°K	kg/m³	1/K	kcal/kg K	kJ/kg K	kcal/hm K	W/m K	m²/s	kps/m²	Ns/m²	m²/s	Pr
0	273.15	999.9	− 0.7	1.0093	4.226	0.480	0.558	0.131	182.9	1793.636	1.789	13.7
5	278.15	1000.0	—	1.0047	4.206	0.488	0.568	0.135	156.5	1534.741	1.535	11.4
10	283.15	999.7	0.95	1.0019	4.195	0.496	0.577	0.137	132.2	1296.439	1.300	9.5
15	288.15	999.1	—	1.0000	4.187	0.505	0.587	0.141	115.8	1135.610	1.146	8.1
20	293.15	998.2	2.1	0.9988	4.182	0.513	0.597	0.143	101.3	993.414	1.006	7.0
25	298.15	997.1	—	0.9980	4.178	0.521	0.606	0.146	89.8	880.637	0.884	6.1
30	303.15	995.7	3.0	0.9975	4.176	0.529	0.615	0.149	80.8	792.377	0.805	5.4
35	308.15	994.1	—	0.9973	4.175	0.537	0.624	0.150	73.4	719.808	0.725	4.8
40	313.15	992.2	3.9	0.9973	4.175	0.544	0.633	0.151	67.1	658.026	0.658	4.3
45	318.15	990.2	—	0.9975	4.176	0.550	0.640	0.155	61.7	605.070	0.611	3.9
50	323.15	988.1	4.6	0.9978	4.178	0.556	0.647	0.157	56.6	555.056	0.556	3.55
55	328.15	985.7	—	0.9982	4.179	0.561	0.652	0.158	52.0	509.946	0.517	3.27
60	333.15	983.2	5.3	0.9987	4.181	0.566	0.658	0.159	48.1	471.670	0.478	3.00
65	338.15	980.6	—	0.9993	4.184	0.570	0.663	0.161	44.4	435.415	0.444	2.76
70	343.15	977.8	5.8	1.0000	4.187	0.574	0.668	0.163	41.2	404.034	0.415	2.55
75	348.15	974.9	—	1.0008	4.190	0.577	0.671	0.164	38.4	376.575	0.366	2.23
80	353.15	971.8	6.3	1.0017	4.194	0.579	0.673	0.165	35.9	352.059	0.364	2.25
85	358.15	968.7	—	1.0026	4.198	0.581	0.676	0.166	33.5	328.523	0.339	2.04
90	363.15	965.3	7.0	1.0036	4.202	0.583	0.678	0.167	31.5	308.909	0.326	1.95
95	368.15	961.9	—	1.0046	4.206	0.585	0.680	0.168	29.8	292.238	0.310	1.84
100	373.15	958.4	7.5	1.0057	4.211	0.586	0.682	0.169	28.3	277.528	0.294	1.75
110	383.15	951.0	8.0	1.0090	4.224	0.588	0.684	0.170	26.0	254.973	0.268	1.57
120	393.15	943.5	8.5	1.0108	4.232	0.589	0.685	0.171	24.0	235.360	0.244	1.43
130	403.15	934.8	9.1	1.0150	4.250	0.590	0.686	0.172	21.6	211.824	0.226	1.32
140	413.15	926.3	9.7	1.0167	4.257	0.588	0.684	0.172	20.5	201.036	0.212	1.23
150	423.15	916.9	10.3	1.0200	4.270	0.588	0.684	0.173	18.9	185.346	0.201	1.17
160	433.15	907.6	10.8	1.0234	4.285	0.585	0.680	0.173	17.5	171.616	0.191	1.10
170	443.15	897.3	11.5	1.050	4.396	0.584	0.679	0.172	16.6	162.290	0.181	1.05
180	453.15	886.6	12.1	1.050	4.396	0.579	0.673	0.172	15.5	152.003	0.173	1.01
190	463.15	876.0	12.8	1.070	4.480	0.576	0.670	0.171	14.8	145.138	0.166	0.97
200	473.15	862.8	13.5	1.075	4.501	0.572	0.665	0.170	14.2	139.254	0.160	0.95
210	483.15	852.8	14.3	1.09	4.560	0.563	0.655	0.168	13.4	131.409	0.154	0.92
220	493.15	837.0	15.2	1.10	4.605	0.561	0.652	0.167	12.7	124.544	0.149	0.90
230	503.15	827.3	16.2	1.12	4.690	0.548	0.637	0.164	12.2	119.641	0.145	0.88
240	513.15	809.0	17.2	1.13	4.731	0.545	0.634	0.162	11.6	113.757	0.141	0.86
250	523.15	799.2	18.6	1.16	4.857	0.531	0.618	0.160	11.2	109.834	0.137	0.86
260	533.15	779.0	20.0	1.19	4.982	0.527	0.613	0.156	10.7	104.931	0.135	0.86
270	543.15	767.9	21.7	1.20	5.030	0.507	0.590	0.152	10.4	101.989	0.133	0.87
280	553.15	750.0	23.8	1.25	5.234	0.506	0.588	0.147	10.0	98.067	0.131	0.89
290	563.15	732.3	26.5	1.30	5.445	0.480	0.558	0.140	9.6	94.144	0.129	0.92
300	573.15	712.5	29.5	1.36	5.694	0.485	0.564	0.132	9.4	92.182	0.128	0.98
310	583.15	690.6	33.5	1.47	6.155	0.446	0.519	0.122	9.0	88.260	0.128	1.05
320	593.15	667.1	38.0	1.58	6.610	0.425	0.494	0.112	8.7	85.318	0.128	1.13
325	598.15	650.0	—	1.60	6.699	0.405	0.471	0.108	8.5	83.357	0.127	1.18
330	603.15	640.2	42.5	1.73	7.245	0.402	0.468	0.101	8.3	81.395	0.127	1.25
340	613.15	609.4	47.5	1.95	8.160	0.376	0.437	0.088	7.9	77.473	0.127	1.45
350	623.15	572.0	—	2.22	9.295	0.344	0.400	0.076	7.4	72.569	0.127	1.67
360	633.15	524.0	—	2.35	9.850	0.306	0.356	0.067	6.8	66.685	0.127	1.91
370	643.15	448.0	—	2.79	11.690	0.252	0.293	0.058	5.8	56.879	0.127	2.18

1 at = 1 kp/cm² = 98066.5 N/m² = 9.80665 N/cm² = 0.980665 bar

1 kcal = 4.1868 kJ
1 kcal/h = 1.163 W

Table 37-1 Thermal Conductivity λ of Water at Various Pressures

Pressure p =		100 at = 98.0665 bar		200 at = 196.1330 bar		300 at = 294.1995 bar		400 at = 392.2660 bar	
Temperature		Thermal conductivity							
t	T	λ		λ		λ		λ	
°C	°K	kcal/hm K	W/m K	kcal/hm K	W/m K	kcal/hm K	W/m K	kcal/hm K	W/m K
0	273.15	0.477	0.555	0.480	0.558	0.484	0.563	0.488	0.568
20	293.15	0.519	0.604	0.523	0.608	0.528	0.614	0.533	0.620
40	313.15	0.549	0.638	0.554	0.644	0.559	0.650	0.564	0.656
60	333.15	0.571	0.664	0.575	0.669	0.580	0.675	0.585	0.680
80	353.15	0.584	0.679	0.589	0.685	0.594	0.691	0.599	0.697
100	373.15	0.593	0.690	0.598	0.695	0.603	0.701	0.608	0.707
120	393.15	0.596	0.693	0.602	0.700	0.608	0.707	0.614	0.714
140	413.15	0.596	0.693	0.602	0.700	0.608	0.707	0.615	0.715
160	433.15	0.593	0.690	0.599	0.697	0.606	0.705	0.613	0.713
180	453.15	0.587	0.683	0.593	0.690	0.600	0.698	0.609	0.708
200	473.15	0.578	0.672	0.584	0.679	0.593	0.690	0.602	0.700
220	493.15	0.564	0.656	0.572	0.665	0.582	0.677	0.591	0.687
240	513.15	0.547	0.636	0.557	0.648	0.567	0.659	0.577	0.671
260	533.15	0.526	0.612	0.536	0.623	0.547	0.636	0.558	0.649
280	553.15	0.500	0.582	0.512	0.595	0.526	0.612	0.538	0.626
300	573.15	0.466	0.542	0.482	0.561	0.499	0.508	0.513	0.597
320	593.15			0.444	0.516	0.463	0.538	0.482	0.561
340	613.15			0.392	0.456	0.422	0.491	0.447	0.520
360	633.15			0.314	0.365	0.364	0.423	0.401	0.466
370	643.15					0.326	0.379	0.376	0.437

1 at = 1 kp/cm² = 98066.5 N/m² = 9.80665 N/cm² = 0.980665 bar   1 kcal/h = 1.163 W

Table 38-1 Composition and Heating Values H_s and H_i of Liquid Fuels

Fuel	Chemical formula	Density at 15°C ρ (kg/m³)	Boiling point t (°C)	Boiling point T (°K)	Characteristic σ	C	H	O + N₂	S	H_s (kcal/kg)	H_s (kJ/kg)	H_i (kcal/kg)	H_i (kJ/kg)	O_{min} (m³ₙ/kg)	L_{min} (m³ₙ/kg)	CO₂ (m³ₙ)	H₂O (m³ₙ)
Alcohol, 100%	C₂H₅OH	794	78.3	351.45	1.500	52	13	—	—	7 100	29 726	6 400	26 796	1.45	7.00	0.97	1.46
95%		809	78.5	351.65	1.500	—	—	—	—	6 730	28 177	6 030	25 246	1.39	6.54	0.92	1.45
90%		823	78.7	351.85	1.500	—	—	—	—	6 350	26 586	5 700	23 865	1.32	6.30	0.88	1.44
85%		836	78.9	352.05	1.500	—	—	—	—	6 020	25 205	5 330	22 316	1.24	5.98	0.83	1.43
Benzene		≈ 760	<120	<393.15	1.530	80.7	14.2	5.1	—	10 800	45 217	10 040	42 035	2.43	11.6	1.59	1.68
Benzene	C₆H₆	875	80.5	353.65	1.250	91.7	7.8	—	0.5	10 100	42 287	9 650	40 403	2.16	10.3	1.72	0.86
— commercial, I (90)		882	—	—	1.260	92.1	7.9	—	—	10 000	41 868	9 600	40 193	2.17	10.4	1.71	0.88
— commercial, II (50)		876	—	—	1.300	91.6	8.4	—	—	10 100	42 287	9 650	40 403	2.20	10.6	1.70	0.98
— motor		870	—	—	—	91.7	8.3	—	—	10 100	42 287	9 650	40 403	—	—	—	—
Gas oil (Diesel oil)		≈ 870	<350	<623.15	1.380	86.6	12.9	0.2	0.3	10 680	44 715	9 994	41 843	2.21	10.54	1.60	1.23
Heptane	C₇H₁₆	683	98	371.15	1.571	83.9	16.1	—	—	11 470	48 023	10 610	44 422	2.46	11.8	1.57	1.79
Hexane	C₆H₁₄	660	65	338.15	1.584	83.6	16.4	—	—	11 500	48 148	10 600	44 380	2.47	11.8	1.56	1.82
Liquid gas		2220	—	—	—	82.5	17.5	—	—	12 160	50 911	11 000	46 055	—	—	—	—
Methane, motor		920	—	—	—	77.4	20.6	2.0	—	12 160	50 911	11 600	48 567	—	—	—	—
Mineral oil		≈ 850	—	—	—	—	—	—	—	10 000	41 868	9 600	40 193	—	—	—	—
Naphthalene (boiling point 80°C)	C₁₀H₈	977 (80°C)	218	491.15	1.200	93.7	6.3	—	—	9 680	40 528	9 340	39 105	2.10	9.99	1.75	0.70
Octane	C₈H₁₈	700	125	398.15	1.562	84.1	15.9	—	—	11 420	47 813	10 570	44 254	2.46	11.8	1.57	1.78
Pentane	C₅H₁₂	626	37	310.15	1.600	83.2	16.8	—	—	11 620	48 651	10 720	44 882	2.49	11.9	1.55	1.87
Petroleum		≈ 810	—	—	—	85	15	—	—	10 000	41 868	9 500	39 775	—	—	—	—
Tetralin	C₁₀H₁₂	975	205	478.15	1.300	90.8	9.2	—	—	10 150	42 496	9 660	40 444	2.20	10.6	1.70	1.02
Toluene	C₇H₈	867	110	383.15	1.285	91.2	8.8	—	—	10 150	42 496	9 680	40 528	2.19	10.5	1.70	0.97
Xylene	C₈H₁₀	863	140	413.15	1.313	90.5	9.5	—	—	10 230	42 831	9 720	40 696	2.22	10.6	1.69	1.06

1 kcal = 4.1868 kJ

Table 38-2 Composition and Heating Values H_s and H_i of Liquid Fuels (Continued)

Fuel	Density ρ (kg/m³)	C	H₂	O₂ +N₂	S	H_s (kcal/kg)	H_s (kJ/kg)	H_i (kcal/kg)	H_i (kJ/kg)	L_{min} (m³ₙ/kg)	$\dfrac{V_{min_{suh}}}{L_{min}}$	CO₂max %
Fuel oils:												
— Californian	950	86.4	11.3	1.1	0.6	10400	43543	9800	41031	—	0.94	16.0
— heavy	950	85.0	11.7	1.2	2.1	10500	43961	9980	41784	10.670	0.939	15.90
— light	900	85.4	12.3	0.7	1.6	10700	44799	10050	42077	10.864	0.937	15.69
— Mexican	910	82.9	12.2	2.1	2.8	10270	42998	9620	40277	—	0.94	15.4
— Pennsylvanian	890	84.9	13.7	1.4	—	10670	44673	9940	41617	—	0.93	15.2
Tar oils:												
— lean tar oil	1120	90.4	6.0	3.2	0.4	9300	38937	8980	37597	9.519	0.966	18.28
— lignite tar oil	925	84.0	11.0	4.3	0.7	10200	42705	9610	40235	10.246	0.942	16.21
— pit coal tar oil	1080	89.5	6.5	3.4	0.6	9380	39272	9000	37681	9.571	0.964	18.06

1 kcal = 4.1868 kJ

VAPORS

IMPORTANT NOTE: Users of the Vapors tables are alerted to the fact that the symbol for pressure, *at*, refers to the **technical atmosphere** and not to the **standard atmosphere**, which is designated by *atm*.

The relationship of conversion factors between the two atmospheres is clearly explained in Table 119 on page 337.

Table 39-1 Properties of Saturated Steam (H₂O) at a Given Temperature (Gravitational Metric System)

Temperature		Pressure	Specific volume		Density		Specific enthalpy		Heat of vapor-ization	Specific entropy	
			Liquid	Vapor	Liquid	Vapor	Liquid	Vapor		Liquid	Vapor
t	T	p	v'	v''	ρ'	ρ''	i'	i''	$r=i''-i'$	s'	s''
°C	K	kp/cm²	m³/kg	m³/kg	kg/m³	kg/m³	kcal/kg	kcal/kg	kcal/kg	kcal/kgK	kcal/kgK
0.01	273.16	0.006228	0.0010002	206.3	999.80	0.004847	0.00	597.3	597.3	0.0000	2.1865
1	274.15	0.006695	0.0010001	192.6	999.90	0.005192	1.01	597.7	596.7	0.0037	2.1802
2	275.15	0.007193	0.0010001	179.9	999.90	0.005559	2.01	598.2	596.2	0.0073	2.1739
3	276.15	0.007724	0.0010001	168.2	999.90	0.005945	3.02	598.6	595.6	0.0109	2.1677
4	277.15	0.008289	0.0010001	157.3	999.90	0.006357	4.02	599.1	595.1	0.0146	2.1615
5	278.15	0.008891	0.0010001	147.2	999.90	0.006793	5.03	599.5	594.5	0.0182	2.1554
6	279.15	0.009532	0.0010001	137.8	999.90	0.007257	6.03	599.9	593.9	0.0218	2.1493
7	280.15	0.010210	0.0010001	129.1	999.90	0.007746	7.03	600.4	593.4	0.0254	2.1433
8	281.15	0.010932	0.0010002	121.0	999.80	0.008264	8.04	600.8	592.8	0.0290	2.1373
9	282.15	0.011699	0.0010003	113.4	999.70	0.008818	9.04	601.3	592.3	0.0326	2.1314
10	283.15	0.012513	0.0010004	106.42	999.60	0.009398	10.04	601.7	591.7	0.0361	2.1256
11	284.15	0.013376	0.0010005	99.91	999.50	0.01001	11.04	602.2	591.2	0.0396	2.1198
12	285.15	0.014292	0.0010006	93.84	999.40	0.01066	12.04	602.6	590.6	0.0431	2.1141
13	286.15	0.015262	0.0010007	88.18	999.30	0.01134	13.04	603.1	590.1	0.0466	2.1084
14	287.15	0.016289	0.0010008	82.90	999.20	0.01206	14.04	603.5	589.5	0.0501	2.1028
15	288.15	0.017377	0.0010010	77.97	999.00	0.01282	15.04	603.9	588.9	0.0536	2.0972
16	289.15	0.018528	0.0010011	73.39	998.90	0.01363	16.04	604.3	588.3	0.0571	2.0916
17	290.15	0.019746	0.0010013	69.10	998.70	0.01447	17.04	604.7	587.7	0.0605	2.0861
18	291.15	0.02103	0.0010015	65.09	998.50	0.01536	18.04	605.1	587.1	0.0640	2.0807
19	292.15	0.02239	0.0010016	61.34	998.40	0.01630	19.04	605.6	586.6	0.0674	2.0753
20	293.15	0.02383	0.0010018	57.84	998.20	0.01729	20.04	606.0	586.0	0.0708	2.0699
21	294.15	0.02535	0.0010021	54.56	997.90	0.01833	21.04	606.4	585.4	0.0742	2.0646
22	295.15	0.02695	0.0010023	51.50	997.71	0.01942	22.04	606.9	584.9	0.0776	2.0593
23	296.15	0.02863	0.0010025	48.62	997.51	0.02057	23.04	607.3	584.3	0.0810	2.0541
24	297.15	0.03041	0.0010028	45.93	997.21	0.02177	24.03	607.8	583.8	0.0843	2.0489
25	298.15	0.03229	0.0010030	43.40	997.01	0.02304	25.03	608.2	583.2	0.0877	2.0438
26	299.15	0.03426	0.0010033	41.04	996.71	0.02437	26.03	608.6	582.6	0.0911	2.0387
27	300.15	0.03634	0.0010036	38.82	996.41	6.02576	27.03	609.1	585.1	0.0944	2.0337
28	301.15	0.03853	0.0010038	36.73	996.21	0.02723	28.03	609.5	581.5	0.0977	2.0287
29	302.15	0.04083	0.0010041	34.77	995.92	0.02876	29.02	610.0	581.0	0.1010	2.0237
30	303.15	0.04325	0.0010044	32.93	995.62	0.03037	30.02	610.4	580.4	0.1043	2.0188
31	304.15	0.04580	0.0010047	31.20	995.32	0.03205	31.02	610.9	579.9	0.1076	2.0139
32	305.15	0.04847	0.0010051	29.57	994.93	0.03382	32.02	611.3	579.3	0.1108	2.0091
33	306.15	0.05128	0.0010054	28.04	994.63	0.03566	33.02	611.7	578.7	0.1141	2.0043
34	307.15	0.05423	0.0010057	26.60	994.33	0.03759	34.02	612.2	578.2	0.1173	1.9995
35	308.15	0.05733	0.0010061	25.24	993.94	0.03962	35.01	612.6	577.6	0.1206	1.9948
36	309.15	0.06057	0.0010064	23.97	993.64	0.04172	36.01	613.0	577.0	0.1239	1.9901
37	310.15	0.06398	0.0010068	22.77	993.25	0.04392	37.01	613.5	576.5	0.1271	1.9855
38	311.15	0.06755	0.0010071	21.03	992.95	0.04623	38.01	613.9	575.9	0.1303	1.9809
39	312.15	0.07129	0.0010075	20.56	992.56	0.04864	39.01	614.3	575.3	0.1335	1.9764
40	313.15	0.07520	0.0010079	19.55	992.16	0.05115	40.01	614.7	574.7	0.1367	1.9719
41	314.15	0.07931	0.0010083	18.59	991.77	0.05379	41.00	615.1	574.1	0.1399	1.9674
42	315.15	0.08360	0.0010087	17.69	991.38	0.05653	42.00	615.5	573.5	0.1430	1.9630
43	316.15	0.08809	0.0010091	16.84	990.98	0.05938	43.00	615.9	572.9	0.1462	1.9586
44	317.15	0.09279	0.0010095	16.04	990.59	0.06234	44.00	616.4	572.4	0.1493	1.9542
45	318.15	0.09771	0.0010099	15.28	990.20	0.06544	45.00	616.8	571.8	0.1525	1.9499
46	319.15	0.10284	0.0010103	14.56	989.81	0.06868	46.00	617.3	571.3	0.1556	1.9456
47	320.15	0.10821	0.0010108	13.88	989.32	0.07205	47.00	617.7	570.7	0.1588	1.9413
48	321.15	0.11382	0.0010112	13.23	988.92	0.07559	47.99	618.1	570.1	0.1619	1.9371
49	322.15	0.11967	0.0010116	12.62	988.53	0.07924	48.99	618.6	569.6	0.1650	1.9329
50	323.15	0.12578	0.0010121	12.04	988.04	0.08306	49.99	619.0	569.0	0.1681	1.9287
51	324.15	0.13216	0.0010126	11.50	987.56	0.08696	50.99	619.4	568.4	0.1712	1.9246
52	325.15	0.13880	0.0010130	10.98	987.17	0.09107	51.99	619.8	567.8	0.1742	1.9205
53	326.15	0.14574	0.0010135	10.49	986.68	0.09533	52.99	620.3	567.3	0.1773	1.9164
54	327.15	0.15297	0.0010140	10.02	986.19	0.09980	53.98	620.7	566.7	0.1804	1.9124

Table 39-2 Properties of Saturated Steam (H₂O) at a Given Temperature (Gravitational Metric System) (*Continued*)

Temperature		Pressure	Specific volume		Density		Specific enthalpy		Heat of vapor-ization	Specific entropy	
			Liquid	Vapor	Liquid	Vapor	Liquid	Vapor		Liquid	Vapor
t	T	p	v'	v''	ρ'	ρ''	i'	i''	$r = i'' - i'$	s'	s''
°C	K	kp/cm²	m³/kg	m³/kg	kg/m³	kg/m³	kcal/kg	kcal/kg	kcal/kg	kcal/kgK	kcal/kgK
55	328.15	0.16050	0.0010145	9.578	985.71	0.1044	54.98	621.1	566.1	0.1834	1.9084
56	329.15	0.16835	0.0010150	9.158	985.22	0.1092	55.98	621.5	565.5	0.1864	1.9045
57	330.15	0.17653	0.0010155	8.757	984.74	0.1142	56.98	622.0	565.0	0.1895	1.9005
58	331.15	0.18504	0.0010160	8.380	984.25	0.1193	57.98	622.4	564.4	0.1925	1.8966
59	332.15	0.19390	0.0010166	8.020	983.67	0.1247	58.98	622.8	563.8	0.1955	1.8928
60	333.15	0.2031	0.0010171	7.678	983.19	0.1302	59.98	623.2	563.2	0.1985	1.8889
61	334.15	0.2127	0.0010177	7.353	982.61	0.1360	60.98	623.6	562.6	0.2015	1.8851
62	335.15	0.2227	0.0010182	7.043	982.13	0.1420	61.98	624.0	562.0	0.2045	1.8813
63	336.15	0.2330	0.0010188	6.749	981.55	0.1482	62.98	624.4	561.4	0.2075	1.8775
64	337.15	0.2438	0.0010193	6.468	981.07	0.1546	63.98	624.8	560.8	0.2104	1.8738
65	338.15	0.2550	0.0010199	6.201	980.49	0.1613	64.98	625.2	560.2	0.2134	1.8701
66	339.15	0.2666	0.0010205	5.947	979.91	0.1681	65.98	625.6	559.6	0.2163	1.8665
67	340.15	0.2787	0.0010210	5.750	979.43	0.1753	66.98	626.1	559.1	0.2193	1.8628
68	341.15	0.2912	0.0010216	5.475	978.86	0.1826	67.98	626.5	558.5	0.2222	1.8592
69	342.15	0.3043	0.0010222	5.255	978.28	0.1903	68.98	626.9	557.9	0.2252	1.8557
70	343.15	0.3178	0.0010228	5.045	977.71	0.1982	69.98	627.3	557.3	0.2281	1.8521
71	344.15	0.3318	0.0010234	4.846	977.14	0.2064	70.98	627.7	556.7	0.2310	1.8485
72	345.15	0.3463	0.0010240	4.655	976.56	0.2148	71.99	628.1	556.1	0.2340	1.8450
73	346.15	0.3613	0.0010246	4.473	975.99	0.2236	72.99	628.5	555.5	0.2369	1.8416
74	347.15	0.3769	0.0010252	4.299	975.42	0.2326	73.99	628.9	554.9	0.2397	1.8381
75	348.15	0.3931	0.0010258	4.133	974.85	0.2420	74.99	629.3	554.3	0.2426	1.8347
76	349.15	0.4098	0.0010264	3.975	974.28	0.2516	75.99	629.7	553.7	0.2454	1.8313
77	350.15	0.4272	0.0010270	3.824	973.71	0.2615	76.99	630.1	553.1	0.2483	1.8280
78	351.15	0.4451	0.0010277	3.679	973.05	0.2718	78.00	630.5	552.5	0.2512	1.8246
79	352.15	0.4637	0.0010283	3.540	972.48	0.2825	79.00	630.9	551.9	0.2540	1.8213
80	353.15	0.4829	0.0010290	3.408	971.82	0.2934	80.00	631.3	551.3	0.2568	1.8180
81	354.15	0.5028	0.0010297	3.282	971.16	0.3047	81.00	631.7	550.7	0.2597	1.8147
82	355.15	0.5234	0.0010304	3.161	970.50	0.3164	82.01	632.1	550.1	0.2625	1.8115
83	556.15	0.5447	0.0010310	3.045	969.93	0.3284	83.01	632.5	549.5	0.2653	1.8082
84	357.15	0.5667	0.0010317	2.934	969.27	0.3408	84.01	632.9	548.9	0.2681	1.8050
85	358.15	0.5894	0.0010324	2.828	968.62	0.3536	85.02	633.3	548.3	0.2709	1.8018
86	359.15	0.6129	0.0010331	2.727	967.96	0.3667	86.02	633.7	547.7	0.2737	1.7986
87	360.15	0.6372	0.0010338	2.629	967.31	0.3804	87.03	634.1	547.1	0.2765	1.7955
88	361.15	0.6623	0.0010345	2.536	966.65	0.3943	88.03	634.4	546.4	0.2792	1.7923
89	362.15	0.6882	0.0010352	2.447	966.00	0.4087	89.03	634.8	545.8	0.2820	1.7893
90	363.15	0.7149	0.0010359	2.361	965.34	0.4235	90.04	635.2	545.2	0.2848	1.7862
91	364.15	0.7424	0.0010366	2.279	964.69	0.4388	91.04	635.6	544.6	0.2876	1.7832
92	365.15	0.7710	0.0010373	2.200	964.04	0.4545	92.05	635.9	543.9	0.2903	1.7802
93	366.15	0.8004	0.0010381	2.124	963.30	0.4708	93.05	636.3	543.3	0.2931	1.7772
94	367.15	0.8307	0.0010388	2.052	962.65	0.4873	94.06	636.8	542.7	0.2959	1.7742
95	368.15	0.8619	0.0010396	1.982	961.91	0.5045	95.07	637.2	542.1	0.2986	1.7712
96	369.15	0.8942	0.0010404	1.915	961.17	0.5222	96.07	637.6	541.5	0.3013	1.7682
97	370.15	0.9274	0.0010412	1.851	960.43	0.5402	97.08	638.0	540.9	0.3041	1.7652
98	371.15	0.9616	0.0010420	1.789	959.69	0.5590	98.09	638.4	540.3	0.3067	1.7623
99	372.15	0.9971	0.0010427	1.730	959.05	0.5780	99.10	638.7	539.6	0.3095	1.7595
100	373.15	1.0332	0.0010435	1.673	958.31	0.5977	100.10	639.1	539.0	0.3122	1.7566
101	374.15	1.0707	0.0010443	1.618	957.58	0.6181	101.11	639.5	538.4	0.3149	1.7538
102	375.15	1.1092	0.0010450	1.566	956.94	0.6386	102.11	639.8	537.7	0.3176	1.7510
103	376.15	1.1489	0.0010458	1.515	956.21	0.6601	103.12	640.2	537.1	0.3203	1.7482
104	377.15	1.1898	0.0010466	1.466	955.47	0.6821	104.13	640.5	536.4	0.3229	1.7454
105	378.15	1.2318	0.0010474	1.419	954.75	0.7047	105.14	640.9	435.8	0.3256	1.7426
106	379.15	1.2751	0.0010482	1.374	954.02	0.7278	106.15	641.3	535.2	0.3283	1.7398
107	380.15	1.3196	0.0010490	1.331	953.29	0.7513	107.16	641.7	534.5	0.3309	1.7370
108	381.15	1.3654	0.0010498	1.289	952.56	0.7758	108.17	642.1	533.9	0.3335	1.7343
109	382.15	1.4125	0.0010507	1.249	951.75	0.8006	109.18	642.4	533.2	0.3362	1.7316

Table 39-3 Properties of Saturated Steam (H$_2$O) at a Given Temperature (Gravitational Metric System) (Continued)

Temperature		Pressure	Specific volume		Density		Specific enthalpy		Heat of vapor-ization	Specific entropy	
			Liquid	Vapor	Liquid	Vapor	Liquid	Vapor		Liquid	Vapor
t	T	p	v'	v''	ρ'	ρ''	i'	i''	r=i''−i'	s'	s''
°C	K	kp/cm²	m³/kg	m³/kg	kg/m³	kg/m³	kcal/kg	kcal/kg	kcal/kg	kcal/kgK	kcal/kgK
110	383.15	1.4609	0.0010515	1.210	951.02	0.8264	110.19	642.8	532.6	0.3388	1.7289
111	384.15	1.5106	0.0010523	1.173	950.30	0.8525	111.20	643.2	532.0	0.3414	1.7262
112	385.15	1.5618	0.0010532	1.137	949.49	0.8795	112.21	643.5	531.3	0.3440	1.7236
113	386.15	1.6144	0.0010540	1.102	948.77	0.9074	113.22	643.9	530.7	0.3467	1.7209
114	387.15	1.6684	0.0010549	1.069	947.96	0.9354	114.23	644.2	530.0	0.3493	1.7183
115	388.15	1.7239	0.0010558	1.036	947.15	0.9652	115.25	644.6	529.4	0.3519	1.7157
116	389.15	1.7809	0.0010567	1.005	946.34	0.9950	116.26	645.0	528.7	0.3545	1.7131
117	390.15	1.8394	0.0010576	0.9754	945.54	1.025	117.27	645.4	528.1	0.3571	1.7105
118	391.15	1.8995	0.0010585	0.9465	944.73	1.056	118.29	645.7	527.4	0.3597	1.7080
119	392.15	1.9612	0.0010594	0.9186	943.93	1.089	119.30	646.0	526.7	0.3623	1.7054
120	393.15	2.0245	0.0010603	0.8917	943.13	1.121	120.3	646.4	526.1	0.3649	1.7029
121	394.15	2.0895	0.0010612	0.8657	942.33	1.155	121.3	646.7	525.4	0.3675	1.7005
122	395.15	2.1561	0.0010621	0.8407	941.53	1.189	122.3	647.0	524.7	0.3700	1.6981
123	396.15	2.2245	0.0010630	0.8164	940.73	1.225	123.4	647.5	524.1	0.3726	1.6954
124	397.15	2.2947	0.0010640	0.7930	939.85	1.261	124.4	647.8	523.4	0.3751	1.6930
125	398.15	2.3666	0.0010649	0.7704	939.06	1.298	125.4	648.1	522.7	0.3777	1.6905
126	399.15	2.4404	0.0010658	0.7486	938.26	1.336	126.4	648.4	522.0	0.3803	1.6880
127	400.15	2.5160	0.0010668	0.7276	937.38	1.374	127.4	648.8	521.4	0.3828	1.6856
128	401.15	2.5935	0.0010677	0.7074	936.59	1.414	128.4	649.1	520.7	0.3854	1.6832
129	402.15	2.6730	0.0010687	0.6880	935.72	1.453	129.5	649.5	520.0	0.3879	1.6808
130	403.15	2.7544	0.0010697	0.6683	934.84	1.496	130.5	649.8	519.3	0.3904	1.6784
131	404.15	2.8378	0.0010707	0.6499	933.97	1.539	131.5	650.1	518.6	0.3929	1.6760
132	405.15	2.9233	0.0010717	0.6321	933.10	1.582	132.5	650.4	517.9	0.3954	1.6737
133	406.15	3.011	0.0010727	0.6148	932.23	1.626	133.5	650.7	517.2	0.3979	1.6713
134	407.15	3.101	0.0010737	0.5981	931.36	1.672	134.6	651.5	516.5	0.4004	1.6690
135	408.15	3.192	0.0010747	0.5820	930.49	1.718	135.6	651.4	515.8	0.4029	1.6667
136	409.15	3.286	0.0010757	0.5664	929.63	1.765	136.6	551.7	515.1	0.4054	1.6644
137	410.15	3.382	0.0010767	0.5512	928.76	1.814	137.6	652.0	514.4	0.4079	1.6621
138	411.15	3.481	0.0010777	0.5366	927.90	1.864	138.7	652.4	513.7	0.4104	1.6598
139	412.15	3.582	0.0010788	0.5224	926.96	1.914	139.7	652.7	513.0	0.4129	1.6575
140	413.15	3.685	0.0010798	0.5087	926.10	1.966	140.7	653.0	512.3	0.4154	1.6553
141	414.15	3.790	0.0010808	0.4953	925.24	2.019	141.7	653.3	511.6	0.4179	1.6531
142	415.15	3.898	0.0010819	0.4824	924.30	2.073	142.8	653.7	510.9	0.4203	1.6508
143	416.15	4.009	0.0010829	0.4699	923.45	2.128	143.8	654.0	510.2	0.4228	1.6486
144	417.15	4.121	0.0010840	0.4579	922.51	2.184	144.8	654.2	509.4	0.4252	1.6464
145	418.15	4.237	0.0010851	0.4461	921.57	2.242	145.8	654.5	508.7	0.4277	1.6642
146	419.15	4.355	0.0010862	0.4347	920.64	2.300	146.9	654.8	507.9	0.4301	1.6420
147	420.15	4.476	0.0010873	0.4237	919.71	2.360	147.9	655.1	507.2	0.4326	1.6398
148	421.15	4.599	0.0010884	0.4130	918.78	2.421	148.9	655.4	506.5	0.4350	1.6376
149	422.15	4.725	0.0010895	0.4026	917.85	2.484	150.0	655.7	505.7	0.4375	1.6355
150	423.15	4.854	0.0010906	0.3926	916.93	2.547	151.0	656.0	505.0	0.4399	1.6333
151	424.15	4.985	0.0010917	0.3828	916.00	2.612	152.0	656.3	504.3	0.4423	1.6311
152	425.15	5.119	0.0010928	0.3733	915.08	2.679	153.1	656.7	503.6	0.4448	1.6290
153	426.15	5.257	0.0010939	0.3641	914.16	2.746	154.1	657.0	502.9	0.4472	1.6269
154	427.15	5.397	0.0010950	0.3552	913.24	2.815	155.1	657.2	502.1	0.4496	1.6248
155	428.15	5.540	0.0010962	0.3466	912.24	2.885	156.2	657.5	501.3	0.4520	1.6220
156	429.15	5.686	0.0010974	0.3381	911.24	2.958	157.2	657.7	500.5	0.4544	1.6207
157	430.15	5.836	0.0010986	0.3299	910.25	3.030	158.2	657.9	499.7	0.4568	1.6186
158	431.15	5.988	0.0010998	0.3220	909.26	3.106	159.3	658.2	498.9	0.4592	1.6165
159	432.15	6.144	0.0011009	0.3143	908.35	3.182	160.3	658.4	498.1	0.4616	1.6145
160	433.15	6.302	0.0011021	0.3068	907.36	3.258	161.3	658.7	497.4	0.4640	1.6124
161	434.15	6.464	0.0011033	0.2996	906.37	3.338	162.4	659.0	496.6	0.4664	1.6103
162	435.15	6.630	0.0011044	0.2925	905.47	4.419	163.4	659.2	495.8	0.4688	1.6083
163	436.15	6.798	0.0011056	0.2856	904.49	3.500	164.5	659.5	496.0	0.4712	1.6062
164	437.15	6.970	0.0011069	0.2790	903.42	3.584	165.5	659.7	494.2	0.4735	1.6042

Temperature		Pressure	Specific volume		Density		Specific enthalpy		Heat of vapor-ization	Specific entropy	
			Liquid	Vapor	Liquid	Vapor	Liquid	Vapor		Liquid	Vapor
t	T	p	v'	v''	ρ'	ρ''	i'	i''	$r=i''-i'$	s'	s''
°C	K	kp/cm²	m³/kg	m³/kg	kg/m³	kg/m³	kcal/kg	kcal/kg	kcal/kg	kcal/kgK	kcal/kgK
165	438.15	7.146	0.0011081	0.2725	902.45	3.670	166.5	660.0	493.5	0.4759	1.6022
166	439.15	7.325	0.0011094	0.2662	901.39	3.757	167.6	660.3	492.7	0.4783	1.6002
167	440.15	7.507	0.0011106	0.2600	900.41	3.846	168.6	660.5	491.9	0.4806	1.5983
168	441.15	7.693	0.0011119	0.2541	899.36	3.935	169.7	660.8	391.1	0.4830	1.5963
169	442.15	7.883	0.0011131	0.2483	898.39	4.027	170.7	661.0	490.3	0.4853	1.5943
170	443.15	8.076	0.0011144	0.2426	897.34	4.122	171.8	661.3	489.5	0.4877	1.5923
171	444.15	8.274	0.0011156	0.2371	896.38	4.218	172.8	661.5	488.7	0.4900	1.5903
172	445.15	8.475	0.0011169	0.2318	895.34	4.314	173.9	661.8	487.9	0.4924	1.5883
173	446.15	8.679	0.0011182	0.2266	894.29	4.413	174.9	662.0	487.1	0.4947	1.5864
174	447.15	8.888	0.0011195	0.2215	893.26	4.515	176.0	662.3	486.3	0.4971	1.5844
175	448.15	9.101	0.0011208	0.2166	892.22	4.617	177.0	662.4	485.4	0.4994	1.5825
176	449.15	9.317	0.0011221	0.2118	891.19	4.721	178.1	662.7	484.6	0.5017	1.5806
177	450.15	9.538	0.0011234	0.2071	890.15	4.829	179.1	662.9	483.8	0.5040	1.5787
178	451.15	9.763	0.0011248	0.2026	889.05	4.936	180.2	663.2	483.0	0.5064	1.5768
179	452.15	9.992	0.0011261	0.1982	888.02	5.045	181.2	663.4	482.2	0.5087	1.5749
180	453.15	10.225	0.0011275	0.1939	886.92	5.157	182.3	663.6	481.3	0.5110	1.5730
181	454.15	10.462	0.0011289	0.1897	885.82	5.271	183.3	663.7	480.4	0.5133	1.5711
182	455.15	10.703	0.0011303	0.1856	884.72	5.388	184.4	663.9	479.5	0.5156	1.5692
183	456.15	10.950	0.0011316	0.1816	883.70	5.507	185.4	664.0	478.6	0.5179	1.5674
184	457.15	11.201	0.0011330	0.1777	882.61	5.627	186.5	664.3	477.8	0.5202	1.5655
185	458.15	11.456	0.0011344	0.1739	881.52	5.750	187.6	664.6	477.0	0.5225	1.5636
186	459.15	11.715	0.0011358	0.1702	880.44	5.875	188.6	664.7	476.1	0.5248	1.5617
187	460.15	11.979	0.0011372	0.1666	879.35	6.002	189.7	664.9	475.2	0.5271	1.5598
188	461.15	12.248	0.0011386	0.1631	878.27	6.131	190.7	665.0	474.3	0.5294	1.5580
189	462.15	12.522	0.0011401	0.1597	877.12	6.262	191.8	665.2	473.4	0.5317	1.5561
190	463.15	12.800	0.0011415	0.1564	876.04	6.394	192.9	665.5	472.6	0.5340	1.5543
191	464.15	13.083	0.0011430	0.1531	874.89	6.532	193.9	665.6	471.7	0.5363	1.5525
192	465.15	13.371	0.0011445	0.1499	873.74	6.671	195.0	665.8	470.8	0.5386	1.5506
193	466.15	13.664	0.0011459	0.1468	872.68	6.812	196.1	666.0	469.9	0.5408	1.5488
194	467.15	13.962	0.0011474	0.1438	871.54	6.954	197.2	666.2	469.0	0.5431	1.5470
195	468.15	14.265	0.0011489	0.1409	870.40	7.097	198.2	666.3	468.1	0.5454	1.5452
196	469.15	14.573	0.0011504	0.1380	869.26	7.246	199.3	666.5	467.2	0.5477	1.5434
197	470.15	14.886	0.0011519	0.1352	868.13	7.396	200.4	666.7	466.3	0.5499	1.5416
198	471.15	15.204	0.0011534	0.1325	867.00	7.547	201.4	666.8	465.4	0.5522	1.5398
199	472.15	15.528	0.0011550	0.1298	865.80	7.704	202.5	667.0	464.5	0.5545	1.5380
200	473.15	15.857	0.0011565	0.1272	864.68	7.862	203.6	667.1	463.5	0.5567	1.5362
201	474.15	16.192	0.0011581	0.1246	863.48	8.026	204.7	667.2	462.5	0.5589	1.5344
202	475.15	16.532	0.0011596	0.1222	862.37	8.183	205.7	667.3	461.5	0.5612	1.5326
203	476.15	16.877	0.0011612	0.1197	861.18	8.354	206.8	667.4	460.6	0.5634	1.5309
204	477.15	17.228	0.0011628	0.1174	859.99	8.518	207.9	667.6	459.7	0.5657	1.5291
205	478.15	17.585	0.0011644	0.1151	858.81	8.688	209.0	667.7	458.7	0.5679	1.5273
206	479.15	17.948	0.0011660	0.1128	857.63	8.865	210.1	667.9	457.8	0.5701	1.5255
207	480.15	18.316	0.0011676	0.1106	856.46	9.042	211.2	668.0	456.8	0.5724	1.5238
208	481.15	18.690	0.0011693	0.1084	855.21	9.225	212.3	668.1	455.8	0.5746	1.5220
209	482.15	19.070	0.0011709	0.1063	854.04	9.407	213.3	668.2	454.8	0.5769	1.5202
210	483.15	19.456	0.0011726	0.1043	852.81	9.588	214.4	668.3	453.9	0.5791	1.5185
211	484.15	19.848	0.0011743	0.1023	851.57	9.775	215.5	668.4	452.9	0.5814	1.5168
212	485.15	20.246	0.0011760	0.1003	850.34	9.970	216.6	668.5	451.9	0.5836	1.5150
213	486.15	20.651	0.0011778	0.09836	849.04	10.17	217.7	668.6	450.9	0.5858	1.5133
214	487.15	21.061	0.0011795	0.09649	847.82	10.36	218.8	668.7	449.9	0.5881	1.5115
215	488.15	21.477	0.0011812	0.09465	846.60	10.56	219.9	668.8	448.9	0.5903	1.5098
216	489.15	21.901	0.0011829	0.09285	845.38	10.77	221.0	668.9	447.9	0.5925	1.5081
217	490.15	22.331	0.0011846	0.09110	844.17	10.98	222.1	669.0	446.9	0.5947	1.5063
218	491.15	22.767	0.0011864	0.08938	842.89	11.19	223.2	669.1	445.9	0.5970	1.5046
219	492.15	23.209	0.0011882	0.08770	841.61	11.40	224.3	669.1	444.8	0.5992	1.5028

Table 39-5 Properties of Saturated Steam (H$_2$O) at a Given Temperature (Gravitational Metric System) (*Continued*)

Temperature		Pressure	Specific volume		Density		Specific enthalpy		Heat of vapor-ization	Specific entropy	
			Liquid	Vapor	Liquid	Vapor	Liquid	Vapor		Liquid	Vapor
t	T	p	v'	v''	ρ'	p''	i'	i''	$r=i''-i'$	s'	s''
°C	K	kp/cm²	m³/kg	m³/kg	kg/m³	kg/m³	kcal/kg	kcal/kg	kcal/kg	kcal/kgK	kcal/kgK
220	493.15	23.659	0.0011900	0.08606	840.34	11.62	225.3	669.1	443.7	0.6014	1.5011
221	494.15	24.115	0.0011918	0.08446	839.07	11.94	226.5	669.2	442.7	0.6036	1.4994
222	495.15	24.577	0.0011937	0.08288	837.73	12.06	227.6	669.3	441.7	0.6058	1.4977
223	496.15	25.047	0.0011955	0.08135	836.47	12.29	228.7	669.3	440.6	0.6080	1.4959
224	497.15	25.523	0.0011973	0.07984	835.21	12.52	229.8	669.3	439.5	0.6102	1.4942
225	498.15	26.007	0.0011992	0.07837	833.89	12.76	230.9	669.3	438.4	0.6124	1.4925
226	499.15	26.497	0.0012011	0.07693	832.57	13.00	232.1	669.4	437.3	0.6146	1.4908
227	500.15	26.995	0.0012029	0.07552	831.32	13.24	233.2	669.4	436.2	0.6168	1.4891
228	501.15	27.499	0.0012048	0.07414	830.01	13.49	234.3	669.4	435.1	0.6190	1.4874
229	502.15	28.011	0.0012068	0.07279	828.64	13.74	235.4	669.5	434.1	0.6212	1.4857
230	503.15	28.531	0.0012087	0.07147	827.34	13.99	236.5	669.5	433.0	0.6234	1.4840
231	504.15	29.057	0.0012107	0.07018	825.97	14.25	237.7	669.7	432.0	0.6256	1.4823
232	505.15	29.591	0.0012126	0.06891	824.67	14.51	238.8	669.7	430.9	0.6278	1.4807
233	506.15	30.133	0.0012146	0.06767	823.32	14.78	239.9	669.7	429.8	0.6300	1.4790
234	507.15	30.682	0.0012167	0.06646	821.90	15.05	241.0	669.7	428.7	0.6322	1.4773
235	508.15	31.239	0.0012187	0.06527	820.55	15.32	242.2	669.7	427.5	0.6344	1.4756
236	509.15	31.803	0.0012208	0.06410	819.13	15.60	243.3	669.6	426.3	0.6366	1.4739
237	510.15	32.376	0.0012228	0.06496	817.80	15.88	244.4	669.6	425.2	0.6388	1.4722
238	511.15	32.955	0.0012249	0.06184	816.39	16.17	245.6	669.7	424.1	0.6410	1.4705
239	512.15	33.544	0.0012270	0.06075	815.00	16.46	246.7	669.6	422.9	0.6432	1.4688
240	513.15	34.140	0.0012291	0.05967	813.60	16.76	247.8	669.5	421.7	0.6454	1.4671
241	514.15	34.745	0.0012312	0.05862	812.22	17.06	249.0	669.6	420.6	0.6476	1.4654
242	515.15	35.357	0.0012334	0.05759	810.77	17.36	250.1	669.5	419.4	0.6498	1.4637
243	516.15	35.978	0.0012355	0.05658	809.39	17.67	251.3	669.5	418.2	0.6520	1.4620
244	517.15	36.607	0.0012377	0.05559	807.95	17.99	252.4	669.4	417.0	0.6541	1.4604
245	518.15	37.244	0.0012399	0.05462	806.52	18.30	253.6	669.4	415.8	0.6563	1.4587
246	519.15	37.890	0.0012421	0.05367	805.09	18.63	254.7	669.3	414.6	0.6585	1.4571
247	520.15	38.545	0.0012443	0.05274	803.66	18.96	255.9	669.3	413.4	0.6606	1.4554
248	521.15	39.208	0.0012466	0.05183	802.18	19.29	257.0	669.2	412.2	0.6628	1.4537
249	522.15	39.880	0.0012489	0.05093	800.70	19.63	258.2	669.2	411.0	0.6650	1.4520
250	523.15	40.56	0.0012512	0.05006	799.23	19.98	259.3	669.0	409.7	0.6672	1.4503
251	524.15	41.25	0.0012536	0.04919	797.70	20.33	260.5	668.9	408.4	0.6694	1.4486
252	525.15	41.95	0.0012559	0.04835	796.24	20.68	261.7	668.9	407.2	0.6716	1.4469
253	526.15	42.66	0.0012583	0.04752	794.72	21.04	262.8	668.7	405.9	0.6738	1.4452
254	527.15	43.37	0.0012607	0.04671	793.21	21.41	264.0	668.6	404.6	0.6760	1.4435
255	528.15	44.10	0.0012631	0.04591	791.70	21.78	265.2	668.5	403.3	0.6782	1.4418
256	529.15	44.83	0.0012655	0.04513	790.20	22.16	266.4	668.4	402.0	0.6804	1.4401
257	530.15	45.58	0.0012680	0.04436	788.64	22.54	267.5	668.2	400.7	0.6826	1.4384
258	531.15	46.33	0.0012705	0.04361	787.09	22.93	268.7	668.1	399.4	0.6847	1.4368
259	532.15	47.09	0.0012730	0.04287	785.55	23.33	269.9	668.0	398.1	0.6869	1.4351
260	533.15	47.87	0.0012755	0.04215	784.01	23.72	271.1	667.9	396.8	0.6891	1.4334
261	534.15	48.65	0.0012781	0.04144	782.41	24.13	272.3	667.8	395.5	0.6913	1.4317
262	535.15	49.44	0.0012807	0.04074	780.82	24.55	273.5	667.7	394.2	0.6935	1.4300
263	536.15	50.24	0.0012833	0.04005	779.24	24.96	274.7	667.6	392.9	0.6957	1.4283
264	537.15	51.05	0.0012859	0.03938	777.67	25.39	275.9	667.5	391.6	0.6978	1.4266
265	538.15	51.87	0.0012886	0.03872	776.04	25.83	277.1	667.3	390.2	0.7000	1.4249
266	539.15	52.71	0.0012913	0.03807	774.41	26.26	278.3	667.0	388.7	0.7021	1.4232
267	540.15	53.55	0.0012940	0.03744	772.80	26.71	279.5	666.8	387.3	0.7043	1.4214
268	541.15	54.40	0.0012967	0.03681	771.19	27.16	280.7	666.6	385.9	0.7065	1.4197
269	542.15	55.26	0.0012995	0.03620	769.53	27.62	281.9	666.4	384.5	0.7087	1.4180
270	543.15	56.14	0.0013023	0.03560	767.87	28.09	283.1	666.3	383.2	0.7109	1.4163
271	544.15	57.02	0.0013051	0.03501	766.22	28.56	284.3	666.1	381.8	0.7131	1.4146
272	545.15	57.91	0.0013080	0.03443	764.53	29.04	285.5	665.9	380.4	0.7153	1.4129
273	546.15	58.82	0.0013109	0.03386	762.83	29.53	286.7	665.7	379.0	0.7155	1.4112
274	547.15	59.73	0.0013138	0.03330	761.15	30.03	288.0	665.5	377.5	0.7197	1.4094

Table 39-6 Properties of Saturated Steam (H_2O) at a Given Temperature (Gravitational Metric System) (Continued)

Temperature		Pressure	Specific volume		Density		Specific enthalpy		Heat of vapor-ization	Specific entropy	
			Liquid	Vapor	Liquid	Vapor	Liquid	Vapor		Liquid	Vapor
t	T	p	v'	v''	ρ'	ρ''	i'	i''	$r=i''-i'$	s'	s''
°C	K	kp/cm²	m³/kg	m³/kg	kg/m³	kg/m³	kcal/kg	kcal/kg	kcal/kg	kcal/kg K	kcal/kgK
275	548.15	60.66	0.0013168	0.03274	759.42	30.53	289.2	665.2	376.0	0.7219	1.4077
276	549.15	61.60	0.0013198	0.03220	757.69	31.06	290.4	664.9	374.5	0.7240	1.4060
277	550.15	62.55	0.0013228	0.03167	755.97	31.58	291.7	664.7	373.0	0.7262	1.4042
278	551.15	63.51	0.0013259	0.03115	754.20	32.10	292.9	664.4	371.5	0.7284	1.4025
279	552.15	64.48	0.0013290	0.03064	752.45	32.64	294.2	664.2	370.0	0.7306	1.4007
280	553.15	65.46	0.0013321	0.03013	750.69	33.19	295.4	663.9	368.5	0.7328	1.3990
281	554.15	66.45	0.0013353	0.02964	748.90	33.74	296.7	663.7	367.0	0.7350	1.3973
282	555.15	67.46	0.0013385	0.02915	747.10	34.30	297.9	663.4	365.5	0.7373	1.3955
283	556.15	68.47	0.0013417	0.02867	745.32	34.88	299.2	663.1	363.9	0.7395	1.3937
284	557.15	69.50	0.0013450	0.02820	743.49	35.46	300.4	662.7	362.3	0.7417	1.3920
285	558.15	70.54	0.0013483	0.02774	741.67	36.05	301.7	662.4	360.7	0.7439	1.3902
286	559.15	71.59	0.0013516	0.02728	739.86	36.66	303.0	662.1	359.1	0.7461	1.3884
287	560.15	72.65	0.0013550	0.02684	738.01	37.26	304.2	661.7	357.5	0.7483	1.3866
288	561.15	73.73	0.0013585	0.02640	736.11	37.88	305.5	661.4	355.9	0.7505	1.3848
289	562.15	74.82	0.0013620	0.02596	734.21	38.52	306.8	661.1	354.3	0.7528	1.3830
290	563.15	75.92	0.0013655	0.02554	732.33	39.15	308.1	660.7	352.6	0.7550	1.3812
291	564.15	77.03	0.0013691	0.02512	730.41	39.81	309.4	660.4	351.0	0.7573	1.3794
292	565.15	78.15	0.0013727	0.02471	728.49	40.47	310.7	660.0	349.3	0.7595	1.3775
293	566.15	79.29	0.0013764	0.02430	726.53	41.15	312.0	659.6	347.6	0.7617	1.3757
294	567.15	80.44	0.0013801	0.02390	724.59	41.84	313.3	659.2	345.9	0.7640	1.3739
295	568.15	81.60	0.0013839	0.02351	722.60	42.53	314.6	658.8	344.2	0.7662	1.3720
296	569.15	82.78	0.0013877	0.02312	720.62	43.23	315.9	658.4	342.5	0.7648	1.3701
297	570.15	83.97	0.0013916	0.02275	718.60	43.96	317.2	658.0	340.8	0.7706	1.3683
298	571.15	85.17	0.0013956	0.02237	716.54	44.70	318.6	657.6	339.0	0.7729	1.3664
299	572.15	86.38	0.0013996	0.02200	714.49	45.43	319.9	657.1	337.2	0.7751	1.3645
300	573.15	87.61	0.0014036	0.02164	712.45	46.21	321.2	656.6	335.4	0.7774	1.3626
301	574.15	88.85	0.001407	0.02129	710.73	46.97	322.6	656.2	333.6	0.7797	1.3607
302	575.15	90.11	0.001412	0.02094	708.22	47.75	323.9	655.7	331.8	0.7820	1.3588
303	576.15	91.38	0.001416	0.02059	706.21	48.57	325.3	655.3	330.0	0.7842	1.3569
304	577.15	92.66	0.001420	0.02025	704.23	49.38	326.6	654.7	328.1	0.7865	1.3549
305	578.15	93.95	0.001425	0.01992	701.75	50.20	328.0	654.2	326.2	0.7888	1.3530
306	579.15	95.26	0.001429	0.01959	699.79	51.05	329.3	653.6	324.3	0.7911	1.3510
307	580.15	96.59	0.001434	0.01926	697.35	51.92	330.7	653.1	322.4	0.7934	1.3491
308	581.15	97.92	0.001438	0.01894	695.41	52.80	332.1	652.6	320.5	0.7957	1.3471
309	582.15	99.28	0.001443	0.01863	693.00	53.68	333.5	652.0	318.5	0.7980	1.3451
310	583.15	100.64	0.001447	0.01832	691.09	54.58	334.9	651.4	316.5	0.8003	1.3431
311	584.15	102.02	0.001452	0.01801	688.71	55.52	336.3	650.8	314.5	0.8026	1.3411
312	585.15	103.42	0.001457	0.01771	686.34	56.46	337.7	650.2	312.5	0.8050	1.3390
313	586.15	104.83	0.001462	0.01741	683.99	57.44	339.1	649.6	310.5	0.8073	1.3370
314	587.15	106.25	0.001467	0.01712	681.66	58.41	340.5	648.9	308.4	0.8096	1.3349
315	588.15	107.69	0.001472	0.01683	679.35	59.42	342.0	648.3	306.3	0.8120	1.3328
316	589.15	109.15	0.001477	0.01655	677.05	60.42	343.4	647.6	304.2	0.8144	1.3307
317	590.15	110.62	0.001483	0.01627	674.31	61.46	344.9	647.0	302.1	0.8167	1.3286
318	591.15	112.11	0.001488	0.01599	672.04	62.54	346.3	646.3	300.0	0.8191	1.3264
319	592.15	113.61	0.001494	0.01572	669.34	63.61	347.8	645.6	297.8	0.8215	1.3243
320	593.15	115.12	0.001499	0.01545	667.11	64.72	349.2	644.9	295.7	0.8239	1.3221
321	594.15	116.66	0.001505	0.01519	664.45	65.83	350.7	644.3	293.6	0.8263	1.3199
322	595.15	118.21	0.001511	0.01493	661.81	66.98	352.2	643.5	291.3	0.8287	1.3177
323	596.15	119.77	0.001517	0.01467	659.20	68.17	353.7	642.7	289.0	0.8311	1.3155
324	597.15	121.35	0.001523	0.01442	656.60	69.35	355.2	641.9	286.7	0.8336	1.3133
325	598.15	122.95	0.001529	0.01417	654.02	70.57	356.7	641.0	284.3	0.8360	1.3111
326	599.15	124.56	0.001535	0.01392	651.47	71.84	358.3	640.2	281.9	0.8384	1.3088
327	600.16	126.19	0.001542	0.01368	648.51	73.10	359.8	639.3	279.5	0.8409	1.3065
328	601.15	127.84	0.001548	0.01344	645.99	74.40	361.3	638.4	277.1	0.8343	1.3043
329	602.15	129.50	0.001555	0.01320	643.09	75.76	362.9	637.6	274.7	0.8459	1.3019

Table 39-7 Properties of Saturated Steam (H₂O) at a Given Temperature (Gravitational Metric System) (Continued)

Temperature		Pressure	Specific volume		Density		Specific enthalpy		Heat of vapor- ization	Specific entropy	
			Liquid	Vapor	Liquid	Vapor	Liquid	Vapor		Liquid	Vapor
t	T	p	v'	v''	ρ'	ρ''	i'	i''	$r = i'' - i'$	s'	s''
°C	K	kp/cm²	m³/kg	m³/kg	kg/m³	kg/m³	kcal/kg	kcal/kg	kcal/kg	kcal/kgK	kcal/kgK
330	603.15	131.18	0.001562	0.01297	640.20	77.10	364.5	636.7	272.2	0.8484	1.2996
331	604.15	132.88	0.001569	0.01274	637.35	78.49	366.1	635.8	269.7	0.8510	1.2972
332	605.15	134.59	0.001577	0.01251	634.12	79.94	367.7	634.9	267.2	0.8535	1.2949
333	606.15	136.33	0.001584	0.01228	631.31	81.43	369.3	633.9	264.6	0.8561	1.2924
334	607.15	138.08	0.001591	0.01206	628.54	82.92	370.9	632.9	262.0	0.8586	1.2900
335	608.15	139.85	0.001599	0.01184	625.39	84.46	372.5	631.8	259.3	0.8612	1.2875
336	609.15	141.63	0.001607	0.01162	622.28	86.06	374.2	630.8	256.6	0.8637	1.2850
337	610.15	143.44	0.001615	0.01141	619.20	87.64	375.8	629.6	253.8	0.8663	1.2824
338	611.15	145.26	0.001623	0.01120	616.14	89.29	377.5	628.5	251.0	0.8689	1.2798
339	612.15	147.10	0.001631	0.01099	613.12	90.99	379.2	627.4	248.2	0.8716	1.2772
340	613.15	148.96	0.001639	0.01078	610.13	92.76	380.9	626.2	245.3	0.8743	1.2745
341	614.15	150.84	0.001648	0.01057	606.80	94.60	382.6	625.0	242.4	0.8770	1.2718
342	615.15	152.73	0.001658	0.01037	603.14	96.43	384.4	623.8	239.4	0.8798	1.2690
343	616.15	154.65	0.001667	0.01017	599.88	98.33	386.1	622.5	236.4	0.8825	1.2662
344	617.15	156.59	0.001676	0.009969	596.66	100.31	387.9	621.2	233.3	0.8853	1.2633
345	618.15	158.54	0.001686	0.009771	593.12	102.34	389.8	619.9	230.1	0.8881	1.2604
346	619.15	160.52	0.001696	0.009574	589.62	104.45	391.6	618.6	227.0	0.8909	1.2574
347	620.15	162.52	0.001707	0.009379	585.82	106.62	393.4	617.2	223.8	0.8937	1.2544
348	621.15	164.53	0.001718	0.009186	582.07	108.86	395.3	615.7	220.4	0.8966	1.2513
349	622.15	166.57	0.001729	0.008995	578.37	111.17	379.2	614.1	216.9	0.8995	1.2481
350	623.15	168.63	0.001741	0.008803	574.38	113.6	399.2	612.5	213.3	0.9025	1.2448
351	624.15	170.71	0.001752	0.008613	570.78	116.1	401.1	610.9	209.8	0.9055	1.2415
352	625.15	172.81	0.001764	0.008425	566.89	118.7	403.1	609.2	206.1	0.9085	1.2381
353	626.15	174.93	0.001777	0.008238	562.75	121.4	405.1	607.4	202.3	0.9116	1.2346
354	627.15	177.07	0.001792	0.008053	558.04	124.2	407.2	605.5	198.3	0.9148	1.2310
355	628.15	179.24	0.001807	0.007869	553.40	127.1	409.4	603.6	194.2	0.9181	1.2273
356	629.15	181.43	0.001823	0.007684	548.55	130.0	411.6	601.5	189.9	0.9214	1.2235
357	630.15	183.64	0.001840	0.007499	543.48	133.2	413.8	599.3	185.5	0.9248	1.2195
358	631.15	185.88	0.001857	0.007314	538.50	136.6	416.1	597.1	181.0	0.9283	1.2154
359	632.15	188.13	0.001875	0.007130	533.33	140.2	418.4	594.9	176.5	0.9318	1.2112
360	633.15	190.42	0.001894	0.006943	527.98	144.0	420.7	592.6	171.9	0.9354	1.2069
361	634.15	192.72	0.001918	0.00675	521.38	148.1	423.3	590.1	166.8	0.9392	1.2024
362	635.15	195.06	0.001943	0.00656	514.67	152.4	425.8	587.5	161.7	0.9430	1.1977
363	636.15	197.41	0.001968	0.00637	508.13	157.0	428.4	584.6	156.2	0.9470	1.1928
364	637.15	199.80	0.00199	0.00618	502.51	161.8	431.2	581.5	150.3	0.9512	1.1872
365	638.15	202.21	0.00202	0.00599	495.05	166.8	434.1	578.2	144.1	0.9556	1.1814
366	639.15	204.64	0.00205	0.00580	487.80	172.5	437.2	574.7	137.5	0.9602	1.1754
367	640.15	207.11	0.00208	0.00559	480.77	178.8	440.4	570.9	130.5	0.9652	1.1689
368	641.15	209.60	0.00212	0.00538	471.70	185.8	443.9	566.6	122.7	0.9704	1.1619
369	642.15	212.12	0.00217	0.00516	460.83	193.6	447.8	562.0	114.2	0.9762	1.1541
370	643.15	214.68	0.00222	0.00493	450.45	203	452.0	556.7	104.7	0.9825	1.1453
371	644.15	217.26	0.00229	0.00468	436.68	214	456.8	550.5	93.7	0.9898	1.1352
372	645.15	219.88	0.00238	0.00440	420.17	227	462.6	542.9	80.3	0.9986	1.1230
373	646.15	222.53	0.00251	0.00405	398.41	247	470.3	532.6	62.4	1.0102	1.1067
374	647.15	225.22	0.00280	0.00347	357.14	228	485.3	512.7	27.4	1.0332	1.0755
374.15	647.30	225.65	0.00326	0.00326	307	307	501.5	501.5	0.0	1.058	1.058

Table 40-1 Properties of Saturated Steam (H_2O) at a Given Pressure (Gravitational Metric System)

Pressure	Temperature		Specific volume		Density		Specific enthalpy		Heat of vapor-ization	Specific entropy	
			Liquid	Vapor	Liquid	Vapor	Liquid	Vapor		Liquid	Vapor
p	t	T	v'	v''	ρ'	ρ''	i'	i''	$r=i''-i'$	s'	s''
kp/cm²	°C	K	m³/kg	m³/kg	kg/m³	kg/m³	kcal/kg	kcal/kg	kcal/kg	kcal/kg K	kcal/kg K
0.010	6.698	279.848	0.0010001	131.6	999.90	0.007599	6.73	600.2	593.5	0.0243	2.1451
0.015	12.737	285.887	0.0010006	89.63	999.40	0.01116	12.78	602.9	590.1	0.0457	2.1100
0.020	17.204	290.354	0.0010013	68.25	998.70	0.01465	17.25	604.9	587.6	0.0612	2.0851
0.025	20.776	293.926	0.0010020	55.27	998.00	0.01809	20.82	606.9	585.6	0.0734	2.0657
0.030	23.772	296.922	0.0010027	46.52	997.31	0.02150	23.81	607.8	584.0	0.0835	2.0501
0.035	26.359	299.509	0.0010034	40.22	996.61	0.02486	26.39	608.9	582.5	0.0922	2.0369
0.040	28.641	301.791	0.0010040	35.46	996.02	0.02820	28.67	609.8	581.1	0.0998	2.0255
0.045	30.69	303.84	0.0010046	31.71	995.42	0.03154	30.71	610.7	580.0	0.1066	2.0154
0.050	32.55	305.70	0.0010052	28.72	994.83	0.03482	32.57	611.5	578.9	0.1126	2.0065
0.055	34.25	307.40	0.0010058	26.26	994.23	0.03808	34.27	612.3	578.0	0.1182	1.9983
0.060	35.82	308.97	0.0010063	24.19	993.74	0.04134	35.83	612.9	577.1	0.1232	1.9909
0.065	37.29	310.44	0.0010069	22.43	993.15	0.04458	37.30	613.6	576.3	0.1280	1.9842
0.070	38.66	311.81	0.0010074	20.91	992.65	0.04782	38.67	614.1	575.4	0.1324	1.9779
0.075	39.95	313.10	0.0010079	19.59	992.16	0.05105	39.96	614.7	574.7	0.1365	1.9721
0.080	41.16	314.31	0.0010084	18.45	991.67	0.05420	41.16	615.2	574.0	0.1404	1.9667
0.085	42.32	315.47	0.0010088	17.41	991.28	0.05744	42.32	615.7	573.4	0.1440	1.9616
0.090	43.41	316.56	0.0010093	16.50	990.79	0.06061	43.41	616.1	572.7	0.1475	1.9568
0.095	44.46	317.61	0.0010097	15.68	990.39	0.06378	44.46	616.6	572.1	0.1508	1.9528
0.10	45.45	318.60	0.0010101	14.95	990.00	0.06689	45.45	617.0	571.6	0.1539	1.9480
0.11	47.33	320.48	0.0010109	13.66	989.22	0.07821	47.32	617.8	570.5	0.1598	1.9400
0.12	49.06	322.21	0.0010117	12.59	988.44	0.07943	49.05	618.6	569.5	0.1652	1.9326
0.13	50.67	323.82	0.0010124	11.67	987.75	0.08562	50.66	619.3	568.6	0.1702	1.9260
0.14	52.18	325.33	0.0010131	10.89	987.07	0.09183	52.17	619.9	567.7	0.1748	1.9197
0.15	53.60	326.75	0.0010138	10.20	986.39	0.09804	53.59	620.5	566.9	0.1791	1.9140
0.16	54.94	328.09	0.0010145	9.603	985.71	0.1041	54.93	621.1	566.2	0.1832	1.9086
0.17	56.21	329.36	0.0010151	9.073	985.12	0.1102	56.19	621.6	565.4	0.1871	1.9036
0.18	57.41	330.56	0.0010157	8.601	984.54	0.1163	57.39	622.1	564.7	0.1907	1.8989
0.19	58.57	331.72	0.0010163	8.172	983.96	0.1224	58.55	622.6	564.0	0.1942	1.8944
0.20	59.67	332.82	0.0010169	7.789	983.38	0.1284	59.65	623.1	563.4	0.1975	1.8902
0.21	60.72	333.87	0.0010175	7.442	982.80	0.1344	60.70	623.5	562 8	0.2006	1.8862
0.22	61.74	334.89	0.0010181	7.122	982.22	0.1404	61.72	623.9	562.2	0.2037	1.8823
0.23	62.71	335.86	0.0010186	6.833	981.74	0.1464	62.69	624.3	561.6	0.2066	1.8786
0.24	63.65	336.80	0.0010191	5.565	981.26	0.1523	63.63	624.6	561.0	0.2094	1.8751
0.25	64.56	337.71	0.0010196	6.318	980.78	0.1583	64.54	625.0	560.5	0.2121	1.8718
0.26	65.44	338.59	0.0010202	6.088	980.20	0.1643	65.42	625.4	560.0	0.2147	1.8685
0.27	66.29	339.44	0.0010206	5.876	979.82	0.1702	66.27	625.7	559.4	0.2172	1.8654
0.28	67.11	340.26	0.0010211	5.679	979.34	0.1761	67.09	626.1	559.0	0.2196	1.8624
0.29	67.91	341.06	0.0010216	5.495	978.86	0.1820	67.89	626.4	558.5	0.2219	1.8595
0.30	68.68	341.83	0.0010220	5.324	978.47	0.1878	68.66	626.8	558.1	0.2242	1.8568
0.32	70.16	343.31	0.0010229	5.013	977.61	0.1995	70.14	627.4	557.3	0.2286	1.8515
0.34	71.57	344.72	0.0010237	4.736	976.85	0.2112	71.56	627.9	556.3	0.2327	1.8465
0.36	72.91	346.06	0.0010245	4.489	976.09	0.2228	72.90	628.5	555.6	0.2365	1.8419
0.38	74.19	347.34	0.0010253	4.267	975.32	0.2344	74.18	629.0	554.8	0.2402	1.8375
0.40	75.42	348.57	0.0010261	4.066	974.56	0.2459	75.41	629.5	554.1	0.2438	1.8333
0.45	78.27	351.42	0.0010279	3.641	972.86	0.2746	78.27	630.6	552.3	0.2519	1.8237
0.50	80.86	354.01	0.0010296	3.299	971.25	0.3031	80.86	631.6	550.7	0.2592	1.8152
0.55	83.25	356.40	0.0010312	3.017	969.74	0.3315	83.26	632.6	549.3	0.2660	1.8074
0.60	85.45	358.60	0.0010327	2.782	968.34	0.3595	85.47	633.5	548.0	0.2722	1.8004
0.65	87.51	360.66	0.0010341	2.581	967.02	0.3875	87.54	634.3	546.8	0.2779	1.7939
0.70	89.45	362.60	0.0010355	2.408	965.72	0.4153	89.49	635.1	545.6	0.2833	1.7879
0.75	91.27	364.42	0.0010368	2.257	964.51	0.4431	91.32	635.8	544.5	0.2883	1.7824
0.80	92.99	366.14	0.0010381	2.125	963.30	0.4706	93.05	636.4	543.3	0.2931	1.7772
0.85	94.62	367.77	0.0010393	2.008	962.19	0.4980	94.69	637.0	542.3	0.2976	1.7723
0.90	96.18	369.33	0.0010405	1.903	961.08	0.5255	96.26	637.6	541.3	0.3018	1.7677
0.95	97.66	370.81	0.0010417	1.810	959.97	0.5525	97.75	638.2	540.4	0.3058	1.7633

Table 40-2 Properties of Saturated Steam (H$_2$O) at a Given Pressure (Gravitational Metric System) (*Continued*)

Pressure	Temperature		Specific volume		Density		Specific enthalpy		Heat of vapor-ization	Specific entropy	
			Liquid	Vapor	Liquid	Vapor	Liquid	Vapor		Liquid	Vapor
p	t	T	v'	v''	ρ'	ρ''	i'	i''	$r=i'-i''$	s'	s''
kp/cm²	°C	K	m³/kg	m³/kg	kg/m³	kg/m³	kcal/kg	kcal/kg	kcal/kg	kcal/kg K	kcal/kg K
1.0	99.09	372.24	0.0010428	1.725	958.96	0.5797	99.19	638.8	539.6	0.3097	1.7593
1.1	101.76	374.91	0.0010448	1.578	957.12	0.6337	101.87	639.8	537.9	0.3169	1.7517
1.2	104.25	377.40	0.0010468	1.455	955.29	0.6873	104.38	640.7	536.3	0.3236	1.7447
1.3	106.56	379.71	0.0010487	1.350	953.56	0.7407	106.72	641.6	534.9	0.3297	1.7382
1.4	108.74	381.89	0.0010505	1.259	951.93	0.7943	108.92	642.3	533.4	0.3355	1.7323
1.5	110.79	383.94	0.0010522	1.181	950.39	0.8467	110.99	643.1	532.1	0.3409	1.7268
1.6	112.73	385.88	0.0010538	1.111	948.95	0.9001	112.95	643.8	530.8	0.3460	1.7217
1.7	114.57	387.72	0.0010554	1.050	947.51	0.9524	114.81	644.5	529.7	0.3508	1.7168
1.8	116.33	389.48	0.0010570	0.9954	946.07	1.0046	116.60	645.1	528.5	0.3554	1.7123
1.9	118.01	391.16	0.0010585	0.9462	994.73	1.0570	118.30	645.7	527.4	0.3597	1.7080
2.0	119.62	392.77	0.0010600	0.9018	943.40	1.109	119.94	646.3	526.4	0.3639	1.7039
2.1	121.16	394.31	0.0010614	0.8616	942.15	1.161	121.5	646.8	525.3	0.3679	1.7000
2.2	122.65	395.80	0.0010627	0.8248	941.00	1.212	123.0	647.3	524.3	0.3717	1.6963
2.3	124.08	397.23	0.0010640	0.7912	939.85	1.264	124.5	647.8	523.3	0.3754	1.6928
2.4	125.46	398.61	0.0010653	0.7603	938.70	1.315	125.9	648.3	522.4	0.3789	1.6894
2.5	126.79	399.94	0.0010666	0.7318	937.56	1.367	127.2	648.7	521.5	0.3822	1.6862
2.6	128.08	401.23	0.0010678	0.7055	936.50	1.417	128.5	649.2	520.7	0.3855	1.6830
2.7	129.34	402.49	0.0010691	0.6808	935.37	1.469	129.8	649.6	519.8	0.3887	1.6800
2.8	130.55	403.70	0.0010703	0.6581	934.32	1.520	131.1	650.0	518.9	0.3918	1.6771
2.9	131.73	404.88	0.0010714	0.6368	933.36	1.570	132.3	650.3	518.0	0.3947	1.6743
3.0	132.88	406.03	0.0010726	0.6169	932.31	1.621	133.4	650.7	517.3	0.3976	1.6717
3.1	134.00	407.15	0.0010737	0.5982	931.36	1.672	134.6	651.7	516.5	0.4004	1.6690
3.2	135.08	408.23	0.0010748	0.5807	930.41	1.722	135.7	651.4	515.7	0.4031	1.6665
3.3	136.14	409.29	0.0010758	0.5645	929.54	1.772	136.8	651.8	515.0	0.4058	1.6641
3.4	137.18	410.33	0.0010759	0.5486	928.59	1.823	137.8	652.1	514.3	0.4084	1.6617
3.5	138.19	411.34	0.0010779	0.5338	927.73	1.873	138.9	652.4	513.5	0.4109	1.6594
3.6	139.18	412.33	0.0010789	0.5199	926.87	1.923	139.9	652.7	512.8	0.4134	1.6572
3.7	140.15	413.20	0.0010800	0.5066	925.93	1.974	140.9	653.0	512.1	0.4158	1.6550
3.8	141.09	414.24	0.0010809	0.4942	925.15	2.024	141.8	653.3	511.5	0.4181	1.6529
3.9	142.02	415.17	0.0010819	0.4822	924.30	2.074	142.8	653.6	510.8	0.4204	1.6508
4.0	142.92	416.07	0.0010829	0.4709	923.45	2.124	143.7	653.9	510.2	0.4226	1.6488
4.1	143.81	416.96	0.0010838	0.4601	922.68	2.173	144.6	654.1	509.5	0.4248	1.6468
4.2	144.68	417.83	0.0010847	0.4498	921.91	2.223	145.5	654.4	508.9	0.4269	1.6449
4.3	145.54	418.69	0.0010857	0.4399	921.06	2.273	146.4	654.7	508.3	0.4290	1.6430
4.4	146.38	419.53	0.0010866	0.4305	920.30	2.323	147.3	654.9	507.6	0.4311	1.6412
4.5	147.20	420.35	0.0010875	0.4215	919.54	2.373	148.1	655.2	507.1	0.4331	1.6494
4.6	148.01	421.16	0.0010884	0.4129	918.78	2.442	149.0	655.4	506.5	0.4351	1.6376
4.7	148.81	421.96	0.0010893	0.4045	918.02	2.472	149.8	655.6	505.8	0.4370	1.6359
4.8	149.59	422.74	0.0010902	0.3966	917.26	2.521	150.6	655.9	505.3	0.4389	1.6342
4.9	150.36	423.51	0.0010910	0.3890	916.59	2.571	151.4	656.1	504.7	0.4408	1.6325
5.0	151.11	424.26	0.0010918	0.3817	915.92	2.620	152.1	656.3	504.2	0.4426	1.6309
5.2	152.59	425.74	0.0010935	0.3679	914.49	2.718	153.7	656.7	503.0	0.4462	1.6278
5.4	154.02	427.17	0.0010951	0.3550	913.16	2.817	155.1	657.1	502.0	0.4496	1.6248
5.6	155.41	428.56	0.0010967	0.3431	911.83	2.915	156.6	657.5	500.9	0.4530	1.6219
5.8	156.76	429.51	0.0010983	0.3319	910.50	3.013	158.0	657.9	499.9	0.4562	1.6191
6.0	158.08	431.23	0.0010998	0.3214	909.26	3.111	159.3	658.3	498.9	0.4594	1.6164
6.2	159.36	432.51	0.0011013	0.3116	908.02	3.209	160.7	658.6	497.9	0.4625	1.6137
6.4	160.61	433.76	0.0011028	0.3024	906.78	3.307	162.0	659.0	497.0	0.4655	1.6112
6.6	161.82	434.97	0.0011043	0.2938	905.55	3.404	163.2	659.3	496.1	0.4683	1.6087
6.8	163.01	436.16	0.0011057	0.2856	904.40	3.501	164.5	659.6	495.1	0.4712	1.6063
7.0	164.17	437.32	0.0011071	0.2778	903.26	3.600	165.7	659.9	494.2	0.4738	1.6039
7.2	165.31	438.46	0.0011085	0.2705	902.12	3.697	166.9	660.2	493.3	0.4766	1.6016
7.4	166.42	439.57	0.0011099	0.2636	900.98	3.794	168.0	660.4	492.4	0.4793	1.5994
7.6	167.51	440.66	0.0011113	0.2570	899.85	3.891	169.2	660.7	491.5	0.4818	1.5972
7.8	168.57	441.72	0.0011126	0.2507	898.80	3.989	170.3	661.0	490.7	0.4843	1.5951

Table 40-3 Properties of Saturated Steam (H₂O) at a Given Pressure (Gravitational Metric System) (*Continued*)

Pressure	Temperature		Specific volume		Density		Specific enthalpy		Heat of vapor- ization	Specific entropy	
			Liquid	Vapor	Liquid	Vapor	Liquid	Vapor		Liquid	Vapor
p	t	T	v'	v''	ρ'	ρ''	i'	i''	$r=i'-i''$	s'	s''
kp/cm²	°C	K	m³/kg	m³/kg	kg/m³	kg/m³	kcal/kg	kcal/kg	kcal/kg	kcal/kg K	kcal/kg K
8.0	169.61	442.76	0.0011139	0.2448	897.75	4.085	171.4	661.2	489.8	0.4868	1.5931
8.2	170.63	443.78	0.0011152	0.2391	896.70	4.182	172.4	661.4	489.0	0.4892	1.5911
8.4	171.63	444.78	0.0011165	0.2337	895.66	4.279	173.4	661.7	488.3	0.4915	1.5891
8.6	172.61	445.76	0.0011177	0.2286	894.69	4.375	174.5	661.9	487.4	0.4938	1.5872
8.8	173.58	446.73	0.0011189	0.2236	893.73	4.472	175.5	662.1	486.6	0.4961	1.5853
9.0	174.53	447.68	0.0011202	0.2189	892.70	4.568	176.5	662.3	485.8	0.4983	1.5834
9.2	175.46	448.61	0.0011214	0.2144	891.74	4.664	177.5	662.5	485.0	0.5005	1.5816
9.4	176.38	449.53	0.0011226	0.2100	890.79	4.762	178.5	662.7	484.2	0.5026	1.5799
9.6	177.28	450.43	0.0011238	0.2058	889.84	4.859	179.4	662.9	483.5	0.5047	1.5782
9.8	178.16	451.31	0.0011250	0.2019	888.89	4.953	180.3	663.1	482.8	0.5067	1.5765
10.0	179.04	452.19	0.0011262	0.1980	887.94	5.051	181.3	663.3	482.1	0.5088	1.5748
10.5	181.16	454.31	0.0011291	0.1890	885.66	5.291	183.5	663.7	480.2	0.5137	1.5708
11.0	183.20	456.35	0.0011319	0.1808	883.47	5.531	185.7	664.1	478.4	0.5184	1.5670
11.5	185.17	458.32	0.0011346	0.1733	881.37	5.770	187.7	664.5	476.8	0.5229	1.5633
12.0	187.08	460.23	0.0011373	0.1663	879.28	6.013	189.8	664.9	475.1	0.5273	1.5597
12.5	188.92	462.07	0.0011399	0.1599	877.27	6.254	191.7	665.3	473.6	0.5315	1.5563
13.0	190.71	463.86	0.0011426	0.1540	875.20	6.494	193.6	665.6	472.0	0.5356	1.5530
13.5	192.45	465.60	0.0011451	0.1485	873.29	6.734	195.5	665.9	470.4	0.5396	1.5498
14.0	194.13	467.28	0.0011476	0.1434	871.38	6.974	197.3	666.2	468.9	0.5434	1.5468
14.5	195.77	468.92	0.0011501	0.1387	869.49	7.210	199.1	666.4	467.4	0.5471	1.5438
15.0	197.36	470.51	0.0011525	0.1342	867.68	7.452	200.7	666.7	465.9	0.5507	1.5410
15.5	198.91	472.06	0.0011548	0.1300	865.95	7.692	202.4	666.9	464.5	0.5542	1.5382
16.0	200.43	473.58	0.0011572	0.1261	864.15	7.930	204.0	667.1	463.1	0.5577	1.5354
16.5	201.91	475.06	0.0011595	0.1224	862.44	8.170	205.6	667.3	461.7	0.5610	1.5328
17.0	203.35	476.50	0.0011618	0.1189	860.73	8.410	207.2	667.5	460.3	0.5642	1.5302
17.5	204.76	477.91	0.0011640	0.1156	859.11	8.651	208.7	667.7	459.0	0.5674	1.5277
18.0	206.14	379.29	0.0011662	0.1125	857.49	8.889	210.2	667.8	457.6	0.5705	1.5253
18.5	207.49	480.64	0.0011684	0.1095	855.87	9.132	211.7	668.0	456.3	0.5753	1.5229
19.0	208.81	481.96	0.0011706	0.1067	854.26	9.372	213.1	668.2	455.1	0.5764	1.5206
19.5	210.11	483.26	0.0011728	0.1040	852.66	9.615	214.5	668.3	453.8	0.5793	1.5183
20.0	211.38	484.53	0.0011749	0.1015	851.14	9.852	215.9	668.5	452.6	0.5822	1.5161
20.5	212.63	485.78	0.0011771	0.09907	849.55	10.09	217.3	668.6	451.3	0.5850	1.5139
21.0	213.85	487.00	0.0011792	0.09676	848.03	10.34	218.6	668.7	450.1	0.5877	1.5118
21.5	215.05	488.20	0.0011813	0.09456	846.53	10.57	220.0	668.8	448.8	0.5904	1.5097
22.0	216.23	489.38	0.0011833	0.09245	845.09	10.82	221.2	668.9	447.7	0.5930	1.5077
22.5	217.39	490.54	0.0011854	0.09042	843.60	11.06	222.5	668.9	446.4	0.5956	1.5056
23.0	218.53	491.68	0.0011874	0.08849	842.18	11.30	223.8	669.0	445.2	0.5981	1.5037
23.5	219.65	492.80	0.0011894	0.08663	840.76	11.54	225.0	669.1	444.1	0.6006	1.5017
24.0	220.75	493.90	0.0011914	0.08486	839.35	11.78	226.2	669.2	443.0	0.6031	1.4998
24.5	221.83	494.98	0.0011933	0.08316	838.01	12.03	227.4	669.2	441.8	0.6054	1.4980
25.0	222.90	496.05	0.0011953	0.08150	836.61	12.27	228.6	669.3	440.7	0.6074	1.4961
25.5	223.95	497.10	0.0011973	0.07991	835.21	12.51	229.8	669.3	439.5	0.6101	1.4943
26.0	224.99	498.14	0.0011992	0.07838	833.89	12.76	230.9	669.4	438.5	0.6124	1.4925
26.5	226.01	499.16	0.0012011	0.07692	832.57	13.00	232.1	669.4	437.3	0.6146	1.4908
27.0	227.01	500.16	0.0012030	0.07551	831.26	13.24	233.2	669.4	436.2	0.6168	1.4891
27.5	228.00	501.15	0.0012048	0.07414	830.01	13.49	234.2	669.5	435.2	0.6190	1.4874
28.0	228.98	502.13	0.0012067	0.07282	828.71	13.73	235.4	669.5	434.1	0.6212	1.4857
28.5	229.94	503.09	0.0012086	0.07155	827.40	13.98	236.5	669.5	433.0	0.6233	1.4841
29.0	230.89	504.04	0.0012105	0.07032	826.10	14.22	237.5	669.5	432.0	0.6254	1.4825
29.5	231.83	504.98	0.0012122	0.06913	824.95	14.47	238.6	669.6	431.0	0.6274	1.4809
30	232.76	505.91	0.0012142	0.06797	823.59	14.71	239.6	669.6	430.0	0.6295	1.4794
31	234.57	507.72	0.0012178	0.06578	821.15	15.20	241.7	669.6	427.9	0.6335	1.4763
32	236.35	509.50	0.0012215	0.06370	818.67	15.70	243.7	669.6	425.9	0.6374	1.4733
33	238.08	511.23	0.0012251	0.06176	816.26	16.19	245.6	669.6	423.9	0.6412	1.4704
34	239.77	512.92	0.0012286	0.05995	813.93	16.68	247.6	669.5	421.9	0.6449	1.4675

Table 40-4 Properties of Saturated Steam (H₂O) at a Given Pressure (Gravitational Metric System) (Continued)

Pressure	Temperature		Specific volume		Density		Specific enthalpy		Heat of vapor- ization	Specific entropy	
			Liquid	Vapor	Liquid	Vapor	Liquid	Vapor		Liquid	Vapor
p	t	T	v'	v''	ρ'	ρ''	i'	i''	$r=i''-i'$	s'	s''
kp/cm²	°C	K	m³/kg	m³/kg	kg/m³	kg/m³	kcal/kg	kcal/kg	kcal/kg	kcal/kg K	kcal/kg K
35	241.42	514.57	0.0012321	0.05819	811.62	17.18	249.5	669.5	420.0	0.6485	1.4647
36	243.04	516.19	0.0012356	0.05654	809.32	17.69	251.3	669.4	418.1	0.6520	1.4620
37	244.62	517.77	0.0012391	0.05499	807.04	18.19	253.1	669.3	416.2	0.6555	1.4593
38	246.17	519.32	0.0012425	0.05352	804.83	18.68	254.9	669.2	414.3	0.6589	1.4567
39	247.69	520.84	0.0012459	0.05211	802.63	19.19	256.7	669.2	412.5	0.6622	1.4542
40	249.18	522.33	0.0012493	0.05077	800.45	19.70	258.4	669.0	410.6	0.6654	1.4517
41	250.64	523.79	0.0012527	0.04950	798.28	20.20	260.1	668.9	408.8	0.6686	1.4492
42	252.07	525.22	0.0012561	0.04829	796.11	20.71	261.8	668.8	407.0	0.6718	1.4468
43	253.48	526.63	0.0012594	0.04713	794.03	21.22	263.4	668.7	405.3	0.6749	1.4444
44	254.87	528.02	0.0012628	0.04601	791.89	21.73	265.0	668.5	403.5	0.6779	1.4420
45	256.23	529.38	0.0012661	0.04495	789.83	22.25	266.6	668.4	401.8	0.6809	1.4397
46	257.56	530.71	0.0012694	0.04394	787.77	22.76	268.2	668.2	400.0	0.6838	1.4375
47	258.88	532.03	0.0012727	0.04296	785.73	23.28	269.8	668.0	398.2	0.6867	1.4353
48	260.17	533.32	0.0012759	0.04203	783.76	23.79	271.3	667.9	396.6	0.6895	1.4331
49	261.45	534.60	0.0012792	0.04112	781.74	24.32	272.8	667.7	394.9	0.6923	1.4309
50	262.70	535.85	0.0012825	0.04026	779.73	24.84	274.3	667.5	393.2	0.6950	1.4288
51	263.93	537.08	0.0012857	0.03943	777.79	25.36	275.8	667.4	391.6	0.6977	1.4267
52	265.15	538.30	0.0012890	0.03863	775.80	25.89	277.2	667.2	390.0	0.7003	1.4246
53	266.35	539.50	0.0012922	0.03785	773.87	26.42	278.7	667.0	388.3	0.7029	1.4226
54	267.53	540.68	0.0012964	0.03711	771.96	26.95	280.1	666.7	386.6	0.7055	1.4205
55	268.69	541.84	0.0012986	0.03639	770.06	27.48	281.5	666.6	385.1	0.7080	1.4186
56	269.84	542.99	0.0013018	0.03569	768.17	28.02	282.9	666.3	383.4	0.7106	1.4166
57	270.98	544.13	0.0013051	0.03502	766.22	28.56	284.3	666.1	381.8	0.7131	1.4146
58	272.10	545.25	0.0013083	0.03437	764.35	29.10	285.6	665.9	380.3	0.7155	1.4127
59	273.20	546.35	0.0013115	0.03374	762.49	29.64	287.0	665.6	378.6	0.7179	1.4108
60	274.29	547.44	0.0013147	0.03313	760.63	30.18	288.3	665.4	377.1	0.7203	1.4089
61	275.37	548.52	0.0013179	0.03255	758.78	30.72	289.7	665.1	375.4	0.7227	1.4071
62	276.43	549.58	0.0013211	0.03197	756.94	31.28	291.0	664.8	373.8	0.7250	1.4052
63	277.48	550.63	0.0013243	0.03142	755.12	31.83	292.3	664.6	372.3	0.7273	1.4034
64	278.51	551.66	0.0013275	0.03089	753.30	32.37	293.6	664.3	370.8	0.7296	1.4016
65	279.54	552.69	0.0013306	0.03036	751.54	32.94	294.8	664.0	369.2	0.7318	1.3998
66	280.55	553.70	0.0013338	0.02986	749.74	33.49	296.1	663.7	367.6	0.7340	1.3980
67	281.55	554.70	0.0013370	0.02937	747.94	34.05	297.3	663.5	366.2	0.7362	1.3963
68	282.54	555.69	0.0013402	0.02889	746.16	34.61	298.6	663.2	364.6	0.7384	1.3945
69	283.52	556.67	0.0013434	0.02843	744.38	35.17	299.8	662.9	363.1	0.7406	1.3928
70	284.48	557.63	0.0013466	0.02798	742.16	35.74	301.0	662.6	361.6	0.7428	1.3911
71	285.44	558.59	0.0013498	0.02754	740.85	36.31	302.3	662.2	359.9	0.7449	1.3894
72	286.39	559.54	0.0013530	0.02711	739.10	36.89	303.5	661.9	358.4	0.7470	1.3877
73	287.32	560.47	0.0013561	0.02669	737.41	37.47	304.6	661.6	356.9	0.7491	1.3860
74	288.25	561.40	0.0013593	0.02629	735.67	38.04	305.8	661.3	355.5	0.7511	1.3843
75	289.17	562.32	0.0013626	0.02589	733.89	38.63	307.0	661.0	354.0	0.7532	1.3827
76	290.08	563.23	0.0013658	0.02550	732.17	39.22	308.2	660.7	352.5	0.7552	1.3811
77	290.97	564.12	0.0013690	0.02513	730.46	39.79	309.4	660.3	351.0	0.7572	1.3794
78	291.86	565.01	0.0013722	0.02476	728.76	40.39	310.5	660.0	349.5	0.7592	1.3778
79	292.75	565.90	0.0013755	0.02440	727.01	40.98	311.7	659.7	348.0	0.7612	1.3761
80	293.62	566.77	0.0012787	0.02405	725.32	41.58	312.9	659.3	346.5	0.7631	1.3745
81	294.48	567.63	0.0013819	0.02372	723.64	42.16	313.9	659.0	345.1	0.7650	1.3730
82	295.34	568.49	0.0013852	0.02338	721.92	42.77	315.0	658.6	343.6	0.7670	1.3714
83	296.19	569.34	0.0013885	0.02305	720.20	43.38	316.2	658.3	342.1	0.7689	1.3698
84	297.03	570.18	0.0013917	0.02273	718.55	43.99	317.3	657.9	340.6	0.7708	1.3682
85	297.86	571.01	0.0013950	0.02243	716.85	44.58	318.4	657.6	339.2	0.7726	1.3666
86	298.69	571.84	0.0013983	0.02212	715.15	45.21	319.5	657.2	337.7	0.7745	1.3651
87	299.51	572.66	0.0014016	0.02182	713.47	45.83	320.6	656.8	336.2	0.7763	1.3635
88	300.32	573.47	0.0014049	0.02153	711.79	46.45	321.6	656.5	334.8	0.7781	1.3620
89	301.12	574.27	0.0014082	0.02124	710.13	47.08	322.7	656.1	333.4	0.7800	1.3604

Table 40-5 Properties of Saturated Steam (H₂O) at a Given Pressure (Gravitational Metric System) (Continued)

Pressure	Temperature		Specific volume		Density		Specific enthalpy		Heat of vapor-ization	Specific entropy	
			Liquid	Vapor	Liquid	Vapor	Liquid	Vapor		Liquid	Vapor
p	t	T	v'	v''	ρ'	ρ''	i'	i''	$r=i''-i'$	s'	s''
kp/cm²	°C	K	m³/kg	m³/kg	kg/m³	kg/m³	kcal/kg	kcal/kg	kcal/kg	kcal/kg K	kcal/kg K
90	301.92	575.07	0.0014115	0.02096	708.47	47.71	323.8	655.7	331.9	0.7818	1.3589
91	302.71	575.86	0.0014148	0.02069	706.81	48.33	324.9	655.3	330.4	0.7836	1.3574
92	303.49	576.64	0.0014181	0.02042	705.17	48.97	325.9	655.0	329.1	0.7854	1.3559
93	304.27	577.42	0.0014215	0.02016	703.48	49.60	327.0	654.6	327.6	0.7871	1.3544
94	305.04	578.19	0.0014249	0.01990	701.80	50.25	328.0	654.2	326.2	0.7889	1.3529
95	305.80	578.95	0,0014282	0.01965	700.18	50.89	329.1	653.8	324.7	0.79C6	1.3514
96	306.56	579.71	0,0014316	0.01940	698.52	51.55	330.1	653.3	323.2	0.7924	1.3499
97	307.31	580.46	0.0014350	0.01916	696.86	52.19	331.2	652.9	321.8	0.7941	1.3484
98	308.06	581.21	0.0014384	0.01892	695.22	52.85	332.2	652.5	320.3	0.7958	1.3469
99	308.80	581.95	0.0014418	0.01869	693.58	53.51	333.2	652.1	318.9	0.7975	1.3455
100	309.53	582.68	0.0014453	0.01846	691.90	54.17	334.2	651.7	317.5	0.7992	1.3440
102	310.98	584.13	0.0014522	0.01802	688.61	55.49	336.3	650.8	314.5	0.8026	1.3411
104	312.41	585.56	0.0014591	0.01759	685.35	56.85	338.3	650.0	311.7	0.8059	1.3381
106	313.82	586.97	0.0014662	0.01717	682.04	58.24	340.3	649.1	308.8	0.8092	1.3352
108	315.21	588.36	0.0014733	0.01677	678.75	59.63	342.3	648.1	305.8	0.8125	1.3324
110	316.58	589.73	0.001480	0.01638	675.68	61.05	344.2	647.2	303.0	0.8158	1.3294
112	317.93	591.08	0.001488	0.01601	672.04	62.46	346.2	646.3	300.1	0.8190	1.3265
114	319.26	592.41	0.001495	0.01565	668.90	63.90	348.2	645.3	297.2	0.8221	1.3237
116	320.57	593.72	0.001502	0.01530	665.78	65.36	350.1	644.4	294.3	0.8253	1.3208
118	321.87	595.02	0.001510	0.01496	662.25	66.85	352.0	643.5	291.5	0.8284	1.3179
120	323.15	596.30	0.001517	0.01463	659.30	68.35	353.9	642.5	288.6	0.8315	1.3151
122	324.41	597.56	0.001525	0.01432	655.74	69.83	355.8	641.4	285.6	0.8346	1.3124
124	325.65	598.80	0.001533	0.01401	652.32	71.38	357.7	640.4	282.7	0.8376	1.3096
126	326.88	600.03	0.001541	0.01371	648.93	72.94	359.6	639.4	279.8	0.8407	1.3067
128	328.10	601.25	0.001549	0.01341	645.58	74.57	361.5	638.3	276.8	0.8437	1.3039
130	329.30	602.45	0.001557	0.01313	642.26	76.16	363.4	637.2	273.8	0.8467	1.3012
132	330.48	603.63	0.001565	0.01286	638.98	77.76	365.2	636.2	271.0	0.8496	1.2984
134	331.65	604.80	0.001574	0.01259	635.32	79.43	367.1	635.1	268.0	0.8526	1.2955
136	332.81	605.96	0.001582	0.01233	632.11	81.10	369.0	634.0	265.0	0.8556	1.2923
138	333.96	607.11	0.001591	0.01207	628.54	82.85	370.8	632.8	262.0	0.8584	1.2899
140	335.09	608.24	0.001600	0.01182	625.00	84.60	372.7	631.7	259.0	0.8614	1.2873
142	336.21	609.36	0.001608	0.01159	621.89	86.28	374.5	630.5	256.0	0.8643	1.2843
144	337.31	610.46	0.001617	0.01134	618.43	88.18	376.4	629.4	253.0	0.8672	1.2814
146	338.40	611.55	0.001625	0.01111	615.38	90.01	378.2	628.1	249.9	0.8700	1.2786
148	339.49	612.64	0.001635	0.01088	611.62	91.91	380.1	626.9	246.8	0.8730	1.2757
150	340.56	613.71	0.001644	0.01066	608.27	93.81	381.9	625.6	243.7	0.8758	1.2728
152	341.61	614.76	0.001653	0.01045	604.96	95.69	383.7	624.3	240.6	0.8786	1.2698
154	342.66	615.81	0.001663	0.01024	601.32	97.66	385.6	623.0	237.4	0.8815	1.2668
156	343.70	616.85	0.001673	0.01003	597.73	99.70	387.4	621.6	234.2	0.8844	1.2640
158	344.72	617.87	0.001683	0.009826	594.18	101.77	389.3	620.3	231.0	0.8873	1.2611
160	345.74	618.89	0.001693	0.009625	590.67	103.9	391.1	618.9	227.8	0.8901	1.2582
162	346.74	619.89	0.001704	0.009431	586.85	106.0	392.9	617.5	224.5	0.8930	1.2552
164	347.74	620.89	0.001715	0.009237	583.09	108.3	394.8	616.0	221.2	0.8959	1.2521
166	348.72	621.87	0.001726	0.009048	579.37	110.5	396.7	614.5	217.8	0.8987	1.2489
168	349.70	622.85	0.001737	0.008862	575.71	112.8	398.6	613.0	214.4	0.9016	1.2456
170	350.66	623.81	0.001748	0.008681	572.08	115.2	400.4	611.5	211.1	0.9045	1.2422
172	351.62	624.77	0.001760	0.008501	568.18	117.6	402.3	609.8	207.5	0.9074	1.2387
174	352.56	625.71	0.001772	0.008325	564.33	120.1	404.2	608.2	204.0	0.9103	1.2353
176	353.50	626.65	0.001785	0.008150	560.22	122.7	406.2	606.5	200.3	0.9132	1.2321
178	354.43	627.58	0.001798	0.007974	556.17	125.4	408.1	604.7	196.6	0.9162	1.2292
180	355.35	628.50	0.001812	0.007803	551.88	128.2	410.1	602.8	192.7	0.9192	1.2258
182	356.26	629.41	0.001826	0.007633	547.65	131.0	412.2	600.9	188.7	0.9222	1.2224
184	357.16	630.31	0.001840	0.007466	543,48	133.9	414.2	599.0	184.8	0.9252	1.2190
186	358.06	631.21	0.001856	0.007300	538.79	137.0	416.2	597.1	180.9	0.9284	1.2154
188	358.94	632.09	0.001873	0.007138	533.90	140.1	418.2	595.1	176.9	0.9315	1.2117

Table 40-6 Properties of Saturated Steam (H$_2$O) at a Given Pressure (Gravitational Metric System) (Continued)

Pressure	Temperature		Specific volume		Density		Specific enthalpy		Heat of vapor-ization	Specific entropy	
			Liquid	Vapor	Liquid	Vapor	Liquid	Vapor		Liquid	Vapor
p	t	T	v'	v''	ρ'	ρ''	i'	i''	$r = i'' - i'$	s'	s''
kp/cm^2	°C	K	m^3/kg	m^3/kg	kg/m^3	kg/m^3	kcal/kg	kcal/kg	kcal/kg	kcal/kg K	kcal/kg K
190	359.82	632.97	0.001890	0.00697	529.10	143.5	420.4	593.0	172.6	0.9347	1.2080
192	360.69	633.84	0.001906	0.00682	524.66	146.6	422.3	591.0	168.7	0.9378	1.2042
194	361.55	634.70	0.001923	0.00666	520.02	150.1	424.4	588.9	164.5	0.9410	1.2002
196	362.40	635.55	0.001942	0.00650	514.93	153.8	426.6	586.6	160.0	0.9443	1.1960
198	363.25	636.40	0.001963	0.00634	509.42	157.7	428.9	584.0	155.1	0.9477	1.1914
200	364.08	637.23	0.001987	0.00618	503.27	161.9	431.3	581.4	150.1	0.9514	1.1867
202	364.91	638.06	0.00201	0.00601	497.51	166.3	433.8	578.5	144.7	0.9551	1.1819
204	365.74	638.89	0.00204	0.00585	490.20	171.0	436.4	575.6	139.2	0.9590	1.1770
206	366.55	639.70	0.00207	0.00568	483.09	175.9	438.9	572.6	133.6	0.9630	1.1717
208	367.36	640.51	0.00210	0.00552	476.19	181.2	441.7	569.4	127.7	0.9670	1.1663
210	368.16	641.31	0.00213	0.00535	469.48	186.9	444.5	565.9	121.4	0.9713	1.1606
212	368.95	642.10	0.00217	0.00517	460.83	193.2	447.6	562.2	114.6	0.9759	1.1545
214	369.74	642.89	0.00221	0.00499	452.49	200.2	450.8	558.2	107.6	0.9808	1.1475
216	370.51	643.66	0.00226	0.00481	442.48	208	454.2	553.8	99.6	0.9860	1.1406
218	371.29	644.44	0.00231	0.00462	432.90	216	458.6	548.2	89.6	0.9921	1.1315
220	372.1	645.25	0.00238	0.00436	420.17	229	463	542.3	79.3	0.9993	1.1220
222	372.8	645.95	0.00247	0.00412	404.86	242	468	535.4	67.3	1.0070	1.1112
224	373.6	646.75	0.00267	0.00373	374.53	268	479	520.7	45.7	1.0240	1.0880
225.65	374.15	647.30	0.00326	0.00326	307	307	501.5	501.5	0.0	1.058	1.058

Table 40-7 Properties of Saturated Steam (H$_2$O) at a Given Pressure (Gravitational Metric System) (Continued)

Temperature	Pressure	Specific volume	Density	Specific enthalpy	Specific entropy
t_k	p_k	v_k	ρ_k	i_k	s_k
°C	kp/cm^2	m^3/kg	kg/m^3	kcal/kg	kcal/kg K
374.15	225.65	0.00326	307	501.5	1.058

Table 41-1 Properties of Superheated Steam (H₂O) (Gravitational Metric System)

Pressure p				0.01 at			0.04 at			0.05 at			0.06 at		
Temperature				$t_s = 6.698\ °C$; $i'' = 600.2$ kcal/kg $v'' = 131.6$ m³/kg $s'' = 2.1451$ kcal/kg K			$t_s = 28.641\ °C$; $i'' = 609.8$ kcal/kg $v'' = 35.46$ m³/kg $s'' = 2.0255$ kcal/kg K			$t_s = 32.55\ °C$; $i'' = 611.5$ kcal/kg $v'' = 28.72$ m³/kg $s'' = 2.0065$ kcal/kg K			$t_s = 35.82\ °C$; $i'' = 612.9$ kcal/kg $v'' = 24.19$ m³/kg $s'' = 1.9909$ kcal/kg K		
t	T			v	i	s	v	i	s	v	i	s	v	i	s
°C	K			$\frac{m^3}{kg}$	$\frac{kcal}{kg}$	$\frac{kcal}{kg\ K}$	$\frac{m^3}{kg}$	$\frac{kcal}{kg}$	$\frac{kcal}{kg\ K}$	$\frac{m^3}{kg}$	$\frac{kcal}{kg}$	$\frac{kcal}{kg\ K}$	$\frac{m^3}{kg}$	$\frac{kcal}{kg}$	$\frac{kcal}{kg\ K}$
0	273.15			0.0010002	0.0	0.000	0.0010002	0.0	0.0000	0.0010002	0.0	0.0000	0.0010002	0.0	0.0000
10	283.15			133.2	601.7	2.1500	0.0010003	10.0	0.0361	0.0010003	10.0	0.0361	0.0010003	10.0	0.0361
20	293.15			137.9	606.1	2.1648	0.0010018	20.0	0.0708	0.0010018	20.0	0.0708	0.0010018	20.0	0.0708
30	303.15			142.6	610.6	2.1795	35.61	610.4	2.0272	0.0010044	30.0	0.1042	0.0010044	30.0	0.1042
40	313.15			147.3	615.1	2.1941	36.79	614.9	0.0412	29.42	614.8	2.0164	24.50	614.8	1.9965
50	323.15			152.0	619.6	2.2084	37.98	619.4	2.0549	30.36	619.4	2.0303	25.29	619.4	2.0102
60	333.15			156.8	624.2	2.2221	39.16	623.9	2.0648	31.31	623.9	2.0439	26.08	623.9	2.0237
70	343.15			161.5	628.6	2.2354	40.34	628.4	2.0818	32.25	628.4	2.0573	26.87	628.4	2.0370
80	353.15			166.2	633.1	2.2484	41.51	633.0	2.0949	33.19	633.0	2.0703	27.65	632.9	2.0500
90	363.15			170.9	637.6	2.2610	42.69	637.6	2.1077	34.14	637.6	2.0830	28.44	637.5	2.0628
100	373.15			175.6	642.1	2.2732	43.87	642.1	2.1199	35.08	642.1	2.0953	29.23	642.1	2.0751
110	383.15			180.3	646.6	2.2851	45.05	646.6	2.1319	36.02	646.6	2.1073	30.01	646.6	2,0871
120	393.15			185.0	651.1	2.2968	46.23	651.1	2.1436	36.96	651.1	2.1190	30.80	651.1	2.0988
130	403.15			189.7	655.6	2.3082	47.40	655.6	2.1550	37.91	655.6	2.1304	31.58	655.6	2.1102
140	413.15			194.4	660.2	2.3193	48.58	660.2	1.1661	38.85	660.2	2.1415	32.37	660.2	2.1214
150	423.15			199.1	664.8	2.3302	49.76	664.8	2.1770	39.79	664.8	2.1524	33.15	664.8	2.1323
160	433.15			203.8	669.4	2.3409	50.94	669.4	2.1877	40.73	669.4	2.1631	33.94	669.4	2.1430
170	443.15			208.5	674.0	2.3514	52.11	674.0	2.1982	41.68	674.0	2.1736	34.73	673.9	2.1535
180	453.15			213.2	678.6	2.3616	53.28	678.6	2.2085	42.62	678.6	2.1839	35.51	678.5	2.1638
190	463.15			218.0	683.2	2.3716	54.46	683.2	2.2186	43.56	683.2	2.1940	36.30	683.1	2.1739
200	473.15			222.7	687.8	2.3815	55.64	687.8	2.2284	44.50	687.8	2.2039	37.08	687.8	2.1838
210	483.15			227.4	692.4	2.3912	56.81	692.4	2.2381	45.44	692.4	2.3136	37.87	692.4	2.1935
220	493.16			232.1	697.0	2.4007	57.99	697.0	2.2476	46.39	697.0	2.2231	38.65	697.0	2.2030
230	503.15			236.8	701.7	2.4100	59.17	701.7	2.2569	47.33	701.7	2.2324	39.40	701.7	2.2123
240	513.15			241.5	706.4	2.4192	60.34	706.4	2.2661	48.27	706.4	2.2415	40.22	706.4	2.2214
250	523.15			246.2	711.1	2.4282	61.52	711.1	2.2752	49.21	711.1	2.2505	41.00	711.1	2.2304
260	533.15			250.9	715.8	2.4371	62.70	715.8	2.2841	50.15	715.8	2.2595	41.79	715.8	2.2393
280	553,15			260.3	725.3	2.4545	65.05	725.3	2.3015	52.04	725.2	2.2769	43.36	725.2	2.2568
300	573.15			269.7	734.9	2.4714	67.41	734.9	2.3184	53.92	734.8	2.2938	44.93	734.8	2.2737
310	583.15			274.4	739.7	2.4796	68.58	739.7	2.3266	54.86	739.6	2.3021	45.71	739.6	2.2819
320	593.15			279.1	744.5	2.4877	69.76	744.5	2.3347	55.80	744.4	2.3102	46.50	744.4	2.2900
330	603.15			283.8	749.3	2.4957	70.94	749.3	2.3427	56.74	749.2	2.3182	47.28	749.2	2.2980
340	613.15			288.6	754.1	2.5036	72.12	754.1	2.3506	57.69	754.0	2.3261	48.07	754.0	2.3059
350	623.15			293.3	759.0	2.5114	73.30	759.0	2.3585	58.63	758.9	2.3340	48.85	758.9	2.3138
360	633.15			298.0	763.9	2.5191	74.47	763.9	2.3663	59.57	763.8	2.3418	49.64	763.8	2.3216
380	653.15			307.4	773.7	2.5344	76.82	773.7	2.3815	61.46	773.6	2.3569	51.20	773.6	2.3368
400	673.15			316.8	783.5	2.5491	79.18	783.5	2.3963	63.33	783.4	2.3717	52.78	783.4	2.3516
410	683.15			321.5	788.4	2.5564	80.35	788.4	2.4036	64.27	788.3	2.3790	53.56	788.3	2.3589
420	693.15			326.2	793.4	2.5636	81.53	793.3	2.4109	65.21	793.3	2.3863	54.35	793.3	2.3662
430	703.15			330.9	798.4	2.5708	82.71	798.3	2.4181	66.15	798.3	2.3935	55.13	798.3	2.3734
440	713.15			335.6	803.4	2.5779	83.88	803.3	2.4252	67.09	803.3	2.4006	55.92	803.3	2.3805
450	723.15			340.3	808.4	2.5849	85.06	808.3	2.4322	68.03	808.3	2.4076	56.70	808.3	2.3875
460	733.15			345.0	813.4	2.5919	86.23	813.3	2.4391	68.98	813.3	2.4146	57.48	813.3	2.3945
480	753.16			354.4	823.5	2.6056	88.59	823.4	2.4529	70.86	823.4	2.4283	59.05	823.4	2.4082
500	773.15			363.8	833.7	2.6191	90.94	833.6	2.4664	72.74	833.6	2.4417	60.62	833.6	2.4217
510	783.15			368.5	838.8	2.6257	92.12	838.7	2.4730	73.68	838.7	2.4484	61.40	838.7	2.4283
520	793.15			373.2	843.9	2.6324	93.30	843.8	2.4796	74.62	843.8	2.4550	62.18	843.8	2.4349
530	803.15			377.9	849.0	2.6390	94.48	848.9	2.4862	75.56	848.9	2.4616	62.97	848.9	2.4415
540	813.15			382.6	854.1	2.6455	95.65	854.1	2.4927	76.51	854.1	2.4681	63.76	854.1	2.4480
550	823.15			387.3	859.3	2.6519	96.83	859.3	2.4991	77.45	859.3	2.4745	64.55	859.3	2.4544
560	833.15			392.0	864.5	2.6582	98.0	864.5	2.5055	78.40	864.5	2.4808	65.34	864.5	2.4607
580	853.15			401.4	875.0	2.6707	100.36	875.0	2.5180	80.28	875.0	2.4934	66.90	875.0	2.4733
600	873.15			410.8	885.5	2.6830	102.71	885.5	2.5303	82.17	885.5	2.5057	68.47	885.5	2.4856
610	883.15			415.5	890.7	2.6890	103.89	890.7	2.5363	83.11	890.7	2.5117	69.26	890.7	2.4916
620	893.15			420.3	896.0	2.6950	105.06	896.0	2.5422	84.05	896.0	2.5177	70.04	896.0	2.4976
630	903.15			425.0	901.3	2.7009	106.24	901.3	2.5481	84.99	901.3	2.5236	70.83	901.3	2.5035
640	913.15			429.7	906.6	2.7068	107.41	906.6	2.5540	85.93	906.6	2.5294	71.61	906.6	2.5093
650	923.15			434.4	912.0	2.7126	108.59	912.0	2.5598	86.88	912.0	2.5352	72.40	912.0	2.5151
660	933.15			439.1	917.3	2.7183	109.76	917.3	2.5656	87.82	917.3	2.5410	73.18	917.3	2.5209
680	853.15			448.5	928.1	2.7297	112.11	928.1	2.5770	89.70	928.1	2.5524	74.75	928.1	2.5323
700	973.15			457.9	938.9	2.7410	114.47	938.9	2.5882	91.58	938.9	2.5637	76.32	938.9	2.5436

All values above the solid horizontal lines refer to the saturated liquid state.

Table 41-2 Properties of Superheated Steam (H₂O) (Gravitational Metric System) (Continued)

Pressure p		0.07 at			0.08 at			0.09 at			0.10 at		
Temperature		$t_s = 38.66$ °C; $i'' = 614.1$ kcal/kg $v'' = 20.91$ m³/kg $s'' = 1.9779$ kcal/kg K			$t_s = 41.16$ °C; $i'' = 615.2$ kcal/kg $v'' = 18.45$ m³/kg $s'' = 1.9667$ kcal/kg K			$t_s = 43.41$ °C; $i'' = 616.1$ kcal/kg $v'' = 16.50$ m³/kg $s'' = 1.9568$ kcal kg K			$t_s = 45.45$ °C; $i'' = 617.0$ kcal/kg $v'' = 14.95$ m³/kg $s'' = 1.9480$ kcal/kg K		
t	T	v	i	s	v	i	s	v	i	s	v	i	s
°C	K	$\frac{m^3}{kg}$	$\frac{kcal}{kg}$	$\frac{kcal}{kg\ K}$	$\frac{m^3}{kg}$	$\frac{kcal}{kg}$	$\frac{kcal}{kg\ K}$	$\frac{m^3}{kg}$	$\frac{kcal}{kg}$	$\frac{kcal}{kg\ K}$	$\frac{m^3}{kg}$	$\frac{kcal}{kg}$	$\frac{kcal}{kg\ K}$
0	273.15	0.0010002	0.0	0.0000	0.0010002	0.0	0.0000	0.0010002	0.0	0.0000	0.0010002	0.0	0.0000
10	283.15	0.0010003	10.0	0.0361	0.0010003	10.0	0.0361	0.0010003	10.0	0.0361	0.0010003	10.0	0,0361
20	293.15	0.0010018	20.0	0.0708	0.0010018	20.0	0.0708	0.0010018	20.0	0.0708	0.0010018	20.0	0.0708
30	303.15	0.0010044	30.0	0.1042	0.0010044	30.0	0.1042	0.0010044	30.0	0.1042	0.0010044	30.0	0.1042
40	313.15	21.00	614.7	1.9797	0.0010079	40.0	0.1365	0.0010079	40.0	0.1365	0.0010079	40.0	0.1365
50	323.15	21.68	619.3	1.9934	18.97	619.2	1.9783	16.85	619.2	1.9653	15.16	619.1	1.9537
60	333.15	22.36	623.9	2.0069	19.56	623.8	1.9919	17.38	623.8	1.9789	15.64	623.7	1.9672
70	343.15	23.03	628.4	2.0201	20.15	628.3	2.0053	17.90	628.3	1.9922	16.11	628.2	1.9805
80	353.15	23.71	632.9	2.0331	20.74	632.9	2.0183	18.43	632.9	2.0052	16.58	632.8	1.9935
90	363.15	24.38	637.5	2.0458	21.33	637.5	2.0309	18.96	637.4	2.0179	17.06	637.4	2.0062
100	373.15	25.06	642.1	2.0581	21.92	642.1	2.0432	19.48	642.0	2.0302	17.53	642.0	2.0186
110	383.15	25.73	646.6	2.0701	22.51	646.6	2.0552	20.00	646.5	2.0422	18.00	646.5	2.0306
120	393.15	26.40	651.1	2.0818	23.10	651.1	2.0669	20.53	651.0	2.0539	18.47	651.0	2.0423
130	403.15	27.08	655.6	2.0933	23.69	655.6	2.0784	21.05	655.5	2.0654	18.94	655.5	2.0537
140	413.15	27.75	660.2	2.1045	24.28	660.1	2.0896	21.58	660.1	2.0766	19.42	660.1	2.0649
150	423.15	28.42	664.8	2.1154	24.87	664.7	2.1005	22.10	664.7	2.0875	19.89	664.7	2.0758
160	433.15	29.09	669.3	2.1261	25.46	669.3	2.1112	22.62	669.3	2.0982	20.36	669.3	2.0865
170	443.15	29.76	673.9	2.1366	26.05	673.9	2.1217	23.15	673.9	2.1087	20.83	673.9	2.0970
180	453.15	30.43	678.5	2.1468	26.64	678.5	2.1319	23.67	678.5	2.1189	21.30	678.5	2.1073
190	463.15	31.11	683.1	2.1568	27.23	683.1	2.1419	24.20	683.1	2.1289	21.77	683.1	2.1174
200	473.15	31.78	687.8	2.1667	27.82	687.8	2.1518	24.72	687.7	2.1388	22.24	687.7	2,1273
210	483.15	32.45	692.4	2.1764	28.40	692.4	2.1615	25.24	692.3	2.1485	22.72	692.3	2.1370
220	493.15	33.12	697.0	2.1859	28.99	697.0	2.1711	25.77	697.0	2.1581	23.19	697.0	2.1465
230	503.15	33.80	701.6	2.1953	29.58	701.6	2.1805	26.29	701.6	2.1675	23.66	701.6	2.1558
240	513.15	34.47	706.3	2.2045	30.17	706.3	2.1897	26.81	706.3	2.1767	24.13	706.3	2.1650
250	523.15	35.14	711.0	2.2135	30.76	711.0	2.1987	27.34	711.0	2,1857	24.60	711.0	2.1741
260	533.15	35.82	715.7	2.2224	31.35	715.7	2.2076	27.86	715.7	2.1946	25.07	715.7	2.1830
280	553.15	37.16	725.2	2.2398	32.52	725.2	2.2249	28.91	725.2	2.2119	26.02	725.2	2.2003
300	573.15	38.51	734.8	2.2567	33.70	734.8	2.2418	29.95	734.8	2.2288	26.96	734.8	2.2172
310	583.15	39.18	739.6	2.2650	34.29	739.6	2.2501	30.48	739.6	2.2371	27.43	739.6	2.2255
320	593.15	39.85	744.4	2.2731	34.88	744.4	2.2582	31.00	744.4	2.2453	27.90	744.4	2.2337
330	603.15	40.52	749.2	2.2811	35.47	749.2	2.2662	31.52	749.2	2.2533	28.37	749.2	2.2418
340	613.15	41.20	754.0	2.2890	36.05	754.0	2.2742	32.05	754.0	2.2612	28.84	754.0	2.2497
350	623.15	41.87	758.9	2.2968	36.64	758.9	2.2821	32.57	758.9	2.2690	29.31	758.9	2.2575
360	633.15	42.54	763.8	2.3046	37.23	763.8	2.2898	33.09	763.8	2.2767	29.78	763.8	2.2652
380	653.15	43.89	773.6	2.3198	38.41	773.6	2.3049	34.14	773.6	2.2919	30.72	773.6	2.2804
400	673.15	45.23	783.4	2.3346	39.59	783.4	2.3198	35.19	783.4	2.3068	31.67	783.4	2.2953
410	683.15	45.90	788.3	2.3419	40.17	788.3	2.3271	35.71	788.3	2.3141	32.14	788.3	2.3026
420	693.15	46.58	793.3	2.3491	40.76	793.3	2.3343	36.23	793.3	2.3214	32.61	793.3	2,3098
430	703.15	47.25	798.3	2.3563	41.35	798.3	2.3415	36.75	798.3	2.3286	33.08	798.3	2,3169
440	713.15	47.92	803.3	2.3634	41.94	803.3	2.3486	37.28	803.3	2.3357	33.55	803.3	2.3240
450	723.15	48.59	808.3	2.3705	42.53	808.3	2.3556	37.80	808.3	2.3427	34.02	808.3	2.3311
460	733.15	49.26	813.3	2.3775	43.12	813.3	2.3626	38.32	813.3	2.3497	34.49	813.3	2.3381
480	753.15	50.61	823.4	2.3912	44.30	823.4	2.3764	39.37	823.4	2.3634	35.43	823.4	2.3519
500	773.15	51.96	833.6	2.4047	45.47	833.6	2.3899	40.42	833.6	2.3769	36.38	833.6	2.3654
510	783.15	52.63	838.7	2.4113	46.06	838.7	2.3965	40.94	838.7	2.3835	36.85	838.7	2.3720
520	793.15	53.30	843.8	2.4179	46.65	843.8	2.4031	41.46	843.8	2.3901	37.32	843.8	2.3786
530	803.15	53.97	848.9	2.4245	47.24	848.9	2.4097	41.99	848.9	2.3967	37.79	848.9	2.3851
540	813.15	54.65	854.1	2.4310	47.82	854.1	2.4162	42.51	854.1	2.4032	38.26	854.1	2.3916
550	823.15	55.32	859.3	2.4374	48.41	859.3	2.4226	43.04	859.3	2.4096	38.73	859.3	2,3980
560	833.15	56.00	864.5	2.4437	49.00	864.5	2.4290	43.56	864.5	2.4160	39.20	864.5	2.4044
580	853.15	57.35	874.9	2.4563	50.18	874.9	2.4416	44.60	874.9	2.4286	40.14	874.9	2.4170
600	873.15	58.69	885.4	2.4686	51.35	885.4	2.4539	45.65	885.4	2.4409	41.08	885.4	2.4293
610	883.15	59.37	890.7	2.4746	51.94	890.7	2.4599	46.17	890.7	2.4469	41.55	890.7	2.4353
620	893.15	60.04	896.0	2.4806	52.53	896.0	2.4659	46.69	896.0	2.4529	42.02	896.0	2.4412
630	903.15	60.71	901.3	2.4865	53.12	901.3	2.4718	47.22	901.3	2.4588	42.49	901.3	2.4471
640	913.15	61.38	906.6	2.4923	53.17	906.6	2.4776	47.74	906.6	2.4647	42.96	906.6	2.4530
650	923.15	62.05	912.0	2.4981	54.30	912.0	2.4834	48.27	912.0	2.4705	43.44	912.0	2.4588
660	933.15	62.72	917.3	2.5039	54.88	917.3	2.4892	48.79	917.3	2.4762	43.91	917.3	2.4646
680	953.15	64.07	928.1	2.5153	56.06	928.1	2.5006	49.83	928.1	2.4876	44.85	928.1	2.4760
700	973.15	65.41	938.9	2.5266	57.24	938.9	2.5119	50.88	938.9	2.4989	45.79	938.9	2.4873

Table 41-3 Properties of Superheated Steam (H₂O) (Gravitational Metric System) (Continued)

Pressure p		0.12 at			0.14 at			0.16 at			0.18 at		
Temperature		t_s = 49.06 °C, i'' = 618.6 kcal/kg, v'' = 12.59 m³/kg, s'' = 1.9326 kcal/kg K			t_s = 52.18 °C, i'' = 619.9 kcal/kg, v'' = 10.89 m³/kg, s'' = 1.9197 kcal/kg K			t_s = 54.94 °C, i'' = 621.1 kcal/kg, v'' = 9.603 m³/kg, s'' = 1.9086 kcal/kg K			t_s = 57.41 °C, i'' = 622.1 kcal/kg, v'' = 8.601 m³/kg, s'' = 1.8989 kcal/kg K		
t	T	v	i	s	v	i	s	v	i	s	v	i	s
°C	K	$\frac{m^3}{kg}$	$\frac{kcal}{kg}$	$\frac{kcal}{kg\,K}$	$\frac{m^3}{kg}$	$\frac{kcal}{kg}$	$\frac{kcal}{kg\,K}$	$\frac{m^3}{kg}$	$\frac{kcal}{kg}$	$\frac{kcal}{kg\,K}$	$\frac{m^3}{kg}$	$\frac{kcal}{kg}$	$\frac{kcal}{kg\,K}$
0	273.15	0.0010002	0.0	0.0000	0.0010002	0.0	0.0000	0.0010002	0.0	0.0000	0.0010002	0.0	0.0000
10	283.15	0.0010003	10.0	0.0361	0.0010003	10.0	0.0361	0.0010003	10.0	0.0361	0.0010003	10.0	0.0361
20	293.15	0.0010018	20.0	0.0708	0.0010018	20.0	0.0708	0.0010018	20.0	0.0708	0.0010018	20.0	0.0708
30	303.15	0.0010044	30.0	0.1042	0.0010044	30.0	0.1042	0.0010044	30.0	0.1042	0.0010044	30.0	0.1042
40	313.15	0.0010079	40.0	0.1365	0.0010079	40.0	0.1365	0.0010079	40.0	0.1365	0.0010079	40.0	0.1365
50	323.15	12.62	619.0	1.9335	0.0010121	50.0	0.1679	0.0010121	50.0	0.1679	0.0010121	50.0	0.1679
60	333.15	13.02	623.6	1.9470	11.16	623.5	1.9300	9.759	623.4	1.9150	8.669	623.3	1.9020
70	343.15	13.41	628.2	1.9603	11.50	628.1	1.9433	10.060	628.0	1.9284	8.935	627.9	1.9154
80	353.15	13.81	632.8	1.9733	11.84	632.7	1.9563	10.358	632.6	1.9415	9.200	632.5	1.9284
90	363.15	14.21	637.3	1.9860	12.17	637.3	1.9690	10.655	637.2	1.9542	9.464	637.1	1.9411
100	373.15	14.60	641.9	1.9984	12.51	641.9	1.9813	10.950	641.8	1.9666	9.728	641.7	1.9534
110	383.15	15.00	646.5	2.0104	12.85	646.4	1.9933	11.245	646.4	1.9786	9.991	646.3	1.9654
120	393.15	15.39	651.0	2.0221	13.19	650.9	2.0050	11.540	650.9	1.9903	10.254	650.8	1.9772
130	403.15	15.78	655.5	2.0336	13.52	655.4	2.0165	11.835	655.4	2.0017	10.517	655.4	1.9887
140	413.15	16.18	660.1	2.0448	13.86	660.0	2.0277	12.130	660.0	2.0129	10.780	660.0	1.9999
150	423.15	16.57	664.7	2.0557	14.20	664.6	2.0386	12.425	664.6	2.0239	11.043	664.6	2.0108
160	433.15	16.96	669.3	2.0664	14.54	669.2	2.0493	12.720	669.2	2.0347	11.306	669.2	2.0215
170	443.15	17.36	673.9	2.0769	14.87	673.8	2.0598	13.015	673.8	2.0452	11.568	673.8	2.0320
180	453.15	17.75	678.5	2.0872	15.21	678.4	2.0701	13.310	678.4	2.0554	11.831	678.4	2.0423
190	463.15	18.14	683.1	2.0973	15.55	683.0	2.0802	13.604	683.0	2.0654	12.093	683.0	2.0524
200	473.15	18.54	687.7	2.1072	15.88	687.7	2.0901	13.899	687.7	2.0735	12.356	687.6	2.0623
210	483.15	18.93	692.3	2.1169	16.22	692.3	2.0998	14.194	692.3	2.0850	12.618	692.2	2.0720
220	493.15	19.32	697.0	2.1264	16.56	697.0	2.1093	14.488	696.9	2.0946	12.880	696.9	2.0815
230	503.15	19.71	701.6	2.1358	16.89	701.6	2.1186	14.782	701.5	2.1040	13.143	701.5	2.0909
240	513.15	20.11	706.3	2.1450	17.23	706.3	2.1278	15.076	706.2	2.1132	13.405	706.2	2.1001
250	523.15	20.50	711.0	2.1541	17.57	711.0	2.1369	15.370	711.0	2.1222	13.667	710.9	2.1092
260	533.15	20.89	715.7	2.1630	17.90	715.7	2.1458	15.664	715.7	2.1310	13.929	715.6	2.1181
280	553.15	21.68	725.2	2.1803	18.58	725.2	2.1633	16.25	725.2	2.1484	14.453	725.2	2.1354
300	573.15	22.46	734.8	2.1972	19.25	734.8	2.1802	16.84	734.8	2.1654	14.976	734.8	2.1525
310	583.15	22.85	739.6	2.2054	19.59	739.6	2.1885	17.14	739.6	2.1737	15.238	739.6	2.1608
320	593.15	23.25	744.4	2.2135	19.92	744.4	2.1966	17.43	744.4	2.1819	15.500	744.4	2.1689
330	603.15	23.64	749.2	2.2215	20.26	749.2	2.2046	17.72	749.2	2.1899	15.762	749.2	2.1769
340	613.15	24.03	754.0	2.2295	20.60	754.0	2.2126	18.02	754.0	2.1978	16.02	754.0	2.1848
350	623.15	24.42	758.9	2.2374	20.93	758.9	2.2205	18.31	758.9	2.2056	16.28	758.9	2.1926
360	633.15	24.82	763.8	2.2451	21.27	763.8	2.2282	18.61	763.8	2.2133	16.53	763.8	2.2003
380	653.15	25.60	773.6	2.2603	21.94	773.6	2.2433	19.20	773.6	2.2286	17.07	773.6	2.2156
400	673.15	26.39	783.4	2.2752	22.61	783.4	2.2582	19.79	783.4	2.2435	17.59	783.4	2.2305
410	683.15	26.78	788.3	2.2825	22.95	788.3	2.2655	20.08	788.3	2.2508	17.85	788.3	2.2378
420	693.15	27.17	793.3	2.2897	23.29	793.3	2.2727	20.37	793.3	2.2580	18.11	793.3	2.2450
430	703.15	27.56	798.3	2.2968	23.62	798.3	2.2798	20.67	798.3	2.2651	18.37	798.3	2.2521
440	713.15	27.95	803.3	2.3040	23.96	803.3	2.2869	20.96	803.3	2.2722	18.63	803.3	2.2592
450	723.15	28.35	808.3	2.3111	24.30	808.3	2.2940	21.25	808.3	2.2793	18.89	808.3	2.2663
460	733.15	28.74	813.3	2.3181	24.63	813.3	2.3010	21.54	813.3	2.2863	19.15	813.3	2.2733
480	753.15	29.52	823.4	2.3318	25.30	823.4	2.3148	22.13	823.4	2.3001	19.67	823.4	2.2871
500	773.15	30.31	833.6	2.3453	25.98	833.6	2.3283	22.72	833.6	2.3137	20.19	833.6	2.3007
510	783.15	30.71	838.7	2.3519	26.32	838.7	2.3349	23.02	838.7	2.3204	20.46	838.7	2.3074
520	793.15	31.10	843.8	2.3585	26.65	843.8	2.3415	23.31	843.8	2.3271	20.72	843.8	2.3141
530	803.15	31.49	848.9	2.3651	26.99	848.9	2.3481	23.61	848.9	2.3337	20.98	848.9	2.3207
540	813.15	31.88	854.1	2.3716	27.32	854.1	2.3546	23.90	854.1	2.3402	21.24	854.1	2.3272
550	823.15	32.27	859.3	2.3780	27.66	859.3	2.3610	24.20	859.3	2.3466	21.50	859.3	2.3336
560	833.15	32.67	864.5	2.3844	28.00	864.5	2.3674	24.49	864.5	2.3529	21.77	864.5	2.3399
580	853.15	33.45	874.9	2.3969	28.67	874.9	2.3799	25.09	874.9	2.3653	22.30	874.9	2.3523
600	873.15	34.24	885.4	2.4092	29.34	885.4	2.3922	25.68	885.4	2.3775	22.82	885.4	2.3645
610	883.15	34.63	890.7	2.4152	29.68	890.7	2.3982	25.97	890.7	2.3835	23.08	890.7	2.3705
620	893.15	35.02	896.0	2.4212	30.02	896.0	2.4042	26.26	896.0	2.3895	23.35	896.0	2.3765
630	903.15	35.42	901.3	2.4271	30.35	901.3	2.4101	26.56	901.3	2.3954	23.61	901.3	2.3824
640	913.15	35.81	906.6	2.4330	30.68	906.6	2.4159	26.86	906.6	2.4013	23.87	906.6	2.3883
650	923.15	36.20	912.0	2.4388	31.02	912.0	2.4217	27.15	912.0	2.4071	24.13	912.0	2.3941
660	933.15	36.59	917.3	2.4445	31.36	917.3	2.4275	27.44	917.3	2.4128	24.39	917.3	2.3999
680	953.15	37.38	928.1	2.4559	32.03	928.1	2.4389	28.03	928.1	2.4242	24.92	928.1	2.4113
700	973.15	38.16	938.9	2.4672	32.70	938.9	2.4502	28.62	938.9	2.4355	25.44	938.9	2.4225

Table 41-4 Properties of Superheated Steam (H_2O) (Gravitational Metric System) (Continued)

Pressure p		0.20 at			0.30 at			0.40 at			0.50 at		
Temperature		$t_s = 59.67\ °C$ $i'' = 623.1$ kcal/kg $v'' = 7.789$ m³/kg $s'' = 1.8902$ kcal/kg K			$t_s = 68.68\ °C$ $i'' = 626.8$ kcal/kg $v'' = 5.324$ m³/kg $s'' = 1.8568$ kcal/kg K			$t_s = 75.42\ °C$ $i'' = 629.5$ kcal/kg $v'' = 4.066$ m³/kg $s'' = 1.8333$ kcal/kg K			$t_s = 80.86\ °C$ $i'' = 631.6$ kcal/kg $v'' = 3.299$ m³/kg $s'' = 1.8152$ kcal/kg K		
t	T	v	i	s	v	i	s	v	i	s	v	i	s
°C	K	$\frac{m^3}{kg}$	$\frac{kcal}{kg}$	$\frac{kcal}{kg\ K}$	$\frac{m^3}{kg}$	$\frac{kcal}{kg}$	$\frac{kcal}{kg\ K}$	$\frac{m^3}{kg}$	$\frac{kcal}{kg}$	$\frac{kg}{kg\ K}$	$\frac{m^3}{kg}$	$\frac{kcal}{kg}$	$\frac{kcal}{kg\ K}$
0	273.15	0.0010002	0.0	0.0000	0.0010002	0.0	0.0000	0.0010002	0.0	0.0000	0.0010002	0.0	0.0000
10	283.15	0.0010003	10.0	0.0361	0.0010003	10.0	0.0361	0.0010003	10.0	0.0361	0.0010003	10.1	0.0361
20	293.15	0.0010018	20.0	0.0708	0.0010018	20.0	0.0708	0.0010018	20.0	0.0708	0.0010018	20.0	0.0708
30	303.15	0.0010044	30.0	0.1042	0.0010044	30.0	0.1042	0.0010044	30.0	0.1042	0.0010044	30.0	0.1042
40	313.15	0.0010079	40.0	0.1365	0.0010079	40.0	0.1365	0.0010079	40.0	0.1365	0.0010079	40.0	0.1365
50	323.15	0.0010121	50.0	0.1679	0.0010121	50.0	0.1679	0.0010121	50.0	0.1680	0.0010121	50.0	0.1680
60	333.15	7.797	623.2	1.8903	0.0010171	60.0	0.1984	0.0010171	60.0	0.1984	0.0010171	60.0	0.1984
70	343.15	8.038	627.8	1.9036	5.345	627.4	1.8582	0.0010228	70.0	0.2280	0.0010228	70.0	0.2280
80	353.15	8.277	632.5	1.9166	5.507	632.1	1.8713	4.123	631.7	1.8389	0.0010290	80.0	0.2567
90	363.15	8.515	637.1	1.9293	5.667	636.8	1.8840	4.244	636.4	1.8520	3.390	636.1	1.8270
100	373.15	8.752	641.7	1.9417	5.826	641.4	1.8965	4.365	641.1	1.8646	3.487	640.8	1.8397
110	383.15	8.989	646.2	1.9537	5.984	646.0	1.9086	4.485	645.7	1.8768	3.583	645.5	1.8520
120	393.15	9.226	650.8	1.9655	6.143	650.6	1.9204	4.604	650.3	1.8887	3.679	650.1	1.8639
130	403.15	9.463	655.3	1.9770	6.301	655.2	1.9319	4.723	654.9	1.9003	3.775	654.7	1.8754
140	413.15	9.699	659.9	1.9882	6.459	659.8	1.9431	4.842	659.5	1.9116	3.870	659.3	1.8866
150	423.15	9.936	664.5	1.9991	6.617	664.4	1.9541	4.961	664.2	1.9226	3.965	664.0	1.8976
160	433.15	10.172	669.1	2.0098	6.776	669.0	1.9648	5.079	668.8	1.9334	4.060	668.6	1.9084
170	443.15	10.409	673.7	2.0203	6.934	673.6	1.9753	5.198	673.4	1.9439	4.155	673.3	1.9190
180	453.15	10.645	678.3	2.0306	7.092	678.2	1.9856	5.317	678.0	1.9542	4.250	677.9	1.9294
190	463.15	10.882	682.9	2.0407	7.249	682.8	1.9958	5.435	682.7	1.9643	4.345	682.5	1.9395
200	473.15	11.118	687.6	2.0506	7.407	687.5	2.0058	5.553	687.4	1.9742	4.440	687.2	1.9494
210	483.15	11.355	692.2	2.0603	7.564	692.1	2.0155	5.672	692.0	1.9839	4.535	691.9	1.9591
220	493.15	11.591	696.9	2.0698	7.722	696.8	2.0250	5.790	696.7	1.9934	4.629	696.6	1.9686
230	503.15	11.827	701.5	2.0792	7.880	701.4	2.0344	5.908	701.4	2.0028	4.724	701.2	1.9780
240	513.15	12.063	706.2	2.0885	8.038	706.1	2.0436	6.026	706.1	2.0120	4.819	705.9	1.9873
250	523.15	12.299	710.9	2.0976	8.195	710.8	2.0527	6.145	710.8	2.0211	4.913	710.6	1.9964
260	533.15	12.535	715.6	2.1065	8.352	715.5	2.0616	6.263	715.5	2.0300	5.008	715.4	2.0053
280	553.15	13.007	725.2	2.1238	8.667	725.1	2.0790	6.500	725.1	2.0475	5.197	725.0	2.0228
300	573.15	13.478	734.8	2.1409	8.983	734.7	2.0959	6.736	734.7	2.0644	5.387	734.6	2.0397
310	583.15	13.714	739.6	2.1492	9.141	739.5	2.1042	6.854	739.5	2.0726	5.482	739.4	2.0479
320	593.15	13.949	744.4	2.1573	9.298	744.3	2.1124	6.971	744.3	2.0807	5.577	744.2	2.0560
330	603.15	14.185	749.2	2.1653	9.455	749.1	2.1205	7.089	749.1	2.0888	5.672	749.0	2.0641
340	613.15	14.420	754.0	2.1732	9.612	754.0	2.1285	7.207	753.9	2.0968	5.767	753.9	2.0721
350	623.15	14.656	758.9	2.1810	9.769	758.9	2.1364	7.325	758.8	2.1047	5.861	758.8	2.0799
360	633.15	14.891	763.8	2.1887	9.926	763.8	2.1441	7.443	763.7	2.1124	5.995	763.7	2.0876
380	653.15	15.362	773.6	2.2040	10.240	773.6	2.1592	7.679	773.5	2.1276	6.144	773.5	2.1029
400	673.15	15.833	783.4	2.2189	10.554	783.4	2.1741	7.916	783.3	2.1425	6.333	783.3	2.1177
410	683.15	16.07	788.3	2.2262	10.711	788.3	2.1814	8.033	788.2	2.1498	6.427	788.2	2.1250
420	693.16	16.30	793.3	2.2334	10.868	793.3	2.1887	8.151	793.2	2.1571	6.521	793.2	2.1323
430	703.15	16.54	798.3	2.2405	11.025	798.3	2.1959	8.269	798.2	2.1643	6.615	798.2	2.1396
440	713.15	16.77	803.3	2.2476	11.182	803.3	2.2030	8.387	803.2	2.1714	6.710	803.2	2.1467
450	723.15	17.00	808.3	2.2547	11.339	808.3	2.2101	8.505	808.2	2.1785	6.804	808.2	2.1537
460	733.15	17.23	813.3	2.2617	11.496	813.3	2.2171	8.623	813.2	2.1855	6.898	813.2	2.1607
480	753.15	17.70	823.4	2.2755	11.810	823.4	2.2309	8.858	823.4	2.1993	7.087	823.3	2.1745
500	773.15	18.17	833.6	2.2891	12.124	833.6	2.2445	9.093	833.6	2.2128	7.275	833.5	2 1880
510	783.15	18.41	838.7	2.2958	12.281	838.7	2.2512	9.211	838.7	2.2195	7.370	838.6	2.1947
520	793.15	18.64	843.8	2.3025	12.438	843.8	2.2579	9.329	843.8	2.2261	7.464	843.7	2.2014
530	803.15	18.88	848.9	2.3091	12.595	848.9	2.2644	9.447	848.9	2.2327	7.558	848.9	2.2080
540	813.15	19.11	854.1	2.3156	12.752	854.1	2.2709	9.564	854.1	2.2392	7.652	854.1	2.2145
550	823.15	19.35	859.3	2.3220	12.909	859.3	2.2773	9.682	859.3	2.2456	7.746	859.3	2.2209
560	833.15	19.60	864.5	2.3283	13.066	864.5	2.2836	9.799	864.5	2.2519	7.839	864.5	2.2272
580	853.15	20.07	874.9	2.3407	13.380	874.9	2.2960	10.034	874.9	2.2643	8.027	874.9	2.2397
600	873.15	20.54	885.4	2.3529	13.693	885.4	2.3082	10.269	885.4	2.2765	8.215	885.4	2.2519
610	983.15	20.78	890.7	2.3589	13.850	890.7	2.3142	10.387	890.7	2.2825	8.309	890.7	2.2579
620	893.15	21.01	896.0	2.3648	14.007	896.0	2.3202	10.505	895.9	2.2885	8.403	895.9	2.2639
630	903.15	21.25	901.3	2.3707	14.164	901.3	2.3261	10.622	901.2	2.2944	8.498	901.2	2.2698
640	913.15	21.48	906.6	2.3766	14.321	906.6	2.3320	10.740	906.6	2.3002	8.592	906.6	2.2756
650	923.15	21.72	911.9	2.3824	14.478	911.9	2.3378	10.858	911.9	2.3060	8.686	911.9	2.2814
660	933.15	21.95	917.3	2.3882	14.634	917.3	2.3436	10.975	917.3	2.3118	8.780	917.3	2.2872
680	953.15	22.42	928.1	2.3996	14.948	928.0	2.3550	11.211	928.0	2.3232	8.969	928.0	2.2986
700	973.15	22.89	938.9	2.4109	15.262	938.9	2.3662	11.446	938.9	2.3345	9.157	938.8	2.3099

Table 41-5 Properties of Superheated Steam (H₂O) (Gravitational Metric System) (Continued)

Pressure p		0.60 at			0.70 at			0.80 at			0.90 at		
Temperature		$t_s = 85.45\ °C$ $i'' = 633.5$ kcal/kg $v'' = 2.782$ m³/kg $s'' = 1.8004$ kcal/kg K			$t_s = 89.45\ °C$ $i'' = 635.1$ kcal/kg $v'' = 2.408$ m³/kg $s'' = 1.7879$ kcal/kg K			$t_s = 92.99\ °C$ $i'' = 636.4$ kcal/kg $v'' = 2.125$ m³/kg $s'' = 1.7772$ kcal/kg K			$t_s = 96.18\ °C$ $i'' = 637.6$ kcal/kg $v'' = 1.903$ m³/kg $s'' = 1.7677$ kcal/kg K		
t	T	v	i	s	v	i	s	v	i	s	v	i	s
°C	K	$\frac{m^3}{kg}$	$\frac{kcal}{kg}$	$\frac{kcal}{kg\ K}$	$\frac{m^3}{kg}$	$\frac{kcal}{kg}$	$\frac{kcal}{kg\ K}$	$\frac{m^3}{kg}$	$\frac{kcal}{kg}$	$\frac{kcal}{kg\ K}$	$\frac{m^3}{kg}$	$\frac{kcal}{kg}$	$\frac{kcal}{kg\ K}$
0	273.15	0,0010002	0	0.0000	0.0010002	0,0	0.0000	0.0010002	0.0	0.0000	0.0010002	0.0	0.0000
10	283.15	0,0010003	10.1	0.0361	0,0010003	10.1	0.0361	0.0010003	10.1	0.0361	0.0010003	10.1	0.0361
20	293.15	0.0010018	20.0	0.0708	0,0010018	20.1	0.0708	0.0010018	20.1	0.0708	0.0010018	20.1	0.0708
30	303.15	0.0010044	30.0	0.1042	0.0010044	30.0	0.1042	0.0010044	30.0	0.1042	0.0010044	30.0	0.1042
40	313.15	0.0010079	40.0	0.1365	0.0010079	40.0	0.1365	0.0010079	40.0	0.1365	0.0010079	40.0	0.1365
50	323.15	0.0010121	50.0	0.1680	0.0010121	50.0	0.6180	0.0010121	50.0	0.1680	0.0010121	50.0	0.1680
60	333.15	0.0010171	60.0	0.1984	0.0010171	60.0	0.1984	0.0010171	60.0	0.1984	0.0010171	60.0	0.1984
70	343.15	0.0010228	70.0	0.2280	0.0010228	70.0	0.2280	0.0010228	70.0	0.2280	0.0010228	70.0	0.2280
80	353.15	0.0010290	80.0	0.2567	0.0010290	80.0	0.2567	0.0010289	80.0	0.2567	0.0010289	80.0	0.2567
90	363.15	2.820	635.6	1.8057	2.412	635.2	1.7885	0.0010359	90.0	0.2848	0.0010359	90.0	0,2848
100	373.15	2.902	640.4	1.8186	2.484	640.2	1.8012	2.169	639.9	1.7859	1.925	639.5	1.7722
110	383.15	2.983	645.2	1.8311	2.554	645.0	1.8136	2.231	644.7	1.7985	1.980	644.4	1.7850
120	393.15	3.063	649.9	1.8432	2.623	649.7	1.8257	2.292	649.4	1.8107	2.035	649.2	1.7973
130	403.15	3.143	654.4	1.8549	2.692	654.3	1.8374	2.353	654.1	1.8225	2.089	653.9	1.8092
140	413.15	3.223	659.2	1.8662	2.760	659.0	1.8488	2.413	658.8	1.8339	2.143	658.6	1.8206
150	423.15	3.302	663.9	1.8772	2.828	663.7	1.8598	2.472	663.5	1.8450	2.196	663.3	1.8317
160	433.15	3.382	668.5	1.8880	2.896	668.3	1.8706	2.532	668.2	1.8559	2.249	668.0	1.8426
170	443.15	3.461	673.1	1.8986	2.965	673.0	1.8812	2.592	672.8	1.8665	2.302	672.7	1.8533
180	453.15	3.540	677.7	1.9089	3.033	677.6	1.8916	2.652	677.5	1.8769	2.356	677.3	1.8637
190	463.15	3.620	682.4	1.9190	3.101	682.3	1.9018	2.711	682.2	1.8871	2.409	682.0	1.8739
200	473.15	3.700	687.1	1.9289	3.169	687.0	1.9118	2.771	686.9	1.8971	2.462	686.7	1.8839
210	483.15	3.779	691.8	1.9387	3.236	691.7	1.9215	2.830	691.6	1.9069	2.515	691.4	1.8937
220	493.15	3.858	696.5	1.9483	3.304	696.4	1.9311	2.890	696.3	1.9165	2.568	696.1	1.9033
230	503.15	3.937	701.2	1.9577	3.372	701.1	1.9406	2.949	701.0	1.9259	2.620	700.8	1.9127
240	513.15	4.016	705.9	1.9669	3.440	705.8	1.9499	3.009	705.7	1.9351	2,673	705.6	1.9220
250	523.15	4.095	710.6	1.9760	3.508	710.5	1.9590	3.068	710.4	1.9442	2.726	710.3	1.9311
260	533.15	4.174	715.3	1.9849	3.576	715.3	1.9679	3.127	715.2	1.9531	2.779	715.1	1.9400
280	553.15	4.331	724.9	2.0023	3.711	724.9	1.9853	3.246	724.8	1.9706	2.884	724.7	1.9575
300	573.15	4.489	734.5	2.0193	3.847	734.5	2.0024	3.364	734.4	1.9876	2.989	734.3	1.9745
310	583.15	4.568	739.3	2.0277	3.914	739.3	2.0107	3.423	739.2	1.9958	3.042	739.1	1.9827
320	593.15	4.646	744.1	2.0359	3.982	744.1	2.0189	3.482	744.0	2.0039	3.095	743.9	1.9908
330	603.15	4.725	748.9	2.0440	4.049	748.9	2.0270	3.542	748.8	2.0119	3.147	748.7	1.9989
340	613.15	4.804	753.8	2.0519	4.117	753.8	2.0349	3.601	753.7	2.0199	3.200	753.6	2.0069
350	623.15	4.882	758.7	2.0598	4.184	758.7	2.0427	3.660	758.6	2.0278	3.252	758.5	2.0148
360	633.15	4.961	763.6	2.0675	4.252	763.6	2.0504	3.720	763.5	2.0356	3.305	763.4	2.0226
380	653.15	5.118	773.4	2.0827	4.388	773.4	2.0657	3.838	773.3	2.0508	3.410	773.2	2.0377
400	673.15	5.277	783.3	2.0976	4.522	783.2	2.0806	3.956	783.2	2.0657	3.515	783.1	2.0526
410	683.15	5.356	788.2	2.1049	4.590	788.1	2.0879	4.015	788.1	2.0730	3.567	788.1	2.0599
420	693.15	5.434	793.2	2.1122	4.657	793.1	2.0951	4.074	793.1	2.0803	3.620	793.1	2.0671
430	703.15	5.512	798.2	2,1194	4.725	798.1	2.1023	4.132	798.1	2.0875	3.673	798.1	2,0742
440	713.15	5.591	803.2	2.1265	4.792	803.1	2.1094	4.191	803.1	2.0946	3.725	803.1	2.0813
450	723.15	5.671	808.2	2.1335	4.859	808.1	2.1165	4.250	808.1	2.1016	3.778	808.1	2.0884
460	733.15	5.750	813.2	2.1405	4.927	813.1	2.1235	4.309	813.1	2.1086	3.830	813.1	2,0954
480	753.15	5.906	823.3	2.1542	5.061	823.3	2.1372	4.427	823.3	2.1224	3.936	823.2	2.1092
500	773.15	6.063	833.5	2.1678	5.196	833.5	2.1508	4.545	833.5	2.1360	4.040	833.4	2.1228
510	783.15	6.142	838.6	2.1745	5.264	838.6	2.1575	4.604	338.6	2.1427	4.093	838.5	2.1295
520	793.15	6.220	843.7	2.1811	5.331	843.7	2.1641	4.663	843.7	2.1494	4.145	843.6	2.1362
530	803.15	6.298	848.8	2.1877	5.398	848.8	2.1707	4.722	848.8	2.1560	4.197	848.8	2.1428
540	813.15	6.376	854.0	2.1942	5.465	854.0	2.1722	4.781	854.0	2.1625	4.249	854.0	2.1493
550	823.15	6.454	859.2	2.2007	5.532	859.2	2.1837	4.840	859.2	2.1690	4.302	859.2	2,1558
560	833.15	.6.532	864.4	2.2071	5.598	864.4	2.1901	4.898	864.4	2.1754	4.354	864.4	2.1622
580	853.15	6.689	874.9	2.2196	5.733	874.8	2.2026	5.016	874.8	2.1879	4.458	874.8	2.1748
600	873.15	6.846	885.4	2.2318	5.867	885.3	2.2148	5.134	885.3	2.2001	4.563	885.3	2.1871
610	883.15	.6.924	890.6	2.2378	5.934	890.6	2.2208	5.193	890.6	2.2061	4.615	890.6	2.1931
620	893.15	7.002	895.9	2.2438	6.002	895.9	2.2268	5.252	895.9	2.2121	4.668	895.8	2,1991
630	903.15	7.081	901.2	2.2497	6.069	901.2	2.2327	5.310	901.2	2.2180	4.720	901.1	2.2050
640	913.15	7.160	906.5	2.2555	6.136	906.5	2.2386	5.369	906.5	2.2238	4.772	906.4	2.2108
650	923.15	7.238	911.9	2.2613	6.204	911.9	2.2444	5.428	911.8	2.2296	4.825	911.8	2.2166
660	933.15	7.316	917.2	2.2671	6.271	917.2	2.2502	5.487	917.1	2.2354	4.877	917.1	2.2224
680	953.15	7.473	928.0	2.2785	6.405	928.0	2.2616	5.604	927.9	2.2468	4.981	927.9	2.2338
700	973.15	7.630	938.7	2.2899	6.540	938.7	2.2728	5.722	938.6	2.2581	5.086	938.5	2.2451

Table 41-6 Properties of Superheated Steam (H_2O) (Gravitational Metric System) (*Continued*)

Pressure p		1.0 at			1.2 at			1.3 at			1.4 at		
Temperature		$t_s = 99.09\,°C$ $i'' = 638.8\,kcal/kg$ $v'' = 1.725\,m^3/kg$ $s'' = 1.7593\,kcal/kg\,K$			$t_s = 104.25\,°C$ $i'' = 604.7\,kcal/kg$ $v'' = 1.455\,m^3/kg$ $s'' = 1.7447\,kcal/kg\,K$			$t_s = 106.56\,°C$ $i'' = 641.6\,kcal/kg$ $v'' = 1.350\,m^3/kg$ $s'' = 1.7382\,kcal/kg\,K$			$t_s = 108.74\,°C$ $i'' = 642.3\,kcal/kg$ $v'' = 1.259\,m^3/kg$ $s'' = 1.7323\,kcal/kg\,K$		
t	T	v	i	s	v	i	s	v	i	s	v	i	s
°C	K	$\frac{m^3}{kg}$	$\frac{kcal}{kg}$	$\frac{kcal}{kg\,K}$	$\frac{m^3}{kg}$	$\frac{kcal}{kg}$	$\frac{kcal}{kg\,K}$	$\frac{m^3}{kg}$	$\frac{kcal}{kg}$	$\frac{kcal}{kg\,K}$	$\frac{m^3}{kg}$	$\frac{kcal}{kg}$	$\frac{kcal}{kg\,K}$
0	273.15	0.0010002	0.0	0.0000	0.0010002	0.0	0.0000	0.0010001	0.0	0.0000	0.0010001	0.0	0.0000
20	293.15	0.0010018	20.1	0.0708	0.0010018	20.1	0.0708	0.0010018	20.1	0.0708	0.0010018	20.1	0.0708
40	313.15	0.0010079	40.0	0.1365	0.0010078	40.0	0.1365	0.0010078	40.0	0.1365	0.0010078	40.0	0.1365
50	323.15	0.0010121	50.0	0.1680	0.0010121	50.0	0.1680	0.0010120	50.0	0.1680	0.0010120	50.0	0.1680
60	333.15	0.0010170	60.0	0.1984	0.0010170	60.0	0.1984	0.0010170	60.0	0.1984	0.0010170	60.0	0.1984
80	353.15	0.0010289	80.0	0.2567	0.0010289	80.0	0.2567	0.0010289	80.0	0.2567	0.0010289	80.0	0.2567
100	373.15	1.730	639.2	1.7603	0.0010435	100.1	0.3121	0.0010435	100.1	0.3121	0.0010435	100.1	0.3121
120	393.15	1.830	649.0	1.7851	1.521	648.5	1.7640	1.403	648.2	1.7549	1.300	648.0	1.7461
140	413.15	1.926	658.4	1.8083	1.602	658.0	1.7875	1.478	657.8	1.7785	1.371	657.7	1.7700
150	423.15	1.975	663.1	1.8194	1.643	662.8	1.7987	1.515	662.6	1.7897	1.406	662.5	1.7814
160	433.15	2.023	667.8	1.8303	1.683	667.5	1.8097	1.552	667.4	1.8007	1.440	667.2	1.7924
180	453.15	2.119	677.2	1.8515	1.763	676.9	1.8310	1.626	676.8	1.8220	1.509	676.7	1.8136
200	473.15	2.214	686.6	1.8717	1.843	686.4	1.8514	1.700	686.3	1.8424	1.578	686.1	1.8340
220	493.15	2.310	696.0	1.8913	1.923	695.8	1.8710	1.774	695.7	1.8620	1.647	695.6	1.8536
240	513.15	2.405	705.5	1.9101	2.002	705.3	1.8898	1.848	705.2	1.8808	1.715	705.1	1.8725
250	523.15	2.452	710.2	1.9193	2.042	710.0	1.8989	1.884	709.9	1.8900	1.749	709.8	1.8818
260	533.15	2.500	714.9	1.9284	2.082	714.8	1.9079	1.921	714.7	1.8991	1.783	714.6	1.8909
280	553.15	2.595	724.5	1.9461	2.161	724.3	1.9258	1.994	724.2	1.9168	1.851	724.2	1.9086
300	573.15	2.690	734.1	1.9634	2.240	733.9	1.9431	2.067	733.8	1.9341	1.919	733.8	1.9259
320	593.15	2.784	743.7	1.9800	2.320	743.5	1.9598	2.140	743.4	1.9508	1.987	743.4	1.9426
340	613.15	2.880	753.4	1.9961	2.399	753.2	1.9759	2.214	753.2	1.9670	2.056	753.2	1.9587
350	623.15	2.927	758.3	2.0040	2.439	758.1	1.9838	2.251	758.0	1.9749	2.089	758.0	1.9666
360	633.15	2.975	763.2	2.0118	2.478	763.0	1.9916	2.287	762.9	1.9827	2.123	762.9	1.9744
380	653.15	3.068	773.0	2.0271	2.556	772.8	2.0069	2.359	772.7	1.9980	2.190	772.7	1.9898
400	673.15	3.163	782.9	2.0421	2.635	782.7	2.0218	2.432	782.6	2.0129	2.258	782.6	2.0048
420	693.15	3.257	792.9	2.0567	2.713	792.7	2.0364	2.504	792.6	2.0276	2.325	792.6	2.0194
440	713.15	3.352	802.9	2.0709	2.792	802.7	2.0507	2.577	802.6	2.0419	2.393	802.6	2.0337
450	723.15	3.399	807.9	2.0779	2.832	807.7	2.0577	2.613	807.6	2.0489	2.426	807.6	2.0407
460	733.15	3.446	812.9	2.0848	2.871	812.7	2.0646	2.650	812.6	2.0558	2.460	812.6	2.0476
480	753.15	3.540	823.0	2.0985	2.950	822.8	2.0783	2.722	822.8	2.0695	2.528	822.8	2.0613
500	773.15	3.635	833.2	2.1119	3.028	833.0	2.0918	2.795	833.0	2.0829	2.595	833.0	2.0748
520	793.15	3.729	843.4	2.1250	3.107	843.2	2.1049	2.868	843.2	2.0960	2.662	843.2	2.0879
540	813.15	3.824	853.8	2.1379	3.186	853.6	2.1178	2.941	853.6	2.1089	2.731	853.6	2.1008
550	823.15	3.871	859.0	2.1443	3.225	858.8	2.1242	2.977	858.8	2.1153	2.765	858.8	2.1072
560	833.15	3.918	864.2	2.1506	3.265	864.0	2.1305	3.013	864.0	2.1216	2.798	864.0	2.1135
580	853.15	4.012	874.6	2.1629	3.343	874.4	2.1428	3.086	874.4	2.1339	2.865	874.4	2.1258
600	873.15	4.107	885.1	2.1749	3.422	884.9	2.1548	3.158	884.8	2.1459	2.933	884.8	2.1378
620	893.15	4.202	895.6	2.1866	3.502	895.4	2.1666	3.232	895.3	2.1577	3.001	895.2	2.1495
640	913.15	4.296	906.2	2.1982	3.581	906.0	2.1782	3.306	905.9	2.1693	3.069	905.8	2.1611
650	923.15	4.343	911.6	2.2040	3.620	911.4	2.1840	3.342	911.3	2.1751	3.103	911.2	2.1669
660	933.15	4.390	917.0	2.2098	3.659	916.8	2.1898	3.378	916.7	2.1809	3.137	916.6	2.1727
680	953.15	4.484	927.8	2.2212	3.737	927.6	2.2012	3.450	927.5	2.1923	3.203	927.4	2.1841
700	973.15	4.578	938.4	2.2325	3.815	938.2	2.2126	3.522	938.2	2.2037	3.269	938.1	2.1954
720	993.15	4.673	949.0	2.2435	3.895	948.9	2.2237	3.595	948.9	2.2148	3.337	948.9	2.2065
740	1013.15	4.767	959.7	2.2545	3.974	959.8	2.2347	3.667	959.8	2.2258	3.405	959.8	2.2175
750	1023.15	4.814	965.2	2.2599	4.013	965.3	2.2401	3.703	965.3	2.2312	3.439	965.3	2.2229
760	1033.15	4.861	970.7	2.2653	4.052	970.8	2.2455	3.739	970.8	2.2366	3.473	970.8	2.2283
780	1053.15	4.955	981.8	2.2759	4.130	981.8	2.2560	3.811	981.8	2.2471	3.540	981.8	2.2388
800	1073.15	5.049	993.0	2.2863	4.208	993.0	2.2664	3.883	993.0	2.2575	3.606	993.0	2.2492
820	1093.15	5.143	1004.2	2.2967	4.286	1004.2	2.2768	3.957	1004.2	2.2679	3.674	1004.2	2.2596
840	1113.15	5.237	1015.6	2.3069	4.364	1015.6	2.2871	4.029	1015.6	2.2782	3.741	1015.6	2.2699
850	1123.15	5.284	1021.3	2.3120	4.403	1021.3	2.2922	4.065	1021.3	2.2833	3.774	1021.3	2.2750
860	1133.15	5.331	1027.0	2.3170	4.443	1027.0	2.2972	4.101	1027.0	2.2883	3.808	1027.0	2.2800
880	1153.15	5.425	1038.4	2.3269	4.522	1038.4	2.3071	4.173	1038.4	2.2982	3.876	1038.4	2.2899
900	1173.15	5.519	1049.8	2.3367	4.600	1049.8	2.3169	4.245	1049.8	2.3080	3.942	1049.8	2.2997
920	1193.15	5.613	1061.2	2.3465	4.680	1061.2	2.3267	4.319	1061.2	2.3178	4.010	1061.2	2.3095
940	1213.15	5.707	1072.6	2.3561	4.758	1072.6	2.3363	4.391	1072.6	2.3274	4.078	1072.6	2.3191
950	1223.15	5.754	1078.3	2.3609	4.797	1078.3	2.3411	4.427	1078.3	2.3322	4.112	1078.3	2.3239
960	1233.15	5.801	1084.0	2.3656	4.836	1084.0	2.3458	4.463	1084.0	2.3369	4.146	1084.0	2.3286
980	1253.15	5.895	1095.6	2.3750	4.914	1095.6	2.3552	4.535	1095.6	2.3463	4.212	1095.6	2.3380
1000	1273.15	5.989	1107.2	2.3842	4.992	1107.2	2.3644	4.607	1107.2	2.3555	4.278	1107.2	2.3472

Table 41-7 Properties of Superheated Steam (H₂O) (Gravitational Metric System) (Continued)

Pressure p		1.5 at			1.6 at			1.8 at			2.0 at		
Temperature		$t_s = 110.79\,°C$ $i'' = 643.1\,kcal/kg$ $v'' = 1.181\,m^3/kg$ $s'' = 1.7268\,kcal/kg\,K$			$t_s = 112.73\,°C$ $i'' = 643.8\,kcal/kg$ $v'' = 1.111\,m^3/kg$ $s'' = 1.7217\,kcal/kg\,K$			$t_s = 116.33\,°C$ $i'' = 645.1\,kcal/kg$ $v'' = 0.9954\,m^3/kg$ $s'' = 1.7123\,kcal/kg\,K$			$t_s = 119.62\,°C$ $i'' = 646.3\,kcal/kg$ $v'' = 0.9018\,m^3/kg$ $s'' = 1.7039\,kcal/kg\,K$		
t	T	v	i	s	v	i	s	v	i	s	v	i	s
°C	K	$\frac{m^3}{kg}$	$\frac{kcal}{kg}$	$\frac{kcal}{kg\,K}$	$\frac{m^3}{kg}$	$\frac{kcal}{kg}$	$\frac{kcal}{kg\,K}$	$\frac{m^3}{kg}$	$\frac{kcal}{kg}$	$\frac{kcal}{kg\,K}$	$\frac{m^3}{kg}$	$\frac{kcal}{kg}$	$\frac{kcal}{kg\,K}$
0	273.15	0.0010001	0.0	0.0000	0.0010001	0.0	0.0000	0.0010001	0.0	0.0000	0.0010001	0.0	0.0000
20	293.15	0.0010018	20.1	0.0708	0.0010018	20.1	0.0708	0.0010018	20.1	0.0708	0.0010018	20.1	0.0708
40	313.15	0.0010078	40.0	0.1365	0.0010078	40.0	0.1365	0.0010078	40.0	0.1365	0.0010078	40.0	0.1365
50	323.15	0.0010120	50.0	0.1680	0.0010120	50.0	0.1680	0.0010120	50.0	0.1680	0.0010120	50.0	0.1680
60	333.15	0.0010170	60.0	0.1984	0.0010170	60.0	0.1984	0.0010170	60.0	0.1984	0.0010170	60.0	0.1984
80	353.15	0.0010289	80.0	0.2567	0.0010289	80.0	0.2567	0.0010289	80.0	0.2567	0.0010289	80.0	0.2567
100	373.15	0.0010435	100.1	0.3121	0.0010435	100.1	0.3121	0.0010435	100.1	0.3121	0.0010435	100.1	0.3121
120	393.15	1.212	647.8	1.7381	1.135	647.5	1.7306	1.006	647.0	1.7166	0.9027	646.5	1.7043
140	413.15	1.278	657.5	1.7622	1.197	657.3	1.7548	1.062	656.9	1.7409	0.9545	656.5	1.7284
150	423.15	1.311	662.3	1.7736	1.228	662.1	1.7661	1.089	661.8	1.7525	0.9795	661.5	1.7401
160	433.15	1.343	667.1	1.7847	1.258	666.9	1.7771	1.117	666.6	1.7637	1.003	666.4	1.7515
180	453.15	1.408	676.5	1.8060	1.319	676.4	1.7984	1.171	676.1	1.7852	1.052	675.9	1.7732
200	473.15	1.472	686.0	1.8262	1.379	685.9	1.8188	1.225	685.7	1.8056	1.101	685.4	1.7937
220	493.15	1.536	695.5	1.8459	1.439	695.4	1.8386	1.278	695.2	1.8252	1.149	695.0	1.8133
240	513.15	1.600	705.0	1.8649	1.499	704.9	1.8576	1.332	704.7	1.8442	1.197	704.5	1.8324
250	523.15	1.632	709.7	1.8741	1.529	709.6	1.8668	1.358	709.5	1.8535	1.221	709.3	1.8417
260	533.15	1.664	714.5	1.8832	1.559	714.4	1.8759	1.385	714.3	1.8626	1.245	714.1	1.8509
280	553.15	1.727	724.1	1.9009	1.619	724.0	1.8936	1.438	723.9	1.8804	1.293	723.8	1.8687
300	573.15	1.791	733.7	1.9182	1.678	733.6	1.9109	1.491	733.5	1.8977	1.341	733.4	1.8858
320	593.15	1.855	743.3	1.9348	1.738	743.3	1.9276	1.545	743.2	1.9145	1.389	743.1	1.9027
340	613.15	1.918	753.2	1.9509	1.798	753.1	1.9439	1.598	753.0	1.9308	1.437	753.0	1.9190
350	623.15	1.950	758.0	1.9588	1.828	758.0	1.9518	1.624	757.9	1.9387	1.461	757.9	1.9270
360	633.15	1.981	762.9	1.9667	1.857	762.9	1.9596	1.650	762.8	1.9465	1.485	762.8	1.9349
380	653.15	2.044	772.7	1.9822	1.916	772.7	1.9749	1.702	772.7	1.9618	1.532	772.7	1.9503
400	673.15	2.107	782.6	1.9972	1.975	782.6	1.9899	1.755	782.6	1.9768	1.579	782.6	1.9652
420	693.15	2.170	792.6	2.0117	2.034	792.6	2.0046	1.807	792.6	1.9915	1.626	792.6	1.9799
440	713.15	2.233	802.6	2.0260	2.093	802.6	2.0189	1.860	802.6	2.0058	1.673	802.6	1.9942
450	723.15	2.264	807.6	2.0330	2.123	807.6	2.0259	1.886	807.6	2.0128	1.697	807.6	2.0012
460	733.15	2.296	812.6	2.0399	2.152	812.6	2.0328	1.913	812.6	2.0198	1.721	812.6	2.0081
480	753.15	2.359	822.8	2.0536	2.211	822.8	2.0464	1.965	822.8	2.0335	1.768	822.8	2.0218
500	773.15	2.422	833.0	2.0671	2.270	833.0	2.0599	2.018	833.0	2.0469	1.815	833.0	2.0353
520	793.15	2.486	843.2	2.0802	2.330	843.2	2.0730	2.071	843.2	2.0600	1.864	843.2	2.0484
540	813.15	2.549	853.6	2.0931	2.389	853.6	2.0859	2.124	853.6	2.0729	1.911	853.6	2.0613
550	823.15	2.580	858.8	2.0995	2.418	858.8	2.0923	2.150	858.8	2.0793	1.935	858.8	2.0677
560	833.15	2.611	864.0	2.1058	2.448	864.0	2.0986	2.176	864.0	2.0856	1.958	864.0	2.0740
580	853.15	2.674	874.4	2.1181	2.507	874.4	2.1109	2.228	874.4	2.0979	2.005	874.4	2.0863
600	873.15	2.737	884.8	2.1301	2.566	884.8	2.1229	2.281	884.8	2.1099	2.052	884.8	2.0983
620	893.15	2.801	895.2	2.1418	2.626	895.2	2.1346	2.335	895.2	2.1216	2.100	895.2	2.1101
640	913.15	2.865	905.8	2.1534	2.686	905.8	2.1462	2.387	905.8	2.1332	2.148	905.8	2.1217
650	923.15	2.896	911.1	2.1592	2.716	911.1	2.1520	2.413	911.1	2.1390	2.172	911.1	2.1275
660	933.15	2.927	916.5	2.1650	2.745	916.4	2.1578	2.439	916.4	2.1448	2.196	916.4	2.1333
680	953.15	2.989	927.3	2.1764	2.803	927.2	2.1692	2.491	927.2	2.1562	2.242	927.1	2.1447
700	973.15	3.051	938.1	2.1877	2.861	938.0	2.1805	2.543	938.0	2.1675	2.288	937.9	2.1560
720	993.15	3.115	948.9	2.1988	2.921	948.8	2.1916	2.596	948.8	2.1785	2.336	948.8	2.1670
740	1013.15	3.179	959.8	2.2098	2.981	959.8	2.2026	2.648	959.8	2.1895	2.384	959.8	2.1780
750	1023.15	3.210	965.3	2.2152	3.010	965.3	2.2080	2.674	965.3	2.1949	2.408	965.3	2.1834
760	1033.15	3.241	970.8	2.2206	3.039	970.8	2.2134	2.700	970.8	2.2003	2.432	970.8	2.1887
780	1053.15	3.303	981.8	2.2311	3.097	981.8	2.2239	2.752	981.8	2.2108	2.478	981.8	2.1992
800	1073.15	3.365	993.0	2.2415	3.155	993.0	2.2343	2.804	993.0	2.2212	2.524	993.0	2.2096
820	1093.15	3.428	1004.2	2.2519	3.215	1004.2	2.2447	2.858	1004.2	2.2316	2.572	1004.2	2.2200
840	1113.15	3.490	1015.6	2.2622	3.273	1015.6	2.2550	2.910	1015.6	2.2420	2.619	1015.6	2.2304
850	1123.15	3.521	1021.3	2.2673	3.302	1021.3	2.2601	2.936	1021.3	2.2471	2.642	1021.3	2.2356
860	1133.15	3.553	1027.0	2.2723	3.332	1027.0	2.2651	2.962	1027.0	2.2521	2.666	1027.0	2.2407
880	1153.15	3.617	1038.4	2.2822	3.391	1038.4	2.2750	3.014	1038.4	2.2620	2.713	1038.4	2.2506
900	1173.15	3.679	1049.8	2.2920	3.449	1049.8	2.2848	3.066	1049.8	2.2718	2.759	1049.8	2.2603
920	1193.15	3.743	1061.2	2.3018	3.509	1061.2	2.2946	3.120	1061.2	2.2816	2.807	1061.2	2.2701
940	1213.15	3.807	1072.6	2.3114	3.569	1072.6	2.3042	3.172	1072.6	2.2912	2.855	1072.6	2.2797
950	1223.15	3.838	1078.3	2.3162	3.598	1078.3	2.3090	3.198	1078.3	2.2960	2.879	1078.3	2.2845
960	1233.15	3.869	1084.0	2.3209	3.627	1084.0	2.3137	3.224	1084.0	2.3007	2.902	1084.0	2.2892
980	1253.15	3.931	1095.6	2.3303	3.685	1095.6	2.3231	3.276	1095.6	2.3101	2.948	1095.6	2.2986
1000	1273.15	3.993	1107.2	2.3395	3.743	1107.2	2.3323	3.328	1107.2	2.3193	2.994	1107.2	2.3078

Pressure p		2.5 at			3.0 at			4.0 at			5.0 at		
Temperature		$t_s = 126.79\,°C$ $i'' = 648.7\,kcal/kg$ $v'' = 0.7318\,m^3/kg$ $s'' = 1.6862\,kcal/kg\,K$			$t_s = 132.88\,°C$ $i'' = 650.7\,kcal/kg$ $v'' = 0.6169\,m^3/kg$ $s'' = 1.6717\,kcal/kg\,K$			$t_s = 142.92\,°C$ $i'' = 653.9\,kcal/kg$ $v'' = 0.4709\,m^3/kg$ $s'' = 1.6488\,kcal/kg\,K$			$t_s = 151.11\,°C$ $i'' = 656.3\,kcal/kg$ $v'' = 0.3817\,m^3/kg$ $s'' = 1.6309\,kcal/kg\,K$		
t	T	v	i	s	v	i	s	v	i	s	v	i	s
°C	K	$\frac{m^3}{kg}$	$\frac{kcal}{kg}$	$\frac{kcal}{kg\,K}$	$\frac{m^3}{kg}$	$\frac{kcal}{kg}$	$\frac{kcal}{kg\,K}$	$\frac{m^3}{kg}$	$\frac{kcal}{kg}$	$\frac{kcal}{kg\,K}$	$\frac{m^3}{kg}$	$\frac{kcal}{kg}$	$\frac{kcal}{kg\,K}$
0	273.15	0.0010001	0.1	0.0000	0.0010001	0.1	0.0000	0.0010000	0.1	0.0000	0.0009999	0.1	0.0000
20	293.15	0.0010017	20.1	0.0708	0.0010017	20.1	0.0708	0.0010017	20.1	0.0708	0.0010016	20.1	0.0708
40	313.15	0.0010078	40.0	0.1365	0.0010078	40.0	0.1365	0.0010077	40.1	0.1365	0.0010077	40.1	0.1365
50	323.15	0.0010120	50.0	0.1680	0.0010120	50.0	0.1679	0.0010119	50.0	0.1679	0.0010119	50.0	0.1679
60	333.15	0.0010170	60.0	0.1984	0.0010170	60.0	0.1983	0.0010169	60.0	0.1983	0.0010168	60.0	0.1983
80	353.15	0.0010289	80.0	0.2567	0.0010288	80.0	0.2567	0.0010288	80.0	0.2566	0.0010287	80.0	0.2566
100	373.15	0.0010434	100.1	0.3121	0.0010434	100.1	0.3121	0.0010433	100.1	0.3120	0.0010433	100.1	0.3120
120	393.15	0.0010603	120.3	0.3647	0.0010602	120.3	0.3647	0.0010602	120.3	0.3646	0.0010601	120.3	0.3646
140	413.15	0.7597	655.6	1.7024	0.6296	654.5	1.6802	0.0010798	140.7	0.4150	0.0010797	140.7	0.4150
150	423.15	0.7802	660.6	1.7143	0.6472	659.7	1.6926	0.4806	657.9	1.6573	0.0010906	151.0	0.4395
160	433.15	0.8003	665.5	1.7258	0.6643	664.7	1.7044	0.4940	663.1	1.6697	0.3917	661.3	1.6420
180	453.15	0.8399	675.2	1.7475	0.6975	674.5	1.7263	0.5197	673.2	1.6927	0.4129	671.7	1.6659
200	473.15	0.8790	684.8	1.7683	0.7304	684.2	1.7471	0.5448	683.0	1.7139	0.4334	681.7	1.6875
220	493.15	0.9179	694.4	1.7882	0.7631	693.9	1.7671	0.5697	692.8	1.7341	0.4537	691.7	1.7079
240	513.15	0.9567	704.0	1.8072	0.7956	703.6	1.7864	0.5944	702.6	1.7535	0.4736	701.6	1.7277
250	523.15	0.9760	708.8	1.8165	0.8119	708.4	1.7957	0.6067	707.5	1.7630	0.4836	706.6	1.7373
260	533.15	0.9953	713.7	1.8256	0.8281	713.2	1.8048	0.6190	712.4	1.7723	0.4935	711.5	1.7467
280	553.15	1.033	723.4	1.8435	0.8603	723.0	1.8227	0.6433	722.2	1.7904	0.5131	721.5	1.7649
300	573.15	1.071	733.0	1.8609	0.8923	732.7	1.8402	0.6676	732.1	1.8079	0.5327	731.4	1.7826
320	593.15	1.111	742.8	1.8778	0.9243	742.5	1.8572	0.6917	742.0	1.8248	0.5521	741.4	1.7996
340	613.15	1.149	752.7	1.8941	0.9562	752.5	1.8737	0.7158	751.9	1.8413	0.5715	751.4	1.8161
350	623.15	1.168	757.6	1.9021	0.9722	757.4	1.8817	0.7278	756.9	1.8494	0.5812	756.4	1.8242
360	633.15	1.187	762.5	1.9100	0.9881	762.3	1.8896	0.7398	761.8	1.8573	0.5908	761.4	1.8322
380	653.15	1.225	772.5	1.9255	1.020	772.2	1.9051	0.7637	771.8	1.8728	0.6101	771.4	1.8478
400	673.15	1.262	782.4	1.9404	1.052	782.2	1.9200	0.7875	781.8	1.8879	0.6294	781.5	1.8630
420	693.15	1.300	792.4	1.9551	1.083	792.2	1.9347	0.8114	791.8	1.9026	0.6485	791.5	1.8778
440	713.15	1.338	802.5	1.9694	1.115	802.3	1.9490	0.8352	802.0	1.9170	0.6676	801.7	1.8921
450	723.15	1.357	807.5	1.9764	1.131	807.3	1.9561	0.8471	807.1	1.9241	0.6772	806.7	1.8992
460	733.15	1.376	812.5	1.9834	1.147	812.3	1.9631	0.8590	812.1	1.9311	0.6867	811.7	1.9062
480	753.15	1.414	822.7	1.9971	1.179	822.5	1.9768	0.8828	822.3	1.9448	0.7058	821.9	1.9200
500	773.15	1.453	832.9	2.0105	1.210	832.7	1.9903	0.9066	832.5	1.9583	0.7248	832.1	1.9335
520	793.15	1.490	843.1	2.0236	1.242	842.9	2.0035	0.9304	842.8	1.9715	0.7439	842.4	1.9467
540	813.15	1.528	853.5	2.0365	1.273	853.3	2.0164	0.9542	853.2	1.9844	0.7629	852.8	1.9597
550	823.15	1.547	858.7	2.0429	1.289	858.5	2.0228	0.9660	858.4	1.9908	0.7724	858.0	1.9661
560	833.15	1.566	863.9	2.0492	1.304	863.7	2.0291	0.9779	863.6	1.9971	0.7819	863.2	1.9724
580	853.15	1.604	874.3	2.0615	1.336	874.1	2.0414	1.0016	874.0	2.0094	0.8009	873.6	1.9847
600	873.15	1.642	884.7	2.0735	1.368	884.5	2.0534	1.0252	884.4	2.0215	0.8198	884.1	1.9968
620	893.15	1.680	895.1	2.0854	1.399	894.9	2.0653	1.0489	894.8	2.0334	0.8388	894.7	2.0087
640	913.15	1.718	905.7	2.0970	1.430	905.5	2.0769	1.0726	905.4	2.0451	0.8578	905.3	2.0204
650	923.15	1.737	911.0	2.1028	1.446	910.8	2.0827	1.0845	910.7	2.0509	0.8673	910.6	2.0262
660	933.15	1.756	916.3	2.1086	1.462	916.1	2.0885	1.0963	916.0	2.0567	0.8768	915.9	2.0320
680	953.15	1.794	927.0	2.1200	1.493	926.9	2.0999	1.1199	926.8	2.0681	0.8956	926.7	2.0434
700	973.15	1.831	937.8	2.1313	1.525	937.7	2.1112	1.1435	937.6	2.0794	0.9144	937.5	2.0547
720	993.15	1.869	948.2	2.1424	1.556	948.6	2.1222	1.1671	948.5	2.0905	0.9334	948.4	2.0658
740	1013.15	1.907	959.7	2.1534	1.588	959.6	2.1332	1.1907	959.5	2.1015	0.9524	959.4	2.0768
750	1023.15	1.926	965.2	2.1588	1.604	965.1	2.1386	1.2025	965.0	2.1069	0.9618	964.9	2.0822
760	1033.15	1.945	970.7	2.1641	1.620	970.6	2.1440	1.2143	970.5	2.1123	0.9712	970.4	2.0876
780	1053.15	1.983	981.7	2.1747	1.652	981.6	2.1546	1.2379	981.6	2.1229	0.9900	981.5	2.0982
800	1073.15	2.019	992.9	2.1852	1.682	992.8	2.1651	1.2615	992.8	2.1333	1.0088	992.7	2.1086
820	1093.15	2.057	1004.1	2.1956	1.714	1004.0	2.1755	1.2851	1004.0	2.1437	1.0277	1003.9	2.1190
840	1113.15	2.095	1015.5	2.2058	1.746	1015.4	2.1857	1.3087	1015.4	2.1539	1.0466	1015.3	2.1292
850	1123.15	2.113	1021.2	2.2109	1.761	1021.1	2.1908	1.3205	1021.1	2.1590	1.0560	1021.0	2.1343
860	1133.15	2.132	1026.9	2.2159	1.777	1026.8	2.1958	1.3322	1026.8	2.1640	1.0655	1026.7	2.1393
880	1153.15	2.170	1038.3	2.2258	1.809	1038.2	2.2057	1.3556	1038.2	2.1740	1.0844	1038.1	2.1493
900	1173.15	2.207	1049.7	2.2356	1.839	1049.6	2.2155	1.3790	1049.6	2.1838	1.1032	1049.5	2.1592
920	1193.15	2.245	1061.1	2.2454	1.871	1061.0	2.2253	1.4025	1061.0	2.1936	1.1221	1060.9	2.1690
940	1213.15	2.283	1072.5	2.2551	1.903	1072.4	2.2350	1.4260	1072.4	2.2032	1.1410	1072.3	2.1786
950	1223.15	2.302	1078.2	2.2599	1.919	1078.1	2.2398	1.4378	1078.1	2.2080	1.1504	1078.0	2.1834
960	1233.15	2.321	1084.0	2.2646	1.935	1083.9	2.2445	1.4495	1083.9	2.2137	1.1599	1083.8	2.1881
980	1253.15	2.359	1095.6	2.2740	1.966	1095.5	2.2539	1.4730	1095.5	2.2221	1.1788	1095.4	2.1975
1000	1273.15	2.395	1107.2	2.2832	1.996	1107.1	2.2631	1.4965	1107.1	2.2313	1.1976	1107.0	2.2068

Table 41-9 Properties of Superheated Steam (H_2O) (Gravitational Metric System) (*Continued*)

Pressure p		6 at			7 at			8 at			9 at		
Temperature		$t_s = 158.08\,°C$ $i'' = 658.3\,kcal/kg$ $v'' = 0.3214\,m^3/kg$ $s'' = 1.6164\,kcal/kg\,K$			$t_s = 164.17\,°C$ $i'' = 659.9\,kcal/kg$ $v'' = 0.2778\,m^3/kg$ $s'' = 1.6039\,kcal/kg\,K$			$t_s = 169.61\,°C$ $i'' = 661.2\,kcal/kg$ $v'' = 0.2448\,m^3/kg$ $s'' = 1.5931\,kcal/kg\,K$			$t_s = 174.53\,°C$ $i'' = 662.3\,kcal/kg$ $v'' = 0.2189\,m^3/kg$ $s'' = 1.5834\,kcal/kg\,K$		
t	T	v	i	s	v	i	s	v	i	s	v	i	s
°C	K	$\frac{m^3}{kg}$	$\frac{kcal}{kg}$	$\frac{kcal}{kg\,K}$	$\frac{m^3}{kg}$	$\frac{kcal}{kg}$	$\frac{kcal}{kg\,K}$	$\frac{m^3}{kg}$	$\frac{kcal}{kg}$	$\frac{kcal}{kg\,K}$	$\frac{m^3}{kg}$	$\frac{kcal}{kg}$	$\frac{kcal}{kg\,K}$
0	273.15	0.0009999	0.1	0.0000	0.0009999	0.2	0.0000	0.0009998	0.2	0.0000	0.0009997	0.2	0.0000
20	293.15	0.0010016	20.2	0.0708	0.0010015	20.2	0.0708	0.0010015	20.2	0.0708	0.0010015	20.2	0.0707
40	313.15	0.0010077	40.1	0.1356	0.0010076	40.1	0.1365	0.0010076	40.2	0.1365	0.0010075	40.2	0.1364
50	323.15	0.0010118	50.1	0.1679	0.0010118	50.1	0.1679	0.0010118	50.1	0.1679	0.0010117	50.1	0.1679
60	333.15	0.0010168	60.1	0.1983	0.0010168	60.1	0.1983	0.0010167	60.1	0.1983	0.0010167	60.1	0.1983
80	353.15	0.0010287	80.1	0.2566	0.0010286	80.1	0.2566	0.0010286	80.1	0.2566	0.0010285	80.1	0.2565
100	373.15	0.0010432	100.1	0.3120	0.0010432	100.1	0.3120	0.0010431	100.2	0.3119	0.0010431	100.2	0.3119
120	393.15	0.0010601	120.3	0.3646	0.0010600	120.3	0.3646	0.0010600	120.3	0.3646	0.0010599	120.4	0.3645
140	413.15	0.0010797	140.7	0.4150	0.0010796	140.7	0.4150	0.0010795	140.7	0.4149	0.0010795	140.7	0.4149
150	423.15	0.0010906	151.0	0.4395	0.0010904	151.0	0.4394	0.0010904	151.0	0.4394	0.0010903	151.0	0.4394
160	433.15	0.3232	659.4	1.6186	0.0011020	161.3	0.4637	0.0011020	161.3	0.4636	0.0011019	161.3	0.4636
180	453.15	0.3416	670.1	1.6431	0.2906	668.8	1.6235	0.2524	667.3	1.6063	0.2226	665.5	1.5905
200	473.15	0.3591	680.6	1.6655	0.3059	679.4	1.6467	0.2662	678.2	1.6300	0.2353	676.8	1.6147
220	493.15	0.3763	690.7	1.6864	0.3209	689.7	1.6680	0.2795	688.7	1.6517	0.2472	687.5	1.6369
240	513.15	0.3932	700.7	1.7064	0.3356	699.8	1.6882	0.2925	699.0	1.6722	0.2589	698.0	1.6577
250	523.15	0.4016	705.7	1.7160	0.3429	704.9	1.6980	0.2990	704.1	1.6820	0.2647	703.2	1.6678
260	533.15	0.4099	710.7	1.7254	0.3501	709.9	1.7075	0.3054	709.2	1.6916	0.2704	708.4	1.6776
280	553.15	0.4264	720.7	1.7438	0.3644	720.0	1.7260	0.3180	719.4	1.7102	0.2818	718.6	1.6964
300	573.15	0.4428	730.7	1.7615	0.3785	730.1	1.7438	0.3305	729.4	1.7282	0.2930	728.7	1.7144
320	593.15	0.4591	740.8	1.7786	0.3926	740.2	1.7609	0.3429	739.5	1.7455	0.3040	739.0	1.7318
340	613.15	0.4753	750.9	1.7953	0.4066	750.3	1.7776	0.3552	749.8	1.7623	0.3150	749.2	1.7487
350	623.15	0.4834	755.9	1.8035	0.4136	755.3	1.7858	0.3613	754.9	1.7705	0.3205	754.3	1.7570
360	633.15	0.4915	760.9	1.8115	0.4206	760.3	1.7939	0.3674	760.0	1.7786	0.3260	759.4	1.7651
380	653.15	0.5077	771.0	1.8271	0.4345	770.5	1.8097	0.3796	770.1	1.7945	0.3369	769.7	1.7810
400	673.15	0.5237	781.1	1.8422	0.4483	780.6	1.8250	0.3918	780.3	1.8099	0.3477	779.9	1.7963
420	693.15	0.5398	791.1	1.8571	0.4621	790.8	1.8399	0.4039	790.5	1.8248	0.3586	790.1	1.8114
440	713.15	0.5558	801.3	1.8716	0.4759	801.0	1.8544	0.4159	800.6	1.8393	0.3693	800.3	1.8259
450	723.15	0.5637	806.4	1.8787	0.4827	806.1	1.8615	0.4219	805.7	1.8464	0.3747	805.4	1.8331
460	733.15	0.5717	811.5	1.8857	0.4896	811.2	1.8685	0.4280	810.8	1.8534	0.3800	810.5	1.8402
480	753.15	0.5876	821.7	1.8996	0.5033	821.4	1.8824	0.4400	821.0	1.8673	0.3907	820.8	1.8541
500	773.15	0.6036	831.9	1.9131	0.5169	831.7	1.8959	0.4519	831.4	1.8808	0.4014	831.2	1.8677
520	793.15	0.6194	842.2	1.9263	0.5306	842.0	1.9091	0.4639	841.8	1.8941	0.4121	841.6	1.8810
540	813.15	0.6352	852.6	1.9393	0.5442	852.4	1.9221	0.4759	852.2	1.9071	0.4227	852.0	1.8940
550	823.15	0.6432	857.8	1.9457	0.5510	857.6	1.9285	0.4819	857.4	1.9135	0.4280	857.2	1.9004
560	833.15	0.6512	863.0	1.9520	0.5579	862.8	1.9348	0.4879	862.2	1.9198	0.4334	862.4	1.9067
580	853.15	0.6671	873.4	1.9643	0.5715	873.2	1.9473	0.4998	873.0	1.9323	0.4441	872.8	1.9192
600	873.15	0.6829	884.0	1.9764	0.5851	883.8	1.9594	0.5117	883.6	1.9445	0.4546	883.4	1.9314
620	893.15	0.6987	894.6	1.9883	0.5987	894.4	1.9713	0.5237	894.2	1.9565	0.4652	894.0	1.9434
640	913.15	0.7145	905.2	2.0000	0.6123	905.0	1.9830	0.5357	904.8	1.9683	0.4758	904.6	1.9552
650	923.15	0.7224	910.5	2.0058	0.6191	910.3	1.9888	0.5416	910.1	1.9741	0.4811	909.9	1.9610
660	933.15	0.7303	915.8	2.0116	0.6259	915.7	1.9946	0.5475	915.5	1.9799	0.4864	915.3	1.9668
680	953.15	0.7461	926.6	2.0230	0.6394	926.5	2.0060	0.5593	926.3	1.9913	0.4970	926.1	1.9782
700	973.15	0.7618	937.4	2.0343	0.6528	937.3	2.0173	0.5711	937.1	2.0026	0.5074	936.9	1.9895
720	993.15	0.7776	948.3	2.0455	0.6664	948.2	2.0284	0.5831	948.0	2.0137	0.5180	947.9	2.0006
740	1013.15	0.7934	959.3	2.0565	0.6800	959.2	2.0394	0.5949	959.0	2.0247	0.5286	958.9	2.0116
750	1023.15	0.8013	964.8	2.0619	0.6868	964.7	2.0448	0.6008	964.5	2.0301	0.5339	964.4	2.0170
760	1033.15	0.8092	970.3	2.0673	0.6936	970.2	2.0502	0.6067	970.0	2.0355	0.5392	969.9	2.0224
780	1053.15	0.8250	981.4	2.0778	0.7070	981.3	2.0608	0.6185	981.1	2.0461	0.5497	981.0	2.0330
800	1073.15	0.8407	992.6	2.0882	0.7204	992.5	2.0712	0.6303	992.3	2.0567	0.5601	992.2	2.0436
820	1093.15	0.8565	1003.8	2.0986	0.7340	1003.7	2.0816	0.6421	1003.5	2.0671	0.5707	1003.4	2.0540
840	1113.15	0.8722	1015.2	2.1089	0.7474	1015.1	2.0920	0.6539	1014.9	2.0773	0.5811	1014.8	2.0643
850	1123.15	0.8800	1020.9	2.1140	0.7541	1020.8	2.0971	0.6598	1020.6	2.0824	0.5863	1020.5	2.0694
860	1133.15	0.8879	1026.6	2.1190	0.7609	1026.5	2.1021	0.6658	1026.3	2.0874	0.5916	1026.2	2.0744
880	1153.15	0.9036	1038.0	2.1290	0.7744	1037.9	2.1121	0.6776	1037.7	2.0974	0.6022	1037.6	2.0844
900	1173.15	0.9192	1049.4	2.1389	0.7878	1049.3	2.1220	0.6894	1049.1	2.1073	0.6127	1049.0	2.0943
920	1193.15	0.9350	1060.8	2.1487	0.8014	1060.7	2.1318	0.7012	1060.5	2.1171	0.6233	1060.4	2.1041
940	1213.15	0.9508	1072.2	2.1583	0.8150	1072.1	2.1414	0.7130	1071.9	2.1267	0.6339	1071.8	2.1137
950	1223.15	0.9587	1077.9	2.1631	0.8217	1077.8	2.1462	0.7189	1077.7	2.1314	0.6391	1077.6	2.1184
960	1233.15	0.9665	1083.7	2.1678	0.8284	1083.6	2.1509	0.7248	1083.5	2.1361	0.6443	1083.4	2.1231
980	1253.15	0.9821	1095.3	2.1772	0.8418	1095.2	2.1603	0.7366	1095.1	2.1455	0.6547	1095.0	2.1325
1000	1273.15	0.9977	1106.9	2.1865	0.8552	1106.8	2.1695	0.7483	1106.7	2.1548	0.6651	1106.6	2.1418

Table 41-10 Properties of Superheated Steam (H_2O) (Gravitational Metric System) (Continued)

Pressure p		10 at			12 at			14 at			16 at		
Temperature		$t_s = 179.04\ °C$ $i'' = 663.3\ kcal/kg$ $v'' = 0.1980\ m^3/kg$ $s'' = 1.5748\ kcal/kg\ K$			$t_s = 187.08\ °C$ $i'' = 664.9\ kcal/kg$ $v'' = 0.1663\ m^3/kg$ $s'' = 1.5597\ kcal/kg\ K$			$t_s = 194.13\ °C$ $i'' = 666.2\ kcal/kg$ $v'' = 0.1434\ m^3/kg$ $s'' = 1.5468\ kcal/kg\ K$			$t_s' = 200.43\ °C$ $i'' = 667.1\ kcal/kg$ $v'' = 0.1261\ m^3/kg$ $s'' = 1.5354\ kcal/kg\ K$		
t	T	v	i	s	v	i	s	v	i	s	v	i	s
°C	K	$\dfrac{m^3}{kg}$	$\dfrac{kcal}{kg}$	$\dfrac{kcal}{kg\ K}$	$\dfrac{m^3}{kg}$	$\dfrac{kcal}{kg}$	$\dfrac{kcal}{kg\ K}$	$\dfrac{m^3}{kg}$	$\dfrac{kcal}{kg}$	$\dfrac{kcal}{kg\ K}$	$\dfrac{m^3}{kg}$	$\dfrac{kcal}{kg}$	$\dfrac{kcal}{kg\ K}$
0	273.15	0.0009997	0.2	0.0000	0.0009996	0.3	0.0000	0.0009995	0.3	0.0000	0.0009994	0.4	0.0000
20	293.15	0.0010014	20.3	0.0707	0.0010013	20.3	0.0707	0.0010012	20.3	0.0707	0.0010011	20.4	0.0707
40	313.15	0.0010075	40.2	0.1364	0.0010074	40.2	0.1364	0.0010073	40.3	0.1364	0.0010072	40.3	0.1364
50	323.15	0.0010117	50.2	0.1678	0.0010116	50.2	0.1678	0.0010115	50.2	0.1678	0.0010114	50.3	0.1678
60	333.15	0.0010166	60.1	0.1982	0.0010165	60.2	0.1982	0.0010164	60.2	0.1982	0.0010163	60.2	0.1982
80	353.15	0.0010285	80.1	0.2565	0.0010284	80.2	0.2565	0.0010283	80.2	0.2564	0.0010282	80.2	0.2564
100	373.15	0.0010430	100.2	0.3119	0.0010429	100.2	0.3119	0.0010428	100.3	0.3118	0.0010427	100.3	0.3118
120	393.15	0.0010599	120.4	0.3645	0.0010598	120.4	0.3645	0.0010596	120.4	0.3644	0.0010595	120.5	0.3644
140	413.15	0.0010794	140.7	0.4149	0.0010793	140.8	0.4148	0.0010792	140.8	0.4148	0.0010791	140.8	0.4147
150	423.15	0.0010902	151.0	0.4394	0.0010901	151.0	0.4393	0.0010900	151.0	0.4392	0.0010899	151.1	0.4392
160	433.15	0.0011018	161.3	0.4635	0.0011017	161.3	0.4635	0.0011015	161.4	0.4634	0.0011014	161.4	0.4633
180	453.15	0.1987	663.8	1.5760	0.0011273	182.3	0.5106	0.0011272	182.3	0.5105	0.0011270	182.3	0.5104
200	473.15	0.2103	675.4	1.6008	0.1728	672.9	1.5762	0.1460	670.0	1.5545	0.0011565	203.6	0.5562
220	493.15	0.2214	686.5	1.6236	0.1825	684.5	1.6000	0.1547	682.3	1.5796	0.1338	679.9	1.5610
240	513.15	0.2321	697.2	1.6449	0.1918	695.3	1.6220	0.1629	693.5	1.6020	0.1411	691.4	1.5843
250	523.15	0.2374	702.4	1.6551	0.1963	700.6	1.6324	0.1669	698.9	1.6126	0.1447	696.9	1.5951
260	533.15	0.2426	707.6	1.6650	0.2007	705.9	1.6425	0.1708	704.2	1.6229	0.1482	702.3	1.6056
280	553.15	0.2529	717.8	1.6839	0.2095	716.4	1.6618	0.1784	714.6	1.6426	0.1551	713.2	1.6257
300	573.15	0.2630	728.0	1.7019	0.2181	726.7	1.6802	0.1859	725.1	1.6612	0.1618	724.0	1.6447
320	593.15	0.2731	738.4	1.7194	0.2265	737.1	1.6979	0.1933	735.8	1.6792	0.1683	734.7	1.6630
340	613.15	0.2829	748.7	1.7365	0.2348	747.6	1.7151	0.2005	746.4	1.6967	0.7447	745.4	1.6807
350	623.15	0.2879	753.8	1.7448	0.2390	752.8	1.7235	0.2041	751.7	1.7052	0.1779	750.7	1.6893
360	633.15	0.2929	758.9	1.7530	0.2432	757.9	1.7317	0.2077	756.9	1.7135	0.1811	756.0	1.6977
380	653.15	0.3028	769.2	1.7690	0.2515	768.3	1.7478	0.2150	767.4	1.7297	0.1875	766.6	1.7140
400	673.15	0.3126	779.5	1.7843	0.2598	778.7	1.7633	0.2220	777.9	1.7454	0.1937	777.1	1.7299
420	693.15	0.3223	789.7	1.7994	0.2679	788.9	1.7784	0.2291	788.3	1.7607	0.2000	787.5	1.7453
440	713.15	0.3320	799.9	1.8139	0.2761	799.3	1.7933	0.2361	798.7	1.7755	0.2062	798.0	1.7600
450	723.15	0.3369	805.1	1.8211	0.2801	804.5	1.8005	0.2396	803.9	1.7827	0.2092	803.2	1.7673
460	733.15	0.3417	810.2	1.8282	0.2842	809.7	1.8076	0.2431	809.1	1.7898	0.2123	808.4	1.7745
480	753.15	0.3513	820.5	1.8421	0.2922	820.1	1.8215	0.2501	819.5	1.8039	0.2184	818.8	1.7886
500	773.15	0.3609	830.9	1.8558	0.3003	830.5	1.8351	0.2570	829.9	1.8177	0.2245	829.4	1.8024
520	793.15	0.3706	841.3	1.8691	0.3084	840.9	1.8484	0.2639	840.3	1.8311	0.2306	839.8	1.8158
540	813.15	0.3802	851.7	1.8821	0.3164	851.3	1.8615	0.2709	850.7	1.8442	0.2367	850.3	1.8289
550	823.15	0.3851	856.9	1.8885	0.3205	856.5	1.8679	0.2744	856.0	1.8506	0.2398	855.6	1.8353
560	833.15	0.3899	862.1	1.8948	0.3245	861.7	1.8742	0.2779	861.3	1.8569	0.2429	860.9	1.8417
580	853.15	0.3994	872.6	1.9073	0.3326	872.2	1.8867	0.2848	871.9	1.8694	0.2489	871.5	1.8543
600	873.15	0.4088	883.2	1.9195	0.3405	882.8	1.8990	0.2915	882.5	1.8817	0.2548	882.1	1.8667
620	893.15	0.4184	893.8	1.9315	0.3485	893.4	1.9111	0.2984	893.1	1.8938	0.2608	892.7	1.8788
640	913.15	0.4280	904.4	1.9433	0.3565	904.0	1.9229	0.3053	903.8	1.9056	0.2668	903.5	1.8907
650	923.15	0.4328	909.8	1.9492	0.3605	909.4	1.9288	0.3087	909.2	1.9115	0.2698	908.9	1.8966
660	933.15	0.4375	915.2	1.9550	0.3645	914.8	1.9346	0.3121	914.6	1.9173	0.2728	914.3	1.9024
680	953.15	0.4470	926.0	1.9664	0.3724	925.6	1.9461	0.3189	925.4	1.9288	0.2788	925.1	1.9130
700	973.15	0.4565	936.8	1.9777	0.3802	936.5	1.9574	0.3257	936.3	1.9402	0.2848	936.0	1.9253
720	993.15	0.4661	947.8	1.9889	0.3882	947.5	1.9686	0.3325	947.3	1.9514	0.2908	947.0	1.9364
740	1013.15	0.4757	958.8	1.9999	0.3962	958.5	1.9796	0.3393	958.3	1.9624	0.2968	958.0	1.9474
750	1023.15	0.4805	964.3	2.0053	0.4002	964.0	1.9850	0.3427	963.8	1.9678	0.2998	963.6	1.9528
760	1033.15	0.4852	969.8	2.0107	0.4042	969.5	1.9904	0.3461	969.3	1.9732	0.3028	969.1	1.9582
780	1053.15	0.4946	981.0	2.0213	0.4120	980.7	2.0010	0.3529	980.5	1.9839	0.3088	980.3	1.9690
800	1073.15	0.5040	992.2	2.0319	0.4198	991.9	2.0116	0.3597	991.7	1.9945	0.3146	991.5	1.9796
820	1093.15	0.5135	1003.4	2.0423	0.4278	1003.2	2.0219	0.3665	1003.0	2.0049	0.3206	1002.8	1.9900
840	1113.15	0.5229	1014.8	2.0525	0.4356	1014.6	2.0321	0.3733	1014.4	2.0152	0.3265	1014.2	2.0003
850	1123.15	0.5276	1020.5	2.0576	0.4395	1020.3	2.0372	0.3766	1020.1	2.0203	0.3294	1019.9	2.0054
860	1133.15	0.5324	1026.2	2.0626	0.4435	1026.0	2.0423	0.3800	1025.8	2.0253	0.3324	1025.6	2.0105
880	1153.15	0.5419	1037.6	2.0726	0.4515	1037.4	2.0524	0.3868	1037.2	2.0353	0.3384	1037.0	2.0205
900	1173.15	0.5513	1049.0	2.0826	0.4593	1048.8	2.0624	0.3936	1048.7	2.0453	0.3443	1048.5	2.0305
920	1193.15	0.5609	1060.4	2.0924	0.4673	1060.2	2.0722	0.4004	1060.1	2.0551	0.3503	1059.9	2.0403
940	1213.15	0.5703	1071.8	2.1020	0.4753	1071.7	2.0818	0.4072	1071.6	2.0647	0.3563	1071.4	2.0499
950	1223.15	0.5750	1077.6	2.1067	0.4792	1077.5	2.0865	0.4106	1077.4	2.0695	0.3593	1077.2	2.0547
960	1233.15	0.5797	1083.4	2.1114	0.4831	1083.3	2.0912	0.4140	1083.2	2.0742	0.3623	1083.0	2.0594
980	1253.15	0.5891	1095.0	2.1208	0.4909	1094.9	2.1006	0.4208	1094.8	2.0836	0.3681	1094.6	2.0688
1000	1273.15	0.5985	1106.6	2.1301	0.4987	1106.5	2.1099	0.4274	1106.4	2.0928	0.3739	1106.2	2.0781

Table 41-11 Properties of Superheated Steam (H₂O) (Gravitational Metric System) (Continued)

Pressure p		18 at			20 at			25 at			30 at		
Temperature		$t_s = 206.14\ °C$ $i'' = 667.8\ kcal/kg$ $v'' = 0.1125\ m^3/kg$ $s'' = 1.5253\ kcal/kg\ K$			$t_s = 211.38\ °C$ $i'' = 668.5\ kcal/kg$ $v'' = 0.1015\ m^3/kg$ $s'' = 1.5161\ kcal/kg\ K$			$t_s = 222.90\ °C$ $i'' = 669.3\ kcal/kg$ $v'' = 0.08150\ m^3/kg$ $s'' = 1.4961\ kcal/kg\ K$			$t_s = 232.76\ °C$ $i'' = 669.6\ kcal/kg$ $v'' = 0.06797\ m^3/kg$ $s'' = 1.4794\ kcal/kg\ K$		
t	T	v	i	s	v	i	s	v	i	s	v	i	s
°C	K	$\frac{m^3}{kg}$	$\frac{kcal}{kg}$	$\frac{kcal}{kg\ K}$	$\frac{m^3}{kg}$	$\frac{kg}{kg}$	$\frac{kcal}{kg\ K}$	$\frac{m^3}{kg}$	$\frac{kval}{kg}$	$\frac{kcal}{kg\ K}$	$\frac{m^3}{kg}$	$\frac{kcal}{kg}$	$\frac{kcal}{kg\ K}$
0	273.15	0.0009993	0.4	0.0000	0.0009992	0.5	0.0000	0.0009989	0.6	0.0000	0.0009987	0.7	0.0000
20	293.15	0.0010010	20.4	0.0707	0.0010010	20.5	0.0707	0.0010007	20.6	0.0706	0.0010005	20.7	0.0706
40	313.15	0.0010071	40.4	0.1364	0.0010070	40.4	0.1364	0.0010068	40.5	0.1363	0.0010066	40.6	0.1362
50	323.15	0.0010113	50.3	0.1678	0.0010112	50.4	0.1677	0.0010110	50.4	0.1677	0.0010108	50.6	0.1676
60	333.15	0.0010162	60.3	0.1982	0.0010161	60.3	0.1981	0.0010159	60.4	0.1980	0.0010157	60.5	0.1980
80	353.15	0.0010281	80.3	0.2564	0.0010280	80.3	0.2563	0.0010278	80.4	0.2562	0.0010275	80.5	0.2561
100	373.15	0.0010425	100.3	0.3117	0.0010425	100.4	0.3117	0.0010422	100.5	0.3116	0.0010419	100.5	0.3115
120	393.15	0.0010594	120.5	0.3644	0.0010593	120.5	0.3643	0.0010591	120.6	0.3642	0.0010588	120.7	0.3641
140	413.15	0.0010789	140.9	0.4147	0.0010788	140.9	0.4146	0.0010785	141.0	0.4145	0.0010782	141.1	0.4144
150	423.15	0.0010897	151.1	0.4391	0.0010896	151.1	0.4391	0.0010893	151.2	0.4390	0.0010890	151.3	0.4388
160	433.15	0.0011013	161.4	0.4633	0.0011011	161.4	0.4632	0.0011008	161.5	0.4630	0.0011004	161.6	0.4629
180	453.15	0.0011268	182.3	0.5103	0.0011267	182.3	0.5102	0.0011263	182.3	0.5100	0.0011259	182.4	0.5098
200	473.15	0.0011563	203.6	0.5561	0.0011561	203.6	0.5560	0.0011556	203.6	0.5558	0.0011552	203.6	0.5556
220	493.15	0.1175	677.0	1.5438	0.1043	674.4	1.5280	0.0011899	225.4	0.6009	0.0011892	225.4	0.6006
240	513.15	0.1242	689.3	1.5681	0.1108	687.2	1.5530	0.08643	681.4	1.5200	0.06987	675.0	1.4900
250	523.15	0.1275	695.0	1.5793	0.1138	693.1	1.5645	0.08906	687.9	1.5325	0.07230	682.1	1.5038
260	533.15	0.1307	700.6	1.5900	0.1168	698.9	1.5756	0.09158	694.1	1.5443	0.07459	688.9	1.5167
280	553.15	0.1369	711.8	1.6104	0.1225	710.2	1.5967	0.09640	706.2	1.5665	0.07889	701.9	1.5405
300	573.15	0.1430	722.8	1.6298	0.1281	721.3	1.6166	0.1010	717.8	1.5874	0.08294	714.2	1.5624
320	593.15	0.1490	733.6	1.6484	0.1334	732.4	1.6354	0.1055	729.1	1.6071	0.08680	726.1	1.5827
340	613.15	0.1548	744.3	1.6663	0.1386	743.4	1.6534	0.1098	740.5	1.6256	0.09055	737.7	1.6019
350	623.15	0.1577	749.7	1.6750	0.1412	748.8	1.6621	0.1120	746.1	1.6345	0.09239	743.5	1.6112
360	633.15	0.1605	755.0	1.6835	0.1438	754.1	1.6707	0.1141	751.6	1.6432	0.09421	749.2	1.6202
380	653.15	0.1661	765.6	1.7000	0.1491	764.8	1.6874	0.1183	762.6	1.6603	0.09780	760.4	1.6376
400	673.15	0.1717	776.2	1.7160	0.1542	775.5	1.7035	0.1225	773.4	1.6767	0.1013	771.4	1.6542
420	693.15	0.1773	786.7	1.7315	0.1592	786.1	1.7190	0.1266	784.2	1.6924	0.1048	782.4	1.6703
440	713.15	0.1829	797.3	1.7465	0.1642	796.6	1.7341	0.1308	795.1	1.7077	0.1084	793.3	1.6858
450	723.15	0.1856	802.5	1.7538	0.1667	801.9	1.7414	0.1328	800.4	1.7152	0.1101	798.7	1.6934
460	733.15	0.1884	807.7	1.7610	0.1692	807.1	1.7486	0.1347	805.7	1.7225	0.1118	804.1	1.7008
480	753.15	0.1938	818.2	1.7751	0.1741	817.7	1.7628	0.1387	816.3	1.7368	0.1151	814.9	1.7153
500	773.15	0.1992	828.8	1.7889	0.1790	828.3	1.7767	0.1426	827.0	1.7509	0.1185	825.7	1.7295
520	793.15	0.2047	839.3	1.8024	0.1840	838.8	1.7902	0.1466	837.6	1.7646	0.1218	836.5	1.7432
540	813.15	0.2101	849.9	1.8155	0.1888	849.4	1.8035	0.1506	848.3	1.7779	0.1252	847.3	1.7566
550	823.15	0.2129	855.2	1.8220	0.1913	854.7	1.8100	0.1527	853.7	1.7844	0.1269	852.6	1.7632
560	833.15	0.2156	860.5	1.8284	0.1938	860.0	1.8164	0.1547	859.0	1.7908	0.1285	858.0	1.7697
580	853.15	0.2210	871.1	1.8410	0.1987	870.6	1.8291	0.1586	869.7	1.8035	0.1318	868.8	1.7825
600	873.15	0.2264	881.7	1.8534	0.2035	881.3	1.8415	0.1624	880.5	1.8160	0.1350	879.6	1.7950
620	893.15	0.2318	892.3	1.8655	0.2084	892.0	1.8537	0.1664	891.3	1.8283	0.1384	890.5	1.8073
640	913.15	0.2372	903.1	1.8774	0.2132	902.8	1.8656	0.1703	902.1	1.8403	0.1416	901.3	1.8194
650	923.15	0.2398	908.5	1.8833	0.2156	908.2	1.8715	0.1722	907.5	1.8462	0.1432	906.7	1.8254
660	933.15	0.2425	913.9	1.8891	0.2180	913.6	1.8773	0.1741	912.9	1.8521	0.1448	912.1	1.8313
680	953.15	0.2477	924.7	1.9007	0.2228	924.5	1.8889	0.1780	923.8	1.8637	0.1480	923.1	1.8430
700	973.15	0.2529	935.7	1.9121	0.2276	935.5	1.9003	0.1818	934.8	1.8752	0.1512	934.1	1.8545
720	993.15	0.2583	946.7	1.9233	0.2324	946.5	1.9115	0.1858	945.8	1.8864	0.1545	945.1	1.8658
740	1013.15	0.2637	957.7	1.9343	0.2372	957.5	1.9225	0.1896	956.9	1.8975	0.1577	956.3	1.8769
750	1023.15	0.2664	963.3	1.9398	0.2396	963.1	1.9280	0.1915	962.5	1.9030	0.1593	961.9	1.8824
760	1033.15	0.2691	968.9	1.9452	0.2420	968.7	1.9334	0.1934	968.1	1.9084	0.1609	967.5	1.8879
780	1053.15	0.2743	980.1	1.9559	0.2468	979.9	1.9441	0.1972	979.3	1.9192	0.1641	978.7	1.8987
800	1073.15	0.2796	991.3	1.9665	0.2515	991.1	1.9547	0.2010	990.6	1.9298	0.1673	990.1	1.9093
820	1093.15	0.2849	1002.7	1.9769	0.2563	1002.5	1.9652	0.2049	1002.0	1.9403	0.1705	1001.5	1.9198
840	1113.15	0.2901	1014.1	1.9872	0.2611	1013.9	1.9755	0.2087	1013.4	1.9506	0.1737	1012.9	1.9302
850	1123.15	0.2927	1019.8	1.9923	0.2634	1019.6	1.9806	0.2106	1019.2	1.9557	0.1753	1018.7	1.9353
860	1133.15	0.2954	1025.5	1.9974	0.2658	1025.3	1.9857	0.2125	1024.9	1.9608	0.1769	1024.4	1.9404
880	1153.15	0.3007	1036.9	2.0074	0.2706	1036.7	1.9957	0.2163	1036.4	1.9708	0.1801	1036.0	1.9504
900	1173.15	0.3059	1048.4	2.0174	0.2753	1048.3	2.0057	0.2201	1048.0	1.9808	0.1833	1047.6	1.9604
920	1193.15	0.3113	1059.8	2.0272	0.2801	1059.7	2.0155	0.2239	1059.4	1.9906	0.1865	1059.0	1.9703
940	1213.15	0.3166	1071.3	2.0368	0.2849	1071.2	2.0251	0.2277	1070.9	2.0002	0.1897	1070.6	1.9799
950	1223.15	0.3192	1077.1	2.0416	0.2873	1077.0	2.0299	0.2296	1076.7	2.0050	0.1913	1076.4	1.9847
960	1233.15	0.3218	1082.9	2.0464	0.2897	1082.8	2.0347	0.2315	1082.5	2.0098	0.1929	1082.2	1.9895
980	1253.15	0.3270	1094.5	2.0558	0.2944	1094.4	2.0441	0.2353	1094.1	2.0192	0.1961	1093.8	1.9989
1000	1273.15	0.3322	1106.1	2.0651	0.2990	1106.0	2.0533	0.2391	1105.7	2.0285	0.1922	1105.4	2.0082

Pressure p		35 at			40 at			45 at			50 at		
Temperature		$t_s = 241.42\,°C$ $i'' = 669.5\ kcal/kg$ $v'' = 0.05819\ m^3/kg$ $s'' = 1.4647\ kcal/kg\ K$			$t_s = 249.18\,°C$ $i'' = 669.0\ kcal/kg$ $v'' = 0.05077\ m^3/kg$ $s'' = 1.4517\ kcal/kg\ K$			$t_s = 256.23\,°C$ $i'' = 668.4\ kcal/kg$ $v'' = 0.04495\ m^3/kg$ $s'' = 1.4397\ kcal/kg\ K$			$t_s = 262.70\,°C$ $i'' = 667.5\ kcal/kg$ $v'' = 0.04026\ m^3/kg$ $s'' = 1.4288\ kcal/kg\ K$		
t	T	v	i	s	v	i	s	v	i	s	v	i	s
°C	K	$\dfrac{m^3}{kg}$	$\dfrac{kcal}{kg}$	$\dfrac{kcal}{kg\ K}$	$\dfrac{m^3}{kg}$	$\dfrac{kcal}{kg}$	$\dfrac{kcal}{kg\ K}$	$\dfrac{m^3}{kg}$	$\dfrac{kcal}{kg}$	$\dfrac{kcal}{kg\ K}$	$\dfrac{m^3}{kg}$	$\dfrac{kcal}{kg}$	$\dfrac{kcal}{kg\ K}$
0	273.15	0.0009984	0.8	0.0000	0.0009982	1.0	0.0001	0.0009979	1.1	0.0001	0.0009977	1.2	0.0001
20	293.15	0.0010003	20.8	0.0706	0.0010001	20.9	0.0705	0.0009998	21.0	0.0705	0.0009997	21.1	0.0705
40	313.15	0.0010064	40.7	0.1362	0.0010062	40.8	0.1362	0.0010059	40.9	0.1361	0.0010057	41.0	0.1361
50	323.15	0.0010106	50.7	0.1676	0.0010103	50.8	0.1675	0.0010101	50.9	0.1674	0.0010099	51.0	0.1674
60	333.15	0.0010155	60.6	0.1979	0.0010152	60.7	0.1978	0.0010150	60.8	0.1978	0.0010148	60.9	0.1977
80	353.15	0.0010273	80.6	0.2561	0.0010271	80.7	0.2560	0.0010269	80.8	0.2559	0.0010266	80.9	0.2558
100	373.15	0.0010417	100.6	0.3114	0.0010414	100.7	0.3113	0.0010412	100.8	0.3112	0.0010409	100.9	0.3111
120	393.15	0.0010585	120.8	0.3640	0.0010582	120.9	0.3639	0.0010580	121.0	0.3638	0.0010577	121.1	0.3637
140	413.15	0.0010779	141.1	0.4143	0.0010776	141.2	0.4142	0.0010773	141.3	0.4141	0.0010770	141.4	0.4140
150	423.15	0.0010886	151.4	0.4387	0.0010883	151.4	0.4386	0.0010880	151.5	0.4384	0.0010877	151.6	0.4383
160	433.15	0.0011001	161.6	0.4627	0.0010997	161.7	0.4625	0.0010994	161.8	0.4624	0.0010990	161.8	0.4622
180	453.15	0.0011255	182.5	0.5096	0.0011251	182.5	0.5094	0.0011247	182.6	0.5092	0.0011243	182.6	0.5090
200	473.15	0.0011547	203.7	0.5553	0.0011542	203.7	0.5551	0.0011537	203.8	0.5549	0.0011532	203.8	0.5547
220	493.15	0.0011886	225.4	0.6004	0.0011880	225.4	0.6001	0.0011874	225.4	0.5999	0.0011868	225.5	0.5996
240	513.15	0.0012290	247.8	0.6448	0.0012282	247.8	0.6445	0.0012274	247.8	0.6442	0.0012266	247.8	0.6439
250	523.15	0.06019	676.3	1.4779	0.05096	669.7	1.4530	0.0012504	259.3	0.6664	0.0012495	259.3	0.6661
260	533.15	0.06234	683.8	1.4919	0.05302	678.0	1.4684	0.04567	671.6	1.4458	0.0012751	271.1	0.6885
280	553.15	0.06630	697.7	1.5173	0.05679	693.0	1.4957	0.04935	687.9	1.4757	0.04330	682.7	1.4564
300	573.15	0.06998	710.5	1.5404	0.06022	706.6	1.5201	0.05260	702.5	1.5017	0.04646	698.4	1.4842
320	593.15	0.07344	722.8	1.5615	0.06338	719.6	1.5423	0.05556	716.0	1.5250	0.04927	712.5	1.5087
340	613.15	0.07675	734.8	1.5812	0.06636	731.9	1.5631	0.05830	728.7	1.5463	0.05186	725.8	1.5308
350	623.15	0.07837	740.7	1.5907	0.06782	737.9	1.5729	0.05963	735.0	1.5563	0.05310	732.2	1.5412
360	633.15	0.07997	746.5	1.6000	0.06927	743.8	1.5823	0.06095	741.2	1.5660	0.05432	738.5	1.5512
380	653.15	0.08314	758.0	1.6180	0.07212	755.6	1.6003	0.06356	753.3	1.5848	0.05671	750.8	1.5703
400	673.15	0.08624	769.3	1.6350	0.07490	767.2	1.6177	0.06610	765.1	1.6026	0.05904	762.9	1.5885
420	693.15	0.08927	780.4	1.6513	0.07763	778.5	1.6345	0.06856	776.6	1.6194	0.06130	774.7	1.6057
440	713.15	0.09228	791.5	1.6671	0.08030	789.8	1.6505	0.07098	788.1	1.6357	0.06352	786.3	1.6222
450	723.15	0.09377	797.0	1.6747	0.08162	795.4	1.6583	0.07218	793.7	1.6436	0.06462	792.0	1.6302
460	733.15	0.09525	802.5	1.6822	0.08293	801.0	1.6659	0.07337	799.3	1.6513	0.06571	797.7	1.6380
480	753.15	0.09820	813.5	1.6969	0.08555	812.0	1.6808	0.07572	810.5	1.6662	0.06786	809.1	1.6532
500	773.15	0.1011	824.4	1.7112	0.08816	823.0	1.6953	0.07804	821.7	1.6810	0.06999	820.3	1.6681
520	793.15	0.1039	835.3	1.7251	0.09074	834.0	1.7094	0.08036	832.8	1.6952	0.07208	831.5	1.6823
540	813.15	0.1069	846.1	1.7386	0.09330	845.0	1.7230	0.08266	843.8	1.7089	0.07416	842.7	1.6961
550	823.15	0.1083	851.5	1.7452	0.09457	850.4	1.7296	0.08380	849.3	1.7156	0.07519	848.2	1.7029
560	833.15	0.1098	857.0	1.7517	0.09584	855.9	1.7361	0.08494	854.8	1.7222	0.07622	853.8	1.7096
580	853.15	0.1127	867.9	1.7646	0.09835	866.9	1.7490	0.08719	865.8	1.7352	0.07827	864.8	1.7227
600	873.15	0.1154	878.7	1.7772	0.1008	877.7	1.7617	0.08943	876.8	1.7480	0.08029	875.8	1.7356
620	893.15	0.1182	889.7	1.7896	0.1034	888.7	1.7742	0.09163	887.8	1.7605	0.08228	886.9	1.7482
640	913.15	0.1210	900.5	1.8017	0.1059	899.7	1.7864	0.09383	898.8	1.7728	0.08426	897.9	1.7605
650	923.15	0.1224	905.9	1.8077	0.1071	905.1	1.7924	0.09493	904.3	1.7788	0.08526	903.4	1.7666
660	933.15	0.1238	911.4	1.8136	0.1083	910.6	1.7983	0.09602	909.8	1.7848	0.08625	908.9	1.7726
680	953.15	0.1266	922.4	1.8254	0.1107	921.6	1.8101	0.09820	920.8	1.7966	0.08823	920.1	1.7845
700	973.15	0.1294	933.4	1.8370	0.1131	932.7	1.8217	0.1004	932.0	1.8082	0.09020	931.3	1.7961
720	993.15	0.1322	944.5	1.8484	0.1156	943.9	1.8331	0.1026	943.2	1.8197	0.09216	942.5	1.8076
740	1013.15	0.1350	955.7	1.8596	0.1180	955.1	1.8443	0.1048	954.4	1.8309	0.09412	953.8	1.8188
750	1023.15	0.1364	961.3	1.8651	0.1192	960.7	1.8498	0.1059	960.1	1.8365	0.09510	959.5	1.8244
760	1033.15	0.1378	966.9	1.8706	0.1204	966.3	1.8553	0.1070	965.7	1.8420	0.09608	965.2	1.8299
780	1053.15	0.1406	978.2	1.8814	0.1228	977.6	1.8662	0.1092	977.1	1.8529	0.09803	976.6	1.8408
800	1073.15	0.1433	989.6	1.8920	0.1252	989.0	1.8768	0.1112	988.5	1.8635	0.09998	988.0	1.8515
820	1093.15	0.1461	1001.0	1.9025	0.1277	1000.4	1.8873	0.1134	999.9	1.8740	0.1019	999.4	1.8621
840	1113.15	0.1488	1012.4	1.9192	0.1301	1011.9	1.8977	0.1156	1011.5	1.8844	0.1039	1011.0	1.8725
850	1123.15	0.1502	1018.2	1.9180	0.1313	1017.7	1.9029	0.1166	1017.3	1.8896	0.1048	1016.8	1.8777
860	1133.15	0.1516	1023.9	1.9231	0.1325	1023.5	1.9080	0.1177	1023.1	1.8947	0.1058	1022.6	1.8828
880	1153.15	0.1544	1035.5	1.9332	0.1349	1035.1	1.9182	0.1199	1034.7	1.9049	0.1078	1034.2	1.8930
900	1173.15	0.1570	1047.1	1.9432	0.1373	1046.7	1.9282	0.1220	1046.3	1.9149	0.1097	1045.8	1.9030
920	1193.15	0.1598	1058.6	1.9531	0.1397	1058.2	1.9381	0.1242	1057.9	1.9248	0.1116	1057.4	1.9129
940	1213.15	0.1626	1070.2	1.9627	0.1421	1069.8	1.9478	0.1264	1069.5	1.9345	0.1135	1069.0	1.9227
950	1223.15	0.1640	1076.0	1.9675	0.1433	1075.6	1.9526	0.1275	1075.3	1.9393	0.1145	1074.8	1.9275
960	1233.15	0.1654	1081.8	1.9723	0.1445	1081.4	1.9574	0.1286	1081.1	1.9441	0.1155	1080.6	1.9323
980	1253.15	0.1680	1093.4	1.9817	0.1469	1093.0	1.9668	0.1306	1092.7	1.9536	0.1174	1092.2	1.9418
1000	1273.15	0.1706	1105.0	1.9910	0.1493	1104.6	1.9761	0.1326	1104.4	1.9629	0.1193	1104.0	1.9511

Pressure p		60 at $t_s=274.29\,°C$ $i''=665.4$ kcal/kg $v''=0.03313$ m³/kg $s''=1.4089$ kcal/kg K			70 at $t_s=284.48\,°C$ $i''=662.6$ kcal/kg $v''=0.02798$ m³/kg $s''=1.3911$ kcal/kg K			80 at $t_s=293.62\,°C$ $i''=659.3$ kcal/kg $v''=0.02405$ m³/kg $s''=1.3745$ kcal/kg K			90 at $t_s=301.92\,°C$ $i''=655.7$ kcal/kg $v''=0.02096$ m³/kg $s''=1.3587$ kcal/kg K		
t	T	v	i	s	v	i	s	v	i	s	v	i	s
°C	K	m³/kg	kcal/kg	kcal/kg K	m³/kg	kcal/kg	kcal/kg K	m³/kg	kcal/kg	kcal/kg K	m³/kg	kcal/kg	kcal/kg K
0	273.15	0.0009972	1.4	0.0001	0.0009967	1.7	0.0001	0.0009962	1.92	0.0001	0.0009957	2.16	0.0001
20	293.15	0.0009992	21.3	0.0704	0.0009988	21.6	0.0704	0.0009984	21.77	0.0703	0.0009980	22.00	0.0702
40	313.15	0.0010053	41.2	0.1360	0.0010049	41.4	0.1359	0.0010044	41.66	0.1359	0.0010039	41.87	0.1357
50	323.15	0.0010095	51.2	0.1672	0.0010090	51.4	0.1671	0.0010086	51.61	0.1670	0.0010081	51.82	0.1668
60	333.15	0.0010144	61.1	0.1976	0.0010139	61.3	0.1974	0.0010135	61.57	0.1973	0.0010130	61.77	0.1971
80	353.15	0.0010261	81.1	0.2556	0.0010257	81.2	0.2555	0.0010254	81.52	0.2553	0.0010250	81.70	0.2551
100	373.15	0.0010404	101.1	0.3109	0.0010400	101.2	0.3107	0.0010398	101.48	0.3105	0.0010394	101.66	0.3103
120	393.15	0.0010572	121.2	0.3635	0.0010567	121.4	0.3633	0.0010565	121.54	0.3631	0.0010560	121.71	0.3628
140	413.15	0.0010764	141.5	0.4137	0.0010758	141.7	0.4135	0.0010762	141.84	0.4133	0.0010750	141.99	0.4130
150	423.15	0.0010870	151.7	0.4381	0.0010864	151.9	0.4378	0.0010860	152.08	0.4376	0.0010855	152.23	0.4373
160	433.15	0.0010984	162.0	0.4619	0.0010977	162.1	0.4617	0.0010973	162.37	0.4614	0.0010967	162.52	0.4611
180	453.15	0.0011235	182.8	0.5086	0.0011226	182.9	0.5082	0.0011222	183.14	0.5079	0.0011214	183.28	0.5075
200	473.15	0.0011522	203.9	0.5543	0.0011513	204.0	0.5539	0.0011506	204.21	0.5535	0.0011497	204.32	0.5532
220	493.15	0.0011857	225.5	0.5991	0.0011845	225.6	0.5986	0.0011835	225.71	0.5981	0.0011824	225.80	0.5977
240	513.15	0.0012251	247.8	0.6434	0.0012236	247.8	0.6428	0.0012223	247.89	0.6423	0.0012209	247.94	0.6417
250	523.15	0.0012478	259.3	0.6656	0.0012460	259.3	0.6649	0.0012445	259.30	0.6643	0.0012429	259.31	0.6638
260	533.15	0.0012729	270.9	0.6878	0.0012709	270.9	0.6871	0.0012691	270.9	0.6864	0.0012672	270.9	0.6858
280	553.15	0.03405	671.0	1.4188	0.0013308	295.2	0.7317	0.0013280	295.1	0.7308	0.0013250	295.0	0.7300
300	573.15	0.03711	689.0	1.4512	0.03029	678.7	1.4195	0.02503	667.0	1.3875	0.0014024	321.1	0.7764
320	593.15	0.03976	705.1	1.4788	0.03287	697.1	1.4510	0.02757	688.0	1.4239	0.02336	677.7	1.3968
340	613.15	0.04213	719.6	1.5029	0.03512	713.0	1.4773	0.02976	705.8	1.4534	0.02553	698.0	1.4303
350	623.15	0.04324	726.4	1.5140	0.03615	720.3	1.4893	0.03076	713.9	1.4665	0.02652	706.9	1.4446
360	633.15	0.04432	733.1	1.5246	0.03714	727.4	1.5007	0.03171	721.5	1.4787	0.02745	715.1	1.4578
380	653.15	0.04642	746.1	1.5448	0.03903	741.0	1.5220	0.03348	735.9	1.5012	0.02914	730.4	1.4818
400	673.15	0.04845	758.7	1.5635	0.04084	754.1	1.5417	0.03514	749.5	1.5217	0.03070	744.6	1.5031
420	693.15	0.05042	770.9	1.5813	0.04260	766.8	1.5602	0.03674	762.6	1.5408	0.03218	758.3	1.5230
440	713.15	0.05233	782.7	1.5983	0.04430	779.1	1.5776	0.03828	775.3	1.5589	0.03359	771.4	1.5418
450	723.15	0.05327	788.5	1.6065	0.04513	785.1	1.5860	0.03903	781.5	1.5676	0.03428	777.8	1.5508
460	733.15	0.05420	794.3	1.6146	0.04596	791.1	1.5943	0.03977	787.7	1.5761	0.03496	784.2	1.5596
480	753.15	0.05604	805.9	1.6302	0.04759	803.0	1.6103	0.04122	799.8	1.5925	0.03629	796.7	1.5766
500	773.15	0.05785	817.5	1.6453	0.04918	814.7	1.6257	0.04265	811.8	1.6082	0.03758	809.0	1.5925
520	793.15	0.05962	828.9	1.6598	0.05073	826.3	1.6404	0.04405	823.6	1.6233	0.03884	821.0	1.6078
540	813.15	0.06138	840.2	1.6740	0.05227	837.8	1.6547	0.04542	835.3	1.6378	0.04009	832.8	1.6226
550	823.15	0.06227	845.8	1.6809	0.05304	843.5	1.6617	0.04610	841.1	1.6449	0.04071	838.7	1.6298
560	833.15	0.06315	851.4	1.6877	0.05318	849.2	1.6686	0.04678	846.9	1.6519	0.04133	844.6	1.6369
580	853.15	0.06488	862.6	1.7009	0.05532	860.6	1.6821	0.04813	858.4	1.6656	0.04254	856.2	1.6508
600	873.15	0.06658	873.8	1.7138	0.05678	871.8	1.6951	0.04944	869.8	1.6788	0.04372	867.8	1.6642
620	893.15	0.06827	885.0	1.7265	0.05824	883.2	1.7079	0.05074	881.3	1.6917	0.04488	879.4	1.6773
640	913.15	0.06995	896.2	1.7390	0.05970	894.5	1.7205	0.05202	892.7	1.7044	0.04604	891.0	1.6901
650	923.15	0.07078	901.7	1.7451	0.06043	900.1	1.7267	0.05267	898.4	1.7107	0.04663	896.7	1.6964
660	933.15	0.07161	907.3	1.7512	0.06116	905.8	1.7329	0.05331	904.1	1.7169	0.04721	902.5	1.7027
680	953.15	0.07327	918.5	1.7632	0.06260	917.2	1.7450	0.05459	915.5	1.7291	0.04836	914.1	1.7150
700	973.15	0.07493	929.9	1.7750	0.06404	928.6	1.7569	0.05586	927.1	1.7411	0.04950	925.7	1.7270
720	993.15	0.07659	941.2	1.7866	0.06547	940.0	1.7686	0.05712	938.6	1.7528	0.05064	937.3	1.7388
740	1013.15	0.07823	952.6	1.7979	0.06689	951.4	1.7800	0.05838	950.2	1.7643	0.05177	948.9	1.7504
750	1023.15	0.07905	958.3	1.8035	0.06760	957.2	1.7856	0.05901	956.0	1.7700	0.05233	954.8	1.7561
760	1033.15	0.07987	964.0	1.8091	0.06831	962.9	1.7912	0.05964	961.8	1.7756	0.05289	960.6	1.7617
780	1053.15	0.08151	975.4	1.8200	0.06973	974.4	1.8021	0.06089	973.4	1.7866	0.05401	972.2	1.7728
800	1073.15	0.08315	987.0	1.8307	0.07114	986.0	1.8129	0.06214	985.0	1.7975	0.05513	983.9	1.7837
820	1093.15	0.08479	998.5	1.8413	0.07256	997.6	1.8235	0.06338	996.6	1.8082	0.05624	995.5	1.7945
840	1113.15	0.08643	1010.1	1.8518	0.07396	1009.2	1.8341	0.06463	1008.2	1.8188	0.05735	1007.3	1.8051
850	1123.15	0.08725	1015.9	1.8570	0.07467	1015.0	1.8393	0.06525	1014.1	1.8240	0.05791	1013.2	1.8104
860	1133.15	0.08806	1021.7	1.8622	0.07537	1020.8	1.8445	0.06587	1019.9	1.8292	0.05847	1019.0	1.8156
880	1153.15	0.08969	1033.3	1.8724	0.07677	1032.4	1.8548	0.06710	1031.6	1.8395	0.05957	1030.8	1.8259
900	1173.15	0.09131	1045.0	1.8824	0.07817	1044.2	1.8649	0.06833	1043.4	1.8496	0.06067	1042.6	1.8361
920	1193.15	0.09292	1056.6	1.8923	0.07957	1055.8	1.8749	0.06955	1055.0	1.8596	0.06177	1054.2	1.8461
940	1213.15	0.09453	1068.2	1.9021	0.08095	1067.4	1.8847	0.07077	1066.7	1.8694	0.06285	1066.0	1.8560
950	1223.15	0.09533	1074.0	1.9069	0.08164	1073.2	1.8895	0.07138	1072.6	1.8743	0.06339	1071.9	1.8609
960	1233.15	0.09613	1079.8	1.9117	0.08233	1079.1	1.8943	0.07198	1078.5	1.8791	0.06393	1077.8	1.8657
980	1253.15	0.09773	1091.6	1.9212	0.08371	1090.9	1.9038	0.07319	1090.3	1.8887	0.06501	1089.6	1.8753
1000	1273.15	0.09933	1103.4	1.9306	0.08509	1102.7	1.9132	0.07440	1102.1	1.8981	0.06609	1101.4	1.8848

Table 41-14 Properties of Superheated Steam (H$_2$O) (Gravitational Metric System) (Continued)

Pressure p		100 at			110 at			120 at			130 at		
Temperature		$t_s = 309.53\,°C$ $i'' = 651.7\ kcal/kg$ $v'' = 0.01846\ m^3/kg$ $s'' = 1.3440\ kcal/kg\,K$			$t_s = 316.58\,°C$ $i'' = 647.2\ kcal/kg$ $v'' = 0.01638\ m^3/kg$ $s'' = 1.3294\ kcal/kg\,K$			$t_s = 323.15\,°C$ $i'' = 642.5\ kcal/kg$ $v'' = 0.01463\ m^3/kg$ $s'' = 1.3151\ kcal/kg\,K$			$t_s = 329.30\,°C$ $i'' = 637.2\ kcal/kg$ $v'' = 0.01313\ m^3/kg$ $s'' = 1.3012\ kcal/kg\,K$		
t	T	v	i	s	v	i	s	v	i	s	v	i	s
°C	K	$\dfrac{m^3}{kg}$	$\dfrac{kcal}{kg}$	$\dfrac{kcal}{kg\,K}$	$\dfrac{m^3}{kg}$	$\dfrac{kcal}{kg}$	$\dfrac{kcal}{kg\,K}$	$\dfrac{m^3}{kg}$	$\dfrac{kcal}{kg}$	$\dfrac{kcal}{kg\,K}$	$\dfrac{m^3}{kg}$	$\dfrac{kcal}{kg}$	$\dfrac{kcal}{kg\,K}$
0	273.15	0.0009952	2.39	0.0001	0.0009947	2.62	0.0001	0.0009942	2.85	0.0002	0.0009938	3.09	0.0002
20	293.15	0.0009976	22.22	0.0702	0.0009971	22.44	0.0701	0.0009967	22.67	0.0701	0.0009962	22.89	0.0700
40	313.15	0.0010034	42.08	0.1356	0.0010029	42.29	0.1355	0.0010025	42.50	0.1355	0.0010021	42.72	0.1354
50	323.15	0.0010076	52.02	0.1667	0.0010071	52.23	0.1666	0.0010067	52.43	0.1665	0.0010063	52.64	0.1664
60	333.15	0.0010126	61.97	0.1970	0.0010121	62.17	0.1968	0.0010117	62.37	0.1967	0.0010113	62.57	0.1966
80	353.15	0.0010246	81.89	0.2550	0.0010241	82.08	0.2548	0.0010237	82.27	0.2547	0.0010232	82.46	0.2545
100	373.15	0.0010390	101.84	0.3101	0.0010385	102.02	0.3099	0.0010380	102.20	0.3097	0.0010375	102.38	0.3095
120	393.15	0.0010555	121.88	0.3626	0.0010550	122.05	0.3625	0.0010545	122.23	0.3623	0.0010540	122.39	0.3621
140	413.15	0.0010745	142.15	0.4128	0.0010739	142.31	0.4126	0.0010733	142.47	0.4124	0.0010727	142.63	0.4122
150	423.15	0.0010849	152.38	0.4371	0.0010843	152.53	0.4369	0.0010836	152.68	0.4366	0.0010830	152.84	0.4364
160	433.15	0.0010961	162.67	0.4608	0.0010954	162.81	0.4606	0.0010947	162.95	0.4603	0.0010940	163.10	0.4601
180	453.15	0.0011207	183.41	0.5072	0.0011199	183.54	0.5069	0.0011191	183.66	0.5066	0.0011183	183.79	0.5063
200	473.15	0.0011488	204.43	0.5527	0.0011479	204.53	0.5523	0.0011470	204.64	0.5520	0.0011460	204.75	0.5516
220	493.15	0.0011813	225.89	0.5972	0.0011802	225.97	0.5968	0.0011791	226.06	0.5964	0.0011780	226.15	0.5959
240	513.15	0.0012195	247.99	0.6412	0.0012181	248.03	0.6407	0.0012167	248.07	0.6402	0.0012153	248.11	0.6397
250	523.15	0.0012413	259.32	0.6632	0.0012397	259.33	0.6626	0.0012381	259.34	0.6621	0.0012365	259.35	0.6615
260	533.15	0.0012653	270.9	0.6852	0.0012634	270.9	0.6845	0.0012616	270.8	0.6839	0.0012598	270.8	0.6833
280	553.15	0.0013222	294.9	0.7293	0.0013196	294.7	0.7285	0.0013170	294.6	0.7278	0.0013144	294.5	0.7271
300	573.15	0.0013978	320.7	0.7751	0.0013937	320.4	0.7739	0.0013896	320.1	0.7729	0.0013857	319.8	0.7718
320	593.15	0.01988	666.0	1.3688	0.01689	652.5	1.3388	0.001495	348.8	0.8225	0.001487	348.0	0.8204
340	613.15	0.02210	689.2	1.4071	0.01925	679.6	1.3835	0.01679	669.2	1.3594	0.01457	657.2	1.3338
350	623.15	0.02307	699.0	1.4231	0.02023	690.7	1.4017	0.01780	681.5	1.3798	0.01566	671.2	1.3570
360	633.15	0.02397	708.0	1.4376	0.02112	700.8	1.4179	0.01870	692.7	1.3978	0.01659	683.8	1.3772
380	653.15	0.02560	724.6	1.4632	0.02269	718.5	1.4457	0.02024	711.9	1.4282	0.01816	750.2	1.4109
400	673.15	0.02709	739.8	1.4858	0.02414	734.7	1.4695	0.02166	729.5	1.4537	0.01955	724.0	1.4386
420	693.15	0.02848	754.1	1.5065	0.02545	749.8	1.4912	0.02292	745.5	1.4765	0.02078	740.9	1.4626
440	713.15	0.02981	767.8	1.5260	0.02671	763.9	1.5113	0.02412	759.8	1.4975	0.02193	755.7	1.4842
450	723.15	0.03046	774.4	1.5353	0.02732	770.6	1.5209	0.02470	766.7	1.5074	0.02248	762.7	1.4944
460	733.15	0.03109	780.8	1.5444	0.02792	777.2	1.5302	0.02527	773.5	1.5169	0.02302	769.7	1.5042
480	753.15	0.03232	793.5	1.5617	0.02906	790.2	1.5479	0.02635	786.9	1.5351	0.02405	783.5	1.5229
500	773.15	0.03352	806.1	1.5781	0.03018	803.1	1.5647	0.02740	800.1	1.5522	0.02505	797.1	1.5406
520	793.15	0.03469	818.4	1.5938	0.03127	815.8	1.5807	0.02842	813.0	1.5685	0.02602	810.2	1.5572
540	813.15	0.03584	830.3	1.6088	0.03234	827.8	1.5959	0.02942	825.2	1.5841	0.02695	822.6	1.5730
550	823.15	0.03641	836.2	1.6161	0.03286	833.7	1.6033	0.02991	831.2	1.5916	0.02742	828.7	1.5807
560	833.15	0.03697	842.1	1.6233	0.03338	839.7	1.6107	0.03039	837.3	1.5990	0.02787	834.9	1.5882
580	853.15	0.03807	853.7	1.6374	0.03441	851.5	1.6250	0.03135	849.2	1.6134	0.02877	847.0	1.6027
600	873.15	0.03916	865.3	1.6509	0.03541	863.2	1.6387	0.03229	861.1	1.6273	0.02965	859.0	1.6168
620	893.15	0.04022	877.0	1.6641	0.03638	875.0	1.6521	0.03321	873.1	1.6408	0.03050	871.1	1.6305
640	913.15	0.04128	888.6	1.6771	0.03736	886.8	1.6652	0.03411	885.0	1.6541	0.03135	883.2	1.6438
650	923.15	0.04181	894.4	1.6835	0.03785	892.7	1.6716	0.03457	891.0	1.6606	0.03178	889.2	1.6503
660	933.15	0.04234	900.2	1.6898	0.03834	898.6	1.6779	0.03502	896.9	1.6670	0.03220	895.2	1.6568
680	953.15	0.04338	911.9	1.7021	0.03930	910.4	1.6904	0.03591	908.8	1.6795	0.03303	907.2	1.6695
700	973.15	0.04442	923.5	1.7142	0.04025	922.1	1.7025	0.03679	920.6	1.6917	0.03385	919.2	1.6818
720	993.15	0.04546	935.2	1.7261	0.04120	933.9	1.7144	0.03767	932.4	1.7036	0.03467	931.1	1.6938
740	1013.15	0.04648	946.9	1.7377	0.04214	945.6	1.7260	0.03854	944.2	1.7153	0.03548	942.9	1.7056
750	1023.15	0.04699	952.8	1.7434	0.04261	951.5	1.7317	0.03897	950.1	1.7211	0.03588	948.8	1.7114
760	1033.15	0.04750	958.6	1.7491	0.04308	957.3	1.7374	0.03940	956.1	1.7269	0.03629	954.8	1.7172
780	1053.15	0.04849	970.4	1.7603	0.04402	969.2	1.7488	0.04026	968.0	1.7383	0.03709	966.8	1.7287
800	1073.15	0.04953	982.2	1.7713	0.04495	981.0	1.7599	0.04112	979.9	1.7495	0.03789	978.7	1.7400
820	1093.15	0.05054	994.0	1.7821	0.04587	992.8	1.7708	0.04198	991.8	1.7604	0.03868	990.7	1.7509
840	1113.15	0.05155	1005.8	1.7928	0.04679	1004.7	1.7815	0.04283	1003.6	1.7712	0.03947	1002.6	1.7617
850	1123.15	0.05205	1011.7	1.7981	0.04725	1010.6	1.7868	0.04325	1009.6	1.7765	0.03987	1008.5	1.7670
860	1133.15	0.05255	1017.6	1.8033	0.04771	1016.6	1.7921	0.04367	1015.5	1.7818	0.04026	1014.5	1.7723
880	1153.15	0.05354	1029.4	1.8137	0.04863	1028.4	1.8026	0.04451	1027.4	1.7924	0.04104	1026.4	1.7829
900	1173.15	0.05454	1041.3	1.8239	0.04953	1040.3	1.8129	0.04535	1039.4	1.8027	0.04182	1038.5	1.7933
920	1193.15	0.05553	1053.1	1.8339	0.05043	1052.2	1.8229	0.04618	1051.3	1.8128	0.04259	1050.3	1.8036
940	1213.15	0.05651	1065.0	1.8438	0.05133	1064.1	1.8328	0.04701	1063.2	1.8227	0.04335	1062.3	1.8135
950	1223.15	0.05700	1070.9	1.8487	0.05177	1070.0	1.8377	0.04742	1069.1	1.8276	0.04373	1068.2	1.8184
960	1233.15	0.05749	1076.9	1.8535	0.05222	1076.0	1.8426	0.04783	1075.1	1.8325	0.04411	1074.2	1.8233
980	1253.15	0.05847	1088.7	1.8631	0.05311	1087.8	1.8522	0.04865	1087.0	1.8422	0.04487	1086.1	1.8330
1000	1273.15	0.05944	1100.5	1.8726	0.05400	1099.7	1.8618	0.04947	1098.9	1.8518	0.04563	1098.1	1.8426

Table 41-15 Properties of Superheated Steam (H₂O) (Gravitational Metric System) (Continued)

Pressure p		140 at $t_s=335.09$ °C, $i''=631.7$ kcal/kg, $v''=0.01182$ m³/kg, $s''=1.2873$ kcal/kg K			150 at $t_s=340.56$ °C, $i''=625.6$ kcal/kg, $v''=0.01066$ m³/kg, $s''=1.2728$ kcal/kg K			160 at $t_s=345.74$ °C, $i''=618.9$ kcal/kg, $v''=0.009625$ m³/kg, $s''=1.2580$ kcal/kg K			170 at $t_s=350.66$ °C, $i''=611.5$ kcal/kg, $v''=0.008681$ m³/kg, $s''=1.2422$ kcal/kg K		
t	T	v	i	s	v	i	s	v	i	s	v	i	s
°C	K	m³/kg	kcal/kg	kcal/kg K	m³/kg	kcal/kg	kcal/kg K	m³/kg	kcal/kg	kcal/kg K	m³/kg	kcal/kg	kcal/kg K
0	273.15	0.0009933	3.33	0.0002	0.0009928	3.57	0.0002	0.0009923	3.80	0.0002	0.0009918	4.03	0.0002
20	293.15	0.0009958	23.11	0.0700	0.0009954	23.32	0.0700	0.0009949	23.64	0.0699	0.0009944	23.76	0.0698
40	313.15	0.0010016	42.92	0.1352	0.0010012	43.12	0.1352	0.0010008	43.33	0.1351	0.0010004	43.55	0.1350
50	323.15	0.0010059	52.84	0.1663	0.0010055	53.04	0.1662	0.0010051	53.24	0.1661	0.0010046	53.45	0.1659
60	333.15	0.0010105	62.77	0.1965	0.0010105	62.96	0.1963	0.0010101	63.16	0.1962	0.0010096	63.36	0.1961
80	353.15	0.0010227	82.65	0.2544	0.0010223	82.83	0.2543	0.0010219	83.02	0.2541	0.0010214	83.21	0.2540
100	373.15	0.0010370	102.56	0.3094	0.0010365	102.74	0.3092	0.0010360	102.92	0.3090	0.0010355	103.10	0.3089
120	393.15	0.0010535	122.57	0.3619	0.0010529	122.74	0.3617	0.0010524	122.92	0.3615	0.0010519	123.10	0.3613
140	413.15	0.0010721	142.80	0.4119	0.0010715	142.96	0.4117	0.0010709	143.11	0.4115	0.0010703	143.27	0.4113
150	423.15	0.0010823	152.99	0.4361	0.0010817	153.15	0.4359	0.0010811	153.29	0.4357	0.0010804	153.45	0.4354
160	433.15	0.0010934	163.24	0.4598	0.0010927	163.39	0.4595	0.0010920	163.53	0.4593	0.0010913	163.68	0.4591
180	453.15	0.0011176	183.92	0.5060	0.0011168	184.05	0.5057	0.0011160	184.18	0.5054	0.0011152	184.31	0.5051
200	473.15	0.0011451	204.86	0.5513	0.0011442	204.97	0.5509	0.0011433	205.08	0.5506	0.0011423	205.19	0.5502
220	493.15	0.0011769	226.23	0.5955	0.0011758	226.32	0.5951	0.0011747	226.40	0.5946	0.0011736	226.48	0.5942
240	513.15	0.0012140	248.16	0.6392	0.0012126	248.21	0.6387	0.0012113	248.25	0.6382	0.0012100	248.30	0.6377
250	523.15	0.0012350	259.37	0.6610	0.0012335	259.39	0.6605	0.0012320	259.41	0.6599	0.0012306	259.43	0.6594
260	533.15	0.0012580	270.8	0.6827	0.0012562	270.8	0.6822	0.0012544	270.7	0.6816	0.0012527	270.7	0.6810
280	553.15	0.0013118	294.5	0.7263	0.0013093	294.4	0.7256	0.0013069	294.4	0.7250	0.0013045	294.3	0.7242
300	573.15	0.0013819	319.5	0.7709	0.0013782	319.3	0.7699	0.0013746	319.1	0.7690	0.0013711	318.9	0.7680
320	593.15	0.001481	347.5	0.8189	0.001474	346.9	0.8173	0.001468	346.4	0.8159	0.001462	346.0	0.8145
340	613.15	0.01253	642.8	1.3055	0.001639	380.8	0.8737	0.001621	379.4	0.8708	0.001608	378.0	0.8678
350	623.15	0.01374	660.3	1.3338	0.01200	647.6	1.3083	0.01032	630.0	1.2780	0.001742	399.4	0.9022
360	633.15	0.01471	674.4	1.3563	0.01307	664.0	1.3348	0.01154	651.8	1.3112	0.01006	637.8	1.2858
380	653.15	0.01634	698.0	1.3935	0.01474	690.3	1.3759	0.01331	681.9	1.3581	0.01202	672.8	1.3398
400	673.15	0.01772	718.2	1.4234	0.01613	712.1	1.4083	0.01471	705.6	1.3937	0.01345	698.8	1.3794
420	693.15	0.01893	736.1	1.4490	0.01732	731.2	1.4364	0.01590	725.8	1.4234	0.01464	720.1	1.4102
440	713.15	0.02004	751.5	1.4714	0.01840	747.1	1.4589	0.01696	742.6	1.4470	0.01569	738.0	1.4352
450	723.15	0.02057	758.8	1.4820	0.01891	754.7	1.4698	0.01746	750.5	1.4582	0.01618	746.2	1.4470
460	733.15	0.02109	765.9	1.4921	0.01941	762.1	1.4803	0.01794	758.2	1.4691	0.01665	754.1	1.4581
480	753.15	0.02208	780.1	1.5112	0.02037	776.7	1.5003	0.01887	773.2	1.4895	0.01754	769.6	1.4792
500	773.15	0.02303	794.1	1.5292	0.02127	791.0	1.5188	0.01974	787.8	1.5086	0.01839	784.6	1.4987
520	793.15	0.02394	807.4	1.5463	0.02215	804.5	1.5361	0.02058	801.6	1.5262	0.01919	798.6	1.5168
540	813.15	0.02484	820.0	1.5625	0.02300	817.4	1.5525	0.02139	814.6	1.5428	0.01998	811.9	1.5338
550	823.15	0.02527	826.2	1.5702	0.02341	823.7	1.5604	0.02179	821.1	1.5508	0.02035	818.5	1.5420
560	833.15	0.02570	832.4	1.5777	0.02382	830.0	1.5681	0.02218	827.5	1.5587	0.02073	825.0	1.5500
580	853.15	0.02656	844.7	1.5924	0.02464	842.4	1.5831	0.02296	840.1	1.5739	0.02148	837.8	1.5654
600	873.15	0.02739	856.9	1.6067	0.02543	854.8	1.5976	0.02371	852.7	1.5886	0.02220	850.5	1.5802
620	893.15	0.02821	869.1	1.6208	0.02621	867.2	1.6117	0.02445	865.2	1.6029	0.02290	863.2	1.5947
640	913.15	0.02901	881.3	1.6343	0.02697	879.5	1.6253	0.02517	877.7	1.6167	0.02359	875.8	1.6086
650	923.15	0.02940	887.4	1.6409	0.02733	885.6	1.6319	0.02552	883.9	1.6234	0.02393	882.1	1.6154
660	933.15	0.02979	893.4	1.6474	0.02770	891.7	1.6385	0.02587	890.1	1.6301	0.02426	888.4	1.6222
680	953.15	0.03057	905.6	1.6601	0.02844	904.0	1.6514	0.02656	902.4	1.6432	0.02491	900.8	1.6354
700	973.15	0.03134	917.7	1.6726	0.02916	916.2	1.6639	0.02725	914.7	1.6557	0.02557	913.2	1.6480
720	993.15	0.03210	929.7	1.6847	0.02988	928.3	1.6762	0.02793	926.8	1.6680	0.02622	925.4	1.6604
740	1013.15	0.03287	941.6	1.6966	0.03059	940.3	1.6882	0.02861	938.9	1.6801	0.02687	937.6	1.6726
750	1023.15	0.03324	947.6	1.7025	0.03095	946.3	1.6941	0.02895	945.0	1.6861	0.02719	943.7	1.6786
760	1033.16	0.03362	953.5	1.7084	0.03131	952.3	1.7000	0.02929	951.0	1.6920	0.02750	949.7	1.6845
780	1053.15	0.03438	965.6	1.7199	0.03202	964.4	1.7116	0.02995	963.2	1.7037	0.02814	962.0	1.6962
800	1073.15	0.03512	977.6	1.7312	0.03272	976.4	1.7229	0.03062	975.2	1.7151	0.02876	974.1	1.7077
820	1093.15	0.03586	989.5	1.7422	0.03342	988.5	1.7340	0.03128	987.4	1.7263	0.02939	986.3	1.7190
840	1113.15	0.03660	1001.5	1.7530	0.03412	1000.5	1.7448	0.03194	999.4	1.7372	0.03001	998.3	1.7299
850	1123.15	0.03697	1007.5	1.7583	0.03446	1006.5	1.7502	0.03226	1005.5	1.7426	0.03032	1004.4	1.7353
860	1133.15	0.03734	1013.5	1.7636	0.03481	1012.5	1.7555	0.03259	1011.5	1.7479	0.03063	1010.5	1.7407
880	1153.15	0.03807	1025.5	1.7742	0.03549	1024.5	1.7661	0.03323	1023.6	1.7585	0.03124	1022.7	1.7513
900	1173.15	0.03879	1037.5	1.7846	0.03617	1036.5	1.7765	0.03387	1035.6	1.7689	0.03184	1034.7	1.7617
920	1193.15	0.03951	1049.4	1.7949	0.03684	1048.5	1.7867	0.03451	1047.6	1.7792	0.03244	1046.7	1.7719
940	1213.15	0.04023	1061.4	1.8049	0.03751	1060.5	1.7967	0.03514	1059.7	1.7892	0.03304	1058.8	1.7820
950	1223.15	0.04058	1067.3	1.8098	0.03784	1066.5	1.8017	0.03545	1065.7	1.7941	0.03334	1064.9	1.7870
960	1233.15	0.04094	1073.3	1.8147	0.03818	1072.5	1.8066	0.03576	1071.7	1.7990	0.03364	1070.9	1.7919
980	1253.15	0.04164	1085.3	1.8244	0.03884	1084.5	1.8163	0.03638	1083.7	1.8087	0.03423	1082.9	1.8016
1000	1273.15	0.04234	1097.3	1.8340	0.03950	1096.5	1.8259	0.03701	1095.8	1.8183	0.03481	1095.1	1.8112

Table 41-16 Properties of Superheated Steam (H$_2$O) (Gravitational Metric System) (*Continued*)

Pressure p		180 at			190 at			200 at			220 at		
Temperature		$t_s = 355.35\ °C$ $i'' = 602.8$ kcal/kg $v'' = 0.007803$ m³/kg $s'' = 1.2258$ kcal/kg K			$t_s = 359.82\ °C$ $i'' = 593.0$ kcal/kg $v'' = 0.00697$ m³/kg $s'' = 1.2080$ kcal/kg K			$t_s = 364.08\ °C$ $i'' = 581.4$ kcal/kg $v'' = 0.00618$ m³/kg $s'' = 1.867$ kcal/kg K			$t_s = 372.1\ °C$ $i'' = 542.3$ kcal/kg $v'' = 0.00436$ m³/kg $s'' = 1.1220$ kcal/kg K		
t	T	v	i	s	v	i	s	v	i	s	v	i	s
°C	K	$\dfrac{m^3}{kg}$	$\dfrac{kcal}{kg}$	$\dfrac{kcal}{kg\ K}$	$\dfrac{m^3}{kg}$	$\dfrac{kcal}{kg}$	$\dfrac{kcal}{kg\ K}$	$\dfrac{m^3}{kg}$	$\dfrac{kcal}{kg}$	$\dfrac{kcal}{kg\ K}$	$\dfrac{m^3}{kg}$	$\dfrac{kcal}{kg}$	$\dfrac{kcal}{kg\ K}$
0	273.15	0.0009914	4.27	0.0002	0.0009909	4.51	0.0002	0.0009905	4.74	0.0003	0.0009895	5.20	0.0003
20	293.15	0.0009940	23.98	0.0698	0.0009935	24.21	0.0698	0.0009931	24.43	0.0697	0.0009922	24.87	0.0696
40	313.15	0.0010000	43.76	0.1350	0.0009995	43.97	0.1348	0.0009991	44.18	0.1347	0.0009983	44.59	0.1345
50	323.15	0.0010042	53.66	0.1659	0.0010038	53.86	0.1657	0.0010034	54.06	0.1657	0.0010026	54.47	0.1654
60	333.15	0.0010092	63.56	0.1959	0.0010088	63.76	0.1958	0.0010084	63.95	0.1956	0.0010075	64.35	0.1953
80	353.15	0.0010210	83.40	0.2538	0.0010205	83.59	0.2537	0.0010201	83.78	0.2534	0.0010192	84.16	0.2531
100	373.15	0.0010350	103.28	0.3087	0.0010345	103.46	0.3086	0.0010341	103.64	0.3084	0.0010331	104.00	0.3080
120	393.15	0.0010513	123.27	0.3611	0.0010508	123.44	0.3609	0.0010503	123.60	0.3607	0.0010492	123.94	0.3602
140	413.15	0.0010697	143.44	0.4110	0.0010692	143.59	0.4108	0.0010686	143.75	0.4106	0.0010674	144.07	0.4101
150	423.15	0.0010798	153.61	0.4352	0.0010792	153.75	0.4349	0.0010786	153.91	0.4347	0.0010774	154.22	0.4342
160	433.15	0.0010907	163.83	0.4588	0.0010900	163.97	0.4585	0.0010894	164.13	0.4584	0.0010880	164.42	0.4579
180	453.15	0.0011144	184.44	0.5048	0.0011137	184.57	0.5046	0.0011129	184.70	0.5045	0.0011114	184.97	0.5040
200	473.15	0.0011414	205.30	0.5499	0.0011405	205.41	0.5495	0.0011396	205.52	0.5493	0.0011378	205.75	0.5487
220	493.15	0.0011725	226.57	0.5938	0.0011714	226.65	0.5934	0.0011704	226.74	0.5932	0.0011683	226.91	0.5924
240	513.15	0.0012087	248.35	0.6372	0.0012074	248.40	0.6367	0.0012061	248.46	0.6364	0.0012035	248.57	0.6357
250	523.15	0.0012291	259.45	0.6588	0.0012276	259.48	0.6583	0.0012261	259.52	0.6578	0.0012232	259.60	0.6570
260	533.15	0.0012510	270.7	0.6804	0.0012493	270.7	0.6798	0.0012476	270.7	0.6793	0.0012444	270.7	0.6783
280	533.15	0.0013021	294.2	0.7236	0.0012998	294.1	0.7229	0.0012976	294.0	0.7222	0.0012935	293.9	0.7209
300	573.15	0.0013677	318.7	0.7673	0.0013644	318.5	0.7663	0.0013611	318.4	0.7656	0.0013549	318.0	0.7640
320	593.15	0.001457	345.5	0.8133	0.001452	345.1	0.8120	0.001446	344.7	0.8110	0.001446	344.2	0.8088
340	613.15	0.001596	376.9	0.8654	0.001584	375.9	0.8632	0.001573	375.0	0.8612	0.001554	373.5	0.8578
350	623.15	0.001713	396.5	0.8973	0.001690	394.6	0.8936	0.001671	393.1	0.8905	0.001641	390.6	0.8857
360	633.15	0.00862	620.3	1.2543	0.00702	594.5	1.2097	0.001841	416.6	0.9280	0.001768	411.0	0.9183
380	653.15	0.01083	662.8	1.3032	0.00972	651.9	1.2997	0.00868	639.6	1.2770	0.00660	607.8	1.2232
400	673.15	0.01234	691.6	1.3641	0.01127	684.0	1.3485	0.01033	676.0	1.3327	0.00862	657.8	1.2985
420	693.15	0.01351	714.3	1.3970	0.01248	708.4	1.3840	0.01155	702.2	1.3710	0.00991	688.9	1.3448
440	713.15	0.01454	733.2	1.4236	0.01351	728.3	1.4122	0.01259	723.2	1.4008	0.01096	712.6	1.3786
450	723.15	0.01503	741.7	1.4359	0.01399	737.3	1.4248	0.01306	732.6	1.4140	0.01143	722.9	1.3931
460	733.15	0.01549	750.0	1.4475	0.01445	745.9	1.4368	0.01351	741.6	1.4266	0.01188	732.5	1.4064
480	753.15	0.01636	766.0	1.4691	0.01530	762.4	1.4594	0.01436	758.7	1.4498	0.01270	750.6	1.4313
500	773.15	0.01718	781.3	1.4891	0.01610	778.1	1.4799	0.01513	774.8	1.4709	0.01345	767.6	1.4536
520	793.15	0.01796	795.7	1.5075	0.01686	792.7	1.4988	0.01586	789.7	1.4903	0.01414	783.4	1.4738
540	813.15	0.01871	809.2	1.5249	0.01759	806.5	1.5166	0.01657	803.8	1.5084	0.01482	798.2	1.4930
550	823.15	0.01909	815.9	1.5333	0.01795	813.3	1.5251	0.01692	810.7	1.5171	0.01515	805.4	1.5022
560	833.15	0.01945	822.5	1.5415	0.01831	820.0	1.5334	0.01726	817.5	1.5256	0.01547	812.4	1.5110
580	853.15	0.02016	835.5	1.5572	0.01900	833.2	1.5492	0.01793	830.8	1.5419	0.01608	826.0	1.5277
600	873.15	0.02086	848.3	0.5723	0.01966	846.3	1.5646	0.01856	844.1	1.5573	0.01667	839.7	1.5434
620	893.15	0.02153	861.2	1.5868	0.02030	859.2	1.5793	0.01917	857.2	1.5720	0.01725	853.2	1.5587
640	913.15	0.02218	874.0	1.6009	0.02092	872.1	1.5934	0.01978	870.2	1.5863	0.01782	866.5	1.5735
650	923.15	0.02250	880.3	1.6077	0.02123	878.5	1.6003	0.02009	876.7	1.5933	0.01811	873.1	1.5806
660	933.15	0.02282	886.6	1.6145	0.02154	884.9	1.6071	0.02039	883.1	1.6001	0.01839	879.7	1.5875
680	953.15	0.02345	899.2	1.6279	0.02214	897.6	1.6206	0.02097	896.0	1.6136	0.01893	892.8	1.6011
700	973.15	0.02408	911.7	1.6406	0.02274	910.2	1.6335	0.02154	908.7	1.6267	0.01946	905.7	1.6140
720	993.15	0.02470	924.0	1.6531	0.02334	922.6	1.6462	0.02211	921.2	1.6394	0.01999	918.4	1.6268
740	1013.15	0.02532	936.3	1.6654	0.02392	934.9	1.6585	0.02267	933.6	1.6519	0.02051	931.0	1.6396
750	1023.15	0.02562	942.4	1.6714	0.02422	941.1	1.6645	0.02295	939.8	1.6580	0.02076	937.2	1.6458
760	1033.15	0.02592	948.5	1.6773	0.02451	947.2	1.6705	0.02323	946.0	1.6640	0.02102	943.4	1.6519
780	1053.15	0.02652	960.8	1.6892	0.02508	959.6	1.6824	0.02378	958.4	1.6759	0.02153	956.0	1.6639
800	1073.15	0.02712	973.0	1.7007	0.02565	971.9	1.6940	0.02432	970.7	1.6876	0.02204	968.4	1.6757
820	1093.15	0.02772	985.2	1.7120	0.02622	984.1	1.7053	0.02486	983.0	1.6990	0.02254	980.9	1.6874
840	1113.15	0.02830	997.3	1.7230	0.02678	996.3	1.7164	0.02540	995.3	1.7101	0.02303	993.3	1.6985
850	1123.15	0.02860	1003.4	1.7284	0.02706	1002.4	1.7129	0.02567	1001.4	1.7156	0.02327	999.4	1.7040
860	1133.15	0.02889	1009.5	1.7338	0.02734	1008.5	1.7273	0.02593	1007.5	1.7210	0.02352	1005.6	1.7094
880	1153.15	0.02947	1021.7	1.7445	0.02789	1020.8	1.7380	0.02646	1019.8	1.7317	0.02400	1018.0	1.7201
900	1173.15	0.03004	1033.8	1.7549	0.02843	1032.9	1.7484	0.02698	1032.0	1.7422	0.02448	1030.3	1.7307
920	1193.15	0.03061	1045.9	1.7651	0.02897	1045.0	1.7586	0.02750	1044.2	1.7524	0.02496	1042.5	1.7412
940	1213.15	0.03118	1058.0	1.7753	0.02951	1057.2	1.7688	0.02801	1056.4	1.7626	0.02543	1054.8	1.7514
950	1223.15	0.03146	1064.1	1.7803	0.02978	1063.3	1.7738	0.02827	1062.5	1.7676	0.02566	1060.9	1.7564
960	1233.15	0.03174	1070.1	1.7852	0.03005	1069.4	1.7787	0.02852	1068.6	1.7726	0.02590	1067.0	1.7614
980	1253.15	0.03230	1082.2	1.7949	0.03058	1081.5	1.7885	0.02903	1080.8	1.7824	0.02636	1079.3	1.7712
1000	1273.15	0.03286	1094.4	1.8045	0.03111	1093.7	1.7982	0.02953	1093.0	1.7922	0.02682	1091.6	1.7810

Table 41-17 Properties of Superheated Steam (H₂O) (Gravitational Metric System) (Continued)

Pressure p		230 at			240 at			250 at			260 at		
t	T	v	i	s	v	i	s	v	i	s	v	i	s
°C	K	$\frac{m^3}{kg}$	$\frac{kcal}{kg}$	$\frac{kcal}{kg\ K}$	$\frac{m^3}{kg}$	$\frac{kcal}{kg}$	$\frac{kcal}{kg\ K}$	$\frac{m^3}{kg}$	$\frac{kcal}{kg}$	$\frac{kcal}{kg\ K}$	$\frac{m^3}{kg}$	$\frac{kcal}{kg}$	$\frac{kcal}{kg\ K}$
0	273.15	0.0009891	5.44	0.0003	0.0009887	5.67	0.0003	0.0009882	5.90	0.0003	0.0009878	6.13	0.0003
20	293.15	0.0009918	25.09	0.0696	0.0009914	25.31	0.0695	0.0009910	25.53	0.0695	0.0009905	25.75	0.0695
40	313.15	0.0009979	44.80	0.1344	0.0009975	45.00	0.1344	0.0009971	45.21	0.1343	0.0009967	45.42	0.1342
50	323.15	0.0010022	54.67	0.1653	0.0010018	54.87	0.1652	0.0010014	55.07	0.1651	0.0010010	55.27	0.1650
60	333.15	0.0010071	64.55	0.1952	0.0010067	64.74	0.1951	0.0010063	64.94	0.1950	0.0010059	65.14	0.1949
80	353.15	0.0010188	84.34	0.2529	0.0010184	84.53	0.2528	0.0010180	84.72	0.2527	0.0010175	84.91	0.2525
100	373.15	0.0010327	104.18	0.3078	0.0010322	104.36	0.3076	0.0010318	104.54	0.3075	0.0010313	104.72	0.3073
120	393.15	0.0010487	124.12	0.3600	0.0010482	124.28	0.3598	0.0010477	124.45	0.3596	0.0010472	124.62	0.3594
140	413.15	0.0010668	144.23	0.4099	0.0010663	144.39	0.4097	0.0010657	144.55	0.4095	0.0010651	144.72	0.4092
150	423.15	0.0010768	154.37	0.4340	0.0010762	154.53	0.4338	0.0010756	154.68	0.4336	0.0010750	154.84	0.4333
160	433.15	0.0010874	164.56	0.4576	0.0010867	164.71	0.4574	0.0010861	164.86	0.4572	0.0010855	165.01	0.4569
180	453.15	0.0011106	185.09	0.5038	0.0011099	185.22	0.5035	0.0011091	185.35	0.5033	0.0011084	185.48	0.5030
200	473.15	0.0011370	205.86	0.5484	0.0011361	205.97	0.5481	0.0011353	206.08	0.5478	0.0011345	206.19	0.5475
220	493.15	0.0011673	227.00	0.5921	0.0011663	227.08	0.5917	0.0011653	227.17	0.5914	0.0011643	227.26	0.5910
240	513.15	0.0012023	248.63	0.6352	0.0012011	248.68	0.6348	0.0011998	248.74	0.6344	0.0011986	248.80	0.6339
250	523.15	0.0012218	259.64	0.6565	0.0012205	259.68	0.6560	0.0012191	259.72	0.6556	0.0012177	259.77	0.6551
260	533.15	0.0012428	270.7	0.6777	0.0012412	270.7	0.6772	0.0012396	270.8	0.6768	0.0012381	270.9	0.6763
280	553.15	0.0012914	293.9	0.7203	0.0012893	293.9	0.7196	0.0012873	293.8	0.7189	0.0012853	293.8	0.7183
300	573.15	0.0013518	318.0	0.7633	0.0013489	317.9	0.7625	0.001346	317.8	0.7616	0.0013432	317.7	0.7607
320	593.15	0.001432	343.8	0.8078	0.001427	343.6	0.8066	0.001423	343.4	0.8055	0.001418	343.2	0.8045
340	613.15	0.001546	372.8	0.8562	0.001538	372.2	0.8542	0.001530	371.8	0.8527	0.001523	371.3	0.8513
350	623.15	0.001629	389.5	0.8834	0.001617	388.5	0.8808	0.001606	387.9	0.8787	0.001596	386.9	0.8768
360	633.15	0.001743	409.0	0.9144	0.001722	407.3	0.9108	0.001703	405.8	0.9077	0.001688	404.5	0.9047
380	653.15	0.00543	580.0	1.1780	0.00371	526.0	1.0937	0.00235	470.0	1.0066	0.00216	455.6	0.9837
400	673.15	0.00784	647.5	1.2800	0.00710	636.1	1.2606	0.00637	623.0	1.2386	0.00565	608.1	1.2142
420	693.25	0.00918	681.6	1.3311	0.00851	674.1	1.3168	0.00788	666.1	1.3019	0.00728	657.4	1.2869
440	713.15	0.01024	707.1	1.3675	0.00958	701.2	1.3562	0.00897	695.0	1.3444	0.00840	688.3	1.3324
450	723.15	0.01072	717.7	1.3825	0.01006	712.3	1.3718	0.00946	706.7	1.3611	0.00889	700.9	1.3504
460	733.15	0.01117	727.6	1.3964	0.01051	722.7	1.3865	0.00990	717.6	1.3765	0.00934	712.4	1.3665
480	753.15	0.01198	746.4	1.4221	0.01131	742.1	1.4130	0.01071	737.7	1.4040	0.01014	733.4	1.3952
500	773.15	0.01272	763.9	1.4451	0.01204	760.0	1.4367	0.01142	756.1	1.4286	0.01084	752.2	1.4207
520	793.15	0.01340	780.1	1.4660	0.01271	776.8	1.4584	0.01210	773.4	1.4508	0.01150	769.9	1.4434
540	813.14	0.01405	795.3	1.4856	0.01335	792.5	1.4784	0.01272	789.5	1.4713	0.01212	786.4	1.4642
550	823.15	0.01438	802.6	1.4950	0.01366	799.9	1.4879	0.01301	797.1	1.4809	0.01242	794.2	1.4740
560	833.15	0.01469	809.8	1.5039	0.01397	807.1	1.4969	0.01331	804.4	1.4900	0.01271	801.7	1.4833
580'	853.15	0.01529	823.7	1.5208	0.01456	821.2	1.5140	0.01390	818.7	1.5073	0.01328	816.3	1.5009
600	873.15	0.01586	837.5	0.5367	0.01511	835.2	1.5302	0.01444	833.0	1.5237	0.01381	830.8	1.5175
620	893.15	0.01642	851.2	1.5522	0.01565	849.1	1.5458	0.01497	847.0	1.5396	0.01433	844.9	1.5336
640	913.15	0.01697	864.6	1.5672	0.01619	862.7	1.5610	0.01549	860.8	1.5550	0.01483	858.9	1.5491
650	923.15	0.01725	871.3	1.5744	0.01647	869.5	1.5683	0.01575	867.7	1.5624	0.01508	865.9	1.5566
660	933.15	0.01753	878.0	1.5814	0.01674	876.3	1.5754	0.01600	874.5	1.5696	0.01533	872.7	1.5638
680	953.15	0.01805	891.2	1.5952	0.01724	889.6	1.5892	0.01650	887.9	1.5834	0.01581	886.2	1.5779
700	973.15	0.01856	904.2	1.6081	0.01773	902.7	1.6023	0.01698	901.1	1.5966	0.01627	899.5	1.5913
720	993.15	0.01907	917.0	1.6209	0.01823	915.6	1.6153	0.01746	914.1	1.6098	0.01673	912.5	1.6045
740	1013.15	0.01957	929.6	1.6337	0.01871	928.3	1.6281	0.01792	926.9	1.6227	0.01719	925.5	1.6175
750	1023.15	0.01982	935.9	1.6400	0.01895	934.6	1.6344	0.01815	933.3	1.6290	0.01742	931.9	1.6238
760	1033.15	0.02007	942.2	1.6462	0.01919	940.9	1.6407	0.01838	939.7	1.6353	0.01764	938.3	1.6301
780	1053.15	0.02056	954.8	1.6583	0.01967	953.6	1.6529	0.01884	952.4	1.6476	0.01809	951.1	1.6425
800	1073.15	0.02104	967.2	1.6701	0.02013	966.1	1.6647	0.01930	965.1	1.6595	0.01852	963.8	1.6545
820	1093.15	0.02152	979.8	1.6818	0.02059	978.7	1.6764	0.01974	977.7	1.6712	0.01896	976.5	1.6662
840	1113.15	0.02200	992.2	1.6931	0.02105	991.2	1.6877	0.02019	990.3	1.6825	0.01938	989.2	1.6775
850	1123.15	0.02223	998.4	1.6986	0.02128	997.5	1.6932	0.02040	996.5	1.6880	0.01959	995.4	1.6831
860	1133.15	0.02246	1004.6	1.7040	0.02150	1003.7	1.6987	0.02062	1002.8	1.6935	0.01980	1001.7	1.6887
880	1153.15	0.02293	1017.1	1.7147	0.02196	1016.2	1.7095	0.02106	1015.3	1.7044	0.02022	1014.3	1.6996
900	1173.15	0.02339	1029.4	1.7253	0.02240	1028.5	1.7201	0.02148	1027.7	1.7151	0.02063	1026.8	1.7103
920	1193.15	0.02385	1041.7	1.7358	0.02284	1040.9	1.7306	0.02191	1040.1	1.7256	0.02104	1039.2	1.7208
940	1213.15	0.02430	1054.0	1.7461	0.02327	1053.2	1.7410	0.02232	1052.5	1.7360	0.02144	1051.6	1.7312
950	1223.15	0.02452	1060.1	1.7511	0.02348	1059.3	1.7460	0.02253	1058.6	1.7411	0.02164	1057.8	1.7363
960	1233.15	0.02475	1066.3	1.7561	0.02370	1065.5	1.7510	0.02274	1064.8	1.7461	0.02184	1064.0	1.7413
980	1253.15	0.02519	1078.6	1.7659	0.02413	1077.9	1.7608	0.02315	1077.2	1.7559	0.02225	1076.5	1.7512
1000	1273.15	0.02563	1090.9	1.7757	0.02455	1090.2	1.7706	0.02356	1089.5	1.7657	0.02264	1088.8	1.7610

Pressure p		280 at			300 at			350 at			400 at		
t	T	v	i	s	v	i	s	v	i	s	v	i	s
°C	K	$\frac{m^3}{kg}$	$\frac{kcal}{kg}$	$\frac{kcal}{kg\,K}$	$\frac{m^3}{kg}$	$\frac{kcal}{kg}$	$\frac{kcal}{kg\,K}$	$\frac{m^3}{kg}$	$\frac{kcal}{kg}$	$\frac{kcal}{kg\,K}$	$\frac{m^3}{kg}$	$\frac{kcal}{kg}$	$\frac{kcal}{kg\,K}$
0	273.15	0.0009868	6.59	0.0003	0.0009859	7.05	0.0003	0.0009836	8.19	0.0002	0.0009814	9.32	0.0002
20	293.15	0.0009896	26.19	0.0694	0.0009888	26.62	0.0693	0.0009867	27.70	0.0690	0.0009846	28.77	0.0687
40	313.15	0.0009959	45.83	0.1340	0.0009951	46.25	0.1338	0.0009931	47.27	0.1333	0.0009911	48.29	0.1328
50	323.15	0.0010002	55.67	0.1648	0.0009994	56.08	0.1646	0.0009974	57.08	0.1641	0.0009954	58.07	0.1635
60	333.15	0.0010051	65.53	0.1947	0.0010043	65.93	0.1945	0.0010023	66.90	0.1939	0.0010003	67.87	0.1933
80	353.15	0.0010166	85.28	0.2522	0.0010158	85.66	0.2520	0.0010137	86.59	0.2513	0.0010116	87.52	0.2507
100	373.15	0.0010304	105.08	0.3069	0.0010295	105.44	0.3066	0.0010273	106.34	0.3057	0.0010251	107.23	0.3049
120	393.15	0.0010462	124.98	0.3590	0.0010452	125.32	0.3587	0.0010428	126.17	0.3575	0.0010405	127.02	0.3565
140	413.15	0.0010640	145.04	0.4088	0.0010629	145.36	0.4084	0.0010602	146.15	0.4072	0.0010576	146.94	0.4061
150	423.15	0.0010738	155.15	0.4328	0.0010726	155.45	0.4324	0.0010697	156.21	0.4312	0.0010669	156.97	0.4301
160	433.15	0.0010842	165.30	0.4564	0.0010829	165.59	0.4559	0.0010799	166.31	0.4547	0.0010769	167.04	0.4536
180	453.15	0.0011069	185.75	0.5024	0.0011054	186.01	0.5018	0.0011020	186.64	0.5005	0.0010987	187.31	0.4992
200	473.15	0.0011327	206.42	0.5469	0.0011311	206.64	0.5463	0.0011270	207.21	0.5448	0.0011231	207.80	0.5434
220	493.15	0.0011623	227.45	0.5903	0.0011603	227.63	0.5897	0.0011555	228.10	0.5881	0.0011508	228.61	0.5865
240	513.15	0.0011962	248.93	0.6331	0.0011938	249.06	0.6323	0.0011880	249.39	0.6305	0.0011824	249.78	0.6286
250	523.15	0.0012149	259.86	0.6542	0.0012123	259.95	0.6534	0.0012059	260.20	0.6514	0.0011998	260.50	0.6493
260	533.15	0.0012351	271.0	0.6753	0.0012321	271.0	0.6745	0.0012251	271.2	0.6722	0.0012184	271.5	0.6700
280	553.15	0.0012814	293.8	0.7172	0.0012775	293.8	0.7162	0.0012686	293.8	0.7135	0.0012602	293.9	0.7108
300	573.15	0.0013378	317.5	0.7594	0.0013326	317.4	0.7580	0.0013206	317.1	0.7547	0.0013095	316.8	0.7516
320	593.15	0.001410	342.8	0.8027	0.001403	342.4	0.8009	0.001385	341.5	0.7964	0.001369	340.8	0.7926
340	613.15	0.001510	370.4	0.8484	0.001497	369.7	0.8460	0.001470	367.9	0.8402	0.001445	366.4	0.8350
350	623.15	0.001577	385.6	0.8733	0.001561	384.7	0.8703	0.001525	382.0	0.8631	0.001493	380.0	0.8572
360	633.15	0.001659	402.3	0.8998	0.001642	400.7	0.8960	0.001592	396.8	0.8867	0.001550	394.0	0.8797
380	653.15	0.00199	446.2	0.9676	0.00191	440.2	0.9569	0.001778	430.3	0.9385	0.001700	424.2	0.9267
400	673.15	0.00425	572.1	1.1570	0.00306	525.6	1.0859	0.002174	478.3	1.0105	0.001942	463.4	0.9855
420	693.15	0.00618	638.6	1.2558	0.00520	617.6	1.2218	0.00329	559.2	1.1300	0.002432	518.3	1.0651
440	713.15	0.00737	674.3	1.3081	0.00646	659.5	1.2829	0.00461	620.6	1.2179	0.003346	579.1	1.1530
450	723.15	0.00787	688.9	1.3288	0.00697	676.4	1.3071	0.00515	642.8	1.2504	0.003847	606.9	1.1927
460	733.15	0.00832	701.9	1.3473	0.00743	691.1	1.3280	0.00563	662.0	1.2777	0.004316	629.2	1.2243
480	753.15	0.00912	724.7	1.3785	0.00824	715.8	1.3621	0.00646	692.5	1.3203	0.005132	665.8	1.2757
500	773.15	0.00982	744.3	1.4051	0.00893	736.3	1.3895	0.00714	716.1	1.3519	0.005805	695.4	1.3150
520	793.15	0.01046	762.8	1.4289	0.00955	755.7	1.4146	0.00774	738.1	1.3801	0.006393	720.7	1.3475
540	813.15	0.01105	780.0	1.4505	0.01013	773.9	1.4372	0.00829	758.4	1.4054	0.006919	743.0	1.3756
550	823.15	0.01135	788.2	1.4607	0.01042	782.4	1.4476	0.00856	767.8	1.4170	0.007168	753.3	1.3881
560	833.15	0.01163	796.1	1.4703	0.01070	790.6	1.4575	0.00881	776.8	1.4279	0.007407	763.0	1.4000
580	853.15	0.01217	811.3	1.4883	0.01122	806.4	1.4762	0.00928	793.9	1.4483	0.007862	781.4	1.4222
600	873.15	0.01268	826.1	1.5055	0.01171	821.6	1.4940	0.00973	810.3	1.4674	0.008289	799.1	1.4429
620	893.15	0.01318	840.7	1.5219	0.01218	836.5	1.5108	0.01016	826.1	1.4854	0.008694	815.9	1.4619
640	913.15	0.01366	855.1	1.5377	0.01264	851.3	1.5268	0.01058	841.5	1.5022	0.009079	831.9	1.4796
650	923.15	0.01390	862.2	1.5453	0.01287	858.6	1.5345	0.01080	849.1	1.5103	0.009267	839.6	1.4881
660	933.15	0.01413	869.2	1.5527	0.01309	865.7	1.5420	0.01101	856.5	1.5183	0.009450	847.2	1.4964
680	953.16	0.01458	883.0	1.5670	0.01352	879.6	1.5567	0.01140	871.0	1.5338	0.009806	862.3	1.5126
700	973.15	0.01502	896.4	1.5808	0.01394	893.2	1.5709	0.01178	885.1	1.5484	0.01016	877.2	1.5280
720	933.15	0.01546	909.6	1.5944	0.01436	906.5	1.5846	0.01215	899.0	1.5622	0.01050	891.6	1.5426
740	1013.15	0.01589	922.7	1.6075	0.01476	919.9	1.5979	0.01251	912.8	1.5756	0.01082	905.9	1.5567
750	1023.15	0.01610	929.2	1.6139	0.01496	926.5	1.6044	0.01269	919.7	1.5823	0.01098	913.0	1.5636
760	1033.15	0.01631	935.7	1.6202	0.01516	933.1	1.6108	0.01286	926.5	1.5890	0.01114	919.9	1.5703
780	1053.15	0.01673	948.6	1.6327	0.01555	946.2	1.6234	0.01321	940.0	1.6021	0.01145	933.7	1.5836
800	1073.15	0.01714	961.4	1.6448	0.01594	959.1	1.6356	0.01355	953.3	1.6150	0.01176	947.3	1.5965
820	1093.15	0.01755	974.3	1.6565	0.01633	972.1	1.6474	0.01389	966.5	1.6272	0.01207	960.7	1.6089
840	1113.15	0.01795	987.0	1.6679	0.01670	984.9	1.6590	0.01423	979.6	1.6392	0.01237	974.2	1.6210
850	1123.15	0.01815	993.3	1.6736	0.01689	991.3	1.6647	0.01439	986.1	1.6450	0.01252	980.9	1.6269
860	1133.15	0.01835	999.7	1.6792	0.01708	997.7	1.6703	0.01455	992.6	1.6507	0.01267	987.5	1.6327
880	1153.15	0.01874	1012.4	1.6903	0.01745	1010.5	1.6814	0.01487	1005.6	1.6618	0.01295	1000.7	1.6441
900	1173.15	0.01912	1025.0	1.7010	0.01781	1023.2	1.6923	0.01519	1018.4	1.6728	0.01324	1013.8	1.6552
920	1193.15	0.01950	1037.6	1.7115	0.01817	1035.9	1.7029	0.01551	1031.3	1.6835	0.01352	1026.8	1.6660
940	1213.15	0.01989	1050.0	1.7219	0.01853	1048.4	1.7133	0.01583	1044.1	1.6939	0.01380	1039.8	1.6767
950	1223.15	0.02007	1056.2	1.7271	0.01870	1054.7	1.7185	0.01598	1050.4	1.6991	0.01394	1046.3	1.6820
960	1233.15	0.02026	1062.5	1.7322	0.01888	1061.0	1.7236	0.01613	1056.8	1.7043	0.01408	1052.8	1.6872
980	1253.15	0.02063	1075.0	1.7422	0.01923	1073.5	1.7336	0.01644	1069.5	1.7145	0.01436	1065.6	1.6974
1000	1273.15	0.02100	1087.3	1.7521	0.01958	1085.9	1.7436	0.01674	1082.1	1.7246	0.01462	1078.4	1.7075

Table 41-19 Properties of Superheated Steam (H₂O) (Gravitational Metric System) (*Continued*)

Pressure p		450 at			500 at			550 at			600 at		
t	T	v	i	s	v	i	s	v	i	s	v	i	s
°C	K	$\dfrac{m^3}{kg}$	$\dfrac{kcal}{kg}$	$\dfrac{kcal}{kg\ K}$	$\dfrac{m^3}{kg}$	$\dfrac{kcal}{kg}$	$\dfrac{kcal}{kg\ K}$	$\dfrac{m^3}{kg}$	$\dfrac{kcal}{kg}$	$\dfrac{kcal}{kg\ K}$	$\dfrac{m^3}{kg}$	$\dfrac{kcal}{kg}$	$\dfrac{kcal}{kg\ K}$
0	273.15	0.0009792	10.44	0.0001	0.0009770	11.55	0.0000	0.0009749	12.66	-0.0001	0.0009728	13.76	-0.0003
20	293.15	0.0009826	29.84	0.0683	0.0009806	30.89	0.0679	0.0009786	31.95	0.0676	0.0009766	33.00	0.0673
40	313.15	0.0009891	49.30	0.1323	0.0009872	50.31	0.1318	0.0009853	51.32	0.1314	0.0009834	52.32	0.1309
50	323.15	0.0009934	59.06	0.1630	0.0009915	60.05	0.1624	0.0009896	61.03	0.1619	0.0009877	62.01	0.1613
60	333.15	0.0009983	68.84	0.1928	0.0009963	69.81	0.1922	0.0009944	70.76	0.1916	0.0009925	71.71	0.1910
80	353.15	0.0010095	88.44	0.2499	0.0010075	89.37	0.2493	0.0010055	90.28	0.2485	0.0010036	91.19	0.2478
100	373.15	0.0010229	108.12	0.3040	0.0010208	109.00	0.3032	0.0010187	109.87	0.3024	0.0010166	110.74	0.3016
120	393.15	0.0010381	127.89	0.3556	0.0010358	128.69	0.3546	0.0010335	129.53	0.3538	0.0010312	130.36	0.3529
140	413.15	0.0010550	147.76	0.4050	0.0010525	148.53	0.4039	0.0010500	149.33	0.4029	0.0010475	150.11	0.4019
150	423.15	0.0010642	157.74	0.4289	0.0010615	158.51	0.4278	0.0010589	159.28	0.4267	0.0010563	160.04	0.4256
160	433.15	0.0010740	167.75	0.4524	0.0010711	168.52	0.4512	0.0010683	169.27	0.4500	0.0010656	170.00	0.4489
180	453.15	0.0010954	187.95	0.4979	0.0010921	188.67	0.4965	0.0010889	189.35	0.4953	0.0010857	190.03	0.4941
200	473.15	0.0011193	208.40	0.5419	0.0011156	209.00	0.5404	0.0011119	209.60	0.5389	0.0011082	210.20	0.5375
220	493.15	0.0011463	229.13	0.5848	0.0011420	229.60	0.5831	0.0011377	230.12	0.5814	0.0011336	230.64	0.5798
240	513.15	0.0011771	250.17	0.6267	0.0011719	250.52	0.6248	0.0011670	250.96	0.6229	0.0011621	251.40	0.6211
250	523.16	0.0011940	260.80	0.6473	0.0011883	261.10	0.6453	0.0011830	261.50	0.6433	0.0011777	261.90	0.6413
260	533.15	0.0012120	271.7	0.6678	0.0012058	271.9	0.6656	0.0012000	272.3	0.6635	0.0011943	272.5	0.6613
280	553.15	0.0012522	293.9	0.7082	0.0012446	294.0	0.7057	0.0012376	294.2	0.7034	0.0012309	294.3	0.7010
300	573.15	0.0012993	316.8	0.7487	0.0012896	316.7	0.7457	0.0012805	316.7	0.7431	0.0012722	316.7	0.7404
320	593.15	0.001356	340.4	0.7889	0.001343	340.1	0.7856	0.001331	339.9	0.7823	0.001320	339.7	0.7797
340	613.15	0.001425	365.3	0.8302	0.001408	364.5	0.8262	0.001392	364.0	0.8222	0.001379	363.5	0.8190
350	623.15	0.001467	378.4	0.8514	0.001446	377.2	0.8466	0.001428	376.4	0.8423	0.001412	375.7	0.8387
360	633.15	0.001515	391.8	0.8732	0.001490	390.1	0.8676	0.001469	389.0	0.8626	0.001449	388.1	0.8585
380	653.15	0.001644	420.5	0.9177	0.001603	417.6	0.9105	0.001567	415.0	0.9040	0.001536	414.0	0.8990
400	673.15	0.001831	455.5	0.9701	0.001754	449.5	0.9585	0.001693	445.0	0.9488	0.001647	441.8	0.9410
420	693.15	0.002125	498.7	1.0332	0.001964	486.9	1.0127	0.001865	478.4	0.9977	0.001793	472.4	0.9857
440	713.15	0.002673	548.1	1.1039	0.002321	528.2	1.0717	0.002122	515.3	1.0498	0.001991	505.8	1.0333
450	723.15	0.003021	574.2	1.1412	0.002556	550.6	1.1031	0.002291	534.9	1.0773	0.002118	523.6	1.0583
460	733.15	0.003396	598.7	1.1755	0.002824	572.9	1.1343	0.002484	554.9	1.1054	0.002267	542.0	1.0842
480	753.15	0.004151	640.0	1.2333	0.003434	615.2	1.1930	0.002949	596.1	1.1625	0.002627	580.8	1.1378
500	773.15	0.004792	674.6	1.2997	0.004020	654.1	1.2458	0.003451	635.6	1.2156	0.003036	619.5	1.1897
520	793.15	0.005350	703.1	1.3162	0.004548	685.8	1.2870	0.003930	668.7	1.2592	0.003458	653.2	1.2339
540	813.15	0.005857	727.6	1.3469	0.005031	712.3	1.3205	0.004378	697.3	1.2952	0.003864	682.8	1.2716
550	823.15	0.006097	738.7	1.3607	0.005258	724.3	1.3354	0.004590	710.3	1.3113	0.004059	696.5	1.2885
560	833.15	0.006326	749.2	1.3739	0.005474	735.8	1.3497	0.004793	722.6	1.3265	0.004248	709.5	1.3044
580	853.15	0.006758	768.9	1.3978	0.005882	757.1	1.3754	0.005181	745.3	1.3538	0.004610	733.6	1.3332
600	873.15	0.007175	787.6	1.4199	0.006263	776.7	1.3986	0.005544	766.1	1.3783	0.004952	755.4	1.3590
620	893.15	0.007535	805.4	1.4402	0.006620	795.3	1.4199	0.005883	785.6	1.4009	0.005275	775.7	1.3827
640	913.15	0.007895	822.4	1.4589	0.006960	813.1	1.4398	0.006204	804.1	1.4217	0.005578	795.0	1.4044
650	923.15	0.008070	830.6	1.4679	0.007124	821.7	1.4492	0.006358	813.0	1.4316	0.005725	804.3	1.4146
660	933.15	0.008240	838.6	1.4766	0.007284	830.1	1.4583	0.006508	821.6	1.4411	0.005868	813.3	1.4245
680	953.15	0.008573	854.3	1.4934	0.007596	846.4	1.4758	0.006800	838.3	1.4592	0.006145	830.7	1.4434
700	973.15	0.008900	869.6	1.5093	0.007895	862.1	1.4924	0.007084	854.6	1.4763	0.006412	847.3	1.4611
720	993.15	0.009210	884.6	1.5245	0.008188	877.7	1.5082	0.007358	870.6	1.4924	0.006670	863.5	1.4777
740	1013.15	0.009513	899.4	1.5390	0.008471	892.8	1.5232	0.007623	886.1	1.5077	0.006921	879.4	1.4935
750	1023.15	0.009662	906.6	1.5461	0.008609	900.2	1.5304	0.007753	893.7	1.5152	0.007042	887.2	1.5011
760	1033.15	0.009808	913.7	1.5531	0.008744	907.5	1.5375	0.007880	901.2	1.5225	0.007162	894.9	1.5086
780	1053.15	0.01010	927.8	1.5666	0.009011	922.0	1.5513	0.008129	916.1	1.5367	0.007395	910.3	1.5231
800	1073.15	0.01038	941.8	1.5797	0.009273	936.3	1.5646	0.008371	930.7	1.5504	0.007623	925.2	1.5372
820	1093.15	0.01066	955.6	1.5924	0.009538	950.4	1.5775	0.008608	945.2	1.5637	0.007847	939.9	1.5507
840	1113.15	0.01093	969.3	1.6045	0.009780	964.4	1.5900	0.008844	959.4	1.5765	0.008066	954.5	1.5638
850	1123.15	0.01106	976.1	1.6105	0.009906	971.3	1.5961	0.008960	966.4	1.5827	0.008174	961.6	1.5702
860	1133.15	0.01120	982.8	1.6164	0.01003	978.2	1.6021	0.009076	973.4	1.5889	0.008281	968.7	1.5765
880	1153.15	0.01146	996.2	1.6281	0.01027	991.8	1.6140	0.009301	987.2	1.6009	0.008495	982.7	1.5887
900	1173.15	0.01172	1009.6	1.6394	0.01051	1005.4	1.6255	0.009524	1001.0	1.6126	0.008704	996.7	1.6005
920	1193.15	0.01197	1022.8	1.6505	0.01075	1018.8	1.6366	0.009745	1014.6	1.6238	0.008911	1010.5	1.6119
940	1213.15	0.01223	1036.0	1.6613	0.01098	1032.2	1.6475	0.009962	1028.2	1.6348	0.009114	1024.3	1.6230
950	1223.15	0.01236	1042.5	1.6666	0.01110	1038.8	1.6529	0.01007	1034.9	1.6402	0.009214	1031.2	1.6285
960	1233.15	0.01248	1049.1	1.6718	0.01122	1045.5	1.6582	0.01018	1041.6	1.6456	0.009315	1038.0	1.6339
980	1253.15	0.01272	1062.0	1.6822	0.01144	1058.6	1.6686	0.01039	1054.9	1.6562	0.009514	1051.4	1.6445
1000	1273.15	0.01297	1074.9	1.6923	0.01166	1071.6	1.6789	0.01059	1068.1	1.6665	0.009710	1064.9	1.6550

Table 41-20 Properties of Superheated Steam (H_2O) (Gravitational Metric System) (*Continued*)

t	T	650 at			700 at			750 at			800 at		
		v	i	s	v	i	s	v	i	s	v	i	s
°C	K	$\frac{m^3}{kg}$	$\frac{kcal}{kg}$	$\frac{kcal}{kg\,K}$	$\frac{m^3}{kg}$	$\frac{kcal}{kg}$	$\frac{kcal}{kg\,K}$	$\frac{m^3}{kg}$	$\frac{kcal}{kg}$	$\frac{kcal}{kg\,K}$	$\frac{m^3}{kg}$	$\frac{kcal}{kg}$	$\frac{kcal}{kg\,K}$
0	273.15	0.0009707	14.9	-0.0003	0.0009687	16.0	-0.0004	0.0009667	17.1	-0.0004	0.0009647	18.1	-0.0004
20	293.15	0.0009746	34.1	0.0671	0.0009728	35.1	0.0668	0.0009709	36.1	0.0665	0.0009690	37.2	0.0662
40	313.15	0.0009815	53.3	0.1305	0.0009797	54.3	0.1300	0.0009779	55.2	0.1295	0.0009761	56.2	0.1291
50	323.15	0.0009858	63.0	0.1608	0.0009840	63.9	0.1603	0.0009822	64.8	0.1598	0.0009804	65.8	0.1593
60	333.15	0.0009906	72.6	0.1904	0.0009888	73.5	0.1898	0.0009870	74.5	0.1892	0.0009852	75.4	0.1887
80	353.15	0.0010017	92.1	0.2470	0.0009997	93.0	0.2463	0.0009978	93.9	0.2456	0.0009960	94.8	0.2450
100	373.15	0.0010146	111.6	0.3008	0.0010125	112.4	0.3001	0.0010105	113.3	0.2993	0.0010086	114.2	0.2985
120	393.15	0.0010290	131.2	0.3519	0.0010268	132.1	0.3512	0.0010248	132.9	0.3502	0.0010227	133.7	0.3494
140	413.15	0.0010451	150.9	0.4008	0.0010427	151.7	0.3999	0.0010404	152.5	0.3988	0.0010382	153.3	0.3979
150	423.15	0.0010538	160.8	0.4245	0.0010513	161.6	0.4235	0.0010489	162.3	0.4224	0.0010465	163.1	0.4213
160	433.15	0.0010630	170.7	0.4477	0.0010604	171.5	0.4466	0.0010578	172.2	0.4454	0.0010552	173.0	0.4442
180	453.15	0.0010827	190.7	0.4927	0.0010799	191.3	0.4914	0.0010770	192.0	0.4901	0.0010741	192.7	0.4888
200	473.15	0.0011048	210.7	0.5360	0.0011015	211.3	0.5346	0.0010982	212.0	0.5332	0.0010951	212.7	0.5318
220	493.15	0.0011296	231.0	0.5782	0.0011256	231.4	0.5766	0.0011220	232.1	0.5750	0.0011184	232.8	0.5735
240	513.15	0.0011575	251.7	0.6193	0.0011529	252.1	0.6175	0.0011486	252.6	0.6157	0.0011445	253.1	0.6140
250	523.15	0.0011728	262.2	0.6394	0.0011679	262.6	0.6375	0.0011633	263.0	0.6356	0.0011587	263.4	0.6338
260	533.15	0.0011890	272.9	0.6593	0.0011838	273.2	0.6573	0.0011788	273.5	0.6552	0.0011738	273.8	0.6534
280	553.15	0.0012245	294.5	0.6987	0.0012184	294.7	0.6964	0.0012123	294.8	0.6941	0.0012064	294.9	0.6921
300	573.15	0.0012646	316.7	0.7378	0.0012571	316.7	0.7353	0.0012496	316.7	0.7328	0.0012425	316.7	0.7304
320	593.15	0.001311	339.5	0.7767	0.001301	339.4	0.7738	0.001294	339.2	0.7710	0.001284	339.0	0.7683
340	613.15	0.001366	363.0	0.8155	0.001355	362.6	0.8121	0.001345	362.2	0.8088	0.001334	361.8	0.8057
350	623.15	0.001398	375.0	0.8349	0.001386	374.4	0.8312	0.001373	373.9	0.8277	0.001361	373.4	0.8243
360	633.15	0.001433	387.3	0.8543	0.001419	386.5	0.8503	0.001403	385.8	0.8465	0.001390	385.1	0.8428
380	653.15	0.001512	412.7	0.8938	0.001493	411.5	0.8890	0.001472	410.3	0.8845	0.001454	409.3	0.8801
400	673.15	0.001609	439.7	0.9345	0.001579	437.8	0.9284	0.001552	436.1	0.9228	0.001529	434.5	0.9174
420	693.15	0.001734	468.6	0.9767	0.001688	465.4	0.9688	0.001650	462.8	0.9619	0.001619	460.5	0.9553
440	713.15	0.001897	499.3	1.0208	0.001829	494.3	1.0103	0.001773	490.6	1.0015	0.001727	487.4	0.9938
450	723.15	0.001999	515.4	1.0436	0.001914	509.4	1.0318	0.001845	505.0	1.0219	0.001789	501.4	1.0134
460	733.15	0.002118	532.3	1.0673	0.002011	525.2	1.0540	0.001926	520.0	1.0430	0.001859	515.7	1.0335
480	753.15	0.002401	568.7	1.1176	0.002241	559.1	1.1010	0.002119	551.7	1.0874	0.002025	545.8	1.0755
500	773.15	0.002734	605.5	1.1671	0.002515	593.8	1.1479	0.002352	584.3	1.1312	0.002224	576.3	1.1169
520	793.15	0.003090	638.4	1.2103	0.002818	625.8	1.1898	0.002614	615.5	1.1725	0.002449	606.4	1.1565
540	813.15	0.003456	668.5	1.2486	0.003141	655.8	1.2282	0.002893	645.2	1.2102	0.002693	635.4	1.1935
550	823.15	0.003638	682.6	1.2662	0.003303	670.2	1.2461	0.003036	659.3	1.2281	0.002819	649.5	1.2113
560	833.15	0.003814	696.2	1.2829	0.003464	684.0	1.2632	0.003181	673.1	1.2453	0.002947	663.3	1.2286
580	853.15	0.004152	721.7	1.3134	0.003775	710.2	1.2949	0.003464	699.6	1.2777	0.003206	690.0	1.2615
600	873.15	0.004471	744.7	1.3407	0.004071	734.4	1.3234	0.003738	724.5	1.3071	0.003459	715.4	1.2916
620	893.15	0.004774	765.8	1.3656	0.004355	756.5	1.3491	0.004002	747.4	1.3337	0.003705	739.0	1.3190
640	913.15	0.005061	785.9	1.3883	0.004626	777.1	1.3726	0.004258	768.6	1.3581	0.003947	760.8	1.3440
650	923.15	0.005199	795.6	1.3990	0.004757	787.0	1.3836	0.004383	778.7	1.3694	0.004066	771.2	1.3556
660	933.15	0.005335	805.0	1.4093	0.004886	796.6	1.3942	0.004506	788.5	1.3803	0.004181	781.2	1.3667
680	953.15	0.005597	823.0	1.4285	0.005135	815.1	1.4142	0.004741	807.5	1.4007	0.004405	800.4	1.3877
700	973.15	0.005851	840.1	1.4466	0.005374	832.9	1.4329	0.004968	825.7	1.4197	0.004620	819.1	1.4072
720	993.15	0.006095	856.7	1.4638	0.005608	850.0	1.4505	0.005190	843.4	1.4377	0.004830	837.3	1.4256
740	1013.15	0.006331	873.0	1.4801	0.005831	866.7	1.4671	0.005401	860.7	1.4547	0.005030	855.0	1.4430
750	1023.15	0.006446	881.0	1.4879	0.005939	874.9	1.4751	0.005503	869.2	1.4630	0.005127	863.7	1.4514
760	1033.15	0.006559	889.0	1.4956	0.006045	883.0	1.4830	0.005604	877.6	1.4711	0.005223	872.2	1.4596
780	1053.15	0.006780	904.6	1.5105	0.006254	899.1	1.4982	0.005802	894.1	1.4867	0.005413	888.9	1.4756
800	1073.15	0.006994	919.9	1.5247	0.006459	914.7	1.5128	0.005997	910.0	1.5016	0.005597	905.2	1.4908
820	1093.15	0.007206	935.0	1.5385	0.006658	930.0	1.5269	0.006187	925.6	1.5160	0.005779	921.0	1.5056
840	1133.15	0.007412	949.8	1.5519	0.006855	945.2	1.5405	0.006373	940.9	1.5298	0.005957	936.6	1.5197
850	1123.15	0.007514	957.0	1.5585	0.006952	952.6	1.5471	0.006465	948.5	1.5366	0.006044	944.3	1.5265
860	1133.15	0.007615	964.2	1.5649	0.007048	959.9	1.5536	0.006557	955.9	1.5432	0.006132	952.0	1.5332
880	1153.15	0.007817	978.6	1.5773	0.007237	974.5	1.5663	0.006738	970.8	1.5560	0.006303	967.0	1.5463
900	1173.16	0.008014	992.8	1.5892	0.007424	989.0	1.5785	0.006915	985.5	1.5684	0.006472	982.0	1.5588
920	1193.15	0.008209	1006.9	1.6008	0.007609	1003.3	1.5903	0.007091	1000.0	1.5805	0.006640	996.7	1.5710
940	1213.15	0.008400	1020.9	1.6120	0.007790	1017.5	1.6016	0.007263	1014.4	1.5920	0.006803	1011.3	1.5827
950	1223.15	0.008495	1027.8	1.6175	0.007880	1024.5	1.6072	0.007348	1021.5	1.5976	0.006884	1018.4	1.5884
960	1233.15	0.008591	1034.7	1.6229	0.007970	1031.4	1.6127	0.007433	1028.5	1.6031	0.006965	1025.5	1.5940
980	1253.15	0.008778	1048.4	1.6337	0.008148	1045.3	1.6235	0.007601	1042.5	1.6140	0.007125	1039.7	1.6049
1000	1273.15	0.008963	1062.0	1.6443	0.008323	1059.1	1.6341	0.007768	1056.4	1.6246	0.007284	1053.8	1.6156

Table 41-21 Properties of Superheated Steam (H₂O) (Gravitational Metric System) (Continued)

Pressure p		850 at			900 at			950 at			1000 at		
t	$\overline{T}$	v	i	s	v	i	s	v	i	s	v	i	s
°C	K	$\dfrac{m^3}{kg}$	$\dfrac{kcal}{kg}$	$\dfrac{kcal}{kg\ K}$	$\dfrac{m^3}{kg}$	$\dfrac{kcal}{kg}$	$\dfrac{kcal}{kg\ K}$	$\dfrac{m^3}{kg}$	$\dfrac{kcal}{kg}$	$\dfrac{kcal}{kg\ K}$	$\dfrac{m^3}{kg}$	$\dfrac{kcal}{kg}$	$\dfrac{kcal}{kg}$
0	273.15	0.0009627	19.2	-0.0004	0.0009608	20.3	-0.0005	0.0009589	21.4	-0.0006	0.0009571	22.4	-0.0007
20	293.15	0.0009671	38.2	0.0659	0.0009653	39.1	0.0657	0.0009636	40.2	0.0656	0.0009618	41.2	0.0655
40	313.15	0.0009743	57.2	0.1287	0.0009725	58.1	0.1284	0.0009708	59.1	0.1281	0.0009691	60.0	0.1278
50	323.15	0.0009786	66.7	0.1589	0.0009769	67.6	0.1585	0.0009752	68.6	0.1581	0.0009735	69.5	0.1577
60	333.15	0.0009834	76.3	0.1881	0.0009817	77.2	0.1877	0.0009800	78.2	0.1872	0.0009783	79.1	0.1867
80	353.15	0.0009942	95.6	0.2443	0.0009924	96.5	0.2436	0.0009906	97.4	0.2431	0.0009889	98.3	0.2424
100	373.15	0.0010067	115.1	0.2977	0.0010048	115.9	0.2970	0.0010029	116.8	0.2962	0.0010010	117.7	0.2955
120	393.15	0.0010207	134.6	0.3485	0.0010187	135.3	0.3476	0.0010166	136.2	0.3468	0.0010146	137.1	0.3461
140	413.15	0.0010360	154.1	0.3969	0.0010338	154.9	0.3959	0.0010317	155.6	0.3950	0.0010295	156.5	0.3943
150	423.15	0.0010442	163.9	0.4203	0.0010419	164.7	0.4193	0.0010397	165.4	0.4184	0.0010374	166.2	0.4174
160	433.15	0.0010528	173.7	0.4432	0.0010504	174.5	0.4421	0.0010480	175.2	0.4412	0.0010457	175.9	0.4401
180	453.15	0.0010714	193.4	0.4877	0.0010687	194.1	0.4864	0.0010661	194.8	0.4854	0.0010653	195.5	0.4842
200	473.15	0.0010920	213.3	0.5305	0.0010890	213.9	0.5292	0.0010860	214.5	0.5280	0.0010831	215.1	0.5267
220	493.15	0.0011149	233.3	0.5720	0.0011114	233.9	0.5705	0.0011080	234.4	0.5689	0.0011047	234.9	0.5675
240	513.15	0.0011405	253.6	0.6123	0.0011365	254.1	0.6106	0.0011326	254.5	0.6088	0.0011288	254.9	0.6072
250	523.15	0.0011543	263.8	0.6320	0.0011500	264.2	0.6302	0.0011459	264.6	0.6284	0.0011418	264.9	0.6266
260	533.15	0.0011689	274.1	0.6514	0.0011643	274.4	0.6495	0.0011599	274.8	0.6477	0.0011555	275.1	0.6457
280	553.15	0.0012008	295.1	0.6898	0.0011953	295.3	0.6876	0.0011903	295.5	0.6855	0.0011854	295.8	0.6832
300	573.15	0.0012361	316.7	0.7279	0.001230	316.7	0.7253	0.0012242	316.8	0.7227	0.0012190	316.9	0.7202
320	593.15	0.001277	338.8	0.7656	0.001270	338.7	0.7625	0.001263	338.5	0.7594	0.001257	338.4	0.7566
340	613.15	0.001324	361.4	0.8026	0.001316	361.1	0.7990	0.001308	360.8	0.7956	0.001300	360.5	0.7925
350	623.15	0.001350	372.9	0.8210	0.001341	372.4	0.8171	0.001332	372.0	0.8136	0.001323	371.7	0.8103
360	633.15	0.001377	384.5	0.8393	0.001367	383.9	0.8352	0.001357	383.4	0.8316	0.001347	382.9	0.8280
380	653.15	0.001437	408.3	0.8758	0.001424	407.4	0.8713	0.001412	406.5	0.8671	0.001398	405.7	0.8631
400	673.15	0.001507	433.0	0.9124	0.001488	431.6	0.9073	0.001471	430.3	0.9026	0.001455	429.2	0.8980
420	693.15	0.001589	458.4	0.9496	0.001564	456.6	0.9438	0.001540	454.8	0.9383	0.001520	453.1	0.9327
440	713.15	0.001688	484.9	0.9873	0.001653	482.4	0.9807	0.001623	479.9	0.9741	0.001596	477.5	0.9673
450	723.15	0.001744	498.5	1.0064	0.001703	495.6	0.9992	0.001669	492.7	0.9920	0.001638	489.9	0.9846
460	733.15	0.001806	512.3	1.0257	0.001758	509.0	1.0178	0.001719	505.6	1.0098	0.001685	502.3	1.0019
480	753.15	0.001949	540.4	1.0651	0.001886	536.0	1.0550	0.001834	531.6	1.0456	0.001791	527.4	1.0366
500	773.15	0.002122	569.5	1.1043	0.002040	563.4	1.0924	0.001972	557.9	1.0818	0.001914	552.9	1.0717
520	793.15	0.002319	598.5	1.1426	0.002214	591.3	1.1297	0.002127	585.1	1.1184	0.002054	579.3	1.1074
540	813.15	0.002532	627.0	1.1793	0.002402	619.4	1.1660	0.002294	612.8	1.1544	0.002205	606.6	1.1432
550	823.15	0.002643	641.0	1.1970	0.002501	633.3	1.1836	0.002382	626.4	1.1719	0.002284	620.0	1.1605
560	833.15	0.002758	654.7	1.2142	0.002603	646.9	1.2007	0.002474	639.8	1.1888	0.002366	633.2	1.1773
580	853.15	0.002992	681.4	1.2469	0.002815	673.5	1.2334	0.002666	666.0	1.2210	0.002541	659.2	1.2092
600	873.15	0.003228	706.8	1.2770	0.003032	698.8	1.2635	0.002864	691.2	1.2509	0.002723	684.4	1.2390
620	893.15	0.003458	730.6	1.3049	0.003248	722.8	1.2914	0.003062	715.3	1.2791	0.002905	708.5	1.2672
640	913.15	0.003683	753.0	1.3304	0.003457	745.4	1.3173	0.003261	738.2	1.3053	0.003088	731.7	1.2935
650	923.15	0.003794	763.6	1.3423	0.003559	756.2	1.3295	0.003358	749.2	1.3177	0.003179	742.9	1.3060
660	933.15	0.003902	773.9	1.3538	0.003660	766.7	1.3412	0.003453	760.0	1.3297	0.003270	753.9	1.3181
680	953.15	0.004114	793.7	1.3753	0.003861	787.1	1.3633	0.003641	780.8	1.3522	0.003449	775.0	1.3411
700	973.15	0.004319	812.9	1.3953	0.004055	806.7	1.3838	0.003826	800.8	1.3728	0.003623	795.2	1.3625
720	993.15	0.004517	831.5	1.4140	0.004243	825.7	1.4029	0.004003	820.2	1.3923	0.003793	814.8	1.3824
740	1013.15	0.004707	849.4	1.4318	0.004425	843.9	1.4211	0.004176	838.7	1.4111	0.003958	833.7	1.4014
750	1023.15	0.004800	858.2	1.4404	0.004513	852.8	1.4299	0.004260	847.8	1.4201	0.004038	842.9	1.4106
760	1033.15	0.004891	866.9	1.4488	0.004600	861.6	1.4385	0.004344	856.8	1.4289	0.004117	852.0	1.4195
780	1053.15	0.005072	883.9	1.4650	0.004772	878.9	1.4551	0.004506	874.2	1.4458	0.004273	869.6	1.4367
800	1073.15	0.005247	900.4	1.4806	0.004940	895.8	1.4710	0.004667	891.3	1.4618	0.004426	887.0	1.4531
820	1093.15	0.005420	916.5	1.4956	0.005106	912.2	1.4862	0.004825	908.0	1.4771	0.004576	903.8	1.4688
840	1113.15	0.005590	932.4	1.5100	0.005267	928.4	1.5008	0.004980	924.4	1.4919	0.004724	920.4	1.4837
850	1123.15	0.005674	940.3	1.5169	0.005347	936.4	1.5078	0.005056	932.5	1.4991	0.004797	928.7	1.4909
860	1133.15	0.005757	948.0	1.5237	0.005427	944.3	1.5147	0.005132	940.5	1.5061	0.004869	936.8	1.4979
880	1153.15	0.005920	963.4	1.5370	0.005583	959.9	1.5281	0.005281	956.3	1.5196	0.005013	952.9	1.5116
900	1173.15	0.006081	978.6	1.5497	0.005738	975.2	1.5410	0.005430	972.0	1.5327	0.005155	968.7	1.5248
920	1193.15	0.006240	993.5	1.5620	0.005889	990.4	1.5534	0.005576	987.3	1.5452	0.005295	984.3	1.5375
940	1213.15	0.006397	1008.2	1.5738	0.006040	1005.3	1.5653	0.005719	1002.4	1.5573	0.005433	999.6	1.5497
950	1223.15	0.006475	1015.5	1.5796	0.006115	1012.6	1.5712	0.005790	1009.9	1.5632	0.005502	1007.1	1.5556
960	1233.15	0.006552	1022.7	1.5853	0.006189	1019.9	1.5769	0.005861	1017.3	1.5689	0.005570	1014.6	1.5613
980	1253.15	0.006706	1037.0	1.5963	0.006335	1034.3	1.5880	0.006002	1031.8	1.5801	0.005704	1029.3	1.5726
1000	1273.15	0.006858	1051.3	1.6070	0.006480	1048.7	1.5989	0.006141	1046.3	1.5911	0.005837	1043.9	1.5837

Table 42-1 Specific Heats c_p and c_v of Superheated Steam (H$_2$O) at Constant Pressure p and Constant Specific Volume v

Pressure p		1 at = 0.980665 bar				5 at = 4.903325 bar				10 at = 9.806650 bar			
Temperature		Specific heat											
t	T	c_p		c_v		c_p		c_v		c_p		c_v	
°C	K	$\frac{kcal}{kg\,K}$	$\frac{kJ}{kg\,K}$	$\frac{kcal}{kg\,K}$	$\frac{kJ}{kg\,K}$	$\frac{kcal}{kg\,K}$	$\frac{kJ}{kg\,K}$	$\frac{kcal}{kg\,K}$	$\frac{kJ}{kg\,K}$	$\frac{kcal}{kg\,K}$	$\frac{kJ}{kg\,K}$	$\frac{kcal}{kg\,K}$	$\frac{kJ}{kg\,K}$
100	373.15	0.489	2.047	—	—	—	—	—	—	—	—	—	—
120	393.15	0.482	2.018	—	—	—	—	—	—	—	—	—	—
140	413.15	0.477	1.997	0.354	1.482	—	—	—	—	—	—	—	—
160	433.15	0.474	1.985	0.355	1.486	0.534	2.236	0.385	1.612	—	—	—	—
180	453.15	0.472	1.976	0.356	1.491	0.521	2.181	0.380	1.591	0.597	2.500	0.415	1.738
200	473.15	0.473	1.980	0.358	1.499	0.511	2.139	0.377	1.578	0.564	2.361	0.404	1.691
220	493.15	0.474	1.985	0.360	1.507	0.503	2.106	0.375	1.570	0.544	2.278	0.396	1.658
240	513.15	0.475	1.989	0.362	1.516	0.498	2.085	0.374	1.566	0.531	2.223	0.391	1.637
260	533.15	0.477	1.997	0.364	1.524	0.496	2.077	0.374	1.566	0.523	2.190	0.388	1.624
280	553.15	0.479	2.005	0.367	1.537	0.495	2.072	0.375	1.570	0.518	2.169	0.387	1.620
300	573.15	0.481	2.014	0.370	1.549	0.495	2.072	0.377	1.578	0.514	2.152	0.386	1.616
320	593.15	0.484	2.026	0.372	1.557	0.496	2.077	0.379	1.587	0.512	2.144	0.387	1.620
340	613.15	0.487	2.039	0.375	1.570	0.497	2.081	0.381	1.595	0.511	2.139	0.387	1.620
360	633.15	0.489	2.047	0.378	1.583	0.498	2.085	0.383	1.604	0.510	2.135	0.388	1.624
380	653.15	0.492	2.060	0.381	1.595	0.500	2.093	0.385	1.612	0.510	2.135	0.390	1.633
400	673.15	0.495	2.072	0.384	1.608	0.502	2.102	0.387	1.620	0.511	2.139	0.392	1.641
420	693.15	0.498	2.085	0.387	1.620	0.504	2.110	0.390	1.633	0.512	2.144	0.394	1.650
440	713.15	0.501	2.098	0.391	1.637	0.506	2.199	0.393	1.645	0.513	2.148	0.397	1.662
460	733.15	0.504	2.110	0.394	1.650	0.509	2.131	0.396	1.658	0.515	2.156	0.399	1.671
480	753.15	0.507	2.123	0.397	1.662	0.512	2.144	0.399	1.671	0.517	2.165	0.402	1.683
500	773.15	0.510	2.135	0.400	1.675	0.514	2.152	0.402	1.683	0.519	2.173	0.404	1.691
520	793.15	0.513	2.148	0.403	1.687	0.517	2.165	0.405	1.696	0.522	2.186	0.407	1.704
540	813.15	0.517	2.165	0.407	1.704	0.520	2.177	0.408	1.708	0.524	2.194	0.410	1.717
560	833.15	0.520	2.177	0.410	1.717	0.523	2.190	0.411	1.721	0.527	2.206	0.413	1.729
580	853.15	0.523	2.190	0.413	1.729	0.526	2.202	0.414	1.733	0.529	2.215	0.416	1.742
600	873.15	0.526	2.202	0.416	1.742	0.529	2.215	0.417	1.746	0.532	2.227	0.419	1.754
620	893.15	0.529	2.215	0.419	1.754	0.532	2.227	0.421	1.763	0.535	2.240	0.422	1.767
640	913.15	0.532	2.227	0.423	1.771	0.535	2.240	0.424	1.775	0.538	2.252	0.425	1.779
660	933.15	0.536	2.244	0.426	1.784	0.539	2.257	0.427	1.788	0.541	2.265	0.428	1.792
680	953.15	0.539	2.257	0.429	1.796	0.542	2.269	0.430	1.800	0.544	2.278	0.431	1.805
700	973.15	0.542	2.269	0.432	1.809	0.545	2.282	0.433	1.813	0.547	2.290	0.434	1.817

Table 42-2 Specific Heats c_p and c_v of Superheated Steam (H$_2$O) at Constant Pressure p and Constant Specific Volume v (Continued)

Pressure p		20 at = 19.613300 bar				30 at = 29.419950 bar				40 at = 39.226600 bar			
Temperature		Specific heat											
t	T	c_p		c_v		c_p		c_v		c_p		c_v	
°C	K	$\frac{kcal}{kg\,K}$	$\frac{kJ}{kg\,K}$	$\frac{kcal}{kg\,K}$	$\frac{kJ}{kg\,K}$	$\frac{kcal}{kg\,K}$	$\frac{kJ}{kg\,K}$	$\frac{kcal}{kg\,K}$	$\frac{kJ}{kg\,K}$	$\frac{kcal}{kg\,K}$	$\frac{kJ}{kg\,K}$	$\frac{kcal}{kg\,K}$	$\frac{kJ}{kg\,K}$
220	493.15	0.703	2.943	—	—	—	—	—	—	—	—	—	—
240	513.15	0.637	2.667	—	—	0.817	3.421	—	—	—	—	—	—
260	533.15	0.596	2.495	0.421	1.763	0.704	2.948	—	—	0.868	3.634	—	—
280	553.15	0.569	2.382	0.412	1.725	0.643	2.692	0.443	1.855	0.747	3.128	—	—
300	573.15	0.556	2.328	0.407	1.704	0.610	2.554	0.431	1.805	0.680	2.847	0.458	1.918
320	593.15	0.547	2.290	0.404	1.691	0.587	2.458	0.423	1.771	0.638	2.671	0.444	1.859
340	613.15	0.540	2.261	0.402	1.683	0.572	2.395	0.417	1.746	0.609	2.550	0.433	1.813
360	633.15	0.535	2.240	0.400	1.675	0.562	2.353	0.413	1.729	0.592	2.479	0.427	1.788
380	653.15	0.532	2.227	0.401	1.679	0.555	2.324	0.411	1.721	0.580	2.428	0.423	1.771
400	673.15	0.529	2.215	0.401	1.679	0.549	2.299	0.410	1.717	0.570	2.386	0.420	1.758
420	693.15	0.528	2.211	0.402	1.683	0.545	2.282	0.410	1.717	0.563	2.357	0.418	1.750
440	713.15	0.527	2.206	0.404	1.691	0.543	2.273	0.411	1.721	0.558	2.336	0.418	1.750
460	733.15	0.528	2.211	0.405	1.696	0.541	2.265	0.411	1.721	0.554	2.319	0.417	1.746
480	753.15	0.528	2.211	0.407	1.704	0.539	2.257	0.412	1.725	0.551	2.307	0.417	1.746
500	773.15	0.529	2.215	0.409	1.712	0.539	2.257	0.414	1.733	0.550	2.303	0.418	1.750
520	793.15	0.530	2.219	0.411	1.721	0.539	2.257	0.415	1.738	0.549	2.299	0.419	1.754
540	813.15	0.532	2.227	0.414	1.733	0.540	2.261	0.417	1.746	0.548	2.294	0.421	1.763
560	833.15	0.534	2.236	0.416	1.742	0.541	2.265	0.419	1.754	0.549	2.299	0.423	1.771
580	853.15	0.536	2.244	0.419	1.754	0.542	2.269	0.422	1.767	0.549	2.299	0.425	1.779
600	873.15	0.538	2.252	0.421	1.763	0.544	2.278	0.424	1.775	0.550	2.303	0.427	1.788
620	893.15	0.540	2.261	0.424	1.775	0.546	2.286	0.427	1.788	0.552	2.311	0.429	1.796
640	913.15	0.543	2.273	0.427	1.788	0.549	2.299	0.429	1.796	0.554	2.319	0.431	1.805
660	933.15	0.546	2.286	0.430	1.800	0.551	2.307	0.432	1.809	0.556	2.328	0.434	1.817
680	953.15	0.549	2.299	0.433	1.813	0.554	2.319	0.435	1.821	0.559	2.340	0.437	1.830
700	973.15	0.552	2.311	0.436	1.825	0.557	2.332	0.438	1.834	0.562	2.353	0.440	1.842

1 at = 1 kp/cm² = 98 966.5 Pa = 98 066.5 N/m² = 9.80665 N/cm² = 0.980665 bar

1 kcal = 4.1868 kJ

Table 42-3 Specific Heats c_p and c_v of Superheated Steam (H_2O) at a Constant Pressure p and Constant Specific Volume v (*Continued*)

Pressure p		50 at = 49.033 250 bar				60 at = 58.839 900 bar				70 at = 68.646 550 bar			
Temperature		Specific heat											
t	T	c_p		c_v		c_p		c_v		c_p		c_v	
°C	K	$\frac{kcal}{kg\,K}$	$\frac{kJ}{kg\,K}$	$\frac{kcal}{kg\,K}$	$\frac{kJ}{kg\,K}$	$\frac{kcal}{kg\,K}$	$\frac{kJ}{kg\,K}$	$\frac{kcal}{kg\,K}$	$\frac{kJ}{kg\,K}$	$\frac{kcal}{kg\,K}$	$\frac{kJ}{kg\,K}$	$\frac{kcal}{kg\,K}$	$\frac{kJ}{kg\,K}$
280	533.15	0.897	3.756	—	—	1.104	4.622	—	—	—	—	—	—
300	573.15	0.770	3.224	0.487	2.039	0.888	3.718	—	—	1.053	4.409	—	—
320	593.15	0.701	2.935	0.467	1.955	0.776	3.249	0.493	2.064	0.866	3.626	0.522	2.186
340	613.15	0.653	2.734	0.451	1.888	0.706	2.956	0.472	1.976	0.769	3.220	0.495	2.072
360	633.15	0.626	2.621	0.441	1.846	0.664	2.780	0.457	1.913	0.709	2.968	0.475	1.989
380	653.15	0.607	2.541	0.434	1.817	0.637	2.667	0.447	1.871	0.671	2.809	0.460	1.926
400	673.15	0.593	2.483	0.430	1.800	0.618	2.587	0.441	1.846	0.645	2.700	0.452	1.892
420	693.15	0.582	2.437	0.427	1.788	0.603	2.525	0.436	1.825	0.625	2.617	0.445	1.863
440	713.15	0.574	2.403	0.425	1.779	0.591	2.474	0.432	1.809	0.609	2.550	0.440	1.842
460	733.15	0.568	2.378	0.424	1.775	0.583	2.441	0.430	1.800	0.599	2.508	0.437	1.830
480	753.15	0.564	2.361	0.423	1.771	0.577	2.416	0.428	1.792	0.590	2.470	0.434	1.817
500	773.15	0.561	2.349	0.423	1.771	0.572	2.395	0.428	1.792	0.583	2.441	0.433	1.813
520	793.15	0.559	2.340	0.425	1.779	0.569	2.382	0.429	1.796	0.579	2.424	0.433	1.813
540	813.15	0.557	2.332	0.425	1.779	0.566	2.370	0.429	1.796	0.575	2.407	0.433	1.813
560	833.15	0.556	2.328	0.426	1.784	0.564	2.361	0.430	1.800	0.572	2.395	0.433	1.813
580	853.15	0.556	2.328	0.428	1.792	0.563	2.357	0.431	1.805	0.570	2.386	0.434	1.817
600	873.15	0.556	2.328	0.430	1.800	0.562	2.353	0.432	1.809	0.568	2.378	0.435	1.821
620	893.15	0.558	2.336	0.432	1.809	0.564	2.361	0.434	1.817	0.570	2.386	0.437	1.830
640	913.15	0.559	2.340	0.434	1.817	0.565	2.366	0.436	1.825	0.571	2.391	0.438	1.834
660	933.15	0.561	2.349	0.436	1.825	0.567	2.374	0.438	1.834	0.572	2.395	0.440	1.842
680	953.15	0.564	2.361	0.438	1.834	0.569	2.382	0.440	1.842	0.573	2.399	0.442	1.851
700	973.15	0.566	2.370	0.441	1.846	0.571	2.391	0.443	1.855	0.575	2.407	0.444	1.859

Table 42-4 Specific Heats c_p and c_v of Superheated Steam (H_2O) at a Constant Pressure p and Constant Specific Volume v (*Continued*)

Pressure p		80 at = 78.453 200 bar				90 at = 88.259 850 bar				100 at = 98.066 500 bar			
Temperature		Specific heat											
t	T	c_p		c_v		c_p		c_v		c_p		c_v	
°C	K	$\frac{kcal}{kg\,K}$	$\frac{kJ}{kg\,K}$	$\frac{kcal}{kg\,K}$	$\frac{kJ}{kg\,K}$	$\frac{kcal}{kg\,K}$	$\frac{kJ}{kg\,K}$	$\frac{kcal}{kg\,K}$	$\frac{kJ}{kg\,K}$	$\frac{kcal}{kg\,K}$	$\frac{kJ}{kg\,K}$	$\frac{kcal}{kg\,K}$	$\frac{kJ}{kg\,K}$
300	573.15	1.310	5.485	—	—	—	—	—	—	—	—	—	—
320	593.15	0.985	4.124	—	—	1.149	4.811	—	—	1.376	5.761	—	—
340	613.15	0.846	3.542	0.520	2.177	0.941	3.940	0.548	2.294	1.059	4.434	—	—
360	633.15	0.761	3.186	0.495	2.072	0.822	3.442	0.516	2.160	0.894	3.743	0.538	2.252
380	653.15	0.709	2.958	0.475	1.989	0.753	3.153	0.491	2.056	0.803	3.362	0.509	2.131
400	673.15	0.675	2.826	0.463	1.938	0.708	2.964	0.475	1.989	0.744	3.115	0.488	2.043
420	693.15	0.648	2.713	0.454	1.901	0.673	2.818	0.464	1.943	0.701	2.935	0.474	1.985
440	713.15	0.629	2.633	0.448	1.876	0.650	2.721	0.456	1.909	0.672	2.814	0.464	1.943
460	733.15	0.615	2.575	0.443	1.855	0.632	2.646	0.450	1.884	0.650	2.721	0.457	1.913
480	753.15	0.604	2.529	0.440	1.842	0.618	2.587	0.446	1.867	0.634	2.654	0.452	1.892
500	773.15	0.595	2.491	0.438	1.834	0.607	2.541	0.443	1.855	0.621	2.600	0.449	1.880
520	793.15	0.589	2.466	0.437	1.830	0.600	2.512	0.442	1.851	0.611	2.558	0.446	1.867
540	813.15	0.584	2.445	0.437	1.830	0.593	2.483	0.441	1.846	0.603	2.525	0.444	1.859
560	833.15	0.580	2.428	0.436	1.825	0.588	2.462	0.439	1.838	0.596	2.495	0.443	1.855
580	853.15	0.577	2.416	0.437	1.830	0.584	2.445	0.440	1.842	0.591	2.474	0.443	1.855
600	873.15	0.575	2.407	0.438	1.834	0.581	2.433	0.440	1.842	0.588	2.462	0.443	1.855
620	893.15	0.576	2.412	0.439	1.838	0.582	2.437	0.441	1.846	0.588	2.462	0.444	1.859
640	913.15	0.576	2.412	0.440	1.842	0.582	2.437	0.442	1.851	0.598	2.462	0.445	1.863
660	933.15	0.577	2.416	0.442	1.851	0.583	2.441	0.444	1.859	0.588	2.462	0.446	1.867
680	953.15	0.578	2.420	0.444	1.859	0.484	2.445	0.446	1.867	0.589	2.466	0.448	1.876
700	973.15	0.580	2.428	0.446	1.867	0.585	2.449	0.484	1.876	0.590	2.470	0.449	1.880

1 at = 1 kp/cm² = 98 066.5 Pa = 98 066.5 N/m² = 9.806 65 N/cm² = 0.980 665 bar   1 kcal = 4.1868 kJ

Table 42-5 Specific Heats c_p and c_v of Superheated Steam (H_2O) at a Constant Pressure p and Constant Specific Volume v (*Continued*)

Pressure p		110 at = 107.873 150 bar				120 at = 117.679 800 bar				130 at = 127.486 450 bar			
Temperature		Specific heat											
t	T	c_p		c_v		c_p		c_v		c_p		c_v	
°C	K	kcal/kg K	kJ/kg K	kcal/kg K	kJ/kg K	kcal/kg K	kJ/kg K	kcal/kg K	kJ/kg K	kcal/kg K	kJ/kg K	kcal/kg K	kJ/kg K
320	593.15	1.695	7.097	—	—	—	—	—	—	—	—	—	—
340	613.15	1.207	5.053	—	—	1.400	5.862	—	—	1.671	6.996	—	—
360	633.15	0.979	4.099	0.562	2.353	1.080	4.522	0.589	2.466	1.201	5.028	—	—
380	653.15	0.859	3.596	0.529	2.215	0.921	3.856	0.548	2.294	0.990	4.145	0.569	2.382
400	673.16	0.783	3.278	0.502	2.102	0.826	3.458	0.517	2.165	0.873	3.655	0.534	2.236
420	693.15	0.730	3.056	0.484	2.026	0.762	3.190	0.495	2.072	0.798	3.341	0.507	2.123
440	713.15	0.696	2.914	0.473	1.980	0.721	3.019	0.482	2.018	0.747	3.128	0.491	2.056
460	733.15	0.670	2.805	0.464	1.943	0.690	2.889	0.472	1.976	0.711	2.977	0.479	2.005
480	753.15	0.650	2.721	0.458	1.918	0.666	2.788	0.464	1.943	0.684	2.864	0.471	1.972
500	773.15	0.634	2.654	0.454	1.901	0.648	2.713	0.459	1.922	0.663	2.776	0.465	1.947
520	793.15	0.622	2.604	0.451	1.888	0.634	2.654	0.455	1.905	0.647	2.709	0.460	1.926
540	813.15	0.613	2.567	0.448	1.876	0.623	2.608	0.452	1.892	0.634	2.654	0.456	1.909
560	833.15	0.605	2.533	0.447	1.871	0.614	2.571	0.451	1.888	0.623	2.608	0.454	1.901
580	853.15	0.599	2.508	0.446	1.867	0.607	2.541	0.449	1.880	0.615	2.575	0.453	1.897
600	873.15	0.595	2.491	0.446	1.867	0.602	2.520	0.449	1.880	0.610	2.554	0.452	1.892
620	893.15	0.595	2.491	0.447	1.871	0.601	2.516	0.449	1.880	0.608	2.546	0.452	1.892
640	913.15	0.594	2.487	0.447	1.871	0.600	2.512	0.449	1.880	0.607	2.541	0.452	1.892
660	933.15	0.594	2.487	0.448	1.876	0.600	2.512	0.450	1.884	0.606	2.537	0.452	1.892
680	953.15	0.595	2.491	0.449	1.880	0.601	2.516	0.451	1.888	0.607	2.541	0.453	1.897
700	973.15	0.595	2.491	0.451	1.888	0.601	2.516	0.453	1.897	0.606	2.537	0.455	1.905

Table 42-6 Specific Heats c_p and c_v of Superheated Steam (H_2O) at a Constant Pressure p and Constant Specific Volume v (*Continued*)

Pressure p		140 at = 137.293 100 bar				150 at = 147.099 750 bar				160 at = 156.906 400 bar			
Temperature		Specific heat											
t	T	c_p		c_v		c_p		c_v		c_p		c_v	
°C	K	kcal/kg K	kJ/kg K	kcal/kg K	kJ/kg K	kcal/kg K	kJ/kg K	kcal/kg K	kJ/kg K	kcal/kg K	kJ/kg K	kcal/kg K	kJ/kg K
340	613.15	2.091	8.755	—	—	—	—	—	—	—	—	—	—
360	633.15	1.350	5.652	—	—	1.551	6.494	—	—	1.850	7.746	—	—
380	653.15	1.069	4.476	0.592	2.479	1.164	4.873	0.618	2.587	1.282	5.367	0.647	2.709
400	673.15	0.925	3.873	0.552	2.311	0.983	4.116	0.571	2.391	1.049	4.392	0.591	2.474
420	693.15	0.836	3.500	0.519	2.173	0.876	3.668	0.531	2.223	0.919	3.848	0.545	2.282
440	713.15	0.774	3.241	0.500	2.093	0.802	3.358	0.510	2.135	0.832	3.483	0.521	2.181
460	733.15	0.733	3.069	0.487	2.039	0.757	3.169	0.495	2.072	0.783	3.278	0.503	2.106
480	753.15	0.702	2.939	0.477	1.997	0.721	3.019	0.484	2.026	0.741	3.102	0.491	2.056
500	773.15	0.678	2.839	0.470	1.968	0.694	2.906	0.476	1.993	0.710	2.973	0.481	2.014
520	793.15	0.659	2.759	0.465	1.947	0.673	2.818	0.470	1.968	0.686	2.872	0.474	1.985
540	813.15	0.654	2.700	0.461	1.930	0.656	2.747	0.465	1.947	0.668	2.797	0.469	1.964
560	833.15	0.633	2.650	0.458	1.918	0.643	2.692	0.461	1.930	0.653	2.734	0.465	1.947
580	853.15	0.624	2.613	0.456	1.909	0.632	2.646	0.459	1.922	0.641	2.684	0.462	1.934
600	873.15	0.617	2.583	0.455	1.905	0.624	2.613	0.457	1.913	0.632	2.646	0.460	1.926
620	893.15	0.615	2.575	0.455	1.905	0.622	2.604	0.457	1.913	0.629	2.633	0.460	1.926
640	913.15	0.614	2.571	0.454	1.901	0.620	2.596	0.456	1.909	0.626	2.621	0.459	1.922
660	933.15	0.612	2.562	0.454	1.901	0.618	2.587	0.456	1.909	0.624	2.613	0.458	1.918
680	953.15	0.612	2.562	0.455	1.905	0.618	2.587	0.457	1.918	0.624	2.613	0.459	1.922
700	973.15	0.611	2.558	0.456	1.909	0.617	2.583	0.458	1.918	0.623	2.608	0.460	1.926

1 at = 1 kp/cm² = 98 066.5 Pa = 98 066.5 N/m² = 9.806 65 N/cm² = 0.980 665 bar   1 kcal = 4.1868 kJ

Table 42-7 Specific Heats c_p and c_v of Superheated Steam (H_2O) at a Constant Pressure p and Constant Specific Volume v (Continued)

Pressure p		170 at = 166.713 050 bar				180 at = 176.519 700 bar				190 at = 186.326 350 bar			
Temperature		Specific heat											
t	T	c_p		c_v		c_p		c_v		c_p		c_v	
°C	K	$\frac{kcal}{kg\,K}$	$\frac{kJ}{kg\,K}$	$\frac{kcal}{kg\,K}$	$\frac{kJ}{kg\,K}$	$\frac{kcal}{kg\,K}$	$\frac{kJ}{kg\,K}$	$\frac{kcal}{kg\,K}$	$\frac{kJ}{kg\,K}$	$\frac{kcal}{kg\,K}$	$\frac{kJ}{kg\,K}$	$\frac{kcal}{kg\,K}$	$\frac{kJ}{kg\,K}$
360	633.15	2.335	9.776	—	—	—	—	—	—	—	—	—	—
380	653.15	1.431	5.991	0.680	2.847	1.623	6.795	0.719	3.010	1.876	7.854	0.764	3.199
400	673.15	1.126	4.714	0.612	2.562	1.217	5.095	0.635	2.659	1.330	5.568	0.660	2.763
420	693.15	0.965	4.040	0.559	2.340	1.020	4.271	0.574	2.403	1.083	4.534	0.590	2.470
440	713.15	0.866	3.626	0.531	2.223	0.904	3.785	0.542	2.269	0.946	3.961	0.554	2.319
460	733.15	0.811	3.395	0.512	2.144	0.840	3.517	0.520	2.177	0.871	3.647	0.529	2.215
480	753.15	0.762	3.190	0.498	2.085	0.784	3.282	0.505	2.114	0.808	3.383	0.512	2.144
500	773.15	0.727	3.044	0.487	2.039	0.745	3.119	0.493	2.064	0.764	3.199	0.499	2.098
520	793.15	0.700	2.931	0.479	2.005	0.716	2.998	0.484	2.026	0.731	3.061	0.489	2.047
540	813.15	0.680	2.847	0.473	1.980	0.692	2.897	0.477	1.997	0.705	2.952	0.482	2.018
560	833.15	0.663	2.776	0.469	1.964	0.674	2.822	0.472	1.976	0.685	2.868	0.476	1.993
580	853.15	0.649	2.717	0.465	1.947	0.658	2.755	0.469	1.964	0.667	2.793	0.472	1.976
600	873.15	0.640	2.680	0.463	1.938	0.648	2.713	0.466	1.951	0.656	2.747	0.469	1.964
620	893.15	0.636	2.663	0.462	1.934	0.644	2.696	0.465	1.947	0.652	2.730	0.467	1.955
640	913.15	0.633	2.650	0.461	1.930	0.640	2.680	0.464	1.943	0.648	2.713	0.466	1.951
660	933.15	0.631	2.642	0.460	1.926	0.638	2.671	0.463	1.938	0.645	2.700	0.465	1.947
680	853.15	0.630	2.638	0.461	1.930	0.636	2.663	0.462	1.934	0.642	2.688	0.464	1.943
700	973.15	0.629	2.633	0.461	1.930	0.634	2.654	0.462	1.934	0.640	2.680	0.464	1.943

Table 42-8 Specific Heats c_p and c_v of Superheated Steam (H_2O) at a Constant Pressure p and Constant Specific Volume v (Continued)

Pressure p		200 at = 196.133 000 bar				210 at = 205.939 650 bar				220 at = 215.746 300 bar			
Temperature		Specific heat											
t	T	c_p		c_v		c_p		c_v		c_p		c_v	
°C	K	$\frac{kcal}{kg\,K}$	$\frac{kJ}{kg\,K}$	$\frac{kcal}{kg\,K}$	$\frac{kJ}{kg\,K}$	$\frac{kcal}{kg\,K}$	$\frac{kJ}{kg\,K}$	$\frac{kcal}{kg\,K}$	$\frac{kJ}{kg\,K}$	$\frac{kcal}{kg\,K}$	$\frac{kJ}{kg\,K}$	$\frac{kcal}{kg\,K}$	$\frac{kJ}{kg\,K}$
380	653.15	2.255	9.441	0.818	3.425	2.830	11.849	—	—	—	—	—	—
400	673.15	1.469	6.150	0.688	2.881	1.636	6.850	0.719	3.010	1.832	7.670	0.756	3.165
420	693.15	1.154	4.832	0.607	2.541	1.235	5.171	0.625	2.617	1.327	5.556	0.645	2.700
440	713.15	0.993	4.157	0.566	2.370	1.045	4.375	0.579	2.424	1.102	4.614	0.592	2.479
460	733.15	0.904	3.785	0.538	2.252	0.939	3.931	0.548	2.294	0.977	4.091	0.558	2.336
480	753.15	0.833	3.488	0.519	2.173	0.859	3.596	0.526	2.202	0.887	3.714	0.534	2.236
500	773.15	0.783	3.278	0.505	2.144	0.804	3.366	0.511	2.139	0.825	3.454	0.517	2.165
520	793.15	0.746	3.123	0.494	2.068	0.763	3.195	0.499	2.089	0.780	3.266	0.504	2.110
540	813.15	0.718	3.006	0.486	2.035	0.731	3.061	0.490	2.052	0.745	3.119	0.495	2.072
560	833.15	0.696	2.914	0.480	2.010	0.707	2.960	0.484	2.026	0.718	3.006	0.487	2.039
580	853.15	0.676	2.830	0.475	1.989	0.686	2.872	0.479	2.005	0.695	2.910	0.482	2.018
600	873.15	0.664	2.780	0.472	1.976	0.673	2.818	0.475	1.989	0.681	2.851	0.478	2.001
620	893.15	0.659	2.759	0.469	1.964	0.667	2.793	0.472	1.976	0.675	2.826	0.475	1.989
640	913.15	0.655	2.742	0.468	1.959	0.663	2.776	0.471	1.972	0.670	2.805	0.473	1.980
660	933.15	0.652	2.730	0.467	1.955	0.659	2.759	0.469	1.964	0.666	2.788	0.471	1.972
680	953.15	0.649	2.717	0.466	1.951	0.656	2.747	0.468	1.959	0.663	2.776	0.470	1.968
700	973.15	0.647	2.709	0.466	1.951	0.653	2.734	0.467	1.955	0.660	2.763	0.469	1.964

1 at = 1 kp/cm² = 98 066.5 Pa = 98 066.5 N/m² = 9.806 65 N/cm² = 0.980 665 bar   1 kcal = 4.1868 kJ

Table 42-9 Specific Heats c_p and c_v of Superheated Steam (H$_2$O) at a Constant Pressure p and Constant Specific Volume v (Continued)

Pressure p		230 at = 225.552950 bar				240 at = 235.359600 bar				250at=245.166250 bar	
Temperature		Specific heat									
t	T	c_p		c_v		c_p		c_v		c_p	
°C	K	kcal/kg K	kJ/kg K	kcal/kg K	kJ/kg K	kcal/kg K	kJ/kg K	kcal/kg K	kJ/kg K	kcal/kg K	kJ/kg K
400	673.15	2.057	8.612	—	—	2.320	9.713	—	—	2.660	11.137
420	693.15	1.431	5.991	0.667	2.793	1.548	6.481	0.691	2.893	1.679	7.030
440	713.15	1.164	4.873	0.606	2.537	1.231	5.154	0.621	2.600	1.303	5.455
460	733.15	1.017	4.258	0.568	2.378	1.060	4.438	0.578	2.420	1.105	4.626
480	753.15	0.916	3.835	0.542	2.269	0.946	3.961	0.549	2.299	0.977	4.091
500	773.15	0.847	3.546	0.523	2.190	0.870	3.643	0.529	2.215	0.893	3.739
520	793.15	0.797	3.337	0.509	2.131	0.815	3.412	0.514	2.152	0.834	3.492
540	813.15	0.759	3.178	0.499	2.089	0.774	3.241	0.503	2.106	0.789	3.303
560	833.15	0.730	3.056	0.491	2.056	0.742	3.107	0.495	2.072	0.755	3.161
580	853.15	0.705	2.952	0.485	2.031	0.715	2.994	0.488	2.043	0.726	3.040
600	873.15	0.689	2.885	0.480	2.010	0.698	2.922	0.483	2.022	0.707	2.960
620	893.15	0.683	2.860	0.477	1.997	0.691	2.893	0.479	2.005	0.700	2.931
640	913.15	0.678	2.839	0.475	1.989	0.686	2.872	0.477	1.997	0.694	2.906
660	933.15	0.673	2.818	0.473	1.980	0.681	2.851	0.475	1.989	0.688	2.881
680	953.15	0.670	2.805	0.472	1.976	0.677	2.834	0.474	1.985	0.684	2.864
700	973.15	0.666	2.788	0.471	1.972	0.673	2.818	0.473	1.980	0.680	2.847

Table 42-10 Specific Heats c_p and c_v of Superheated Steam (H$_2$O) at a Constant Pressure p and Constant Specific Volume v (Continued)

Pressure p		250at=245.166250 bar		260 at = 254.972900 bar				270 at = 264.779550 bar			
Temperature		Specific heat									
t	T	c_v		c_p		c_v		c_p		c_v	
°C	K	kcal/kg K	kJ/kg K	kcal/kg K	kJ/kg K	kcal/kg K	kJ/kg K	kcal/kg K	kJ/kg K	kcal/kg K	kJ/kg K
400	673.15	—	—	—	—	—	—	—	—	—	—
420	693.15	0.716	2.998	1.825	7.641	0.748	3.132	1.990	8.332	—	—
440	713.15	0.637	2.667	1.380	5.778	0.655	2.742	1.462	6.121	0.673	2.818
460	733.15	0.589	2.466	1.153	4.827	0.600	2.512	1.204	5.041	0.612	2.562
480	753.15	0.557	2.332	1.010	4.229	0.566	2.370	1.045	4.375	0.574	2.403
500	773.15	0.535	2.240	0.916	3.835	0.542	2.269	0.941	3.940	0.548	2.294
520	793.15	0.520	2.177	0.853	3.571	0.525	2.198	0.873	3.655	0.530	2.219
540	813.15	0.508	2.127	0.804	3.366	0.512	2.144	0.819	3.429	0.517	2.165
560	833.15	0.499	2.089	0.767	3.211	0.502	2.102	0.780	3.266	0.506	2.119
580	853.15	0.492	2.060	0.737	3.086	0.495	2.072	0.748	3.132	0.498	2.085
600	873.15	0.486	2.035	0.716	2.998	0.489	2.047	0.725	3.035	0.492	2.060
620	893.15	0.482	2.018	0.708	2.964	0.484	2.026	0.717	3.002	0.487	2.039
640	913.15	0.479	2.005	0.702	2.939	0.481	2.014	0.711	2.977	0.483	2.022
660	933.15	0.477	1.997	0.696	2.914	0.479	2.005	0.704	2.948	0.481	2.014
680	953.15	0.475	1.989	0.691	2.893	0.477	1.997	0.699	2.927	0.479	2.005
700	973.15	0.474	1.985	0.687	2.876	0.476	1.993	0.694	2.906	0.478	2.001

1 at = 1 kp/cm² = 98066.5 Pa = 98066.5 N/m² = 9.80665 N/cm² = 0.980665 bar   1 kcal = 4.1868 kJ

Table 42-11 Specific Heats c_p and c_v of Superheated Steam (H_2O) at a Constant Pressure p and Constant Specific Volume v (Continued)

Pressure p		280 at = 274.586200 bar				290 at = 284.392850 bar				300 at = 294.199500 bar			
Temperature		Specific heat											
t	T	c_p		c_v		c_p		c_v		c_p		c_v	
°C	K	$\frac{kcal}{kg\,K}$	$\frac{kJ}{kg\,K}$	$\frac{kcal}{kg\,K}$	$\frac{kJ}{kg\,K}$	$\frac{kcal}{kg\,K}$	$\frac{kJ}{kg\,K}$	$\frac{kcal}{kg\,K}$	$\frac{kJ}{kg\,K}$	$\frac{kcal}{kg\,K}$	$\frac{kJ}{kg\,K}$	$\frac{kcal}{kg\,K}$	$\frac{kJ}{kg\,K}$
420	693.15	2.178	9.119	—	—	2.400	10.048	—	—	2.670	11.179	—	—
440	713.15	1.549	6.485	0.692	2.897	1.646	6.891	0.711	2.977	1.764	7.386	0.731	3.061
460	733.15	1.256	5.259	0.624	2.613	1.308	5.476	0.636	2.663	1.359	5.690	0.648	2.713
480	753.15	1.081	4.526	0.583	2.441	1.117	4.677	0.591	2.474	1.153	4.827	0.600	2.512
500	773.15	0.966	4.044	0.555	2.324	0.992	4.153	0.561	2.349	1.018	3.262	0.568	2.378
520	793.15	0.893	3.739	0.535	2.240	0.913	3.823	0.541	2.265	0.934	3.910	0.546	2.286
540	813.15	0.835	3.496	0.521	2.181	0.851	3.563	0.525	2.198	0.869	3.638	0.529	2.215
560	833.15	0.794	3.324	0.510	2.135	0.807	3.379	0.513	2.148	0.821	3.437	0.517	2.165
580	853.15	0.759	3.178	0.501	2.098	0.770	3.224	0.504	2.110	0.781	3.270	0.508	2.127
600	873.15	0.734	3.073	0.494	2.068	0.743	3.111	0.497	2.081	0.753	3.153	0.500	2.093
620	893.15	0.725	3.035	0.489	2.047	0.734	3.073	0.491	2.056	0.743	3.111	0.494	2.068
640	913.15	0.719	3.010	0.485	2.031	0.727	3.044	0.487	2.039	0.736	3.081	0.490	2.052
660	933.15	0.712	2.981	0.483	2.022	0.720	3.014	0.485	2.031	0.729	3.052	0.487	2.039
680	953.15	0.707	2.960	0.481	2.014	0.714	2.989	0.483	2.022	0.722	3.023	0.484	2.026
700	973.15	0.702	2.939	0.479	2.005	0.709	2.968	0.481	2.014	0.716	2.998	0.482	2.018

1 at = 1 kp/cm² = 98066.5 Pa = 98066.5 N/m² = 9.80665 N/cm² = 0.980665 bar   1 kcal = 4.1868 kJ

Table 43-1 Mean Specific Heat c_{pm} of Superheated Steam (H_2O) at a Constant Pressure p
(Between the Temperature of Saturation and a Given Temperature)

Pressure p		1 at = 0.980665 bar		5 at = 4.903325 bar		10 at = 9.806650 bar		20 at = 19.613300 bar	
Temperature		Mean specific heat							
t	T	c_{pm}		c_{pm}		c_{pm}		c_{pm}	
°C	K	kcal/kg K	kJ/kg K	kcal/kg K	kJ/kg K	kcal/kg K	kJ/kg K	kcal/kg K	kJ/kg K
120	393.15	0.488	2.043	—	—	—	—	—	—
140	413.15	0.479	2.005	—	—	—	—	—	—
160	433.15	0.476	1.993	0.562	2.353	—	—	—	—
180	453.15	0.475	1.989	0.533	2.232	—	—	—	—
200	473.15	4.474	1.985	0.520	2.177	0.577	2.416	—	—
220	493.15	0.473	1.980	0.514	2.152	0.566	2.370	0.681	2.815
240	513.15	0.473	1.980	0.511	2.139	0.556	2.328	0.653	2.734
260	533.15	0.474	1.985	0.509	2.131	0.548	2.294	0.626	2.621
280	553.15	0.475	1.989	0.508	2.127	0.544	2.278	0.613	2.567
300	573.15	0.475	1.989	0.507	2.123	0.540	2.261	0.603	2.525
320	593.15	0.476	1.993	0.506	2.119	0.536	2.244	0.594	2.487
340	613.15	0.476	1.993	0.504	2.110	0.533	2.232	0.586	2.453
360	633.15	0.477	1.997	0.504	2.110	0.530	2.219	0.579	2.424
380	653.15	0.487	2.001	0.504	2.110	0.528	2.211	0.573	2.399
400	673.15	0.479	2.005	0.503	2.106	0.527	2.206	0.568	2.378
420	693.15	0.481	2.014	0.503	2.106	0.525	2.198	0.564	2.361
440	713.15	0.482	2.018	0.504	2.110	0.524	2.194	0.561	2.349
460	733.15	0.483	2.022	0.504	2.110	0.524	2.194	0.559	2.340
480	753.15	0.484	2.026	0.504	2.110	0.523	2.190	0.557	2.332
500	773.15	0.485	2.031	0.505	2.114	0.523	2.190	0.555	2.324
520	793.15	0.487	2.039	0.505	2.114	0.523	2.190	0.553	2.315
540	813.15	0.488	2.043	0.506	2.119	0.523	2.190	0.552	2.311
560	833.15	0.489	2.047	0.507	2.123	0.523	2.190	0.551	2.307
580	853.15	0.491	2.056	0.508	2.127	0.523	2.190	0.550	2.303
600	873.15	0.492	2.060	0.508	2.127	0.524	2.194	0.549	2.299
620	893.15	0.493	2.064	0.510	2.135	0.524	2.194	0.549	2.299
640	913.15	0.495	2.072	0.511	2.139	0.525	2.198	0.548	2.294
660	933.15	0.496	2.077	0.512	2.144	0.525	2.198	0.548	2.294
680	953.15	0.498	2.085	0.513	2.148	0.526	2.202	0.548	2.294
700	973.15	0.499	2.089	0.514	2.152	0.527	2.206	0.548	2.294

Table 43-2 Mean Specific Heat c_{pm} of Superheated Steam (H_2O) at a Constant Pressure p
(Between the Temperature of Saturation and a Given Temperature) (Continued)

Pressure p		30 at = 29.419950 bar		40 at = 39.226600 bar		50 at = 49.033250 bar		60 at = 58.839900 bar	
Temperature		Mean specific heat							
t	T	c_{pm}		c_{pm}		c_{pm}		c_{pm}	
°C	K	kcal/kg K	kJ/kg K	kcal/kg K	kJ/kg K	kcal/kg K	kJ/kg K	kcal/kg K	kJ/kg K
240	513.15	0.746	3.123	—	—	—	—	—	—
260	533.15	0.709	2.968	0.832	3.483	—	—	—	—
280	553.15	0.678	2.876	0.779	3.262	0.879	3.680	0.981	4.107
300	573.15	0.669	2.801	0.749	3.136	0.828	3.467	0.919	3.848
320	593.15	0.653	2.734	0.723	3.027	0.788	3.299	0.873	3.655
340	613.15	0.640	2.680	0.700	2.931	9.761	3.186	0.838	3.509
360	633.15	0.629	2.633	0.681	2.851	0.738	3.090	0.799	3.345
380	653.15	0.620	2.596	0.677	2.793	0.717	3.002	0.771	3.228
400	673.15	0.612	2.562	0.655	2.742	0.700	2.931	0.747	3.128
420	693.15	0.605	2.533	0.645	2.700	0.686	2.872	0.728	3.048
440	713.15	0.599	2.508	0.636	2.663	0.673	2.818	0.712	2.981
460	733.15	0.594	2.487	0.629	2.633	0.663	2.776	0.699	2.927
480	753.15	0.589	2.466	0.622	2.604	0.654	2.738	0.687	2.876
500	773.15	0.586	2.453	0.616	2.579	0.646	2.705	0.677	2.834
520	793.15	0.583	2.441	0.611	2.558	0.640	2.680	0.669	2.801
540	813.15	0.580	2.428	0.607	2.541	0.634	2.654	0.661	2.767
560	833.15	0.577	2.416	0.603	2.525	0.629	2.633	0.654	2.738
580	853.15	0.575	2.407	0.600	2.512	0.624	2.613	0.648	2.713
600	873.15	0.573	2.399	0.597	2.500	0.620	2.596	0.643	2.692
620	893.15	0.572	2.395	0.594	2.487	0.616	2.579	0.638	2.671
640	913.15	0.571	2.391	0.592	2.479	0.613	2.567	0.634	2.654
660	933.15	0.570	2.386	0.590	2.470	0.610	2.554	0.630	2.638
680	953.15	0.569	2.382	0.589	2.466	0.608	2.546	0.627	2.625
700	973.15	0.568	2.378	0.587	2.458	0.606	2.537	0.624	2.613

1 at = kp/cm² = 98066.5 Pa = 98066.5 N/m² = 9.80665 N/cm² = 0.980665 bar   1 kcal = 4.1868 kJ

Table 43-3 Mean Specific Heat c_{pm} of Superheated Steam (H_2O) at a Constant Pressure p (Between the Temperature of Saturation and a Given Temperature) (Continued)

Pressure p		70 at = 68.646550 bar		80 at = 78.453200 bar		90 at = 88.259850 bar		100 at = 98.066500 bar	
Temperature		Mean specific heat							
t	T	c_{pm}		c_{pm}		c_{pm}		c_{pm}	
°C	K	kcal/kg K	kJ/kg K	kcal/kg K	kJ/kg K	kcal/kg K	kJ/kg K	kcal/kg K	kJ/kg K
300	573.15	1.039	4.350	1.207	5.053	—	—	—	—
320	593.15	0.972	4.070	1.090	4.564	1.217	5.095	1.366	5.719
340	613.15	0.915	3.381	1.008	4.220	1.115	4.668	1.236	5.175
360	633.15	0.866	3.626	0.943	3.948	1.027	4.300	1.127	4.719
380	653.15	0.829	3.471	0.893	3.739	0.964	4.036	1.044	4.371
400	673.15	0.798	3.341	0.855	3.580	0.915	3.831	0.982	4.111
420	693.15	0.773	3.236	0.824	3.450	0.876	3.668	0.934	3.910
440	713.15	0.754	3.157	0.798	3.341	0.845	3.538	0.896	3.751
460	733.15	0.737	3.086	0.777	3.253	0.820	3.433	0.865	3.622
480	753.15	0.722	3.023	0.760	3.182	0.798	3.341	0.839	3.513
500	773.15	0.710	2.973	0.744	3.115	0.779	3.262	0.817	3.421
520	793.15	0.699	2.927	0.731	3.061	0.763	3.195	0.798	3.341
540	813.15	0.690	2.889	0.719	3.010	0.749	3.136	0.781	3.270
560	833.15	0.681	2.851	0.708	2.964	0.737	3.086	0.766	3.207
580	853.15	0.673	2.818	0.699	2.927	0.726	3.040	0.754	3.157
600	873.15	0.667	2.793	0.691	2.893	0.716	2.998	0.743	3.111
620	893.15	0.661	2.767	0.684	2.864	0.708	2.964	0.733	3.069
640	913.15	0.656	2.747	0.678	2.839	0.700	2.931	0.724	3.031
660	933.15	0.651	2.726	0.672	3.814	0.693	2.901	0.716	2.998
680	953.15	0.647	2.709	0.667	2.793	0.687	2.876	0.708	2.964
700	973.15	0.643	2.692	0.662	2.772	0.682	2.855	0.702	2.939

Table 43-4 Mean Specific Heat c_{pm} of Superheated Steam (H_2O) at a Constant Pressure p (Between the Temperature of Saturation and a Given Temperature) (Continued)

Pressure p		110 at = 107.873150 bar		120 at = 117.679800 bar		130 at = 127.486450 bar		140 at = 137.293100 bar	
Temperature		Mean specific heat							
t	T	c_{pm}		c_{pm}		c_{pm}		c_{pm}	
°C	K	kcal/kg K	kJ/kg K	kcal/kg K	kJ/kg K	kcal/kg K	kJ/kg K	kcal/kg K	kJ/kg K
320	593.15	1.550	6.490	—	—	—	—	—	—
340	613.15	1.384	5.795	1.585	6.626	1.869	7.825	2.261	9.466
360	633.15	1.235	5.171	1.360	5.694	1.521	6.368	1.718	7.193
380	653.15	1.136	4.756	1.235	5.171	1.355	5.673	1.492	6.247
400	673.15	1.059	4.434	1.140	4.773	1.236	5.175	1.336	5.594
420	693.15	0.999	4.183	1.068	4.472	1.147	4.802	1.232	5.158
440	713.15	0.952	3.986	1.012	4.237	1.079	4.518	1.151	4.819
460	733.15	0.915	3.831	0.967	4.049	1.025	4.291	1.086	4.547
480	753.15	0.884	3.701	0.930	3.894	0.981	4.107	1.035	4.333
500	773.15	0.857	3.588	0.899	3.764	0.946	3.961	0.994	4.162
520	793.15	0.835	3.496	0.873	3.655	0.915	3.831	0.959	4.015
540	813.15	0.816	3.416	0.851	3.563	0.889	3.772	0.929	3.890
560	833.15	0.799	3.345	0.831	3.479	0.866	3.626	0.903	3.781
580	853.15	0.784	3.282	0.814	3.408	0.847	3.546	0.881	3.689
600	873.15	0.771	3.228	0.799	3.345	0.829	3.471	0.861	3.605
620	893.15	0.759	3.178	0.785	3.287	0.814	3.408	0.843	3.529
640	913.15	0.748	3.132	0.773	3.236	0.800	3.349	0.828	3.467
660	933.15	0.739	3.094	0.763	3.195	0.788	3.299	0.815	3.412
680	953.15	0.731	3.061	0.753	3.153	0.777	3.253	0.802	3.358
700	973.15	0.723	3.027	0.745	3.119	0.767	3.211	0.791	3.312

1 at = 1 kp/cm² = 98066.5 Pa = 98066.5 N/m² = 9.80665 N/cm² = 0.980665 bar

1 kcal = 4.1868 kJ

Table 43-5 Mean Specific Heat c_{pm} of Superheated Steam (H$_2$O) at a Constant Pressure p
(Between the Temperature of Saturation and a Given Temperature) (Continued)

Pressure p		150 at = 147.099750 bar		160 at = 156.906400 bar		170 at = 166.713050 bar		180 at = 176.519700 bar	
Temperature		Mean specific heat							
t	T	c_{pm}		c_{pm}		c_{pm}		c_{pm}	
°C	K	kcal/kg K	kJ/kg K	kcal/kg K	kJ/kg K	kcal/kg K	kJ/kg K	kcal/kg K	kJ/kg K
360	633.15	1.975	8.269	2.307	9.659	2.816	11.790	—	—
380	653.15	1.654	6.925	1.850	7.746	2.113	8.847	2.462	10.308
400	673.15	1.462	6.121	1.604	6.716	1.777	7.440	1.991	8.336
420	693.15	1.330	5.568	1.440	6.029	1.571	6.577	1.729	7.239
440	713.15	1.232	5.158	1.322	5.535	1.427	5.975	1.551	6.494
460	733.15	1.155	4.836	1.232	5.158	1.319	5.522	1.421	5.949
480	753.15	1.095	4.585	1.162	4.865	1.237	5.179	1.323	5.539
500	773.15	1.047	4.384	1.106	4.631	1.171	4.903	1.247	5.221
520	793.15	1.007	4.216	1.059	4.434	1.118	4.681	1.184	4.957
540	813.15	0.972	4.070	1.019	4.226	1.072	4.488	1.132	4.739
560	833.15	0.943	3.948	0.986	4.128	1.034	4.329	1.088	4.555
580	853.15	0.918	3.843	0.957	4.007	1.001	4.191	1.050	4.396
600	873.15	0.895	3.747	0.932	3.902	0.972	4.070	1.018	4.262
620	893.15	0.875	3.663	0.910	3.810	0.947	3.965	0.990	4.145
640	913.15	0.858	3.592	0.890	3.726	0.925	3.873	0.965	4.040
660	933.15	0.842	3.525	0.873	3.655	0.905	3.789	0.943	3.948
680	953.15	0.828	3.467	0.857	3.588	0.888	3.718	0.923	3.864
700	973.15	0.816	3.416	0.843	3.529	0.872	3.651	0.905	3.789

Table 43-6 Mean Specific Heat c_{pm} of Superheated Steam (H$_2$O) at a Constant Pressure p
(Between the Temperature of Saturation and a Given Temperature) (Continued)

Pressure p		190 at = 186.326350 bar		200 at = 196.133000 bar		210 at = 205.939650 bar		220 at = 215.746300 bar	
Temperature		Mean specific heat							
t	T	c_{pm}		c_{pm}		c_{pm}		c_{pm}	
°C	K	kcal/kg K	kJ/kg K	kcal/kg K	kJ/kg K	kcal/kg K	kJ/kg K	kcal/kg K	kJ/kg K
380	653.15	2.965	12.414	3.704	15.508	—	—	—	—
400	673.15	2.264	9.479	2.629	11.007	3.176	13.297	4.111	17.212
420	693.15	1.905	7.976	2.161	9.048	2.502	10.475	3.058	12.803
440	713.15	1.699	7.113	1.880	7.871	2.129	8.914	2.521	10.555
460	733.15	1.541	6.452	1.687	7.063	1.882	7.880	2.183	9.140
480	753.15	1.424	5.962	1.546	6.473	1.707	7.147	1.951	8.168
500	773.15	1.333	5.581	1.437	6.016	1.573	6.586	1.780	7.453
520	793.15	1.260	5.275	1.350	5.652	1.470	6.155	1.648	6.900
540	813.15	1.200	5.024	1.280	5.359	1.385	5.799	1.542	6.456
560	833.15	1.149	4.811	1.221	5.112	1.315	5.506	1.455	6.092
580	853.15	1.106	4.631	1.172	4.907	1.257	5.263	1.383	5.790
600	873.15	1.069	4.476	1.129	4.727	1.207	5.053	1.322	5.535
620	893.15	1.037	4.342	1.093	4.576	1.164	4.873	1.270	5.317
640	913.15	1.009	4.224	1.062	4.446	1.127	4.719	1.225	5.129
660	933.15	0.984	4.120	1.034	4.329	1.094	4.580	1.185	4.961
680	953.15	0.962	4.028	1.008	4.220	1.065	3.459	1.150	4.815
700	973.15	0.942	3.944	0.984	4.120	1.038	4.346	1.118	4.681

1 at = 1 kp/cm² = 98066.5 Pa = 98066.5 N/m² = 9.80665 N/cn.² = 0.980665 bar   1 kcal = 4.1868 kJ

Table 44-1 Properties of Saturated Steam (H₂O) at a Given Temperature (in International System of Units–SI)

Temperature		Pressure	Specific volume		Density		Specific enthalpy		Heat of vaporization	Specific entropy	
			Liquid	Vapor	Liquid	Vapor	Liquid	Vapor		Liquid	Vapor
t	T	p	v'	v''	ρ'	ρ''	i'	i''	$r = i'' - i'$	s'	s''
°C	K	bar	m³/kg	m³/kg	kg/m³	kg/m³	kJ/kg	kJ/kg	kJ/kg	kJ/kg K	kJ/kg K
0.01	273.16	0.006108	0.0010002	206.3	999.80	0.004847	0.00	2501	2501	0.0000	9.1544
1	274.15	0.006566	0.0010001	192.6	999.90	0.005192	4.22	2502	2498	0.0154	9.1281
2	275.15	0.007054	0.0010001	179.9	999.90	0.005559	8.42	2504	2496	0.0306	9.1018
3	276.15	0.007575	0.0010001	168.2	999.90	0.005945	12.63	2506	2493	0.0458	9.0757
4	277.15	0.008129	0.0010001	157.3	999.90	0.006357	16.84	2508	2491	0.0610	9.0498
5	278.15	0.008719	0.0010001	147.2	999.90	0.006793	21.05	2510	2489	0.0762	9.0241
6	279.15	0.009347	0.0010001	137.8	999.90	0.007257	25.25	2512	2489	0.0913	8.9978
7	280.15	0.010013	0.0010001	129.1	999.90	0.007746	29.45	2514	2485	0.1063	8.9736
8	281.15	0.010721	0.0010002	121.0	999.80	0.008264	33.55	2516	2482	0.1212	8.9485
9	282.15	0.011473	0.0010003	113.4	999.70	0.008818	37.85	2517	2479	0.1361	8.9238
10	283.15	0.012277	0.0010004	106.42	999.60	0.009398	42.04	2519	2477	0.1510	8.8994
11	284.15	0.013118	0.0010005	99.91	999.50	0.01001	46.22	2521	2475	0.1658	8.8752
12	285.15	0.014016	0.0010006	93.84	999.40	0.01066	50.41	2523	2473	0.1805	8.8513
13	286.15	0.014967	0.0010007	88.18	999.30	0.01134	54.60	2525	2470	0.1952	8.8276
14	287.15	0.015974	0.0010008	82.90	999.20	0.01206	58.78	2527	2468	0.2098	8.8040
15	288.15	0.017041	0.0010010	77.97	999.00	0.01282	62.97	2528	2465	0.2244	8.7806
16	289.15	0.018170	0.0010011	73.39	998.90	0.01363	67.16	2530	2463	0.2389	8.7574
17	290.14	0.019364	0.0010013	69.10	998.70	0.01447	71.34	2532	2461	0.2534	8.7344
18	291.15	0.02062	0.0010015	65.09	998.50	0.01536	75.53	2534	2458	0.2678	8.7116
19	292.15	0.02196	0.0010016	61.34	998.40	0.01630	79.72	2536	2456	0.2821	8.6890
20	293.15	0.02337	0.0010018	57.84	998.20	0.01729	83.90	2537	2454	0.2964	8.6665
21	294.15	0.02486	0.0010021	54.56	997.90	0.01833	88.09	2539	2451	0.3107	8.6442
22	295.15	0.02643	0.0010023	51.50	997.71	0.01942	92.27	2541	2449	0.3249	8.6220
23	296.15	0.02808	0.0010025	48.62	997.51	0.02057	96.46	2543	2447	0.3391	8.6001
24	297.15	0.02982	0.0010028	45.93	997.21	0.02177	100.63	2545	2444	0.3532	8.5785
25	298.15	0.03166	0.0010030	43.40	997.01	0.02304	104.81	2547	2442	0.3672	8.5570
26	299.15	0.03360	0.0010033	41.04	996.71	0.02437	108.99	2548	2440	0.3812	8.5358
27	300.15	0.03564	0.0010036	38.82	996.41	0.02576	113.17	2550	2437	0.3951	8.5147
28	301.15	0.03779	0.0010038	36.73	996.21	0.02723	117.35	2552	2435	0.4090	8.4938
29	302.15	0.04004	0.0010041	34.77	995.92	0.02876	121.53	2554	2432	0.4228	8.4730
30	303.15	0.04241	0.0010044	32.93	995.62	0.03037	125.71	2556	2430	0.4366	8.4523
31	304.15	0.04491	0.0010047	31.20	995.32	0.03205	129.89	2558	2428	0.4503	8.4319
32	305.15	0.04753	0.0010051	29.57	994.93	0.03382	134.07	2559	2425	0.4640	8.4117
33	306.15	0.05029	0.0010054	28.04	994.63	0.03566	138.25	2561	2423	0.4777	8.3916
34	307.15	0.05318	0.0010057	26.60	994.33	0.03759	142.42	2563	2421	0.4913	8.3716
35	308.15	0.05622	0.0010061	25.24	993.94	0.03962	146.60	2565	2418	0.5049	8.3519
36	309.15	0.05940	0.0010064	23.97	993.64	0.04172	150.78	2567	2416	0.5185	8.3323
37	310.15	0.06274	0.0010068	22.77	993.25	0.04392	154.96	2569	2414	0.5320	8.3129
38	311.15	0.06624	0.0010071	21.63	992.95	0.04623	159.14	2570	2411	0.5455	8.2938
39	312.15	0.06991	0.0010075	20.56	992.56	0.04864	163.32	2572	2409	0.5589	8.2748
40	313.15	0.07375	0.0010079	19.55	992.16	0.05115	167.50	2574	2406	0.5723	8.2559
41	314.15	0.07777	0.0010083	18.59	991.77	0.05379	171.67	2575	2403	0.5856	8.2372
42	315.15	0.08198	0.0010087	17.69	991.38	0.05653	175.86	2577	2401	0.5988	8.2187
43	316.15	0.08639	0.0010091	16.84	990.98	0.05938	180.04	2579	2399	0.6120	8.2003
44	317.15	0.09101	0.0010095	16.04	990.59	0.06234	184.22	2581	2397	0.6252	8.1820
45	318.15	0.09584	0.0010099	15.28	990.20	0.06544	188.40	2582	2394	0.6384	8.1638
46	319.15	0.10088	0.0010103	14.56	989.81	0.06868	192.58	2584	2391	0.6516	8.1458
47	320.15	0.10614	0.0010108	13.88	989.32	0.07205	196.76	2586	2389	0.6647	8.1279
48	321.15	0.11163	0.0010112	13.23	988.92	0.07559	200.93	2588	2387	0.6778	8.1102
49	322.15	0.11736	0.0010116	12.62	988.53	0.07924	205.11	2590	2385	0.6908	8.0927
50	323.15	0.12335	0.0010121	12.04	988.04	0.08306	209.3	2592	2383	0.7038	8.0753
51	324.15	0.12960	0.0010126	11.50	987.56	0.08696	213.5	2593	2380	0.7167	8.0579
52	325.15	0.13612	0.0010130	10.98	987.17	0.09107	217.7	2595	2377	0.7295	8.0407
53	326.15	0.14292	0.0010135	10.49	986.68	0.09533	221.9	2597	2375	0.7423	8.0236
54	327.15	0.15001	0.0010140	10.02	986.19	0.09980	226.0	2599	2373	0.7551	8.0068

Table 44-2 Properties of Saturated Steam (H₂O) at a Given Temperature (in International System of Units—SI) (*Continued*)

Temperature		Pressure	Specific volume		Density		Specific enthalpy		Heat of vaporization	Specific entropy	
			Liquid	Vapor	Liquid	Vapor	Liquid	Vapor		Liquid	Vapor
t	T	p	v'	v''	ρ'	ρ''	i'	i''	$r=i''-i'$	s'	s''
°C	K	bar	m³/kg	m³/kg	kg/m³	kg/m³	kJ/kg	kJ/kg	kJ/kg	kJ/kg K	kJ/kg K
55	328.15	0.15740	0.0010145	9.578	985.71	0.1044	230.2	2600	2370	0.7679	7.9901
56	329.15	0.16510	0.0010150	9.158	985.22	0.1092	234.4	2602	2368	0.7806	7.9736
57	330.15	0.17312	0.0010155	8.757	984.74	0.1142	238.6	2604	2365	0.7933	7.9571
58	331.15	0.18146	0.0010160	8.380	984.25	0.1193	242.8	2606	2363	0.8059	7.9407
59	332.15	0.19014	0.0010166	8.020	983.67	0.1247	246.9	2608	2361	0.8185	7.9245
60	333.15	0.19917	0.0010171	7.678	983.19	0.1302	251.1	2609	2358	0.8311	7.9084
61	334.15	0.2086	0.0010177	7.353	982.61	0.1360	255.3	2611	2355	0.8436	7.8925
62	335.15	0.2184	0.0010182	7.043	982.13	0.1420	259.5	2613	2353	0.8561	7.8767
63	336.15	0.2285	0.0010188	6.749	981.55	0.1482	263.7	2614	2350	0.8686	7.8609
64	337.15	0.2391	0.0010193	6.468	981.07	0.1546	267.9	2616	2348	0.8810	7.8452
65	338.15	0.2501	0.0010199	6.201	980.49	0.1613	272.1	2617	2345	0.8934	7.8297
66	339.15	0.2615	0.0010205	5.947	979.91	0.1681	276.2	2619	2343	0.9057	7.8144
67	340.15	0.2733	0.0010210	5.705	979.43	0.1753	280.4	2621	2341	0.9180	7.7992
68	341.15	0.2856	0.0010216	5.475	978.86	0.1826	284.6	2623	2338	0.9303	7.7841
69	342.15	0.2984	0.0010222	5.255	978.28	0.1903	288.8	2625	2336	0.9426	7.7692
70	343.15	0.3117	0.0010228	5.045	977.71	0.1982	293.0	2626	2333	0.9549	7.7544
71	344.15	0.3254	0.0010234	4.846	977.14	0.2064	297.2	2628	2331	0.9672	7.7396
72	345.15	0.3396	0.0010240	4.655	976.56	0.2148	301.4	2630	2329	0.9794	7.7249
73	346.15	0.3543	0.0010246	4.473	975.99	0.2236	305.6	2631	2326	0.9916	7.7103
74	347.15	0.3696	0.0010252	4.299	975.42	0.2326	309.8	2633	2323	1.0037	7.6958
75	348.15	0.3855	0.0010258	4.133	974.85	0.2420	314.0	2635	2321	1.0157	7.6815
76	349.15	0.4019	0.0010264	3.975	974.28	0.2516	318.2	2636	2318	1.0277	7.6673
77	350.15	0.4189	0.0010270	3.824	973.71	0.2615	322.4	2638	2316	1.0396	7.6533
78	351.15	0.4365	0.0010277	3.679	973.05	0.2718	326.4	2639	2313	1.0515	7.6393
79	352.15	0.4547	0.0010283	3.540	972.48	0.2825	330.6	2641	2310	1.0634	7.6254
80	353.15	0.4736	0.0010290	3.408	971.82	0.2934	334.9	2643	2308	1.0753	7.6116
81	354.15	0.4931	0.0010297	3.282	971.16	0.3047	339.1	2645	2306	1.0872	7.5979
82	355.15	0.5133	0.0010304	3.161	970.50	0.3164	343.3	2646	2303	1.0990	7.5843
83	356.15	0.5342	0.0010310	3.045	969.93	0.3284	347.5	2648	2300	1.1107	7.5707
84	357.15	0.5558	0.0010317	2.934	969.27	0.3408	351.7	2650	2298	1.1225	7.5572
85	358.15	0.5781	0.0010324	2.828	968.62	0.3536	355.9	2651	2295	1.1342	7.5438
86	359.15	0.6011	0.0010331	2.727	967.96	0.3667	360.1	2653	2293	1.1459	7.5305
87	360.15	0.6249	0.0010338	2.629	967.31	0.3804	364.3	2655	2291	1.1576	7.5174
88	361.15	0.6495	0.0010345	2.536	966.65	0.3943	368.5	2656	2288	1.1693	7.5044
89	362.15	0.6749	0.0010352	2.447	966.00	0.4087	372.7	2658	2285	1.1809	7.4915
90	363.15	0.7011	0.0010359	2.361	965.34	0.4235	377.0	2659	2282	1.1925	7.4787
91	364.15	0.7281	0.0010366	2.279	964.69	0.4388	381.2	2661	2280	1.2041	7.4660
92	365.15	0.7560	0.0010373	2.200	964.04	0.4545	385.4	2662	2277	1.2157	7.4533
93	366.15	0.7848	0.0010381	2.124	963.30	0.4708	389.6	2664	2274	1.2272	7.4407
94	367.15	0.8145	0.0010388	2.052	962.65	0.4873	393.8	2666	2272	1.2387	7.4281
95	368.15	0.8451	0.0010396	1.982	961.91	0.5045	398.0	2668	2270	1.2502	7.4155
96	369.15	0.8767	0.0010404	1.915	961.17	0.5222	402.2	2669	2267	1.2617	7.4030
97	370.15	0.9093	0.0010412	1.851	960.43	0.5402	406.4	2671	2265	1.2731	7.3907
98	371.15	0.9429	0.0010420	1.789	959.69	0.5590	410.7	2673	2262	1.2845	7.3786
99	372.15	0.9775	0.0010427	1.730	959.05	0.5780	414.9	2674	2259	1.2958	7.3666
100	373.15	1.0131	0.0010435	1.673	958.31	0.5977	419.1	2676	2257	1.3071	7.3547
101	374.15	1.0498	0.0010443	1.618	957.58	0.6181	423.3	2677	2254	1.3184	7.3429
102	375.15	1.0876	0.0010450	1.566	956.94	0.6386	427.5	2679	2251	1.3297	7.3311
103	376.15	1.1265	0.0010458	1.515	956.21	0.6601	431.7	2680	2248	1.3409	7.3193
104	377.15	1.1666	0.0010466	1.466	955.47	0.6821	436.0	2681	2245	1.3521	7.3076
105	378.15	1.2079	0.0010474	1.419	954.75	0.7047	440.2	2683	2243	1.3632	7.2959
106	379.15	1.2504	0.0010482	1.374	954.02	0.7278	444.4	2685	2241	1.3743	7.2843
107	380.15	1.2941	0.0010490	1.331	953.29	0.7513	448.6	2687	2238	1.3854	7.2728
108	381.15	1.3390	0.0010498	1.289	952.56	0.7758	452.9	2688	2235	1.3964	7.2614
109	382.15	1.3852	0.0010507	1.249	951.75	0.8006	457.1	2689	2232	1.4074	7.2500

Table 44-3 Properties of Saturated Steam (H_2O) at a Given Temperature (in International System of Units—SI) (Continued)

Temperature		Pressure	Specific volume		Density		Specific enthalpy		Heat of vaporization	Specific entropy	
			Liquid	Vapor	Liquid	Vapor	Liquid	Vapor		Liquid	Vapor
t	T	p	v'	v''	ρ'	ρ''	i'	i''	$r=i''-i'$	s'	s''
°C	K	bar	m³/kg	m³/kg	kg/m³	kg/m³	kJ/kg	kJ/kg	kJ/kg	kJ/kg K	kJ/kg K
110	383.15	1.4326	0.0010515	1.210	951.02	0.8264	461.3	2691	2230	1.4184	7.2387
111	384.15	1.4814	0.0010523	1.173	950.30	0.8525	465.6	2693	2272	1.4294	7.2274
112	385.15	1.5316	0.0010532	1.137	949.49	0.8795	469.8	2694	2224	1.4404	7.2162
113	386.15	1.5831	0.0010540	1.102	948.77	0.9074	474.0	2696	2222	1.4514	7.2051
114	387.15	1.6361	0.0010549	1.069	947.96	0.9354	478.2	2697	2219	1.4624	7.1941
115	388.15	1.6905	0.0010559	1.036	947.15	0.9652	482.5	2698	2216	1.4733	7.1832
116	389.15	1.7464	0.0010567	1.005	946.34	0.9950	486.7	2700	2213	1.4842	7.1724
117	390.15	1.8038	0.0010576	0.9754	945.54	1.025	491.0	2702	2211	1.4951	7.1616
118	391.15	1.8628	0.0010585	0.9465	944.73	1.056	495.2	2703	2208	1.5060	7.1509
119	392.15	1.9233	0.0010594	0.9186	943.93	1.089	499.5	2705	2205	1.5169	7.1403
120	393.15	1.9854	0.0010603	0.8917	943.13	1.121	503.7	2706	2202	1.5277	7.1298
121	394.15	2.0491	0.0010612	0.8657	942.33	1.155	507.9	2708	2200	1.5385	7.1193
122	395.15	2.1144	0.0010621	0.8407	941.53	1.189	512.2	2709	2197	1.5492	7.1089
123	396.15	2.1814	0.0010630	0.8164	940.73	1.225	516.5	2710	2194	1.5599	7.0985
124	397.15	2.2502	0.0010640	0.7930	939.85	1.261	520.8	2712	2191	1.5706	7.0881
125	398.15	2.3208	0.0010649	0.7704	939.06	1.298	525.0	2713	2188	1.5814	7.0777
126	399.15	2.3932	0.0010658	0.7486	938.26	1.336	529.2	2715	2186	1.5922	7.0674
127	400.15	2.4674	0.0010668	0.7276	937.38	1.374	533.4	2716	2183	1.6029	7.0573
128	401.15	2.5434	0.0010677	0.7074	936.59	1.414	537.7	2718	2180	1.6135	7.0472
129	402.15	2.6213	0.0010687	0.6880	935.72	1.454	542.0	2719	2177	1.6240	7.0372
130	403.15	2.7011	0.0010697	0.6683	934.84	1.496	546.3	2721	2174	1.6354	7.0272
131	404.15	2.7829	0.0010707	0.6499	933.97	1.539	550.5	2722	2171	1.6450	7.0173
132	504.15	2.8668	0.0010717	0.6321	933.10	1.582	554.8	2723	2168	1.6555	7.0074
133	406.15	2.9528	0.0010727	0.6148	932.23	1.626	559.0	2724	2165	1.6659	6.9976
134	407.15	3.041	0.0010737	0.5981	931.36	1.672	563.2	2725	2162	1.6764	6.9878
135	408.15	3.130	0.0010747	0.5820	930.49	1.718	567.5	2727	2159	1.6869	6.9781
136	409.15	3.222	0.0010757	0.5664	929.63	1.765	571.8	2728	2156	1.6973	6.9685
137	410.15	3.317	0.0010767	0.5512	928.76	1.814	576.1	2730	2154	1.7078	6.9589
138	411.15	3.414	0.0010777	0.5366	927.90	1.864	580.4	2731	2151	1.7183	6.9493
139	412.15	3.513	0.0010788	0.5224	926.96	1.914	584.7	2733	2148	1.7278	6.9398
140	413.15	3.614	0.0010798	0.5087	926.10	1.966	589.0	2734	2145	1.7392	6.9304
141	414.15	3.717	0.0010808	0.4953	925.24	2.019	593.3	2735	2142	1.7496	6.9211
142	415.15	3.823	0.0010819	0.4824	924.30	2.073	597.6	2737	2139	1.7599	6.9117
143	416.15	3.931	0.0010829	0.4699	923.45	2.128	601.9	2738	2136	1.7702	6.9024
144	417.15	4.042	0.0010840	0.4579	922.51	2.184	606.2	2739	2133	1.7804	6.8932
145	418.15	4.155	0.0010851	0.4461	921.57	2.242	610.5	2740	2130	1.7907	6.8839
146	419.15	4.271	0.0010862	0.4347	920.64	2.300	614.8	2742	2127	1.8009	6.8747
147	420.15	4.389	0.0010873	0.4237	919.71	2.360	619.1	2743	2124	1.8112	6.8655
148	421.15	4.510	0.0010884	0.4130	918.78	2.421	623.4	2744	2121	1.8214	6.8564
149	422.15	4.634	0.0010895	0.4026	917.85	2.484	627.8	2745	2117	1.8316	6.8473
150	423.15	4.760	0.0010906	0.3926	916.93	2.547	632.2	2746	2114	1.8418	6.8383
151	424.15	4.889	0.0010917	0.3828	916.00	2.612	636.6	2748	2111	1.8520	6.8293
152	425.15	5.020	0.0010928	0.3733	915.08	2.679	641.0	2749	2108	1.8622	6.8204
153	426.15	5.155	0.0010939	0.3641	914.16	2.746	645.3	2750	2105	1.8723	6.8115
154	427.15	5.293	0.0010950	0.3552	913.24	2.815	649.6	2752	2102	1.8824	6.8027
155	428.15	5.433	0.0010962	0.3466	912.24	2.885	653.9	2753	2099	1.8924	6.7940
156	429.15	5.576	0.0010974	0.3381	911.24	2.958	658.2	2754	2096	1.9025	6.7854
157	430.15	5.723	0.0010986	0.3299	910.25	3.030	662.5	2755	2092	1.9125	6.7768
158	431.15	5.872	0.0010998	0.3220	909.26	3.106	666.9	2756	2089	1.9226	6.7681
159	432.15	6.024	0.0011009	0.3143	908.35	3.182	671.2	2757	2086	1.9326	6.7595
160	433.15	6.180	0.0011021	0.3068	907.36	3.258	675.6	2758	2082	1.9427	6.7508
161	434.15	6.339	0.0011033	0.2996	906.37	3.338	679.9	2759	2079	1.9527	6.7421
162	435.15	6.502	0.0011044	0.2925	905.47	3.419	684.2	2760	2076	1.9627	6.7335
163	436.15	6.667	0.0011056	0.2856	904.49	3.500	688.6	2761	2072	1.9726	6.7250
164	437.15	6.836	0.0011069	0.2790	903.42	3.584	692.9	2762	2069	1.9825	6.7165

Table 44-4 Properties of Saturated Steam (H$_2$O) at a Given Temperature (in International System of Units—SI) (*Continued*)

Temperature		Pressure	Specific volume		Density		Specific enthalpy		Heat of vaporization	Specific entropy	
			Liquid	Vapor	Liquid	Vapor	Liquid	Vapor		Liquid	Vapor
t	T	p	v'	v''	ρ'	ρ''	i'	i''	$r = i'' - i'$	s'	s''
°C	K	bar	m³/kg	m³/kg	kg/m³	kg/m³	kJ/kg	kJ/kg	kJ/kg	kJ/kg K	kJ/kg K
165	438.15	7.008	0.0011081	0.2725	902.45	3.670	697.3	2763	2066	1.9924	6.7081
166	439.15	7.183	0.0011094	0.2662	901.39	3.757	701.7	2764	2062	2.0023	6.6998
167	440.15	7.362	0.0011106	0.2600	900.41	3.846	706.1	2765	2059	2.0122	6.6915
168	441.15	7.545	0.0011119	0.2541	899.36	3.935	710.5	2667	2056	2.0221	6.6832
169	442.15	7.731	0.0011131	0.2483	898.39	4.027	714.8	2768	2053	2.0319	6.6749
170	443.15	7.920	0.0011144	0.2426	897.34	4.122	719.2	2769	2050	2.0417	6.6666
171	444.15	8.114	0.0011156	0.2371	896.38	4.218	723.5	2770	2046	2.0515	6.6583
172	445.14	8.311	0.0011169	0.2318	895.34	4.314	727.9	2771	2043	2.0614	6.6500
173	446.15	8.511	0.0011182	0.2266	894.29	4.413	732.3	2772	2040	2.0712	6.6418
174	447.15	8.176	0.0011195	0.2215	893.26	4.515	736.7	2773	2036	2.0811	6.6336
175	448.15	8.925	0.0011208	0.2166	892.22	4.617	741.1	2773	2032	2.0909	6.6256
176	449.15	9.137	0.0011221	0.2118	891.19	4.721	745.5	2774	2029	2.1006	6.6177
177	450.15	9.354	0.0011234	0.2071	890.15	4.829	749.9	2775	2025	2.1103	6.6097
178	451.15	9.574	0.0011248	0.2026	889.05	4.936	754.3	2776	2022	2.1201	6.6017
179	452.15	9.799	0.0011261	0.1982	888.02	5.045	758.7	2777	2018	2.1298	6.5938
180	453.15	10.027	0.0011275	0.1939	886.92	5.157	763.1	2778	2015	2.1395	6.5858
181	454.15	10.260	0.0011289	0.1897	885.82	5.271	767.5	2779	2011	2.1491	6.5779
182	455.15	10.497	0.0011303	0.1856	884.72	5.388	771.9	2780	2008	2.1587	6.5700
183	456.15	10.738	0.0011316	0.1816	883.70	5.507	776.3	2780	2004	2.1683	6.5622
184	457.15	10.984	0.0011330	0.1777	882.61	5.627	780.7	2781	2000	2.1780	6.5344
185	458.15	11.234	0.0011344	0.1739	881.52	5.750	785.2	2782	1997	2.1876	6.5465
186	459.15	11.488	0.0011358	0.1702	880.44	5.875	789.6	2783	1993	2.1972	6.5386
187	460.15	11.747	0.0011372	0.1666	879.35	6.002	794.0	2784	1990	2.2069	6.5307
188	461.15	12.011	0.0011386	0.1631	878.27	6.131	798.5	2784	1986	2.2165	6.5229
189	462.15	12.280	0.0011401	0.1597	877.12	6.262	803.0	2785	1982	2.2261	6.5151
190	463.15	12.553	0.0011415	0.1564	876.04	6.394	807.5	2786	1979	2.2357	6.5074
191	464.15	12.830	0.0011430	0.1531	874.89	6.532	811.9	2787	1975	2.2453	6.4998
192	465.15	13.112	0.0011445	0.1499	873.74	6.671	816.4	2787	1971	2.2584	6.4921
193	466.15	13.400	0.0011459	0.1468	872.68	6.812	820.9	2788	1967	2.2643	6.4845
194	467.15	13.692	0.0011474	0.1438	871.54	6.954	825.4	2789	1964	2.2739	6.4770
195	468.15	13.989	0.0011489	0.1409	870.40	7.097	829.9	2790	1960	2.2834	6.4694
196	469.15	14.291	0.0011504	0.1380	869.26	7.246	834.4	2790	1956	2.2929	6.4619
197	470.15	14.598	0.0011519	0.1352	868.13	7.396	838.9	2791	1952	2.3024	6.4544
198	471.15	14.910	0.0011534	0.1325	867.00	7.547	843.4	2792	1949	2.3119	6.4468
199	472.15	15.228	0.0011550	0.1298	865.80	7.704	847.9	2793	1945	2.3214	6.4393
200	473.15	15.551	0.0011565	0.1272	864.68	7.862	852.4	2793	1941	2.3308	6.4318
201	474.15	15.879	0.0011581	0.1246	863.48	8.026	856.9	2793	1936	2.3402	6.4243
202	475.15	16.212	0.0011596	0.1221	862.37	8.190	861.5	2794	1932	2.3496	6.4168
203	476.15	16.551	0.0011612	0.1197	861.18	8.354	866.0	2794	1928	2.3590	6.4094
204	677.15	16.895	0.0011628	0.1174	859.99	8.518	870.5	2795	1924	2.3684	6.4020
205	478.15	17.245	0.0011644	0.1151	858.81	8.688	875.0	2796	1921	2.3777	6.3945
206	479.15	17.601	0.0011660	0.1128	857.63	8.865	879.6	2797	1917	2.3870	6.3871
207	480.15	17.962	0.0011676	0.1106	856.46	9.042	884.2	2797	1913	2.3964	6.3797
208	481.15	18.329	0.0011693	0.1084	855.21	9.225	888.7	2797	1908	2.4058	6.3723
209	482.15	18.701	0.0011709	0.1063	854.04	9.407	893.2	2798	1905	2.4152	6.3650
210	483.15	19.080	0.0011726	0.1043	852.81	9.588	897.7	2798	1900	2.4246	6.3577
211	484.15	19.464	0.0011743	0.1023	851.57	9.775	902.3	2798	1896	2.4340	6.3504
212	485.15	19.855	0.0011760	0.1003	850.34	9.970	906.9	2799	1892	2.4434	6.3431
213	486.15	20.252	0.0011778	0.09836	849.04	10.17	911.5	2799	1888	2.4528	6.3358
214	487.15	20.654	0.0011795	0.09649	847.82	10.36	916.1	2800	1884	2.4622	6.3285
215	488.15	21.062	0.0011812	0.09465	846.60	10.56	920.7	2800	1879	2.4751	6.3212
216	489.15	21.477	0.0011829	0.09285	845.38	10.77	925.3	2800	1875	2.4808	6.3140
217	490.15	21.899	0.0011846	0.09110	844.17	10.98	929.9	2801	1871	2.4901	6.3067
218	491.15	22.327	0.0011864	0.08938	842.89	11.19	934.5	2801	1867	2.4994	6.2994
219	492.15	22.761	0.0011882	0.08770	841.61	11.40	939.1	2801	1862	2.5087	6.2921

Table 44-5 Properties of Saturated Steam (H₂O) at a Given Temperature (in International System of Units—SI) *(Continued)*

Temperature		Pressure	Specific volume		Density		Specific enthalpy		Heat of vaporization	Specific entropy	
			Liquid	Vapor	Liquid	Vapor	Liquid	Vapor		Liquid	Vapor
t	T	p	v'	v''	ρ'	ρ''	i'	i''	$r = i'' - i'$	s'	s''
°C	K	bar	m³/kg	m³/kg	kg/m³	kg/m³	kJ/kg	kJ/kg	kJ/kg	kJ/kg K	kJ/kg K
220	493.15	23.201	0.0011900	0.08606	840.34	11.62	943.7	2802	1858	2.5179	6.2849
221	494.15	23.649	0.0011918	0.08446	839.07	11.84	948.3	2802	1854	2.5272	6.2776
222	495.15	24.103	0.0011937	0.08288	837.73	12.06	952.9	2802	1849	2.5364	6.2704
223	496.15	24.563	0.0011955	0.08135	836.47	12.29	957.5	2802	1845	2.5456	6.2632
224	497.15	25.030	0.0011973	0.07984	835.21	12.52	962.2	2802	1840	2.5548	6.2560
225	498.15	25.504	0.0011992	0.07837	833.89	12.76	966.9	2802	1835	2.5640	6.2488
226	499.15	25.984	0.0012011	0.07693	832.57	13.00	971.6	2803	1831	2.5732	6.2417
227	500.15	26.475	0.0012029	0.07552	831.32	13.24	976.3	2803	1826	2.5824	6.2346
228	501.15	26.967	0.0012048	0.07414	830.01	13.49	981.0	2803	1822	2.5916	6.2275
229	502.15	27.469	0.0012068	0.07279	828.64	13.74	985.7	2803	1817	2.6008	6.2204
230	503.15	27.979	0.0012087	0.07147	827.34	13.99	990.4	2803	1813	2.6101	6.2133
231	504.15	28.495	0.0012107	0.07018	825.97	14.25	995.1	2804	1809	2.6193	6.2063
232	505.15	29.019	0.0012126	0.06891	824.67	14.51	999.8	2804	1804	2.6285	6.1993
233	506.15	29.550	0.0012146	0.06767	823.32	14.78	1004.5	2804	1800	2.6377	6.1922
234	507.15	30.089	0.0012167	0.06646	821.90	15.05	1009.2	2804	1795	2.6469	6.1851
235	508.15	30.635	0.0012187	0.06527	820.35	15.32	1013.9	2804	1790	2.6561	6.1780
236	509.15	31.188	0.0012208	0.06410	819.13	15.60	1018.6	2804	1785	2.6653	6.1709
237	510.15	31.749	0,0012228	0.06296	817.80	15.88	1023.3	2804	1781	2.6745	6.1638
238	511.15	32.318	0.0012249	0.06184	816.39	16.17	1028.1	2804	1776	2.6837	6.1567
239	512.15	32.895	0.0012270	0.06075	815.00	16.46	1032.8	2804	1771	2.6929	6.1496
240	513.15	33.480	0.0012291	0.05967	813.60	16.76	1037.5	2803	1766	2.7021	6.1425
241	514.14	34.073	0.0012312	0.05862	812.22	17.06	1042.3	2803	1761	2.7113	6.1354
242	515.15	34.673	0.0012334	0.05759	810.77	17.36	1047.1	2803	1756	2.7205	6.1283
243	516.15	35.282	0.0012355	0.05658	809.39	17.67	1051.9	2803	1751	2.7296	6.1213
244	517.15	35.899	0.0012377	0.05559	807.95	17.99	1056.7	2803	1746	2.7387	6.1143
245	518.15	36.524	0.0012399	0.05462	806.52	18.30	1061.6	2803	1741	2.7478	6.1073
246	519.15	37.157	0.0012421	0.05367	805.09	18.63	1066.4	2802	1736	2.7569	6.1003
247	520.15	37.799	0.0012443	0.05274	803.66	18.96	1071.3	2802	1731	2.7660	6.0933
248	521.15	38.450	0.0012466	0.05183	802.18	19.29	1076.1	2802	1726	2.7751	6.0863
249	522.15	39.109	0.0012489	0.05093	800.70	19.63	1080.9	2802	1721	2.7842	6.0792
250	523.15	39.776	0.0012512	0.05006	799.23	19.28	1085.7	2801	1715	2.7934	6.0721
251	524.15	40.45	0.0012536	0.04919	797.70	20.33	1090.6	2801	1710	2.8026	6.0650
252	525.15	41.14	0.0012559	0.04835	796.24	20.68	1095.5	2801	1705	2.8118	6.0579
253	526.15	41.84	0.0012583	0.04752	794.72	21.04	1100.4	2800	1700	2.8210	6.0508
254	527.15	42.54	0.0012607	0.04671	793.21	21.41	1105.3	2799	1694	2.8302	6.0437
255	528.15	43.25	0.0012631	0.04591	791.70	21.78	1110.2	2799	1689	2.8394	6.0366
256	529.15	43.97	0.0012655	0.04513	790.20	22.16	1115.2	2798	1683	2.8486	6.0295
257	530.15	44.70	0.0012680	0.04436	788.64	22.54	1120.1	2798	1678	2.8577	6.0224
258	531.15	45.43	0.0012705	0.04361	787.09	22.93	1125.1	2797	1672	2.8668	6.0154
259	532.15	46.18	0.0012730	0.04287	785.55	23.33	1130.1	2797	1667	2.8759	6.0084
260	533.15	46.94	0.0012755	0.04215	784.01	23.72	1135.1	2796	1661	2.8851	6.0013
261	534.15	47.71	0.0012781	0.04144	782.41	24.13	1140.1	2796	1656	2.8942	5.9942
262	535.15	48.48	0.0012807	0.04074	780.82	24.55	1145.1	2795	1650	2.9034	5.9871
263	536.51	49.27	0.0012833	0.04005	779.24	24.96	1150.1	2795	1645	2.9125	5.9800
264	537.15	50.06	0.0012859	0.03938	777.67	25.39	1155.1	2795	1640	2.9216	5.9729
265	538.15	50.87	0.0012886	0.03872	776.04	25.83	1160.2	2794	1634	2.9307	5.9657
266	539.15	51.69	0.0012913	0.03807	774.41	26.26	1165.2	2793	1628	2.9398	5.9585
267	540.15	52.51	0.0012940	0.03744	772.80	26.71	1170.2	2792	1622	2.9489	5.9513
268	541.15	53.35	0.0012967	0.03681	771.19	27.16	1175.2	2791	1616	2.9580	5.9441
269	542.15	54.19	0.0012995	0.03620	769.53	27.62	1180.3	2790	1610	2.9672	5.9369
270	543.15	55.05	0.0013023	0.03560	767.87	28.09	1185.3	2790	1605	2.9764	5.9297
271	544.15	55.92	0.0013051	0.03501	766.22	28.56	1190.3	2789	1598,7	2.9856	5.9225
272	545.15	56.79	0.0013080	0.03443	764.53	29.04	1195.4	2788	1592,7	2.9948	5.9153
273	546.15	57.68	0.0013109	0.03386	762.83	29.53	1200.5	2787	1586,6	3.0040	5.9082
274	547.15	58.58	0.0013138	0.03330	761.15	30.03	1205.6	2786	1580,4	3.0132	5.9010

Table 44-6 Properties of Saturated Steam (H₂O) at a Given Temperature (in International System of Units—SI) (*Continued*)

Temperature		Pressure	Specific volume		Density		Specific enthalpy		Heat of vaporization	Specific entropy	
			Liquid	Vapor	Liquid	Vapor	Liquid	Vapor		Liquid	Vapor
t	T	p	v'	v''	ρ'	ρ''	i'	i''	$r=i''-i'$	s'	s''
°C	K	bar	m³/kg	m³/kg	kg/m³	kg/m³	kJ/kg	kJ/kg	kJ/kg	kJ/kg K	kJ/kg K
275	548.15	59.49	0.0013168	0.03274	759.42	30.53	1210.7	2785	1574.2	3.0223	5.8938
276	549.15	60.41	0.0013198	0.03220	757.69	31.06	1215.9	2784	1568.0	3.0314	5.8865
277	550.15	61.34	0.0013228	0.03167	755.97	31.58	1221.1	2783	1561.7	3.0405	5.8792
278	551.15	62.28	0.0013259	0.03115	754.20	32.10	1226.3	2782	1555.5	3.0497	5.8719
279	552.15	63.23	0.0013290	0.03064	752.45	32.64	1231.6	2781	1549.2	3.0589	5.8646
280	553.15	64.19	0.0013321	0.03013	750.69	33.19	1236.9	2780	1542.9	3.0681	5.8573
281	554.15	65.17	0.0013353	0.02964	748.90	33.74	1242.1	2779	1536.5	3.0774	5.8500
282	555.15	66.16	0.0013385	0.02915	747.10	34.30	1247.3	2777	1530.0	3.0867	5.8427
283	556.15	67.15	0.0013417	0.02867	745.32	34.88	1252.5	2776	1523.5	3.0960	5.8353
284	557.15	68.16	0.0013450	0.02820	743.49	35.46	1257.8	2775	1516.9	3.1053	5.8279
285	558.15	69.18	0.0013483	0.02773	741.67	36.05	1263.1	2773	1510.2	3.1146	5.8205
286	559.15	70.21	0.0013516	0.02728	739.86	36.66	1268.4	2772	1503.5	3.1239	5.8130
287	560.15	71.25	0.0013550	0.02684	738.01	37.26	1273.7	2771	1496.8	3.1331	5.8054
288	561.15	72.30	0.0013585	0.02640	736.11	37.88	1279.1	2769	1490.0	3.1424	5.7979
289	562.15	73.37	0.0013620	0.02596	734.12	38.52	1284.5	2768	1483.2	3.1517	5.7903
290	563.15	74.45	0.0013655	0.02554	732.33	39.15	1290.0	2766	1476.3	3.1611	5.7827
291	564.15	75.54	0.0013691	0.02512	730.41	39.81	1295.5	2765	1469.4	3.1705	5.7751
292	565.15	76.64	0.0013727	0.02471	728.49	40.47	1300.9	2763	1462.4	3.1798	5.7674
293	566.15	77.76	0.0013764	0.02430	726.53	41.15	1306.3	2761	1455.3	3.1892	5.7597
294	567.15	78.88	0.0013801	0.02390	724.59	31.84	1311.7	2760	1448.2	3.1986	5.7520
295	568.15	80.02	0.0013839	0.02351	722.60	42.53	1317.2	2758	1441.0	3.2079	5.7443
296	569.15	81.18	0.0013877	0.02312	720.62	43.23	1322.7	2757	1433.8	3.2172	5.7365
297	570.15	82.35	0.0013916	0.02275	718.60	43.96	1328.3	2755	1426.6	3.2265	5.7287
298	571.15	83.52	0.0013956	0.02237	716.54	44.70	1333.8	2753	1419.3	3.2359	5.7208
299	572.15	84.71	0.0013996	0.02200	714.49	45.43	1339.3	2751	1411.8	3.2453	5.7129
300	573.15	85.92	0.0014036	0.02164	712.45	46.21	1344.9	2749	1404.3	3.2548	5.7049
301	574.15	87.14	0.001407	0.02129	710.73	46.97	1350.5	2747	1396.6	3.2644	5.6969
302	575.15	88.37	0.001412	0.02094	708.22	47.75	1356.2	2745	1389.0	3.2739	5.6889
303	576.15	89.61	0.001416	0.02059	706.21	48.57	1361.9	2744	1381.3	3.2834	5.6809
304	577.15	90.87	0.001420	0.02025	704.23	49.38	1367.5	2741	1373.5	3.2930	5.6728
305	578.15	92.14	0.001425	0.01992	701.75	50.20	1373.1	2739	1365.6	3.3026	5.6647
306	579.15	93.42	0.001429	0.01959	699.79	51.05	1378.8	2737	1357.6	3.3122	5.6565
307	580.15	94.72	0.001434	0.01926	697.35	51.92	1384.5	2734	1349.6	3.3218	5.6483
308	581.15	96.03	0.001438	0.01894	695.41	52.80	1390.3	2732	1341.6	3.3314	5.6400
309	582.15	97.36	0.001443	0.01863	693.00	53.68	1396.2	2730	1333.5	3.3411	5.6317
310	583.15	98.70	0.001447	0.01832	691.09	54.58	1402.1	2727	1325.2	3.3508	5.6233
311	584.15	100.05	0.001452	0.01801	688.71	55.52	1408.0	2725	1316.7	3.3605	5.6148
312	585.15	101.42	0.001457	0.01771	686.34	56.46	1413.9	2722	1308.2	3.3702	5.6063
313	586.15	102.80	0.001462	0.01741	683.99	57.44	1419.8	2720	1299.6	3.3800	5.5977
314	587.15	104.20	0.001467	0.01712	681.66	58.41	1425.7	2717	1291.0	3.3898	5.5890
315	588.15	105.61	0.001472	0.01683	679.35	59.42	1431.7	2714	1282.3	3.3996	5.5802
316	589.15	107.04	0.001477	0.01655	677.05	60.42	1437.7	2711	1273.4	3.4095	5.5714
317	590.15	108.48	0.001483	0.01627	674.31	61.46	1443.8	2709	1264.6	3.4194	5.5625
318	591.15	109.94	0.001488	0.01599	672.04	62.54	1449.9	2706	1255.8	3.4294	5.5535
319	592.15	111.41	0.001494	0.01572	669.34	63.61	1456.0	2703	1246.8	3.4395	5.5444
320	593.15	112.90	0.001499	0.01545	667.11	64.72	1462.1	2700	1237.8	3.4495	5.5353
321	594.15	114.40	0.001505	0.01519	664.45	65.83	1468.3	2697	1228.7	3.4596	5.5261
322	595.15	115.92	0.001511	0.01493	661.81	66.98	1474.6	2694	1219.4	3.4697	5.5169
323	596.15	117.15	0.001517	0.01467	659.20	68.17	1480.9	2691	1209.9	3.4798	5.5077
324	597.15	119.00	0.001523	0.01442	656.60	69.35	1487.2	2687	1200.2	3.4900	5.4984
325	598.15	120.57	0.001529	0.01417	654.02	70.57	1493.6	2684	1190.3	3.5002	5.4191
326	599.15	122.15	0.001535	0.01392	651.47	71.84	1500.0	2680	1180.3	3.5104	5.4797
327	600.15	123.75	0.001542	0.01368	648.51	73.10	1506.4	2676	1170.2	3.5207	5.4702
328	601.15	125.37	0.001548	0.01344	645.99	74.40	1512.9	2673	1160.1	3.5311	5.4607
329	602.15	127.00	0.001555	0.01320	643.09	75.76	1519.4	2669	1149.9	3.5416	5.4510

Temperature		Pressure	Specific volume		Density		Specific enthalpy		Heat of vaporization	Specific entropy	
			Liquid	Vapor	Liquid	Vapor	Liquid	Vapor		Liquid	Vapor
t	T	p	v'	v''	ρ'	ρ''	i'	i''	$r=i''-i'$	s'	s''
°C	K	bar	m³/kg	m³/kg	kg/m³	kg/m³	kJ/kg	kJ/kg	kJ/kg	kJ/kg K	kJ/kg K
330	603.15	128.65	0.001562	0.01297	640.20	77.10	1526.1	2666	1139.6	3.5522	5.4412
331	604.15	130.31	0.001569	0.01274	637.35	78.49	1532.8	2662	1129.1	3.5628	5.4313
332	605.15	131.99	0.001577	0.01251	634.12	79.94	1539.5	2658	1118.6	3.5735	5.4213
333	606.15	133.69	0.001584	0.01228	631.31	81.43	1546.2	2654	1107.9	3.5842	5.4112
334	607.15	135.41	0.001591	0.01206	628.54	82.92	1553.0	2650	1096.9	3.5949	5.4009
335	608.15	137.14	0.001599	0.01184	625.39	84.46	1559.8	2646	1085.7	3.6056	5.3905
336	609.15	138.89	0.001607	0.01162	622.28	86.06	1566.6	2641	1074.3	3.6163	5.3799
337	610.15	140.66	0.001615	0.01141	619.20	87.64	1573.4	2636	1062.7	3.6272	5.3692
338	611.15	142.45	0.001623	0.01120	616.14	89.29	1580.4	2631	1050.9	3.6381	5.3583
339	612.15	144.26	0.001631	0.01099	613.12	90.99	1587.6	2627	1039.0	3.6492	5.3473
340	613.15	146.08	0.001639	0.01078	610.13	92.76	1594.7	2622	1027.0	3.6605	5.3361
341	614.15	147.92	0.001648	0.01057	606.80	94.60	1602	2617	1014.8	3.6720	5.3247
342	615.15	149.78	0.001658	0.01037	603.14	96.43	1609	2612	1002.4	3.6835	5.3131
343	616.15	151.66	0.001667	0.01017	599.88	98.33	1616	2606	989.7	3.6950	5.3013
344	617.15	153.56	0.001676	0.09969	596.66	100.31	1624	2601	976.7	3.7067	5.2892
345	618.15	155.48	0.001686	0.009771	593.12	102.34	1632	2595	963.5	3.7184	5.2769
346	619.15	157.42	0.001696	0.009574	589.62	104.45	1640	2590	950.2	3.7302	5.2644
347	620.15	159.38	0.001707	0.009379	585.82	106.62	1647	2584	936.6	3.7420	5.2517
348	621.15	161.35	0.001718	0.009186	582.07	108.86	1655	2578	922.7	3.7540	5.2387
349	622.15	163.35	0.001729	0.008995	578.37	111.17	1663	2571	908.3	3.7662	5.2253
350	623.15	165.37	0.001741	0.008803	574.38	113.6	1671	2565	893.5	3.7786	5.2117
351	624.15	167.41	0.001752	0.008613	570.78	116.1	1679	2558	878.5	3.7912	5.1979
352	625.15	169.47	0.001764	0.008425	566.89	118.7	1688	2551	863.0	3.8040	5.1837
353	626.15	171.55	0.001777	0.008238	562.75	121.4	1696	2543	847.0	3.8170	5.1690
354	627.15	173.65	0.001792	0.008053	558.04	124.2	1705	2535	830.3	3.8303	5.1539
355	628.15	175.77	0.001807	0.007869	553.40	127.1	1714	2527	813.0	3.8439	5.1385
356	629.15	177.92	0.001823	0.007684	548.55	130.0	1723	2518	795.1	3.8577	5.1225
357	630.15	180.09	0.001840	0.007499	543.48	133.2	1732	2509	776.7	3.8718	5.1058
358	631.15	182.28	0.001857	0.007314	538.50	136.6	1742	2500	758.0	3.8862	5.0886
359	632.15	184.50	0.001875	0.007130	533.33	140.2	1752	2491	738.9	3.9010	5.0710
360	633.15	186.74	0.001894	0.006943	527.98	144.0	1762	2481	719.3	3.9162	5.0530
361	634.15	189.00	0.001918	0.00675	521.38	148.1	1772	2471	698.8	3.9320	5.0342
362	635.15	191.29	0.001943	0.00656	514.67	152.4	1783	2460	677.1	3.9482	5.0145
363	636.15	193.60	0.001968	0.00637	508.13	157.0	1794	2448	654.0	3.9649	4.9940
364	637.15	195.94	0.00199	0.00618	502.51	161.8	1805	2435	629.5	3.9825	4.9706
365	638.15	198.30	0.00202	0.00599	495.05	166.8	1817	2421	603.5	4.0009	4.9463
366	639.15	200.69	0.00205	0.00580	487.80	172.5	1830	2406	575.9	4.0202	4.9212
367	640.15	203.11	0.00208	0.00559	480.77	178.8	1844	2390	546.2	4.0413	4.8939
368	641.15	205.56	0.00212	0.00538	471.70	185.8	1859	2372	513.4	4.0629	4.8646
369	642.15	208.03	0.00217	0.00516	460.83	193.6	1875	2353	478.1	4.0870	4.8320
370	643.15	210.53	0.00222	0.00493	450.45	203	1893	2331	438.4	4.1137	4.7951
371	644.15	213.06	0.00229	0.00468	436.68	214	1913	2305	392.3	4.1441	4.7529
372	645.15	215.63	0.00238	0.00440	420.17	227	1937	2273	336.2	4.1809	4.7018
373	646.15	218.23	0.00251	0.00405	398.41	247	1969	2230	261.3	4.2295	4.6335
374	647.15	220.87	0.00280	0.00347	357.14	288	2032	2147	114.7	4.3258	4.5029
374,15	647.30	221.297	0.00326	0.00326	306.75	306.75	2100	2100	0.0	4.4296	4.4296

1 at = 1 kp/cm² = 98 066.5 Pa = 98 066.5 N/m² = 9.806 65 N/cm² = 0.980 665 bar   1 kcal = 4.1868 kJ

Table 45-1 Properties of Superheated Steam (H₂O) (in International System of Units—SI)

Pressure	Temperature		Specific volume		Density		Specific enthalpy		Heat of vaporization	Specific entropy	
			Liquid	Vapor	Liquid	Vapor	Liquid	Vapor		Liquid	Vapor
p	t	T	v'	v''	ρ'	ρ''	i'	i''	$r=i''-i'$	s'	s''
bar	°C	K	m³/kg	m³/kg	kg/m³	kg/m³	kJ/kg	kJ/kg	kJ/kg	kJ/kg K	kJ/kg K
0.010	6.92	280.07	0.0010001	129.9	999.9	0.00770	29.32	2513	2484	0.1054	8.975
0.015	13.038	286.188	0.0010007	87.90	999.3	0.01138	54.75	2525	2470	0.1958	8.827
0.020	17.514	290.664	0.0010014	66.97	998.6	0.01493	73.52	2533	2459	0.2609	8.722
0.025	21.094	294.244	0.0010021	54.24	997.9	0.01843	88.50	2539	2451	0.3124	8.642
0.030	24.097	297.247	0.0010028	45.66	997.2	0.02190	101.04	2545	2444	0.3546	8.576
0.035	26.692	299.842	0.0010035	39.48	996.5	0.02533	111.86	2550	2438	0.3908	8.521
0.040	28.979	302.129	0.0010041	34.81	995.9	0.02873	121.42	2554	2433	0.4225	8.473
0.045	31.033	304.183	0.0010047	31.13	995.3	0.03211	130.00	2557	2427	0.4507	8.431
0.050	32.88	306.03	0.0010053	28.19	994.7	0.03547	137.83	2561	2423	0.4761	8.393
0.055	34.59	307.74	0.0010059	25.77	994.1	0.03880	144.95	2564	2419	0.4993	8.359
0.060	36.18	309.33	0.0010064	23.74	993.6	0.04212	151.50	2567	2415	0.5207	8.328
0.065	37.65	310.80	0.0010070	22.02	993.0	0.04542	157.68	2570	2412	0.5406	8.300
0.070	39.03	312.18	0.0010075	20.53	992.6	0.04871	163.43	2572	2409	0.5591	8.274
0.075	40.32	313.47	0.0010080	19.23	992.1	0.05198	168.8	2574	2405	0.5764	8.250
0.080	41.54	314.69	0.0010085	18.10	991.6	0.05525	173.9	2576	2402	0.5927	8.227
0.085	42.69	315.84	0.0010090	17.10	991.1	0.05849	178.7	2578	2399	0.6080	8.206
0.090	43.79	316.94	0.0010094	16.20	990.7	0.06172	183.3	2580	2397	0.6225	8.186
0.095	44.84	317.99	0.0010098	15.40	990.3	0.06493	187.7	2582	2394	0.6362	8.167
0.10	45.84	318.99	0.0010103	14.68	989.8	0.06812	191.9	2584	2392	0.6492	8.149
0.11	47.72	320.87	0.0010111	13.40	989.0	0.07462	199.7	2588	2388	0.6740	8.116
0.12	49.45	322.60	0.0010119	12.35	988.2	0.08097	207.0	2591	2384	0.6966	8.085
0.13	51.07	324.22	0.0010126	11.46	987.6	0.08726	213.8	2594	2380	0.7174	8.057
0.14	52.58	325.73	0.0010133	10.69	986.9	0.09354	220.1	2596	2376	0.7368	8.031
0.15	54.00	327.15	0.0010140	10.02	986.2	0.09980	226.1	2599	2373	0.7550	8.007
0.16	55.34	328.49	0.0010147	9.429	985.5	0.10600	231.7	2601	2369	0.7722	7.984
0.17	56.61	329.76	0.0010153	8.909	984.9	0.1123	236.9	2603	2366	0.7884	7.963
0.18	57.82	330.97	0.0010159	8.444	984.3	0.1185	241.9	2605	2363	0.8038	7.944
0.19	58.98	332.13	0.0010165	8.025	983.8	0.1247	246.7	2607	2360	0.8183	7.925
0.20	60.08	333.23	0.0010171	7.647	983.2	0.1308	251.4	2609	2358	0.8321	7.907
0.21	61.14	334.29	0.0010177	7.304	982.6	0.1369	255.9	2611	2355	0.8453	7.890
0.22	62.16	335.31	0.0010183	6.992	982.0	0.1430	260.2	2613	2353	0.8581	7.874
0.23	63.14	336.29	0.0010188	6.708	981.5	0.1491	264.3	2614	2350	0.8703	7.859
0.24	64.08	337.23	0.0010193	6.445	981.1	0.1551	268.2	2616	2348	0.8821	7.884
0.25	64.99	338.14	0.0010199	6.202	980.5	0.1612	272.0	2618	2346	0.8934	7.830
0.26	65.88	339.03	0.0010204	5.977	980.0	0.1673	275.7	2620	2344	0.9043	7.816
0.27	66.73	339.88	0.0010209	5.769	979.5	0.1733	279.3	2621	2342	0.9147	7.803
0.28	67.55	340.70	0.0010214	5.576	979.0	0.1793	282.7	2623	2340	0.9248	7.791
0.29	68.35	341.50	0.0010218	5.395	978.7	0.1853	286.0	2624	2338	0.9346	7.779
0.30	69.12	342.27	0.0010222	5.226	978.3	0.1913	289.3	2625	2336	0.9441	7.769
0.32	70.60	343.75	0.0010232	4.922	977.3	0.2032	295.5	2627	2332	0.9625	7.745
0.34	72.02	345.17	0.0010240	4.650	976.6	0.2151	301.5	2630	2328	0.9796	7.724
0.36	73.36	346.51	0.0010248	4.407	975.8	0.2269	307.1	2632	2325	0.9958	7.705
0.38	74.64	347.79	0.0010256	4.189	975.0	0.2387	312.5	2634	2322	1.0113	7.687
0.40	75.88	349.03	0.0010264	3.994	974.3	0.2504	317.7	2636	2318	1.0261	7.670
0.45	78.75	351.90	0.0010282	3.574	972.6	0.2797	329.6	2641	2311	1.0601	7.629
0.50	81.35	354.50	0.0010299	3.239	971.0	0.3087	340.6	2645	2404	1.0910	7.593
0.55	83.74	356.89	0.0010315	2.963	969.5	0.3375	350.7	2649	2298	1.1193	7.561
0.60	85.95	359.10	0.0010330	2.732	968.1	0.3661	360.0	2653	2293	1.1453	7.531
0.65	88.02	361.17	0.0010345	2.534	966.7	0.3946	368.6	2657	2288	1.1693	7.504
0.70	89.97	363.12	0.0010359	2.364	965.3	0.4230	376.8	2660	2283	1.1918	7.479
0.75	91.80	364.95	0.0010372	2.216	964.1	0.4512	384.5	2663	2278	1.2130	7.456
0.80	93.52	366.67	0.0010385	2.087	962.9	0.4792	391.8	2665	2273	1.2330	7.434
0.85	95.16	368.31	0.0010397	1.972	961.8	0.5071	398.7	2668	2269	1.2518	7.414
0.90	96.72	369.87	0.0010409	1.869	960.7	0.5350	405.3	2670	2265	1.2696	7.394
0.95	98.21	371.36	0.0010421	1.777	959.6	0.5627	411.5	2673	2261	1.2865	7.376

Table 45-2 Properties of Superheated Steam (H$_2$O) (in International System of Units—SI) (Continued)

Pressure	Temperature		Specific volume		Density		Specific enthalpy		Heat of vaporization	Specific entropy	
			Liquid	Vapor	Liquid	Vapor	Liquid	Vapor		Liquid	Vapor
p	t	T	v'	v''	ρ'	ρ''	i'	i''	$r=i''-i'$	s'	s''
bar	°C	K	m³/kg	m³/kg	kg/m³	kg/m³	kJ/kg	kJ/kg	kJ/kg	kJ/kg K	kJ/kg K
1.0	99.64	372.79	0.0010432	1.694	958.6	0.5903	417.4	2675	2258	1.3026	7.360
1.1	102.32	375.47	0.0010452	1.550	956.8	0.6453	428.9	2679	2250	1.3327	7.328
1.2	104.81	377.96	0.0010472	1.429	954.9	0.6999	439.4	2683	2244	1.3606	7.298
1.3	107.14	380.29	0.0010492	1.325	953.1	0.7545	449.2	2687	2238	1.3866	7.271
1.4	109.33	382.48	0.0010510	1.236	951.5	0.8088	458.5	2690	2232	1.4109	7.246
1.5	111.38	384.53	0.0010527	1.159	949.9	0.8627	467.2	2693	2226	1.4336	7.223
1.6	113.32	386.47	0.0010543	1.091	948.5	0.9164	475.4	2696	2221	1.4550	7.202
1.7	115.17	388.32	0.0010559	1.031	947.1	0.9699	483.2	2699	2216	1.4752	7.182
1.8	116.94	390.09	0.0010575	0.9773	945.6	1.023	490.7	2702	2211	1.4943	7.163
1.9	118.62	391.77	0.0010591	0.9290	944.2	1.076	497.9	2704	2206	1.5126	7.145
2.0	120.23	393.38	0.0010605	0.8854	943.0	1.129	504.8	2707	2202	1.5302	7.127
2.1	121.78	394.93	0.0010619	0.8459	941.7	1.182	511.4	2709	2198	1.5470	7.111
2.2	123.27	396.42	0.0010633	0.8098	940.5	1.235	517.8	2711	2193	1.5630	7.096
2.3	124.71	397.86	0.0010646	0.7768	939.3	1.287	524.0	2713	2189	1.5783	7.081
2.4	126.09	399.24	0.0010659	0.7465	938.2	1.340	529.8	2715	2185	1.5929	7.067
2.5	127.43	400.58	0.0010672	0.7185	937.0	1.393	535.4	2717	2182	1.6071	7.053
2.6	128.73	401.88	0.0010685	0.6925	935.9	1.444	540.9	2719	2178	1.621	7.040
2.7	129.98	403.13	0.0010697	0.6684	934.8	1.496	546.2	2721	2175	1.634	7.027
2.8	131.20	404.35	0.0010709	0.6461	933.8	1.548	551.4	2722	2171	1.647	7.015
2.9	132.39	405.54	0.0010721	0.6253	932.7	1.599	556.5	2724	2167	1.660	7.003
3.0	133.54	406.69	0.0010733	0.6057	931.7	1.651	561.4	2725	2164	1.672	6.992
3.1	134.66	410.81	0.0010744	0.5873	930.8	1.703	566.3	2727	2161	1.683	6.981
3.2	135.75	408.90	0.0010754	0.5701	929.9	1.754	571.1	2728	2157	1.695	6.971
3.3	136.82	409.97	0.0010765	0.5539	928.9	1.805	575.7	2730	2154	1.706	6.961
3.4	137.86	411.01	0.0010776	0.5386	928.0	1.857	580.2	2731	2151	1.717	6.951
3.5	138.88	412.03	0.0010786	0.5241	927.1	1.908	584.5	2732	2148	1.728	6.941
3.6	139.87	413.02	0.0010797	0.5104	926.2	1.959	588.7	2734	2145	1.738	6.932
3.7	140.84	314.99	0.0010807	0.4975	925.3	2.010	592.8	2735	2142	1.748	6.923
3.8	141.79	414.94	0.0010817	0.4852	924.5	2.061	596.8	2736	2139	1.758	6.914
3.9	142.71	415.86	0.0010827	0.4735	923.6	2.112	600.8	2737	2136	1.768	6.905
4.0	143.62	416.77	0.0010836	0.4624	922.8	2.163	604.7	2738	2133	1.777	6.897
4.1	144.51	417.66	0.0010845	0.4518	922.1	2.213	608.5	2740	2131	1.786	6.889
4.2	145.39	418.54	0.0010855	0.4416	921.2	2.264	612.3	2741	2129	1.795	6.881
4.3	146.25	419.40	0.0010865	0.4391	920.4	2.315	616.1	2742	2126	1.804	6.873
4.4	147.09	420.24	0.0010874	0.4227	919.6	2.366	619.8	2743	2123	1.812	6.865
4.5	147.92	421.07	0.0010883	0.4139	918.9	2.416	623.4	2744	2121	1.821	6.857
4.6	148.73	421.88	0.0010892	0.4054	918.1	2.467	626.9	2745	2118	1.829	6.850
4.7	149.53	422.68	0.0010901	0.3973	917.3	2.517	630.3	2746	2116	1.837	6.843
4.8	150.41	423.46	0.0010190	0.3895	916.6	2.568	633.7	2747	2113	1.845	6.835
4.9	151.08	424.23	0.0010918	0.3819	915.9	2.618	639.9	2748	2111	1.853	6.828
5.0	151.84	424.99	0.0010927	0.3747	915.2	2.669	640.1	2749	2109	1.860	6.822
5.2	153.32	426.47	0.0010943	0.3612	913.8	2.769	646.5	2750	2104	1.875	6.809
5.4	154.76	427.91	0.0010960	0.3485	912.4	2.869	652.7	2752	2099	1.890	6.796
5.6	156.16	429.31	0.0010976	0.3368	911.1	2.969	658.8	2754	2095	1.904	6.784
5.8	157.52	430.67	0.0010992	0.3258	909.8	3.069	664.7	2755	2090	1.918	6.772
6.0	158.84	431.99	0.0011007	0.3156	908.5	3.169	670.5	2757	2086	1.931	6.761
6.2	160.12	433.27	0.0011022	0.3060	907.3	3.268	676.0	2758	2082	1.944	6.750
6.4	161.37	434.52	0.0011037	0.2970	906.0	3.367	681.5	2760	2078	1.956	6.739
6.6	162.59	435.74	0.0011052	0.2885	904.8	3.467	686.9	2761	2074	1.968	6.729
6.8	163.79	436.94	0.0011066	0.2804	903.7	3.566	692.1	2762	2070	1.980	6.719
7.0	164.96	438.11	0.0011081	0.2728	902.4	3.666	697.2	2764	2067	1.992	6.709
7.2	166.10	439.25	0.0011095	0.2656	901.3	3.765	702.2	2765	2063	2.003	6.699
7.4	167.21	440.36	0.0011109	0.2588	900.2	3.864	707.1	2766	2059	2.014	6.690
7.6	168.30	441.45	0.0011123	0.2523	899.0	3.963	711.8	2767	2055	2.025	6.681
7.8	169.37	442.52	0.0011136	0.2462	898.0	4.062	716.4	2768	2052	2.036	6.672

Table 45-3 Properties of Superheated Steam (H₂O) (in International System of Units—SI) (Continued)

Pressure	Temperature		Specific volume		Density		Specific enthalpy		Heat of vaporization	Specific entropy	
			Liquid	Vapor	Liquid	Vapor	Liquid	Vapor		Liquid	Vapor
p	t	T	v'	v''	ρ'	ρ''	i'	i''	$r=i''-i'$	s'	s''
bar	°C	K	m³/kg	m³/kg	kg/m³	kg/m³	kJ/kg	kJ/kg	kJ/kg	kJ/kg K	kJ/kgK
8.0	170.42	443.57	0.0011149	0.2403	896.9	4.161	720.9	2769	2048	2.046	6.663
8.2	171.44	444.59	0.0011162	0.2347	895.9	4.260	725.4	2770	2045	2.056	6.655
8.4	172.44	445.59	0.0011175	0.2294	894.9	4.359	729.8	2771	2041	2.066	6.647
8.6	173.43	446.58	0.0011187	0.2243	893.9	4.458	734.2	2772	2038	2.076	6.639
8.8	174.40	447.55	0.0011200	0.2195	892.9	4.556	738.6	2773	2034	2.085	6.631
9.0	175.35	448.50	0.0011213	0.2149	891.8	4.654	742.8	2774	2031	2.094	6.623
9.2	176.29	449.44	0.0011225	0.2104	890.9	4.753	746.9	2775	2028	2.103	6.615
9.4	177.21	450.36	0.0011237	0.2061	889.9	4.852	750.9	2776	2025	2.112	6.608
9.6	178.12	451.27	0.0011249	0.2020	889.0	4.949	754.8	2777	2022	2.121	6.601
9.8	179.01	452.16	0.0011261	0.1982	888.0	5.045	758.8	2778	2019	2.130	6.594
10.0	179.88	453.03	0.0011273	0.1946	887.1	5.139	762.7	2778	2015	2.138	6.587
10.5	182.00	455.15	0.0011303	0.1856	884.7	5.388	772.1	2779	2007	2.159	6.570
11.0	184.05	457.20	0.0011331	0.1775	882.5	5.634	781.1	2781	2000	2.179	6.554
11.5	186.04	459.19	0.0011358	0.1701	880.4	5.879	789.8	2783	1993	2.198	6.538
12.0	187.95	461.10	0.0011385	0.1633	878.3	6.124	798.3	2785	1987	2.216	6.523
12.5	189.80	462.95	0.0011412	0.1570	876.3	6.369	806.5	2786	1980	2.234	6.509
13.0	191.60	464.75	0.0011438	0.1512	874.3	6.614	814.5	2787	1973	2.251	6.495
13.5	193.34	466.49	0.0011464	0.1458	872.3	6.859	822.3	2789	1967	2.268	6.482
14.0	195.04	468.19	0.0011490	0.1408	870.3	7.103	830.0	2790	1960	2.284	6.469
14.5	196.68	469.83	0.0011515	0.1361	868.4	7.348	837.4	2791	1954	2.299	6.457
15.0	198.28	471.43	0.0011539	0.1317	866.6	7.593	844.6	2792	1947	2.314	6.445
15.5	199.84	472.99	0.0011563	0.1276	864.8	7.837	851.5	2793	1941	2.329	6.433
16.0	201.36	474.51	0.0011586	0.1238	863.1	8.080	858.3	2793	1935	2.344	6.422
16.5	202.85	476.00	0.0011609	0.1201	861.4	8.325	865.0	2794	1929	2.358	6.411
17.0	204.30	477.45	0.0011632	0.1167	859.7	8.569	871.6	2795	1923	2.371	6.400
17.5	205.72	478.87	0.0011655	0.1135	858.0	8.812	878.1	2796	1918	2.384	6.389
18.0	207.10	480.25	0.0011678	0.1104	856.3	9.058	884.4	2796	1912	2.397	6.379
18.5	208.45	481.60	0.0011700	0.1075	854.7	9.303	890.6	2797	1907	2.410	6.369
19.0	209.78	482.93	0.0011722	0.1047	853.1	9.549	896.6	2798	1901	2.422	6.359
19.5	211.09	484.24	0.0011744	0.1021	851.5	9.795	902.6	2799	1896	2.435	6.350
20.0	212.37	485.52	0.0011766	0.09958	849.9	10.041	908.5	2799	1891	2.447	6.340
20.5	213.62	486.77	0.0011788	0.09719	848.3	10.29	914.2	2800	1886	2.458	6.331
21.0	214.84	487.99	0.0011809	0.09492	846.8	10.54	919.8	2800	1880	2.470	6.322
21.5	216.05	489.20	0.0011830	0.09276	845.3	10.78	925.4	2800	1875	2.481	6.314
22.0	217.24	490.39	0.0011851	0.09068	843.8	11.03	930.9	2801	1870	2.492	6.305
22.5	218.41	491.56	0.0011872	0.08869	842.3	11.28	936.4	2801	1865	2.503	6.297
23.0	219.55	492.70	0.0011892	0.08679	840.9	11.52	941.5	2801	1860	2.514	6.288
23.5	220.67	493.82	0.0011912	0.08498	839.5	11.77	946.7	2802	1855	2.524	6.280
24.0	221.77	494.92	0.0011932	0.08324	838.1	12.01	951.8	2802	1850	2.534	6.272
24.5	222.85	496.00	0.0011952	0.08156	836.7	12.26	956.8	2802	1845	2.544	6.264
25.0	223.93	497.08	0.0011972	0.07993	835.3	12.51	961.8	2802	1840	2.554	6.256
25.5	224.99	498.14	0.0011992	0.07837	833.9	12.76	966.8	2803	1836	2.564	6.249
26.0	226.03	499.18	0.0012012	0.07688	835.2	13.01	971.7	2803	1831	2.573	6.242
26.5	227.05	500.20	0.0012031	0.07545	831.2	13.25	976.6	2803	1826	2.582	6.234
27.0	228.06	501.21	0.0012050	0.07406	829.9	13.50	981.3	2803	1822	2.592	6.227
27.5	239.06	502.21	0.0012069	0.07271	828.6	13.75	985.9	2803	1817	2.602	6.220
28.0	230.04	503.19	0.0012088	0.07141	827.3	14.00	990.4	2803	1813	2.611	6.213
28.5	231.01	504.16	0.0012107	0.07016	826.0	14.25	994.9	2803	1808	2.620	6.206
29.0	231.96	505.11	0.0012126	0.06895	824.7	14.50	999.4	2803	1804	2.628	6.199
29.5	232.90	506.05	0.0012145	0.06778	823.4	14.75	1003.8	2804	1800	2.637	6.193
30	233.83	506.98	0.0012163	0.06665	822.2	15.00	1008.3	2804	1796	2.646	6.186
31	235.66	508.81	0.0012201	0.06450	819.6	15.50	1016.9	2804	1787	2.662	6.173
32	237.44	510.59	0.0012238	0.06246	817.1	16.01	1025.3	2803	1778	2.679	6.161
33	239.18	512.33	0.0012274	0.06055	814.7	16.52	1033.7	2803	1769	2.695	6.149
34	240.88	514.03	0.0012310	0.05875	812.3	17.02	1041.9	2803	1761	2.710	6.137

Table 45-4 Properties of Superheated Steam (H₂O) (in International System of Units—SI) (Continued)

Pressure	Temperature		Specific volume		Density		Specific enthalpy		Heat of vaporization	Specific entropy	
			Liquid	Vapor	Liquid	Vapor	Liquid	Vapor		Liquid	Vapor
p	t	T	v'	v''	ρ'	ρ''	i'	i''	$r=i''-i'$	s'	s''
bar	°C	K	m³/kg	m³/kg	kg/m³	kg/m³	kJ/kg	kJ/kg	kJ/kg	kJ/kg K	kJ/kg K
35	242.54	515.69	0.0012345	0.05704	810.0	17.53	1049.8	2803	1753	2.725	6.125
36	244.16	517.31	0.0012380	0.05543	807.8	18.04	1057.5	2802	1745	2.740	6.113
37	245.75	518.90	0.0012415	0.05391	805.5	18.55	1065.2	2802	1737	2.755	6.102
38	247.31	520.46	0.0012450	0.05246	803.2	19.06	1072.7	2802	1729	2.769	6.091
39	248.84	521.99	0.0012485	0.05108	801.0	19.58	1080.2	2801	1721	2.783	6.081
40	250.33	523.48	0.0012520	0.04977	798.7	20.09	1087.5	2801	1713	2.796	6.070
41	251.80	524.95	0.0012554	0.04852	796.6	20.61	1094.7	2800	1705	2.810	6.059
42	253.24	526.39	0.0012588	0.04732	794.4	21.13	1101.7	2800	1698	2.823	6.049
43	254.66	527.81	0.0012622	0.04617	792.3	21.66	1108.5	2799	1691	2.836	6.039
44	256.05	529.20	0.0012656	0.04508	790.1	22.18	1115.3	2798	1683	2.849	6.029
45	257.41	530.56	0.0012690	0.04404	788.0	22.71	1122.1	2798	1676	2.862	6.020
46	258.75	531.90	0.0012724	0.04305	785.9	23.23	1128.8	2797	1668	2.874	6.010
47	260.07	533.22	0.0012757	0.04210	783.9	23.76	1135.4	2796	1661	2.886	6.001
48	261.37	534.52	0.0012790	0.04118	781.9	24.29	1141.8	2796	1654	2.898	5.991
49	262.65	535.80	0.0012824	0.04029	779.8	24.82	1148.2	2795	1647	2.909	5.982
50	263.91	537.06	0.0012857	0.03944	777.8	25.35	1154.4	2794	1640	2.921	5.973
51	265.15	538.30	0.0012890	0.03863	775.8	25.89	1160.6	2793	1632	2.932	5.964
52	266.38	539.53	0.0012923	0.03784	773.8	26.43	1166.8	2792	1625	2.943	5.956
53	267.58	540.73	0.0012955	0.03708	771.9	26.97	1172.9	2791	1618	2.954	5.947
54	268.77	541.92	0.0012988	0.03635	769.9	27.51	1179.0	2791	1612	2.965	5.939
55	269.94	543.09	0.0013021	0.03564	768.0	28.06	1184.9	2790	1604.6	2.976	5.930
56	271.10	544.25	0.0013054	0.03495	766.0	28.61	1190.8	2789	1597.7	2.987	5.922
57	272.24	545.39	0.0013087	0.03429	764.1	29.16	1196.6	2788	1591.0	2.997	5.914
58	273.36	546.51	0.0013120	0.03365	762.2	29.72	1202.4	2786	1584.3	3.007	5.906
59	274.47	547.62	0.0013152	0.03303	760.3	30.28	1208.2	2786	1577.6	3.017	5.898
60	275.56	548.71	0.0013185	0.03243	758.4	30.84	1213.9	2785	1570.8	3.027	5.890
61	276.64	549.79	0.0013217	0.03185	756.6	31.40	1219.6	2784	1564.1	3.037	5.882
62	277.71	550.86	0.0013250	0.03130	754.7	31.95	1225.1	2782	1557.4	3.047	5.874
63	278.76	551.91	0.0013282	0.03076	752.9	32.51	1230.6	2781	1550.7	3.057	5.866
64	279.80	552.95	0.0013314	0.03024	751.1	33.07	1236.0	2780	1544.1	3.066	5.859
65	280.83	553.98	0.0013347	0.02973	749.2	33.64	1241.3	2779	1537.5	3.076	5.851
66	281.85	555.00	0.0013380	0.02923	747.4	34.21	1246.6	2778	1530.9	3.085	5.844
67	282.86	556.01	0.0013412	0.02874	745.6	34.79	1251.8	2776	1524.4	3.095	5.836
68	283.85	557.00	0.0013445	0.02827	743.8	35.37	1257.0	2775	1517.9	3.104	5.829
69	284.83	557.98	0.0013478	0.02782	741.9	35.95	1262.2	2773	1511.4	3.113	5.822
70	285.80	558.95	0.0013510	0.02737	740.2	36.54	1267.4	2772	1504.9	3.122	5.814
71	286.76	559.91	0.0013542	0.02694	738.4	37.12	1272.5	2771	1498.4	3.131	5.807
72	287.71	560.86	0.0013574	0.02652	736.7	37.71	1277.6	2769	1492.0	3.140	5.800
73	288.65	561.80	0.0013607	0.02611	734.9	38.30	1282.6	2768	1485.6	3.149	5.793
74	289.58	562.73	0.0013640	0.02571	733.1	38.89	1287.6	2767	1479.2	3.158	5.786
75	290.50	563.65	0.0013673	0.02532	713.4	39.49	1292.7	2766	1472.8	3.160	5.779
76	291.41	564.56	0.0013706	0.02494	729.6	40.09	1297.7	2764	1466.4	3.174	5.772
77	292.32	565.47	0.0013739	0.02457	727.8	40.70	1302.6	2763	1460.0	3.183	5.765
78	293.22	566.37	0.0013772	0.02421	726.1	41.30	1307.4	2761	1453.7	3.192	5.758
79	294.10	567.25	0.0013805	0.02386	724.4	41.91	1312.2	2759	1447.4	3.200	5.751
80	294.98	568.13	0.0013838	0.02352	722.6	42.52	1317.0	2758	1441.1	3.208	5.745
81	295.85	569.00	0.0013872	0.02318	720.9	43.14	1321.8	2757	1434.8	3.216	5.738
82	296.71	569.86	0.0013905	0.02285	719.2	43.76	1326.6	2755	1428.5	3.224	5.731
83	297.56	570.71	0.0013938	0.02253	717.5	44.38	1331.4	2753	1422.2	3.232	5.724
84	298.40	571.55	0.0013972	0.02222	715.7	45.00	1336.1	2752	1416.0	3.240	5.717
85	299.24	572.39	0.0014005	0.02192	714.0	45.62	1340.8	2751	1409.8	3.248	5.711
86	300.07	573.22	0.0014039	0.02162	712.3	46.25	1345.4	2749	1403.7	3.255	5.704
87	300.89	574.04	0.0014073	0.02132	710.6	46.90	1350.1	2747	1397.6	3.263	5.698
88	301.71	574.86	0.0014106	0.02103	708.9	47.55	1354.7	2746	1391.5	3.271	5.691
89	302.52	575.67	0.0014140	0.02075	707.2	48.19	1359.2	2744	1385.4	3.279	5.685

Table 45-5   Properties of Superheated Steam (H$_2$O) (in International System of Units—SI) (*Continued*)

Pressure	Temperature		Specific volume		Density		Specific enthalpy		Heat of vaporization	Specific entropy	
			Liquid	Vapor	Liquid	Vapor	Liquid	Vapor		Liquid	Vapor
p	t	T	v'	v''	ρ'	ρ''	i'	i''	$r=i''-i'$	s'	s''
bar	°C	K	m³/kg	m³/kg	kg/m³	kg/m³	kJ/kg	kJ/kg	kJ/kg	kJ/kg K	kJ/kg K
90	303.32	376.47	0.0014174	0.02048	705.5	48.83	1363.7	2743	1379.3	3.287	5.678
91	304.11	577.26	0.0014208	0.02021	703.8	49.48	1368.2	2741	1373.2	3.294	5.672
92	304.90	578.05	0.0014242	0.01995	702.1	50.13	1372.7	2740	1367.0	3.301	5.665
93	305.67	578.82	0.0014276	0.01969	700.5	50.79	1377.1	2738	1360.9	3.309	5.659
94	306.45	579.60	0.0014310	0.01944	698.8	51.45	1381.5	2736	1354.7	3.316	5.653
95	307.22	580.37	0.0014345	0.01919	697.1	52.11	1385.9	2734	1348.4	3.324	5.646
96	307.98	581.13	0.0014380	0.01895	695.4	52.77	1390.2	2732	1342.1	3.331	5.640
97	308.74	581.89	0.0014415	0.01871	693.7	53.44	1394.2	2730	1335.8	3.338	5.634
98	309.49	582.64	0.0014450	0.01848	692.0	54.11	1398.9	2728	1329.5	3.346	5.628
99	310.23	583.38	0.0014486	0.01825	690.3	54.79	1403.3	2726	1323.2	3.353	5.621
100	310.96	584.11	0.0014521	0.01803	688.7	55.46	1407.7	2725	1317.0	3.360	5.515
102	312.42	585.57	0.0014592	0.01759	685.3	56.85	1416.4	2721	1304.6	3.374	5.602
104	313.86	587.01	0.0014664	0.01716	681.9	58.27	1425.0	2717	1292.3	3.388	5.590
106	315.28	588.43	0.0014736	0.01675	678.6	59.70	1433.5	2713	1280.0	3.402	5.578
108	316.67	589.82	0.0014808	0.01636	675.3	61.13	1441.9	2709	1267.3	3.416	5.565
110	318.04	591.19	0.001489	0.01598	671.6	62.58	1450.2	2705	1255.4	3.430	5.553
112	319.39	592.54	0.001496	0.01561	668.4	64.05	1458.4	2701	1243.0	3.443	5.541
114	320.73	593.88	0.001503	0.01526	665.3	65.54	1466.6	2697	1230.6	3.457	5.528
116	322.05	595.20	0.001511	0.01491	661.8	67.06	1474.8	2693	1218.3	3.470	5.516
118	323.35	596.50	0.001519	0.01458	658.3	68.59	1483.0	2689	1205.9	3.483	5.504
120	324.63	597.78	0.001527	0.01426	654.9	70.13	1491.1	2685	1193.5	3.496	5.492
122	325.90	599.05	0.001535	0.01395	651.5	71.70	1499.2	2680	1181.0	3.509	5.480
124	327.15	600.30	0.001543	0.01364	648.1	73.30	1507.3	2676	1168.5	3.522	5.468
126	328.39	601.54	0.001551	0.01334	644.7	74.94	1515.4	2671	1156.0	3.535	5.456
128	329.61	602.76	0.001559	0.01305	641.4	76.61	1523.5	2667	1143.4	3.548	5.444
130	330.81	603.96	0.001567	0.01277	638.2	78.30	1531.5	2662	1130.8	3.561	5.432
132	332.00	605.15	0.001576	0.01250	634.5	80.00	1539.5	2658	1118.2	3.573	5.420
134	333.18	606.33	0.001585	0.01224	630.9	81.72	1547.3	2653	1105.5	3.586	5.408
136	334.34	607.49	0.001594	0.01198	627.4	83.47	1555.1	2648	1092.7	3.598	5.396
138	335.49	608.64	0.001602	0.01173	624.2	85.25	1562.9	2643	1079.9	3.610	5.384
140	336.63	609.78	0.001611	0.01149	620.7	87.03	1570.8	2638	1066.9	3.623	5.372
142	337.75	610.90	0.001620	0.01125	617.3	88.89	1578.7	2633	1053.8	3.636	5.360
144	338.86	612.01	0.001629	0.01101	613.9	90.83	1586.6	2628	1040.7	3.648	5.348
146	339.96	613.11	0.001638	0.01078	610.5	92.76	1594.5	2622	1027.6	3.660	5.316
148	341.04	614.19	0.001648	0.01056	606.8	94.69	1602	2617	1014.5	3.672	5.323
150	342.11	615.26	0.001658	0.01035	603.1	92.62	1610	2611	1001.1	3.684	5.310
152	343.18	616.33	0.001668	0.01014	599.5	98.62	1618	2606	987.5	3.697	5.297
154	344.23	617.38	0.001678	0.009928	595.9	100.72	1626	2600	973.8	3.709	5.285
156	345.27	618.42	0.001688	0.009720	592.4	102.9	1634	2594	960.0	3.721	5.273
158	346.30	619.15	0.001699	0.009517	588.6	105.1	1642	2588	946.1	3.733	5.260
160	347.32	620.47	0.001710	0.009318	584.8	107.3	1650	2582	932.0	3.746	5.247
162	348.33	621.48	0.001721	0.009124	581.1	109.6	1658	2576	917.7	3.758	5.233
164	349.32	622.47	0.001732	0.008934	577.4	111.9	1666	2569	903.2	3.770	5.219
166	350.31	623.46	0.001744	0.008747	573.4	114.3	1674	2562	888.4	3.783	5.205
168	351.29	624.44	0.001756	0.008563	569.5	116.8	1682	2555	873.4	3.795	5.191
170	352.26	625.41	0.001768	0.008382	565.6	119.3	1690	2548	848.3	3.807	5.177
172	353.21	626.36	0.001781	0.008203	561.5	121.9	1698	2541	843.0	3.820	5.163
174	354.17	627.32	0.001794	0.008025	557.4	124.6	1707	2534	827.4	3.832	5.149
176	355.11	628.26	0.001808	0.007848	553.1	127.4	1715	2526	811.4	3.845	5.135
178	356.04	629.19	0.001822	0.007674	548.8	130.3	1723	2518	795.0	3.858	5.121
180	356.96	630.11	0.001837	0.007504	544.4	133.2	1732	2510	778.2	3.871	5.107
182	357.87	631.02	0.001853	0.007336	539.7	136.3	1741	2502	761.2	3.884	5.092
184	358.78	631.93	0.001870	0.007169	534.8	139.5	1749	2493	743.9	3.898	5.076
186	359.67	632.82	0.001887	0.007003	529.9	142.8	1758	2484	726.4	3.911	5.060
188	360.56	633.71	0.001904	0.00684	525.2	146.2	1767	2475	708.5	3.925	5.044

Table 45-6   Properties of Superheated Steam (H_2O) (in International System of Units—SI) (*Continued*)

Pressure	Temperature		Specific volume		Density		Specific enthalpy		Heat of vaporization	Specific entropy	
			Liquid	Vapor	Liquid	Vapor	Liquid	Vapor		Liquid	Vapor
p	t	T	v'	v''	ρ'	ρ''	i'	i''	$r = i'' - i'$	s'	s''
bar	°C	K	m³/kg	m³/kg	kg/m³	kg/m³	kJ/kg	kJ/kg	kJ/kg	kJ/kg	kJ/kg K
190	361.44	634.59	0.001921	0.00668	520.6	149.7	1776	2466	690	3.938	5.027
192	362.31	635.46	0.001940	0.00652	515.5	153.4	1785	2456	671	3.952	5.009
194	363.17	636.32	0.001961	0.00636	509.9	157.3	1795	2446	651	3.967	4.990
196	364.02	637.17	0.001985	0.00619	503.8	161.6	1805	2435	630	3.982	4.970
198	364.87	638.02	0.00201	0.00602	497.5	166.1	1816	2423	607	3.998	4.949
200	365.71	638.86	0.00204	0.00585	490.2	170.9	1827	2410	583	4.015	4.928
202	366.54	639.69	0.00207	0.00568	483.1	176.0	1838	2397	559	4.032	4.906
204	367.37	640.52	0.00210	0.00551	476.2	181.4	1849	2383	534	4.049	4.883
206	368.18	641.33	0.00213	0.00534	469.5	187.2	1861	2369	508	4.067	4.858
208	368.99	642.14	0.00217	0.00516	460.8	193.6	1874	2353	479	4.087	4.832
210	369.79	642.94	0.00221	0.00498	452.5	200.7	1888	2336	448	4.108	4.803
212	370.58	643.73	0.00226	0.00480	442.5	208.5	1903	2316	413	4.131	4.771
214	371.4	644.55	0.00232	0.00460	431.0	217.4	1920	2294	374	4.157	4.734
216	372.2	645.35	0.00239	0.00436	418.4	229.3	1940	2269	329	4.188	4.692
218	372.9	646.05	0.00249	0.00402	401.6	248.7	1965	2233	268	4.223	4.645
220	373.7	646.85	0.00273	0.00367	366.3	272.5	2016	2168	152	4.303	4.591
221.29	374.15	647.30	0.00326	0.00326	306.75	306.75	2100	2100	0	4.430	4.430

1at = 1kp/cm² = 98 966.5 Pa = 98 066.5 N/m² = 9.80665 N/cm² = 0.980665 bar   1 kcal = 4.1868 kJ

Table 45-7   Critical Constants of Water (in International System of Units—SI)

Temperature		Pressure	Specific volume	Density	Specific enthalpy	Specific entropy
t_k	T_k	p_k	v_k	ρ_k	i_k	s_k
°C	K	bar	m³/kg	kg/m³	kJ/kg	kJ/kg K
374.15	647.30	221.29	0.00326	306.75	2100	4.430

Table 46-1 Properties of Superheated Steam (H₂O) in International System of Units—SI

Pressure p		0.01 bar			0.04 bar			0.05 bar			0.06 bar		
Temperature		$t_s = 6.92\ °C$; $i'' = 2\,513\ kJ/kg$ $v'' = 129.9\ m^3/kg$ $s'' = 8.975\ kJ/kg\ K$			$t_s = 28.979\ °C$ $i'' = 2\,554\ kJ/kg$ $v'' = 34.81\ m^3/kg$ $s'' = 8.473\ kJ/kg\ K$			$t_s = 32.88\ °C$ $i'' = 2\,561\ kJ/kg$ $v'' = 28.19\ m^3/kg$ $s'' = 8.393\ kJ/kg\ K$			$t_s = 36.18\ °C$ $i'' = 2\,567\ kJ/kg$ $v'' = 23.74\ m^3/kg$ $s'' = 8.328\ kJ/kg\ K$		
t	T	v	i	s	v	i	s	v	i	s	v	i	s
°C	K	m³/kg	kJ/kg	kJ/kg K	m³/kg	kJ/kg	kJ/kg K	m³/kg	kJ/kg	kJ/kg K	m³/kg	kJ/kg	kJ/kg K
0	273.15	0.0010002	0.0	0.0000	0.0010002	0.0	0.0000	0.0010002	0.0	0.0000	0.0010002	0.0	0.0000
10	283.15	131.3	2518	8.995	0.0010003	41.9	0.1511	0.0010003	41.9	0.1511	0.0010003	41.9	0.1511
20	293.15	136.0	2537	9.056	0.0010018	83.7	0.2964	0.0010018	83.7	0.2964	0.0010018	83.7	0.2964
30	303.15	140.7	2556	9.117	34.95	2556	8.470	0.0010044	125.6	0.4363	0.0010044	125.6	0.4363
40	313.15	145.4	2575	9.178	36.12	2574	8.537	28.87	2574	8.434	24.72	2574	8.350
50	323.15	150.0	2594	9.238	37.29	2593	8.595	29.80	2593	8.492	24.82	2593	8.407
60	333.15	154.7	2613	9.296	38.45	2612	8.651	30.73	2612	8.549	25.59	2612	8.464
70	343.15	159.4	2632	9.352	39.60	2631	8.707	31.65	2631	8.605	26.36	2631	8.520
80	353.15	164.0	2651	9.406	40.75	2650	8.762	32.58	2650	8.659	27.13	2650	8.574
90	363.15	168.7	2669	9.459	41.91	2669	8.815	33.50	2669	8.712	27.91	2669	8.627
100	373.15	173.3	2688	9.510	43.07	2688	8.867	34.43	2688	8.764	28.68	2688	8.679
110	383.15	177.9	2707	9.560	44.23	2707	8.917	35.35	2707	8.814	29.45	2707	8.729
120	393.15	182.6	2726	9.609	45.39	2726	8.966	36.28	2726	8.863	30.22	2726	8.778
130	403.15	187.2	2745	9.656	46.54	2745	9.014	37.20	2745	8.911	30.99	2745	8.826
140	413.15	191.9	2764	9.703	47.69	2764	9.060	38.13	2764	8.957	31.76	2764	8.873
150	423.15	196.5	2783	9.748	48.85	2783	9.105	39.05	2783	9.002	32.53	2783	8.918
160	433.15	201.1	2803	9.793	50.01	2803	9.150	39.98	2803	9.047	33.30	2803	8.963
170	443.15	205.8	2822	9.837	51.16	2822	9.195	40.90	2822	9.092	34.07	2822	9.007
180	453.15	210.4	2841	9.880	52.31	2841	9.238	41.83	2841	9.135	34.84	2841	9.050
190	463.15	215.1	2860	9.922	53.47	2860	9.280	42.75	2860	9.177	35.61	2860	9.092
200	473.15	219.8	2880	9.963	54.63	2880	9.321	43.68	2880	9.219	36.38	2880	9.134
210	483.15	224.4	2899	10.004	55.78	2899	9.362	44.60	2899	9.259	37.15	2899	9.175
220	493.15	229.1	2918	10.044	56.93	2918	9.402	45.53	2918	9.299	37.92	2918	9.215
230	503.15	233.7	2938	10.083	58.09	2938	9.441	46.45	2938	9.338	38.69	2938	9.254
240	513.15	238.3	2958	10.121	59.24	2958	9.479	47.37	2958	9.376	39.46	2958	9.292
250	523.15	243.0	2977	10.159	60.40	2977	9.517	48.30	2977	9.414	40.23	2977	9.330
260	533.15	247.6	2997	10.196	61.56	2997	9.554	49.22	2997	9.451	41.00	2997	9.367
280	553.15	256.9	3037	10.269	63.87	3037	9.627	51.07	3037	9.524	42.54	3037	9.440
300	573.15	266.2	3077	10.340	66.18	3077	9.698	52.92	3077	9.595	44.08	3077	9.511
310	583.15	270.8	3097	10.374	67.33	3097	9.732	53.84	3097	9.630	44.85	3097	9.545
320	593.15	275.4	3117	10.408	68.49	3117	9.766	54.77	3117	9.664	45.62	3117	9.579
330	603.15	280.1	3137	10.441	69.64	3137	9.800	55.69	3137	9.697	46.39	3137	9.613
340	513.15	284.8	3157	10.474	70.80	3157	9.833	56.62	3157	9.730	47.16	3157	9.646
350	623.15	289.5	3177	10.507	71.96	3177	9.866	57.74	3177	9.763	47.93	3177	9.679
360	633.15	294.1	3198	10.539	73.11	3198	9.899	58.47	3198	9.796	48.70	3198	9.711
380	653.15	303.4	3238	10.603	75.42	3238	9.962	60.32	3238	9.859	50.24	3238	9.775
400	673.15	312.6	3280	10.665	77.73	3280	10.024	62.16	3280	9.921	51.78	3280	9.837
410	683.15	317.3	3301	10.696	78.89	3301	10.055	63.08	3301	9.952	52.55	3301	9.868
420	693.15	321.9	3321	10.726	80.04	3321	10.085	64.00	3321	9.982	53.32	3321	9.898
430	703.15	326.6	3342	10.756	81.20	3342	10.115	64.92	3342	10.012	54.09	3342	9.928
440	713.15	331.2	3363	10.786	82.35	3363	10.145	65.85	3363	10.042	54.86	3363	9.958
450	723.15	335.8	3384	10.815	83.51	3384	10.174	66.77	3384	10.071	55.63	3384	9.987
460	733.15	340.5	3405	10.844	84.66	3405	10.203	67.70	3405	10.100	56.40	3405	10.016
480	753.15	349.8	3448	10.902	86.97	3448	10.261	69.54	3448	10.158	57.94	3448	10.074
500	773.15	359.0	3490	10.958	89.28	3490	10.317	71.39	3490	10.214	59.84	3490	10.130
510	783.15	363.7	3512	10.986	90.44	3512	10.345	72.31	3512	10.242	60.25	3512	10.158
520	793.15	368.3	3533	11.014	91.59	3533	10.373	73.24	3533	10.270	61.02	3533	10.186
530	803.15	372.9	3555	11.041	92.75	3555	10.400	74.16	3555	10.297	61.79	3555	10.213
540	813.15	377.6	3576	11.068	93.90	3576	10.427	75.09	3576	10.324	62.56	3576	10.240
550	823.15	382.2	3598	11.095	95.06	3598	10.454	76.01	3598	10.351	63.34	3598	10.267
560	833.15	386.9	3619	11.122	96.22	3619	10.481	76.94	3619	10.378	64.11	3619	10.294
580	853.15	396.2	3663	11.174	98.53	3663	10.533	78.79	3663	10.430	65.65	3663	10.346
600	873.15	405.6	3707	11.226	100.84	3707	10.585	80.64	3707	10.482	67.19	3707	10.398
610	883.15	410.2	3729	11.251	102.00	3729	10.610	81.57	3729	10.507	67.96	3729	10.423
620	893.15	414.8	3751	11.276	103.15	3751	10.635	82.49	3751	10.532	68.73	3751	10.448
630	903. 5	419.4	3773	11.301	104.31	3773	10.660	83.42	3773	10.557	69.50	3773	10.473
640	913.15	424.1	3796	11.325	105.46	3796	10.684	84.34	3796	10.581	70.27	3796	10.497
650	923.15	428.7	3818	11.349	106.62	3818	10.709	85.27	3818	10.605	71.04	3818	20.521
660	933.15	433.4	3841	11.373	107.77	3841	10.733	86.19	3841	10.629	71.81	3841	10.545
680	953.15	442.6	3886	11.421	110.08	3886	10.781	88.04	3886	10.677	73.35	3886	10.593
700	973.15	451.9	3931	11.468	112.39	3931	10.828	89.88	3931	10.725	74.89	3931	10.640

All values above the solid horizontal lines refer to the saturated liquid state.

Table 46-2 Properties of Superheated Steam (H_2O) in International System of Units—SI (*Continued*)

Pressure p		0.07 bar			0.08 bar			0.09 bar			0.10 bar		
Temperature		$t_s = 39.03$ °C $i'' = 2\,572$ kJ/kg $v'' = 20.53$ m³/kg $s'' = 8.274$ kJ/kg K			$t_s = 41.54$ °C $i'' = 2\,576$ kJ/kg $v'' = 18.10$ m³/kg $s'' = 8.227$ kJ/kg K			$t_s = 43.79$ °C $i'' = 2\,580$ kJ/kg $v'' = 16.20$ m³/kg $s'' = 8.186$ kJ/kg K			$t_s = 45.84$ °C $i'' = 2\,584$ kJ/kg $v'' = 14.68$ m³/kg $s'' = 8.149$ kJ/kg K		
t	T	v	i	s	v	i	s	v	i	s	v	i	s
°C	K	m³/kg	kJ/kg	kJ/kg K	m³/kg	kJ/kg	kJ/kg K	m³/kg	kJ/kg	kJ/kg K	m³/kg	kJ/kg	kJ/kg K
0	273.15	0.0010002	0.0	0.0000	0.0010002	0.0	0.0000	0.0010002	0.0	0.0000	0.0010002	0.0	0.0000
10	283.15	0.0010003	41.9	0.1511	0.0010003	41.9	0.1511	0.0010003	41.9	0.1511	0.0010003	41.9	0.1511
20	293.15	0.0010018	83.7	0.2964	0.0010018	83.7	0.2964	0.0010018	83.7	0.2964	0.0010018	83.7	0.2964
30	303.15	0.0010044	125.6	0.4363	0.0010044	125.6	0.4363	0.0010044	125.6	0.4363	0.0010044	125.6	0.4363
40	313.15	20.94	2574	8.279	0.0010079	167.5	0.5715	0.0010079	167.5	0.5715	0.0010079	167.5	0.5715
50	323.15	21.27	2593	8.336	18.61	2593	8.274	16.57	2593	8.220	15.00	2592	8.170
60	333.15	21.94	2612	8.393	19.19	2612	8.331	17.09	2612	8.277	15.35	2611	8.227
70	343.15	22.60	2631	8.449	19.76	2631	8.387	17.61	2631	8.333	15.81	2630	8.283
80	353.15	23.26	2650	8.503	20.34	2650	8.441	18.12	2650	8.387	16.27	2649	8.337
90	363.15	23.92	2669	8.556	20.92	2669	8.494	18.64	2669	8.440	16.74	2669	8.390
100	373.15	24.58	2688	8.608	21.50	2688	8.546	19.16	2688	8.492	17.20	2688	8.442
110	383.15	25.24	2707	8.658	22.08	2707	8.596	19.67	2707	8.542	17.67	2707	8.493
120	393.15	25.90	2726	8.707	22.26	2726	8.645	20.19	2726	8.591	18.13	2726	8.542
130	403.15	26.56	2745	8.755	23.24	2745	8.693	20.70	2745	8.639	18.59	2745	8.589
140	413.15	27.22	2764	8.802	23.82	2764	8.740	21.22	2764	8.686	19.06	2764	8.636
150	423.15	27.88	2783	8.847	24.40	2783	8.785	21.73	2783	8.731	19.52	2783	8.682
160	433.15	28.54	2802	8.892	24.97	2802	8.830	22.25	2802	8.776	19.98	2802	8.727
170	443.15	29.20	2822	8.936	25.55	2822	8.874	22.76	2822	8.820	20.44	2822	8.771
180	453.15	29.86	2841	8.979	26.13	2841	8.917	23.28	2841	8.863	20.90	2841	8.814
190	463.15	30.52	2860	9.021	26.71	2860	8.959	23.79	2860	8.905	21.36	2860	8.856
200	473.15	31.18	2880	9.062	27.29	2880	9.000	24.31	2879	8.946	21.83	2879	8.897
210	483.15	31.84	2899	9.103	27.86	2899	9.041	24.83	2899	8.987	22.30	2899	8.938
220	493.15	32.50	2918	9.143	28.44	2918	9.081	25.34	2918	9.027	22.76	2918	8.978
230	503.15	33.16	2938	9.182	29.02	2938	9.120	25.86	2938	9.066	23.22	2938	9.017
240	513.15	33.82	2957	9.221	29.60	2957	9.159	26.37	2957	9.105	23.68	2957	9.056
250	523.15	34.48	2977	9.258	30.18	2977	9.197	26.89	2977	9.143	24.14	2977	9.094
260	533.15	35.14	2997	9.295	30.75	2997	9.234	27.40	2997	9.180	24.60	2997	9.131
280	553.15	36.46	3037	9.368	31.90	3037	9.306	28.43	3037	9.252	25.53	3037	9.203
300	573.15	37.78	3077	9.439	33.06	3077	9.377	29.46	3077	9.323	26.46	3077	9.274
310	583.15	38.44	3097	9.474	33.64	3097	9.412	29.98	3097	9.358	26.92	3097	9.309
320	593.15	39.10	3117	9.508	34.22	3117	9.446	30.49	3117	9.392	27.38	3117	9.343
330	603.15	39.76	3137	9.541	34.79	3137	9.480	31.00	3137	9.426	27.84	3137	9.377
340	613.15	40.42	3157	9.574	35.37	3157	9.513	31.51	3157	9.459	28.30	3157	9.410
350	623.15	41.08	3177	9.607	35.94	3177	9.546	32.03	3177	9.492	28.76	3177	9.443
360	633.15	41.74	3198	9.640	36.52	3198	9.578	32.54	3198	9.524	29.23	3198	9.475
380	653.15	43.06	3238	9.703	37.68	3238	9.641	33.57	3238	9.587	30.15	3238	9.539
400	673.15	44.38	3280	9.765	38.84	3280	9.704	34.60	3280	9.650	31.08	3280	9.601
410	683.15	45.04	3301	9.796	39.41	3301	9.375	35.12	3301	9.681	31.54	3301	9.632
420	693.15	45.70	3321	9.826	39.98	3321	9.765	35.63	3321	9.711	32.00	3321	9.662
430	703.15	46.36	3342	9.856	40.56	3342	9.795	36.15	3342	9.741	32.46	3342	9.692
440	713.15	47.02	3363	9.886	41.14	3363	9.825	36.66	3363	9.771	32.93	3363	9.722
450	723.15	47.68	3384	9.916	41.72	3384	9.854	37.18	3384	9.800	33.39	3384	9.751
460	733.15	48.34	3405	9.945	42.30	3405	9.883	37.69	3405	9.829	33.85	3405	9.780
480	753.15	49.66	3448	10.002	43.46	3448	9.941	38.72	3448	9.887	34.77	3448	9.838
500	773.15	50.98	3490	10.059	44.61	3490	9.997	39.75	3490	9.943	35.70	3490	9.895
510	783.15	51.64	3512	10.086	45.18	3512	10.025	40.27	3512	9.971	36.16	3512	9.922
520	793.15	52.30	3533	10.114	45.76	3533	10.053	40.78	3533	9.999	36.63	3533	9.950
530	803.15	52.96	3555	10.142	46.34	3555	10.080	41.30	3555	10.026	37.09	3555	9.977
540	8131.5	53.62	3576	10.169	46.91	3576	10.107	41.81	3576	10.053	37.55	3576	10.004
550	823.15	54.28	3598	10.196	47.49	3598	10.134	42.32	3598	10.080	38.01	3598	10.031
560	833.15	54.94	3619	10.223	48.07	3619	10.161	42.83	3619	10.107	38.47	3619	10.058
580	853.15	56.26	3663	10.275	49.23	3663	10.213	43.86	3663	10.159	39.40	3663	10.110
600	873.15	57.58	3707	10.327	50.38	3707	10.265	44.89	3707	10.211	40.32	3707	10.162
610	883.15	58.24	2729	10.352	50.95	3729	10.290	45.40	3729	10.236	40.78	3729	10.187
620	893.15	58.90	3751	10.377	51.53	3751	10.315	45.92	3751	10.261	41.24	3751	10.212
630	903.15	59.56	3773	10.402	52.11	3773	10.340	46.43	3773	10.286	41.70	3773	10.237
640	913.15	60.22	3796	10.426	52.69	3796	10.365	46.95	3796	10.311	42.17	3796	10.262
650	923.15	60.88	3818	10.450	53.27	3818	10.389	47.46	3818	10.355	42.63	3818	10.286
660	933.15	61.54	3841	10.474	53.84	3841	10.413	47.98	3841	10.359	43.10	3841	10.310
680	953.15	62.86	3886	10.522	54.99	3886	10.461	49.01	3886	10.407	44.02	3886	10.358
700	973.15	64.17	3931	10.569	56.15	3931	10.508	50.04	3931	10.454	44.94	3931	10.405

Table 46-3 Properties of Superheated Steam (H₂O) in International System of Units—SI (Continued)

Pressure p		0.12 bar			0.14 bar			0.16 bar			0.18 bar		
Temperature		t_s =49.45 °C i'' =2 591 kJ/kg v'' =12.35 m³/kg s'' =8.085 kJ/kcal K			t_s =52.58 °C i'' =2 596 kJ/kg v'' =10.69 m³/kg s'' =8.031 kJ/kg K			t_s =55.34° C i'' =2 601 kJ/kg v'' =9.429 m³/kg s'' =7.984 kJ/kg K			t_s =57.82 °C i'' =2 605 kJ/kg v'' =8.444 m³/kg s'' =7.944 kJ/kg K		
t	T	v	i	s	v	i	s	v	i	s	v	i	s
°C	K	m³/kg	kJ/kg	kJ/kgK	m³/kg	kJ/kg	kJ/kg K	m³/kg	kJ/kg	kJ/kg K	m³/kg	kJ/kg	kJ/kg K
0	273.15	0.0010002	0.0	0.0000	0.0010002	0.0	0.0000	0.0010002	0.0	0.0000	0.0010002	0.0	0.0000
10	283.15	0.0010003	41.9	0.1511	0.0010003	41.9	0.1511	0.0010003	41.9	0.1511	0.0010003	41.9	0.1511
20	293.15	0.0010018	83.7	0.2964	0.0010018	83.7	0.2964	0.0010018	83.7	0.2964	0.0010018	83.7	0.2964
30	303.15	0.0010044	125.6	0.4363	0.0010044	125.6	0.4363	0.0010044	125.6	0.4363	0.0010044	125.6	0.4363
40	313.15	0.0010079	167.5	0.5715	0.0010079	167.5	0.5715	0.0010079	167.5	0.5715	0.0010079	167.5	0.5715
50	323.15	12.44	2592	8.086	0.0010121	209.3	0.7030	0.0010121	209.3	0.7030	0.0010121	209.3	0.7030
60	333.15	12.78	2611	8.143	10.95	2611	8.071	9.573	2610	8.009	8.497	2610	7.954
70	343.15	13.17	2630	8.199	11.28	2630	8.127	9.867	2629	8.065	8.764	2629	8.010
80	353.15	13.55	2649	8.253	11.61	2649	8.181	10.160	2649	8.120	9.024	2648	8.064
90	363.15	13.94	2668	8.306	11.94	2668	8.235	10.450	2668	8.173	9.283	2667	8.117
100	373.15	14.33	2687	8.358	12.27	2687	8.287	10.740	2687	8.225	9.542	2687	8.169
110	383.15	14.72	2706	8.408	12.61	2706	8.337	11.030	2706	8.275	9.800	2706	8.220
120	393.15	15.10	2725	8.457	12.94	2725	8.386	11.320	2725	8.324	10.058	2725	8.269
130	403.15	15.49	2744	8.505	13.27	2744	8.434	11.610	2744	8.372	10.316	2744	8.317
140	413.15	15.87	2764	8.552	13.60	2763	8.481	11.899	2763	8.419	10.574	2763	8.364
150	423.15	16.26	2783	8.598	13.93	2782	8.527	12.189	2782	8.465	10.832	2782	8.410
160	433.15	16.64	2802	8.643	14.26	2802	8.572	12.478	2802	8.510	11.090	2802	8.455
170	443.15	17.03	2822	8.687	14.59	2821	8.616	12.768	2821	8.554	11.347	2821	8.499
180	453.15	17.42	2841	8.730	14.92	2840	8.659	13.057	2840	8.597	11.605	2840	8.542
190	463.15	17.80	2860	8.772	15.25	2860	8.701	13.346	2860	8.639	11.862	2860	8.584
200	473.15	18.19	2879	8.813	15.58	2879	8.742	13.635	2879	8.680	12.120	2879	8.625
210	483.15	18.57	2898	8.854	15.91	2898	8.783	13.924	2898	8.721	12.377	2898	8.666
220	493.15	18.96	2918	8.894	16.24	2918	8.823	14.213	2918	8.761	12.634	2918	8.706
230	503.15	19.34	2937	8.933	16.57	2937	8.862	14.502	2937	8.800	12.892	2937	8.745
240	513.15	19.73	2957	8.972	16.90	2957	8.900	14.790	2957	8.838	13.149	2957	8.784
250	523.15	20.11	2977	9.010	17.23	2977	8.938	15.079	2977	8.876	13.406	2976	8.822
260	533.15	20.50	2996	9.047	17.56	2997	8.975	15.367	2997	8.913	13.663	2997	8.859
280	553.15	21.27	3036	9.119	18.22	3037	9.048	15.943	3037	8.986	14.177	3037	8.932
300	573.15	22.04	3077	9.190	18.88	3077	9.119	16.52	3077	9.057	14.690	3077	9.003
310	583.15	22.42	3097	9.225	19.21	3097	9.154	16.81	3097	9.092	14.947	3097	9.038
320	593.15	22.81	3117	9.259	19.54	3117	9.188	17.10	3117	9.126	15.204	3117	9.072
330	603.15	23.19	3137	9.292	19.87	3137	9.221	17.39	3137	9.160	15.460	3137	9.106
340	613.15	23.58	3157	9.326	20.20	3157	9.255	17.68	3157	9.193	15.716	3157	9.139
350	623.15	23.96	3177	9.359	20.53	3177	9.288	17.96	3177	9.226	15.971	3177	9.171
360	633.15	24.35	3198	9.391	20.86	3198	9.320	18.25	3198	9.258	16.23	3198	9.204
380	653.15	25.12	3238	9.455	21.52	3238	9.383	18.83	3238	9.322	16.74	3238	9.268
400	673.15	25.89	3280	9.517	22.18	3280	9.446	19.41	3280	9.384	17.26	3280	9.330
410	683.15	26.28	3301	9.548	22.51	3301	9.476	19.70	3301	9.415	17.51	3301	9.361
420	693.15	26.66	3321	9.578	22.84	3321	9.506	19.99	3321	9.445	17.76	3321	9.391
430	703.15	27.04	3342	9.608	23.17	3342	9.536	20.28	3342	9.475	18.02	3342	9.421
440	713.15	27.43	3363	9.638	23.50	3363	9.566	20.56	3363	9.504	18.27	3363	9.450
450	723.15	27.82	3384	9.667	23.83	3384	9.596	20.85	3384	9.534	18.53	3384	9.480
460	733.15	28.20	3405	9.696	24.16	3405	9.625	21.13	3405	9.563	18.78	3405	9.509
480	753.15	28.96	3448	9.754	24.82	3448	9.683	21.71	3448	9.621	19.29	3448	9.567
500	773.15	29.74	3490	9.810	25.49	3490	9.739	22.29	3490	9.678	19.80	3490	9.624
510	783.15	30.13	3512	9.838	25.82	3512	9.767	22.58	3512	9.706	20.06	3512	9.652
520	793.15	30.52	3533	9.866	26.15	3533	9.795	22.87	3533	9.734	20.32	3533	9.680
530	803.15	30.90	3555	8.894	26.48	3555	9.823	23.16	3555	9.762	20.58	3555	9.708
540	813.15	31.29	3576	9.921	26.81	3576	9.850	23.45	3576	9.789	20.83	3576	9.735
550	823.15	31.67	3598	9.948	27.14	3598	9.877	23.74	3598	9.816	21.09	3598	9.762
560	833.15	32.06	3619	9.974	27.47	3619	9.903	24.03	3619	9.842	21.35	3619	9.788
580	853.15	32.83	3663	10.026	28.13	3663	9.955	24.60	3663	9.894	21.87	3663	9.840
600	873.15	33.60	3707	10.078	28.79	3707	10.007	25.18	3707	9.945	22.39	3707	9.891
610	883.15	33.98	3729	10.103	29.12	3729	10.032	25.47	3729	9.970	22.64	3729	9.916
620	893.15	34.37	3751	10.128	29.45	3751	10.057	25.76	3751	9.995	22.90	3751	9.941
630	903.15	34.75	3773	10.153	29.78	3773	10.082	26.05	3773	10.020	23.16	3773	9.966
640	913.15	35.14	3796	10.178	30.11	3796	10.106	26.34	3796	10.045	23.41	3796	9.991
650	923.15	35.52	3818	10.202	30.44	3818	10.130	26.63	3818	10.069	23.67	3818	10.015
660	933.15	35.91	3841	10.226	30.77	3841	10.154	26.92	3841	10.093	23.92	3841	10.039
680	953.15	36.68	3886	10.274	31.42	3886	10.202	27.50	3886	10.141	24.44	3886	10.087
700	973.15	37.44	3931	10.321	32.08	3931	10.249	28.08	3931	10.188	24.95	3931	10.133

Table 46-4 Properties of Superheated Steam (H$_2$O) in International System of Units—SI (Continued)

Pressure p		0.20 bar			0.30 bar			0.40 bar			0.50 bar		
Temperature		$t_s = 60.08\ °C$ $i'' = 2\ 609\ kJ/kg$ $v'' = 7.647\ m^3/kg$ $s'' = 7.907\ kJ/kg\ K$			$t_s = 69.12\ °C$ $i'' = 2\ 625\ kJ/kg$ $v'' = 5.226\ m^3/kg$ $s'' = 7.769\ kJ/kg\ K$			$t_s = 75.88\ °C$ $i'' = 2\ 636\ kJ/kg$ $v'' = 3.994\ m^3/kg$ $s'' = 7.760\ kJ/kg\ K$			$t_s = 81.35\ °C$ $i'' = 2\ 645\ kJ/kg$ $v'' = 3.239\ m^3/kg$ $s'' = 7.593\ kJ/kg\ K$		
t	T	v	i	s	v	i	s	v	i	s	v	i	s
°C	K	m³/kg	kJ/kg	kJ/kg K	m³/kg	kJ/kg	kJ/kg K	m³/kg	kJ/kg	kJ/kg K	m³/kg	kj/kg	kJ/kg K
0	273.15	0.0010002	0.0	0.0000	0.0010002	0.0	0.0000	0.0010002	0.0	0.0000	0.0010002	0.1	0.0000
10	283.15	0.0010003	41.9	0.1511	0.0010003	41.9	0.1511	0.0010003	41.9	0.1511	0.0010003	42.0	0.1511
20	293.15	0.0010018	83.7	0.2964	0.0010018	83.7	0.2964	0.0010018	83.7	0.2964	0.0010018	83.8	0.2964
30	303.15	0.0010044	125.6	0.4363	0.0010044	125.6	0.4363	0.0010044	125.6	0.4363	0.0010044	125.6	0.4363
40	313.15	0.0010079	167.5	0.5715	0.0010079	167.5	0.5715	0.0010079	167.5	0.5715	0.0010079	167.5	0.5715
50	323.15	0.0010121	209.3	0.7030	0.0010121	209.3	0.7030	0.0010121	209.3	0.7031	0.0010121	209.3	0.7031
60	333.15	0.0010171	251.1	0.8307	0.0010171	251.1	0.8307	0.0010171	251.1	0.8307	0.0010171	251.1	0.8307
70	343.15	7.887	2629	7.961	5.268	2627	7.770	0.0010228	293.0	0.9546	0.0010228	293.0	0.9546
80	353.15	8.119	2648	8.015	5.400	2646	7.825	4.088	2645	7.690	0.0010290	334.9	1.0748
90	363.15	8.351	2667	8.068	5.557	2666	7.879	4.163	2665	7.745	3.324	2663	7.640
100	373.15	8.584	2687	8.120	5.713	2685	7.931	4.282	2684	7.798	3.420	2683	7.693
110	383.15	8.816	2706	8.171	5.869	2705	7.982	4.399	2703	7.849	3.514	2703	7.745
120	393.15	9.049	2725	8.220	6.025	2724	8.031	4.516	2723	7.899	3.608	2722	7.795
130	403.15	9.281	2744	8.268	6.180	2743	8.079	4.633	2742	7.947	3.702	2741	7.843
140	413.15	9.513	2763	8.315	6.335	2762	8.126	4.750	2761	7.995	3.795	2761	7.890
150	423.15	9.745	2782	8.361	6.490	2782	8.172	4.866	2781	8.041	3.889	2780	7.936
160	433.15	9.977	2801	8.406	6.645	2801	8.217	4.982	2800	8.086	3.982	2799	7.981
170	443.15	10.209	2821	8.450	6.800	2820	8.261	5.099	2819	8.130	4.075	2819	8.025
180	453.15	10.441	2840	8.493	6.955	2839	8.304	5.215	2838	8.173	4.169	2838	8.069
190	463.15	10.673	2859	8.535	7.110	2859	8.346	5.331	2858	8.215	4.262	2858	8.111
200	473.15	10.905	2879	8.576	7.264	2878	8.388	5.447	2878	8.256	4.355	2877	8.152
210	483.15	11.137	2898	8.617	7.419	2898	8.429	5.564	2897	8.297	4.448	2897	8.193
220	493.15	11.369	2918	8.657	7.573	2917	8.469	5.680	2917	8.337	4.540	2916	8.233
230	503.15	11.600	2937	8.696	7.728	2937	8.508	5.796	2937	8.376	4.633	2936	8.272
240	513.15	11.832	2957	8.735	7.882	2956	8.547	5.912	2956	8.415	4.726	2956	8.311
250	523.15	12.064	2976	8.773	8.037	2976	8.585	6.028	2976	8.453	4.819	2975	8.349
260	533.15	12.295	2996	8.810	8.191	2996	8.622	6.144	2995	8.490	4.912	2995	8.386
280	553.15	12.758	3036	8.883	8.500	3036	8.695	6.376	3035	8.564	5.098	3035	8.460
300	573.15	13.220	3077	8.954	8.809	3076	8.766	6.608	3076	8.635	5.284	3076	8.531
310	583.15	13.452	3097	8.989	8.964	3096	8.801	6.723	3096	8.669	5.377	3096	8.565
320	593.15	13.683	3117	9.023	9.118	3116	8.835	6.839	3116	8.703	5.470	3116	8.599
330	603.15	13.914	3137	9.057	9.272	3136	8.869	6.954	3136	8.737	5.563	3136	8.633
340	613.15	14.145	3157	9.090	9.426	3157	8.902	7.070	3156	8.770	5.656	3156	8.666
350	623.15	14.376	3177	9.123	9.580	3177	8.935	7.186	3177	8.803	5.749	3176	8.699
360	633.15	14.606	3198	9.155	9.734	3198	8.967	7.301	3197	8.835	5.841	3197	8.731
380	653.15	15.068	3238	9.219	10.042	3238	9.031	7.533	3238	8.899	6.027	3237	8.795
400	673.15	15.530	3280	9.281	10.351	3280	9.093	7.765	3279	8.962	6.212	3279	8.858
410	683.15	15.761	3301	9.312	10.505	3300	9.124	7.880	3300	8.992	6.304	3300	8.889
420	693.15	15.992	3321	9.342	10.659	3321	9.155	7.996	3321	9.022	6.397	3320	8.919
430	703.15	16.220	3342	9.372	10.813	3342	9.185	8.112	3341	9.053	6.489	3341	8.949
440	713.15	16.45	3363	9.402	10.967	3363	9.215	8.228	3362	9.083	6.582	3362	8.979
450	723.15	16.68	3384	9.431	11.121	3384	9.244	8.343	3383	9.112	6.674	3383	9.008
460	733.15	16.90	3405	9.460	11.275	3405	9.273	8.459	3404	9.141	6.766	3404	9.037
480	753.15	17.36	3448	9.518	11.583	3447	9.331	8.690	3447	9.199	6.951	3447	9.095
500	773.15	17.82	3490	9.575	11.891	3490	9.388	8.921	3490	9.256	7.136	3489	9.152
510	783.15	18.05	3512	9.603	12.045	3512	9.416	9.036	3511	9.284	7.229	3511	9.180
520	793.51	18.28	3533	9.631	12.199	3533	9.444	9.152	3532	9.311	7.321	3532	9.208
530	803.15	18.52	3555	9.659	12.353	3555	9.472	9.267	3554	9.339	7.413	3554	9.236
540	813.15	18.75	3576	9.686	12.507	3576	9.499	9.382	3576	9.366	7.506	3576	9.263
550	823.15	18.99	3598	9.713	12.661	3598	9.526	9.498	3597	9.393	7.598	3597	9.290
560	833.15	19.22	3619	9.739	12.815	3619	9.552	9.613	3619	9.419	7.690	3619	9.316
580	853.15	19.69	3663	9.791	13.123	3663	9.604	9.843	3663	9.471	7.874	3663	9.368
600	873.15	20.15	3707	9.842	13.430	3707	9.655	10.074	3707	9.522	8.058	3707	9.419
610	883.15	20.38	3729	9.867	13.584	3729	9.680	10.190	3729	9.547	8.150	3729	9.444
620	893.15	20.61	3751	9.892	13.738	3751	9.705	10.305	3751	9.572	8.242	3741	9.469
630	903.15	20.84	3773	9.917	13.892	3773	9.730	10.420	3773	9.597	8.335	3773	9.494
640	913.15	21.07	3796	9.942	14.046	3796	9.755	10.536	3795	9.622	8.427	3795	9.519
650	923.15	21.30	3818	9.966	14.200	3818	9.779	10.651	3818	9.646	8.520	3818	9.543
660	933.15	21.53	3841	9.990	14.353	3841	9.803	10.767	3840	9.670	8.612	3840	9.567
680	953.15	21.99	3886	10.038	14.661	3886	9.851	10.998	3885	9.718	8.797	3885	9.615
700	973.15	22.45	3931	10.085	14.969	3931	9.898	11.228	3930	9.765	8.982	3930	9.662

Table 46-5 Properties of Superheated Steam (H_2O) in International System of Units—SI (*Continued*)

Pressure p		0.60 bar			0.70 bar			0.80 bar			0.90 bar		
Temperature		$t_s = 85.95\,°C$ $i'' = 2\,653$ kJ/kg $v'' = 2.732$ m³/kg $s'' = 7.531$ kJ/kg K			$t_s = 89.97\,°C$ $i'' = 2\,660$ kJ/kg $v'' = 2.364$ m³/kg $s'' = 7.479$ kJ/kg K			$t_s = 93.52\,°C$ $i'' = 2\,665$ kJ/kg $v'' = 2.087$ m³/kg $s'' = 7.434$ kJ/kg K			$t_s = 96.72\,°C$ $i'' = 2\,670$ kJ/kg $v'' = 1.869$ m³/kg $s'' = 7.394$ kJ/kg K		
t	T	v	i	s	v	i	s	v	i	s	v	i	s
°C	K	m³/kg	kJ/kg	kJ/kg K	m³/kg	kJ/kg	kJ/kg K	m³/kg	kJ/kg	kJ/kg K	m³/kg	kJ/kg	kJ/kg K
0	273.15	0.0010002	0.1	0.0000	0.0010002	0.1	0.0000	0.0010002	0.1	0.0000	0.0010001	0.1	0.0000
10	283.15	0.0010003	42.0	0.1511	0.0010003	42.0	0.1511	0.0010003	42.0	0.1511	0.0010003	42.0	0.1511
20	293.15	0.0010018	83.8	0.2964	0.0010018	83.9	0.2964	0.0010018	83.9	0.2964	0.0010018	83.9	0.2964
30	303.15	0.0010044	125.6	0.4363	0.0010044	125.7	0.4363	0.0010044	125.7	0.4363	0.0010044	125.7	0.4363
40	313.15	0.0010079	167.5	0.5715	0.0010079	167.5	0.5715	0.0010079	167.5	0.5715	0.0010079	167.5	0.5715
50	323.15	0.0010121	209.3	0.7031	0.0010121	209.3	0.7031	0.0010121	209.3	0.7031	0.0010121	209.3	0.7031
60	333.15	0.0010171	251.1	0.8307	0.0010171	251.1	0.8307	0.0010171	251.1	0.8307	0.0010171	251.1	0.8307
70	343.15	0.0010228	293.0	0.9546	0.0010228	293.0	0.9546	0.0010227	293.0	0.9546	0.0010227	293.0	0.9546
80	353.15	0.0010289	334.9	1.0748	0.0010290	334.9	1.0748	0.0010289	334.9	1.0748	0.0010289	334.9	1.0748
90	363.15	2.765	2661	7.551	2.366	2659	7.479	0.0010359	376.9	1.1924	0.0010359	376.8	1.1924
100	373.15	2.846	2681	7.604	2.436	2680	7.532	2.127	2679	7.468	1.887	2677	7.411
110	383.15	2.925	2701	7.657	2.504	2700	7.584	2.187	2699	7.520	1.941	2698	7.464
120	393.15	3.004	2721	7.708	2.572	2720	7.635	2.247	2719	7.572	1.995	2718	7.515
130	403.15	3.082	2740	7.757	2.640	2739	7.684	2.307	2739	7.621	2.048	2738	7.565
140	413.15	3.160	2760	7.804	2.707	2759	7.731	2.366	2758	7.669	2.101	2757	7.613
150	423.15	3.238	2780	7.850	2.773	2779	7.778	2.424	2778	7.715	2.153	2777	7.660
160	433.15	3.316	2799	7.895	2.840	2798	7.823	2.483	2797	7.761	2.205	2797	7.705
170	443.15	3.394	2818	7.939	2.907	2818	7.867	2.542	2817	7.805	2.258	2816	7.750
180	453.15	3.472	2837	7.983	2.974	2837	7.911	2.600	2836	7.849	2.310	2836	7.794
190	463.15	3.550	2857	8.025	3.041	2857	7.954	2.659	2856	7.892	2.362	2855	7.836
200	473.15	3.628	2877	8.067	3.107	2876	7.995	2.717	2876	7.934	2.414	2875	7.878
210	483.15	3.706	2897	8.108	3.174	2896	8.036	2.775	2896	7.975	2.466	2895	7.919
220	493.15	3.784	2916	8.148	3.240	2916	8.076	2.834	2915	8.015	2.518	2914	7.960
230	503.15	3.861	2936	8.187	3.307	2935	8.116	2.892	2935	8.054	2.570	2934	7.999
240	513.15	3.939	2956	8.226	3.374	2955	8.155	2.951	2955	8.093	2.621	2954	8.037
250	523.15	4.016	2975	8.264	3.440	2975	8.193	3.009	2974	8.131	2.673	2974	8.075
260	533.15	4.093	2995	8.302	3.507	2995	8.230	3.067	2994	8.168	2.725	2994	8.133
280	553.15	4.247	3035	8.375	3.639	3035	8.303	3.183	3034	8.242	2.828	3034	8.186
300	573.15	4.401	3075	8.446	3.772	3075	8.374	3.299	3075	8.313	2.931	3074	8.257
310	583.15	4.479	3095	8.481	3.839	3095	8.409	3.357	3095	8.347	2.983	3094	8.292
320	593.15	4.556	3115	8.515	3.905	3115	8.443	3.415	3115	8.381	3.035	3114	8.326
330	603.15	4.634	3136	8.549	3.971	3135	8.477	3.474	3135	8.415	3.087	3134	8.360
340	613.15	4.711	3156	8.582	4.038	3156	8.510	3.532	3156	8.448	3.138	3155	8.393
350	623.15	4.788	3176	8.615	4.104	3176	8.543	3.590	3176	8.481	3.190	3175	8.426
360	633.15	4.865	3197	8.647	4.170	3197	8.575	3.648	3196	8.513	3.242	3196	8.458
380	653.15	5.020	3237	8.711	4.303	3237	8.639	3.764	3237	8.577	3.345	3237	8.522
400	673.15	5.175	3279	8.773	4.435	3279	8.702	3.880	3279	8.640	3.448	3278	8.585
410	683.15	5.252	3300	8.804	4.501	3300	8.733	3.937	3299	8.671	3.499	3299	8.616
420	693.15	5.330	3320	8.834	4.567	3320	8.763	3.995	3320	8.701	3.551	3320	8.646
430	703.15	5.407	3341	8.864	4.634	3341	8.793	4.052	3341	8.731	3.602	3341	8.675
440	713.15	5.484	3362	8.894	4.700	3362	8.823	4.110	3362	8.761	3.654	3362	8.705
450	723.15	5.562	3383	8.924	4.766	3383	8.852	4.168	3383	8.790	3.705	3383	8.734
460	733.15	5.639	3404	8.953	4.832	3404	8.881	4.226	3404	8.819	3.757	3404	8.763
480	753.15	5.793	3447	9.011	4.964	3447	8.939	4.342	3446	8.877	3.860	3446	8.821
500	773.15	5.947	3489	9.067	5.096	3489	8.996	4.458	3489	8.934	3.962	3489	8.879
510	783.15	6.024	3511	9.095	5.162	3511	9.024	4.515	3510	8.962	4.013	3510	8.907
520	793.15	6.101	3532	9.123	5.228	3532	9.052	4.573	3532	8.990	4.065	3532	8.935
530	803.15	6.177	3554	9.151	5.294	3554	9.079	4.631	3554	9.018	4.116	3553	8.963
540	813.15	6.254	2576	9.178	5.360	3576	9.106	4.689	3576	9.045	4.167	3575	8.990
550	823.15	6.331	3597	9.205	5.246	3597	9.133	4.747	3597	9.072	4.219	3597	9.017
560	833.15	6.407	3619	9.232	5.491	3619	9.160	4.804	3619	9.099	4.270	3619	9.044
580	853.15	6.560	3663	9.284	5.623	3663	9.213	4.919	3663	9.151	4.372	3662	9.097
600	873.15	6.714	3707	9.335	5.754	3707	9.264	5.035	3707	9.202	4.475	3706	9.147
610	883.15	6.791	3729	9.360	5.820	3729	9.289	5.093	3729	9.227	4.526	3728	9.172
620	893.15	6.868	3751	9.385	5.886	3751	9.314	5.150	3751	9.252	4.578	3750	9.197
630	903.15	6.945	3773	9.410	5.952	3773	9.339	5.208	3773	9.277	4.629	3772	9.222
640	913.15	7.022	3795	9.435	6.018	3795	9.364	5.266	3795	9.302	4.680	3795	9.274
650	923.15	7.099	3818	9.459	6.084	3818	9.388	5.324	3818	9.326	4.732	3817	9.271
660	933.15	7.176	3840	9.483	6.150	3840	9.412	5.381	3840	9.350	4.783	3840	9.295
680	953.15	7.330	3885	9.531	6.282	3885	9.460	5.496	3885	9.398	4.885	3885	9.343
700	973.15	7.484	3930	9.578	6.414	3930	9.507	5.612	3930	9.445	4.988	3930	9.390

Table 46-6 Properties of Superheated Steam (H₂O) in International System of Units—SI (Continued)

Pressure p		1.0 bar			1.2 bar			1.3 bar			1.4 bar		
Temperature		$t_s=99.64\,°C$ $i''=2\,675\ kJ/kg$ $v''=1.694\ m^3/kg$ $s''=7.360\ kJ/kg\,K$			$t_s=104.81\,°C$ $i''=2\,683\ kJ/kg$ $v''=1.429\ m^3/kg$ $s''=7.298\ kJ/kg\,K$			$t_s=107.14\,°C$ $i''=2\,687\ kJ/kg$ $v''=1.325\ m^3/kg$ $s''=7.271\,kJ/kg\,K$			$t_s=109.33\,°C$ $i''=2\,690\ kJ/kg$ $v''=1.236\ m^3/kg$ $s''=7.246\ kJ/kg\,K$		
t	T	v	i	s	v	i	s	v	i	s	v	i	s
°C	K	m³/kg	kJ/kg	kJ/kg K	m³/kg	kJ/kg	kJ/kg K	m³/kg	kJ/kg	kJ/kg K	m³/kg	kJ/kg	kJ/kg K
0	273.15	0.0010001	0.1	0.0000	0.0010001	0.1	0.0000	0.0010001	0.1	0.0000	0.0010000	0.1	0.0000
20	293.15	0.0010018	83.9	0.2964	0.0010018	83.9	0.2964	0.0010017	83.9	0.2964	0.0010017	83.9	0.2964
40	313.15	0.0010079	167.5	0.5715	0.0010079	167.5	0.5715	0.0010078	167.5	0.5715	0.0010078	167.5	0.5715
50	323.15	0.0010121	209.3	0.7031	0.0010121	209.3	0.7031	0.0010120	209.3	0.7032	0.0010120	209.3	0.7032
60	333.15	0.0010171	251.1	0.8307	0.0010171	251.1	0.8307	0.0010170	251.1	0.8307	0.0010170	251.1	0.8307
80	353.15	0.0010289	334.9	1.0748	0.0010289	334.9	1.0748	0.0010289	334.9	1.0748	0.0010289	334.9	1.0748
100	373.15	1.695	2676	7.361	0.0010434	419.0	1.3067	0.0010434	419.0	1.3067	0.0010434	419.0	1.3067
120	393.15	1.795	2717	7.465	1.491	2715	7.376	1.376	2714	7.337	1.275	2713	7.301
140	413.15	1.889	2757	7.562	1.572	2755	7.475	1.449	2754	7.437	1.344	2753	7.401
150	423.15	1.937	2776	7.608	1.611	2775	7.522	1.486	2774	7.484	1.378	2773	7.449
160	433.15	1.984	2796	7.654	1.650	2795	7.568	1.522	2794	7.530	1.412	2793	7.495
180	453.15	2.078	2835	7.743	1.729	2834	7.657	1.594	2834	7.619	1.480	2833	7.584
200	473.15	2.172	2875	7.828	1.807	2874	7.742	1.667	2873	7.704	1.547	2873	7.669
220	493.15	2.266	2914	7.910	1.886	2913	7.824	1.740	2913	7.787	1.615	2912	7.752
240	513.15	2.359	2954	7.988	1.964	2953	7.903	1.812	2952	7.865	1.682	2952	7.831
250	523.15	2.405	2974	8.026	2.003	2973	7.941	1.848	2972	7.904	1.715	2972	7.870
260	533.15	2.452	2993	8.064	2.042	2993	7.979	1.884	2992	7.942	1.749	2992	7.908
280	553.15	2.545	3033	8.139	2.120	3033	8.053	1.956	3032	8.016	1.815	3032	7.982
300	573.15	2.638	3074	8.211	2.197	3073	8.126	2.028	3072	8.088	1.882	3072	8.054
320	593.15	2.731	3114	8.281	2.275	3113	8.196	2.100	3113	8.159	1.949	3113	8.124
340	613.15	2.825	3155	8.348	2.352	3154	8.263	2.172	3154	8.226	2.016	3154	8.191
350	623.15	2.871	3175	8.381	2.391	3174	8.296	2.208	3174	8.259	2.049	3174	8.224
360	633.15	2.918	3195	8.414	2.430	3195	8.329	2.243	3195	8.292	2.082	3194	8.257
380	653.15	3.010	3236	8.478	2.507	3236	8.393	2.314	3236	8.356	2.148	3235	8.322
400	673.15	3.102	3278	8.541	2.584	3278	8.456	2.385	3278	8.419	2.214	3277	8.385
420	693.15	3.195	3319	8.602	2.661	3319	8.517	2.456	3319	8.480	2.280	3318	8.446
440	713.15	3.288	3361	8.661	2.738	3361	8.577	2.527	3361	8.540	2.346	3360	8.506
450	723.15	3.334	3382	8.690	2.777	3382	8.606	2.563	3382	8.569	2.379	3381	8.535
460	733.15	3.380	3403	8.719	2.816	3403	8.635	2.599	3403	8.598	2.413	3402	8.564
480	753.15	3.472	3446	8.777	2.893	3445	8.693	2.670	3445	8.656	2.479	3445	8.621
500	773.15	3.565	3488	8.833	2.970	3488	8.749	2.741	3488	8.712	2.545	3488	8.677
520	793.15	3.658	3531	8.888	3.048	3531	8.804	2.813	3531	8.767	2.611	3531	8.732
540	813.15	3.751	3575	8.942	3.125	3574	8.858	2.884	3574	8.821	2.678	3574	8.787
550	823.15	3.797	3596	8.969	3.164	3596	8.885	2.920	3596	8.848	2.711	3596	8.814
560	833.15	3.843	3618	8.995	3.202	3618	8.911	2.955	3618	8.874	2.744	3618	8.840
580	853.15	3.935	3662	9.047	3.279	3662	8.963	3.026	3661	8.926	2.810	3661	8.891
600	873.15	4.028	3706	9.097	3.357	3705	9.013	3.098	3705	8.976	2.876	3705	8.941
620	893.15	4.121	3750	9.146	3.435	3749	9.062	3.170	3749	9.025	2.943	3749	8.990
640	913.15	4.214	3795	9.195	3.512	3794	9.111	3.242	3794	9.074	3.010	3794	9.039
650	923.15	4.260	3817	9.219	3.551	3816	9.135	3.278	3816	9.098	3.043	3816	9.063
660	933.15	4.306	3840	9.243	3.589	3839	9.159	3.313	3839	9.122	3.076	3839	9.087
680	953.15	4.398	3885	9.291	3.665	3884	9.207	3.384	3884	9.170	3.142	3884	9.135
700	973.15	4.491	3929	9.338	3.742	3928	9.255	3.454	3928	9.217	3.207	3928	9.182
720	993.15	4.583	3974	9.384	3.820	3973	9.301	3.525	3973	9.264	3.273	3973	9.229
740	1013.15	4.676	4019	9.430	3.897	4019	9.347	3.596	4019	9.310	3.340	4018	9.275
750	1023.15	4.722	4042	9.453	3.935	4042	9.370	3.632	4042	9.333	3.373	4041	9.298
760	1033.15	4.768	4065	9.475	3.974	4065	9.392	3.667	4065	9.355	3.406	4064	9.320
780	1053.15	4.860	4111	9.519	4.050	4111	9.436	3.738	4111	9.399	3.471	4110	9.364
800	1073.15	4.952	4175	9.563	4.127	4157	9.480	3.809	4157	9.443	3.537	4157	9.408
820	1093.15	5.045	4205	9.606	4.204	4205	9.524	3.881	4204	9.486	3.603	4204	9.451
840	1113.15	5.137	4252	9.649	4.280	4252	9.567	3.952	4252	9.529	3.668	4252	9.494
850	1123.15	5.183	4276	9.670	4.319	4276	9.588	3.987	4276	9.550	3.701	4276	9.515
860	1133.15	5.229	4300	9.691	4.358	4300	9.606	4.022	4300	9.571	3.734	4300	9.536
880	1153.15	5.321	4348	9.733	4.435	4348	9.650	4.093	4348	9.613	3.801	4347	9.578
900	1173.15	5.413	4395	9.774	4.511	4395	9.691	4.164	4395	9.654	3.867	4395	9.619
920	1193.15	5.505	4443	9.815	4.589	4443	9.732	4.236	4443	9.695	3.933	4443	9.660
940	1213.51	5.598	4491	9.855	4.666	4491	9.773	4.307	4491	9.375	4.000	4491	9.700
950	1223.15	5.644	4515	9.875	4.704	4515	9.793	4.342	4515	9.755	4.033	4515	9.720
960	1233.15	5.690	4539	9.895	4.743	4539	9.812	4.378	4539	9.775	4.066	4539	9.740
980	1253.15	5.782	4587	9.934	4.819	4587	9.852	4.448	4587	9.814	4.131	4587	9.779
1000	1273.15	5.874	4636	9.973	4.896	4636	9.890	4.518	4636	9.853	4.196	4636	9.818

Table 46-7 Properties of Superheated Steam (H₂O) in International System of Units—SI (Continued)

Pressure p		1.5 bar			1.6 bar			1.8			2.0 bar		
Temperature		t_s = 111.38 °C i'' = 2 693 kJ/kg v'' = 1.159 m³/kg s'' = 7.223 kJ/kg K			t_s = 113.32 °C i'' = 2 696 kJ/kg v'' = 1.091 m³/kg s'' = 7.202 kJ/kg K			t_s = 116.94 °C i'' = 2 702 kJ/kg v'' = 0.9773 m³/kg s'' = 7.163 kJ/kg K			t_s = 120.23 °C i'' = 2 707 kJ/kg v'' = 0.8854 m³/kg s'' = 7.127 kJ/kg K		
t	T	v	i	s	v	i	s	v	i	s	v	i	s
°C	K	m³/kg	kJ/kg	kJ/kg K	m³/kg	kJ/kg	kJ/kg K	m³/kg	kJ/kg	kJ/kg K	m³/kg	kJ/kg	kJ/kg K
0	273.15	0.0010000	0.1	0.0000	0.0010000	0.2	0.0000	0.0010000	0.2	0.0000	0.0010000	0.2	0.0000
20	293.15	0.0010017	83.9	0.2964	0.0010017	84.0	0.2964	0.0010017	84.0	0.2964	0.0010017	84.0	0.2964
40	313.15	0.0010078	167.5	0.5715	0.0010078	167.6	0.5715	0.0010078	167.6	0.5716	0.0010078	167.6	0.5716
50	323.15	0.0010120	209.3	0.7032	0.0010120	209.4	0.7032	0.0010120	209.4	0.7033	0.0010120	209.4	0.7033
60	333.15	0.0010170	251.1	0.8307	0.0010170	251.2	0.8307	0.0010170	251.2	0.8307	0.0010170	251.2	0.8307
80	353.15	0.0010289	334.9	1.0748	0.0010289	335.0	1.0748	0.0010289	335.0	1.0748	0.0010289	335.0	1.0748
100	373.15	0.0010434	419.0	1.3067	0.0010434	419.0	1.3067	0.0010434	419.0	1.3067	0.0010434	419.0	1.3067
120	393.15	1.188	2712	7.268	1.113	2711	7.236	0.986	2708	7.177	0.0010603	503.7	1.5269
140	413.15	1.253	2753	7.368	1.173	2752	7.337	1.041	2750	7.279	0.9357	2749	7.227
150	423.15	1.285	2773	7.416	1.203	2772	7.385	1.068	2771	7.328	0.9603	2769	7.276
160	433.15	1.317	2793	7.462	1.233	2792	7.431	1.095	2791	7.375	0.9840	2790	7.324
180	453.15	1.380	2833	7.551	1.293	2832	7.520	1.148	2831	7.465	1.032	2830	7.415
200	473.15	1.443	2872	7.636	1.352	2872	7.606	1.201	2871	7.550	1.080	2870	7.501
220	493.15	1.506	2912	7.719	1.411	2912	7.689	1.253	2911	7.633	1.128	2910	7.583
240	513.15	1.569	2952	7.799	1.470	2951	7.768	1.306	2950	7.712	1.175	2950	7.663
250	523.15	1.600	2972	7.837	1.500	2971	7.807	1.332	2970	7.751	1.198	2970	7.702
260	533.15	1.632	2992	7.875	1.529	2991	7.845	1.358	2990	7.789	1.222	2990	7.740
280	553.15	1.694	3032	7.950	1.587	3031	7.919	1.410	3030	7.864	1.269	3030	7.815
300	573.15	1.756	3072	8.022	1.646	3071	7.991	1.462	3071	7.936	1.316	3071	7.887
320	593.15	1.819	3112	8.092	1.705	3112	8.061	1.514	3112	8.006	1.363	3111	7.957
340	613.15	1.881	3154	8.159	1.763	3153	8.129	1.566	3153	8.075	1.410	3153	8.025
350	623.15	1.912	3174	8.192	1.792	3174	8.162	1.592	3173	8.108	1.433	3173	8.059
360	633.15	1.943	3194	8.225	1.821	3194	8.195	1.618	3194	8.141	1.457	3194	8.092
380	653.15	2.004	3235	8.290	1.879	3235	8.259	1.669	3235	8.205	1.503	3235	8.156
400	673.15	2.066	3277	8.353	1.937	3277	8.322	1.721	3277	8.268	1.549	3276	8.219
420	693.15	2.128	3318	8.414	1.995	3318	8.384	1.772	3318	8.329	1.595	3318	8.280
440	713.15	2.190	3360	8.474	2.054	3360	8.444	1.824	3360	8.389	1.641	3360	8.340
450	723.15	2.221	3381	8.503	2.082	3381	8.473	1.850	3381	8.418	1.664	3381	8.369
460	733.15	2.252	3402	8.532	2.111	3402	8.502	1.876	3402	8.447	1.687	3402	8.398
480	753.15	2.313	3445	8.589	2.169	3445	8.559	1.928	3445	8.505	1.734	3445	8.456
500	773.15	2.375	3488	8.645	2.227	3488	8.615	1.979	3488	8.561	1.781	3487	8.512
520	793.15	2.437	3530	8.700	2.285	3530	8.670	2.031	3530	8.616	1.828	3530	8.567
540	813.15	2.499	3574	8.754	2.343	3574	8.724	2.083	3574	8.670	1.874	3574	8.621
550	823.15	2.530	3596	8.781	2.372	3596	8.751	2.109	3596	8.697	1.897	3595	8.648
560	833.15	2.561	3617	8.807	2.401	3617	8.777	2.134	3617	8.723	1.920	3617	8.674
580	853.15	2.623	3661	8.859	2.459	3661	8.829	2.185	3661	8.775	1.967	3661	8.726
600	873.15	2.685	3705	8.909	2.517	3705	8.879	2.237	3705	8.825	2.013	3705	8.776
620	893.15	2.747	3749	8.958	2.575	3749	8.928	2.289	3749	8.874	2.060	3749	8.825
640	913.15	2.810	3794	9.007	2.633	3794	8.977	2.340	3794	8.922	2.107	3794	8.874
650	923.15	2.841	3816	9.031	2.662	3816	9.001	2.366	3816	8.947	2.130	3816	8.898
660	933.15	2.871	3839	9.055	2.691	3839	9.025	2.391	3839	8.971	2.153	3839	8.922
680	953.15	2.932	3884	9.103	2.749	3884	9.073	2.443	3884	9.019	2.199	3884	8.970
700	973.15	2.993	3928	9.150	2.807	3928	9.120	2.494	3928	9.066	2.245	3928	9.018
720	993.15	3.055	3973	9.197	2.865	3973	9.167	2.545	3973	9.112	2.291	3973	9.064
740	1013.15	3.118	4018	9.243	2.923	4018	9.213	2.596	4018	9.158	2.338	4018	9.110
750	1023.15	3.149	4041	9.266	2.952	4041	9.235	2.622	4041	9.181	2.361	4041	9.133
760	1033.15	3.179	4064	9.288	2.981	4064	9.258	2.648	4064	9.203	2.384	4064	9.155
780	1053.15	3.240	4111	9.332	3.038	4110	9.302	2.699	4110	9.247	2.430	4110	9.199
800	1073.15	3.301	4157	9.376	3.094	4157	9.346	2.750	4157	9.291	2.476	4157	9.242
820	1093.15	3.363	4204	9.419	3.152	4204	9.389	2.802	4204	9.334	2.522	4204	9.286
840	1113.15	3.423	4252	9.462	3.210	4252	9.432	2.854	4252	9.378	2.568	4252	9.329
850	1123.15	3.454	4276	9.483	3.239	4276	9.453	2.879	4276	9.399	2.591	4276	9.351
860	1133.15	3.485	4300	9.504	3.268	4300	9.474	2.905	4300	9.420	2.614	4299	9.372
880	1153.15	3.547	4347	9.546	3.326	4347	9.516	2.956	4347	9.462	2.661	4347	9.414
900	1173.15	3.609	4395	9.587	3.383	4395	9.557	3.007	4395	9.503	2.706	4395	9.455
920	1193.15	3.671	4443	9.628	3.441	4443	9.598	3.059	4443	9.544	2.753	4443	9.495
940	1213.15	3.733	4491	9.668	3.499	4491	9.638	3.110	4491	9.584	2.800	4491	9.535
950	1223.5	3.764	4515	9.688	3.528	4515	9.658	3.136	4515	9.604	2.823	4515	9.555
960	1233.15	3.794	4539	9.708	3.557	4539	9.678	3.162	4539	9.624	2.846	4539	9.575
980	1253.15	3.855	4587	9.747	3.614	4587	9.717	3.214	4587	9.663	2.891	4587	9.615
1000	1273.15	3.916	4636	9.786	3.671	4636	9.756	3.264	4636	9.702	2.937	4635	9.653

Pressure p		2.5 bar			3.0 bar			4.0 bar			5.0 bar		
Temperature		$t_s = 127.43\,°C$ $i'' = 2\,717\,kJ/kg$ $v'' = 0.7185\,m^3/kg$ $s'' = 7.053\,kJ/kg\,K$			$t_s = 133.54\,°C$ $i'' = 2\,725\,kJ/kg$ $v'' = 0.6057\,m^3/kg$ $s'' = 6.992\,kJ/kg\,K$			$t_s = 143.62\,°C$ $i'' = 2\,738\,kJ/kg$ $v'' = 0.4624\,m^3/kg$ $s'' = 6.897\,kJ/kg\,K$			$t_s = 151.84\,°C$ $i'' = 2\,749\,kJ/kg$ $v'' = 0.3747\,m^3/kg$ $s'' = 6.822\,kJ/kg\,K$		
t	T	v	i	s	v	i	s	v	i	s	v	i	s
°C	K	m^3/kg	kJ/kg	kJ/kg K	m^3/kg	kJ/kg	kJ/kgK	m^3/kg	kJ/kg	kJ/kg K	m^3/kg	kJ/kg	kJ/kg K
0	273.15	0.0010000	0.2	0.0000	0.0010000	0.3	0.0000	0.0010000	0.5	0.0000	0.0009999	0.6	0.0000
20	293.15	0.0010017	84.0	0.2964	0.0010017	84.1	0.2964	0.0010017	84.1	0.2964	0.0010016	84.2	0.2964
40	313.15	0.0010078	167.6	0.5716	0.0010078	167.7	0.5716	0.0010078	167.7	0.5716	0.0010077	167.8	0.5716
50	323.15	0.0010120	209.4	0.7032	0.0010120	209.5	0.7031	0.0010120	209.5	0.7030	0.0010119	209.6	0.7029
60	333.15	0.0010170	251.2	0.8305	0.0010170	251.3	0.8304	0.0010170	251.3	0.8303	0.0010169	251.4	0.8302
80	353.15	0.0010289	335.0	1.0747	0.0010288	335.1	1.0746	0.0010288	335.1	1.0745	0.0010287	335.1	1.0744
100	373.15	0.0010434	419.0	1.3067	0.0010434	419.1	1.3066	0.0010433	419.1	1.3063	0.0010433	419.1	1.3063
120	393.15	0.0010603	503.7	1.5269	0.0010602	503.7	1.5268	0.0010602	503.7	1.5265	0.0010601	503.7	1.5265
140	413.15	0.7445	2745	7.118	0.6171	2740	7.025	0.0010798	589.1	1.738	0.0010797	589.1	1.738
150	423.15	0.7647	2766	7.168	0.6344	2762	7.077	0.4709	2754	6.928	0.0010906	632.1	1.840
160	433.15	0.7845	2787	7.216	0.6512	2783	7.126	0.4840	2776	6.980	0.3839	2767	6.864
180	453.15	0.8234	2827	7.308	0.6838	2824	7.218	0.5094	2818	7.077	0.4047	2812	6.965
200	473.15	0.8618	2867	7.395	0.7161	2864	7.306	0.5341	2859	7.166	0.4249	2854	7.056
220	493.15	0.9000	2908	7.478	0.7482	2905	7.389	0.5585	2900	7.251	0.4448	2896	7.141
240	513.15	0.9380	2948	7.557	0.7802	2946	7.470	0.5827	2941	7.332	0.4644	2937	7.224
250	523.15	0.9570	2968	7.596	0.7961	2966	7.509	0.5948	2962	7.371	0.4742	2958	7.264
260	533.15	0.9758	2988	7.634	0.8120	2986	7.547	0.6068	2982	7.410	0.4839	2979	7.304
280	553.15	1.0133	3029	7.709	0.8436	3027	7.623	0.6307	3023	7.486	0.5031	3020	7.380
300	573.15	1.051	3069	7.781	0.8750	3068	7.695	0.6545	3065	7.560	0.5224	3062	7.454
320	593.15	1.089	3110	7.852	0.9064	3109	7.766	0.6782	3106	7.631	0.5414	3104	7.525
340	613.15	1.126	3152	7.920	0.9377	3150	7.835	0.7019	3148	7.700	0.5605	3146	7.595
350	623.15	1.145	3172	7.954	0.9534	3171	7.869	0.7137	3169	7.734	0.5700	3167	7.629
360	633.15	1.164	3193	7.987	0.9690	3192	7.902	0.7254	3190	7.767	0.5794	3188	7.662
380	653.15	1.201	3234	8.052	1.000	3233	7.967	0.7488	3231	7.832	0.5984	3230	7.727
400	673.15	1.238	3276	8.115	1.032	3275	8.030	0.7723	3273	7.895	0.6173	3272	7.791
420	693.15	1.275	3318	8.176	1.063	3317	8.091	0.7957	3315	7.957	0.6361	3314	7.853
440	713.15	1.312	3360	8.236	1.094	3359	8.151	0.8190	3358	8.017	0.6548	3356	7.913
450	723.15	1.330	3381	8.266	1.110	3380	8.181	0.8307	3379	8.047	0.6642	3377	7.943
460	733.15	1.349	3402	8.295	1.125	3401	8.210	0.8424	3400	8.076	0.6735	3398	7.792
480	753.15	1.386	3444	8.353	1.156	3444	8.268	0.8657	3443	8.134	0.6922	3441	8.030
500	773.15	1.424	3487	8.409	1.187	3486	8.324	0.8890	3485	8.190	0.7109	3484	8.086
520	793.15	1.461	3529	8.464	1.218	3529	8.379	0.9123	3528	8.245	0.7296	3527	8.141
540	813.15	1.498	3573	8.518	1.248	3573	8.433	0.9357	3572	8.299	0.7483	3571	8.196
550	823.15	1.517	3594	8.545	1.264	3594	8.460	0.9473	3593	8.326	0.7576	3592	8.223
560	833.15	1.536	3616	8.571	1.279	3616	8.486	0.9590	3615	8.352	0.7669	3614	8.249
580	853.15	1.573	3660	8.622	1.310	3660	8.538	0.9822	3659	8.404	0.7855	3658	8.301
600	873.15	1.610	3704	8.672	1.341	3704	8.588	1.0054	3703	8.455	0.8041	3702	8.351
620	893.15	1.648	3748	8.722	1.372	3748	8.638	1.0287	3747	8.504	0.8228	3746	8.401
640	913.15	1.685	3793	8.771	1.403	3793	8.686	1.0519	3792	8.553	0.8414	3791	8.450
650	923.15	1.703	3815	8.795	1.418	3815	8.711	1.0636	3814	8.578	0.8507	3813	8.474
660	933.15	1.722	3838	8.819	1.434	3838	8.735	1.0752	3837	8.602	0.8600	3836	8.498
680	953.15	1.759	3883	8.867	1.465	3883	8.783	1.0983	3882	8.650	0.8785	3881	8.546
700	973.15	1.795	3927	8.914	1.496	3927	8.830	1.1214	3926	8.697	0.8969	3925	8.594
720	993.15	1.833	3972	8.961	1.527	3972	8.876	1.1446	3971	8.744	0.9155	3971	8.640
740	1013.15	1.870	4018	9.007	1.558	4018	8.922	1.1677	4017	8.790	0.9342	4017	8.686
750	1023.15	1.888	4041	9.030	1.573	4041	8.945	1.1793	4040	8.812	0.9434	4040	8.709
760	1033.15	1.907	4064	9.052	1.588	4064	8.968	1.1909	4063	8.835	0.9526	4063	8.731
780	1053.15	1.944	4110	9.096	1.619	4110	9.012	1.2141	4109	8.879	0.9711	4109	8.775
800	1073.15	1.980	4157	9.140	1.650	4157	9.056	1.2372	4156	8.923	0.9895	4156	8.819
820	1093.15	2.017	4204	9.183	1.681	4204	9.099	1.2604	4203	8.966	1.0080	4203	8.862
840	1113.15	2.054	4252	9.226	1.713	4251	9.142	1.2835	4251	9.009	1.0266	4250	8.905
850	1123.15	2.072	4275	9.247	1.728	4275	9.163	1.2950	4275	9.030	1.0358	4274	8.926
860	1133.15	2.091	4299	9.268	1.743	4299	9.184	1.3065	4299	9.051	1.0451	4298	8.948
880	1153.15	2.128	4347	9.310	1.774	4347	9.226	1.3295	4346	9.093	1.0636	4346	8.990
900	1173.15	2.164	4395	9.351	1.804	4395	9.267	1.3525	4394	9.134	1.0821	4394	9.031
920	1193.15	2.202	4442	9.392	1.835	4442	9.308	1.3756	4442	9.175	1.1006	4441	9.072
940	1213.15	2.239	4490	9.432	1.866	4490	9.348	1.3987	4490	9.215	1.1192	4489	9.112
950	1223.15	2.257	4514	9.542	1.882	4514	9.368	1.4103	4514	9.235	1.1284	4513	9.132
960	1233.15	2.276	4538	9.472	1.898	4538	9.388	1.4218	4538	9.255	1.1377	4537	9.152
980	1253.15	2.313	4587	9.511	1.928	4587	9.427	1.4448	4586	9.294	1.1562	4586	9.192
1000	1273.15	2.349	4635	9.550	1.958	4635	9.466	1.4679	4635	9.333	1.1746	4635	9.230

Table 46-9 Properties of Superheated Steam (H₂O) in International System of Units—SI (Continued)

Pressure p		6 bar $t_s=158.84\,°C$ $i''=2\,757\ kJ/kg$ $v''=0.3156\ m^3/kg$ $s''=6.761\ kJ/kg\ K$			7 bar $t_s=164.96\,°C$ $i''=2\,764\ kJ/kg$ $v''=0.2728\ m^3/kg$ $s''=6.709\ kJ/kg\ K$			8 bar $t_s=170.42\,°C$ $i''=2\,769\ kJ/kg$ $v''=0.2403\ m^3/kg$ $s''=6.663\ kJ/kg\ K$			9 bar $t_s=175.35\,°C$ $i''=2\,774\ kJ/kg$ $v''=0.2149\ m^3/kg$ $s''=6.623\ kJ/kg\ K$		
t	T	v	i	s	v	i	s	v	i	s	v	i	s
°C	K	m³/kg	kJ/kg	kJ/kg	m³/kg K	kJ/kg	kJ/kg K	m³/kg	kJ/kg	kJ/kg K	m³/kg	kJ/kg	kJ/kg K
0	273.15	0.0009998	0.7	0.0000	0.0009998	0.8	0.0000	0.0009997	0.9	0.0000	0.0009997	1.0	0.0000
20	293.15	0.0010015	84.3	0.2964	0.0010015	84.4	0.2963	0.0010015	84.5	0.2962	0.0010015	84.6	0.2961
40	313.15	0.0010076	167.9	0.5716	0.0010076	168.0	0.5715	0.0010076	168.1	0.5714	0.0010076	168.2	0.5713
50	323.15	0.0010118	209.7	0.7028	0.0010118	209.8	0.7027	0.0010118	209.9	0.7026	0.0010118	210.0	0.7025
60	333.15	0.0010168	251.5	0.8302	0.0010168	251.6	0.8301	0.0010167	251.7	0.8300	0.0010167	251.8	0.8299
80	353.15	0.0010287	335.2	1.0744	0.0010286	335.2	1.0743	0.0010286	335.3	1.0742	0.0010285	335.4	1.0741
100	373.15	0.0010432	419.1	1.3062	0.0010432	419.1	1.3061	0.0010431	419.2	1.3060	0.0010431	419.3	1.3059
120	393.15	0.0010601	503.7	1.5265	0.0010600	503.7	1.5264	0.0010600	503.8	1.5263	0.0010599	503.9	1.5262
140	413.15	0.0010797	589.1	1.738	0.0010796	589.1	1.738	0.0010795	589.1	1.737	0.0010795	589.2	1.737
150	423.15	0.0010906	632.1	1.840	0.0010904	632.1	1.840	0.0010904	632.1	1.840	0.0010903	632.1	1.140
160	433.15	0.3167	2759	6.767	0.0011020	675.3	1.941	0.0011020	675.3	1.941	0.0011019	675.7	1.941
180	453.15	0.3348	2805	6.869	0.2847	2799	6.787	0.2473	2792	6.715	0.2180	2785	6.648
200	473.15	0.3520	2849	6.963	0.2998	2844	6.884	0.2609	2839	6.814	0.2304	2833	6.750
220	493.15	0.3688	2891	7.051	0.3145	2887	6.973	0.2739	2883	6.905	0.2422	2878	6.844
240	513.15	0.3855	2933	7.135	0.3290	2929	7.058	0.2867	2926	6.991	0.2537	2922	6.931
250	523.15	0.3937	2954	7.175	0.3361	2951	7.099	0.2930	2947	7.032	0.2594	2944	6.973
260	533.15	0.4019	2975	7.215	0.3432	2972	7.139	0.2993	2969	7.073	0.2651	2965	7.014
280	553.15	0.4181	3017	7.292	0.3572	3014	7.216	0.3118	3011	7.151	0.2762	3008	7,093
300	573.15	0.4342	3059	7.366	0.3711	3056	7.291	0.3240	3054	7.226	0.2872	3051	7.168
320	593.15	0.4502	3101	7.437	0.3850	3099	7.363	0.3362	3096	7.299	0.2980	3093	7.241
340	613.15	0.4661	3143	7.507	0.3987	3141	7.433	0.3482	3139	7.369	0.3088	3136	7.312
350	623.15	0.4741	3164	7.451	0.4055	3162	7.468	0.3542	3160	7.404	0.3142	3158	7.347
360	633.15	0.4820	3185	7.575	0.4124	3183	7.502	0.3602	3181	7.438	0.3196	3179	7.381
380	653.15	0.4979	3228	7.640	0.4261	3226	7.568	0.3722	3224	7.504	0.3303	3222	7.447
400	673.15	0.5136	3270	7.704	0.4396	3268	7.632	0.3842	3267	7.568	0.3409	3265	7.511
420	693.15	0.5293	3312	7.766	0.4531	3311	7.694	0.3960	3309	7.631	0.3516	3308	7.574
440	713.15	0.5450	3355	7.827	0.4667	3353	7.755	0.4079	3352	7.692	0.3621	3351	7.635
450	723.15	0.5528	3376	7.857	0.4734	3375	7.785	0.4137	3373	7.722	0.3674	3372	7.665
460	733.15	0.5607	3397	7.886	0.4801	3396	7.814	0.4196	3395	7.751	0.3726	3393	7.695
480	753.15	0.5763	3440	7.944	0.4935	3439	7.872	0.4315	3437	7.809	0.3831	3436	7.753
500	773.15	0.5919	3483	8.001	0.5069	3482	7.929	0.4432	3481	7.866	0.3936	3480	7.810
520	793.15	0.6075	3526	8.056	0.5203	3525	7.984	0.4549	3524	7.921	0.4041	3523	7.866
540	813.15	0.6230	3570	8.110	0.5337	3569	8.038	0.4667	3568	7.975	0.4145	3567	7.920
550	823.15	0.6308	3592	8.137	0.5403	3591	8.065	0.4725	3590	8.002	0.4197	3589	7.947
560	833.15	0.6387	3613	8.163	0.5470	3612	8.092	0.4784	3611	8.029	0.4250	3610	7.974
580	853.15	0.6542	3657	8.215	0.5605	3656	8.144	0.4901	3655	8.081	0.4355	3654	8.026
600	973.15	0.6697	3701	8.266	0.5738	3700	8.195	0.5018	3699	8.132	0.4458	3698	8.077
620	893.15	0.6852	3745	8.316	0.5872	3745	8.245	0.5135	3744	8.182	0.4561	3743	8.127
640	913.15	0.7007	3790	8.365	0.6005	3789	8.294	0.5253	3788	8.232	0.4665	3787	8.177
650	923.15	0.7085	3812	8.389	0.6071	3811	8.318	0.5311	3810	8.256	0.4717	3809	8.201
660	933.15	0.7162	3835	8.413	0.6138	3834	8.342	0.5369	3833	8.280	0.4769	3832	8.225
680	953.15	0.7317	3880	8.461	0.6270	3879	8.390	0.5485	3878	8.328	0.4873	3877	8.273
700	973.15	0.7472	3925	8.508	0.6402	3924	8.437	0.5601	3924	8.375	0.4976	3923	8.321
720	993.15	0.7627	3970	8.555	0.6535	3970	8.484	0.5718	3969	8.422	0.5079	3969	8.367
740	1013.15	0.7782	4016	8.601	0.6668	4016	8.530	0.5834	4015	8.468	0.5183	4015	8.413
750	1023.15	0.7859	4039	8.624	0.6735	4039	8.552	0.5893	4038	8.490	0.5235	4038	8.436
760	1033.15	0.7937	4062	8.646	0.6802	4062	8.575	0.5951	4061	8.513	0.5287	4061	8.458
780	1053.15	0.8091	4108	8.690	0.6933	4108	8.619	0.6066	4108	8.557	0.5390	4107	8.503
800	1073.15	0.8245	4155	8.734	0.7065	4155	8.663	0.6182	4155	8.601	0.5493	4154	8.547
820	1093.15	0.8400	4202	8.777	0.7199	4202	8.706	0.6298	4202	8.645	0.5597	4201	8.590
840	1113.15	0.8554	4250	8.820	0.7331	4249	8.749	0.6413	4249	8.688	0.5700	4249	8.633
850	1123.15	0.8631	4274	8.842	0.7397	4273	8.771	0.6471	4273	8.709	0.5751	4273	8.655
860	1133.15	0.8708	4298	8.863	0.7463	4297	8.792	0.6529	4297	8.730	0.5802	4297	8.676
880	1153.15	0.8862	4345	8.905	0.7595	4345	8.834	0.6645	4345	8.772	0.5906	4344	8.718
900	1173.15	0.9016	4393	8.946	0.7727	4393	8.875	0.6761	4392	8.814	0.6009	4392	8.759
920	1193.15	0.9171	4441	8.987	0.7860	4441	8.916	0.6877	4440	8.855	0.6112	4440	8.800
940	1213.15	0.9326	4489	9.028	0.7992	4489	8.957	0.6993	4488	8.895	0.6216	4488	8.840
950	1223.15	0.9403	4513	9.048	0.8058	4513	8.977	0.7051	4512	8.915	0.6268	4512	8.860
960	1233.15	0.9480	4537	9.067	0.8124	4537	8.996	0.7109	4536	8.935	0.6319	4536	8.880
980	1253.15	0.9633	4586	9.107	0.8256	4585	9.036	0.7224	4585	8.974	0.6421	4585	8.919
1000	1273.15	0.9786	4634	9.145	0.8388	4634	9.074	0.7338	4634	9.013	0.6523	4633	8.958

Table 46-10 Properties of Superheated Steam (H$_2$O) in International System of Units—SI (Continued)

Pressure p		10 bar			12 bar			14 bar			16 bar		
Temperature		$t_s = 187.95\,°C$ $i'' = 2.785\ kJ/kg$ $v'' = 0.1633\ m^3/kg$ $s'' = 6.523\ kJ/kg\ K$			$t_s = 179.88\,°C$ $i'' = 2.778\ kJ/kg$ $v'' = 0.1946\ m^3/kg$ $s'' = 6.587\ kJ/kg\ K$			$t_s = 201.36\,°C$ $i'' = 2.793\ kJ/kg$ $v'' = 0.1238\ m^3/kg$ $s'' = 6.422\ kJ/kg\ K$			$t_s = 195.04\,°C$ $i'' = 2.790\ kJ/kg$ $v'' = 0.1408\ m^3/kg$ $s'' = 6.469\ kJ/kg\ K$		
t	T	v	i	s	v	i	s	v	i	s	v	i	s
°C	K	m³/kg	kJ/kg	kJ/kg K	m³/kg	kJ/kg	kJ/kg K	m³/kg	kJ/kg	kJ/kg K	m³/kg	kJ/kg	kJ/kg K
0	273.15	0.0009996	1.1	0.0000	0.0009995	1.3	0.0000	0.0009994	1.5	0.0000	0.0009994	1.7	0.0000
20	293.15	0.0010014	84.7	0.2960	0.0010013	84.9	0.2959	0.0010012	85.1	0.2958	0.0010011	85.3	0.2958
40	313.15	0.0010075	168.3	0.5712	0.0010074	168.5	0.5711	0.0010073	168.7	0.5711	0.0010072	168.8	0.5710
50	323.15	0.0010117	210.1	0.7024	0.0010116	210.2	0.7023	0.0010115	210.4	0.7022	0.0010114	210.5	0.7022
60	333.15	0.0010166	251.8	0.8298	0.0010165	251.9	0.8297	0.0010164	252.1	0.8296	0.0010163	252.2	0.8296
80	353.15	0.0010285	335.4	1.0740	0.0010284	335.5	1.0738	0.0010282	335.7	1.0736	0.0010282	335.8	1.0735
100	373.15	0.0010430	419.3	1.3058	0.0010429	419.4	1.3056	0.0010427	419.6	1.3054	0.0010426	419.7	1.3052
120	393.10	0.0010598	503.9	1.5261	0.0010597	504.0	1.5259	0.0010596	504.2	1.5257	0.0010595	504.3	1.5256
140	413.15	0.0010794	589.2	1.737	0.0010793	589.3	1.737	0.0010792	589.5	1.736	0.0010790	589.6	1.736
150	423.15	0.0010902	632.1	1.840	0.0010901	632.2	1.839	0.0010900	632.4	1.839	0.0010898	632.5	1.839
160	433.15	0.0011018	675.4	1.941	0.0011016	675.5	1.940	0.0011015	675.7	1.940	0.0011013	675.7	1.940
180	453.15	0.1949	2778	6.588	0.0011273	763.2	2.138	0.0011271	763.2	1.137	0.0011270	763.2	2.137
200	473.15	0.2060	2827	6.692	0.1693	2816	6.588	0.1429	2803	6.497	0.0011565	852.4	2.329
220	493.15	0.2169	2874	6.788	0.1788	2865	6.688	0.1515	2855	6.602	0.1309	2844	6.524
240	513.15	0.2274	2918	6.877	0.1879	2911	6.780	0.1596	2902	6.697	0.1382	2893	6.622
250	523.15	0.2326	2940	6.920	0.1924	2933	6.824	0.1635	2925	6.741	0.1417	2917	6.667
260	533.15	0.2377	2962	6.961	0.1967	2955	6.866	0.1673	2948	6.784	0.1452	2940	6.711
280	553.15	0.2478	3005	7.040	0.2054	2999	6.947	0.1748	2992	6.867	0.1519	2986	6.796
300	573.15	0.2578	3048	7.116	0.2139	3042	7.025	0.1823	3036	6.945	0.1585	3030	6.877
320	593.15	0.2677	3091	7.189	0.2221	3086	7.099	0.1894	3080	7.021	0.1649	3075	6.953
340	613.15	0.2774	3134	7.261	0.2302	3129	7.171	0.1965	3125	7.094	0.1712	3120	7.027
350	623.15	0.2822	3156	7.296	0.2343	3151	7.206	0.2001	3147	7.130	0.1743	3142	7.063
360	633.15	0.2871	3177	7.330	0.2384	3173	7.241	0.2036	3169	7.164	0.1775	3164	7.098
380	653.15	0.2968	3220	7.397	0.2466	3216	7.308	0.2107	3213	7.232	0.1838	3209	7.166
400	673.15	0.3065	3263	7.461	0.2547	3260	7.373	0.2176	3256	7.299	0.1899	3253	7.233
420	693.15	0.3160	3306	7.524	0.2627	3302	7.437	0.2246	3300	7.363	0.1960	3297	7.298
440	713.15	0.3255	3349	7.585	0.2707	3346	7.499	0.2315	3344	7.425	0.2021	3341	7.360
450	723.15	0.3303	3370	7.615	0.2747	3368	7.529	0.2349	3365	7.455	0.2051	3363	7.390
460	733.15	0.3351	3392	7.645	0.2786	3390	7.559	0.2383	3387	7.485	0.2082	3384	7.420
480	753.15	0.3445	3435	7.703	0.2865	3433	7.617	0.2452	3431	7.543	0.2141	3428	7.479
500	773.15	0.3539	3479	7.761	0.2944	3477	6.674	0.2520	3474	7.601	0.2201	3472	7.537
520	793.15	0.3634	3522	7.817	0.3023	3520	7.730	0.2588	3518	7.657	0.2261	3516	7.593
540	813.15	0.3728	3566	7.871	0.3103	3564	7.784	0.2656	3561	7.712	0.2320	3560	7.648
550	823.15	0.3776	3588	7.898	0.3143	3586	7.811	0.2690	3584	7.739	0.2350	3582	7.675
560	833.15	0.3824	3609	7.924	0.3182	3608	7.838	0.2725	3606	7.766	0.2381	3604	7.702
580	853.15	0.3917	3653	7.976	0.3261	3652	7.890	0.2793	3650	7.818	0.2441	3648	7.754
600	873.15	0.4010	3698	8.027	0.3339	3696	7.942	0.2858	3695	7.870	0.2490	3693	7.806
620	893.15	0.4104	3742	8.077	0.3417	3740	7.992	0.2925	3739	7.920	0.2558	3737	7.857
640	913.15	0.4199	3787	8.127	0.3495	3785	8.042	0.2993	3784	7.970	0.2617	3782	7.907
650	923.15	0.4246	3809	8.152	0.3534	3808	8.067	0.3026	3806	7.994	0.2646	3805	7.932
660	933.15	0.4292	3832	8.176	0.3573	3830	8.091	0.3060	3829	8.018	0.2676	3828	7.956
680	953.15	0.4384	3877	8.224	0.3651	3875	8.139	0.3127	3874	8.066	0.2735	3873	8.004
700	973.15	0.4477	3923	8.272	0.3728	3921	8.187	0.3194	3920	8.114	0.2783	3919	8.052
720	993.15	0.4571	3968	9.318	0.3807	3967	8.233	0.3261	3966	8.161	0.2852	3965	8.099
740	1013.15	0.4664	4014	8.364	0.3884	4013	8.279	0.3327	4012	8.207	0.2911	4011	8.145
750	1023.15	0.4711	4037	8.387	0.3923	4036	8.302	0.3361	4035	8.230	0.2940	4034	8.168
760	1033.15	0.4757	4060	8.409	0.3962	4059	8.325	0.3394	4058	8.253	0.2969	4057	8.191
780	1053.15	0.4850	4107	8.454	0.4040	4106	8.369	0.3461	4105	8.297	0.3028	4105	8.235
800	1073.15	0.4942	4154	8.498	0.4117	4153	8.413	0.3527	4152	8.341	0.3086	4151	8.279
820	1093.15	0.5036	4201	8.542	0.4195	4200	8.457	0.3594	4199	8.385	0.3144	4199	8.323
840	1113.15	0.5128	4249	9.584	0.4272	4248	8.499	0.3660	4247	8.428	0.3202	4246	8.366
850	1123.15	0.5174	4272	8.606	0.4310	4272	8.521	0.3693	4271	8.449	0.3231	4270	8.387
860	1133.15	0.5221	4296	8.627	0.4349	4296	8.542	0.3727	4295	8.470	0.3260	4294	8.408
880	1153.15	0.5314	4344	8.669	0.4427	4343	8.584	0.3793	4343	8.512	0.3318	4342	8.451
900	1173.15	0.5406	4392	8.710	0.4504	4391	8.626	0.3860	4391	8.554	0.3377	4390	8.492
920	1193.15	0.5500	4440	8.751	0.4582	4432	8.667	0.3927	4439	8.595	0.3435	4438	8.533
940	1213.15	0.5592	4488	8.791	0.4660	4487	8.707	0.3994	4487	8.635	0.3493	4486	8.574
950	1223.15	0.5638	4512	8.811	0.4699	4511	8.727	0.4027	4511	8.655	0.3522	4510	8.594
960	1233.15	0.5685	4536	8.831	0.4737	4535	8.747	0.4061	4535	8.675	0.3551	4534	8.614
980	1253.15	0.5777	4584	8.870	0.4814	4584	8.786	0.4127	4584	8.714	0.3609	4583	8.653
1000	1273.15	0.5870	4633	8.909	0.4890	4632	8.825	0.4192	4632	8.753	0.3666	4632	8.691

Table 46-11 Properties of Superheated Steam (H$_2$O) in International System of Units—SI (*Continued*)

Pressure p		18 bar			20 bar			25 bar			30 bar		
Temperature		$t_s = 207.10$ °C $i'' = 2.796$ kJ/kg $v'' = 0.1104$ m³/kg $s'' = 6.379$ kJ/kg K			$t_s = 212.37$ °C $i'' = 2.799$ kJ/kg $v'' = 0.09958$ m³/kg $s'' = 6.340$ kJ/kg K			$t_s = 223.93$ °C $i'' = 2.802$ kJ/kg $v'' = 0.07993$ m³/kg $s'' = 6.256$ kJ/kg K			$t_s = 233.83$ °C $i'' = 2.804$ kJ/kg $v'' = 0.06665$ m³/kg $s'' = 6.186$ kJ/kg K		
t	T	v	i	s	v	i	s	v	i	s	v	i	s
°C	K	m³/kg	kJ/kg	kJ/kg K	m³/kg	kJ/kg	kJ/kg K	m³/kg	kJ/kg	kJ/kg K	m³/kg	kJ/kg	kJ/kg K
0	273.15	0.0009992	1.9	0.0000	0.0009991	2.1	0.0000	0.0009989	2.6	0.0000	0.0009986	3.1	0.0000
20	293.15	0.0010010	85.5	0.2957	0.0010009	85.7	0.2957	0.0010007	86.2	0.2975	0.0010004	86.7	0.2956
40	313.15	0.0010071	169.0	0.5709	0.0010070	169.2	0.5708	0.0010068	169.7	0.5708	0.0010065	170.1	0.5707
50	323.15	0.0010113	210.7	0.7021	0.0010112	210.9	0.7020	0.0010110	211.4	0.7019	0.0010107	211.8	0.7018
60	333.15	0.0010162	252.4	0.8295	0.0010161	252.6	0.8294	0.0010159	253.1	0.8292	0.0010157	253.5	0.8290
80	353.15	0.0010281	336.0	1.0733	0.0010280	336.2	1.0731	0.0010277	336.6	1.0728	0.0010275	337.0	1.0726
100	373.15	0.0010425	419.9	1.3050	0.0010424	420.1	1.3048	0.0010422	420.5	1.3043	0.0010419	420.9	1.3038
120	393.15	0.0010594	504.5	1.5254	0.0010593	504.7	1.5252	0.0010590	505.1	1.5247	0.0010587	505.4	1.5244
140	413.15	0.0010789	589.8	1.736	0.0010787	589.9	1.736	0.0010785	590.3	1.735	0.0010782	590.6	1.735
150	423.15	0.0010897	632.7	1.838	0.0010895	632.8	1.838	0.0010892	633.1	1.838	0.0010889	633.4	1.837
160	433.15	0.0011012	675.8	1.939	0.0011011	675.9	1.939	0.0011007	676.2	1.938	0.0011004	676.4	1.938
180	453.15	0.0011268	763.2	2.136	0.0011267	763.2	2.136	0.0011262	763.5	2.135	0.0011258	763.7	2.134
200	473.15	0.0011562	852.4	2.328	0.0011561	852.4	2.328	0.0011556	852.5	2.327	0.0011551	852.6	2.326
220	493.15	0.1149	2833	6.452	0.1021	2821	6.385	0.0011898	943.6	2.516	0.0011891	943.5	2.514
240	513.15	0.1216	2884	6.554	0.1084	2875	6.491	0.08453	2850	6.351	0.06826	2823	6.225
250	523.15	0.1248	2908	6.601	0.1114	2900	6.539	0.08713	2878	6.404	0.07067	2853	6.283
260	533.15	0.1280	2932	6.646	0.1143	2924	6.585	0.08962	2904	6.454	0.07294	2882	6.337
280	553.15	0.1341	2979	6.732	0.1200	2972	6.674	0.09437	2955	6.547	0.07720	2937	6.438
300	573.15	0.1401	3025	6.814	0.1255	3019	6.757	0.09891	3004	6.635	0.08119	2988	6.530
320	593.15	0.1460	3071	6.892	0.1308	3065	6.837	0.1033	3052	6.717	0.08500	3038	6.615
340	613.15	0.1516	3116	6.966	0.1358	3111	6.913	0.1075	3099	6.795	0.08870	3087	6.696
350	623.15	0.1545	3138	7.003	0.1384	3134	6.949	0.1096	3123	6.833	0.09051	3111	6.735
360	633.15	0.1573	3160	7.039	0.1410	3156	6.985	0.1117	3146	6.870	0.09230	3135	6.773
380	653.15	0.1629	3205	7.108	0.1461	3201	7.055	0.1159	3192	6.941	0.09582	3182	6.847
400	673.15	0.1683	3249	7.175	0.1511	3246	7.122	0.1201	3238	7.010	0.09929	3229	6.916
420	693.15	0.1738	3294	7.240	0.1560	3291	7.187	0.1241	3283	7.076	0.1027	3275	6.984
440	713.15	0.1792	3338	7.303	0.1609	3335	7.251	0.1281	3328	7.140	0.1061	3321	7.048
450	723.15	0.1819	3360	7.333	0.1634	3357	7.282	0.1301	3350	7.172	0.1078	3343	7.080
460	733.15	0.1847	3381	7.363	0.1659	3379	7.312	0.1321	3373	7.202	0.1095	3366	7.111
480	753.15	0.1900	3425	7.422	0.1707	3423	7.371	0.1360	3417	7.262	0.1128	3411	7.172
500	773.15	0.1953	3470	7.480	0.1755	3468	7.429	0.1399	3462	7.321	0.1161	3456	7.231
520	793.15	0.2007	3514	7.537	0.1804	3512	7.486	0.1438	3507	7.378	0.1194	3501	7.289
540	813.15	0.2061	3558	7.592	0.1851	3556	7.542	0.1477	3552	7.434	0.1227	3547	7.345
550	823.15	0.2088	3580	7.619	0.1875	3578	7.569	0.1497	3574	7.461	0.1243	3569	7.373
560	833.15	0.2115	3602	7.646	0.1900	3600	7.596	0.1516	3597	7.488	0.1260	3592	7.400
580	853.15	0.2167	3647	7.698	0.1948	3645	7.649	0.1555	3641	7.542	0.1292	3637	7.454
600	873.15	0.2219	3691	7.750	0.1995	3690	7.701	0.1593	3686	7.594	0.1325	3682	7.506
620	893.15	0.2272	3736	7.801	0.2043	3735	7.752	0.1631	3732	7.646	0.1357	3728	7.558
640	913.15	0.2325	3781	7.851	0.2090	3780	7.802	0.1670	3777	7.696	0.1389	3773	7.608
650	923.15	0.2351	3803	7.876	0.2114	3802	7.827	0.1689	3799	7.721	0.1405	3796	7.633
660	933.15	0.2377	3826	7.900	0.2137	3825	7.851	0.1708	3822	7.746	0.1421	3819	7.658
680	953.15	0.2429	3871	7.948	0.2185	3871	7.899	0.1746	3868	7.794	0.1453	3865	7.707
700	973.15	0.2481	3917	7.996	0.2232	3917	7.947	0.1784	3914	7.842	0.1484	3911	7.755
720	993.15	0.2534	3963	8.043	0.2279	3963	7.994	0.1822	3960	7.889	0.1516	3957	7.803
740	1013.15	0.2586	4009	8.089	0.2326	4009	8.040	0.1860	4006	7.935	0.1548	4004	7.849
750	1023.15	0.2613	4033	8.112	0.2350	4032	8.063	0.1879	4030	7.958	0.1564	4027	7.872
760	1033.15	0.2639	4056	8.135	0.2373	4055	8.086	0.1897	4053	7.981	0.1580	4050	7.895
780	1053.15	0.2691	4103	8.180	0.2421	4102	8.130	0.1934	4100	8.026	0.1610	4097	7.940
800	1073.15	0.2742	4150	8.224	0.2467	4150	8.174	0.1971	4147	8.070	0.1641	4145	7.984
820	1093.15	0.2794	4198	8.268	0.2514	4197	8.218	0.2009	4195	8.114	0.1673	4193	8.028
840	1113.15	0.2845	4245	8.311	0.2560	4245	8.262	0.2046	4243	8.158	0.1704	4241	8.072
850	1123.15	0.2871	4269	8.332	0.2583	4269	8.283	0.2065	4267	8.179	0.1720	4265	8.093
860	1133.15	0.2897	4293	8.354	0.2606	4293	8.304	0.2084	4291	8.200	0.1735	4289	8.115
880	1153.15	0.2949	4341	8.396	0.2654	4340	8.346	0.2121	4339	8.242	0.1767	4337	8.157
900	1173.15	0.3001	4389	8.438	0.2700	4388	8.388	0.2158	4387	8.284	0.1798	4386	8.199
920	1193.15	0.3053	4437	8.478	0.2747	4437	8.429	0.2196	4435	8.325	0.1830	4434	8.240
940	1213.15	0.3105	4485	8.519	0.2794	4485	8.469	0.2234	4484	8.365	0.1861	4482	8.280
950	1223.15	0.3131	4510	8.539	0.2817	4509	8.489	0.2252	4508	8.385	0.1877	4506	8.300
960	1233.15	0.3157	4534	8.559	0.2841	4533	8.509	0.2271	4532	8.405	0.1892	4531	8.320
980	1253.15	0.3208	4582	8.598	0.2887	4582	8.549	0.2308	4581	8.445	0.1923	4579	8.360
1000	1273.15	0.3258	4631	8.637	0.2933	4630	8.588	0.2345	4629	8.484	0.1953	4628	8.399

Table 46-12 **Properties of Superheated Steam (H_2O) in International System of Units—SI** (*Continued*)

Pressure p		35 bar			40 bar			45 bar			50 bar		
Temperature		$t_s = 242.54\,°C$ $i'' = 2.803\,kJ/kg$ $v'' = 0.05704\,m^3/kg$ $s'' = 6.125\,kJ/kg\,K$			$t_s = 250.33\,°C$ $i'' = 2.801\,kJ/kg$ $v'' = 0.04977\,m^3/kg$ $s'' = 6.070\,kJ/kg\,K$			$t_s = 257.41\,°C$ $i'' = 2.798\,kJ/kg$ $v'' = 0.04404\,m^3/kg$ $s'' = 6.020\,kJ/kg\,K$			$t_s = 263.91\,°C$ $i'' = 2.794\,kJ/kg$ $v'' = 0.03944\,m^3/kg$ $s'' = 5.973\,kJ/kg\,K$		
t	T	v	i	s	v	i	s	v	i	s	v	i	s
°C	K	m^3/kg	kJ/kg	kJ/kg K	m^3/kg	kJ/kg	kJ/kg K	m^3/kg	kJ/kg	kJ/kg K	m^3/kg	kJ/kg	kJ/kg K
0	273.15	0.0009983	3.7	0.0001	0.0009981	4.2	0.0002	0.0009978	4.7	0.0003	0.0009976	5.2	0.0004
20	293.15	0.0001002	87.2	0.2955	0.0010000	87.6	0.2953	0.0009998	88.1	0.2952	0.0009995	88.5	0.2951
40	313.15	0.0010063	170.6	0.5706	0.0010061	171.0	0.5704	0.0010059	171.5	0.5702	0.0010056	171.9	0.5699
50	323.15	0.0010105	212.3	0.7015	0.0010103	212.7	0.7012	0.0010101	213.2	0.7008	0.0010098	213.6	0.7005
60	333.15	0.0010154	254.0	0.8286	0.0010152	254.4	0.8282	0.0010150	254.9	0.8277	0.0010147	255.3	0.8273
80	353.15	0.0010272	337.4	1.0722	0.0010270	337.8	1.0718	0.0010268	338.3	1.0713	0.0010265	338.7	1.0709
100	373.15	0.0010416	421.3	1.3034	0.0010414	421.7	1.3030	0.0010411	422.1	1.3025	0.0010408	422.5	1.3020
120	393.15	0.0010584	505.8	1.5240	0.0010582	506.2	1.5236	0.0010579	506.6	1.5230	0.0010576	506.9	1.5223
140	413.15	0.0010778	590.9	1.735	0.0010776	591.2	1.734	0.0010772	591.6	1.734	0.0010769	591.9	1.733
150	423.15	0.0010885	633.7	1.837	0.0010883	634.0	1.836	0.0010879	634.4	1.836	0.0010876	634.7	1.835
160	433.15	0.0011000	676.7	1.937	0.0010997	677.0	1.936	0.0010993	677.4	1.936	0.0010990	677.7	1.935
180	453.15	0.0011254	764.0	2.133	0.0011250	764.2	2.133	0.0011246	764.6	2.132	0.0011242	764.9	2.131
200	473.15	0.0011546	852.9	2.325	0.0011541	853.0	2.324	0.0011536	853.4	2.323	0.0011530	853.6	2.322
220	493.15	0.0011885	943.7	2.513	0.0011879	943.8	2.512	0.0011873	944.0	2.511	0.0011867	944.3	2.510
240	513.15	0.0012288	1037.4	2.699	0.0012280	1037.4	2.698	0.0012273	1037.4	2.697	0.0012264	1037.4	2.696
250	523.15	0.05877	2828	6.173	0.0012511	1085.7	2.791	0.0012502	1085.7	2.790	0.0012492	1085.7	2.789
260	533.15	0.06089	2859	6.232	0.05174	2834	6.133	0.04451	2808	6.038	0.0012749	1135.1	2.882
280	553.15	0.06482	2918	6.339	0.05550	2898	6.249	0.04818	2876	6.164	0.04224	2854	6.083
300	573.15	0.06847	2972	6.437	0.05888	2955	6.352	0.05142	2938	6.273	0.04539	2920	6.200
320	593.15	0.07188	3024	6.526	0.06201	3010	6.446	0.05434	2995	6.372	0.04817	2980	6.304
340	613.15	0.07514	3075	6.610	0.06496	3062	6.532	0.05705	3049	6.462	0.05071	3036	6.397
350	623.15	0.07674	3100	6.649	0.06639	3087	6.573	0.05837	3075	6.504	0.05195	3063	6.440
360	633.15	0.07832	3124	6.688	0.06781	3113	6.613	0.05967	3101	6.545	0.05316	3090	6.483
380	653.15	0.08143	3172	6.763	0.07062	3162	6.690	0.06223	3152	6.624	0.05553	3142	5.564
400	673.15	0.08448	3220	6.835	0.07337	3211	6.762	0.06474	3202	6.699	0.05781	3193	6.640
420	693.15	0.08748	3267	6.904	0.07606	3259	6.832	0.06717	3250	6.770	0.06004	3242	6.712
440	713.15	0.09043	3313	6.970	0.07870	3306	6.900	0.06955	3298	6.838	0.06224	3291	6.781
450	723.15	0.09190	3336	7.002	0.08001	3329	6.933	0.07073	3322	6.871	0.06332	3315	6.815
460	733.15	0.09336	3559	7.033	0.08130	3353	6.965	0.07190	3346	6.903	0.06439	3339	6.848
480	753.15	0.09625	3405	7.095	0.08388	3399	7.027	0.07421	3393	6.966	0.06650	3386	6.912
500	773.15	0.09910	3451	7.155	0.08642	3445	7.087	0.07649	3439	7.028	0.06858	3433	6.974
520	793.15	0.1019	3496	7.213	0.08895	3491	7.146	0.07877	3486	7.088	0.07064	3480	7.033
540	813.15	0.1048	3542	7.270	0.09145	3537	7.203	0.08103	3532	7.145	0.07268	3527	7.091
550	823.15	0.1062	3564	7.298	0.09270	3560	7.231	0.08215	3555	7.173	0.07370	3550	7.120
560	833.15	0.1076	3587	7.325	0.09394	3583	7.259	0.08326	3578	7.201	0.07471	3574	7.148
580	853.15	0.1105	3633	7.379	0.09640	3629	7.313	0.08546	3624	7.255	0.07672	3620	7.203
600	873.15	0.1132	3678	7.432	0.09885	3674	7.367	0.08766	3670	7.309	0.07870	3666	7.257
620	893.15	0.1160	3724	7.484	0.1013	3720	7.419	0.08982	3716	7.361	0.08065	3713	7.310
640	913.15	0.1188	3770	7.534	0.1037	3766	7.470	0.09198	3762	7.413	0.08260	3759	7.362
650	923.15	0.1202	3793	7.559	0.1049	3789	7.495	0.09305	3785	7.438	0.08357	3782	7.387
660	933.15	0.1215	3816	7.584	0.1061	3812	7.520	0.09412	3808	7.463	0.08454	3805	7.412
680	953.15	0.1243	3862	7.633	0.1085	3858	7.570	0.09626	3855	7.513	0.08648	3852	7.462
700	973.15	0.1269	3908	7.681	0.1109	3905	7.618	0.09841	3902	7.561	0.08842	3899	7.510
720	993.15	0.1297	3954	7.729	0.1133	3952	7.666	0.1006	3949	7.609	0.09035	3946	7.558
740	1013.15	0.1325	4001	7.776	0.1157	3998	7.712	0.1027	3996	7.656	0.09227	3993	7.606
750	1023.15	0.1339	4024	7.799	0.1169	4022	7.735	0.1038	4019	7.679	0.09323	4017	7.629
760	1033.15	0.1352	4048	7.822	0.1181	4045	7.758	0.1048	4043	7.702	0.09419	4041	7.652
780	1053.15	0.1379	4095	7.868	0.1205	4093	7.804	0.1070	4091	7.748	0.09612	4088	7.698
800	1073.15	0.1405	4143	7.912	0.1228	4141	7.848	0.1091	4138	7.792	0.09803	4136	7.742
820	1093.15	0.1432	4191	7.956	0.1252	4188	7.892	0.1113	4186	7.836	0.09993	4184	7.786
840	1113.15	0.1459	4239	8.000	0.1276	4236	7.936	0.1134	4235	7.880	0.1018	4232	7.830
850	1123.15	0.1473	4263	8.021	0.1288	4261	7.958	0.1145	4259	7.902	0.1027	4257	7.852
860	1133.15	0.1487	4287	8.042	0.1300	4285	7.979	0.1156	4283	7.924	0.1037	4281	7.874
880	1153.15	0.1513	4335	8.085	0.1324	4333	8.022	0.1176	4332	7.966	0.1057	4329	7.916
900	1173.15	0.1540	4384	8.127	0.1346	4382	8.064	0.1196	4380	8.008	0.1075	4387	7.958
920	1193.15	0.1568	4432	8.168	0.1370	4430	8.105	0.1218	4429	8.050	0.1094	4427	8.000
940	1213.15	0.1595	4480	8.208	0.1394	4479	8.146	0.1240	4477	8.090	0.1114	4475	8.040
950	1223.15	0.1608	4505	8.228	0.1405	4503	8.166	0.1250	4502	8.110	0.1123	4500	8.060
960	1233.15	0.1621	4529	8.248	0.1417	4527	8.186	0.1260	4526	8.130	0.1132	4524	8.080
980	1253.15	0.1647	4578	8.288	0.1441	4576	8.226	0.1280	4575	8.170	0.1150	4573	8.120
1000	1273.15	0.1673	4626	8.327	0.1464	4625	8.265	0.1300	4623	8.208	0.1170	4622	8.159

Table 46-13 Properties of Superheated Steam (H$_2$O) in International System of Units—SI (Continued)

Pressure p		60 bar			70 bar			80 bar			90 bar		
Temperature		$t_s = 275.56\ °C$ $i'' = 2.785\ kJ/kg$ $v'' = 0.03243\ m^3/kg$ $s'' = 5.890\ kJ/kg\ K$			$t_s = 285.80\ °C$ $i'' = 2.772\ kJ/kg$ $v'' = 0.02737\ m^3/kg$ $s'' = 5.814\ kJ/kg\ K$			$t_s = 294.98\ °C$ $i'' = 2.758\ kJ/kg$ $v'' = 0.02352\ m^3/kg$ $s'' = 5.745\ kJ/kg\ K$			$t_s = 303.32\ °C$ $i'' = 2.743\ kJ/kg$ $v'' = 0.02048\ m^3/kg$ $s'' = 5.678\ kJ/kg\ K$		
t	T	v	i	s	v	i	s	v	i	s	v	i	s
°C	K	m³/kg	kJ/kg	kJ/kg K	m³/kg	kJ/kg	kJ/kg K	m³/kg	kJ/kg	kJ/kg K	m³/kg	kJ/kg	kJ/kg K
0	273.15	0.0009971	6.2	0.0004	0.0009966	7.2	0.0004	0.0009961	8.2	0.0004	0.0009956	9.2	0.0004
20	293.15	0.0009991	89.4	0.2948	0.0009987	90.4	0.2945	0.0009983	91.3	0.2943	0.0009978	92.3	0.2941
40	313.15	0.0010052	172.8	0.5694	0.0010048	173.7	0.5689	0.0010043	174.6	0.5686	0.0010038	175.5	0.5681
50	323.15	0.0010094	214.4	0.7000	0.0010090	215.3	0.6995	0.0010085	216.2	0.6992	0.0010080	217.1	0.6986
60	333.15	0.0010143	256.1	0.8268	0.0010139	256.9	0.8263	0.0010134	257.8	0.8260	0.0010129	258.7	0.8253
80	353.15	0.0010261	339.5	1.0702	0.0010257	340.3	1.0694	0.0010254	341.2	1.0689	0.0010249	342.1	1.0682
100	373.15	0.0010403	423.3	1.3012	0.0010400	424.1	1.3003	0.0010398	424.9	1.2996	0.0010393	425.7	1.2988
120	393.15	0.0010571	507.7	1.5215	0.0010567	508.4	1.5205	0.0010564	509.1	1.5198	0.0010559	509.8	1.5189
140	413.15	0.0010763	592.6	1.732	0.0010758	593.2	1.731	0.0010754	593.9	1.730	0.0010749	594.6	1.729
150	423.15	0.0010869	635.4	1.834	0.0010864	636.0	1.833	0.0010859	636.6	1.832	0.0010854	637.3	1.831
160	433.15	0.0010938	678.4	1.934	0.0010977	679.0	1.933	0.0010972	679.6	1.931	0.0010966	680.3	1.930
180	453.15	0.0011234	765.5	2.119	0.0011226	766.1	2.128	0.0011220	766.7	2.126	0.0011213	767.4	2.125
200	473.15	0.0011522	854.0	2.320	0.0011512	854.5	2.319	0.0011504	855.0	2.317	0.0011496	855.5	2.316
220	493.15	0.0011855	944.5	2.508	0.0011845	944.8	2.506	0.0011833	945.1	2.504	0.0011822	945.5	2.502
240	513.15	0.0012249	1037.6	2.693	0.0012235	1037.8	2.691	0.0012221	1037.9	2.688	0.0012207	1038.1	2.686
250	523.15	0.0012476	1085.7	2.786	0.0012459	1085.7	2.783	0.0012443	1085.7	2.781	0.0012427	1085.7	2.778
260	533.15	0.0012727	1134.8	2.879	0.0012706	1134.6	2.876	0.0012689	1134.4	2.873	0.0012669	1134.2	2.870
280	553.15	0.03315	2803	5.923	0.0013304	1235.9	3.063	0.0013275	1235.4	3.059	0.0013246	1234.9	3.056
300	573.15	0.03620	2880	6.060	0.02948	2835	5.925	0.02429	2784	5.788	0.0014016	1344.3	3.249
320	593.15	0.03884	2948	6.177	0.03206	2913	6.058	0.02687	2874	5.943	0.02272	2829	5.827
340	613.15	0.04118	3010	6.279	0.03430	2981	6.171	0.02904	2951	6.070	0.02488	2916	5.972
350	623.15	0.04227	3039	6.326	0.03532	3012	6.222	0.03003	2985	6.126	0.02586	2954	6.033
360	633.15	0.04334	3067	6.371	0.03630	3042	6.270	0.03098	3017	6.177	0.02678	2989	6.089
380	653.15	0.04542	3121	6.456	0.03819	3099	6.360	0.03274	3077	6.272	0.02847	3054	6.189
400	673.15	0.04742	3174	6.535	0.03997	3155	6.442	0.03438	3135	6.358	0.03001	3114	6.280
420	693.15	0.04935	3225	6.610	0.04170	3208	6.520	0.03595	3190	6.439	0.03147	3172	6.364
440	713.15	0.05124	3275	6.681	0.04338	3259	6.593	0.03746	3244	6.515	0.03286	3227	6.443
450	723.15	0.05217	3299	6.716	0.04420	3285	6.628	0.03821	3270	6.552	0.03354	3254	6.481
460	733.15	0.05309	3324	6.750	0.04501	3310	6.663	0.03894	3296	6.588	0.03421	3281	6.518
480	753.15	0.05490	3373	6.815	0.04661	3360	6.731	0.04037	3347	6.657	0.03552	3334	6.589
500	773.15	0.05667	3421	6.878	0.04817	3409	6.795	0.04177	3397	6.722	0.03680	3386	6.656
520	793.15	0.05842	3469	6.939	0.04970	3458	6.858	0.04315	3447	6.785	0.03805	3436	6.720
540	813.15	0.06016	3517	6.999	0.05122	3506	6.918	0.04449	3496	6.846	0.03927	3485	6.783
550	823.15	0.06103	3540	7.028	0.05197	3530	6.947	0.04516	3520	6.876	0.03988	3510	6.813
560	833.15	0.06189	3564	7.056	00.5272	3554	6.976	0.04583	3544	6.905	0.04049	3534	6.843
580	853.15	0.06358	3611	7.111	0.05421	3602	7.032	0.04716	3592	6.963	0.04169	3582	6.901
600	873.15	0.06525	3658	7.165	0.05565	3649	7.087	0.04844	3640	7.019	0.04285	3631	6.957
620	893.15	0.06691	3705	7.219	0.05708	3697	7.141	0.04972	3688	7.073	0.04399	3679	7.012
640	913.15	0.06855	3751	7.271	0.05851	3744	7.194	0.05098	3736	7.126	0.04513	3728	7.066
650	923.15	0.06937	3775	7.297	0.05923	3768	7.220	0.05161	3760	7.152	0.04570	3752	7.092
660	933.15	0.07019	3798	7.322	0.05995	3791	7.246	0.05225	3784	7.178	0.04627	3776	7.118
680	953.15	0.07183	3846	7.372	0.06137	3839	7.296	0.05350	3832	7.230	0.04739	3825	7.170
700	973.15	0.07347	3893	7.422	0.06277	3887	7.346	0.05475	3881	7.280	0.04851	3874	7.220
720	993.15	0.07509	3940	7.471	0.06417	3935	7.395	0.05599	3929	7.329	0.04963	3922	7.270
740	1013.15	0.07670	3988	7.519	0.06557	3983	7.443	0.05723	3977	7.377	0.05075	3971	7.319
750	1023.15	0.07751	4012	7.542	0.06627	4007	7.467	0.05785	4002	7.401	0.05130	3996	7.343
760	1033.15	0.07831	4036	7.565	0.06697	4031	7.490	0.05847	4026	7.425	0.05185	4020	7.367
780	1053.15	0.07992	4084	7.610	0.06836	4079	7.535	0.05970	4074	7.470	0.05295	4069	7.413
800	1073.15	0.08153	4132	7.655	0.06975	4127	7.580	0.06092	4122	7.515	0.05405	4117	7.459
820	1093.15	0.08314	4180	7.699	0.07114	4176	7.624	0.06214	4171	7.559	0.05515	4166	7.504
840	1113.15	0.08474	4229	7.743	0.07252	4225	7.668	0.06337	4221	7.603	0.05623	4216	7.548
850	1123.15	0.08554	4253	7.765	0.07321	4249	7.690	0.06398	4245	7.625	0.05677	4240	7.570
860	1133.15	0.08634	4277	7.787	0.07390	4273	7.712	0.06459	4269	7.647	0.05732	4264	7.592
880	1153.15	0.08794	4326	7.830	0.07527	4322	7.755	0.06579	4318	7.690	0.05840	4314	7.635
900	1173.15	0.08953	4375	7.872	0.07665	4371	7.797	0.06700	4367	7.732	0.05949	4363	7.677
920	1193.15	0.09111	4424	7.914	0.07802	4420	7.839	0.06819	4416	7.774	0.06056	4413	7.719
940	1213.15	0.09269	4472	7.954	0.07938	4469	7.879	0.06939	4466	7.816	0.06162	4462	7.761
950	1223.15	0.09348	4496	7.974	0.08005	4493	7.899	0.06998	4490	7.836	0.06215	4487	7.781
960	1233.15	0.09427	4521	7.994	0.08073	4518	7.919	0.07058	4515	7.856	0.06268	4512	7.801
980	1253.15	0.09584	4570	8.034	0.08208	4567	7.959	0.07176	4565	7.896	0.06374	4561	7.841
1000	1273.15	0.09740	4619	8.073	0.08343	4617	7.999	0.07295	4614	7.936	0.06480	4611	7.881

Table 46-14 Properties of Superheated Steam (H$_2$O) in International System of Units—SI (*Continued*)

Pressure p		100 bar			110 bar			120 bar			130 bar		
Temperature		$t_s = 318.04$ °C $i'' = 2.705$ kJ/kg $v'' = 0.01598$ m³/kg $s'' = 5.553$ kJ/kg K			$t_s = 310.96$ °C $i'' = 2.725$ kJ/kg $v'' = 0.01803$ m³/kg $s'' = 5.615$ kJ/kg K			$t_s = 324.63$ °C $i'' = 2.685$ kJ/kg $v'' = 0.01426$ m³/kg $s'' = 5.492$ kJ/kg K			$t_s = 330.81$ °C $i'' = 2.662$ kJ/kg $v'' = 0.01277$ m³/kg $s'' = 5.432$ kJ/kg K		
t	T	v	i	s	v	i	s	v	i	s	v	i	s
°C	K	m³/kg	kJ/kg	kJ/kg K	m³/kg	kJ/kg	kJ/kg K	m³/kg	kJ/kg	kJ/kg K	m³/kg	kJ/kg	kJ/kg K
0	273.15	0.0009951	10.2	0.0004	0.0009946	11.2	0.0005	0.0009941	12.2	0.0006	0.0009936	13.2	0.0007
20	293.15	0.0009975	93.2	0.2939	0.0009970	94.1	0.2937	0.0009965	95.1	0.2935	0.0009961	96.0	0.2931
40	313.15	0.0010033	176.4	0.5677	0.0010028	177.3	0.5672	0.0010024	178.2	0.5668	0.0010020	179.0	0.5664
50	323.15	0.0010075	218.0	0.6980	0.0010070	218.9	0.6975	0.0010066	219.8	0.6970	0.0010062	220.6	0.6965
60	333.15	0.0010125	259.6	0.8247	0.0010120	260.5	0.8241	0.0010116	216.4	0.8236	0.0010112	262.2	0.8230
80	353.15	0.0010245	342.9	1.0676	0.0010240	343.8	1.0669	0.0010236	344.6	1.0662	0.0010231	345.4	1.0655
100	373.15	0.0010386	426.5	1.2982	0.0010384	427.3	1.2974	0.0010379	428.1	1.2967	0.0010373	428.9	1.2959
120	393.15	0.0010552	510.5	1.5182	0.0010549	511.3	1.5173	0.0010544	512.0	1.5165	0.0010538	512.7	1.5156
140	413.15	0.0010741	595.3	1.728	0.0010738	596.0	1.728	0.0010732	596.7	1.727	0.0010725	597.4	1.726
150	423.15	0.0010845	638.0	1.830	0.0010841	638.7	1.829	0.0010835	639.4	1.828	0.0010828	640.1	1.827
160	433.15	0.0010956	681.0	1.929	0.0010952	681.7	1.928	0.0010946	682.4	1.927	0.0010939	683.0	1.926
180	453.15	0.0011201	768.0	2.123	0.0011197	768.6	2.123	0.0011189	769.1	2.121	0.0011182	769.7	2.119
200	473.15	0.0011482	856.0	2.314	0.0011477	856.5	2.312	0.0011468	857.0	2.311	0.0011458	857.4	2.309
220	493.15	0.0011805	945.8	2.500	0.0011799	946.2	2.498	0.0011788	946.6	2.497	0.0011777	946.9	2.495
240	513.15	0.0012185	1038.3	2.684	0.0012178	1038.5	2.682	0.0012164	1038.7	2.680	0.0012150	1038.9	2.678
250	523.15	0.0012402	1085.7	2.776	0.0012394	1085.8	2.774	0.0012377	1085.8	2.772	0.0012361	1085.9	2.769
260	533.15	0.0012650	1134.1	2.868	0.0012630	1134.0	2.865	0.0012612	1133.9	2.863	0.0012593	1133.8	2.860
280	553.15	0.0013217	1234.5	3.053	0.0013190	1234.1	3.049	0.0013164	1233.7	3.046	0.0013137	1233.3	3.043
300	573.15	0.0013970	1342.2	3.244	0.0013928	1341.1	3.239	0.0013886	1340.0	3.235	0.0013847	1339.0	3.230
320	593.15	0.01926	2778	5.705	0.01629	2719	5.579	0.001493	1459.3	3.441	0.001485	1456.5	3.433
340	613.15	0.02150	2878	5.872	0.01868	2836	5.770	0.01624	2789	5.667	0.01403	2737	5.555
350	623.15	0.02247	2920	5.940	0.01967	2884	5.849	0.01726	2844	5.755	0.01514	2799	5.657
360	633.15	0.02337	2958	6.002	0.02056	2927	5.918	0.01816	2892	5.832	0.01610	2853	5.744
380	653.15	0.02498	3028	6.111	0.02214	3002	6.037	0.01973	2974	5.963	0.01767	2945	5.888
400	673.15	0.02646	3093	6.207	0.02356	3071	6.138	0.02113	3049	6.071	0.01905	3026	6.006
420	693.15	0.02784	3154	6.294	0.02487	3135	6.230	0.02239	3116	6.168	0.02028	3097	6.108
440	713.15	0.02915	3211	6.377	0.02612	3194	6.315	0.02357	3177	6.256	0.02143	3159	6.200
450	723.15	0.02979	3239	6.416	0.02672	3222	6.355	0.02414	3206	6.298	0.02197	3189	6.243
460	733.15	0.03042	3266	6.454	0.02731	3250	6.394	0.02471	3235	6.338	0.02250	3219	6.285
480	753.15	0.03163	3320	6.527	0.02844	3305	6.469	0.02578	3291	6.415	0.02352	3277	6.364
500	773.15	0.03281	3372	6.596	0.02954	3360	6.540	0.02681	3347	6.487	0.02450	3334	6.438
520	793.15	0.03397	3424	6.662	0.03061	3412	6.607	0.02782	3400	6.556	0.02546	3388	6.507
540	813.15	0.03510	3474	6.725	0.03167	3463	6.671	0.02880	3452	6.621	0.02638	3441	6.574
550	823.15	0.03566	3499	6.756	0.03218	3488	6.702	0.02928	3478	6.653	0.02683	3467	6.606
560	833.15	0.03621	3524	6.786	0.03269	3513	6.733	0.02976	3503	6.684	0.02728	3493	6.638
580	853.15	0.03730	3572	6.845	0.03370	3563	6.793	0.03070	3553	6.744	0.02817	3543	6.698
600	873.15	0.03837	3621	6.901	0.03469	3612	6.850	0.03163	3603	6.803	0.02903	3594	6.758
620	893.15	0.03941	3670	6.957	0.03565	3662	6.906	0.03253	3653	6.859	0.02988	3645	6.816
640	913.15	0.04045	3719	7.011	0.03661	3711	6.961	0.03342	3703	6.915	0.03071	3696	6.872
650	923.15	0.04097	3744	7.038	0.03709	3736	6.988	0.03387	3728	6.942	0.03113	3721	6.899
660	933.15	0.04149	3768	7.064	0.03757	3761	7.015	0.03431	3753	6.969	0.03155	3746	6.926
680	953.15	0.04252	3818	7.116	0.03851	3811	7.067	0.03519	3804	7.021	0.03236	3797	6.980
700	973.15	0.04354	3867	7.167	0.03945	3860	7.118	0.03605	3853	7.073	0.03317	3847	7.032
720	993.15	0.04456	3915	7.217	0.04038	3909	7.168	0.03691	3903	7.123	0.03398	3897	7.082
740	1013.15	0.04556	3964	7.265	0.04130	3958	7.216	0.03777	3952	7.172	0.03478	3947	7.132
750	1023.15	0.04606	3989	7.289	0.04176	3983	7.240	0.03820	3977	7.196	0.03517	3972	7.156
760	1033.15	0.04656	4013	7.313	0.04222	4007	7.264	0.03863	4002	7.220	0.03557	3997	7.180
780	1053.15	0.04756	4062	7.360	0.04314	4057	7.311	0.03947	4052	7.268	0.03635	4047	7.228
800	1073.15	0.04856	4111	7.406	0.04406	4106	7.357	0.04031	4102	7.314	0.03714	4097	7.274
820	1093.15	0.04956	4160	7.452	0.04497	4156	7.403	0.04115	4151	7.360	0.03792	4147	7.320
840	1113.15	0.05054	4210	7.497	0.04587	4205	7.448	0.04199	4201	7.405	0.03969	4197	7.365
850	1123.15	0.05103	4235	7.519	0.04632	4230	7.470	0.04240	4226	7.427	0.03908	4222	7.388
860	1133.15	0.05152	4259	7.541	0.04677	4255	7.492	0.04282	4251	7.449	0.03947	4247	7.410
880	1153.15	0.05250	4309	7.585	0.04767	4305	7.536	0.04364	4301	7.493	0.04023	4297	7.454
900	1173.15	0.05347	4358	7.627	0.04856	4354	7.579	0.04446	4351	7.536	0.04100	4347	7.498
920	1193.15	0.05445	4408	7.669	0.04944	4404	7.621	0.04528	4400	7.578	0.04175	4397	7.540
940	1213.15	0.05541	4457	7.711	0.05032	4454	7.663	0.04609	4450	7.620	0.04250	4447	7.582
950	1223.15	0.05589	4482	7.732	0.05076	4479	7.684	0.04649	4475	7.641	0.04287	4472	7.603
960	1233.15	0.05637	4507	7.752	0.05120	4504	7.704	0.04690	4500	7.661	0.04324	4497	7.623
980	1253.15	0.05733	4557	7.792	0.05206	4553	7.744	0.04770	4550	7.701	0.04400	4547	7.663
1000	1273.15	0.05829	4606	7.832	0.05294	4603	7.784	0.04850	4600	7.741	0.04475	4597	7.703

Table 46-15 Properties of Superheated Steam (H_2O) in International System of Units—SI (Continued)

Pressure p		140 bar			150 bar			160 bar			170 bar		
Temperature		$t_s = 336.63$ °C $i'' = 2.638$ kJ/kg $v'' = 0.01149$ m³/kg $s'' = 5.372$ kJ/kg K			$t_s = 342.11$ °C $i'' = 2.611$ kJ/kg $v'' = 0.01035$ m³/kg $s'' = 5.310$ kJ/kg K			$t_s = 347.32$ °C $i'' = 2.582$ kJ/kg $v'' = 0.009318$ m³/kg $s'' = 5.247$ kJ/kg K			$t_s = 352.26$ °C $i''' = 2.548$ kJ/kg $v'' = 0.008382$ m³/kg $s'' = 5.177$ kJ/kg K		
t	T	v	i	s	v	i	s	v	i	s	v	i	s
°C	K	m³/kg	kJ/kg	kJ/kg K	m³/kg	kJ/kg	kJ/kg K	m³/kg	kJ/kg	kJ/kg K	m³/kg	kJ/kg	kJ/kg K
0	273.15	0.0009931	14.2	0.0008	0.0009927	15.2	0.0008	0.0009922	16.2	0.0009	0.0009917	17.2	0.0010
20	293.15	0.0009957	96.9	0.2930	0.0009953	97.9	0.2927	0.0009948	98.9	0.2925	0.0009943	99.8	0.2923
40	313.15	0.0010016	179.9	0.5660	0.0010012	180.8	0.5656	0.0010007	181.7	0.5653	0.0010003	182.6	0.5650
50	323.15	0.0010058	221.4	0.6960	0.0010054	222.3	0.6955	0.0010049	223.2	0.6951	0.0010045	224.1	0.6947
60	333.15	0.0010108	263.0	0.8224	0.0010104	263.8	0.8218	0.0010099	264.7	0.8212	0.0010095	265.6	0.8206
80	353.15	0.0010226	346.2	1.0648	0.0010222	347.0	1.0641	0.0010217	347.9	1.0634	0.0010213	348.7	1.0627
100	373.15	0.0010368	429.6	1.2951	0.0010363	430.4	1.2944	0.0010359	431.2	1.2937	0.0010354	431.9	1.2930
120	393.15	0.0010533	513.4	1.5148	0.0010527	514.1	1.5139	0.0010522	514.9	1.5131	0.0010517	515.6	1.5123
140	413.15	0.0010719	598.0	1.724	0.0010713	598.7	1.723	0.0010707	599.4	1.722	0.0010701	600.1	1.722
150	423.15	0.0010822	640.7	1.826	0.0010815	641.3	1.824	0.0010809	642.0	1.823	0.0010802	642.7	1.823
160	433.15	0.0010932	683.6	1.925	0.0010925	684.2	1.923	0.0010918	684.9	1.922	0.0010911	685.5	1.922
180	453.15	0.0011174	770.2	2.118	0.0011166	770.8	2.117	0.0011157	771.3	2.116	0.0011149	771.9	2.115
200	473.15	0.0011448	857.9	2.308	0.0011439	858.3	2.306	0.0011430	858.8	2.305	0.0011420	859.3	2.303
220	493.15	0.0011766	947.3	2.493	0.0011755	947.6	2.491	0.0011744	948.0	2.489	0.0011732	948.4	2.488
240	513.15	0.0012136	1039.1	2.676	0.0012123	1039.2	2.674	0.0012109	1039.5	2.672	0.0012095	1039.7	2.670
250	523.15	0.0012345	1086.0	2.767	0.0012330	1086.0	2.775	0.0012316	1086.2	2.762	0.0012301	1086.3	2.760
260	533.15	0.0012575	1133.8	2.858	0.0012557	1133.7	2.885	0.0012539	1133.7	2.853	0.0012521	1133.7	2.850
280	553.15	0.0013111	1232.9	3.040	0.0013086	1232.5	3.007	0.0013061	1232.2	3.035	0.0013037	1231.9	3.031
300	573.15	0.0013808	1338.0	3.226	0.0013771	1337.0	3.222	0.0013735	1336.2	3.218	0.001370	1335.4	3.214
320	593.15	0.001479	1454.1	3.427	0.001472	1451.7	3.420	0.001466	1449.8	3.414	0.001461	1448.1	3.409
340	613.15	0.01197	2672	5.436	0.001633	1592.2	3.654	0.001616	1586.3	3.642	0.001604	1581.1	3.631
350	623.15	0.01325	2750	5.556	0.01150	2690	5.442	0.00978	2612	5.302	0.001732	1668	3.770
360	633.15	0.01425	2812	5.654	0.01260	2765	5.559	0.01106	2711	5.457	0.00956	2649	5.342
380	653.15	0.01588	2914	5.813	0.01430	2880	5.739	0.01289	2843	5.662	0.01161	2803	5.582
400	673.15	0.01726	3000	5.942	0.01568	2973	5.878	0.01429	2945	5.816	0.01306	2915	5.753
420	693.15	0.01847	3077	6.051	0.01688	3055	5.997	0.01549	3031	5.941	0.01426	3006	5.885
440	713.15	0.01957	3141	6.146	0.01796	3123	6.093	0.01655	3103	6.042	0.01531	3083	5.992
450	723.15	0.02010	3172	6.190	0.01847	3155	6.139	0.01704	3137	6.090	0.01579	3118	6.042
460	733.15	0.02061	3203	6.233	0.01896	3186	6.183	0.01752	3169	6.136	0.01625	3152	6.090
480	753.15	0.02158	3262	6.314	0.01991	3248	6.268	0.01844	3233	6.223	0.01714	3218	6.179
500	773.15	0.02252	3321	6.390	0.02080	3308	6.346	0.01930	3294	6.303	0.01798	3281	6.261
520	793.15	0.02342	3376	6.461	0.02166	3364	6.419	0.02012	3352	6.377	0.01876	3340	6.337
540	813.15	0.02431	3430	6.529	0.02250	3418	6.488	0.02092	3407	6.448	0.01952	3396	6.410
550	823.15	0.02474	3456	6.562	0.02291	3445	6.521	0.02132	3434	6.482	0.01991	3423	6.444
560	833.15	0.02516	3482	6.594	0.02331	3472	6.554	0.02171	3461	6.515	0.02029	3450	6.477
580	853.15	0.02600	3534	6.656	0.02411	3524	6.616	0.02249	3514	6.578	0.02104	3504	6.542
600	873.15	0.02683	3585	6.716	0.02490	3576	6.677	0.02322	3567	6.640	0.02174	3558	6.604
620	893.15	0.02763	3637	6.775	0.02566	3628	6.737	0.02394	3620	6.701	0.02242	3611	6.665
640	913.15	0.02842	3688	6.832	0.02641	3680	6.794	0.02465	3672	6.758	0.02310	3664	6.725
650	923.15	0.02881	3713	6.859	0.02677	3706	6.822	0.02500	3698	6.786	0.02343	3691	6.752
660	933.15	0.02919	3739	6.886	0.02714	3732	6.849	0.02535	3724	6.814	0.02376	3717	6.780
680	953.15	0.02995	3790	6.939	0.02786	3784	6.903	0.02603	3777	6.869	0.02440	3770	6.836
700	973.15	0.03071	3841	6.992	0.02857	3835	6.956	0.02671	3829	6.922	0.02505	3822	6.890
720	993.15	0.03146	3891	7.043	0.02927	3885	7.008	0.02739	3879	6.974	0.02569	3873	6.942
740	1013.15	0.03221	3941	7.093	0.02998	3936	7.058	0.02806	3930	7.024	0.02633	3924	6.992
750	1023.15	0.03258	3966	7.118	0.03033	3961	7.083	0.02839	3955	7.049	0.02664	3949	7.017
760	1033.15	0.03295	3991	7.143	0.03069	3986	7.108	0.02872	3980	7.074	0.02696	3974	7.042
780	1053.15	0.03370	4042	7.191	0.03138	4036	7.156	0.02936	4031	7.123	0.02758	4026	7.092
800	1073.15	0.03442	4092	7.238	0.03206	4087	7.204	0.03001	4082	7.171	0.02819	4077	7.140
820	1093.15	0.03514	4142	7.284	0.03275	4138	7.250	0.03066	4133	7.218	0.02881	4128	7.187
840	1113.15	0.03588	4192	7.330	0.03344	4188	7.296	0.03131	4183	7.264	0.02941	4179	7.233
850	1123.15	0.03624	4217	7.352	0.03378	4213	7.319	0.03162	4209	7.287	0.02971	4204	7.256
860	1133.15	0.03660	4243	7.374	0.03412	4238	7.341	0.03194	4234	7.309	0.03001	4230	7.279
880	1153.15	0.03732	4293	7.418	0.03479	4289	7.385	0.03257	4284	7.353	0.03061	4280	7.323
900	1173.15	0.03803	4343	7.462	0.03545	4339	7.429	0.03320	4335	7.397	0.03121	4331	7.367
920	1193.15	0.03873	4393	7.504	0.03612	4389	7.471	0.03382	4385	7.439	0.03181	4381	7.409
940	1213.15	0.03943	4443	7.546	0.03678	4439	7.513	0.03445	4436	7.481	0.03240	4432	7.451
950	1223.15	0.03978	4468	7.567	0.03710	4464	7.534	0.03475	4461	7.502	0.03269	4457	7.472
960	1233.15	0.04013	4493	7.587	0.03743	4490	7.554	0.03506	4486	7.523	0.03298	4482	7.493
980	1253.15	0.04083	4543	7.627	0.03809	4540	7.594	0.03567	4536	7.563	0.03356	4533	7.533
1000	1273.15	0.04152	4593	7.667	0.03873	4590	7.634	0.03628	4587	7.603	0.03412	4584	7.573

Table 46-16 Properties of Superheated Steam (H_2O) in International System of Units—SI (Continued)

Pressure p		180 bar			190 bar			200 bar			220 bar		
Temperature		$t_s=356.96\ °C$ $i''=2.510\ kJ/kg$ $v''=0.007504\ m^3/kg$ $s''=5.107\ kJ/kg\ K$			$t_s=361.44\ °C$ $i''=2.466\ kJ/kg$ $v''=0.00668\ m^3/kg$ $s''=5.027\ kJ/kg\ K$			$t_s=365.71\ °C$ $i''=2.410\ kJ/kg$ $v''=0.00585\ m^3/kg$ $s''=4.928\ kJ/kg\ K$			$t_s=373.7\ °C$ $i''=2.168\ kJ/kg$ $v''=0.00367\ m^3/kg$ $s''=4.591\ kJ/kg\ K$		
t	T	v	i	s	v	i	s	v	i	s	v	i	s
°C	K	m^3/kg	kJ/kg	kJ/kg K	m^3/kg	kJ/kg	kJ/kg K	m^3/kg	kJ/kg	kJ/kg K	m^3/kg	kJ/kg	skJ/kg K
0	273.15	0.0009913	18.2	0.0011	0.009908	19.2	0.0012	0.0009904	20.2	0.0013	0.0009893	22.2	0.0013
20	293.15	0.0009939	100.7	0.2921	0.0009934	101.7	0.2919	0.0009930	102.6	0.2918	0.0009920	104.5	0.2915
40	313.15	0.0009999	183.5	0.5647	0.0009994	184.4	0.5643	0.0009990	185.3	0.5640	0.0009981	187.1	0.5634
50	323.15	0.0010041	225.0	0.6942	0.0010037	225.8	0.6937	0.0010033	226.7	0.6933	0.0010024	228.4	0.6927
60	333.15	0.0010091	266.5	0.8200	0.0010087	267.3	0.8194	0.0010083	268.1	0.8188	0.0010073	269.8	0.8181
80	353.15	0.0010209	349.5	1.0620	0.0010204	350.3	1.0613	0.0010200	351.1	1.0605	0.0010190	352.7	1.0596
100	373.15	0.0010349	432.7	1.2923	0.0010344	433.4	1.2916	0.0010339	434.2	1.2909	0.0010329	435.7	1.2899
120	393.15	0.0010512	516.4	1.5115	0.0010506	517.1	1.5106	0.0010501	517.8	1.5098	0.0010490	519.3	1.5084
140	413.15	0.0010695	600.8	1.721	0.0010690	601.5	1.720	0.0010684	602.1	1.719	0.0010671	603.5	1.717
150	423.15	0.0010796	643.4	1.822	0.0010790	644.0	1.821	0.0010784	644.6	1.820	0.0010771	646.0	1.818
160	433.15	0.0010905	686.2	1.921	0.0010898	686.8	1.920	0.0010891	687.4	1.919	0.0010877	688.7	1.917
180	453.15	0.0011142	772.4	2.114	0.0011134	773.0	2.113	0.0011126	773.5	2.112	0.0011110	774.7	2.110
200	473.15	0.0011411	859.7	2.302	0.0011402	860.2	2.301	0.0011393	860.6	2.299	0.0011375	861.6	2.297
220	493.15	0.0011721	948.7	2.486	0.0011711	949.1	2.485	0.0011700	949.4	2.483	0.0011679	950.2	2.480
240	513.15	0.0012082	1039.9	2.668	0.0012069	1040.1	2.666	0.0012056	1040.3	2.664	0.0012030	1040.9	2.661
250	523.15	0.0012286	1086.4	2.758	0.0012271	1086.5	2.756	0.0012256	1086.6	2.754	0.0012226	1087.0	2.750
260	533.15	0.0012504	1133.7	2.848	0.0012487	1133.7	2.845	0.0012470	1133.6	2.843	0.0012437	1133.8	2.839
280	553.15	0.0013013	1231.6	3.028	0.0012990	1231.3	3.025	0.0012968	1230.9	3.023	0.0012923	1230.6	3.017
300	573.15	0.0013665	1334.6	3.211	0.0013631	1333.9	3.207	0.0013598	1333.2	3.204	0.0013535	1332.2	3.197
320	593.15	0.001455	1446.3	3.403	0.001450	1444.9	3.399	0.001444	1442.9	3.394	0.001434	1440.5	3.384
340	613.15	0.001592	1576.6	3.620	0.001580	1572.7	3.611	0.001569	1569.1	3.603	0.001551	1562.6	3.589
350	623.15	0.001704	1657	3.751	0.001684	1650	3.736	0.001665	1644	3.724	0.001636	1633	3.704
360	633.15	0.00810	2563	5.194	0.001874	1755	3.905	0.001824	1739	3.876	0.001757	1717	3.837
380	653.15	0.01042	2759	5.498	0.00932	2711	5.408	0.00828	2655	5.309	0.00610	2503	5.052
400	673.15	0.01194	2884	5.688	0.01092	2851	5.622	0.00998	2816	5.553	0.00828	2736	5.406
420	693.15	0.01314	2981	5.830	0.01212	2955	5.774	0.01119	2928	5.719	0.00959	2871	5.606
440	713.15	0.01419	3062	5.943	0.01317	3041	5.895	0.01224	3019	5.847	0.01064	2974	5.752
450	723.15	0.01467	3100	5.995	0.01365	3080	5.949	0.01272	3060	5.903	0.01112	3017	5.813
460	733.15	0.01512	3136	6.044	0.01410	3118	6.000	0.01317	3098	5.956	0.01157	3058	5.870
480	753.15	0.01597	3203	6.137	0.01494	3188	6.095	0.01401	3170	6.055	0.01239	3135	5.975
500	773.15	0.01678	3267	6.221	0.01573	3253	6.182	0.01478	3238	6.144	0.01312	3207	6.070
520	793.15	0.01755	3327	6.300	0.01648	3315	6.263	0.01550	3301	6.227	0.01381	3274	6.159
540	813.15	0.01830	3384	6.373	0.01720	3373	6.338	0.01619	3361	6.304	0.01448	3337	6.239
550	823.15	0.01867	3412	6.407	0.01755	3401	6.373	0.01653	3390	6.339	0.01481	3367	6.276
560	833.15	0.01903	3440	6.441	0.01790	3429	6.407	0.01687	3418	6.374	0.01511	3396	6.312
580	853.15	0.01975	3494	6.507	0.01858	3484	6.474	0.01752	3474	6.442	0.01571	3454	6.382
600	873.15	0.02043	3549	6.572	0.01924	3540	6.540	0.01816	3530	6.508	0.01631	3512	6.449
620	893.15	0.02108	3603	6.633	0.01987	3594	6.601	0.01877	3586	6.571	0.01688	3568	6.513
640	913.15	0.02172	3656	6.691	0.02049	3648	6.660	0.01937	3640	6.631	0.01745	3624	6.576
650	923.15	0.02204	3683	6.720	0.02079	3675	6.689	0.01967	3667	6.660	0.01773	3653	6.606
660	933.15	0.02235	3709	6.748	0.02109	3702	6.718	0.01996	3695	6.689	0.01801	3681	6.635
680	953.15	0.02296	3763	6.804	0.02168	3756	6.774	0.02053	3749	6.745	0.01855	3735	6.692
700	973.15	0.02357	3815	6.858	0.02227	3809	6.828	0.02109	3803	6.800	0.01907	3789	6.747
720	993.15	0.02418	3867	6.911	0.02286	3861	6.881	0.02165	3855	6.853	0.01958	3843	6.801
740	1013.15	0.02480	3919	6.962	0.02345	3913	6.933	0.02221	3907	6.905	0.02008	3896	6.854
750	1023.15	0.02511	3944	6.987	0.02374	3938	6.959	0.02249	3933	6.931	0.02034	3922	6.880
760	1033.15	0.02541	3969	7.012	0.02402	3964	6.984	0.02276	3959	6.957	0.02059	3948	6.906
780	1053.15	0.02599	4021	7.062	0.02458	4016	7.034	0.02330	4011	7.007	0.02109	4000	6.957
800	1073.15	0.02658	4072	7.110	0.02514	4068	7.082	0.02383	4063	7.056	0.02160	4053	7.007
820	1093.15	0.02716	4123	7.158	0.02569	4119	7.130	0.02436	4114	7.104	0.02209	4105	7.055
840	1113.15	0.02774	4174	7.204	0.02624	4170	7.176	0.02489	4165	7.150	0.02257	4157	7.101
850	1123.15	0.02803	4200	7.227	0.02652	4195	7.199	0.02515	4191	7.172	0.02281	4182	7.124
860	1133.15	0.02832	4225	7.250	0.02679	4221	7.222	0.02541	4217	7.196	0.02305	4208	7.147
880	1153.15	0.02889	4276	7.294	0.02733	4272	7.266	0.02594	4268	7.240	0.02352	4260	7.192
900	1173.15	0.02945	4327	7.338	0.02787	4323	7.310	0.02645	4319	7.284	0.02399	4312	7.236
920	1193.12	0.03001	4378	7.381	0.02841	4374	7.353	0.02696	4371	7.328	0.02446	4364	7.280
940	1213.15	0.03056	4429	7.423	0.02893	4425	7.395	0.02746	4422	7.370	0.02492	4415	7.323
950	1223.15	0.03084	4454	7.444	0.02919	4451	7.416	0.02771	4447	7.391	0.02515	4441	7.344
960	1233.15	0.03112	4479	7.465	0.02945	4476	7.437	0.02796	4473	7.412	0.02538	4466	7.365
980	1253.15	0.03167	4530	7.505	0.02997	4527	7.479	0.02846	4524	7.454	0.02584	4518	7.407
1000	1273.15	0.03221	4581	7.545	0.03049	4578	7.519	0.02894	4575	7.494	0.02629	4569	7.447

Pressure p		230 bar			240 bar			250 bar			260 bar		
t	T	v	i	s	v	i	s	v	i	s	v	i	s
°C	K	m³/kg	kJ/kg	kJ/kg K	m³/kg	kJ/kg	kJ/kg K	m³/kg	kJ/kg	kJ/kg K	m³/kg	kJ/kg	kJ/kg K
0	273.15	0.0009889	23.2	0.0013	0.0009884	24.2	0.0013	0.0009880	25.2	0.0013	0.0009875	26.2	0.0013
20	293.15	0.0009916	105.5	0.2912	0.0009912	106.4	0.2911	0.0009908	107.3	0.2909	0.0009903	108.3	0.2908
40	313.15	0.0009977	188.0	0.5628	0.0009973	188.8	0.5625	0.0009969	189.7	0.5621	0.0009965	190.6	0.5618
50	323.15	0.0010020	229.3	0.6920	0.0010016	230.1	0.6916	0.0010012	231.0	0.6911	0.0010008	231.9	0.6907
60	333.15	0.0010069	270.7	0.8174	0.0010065	271.5	0.8169	0.0010061	272.3	0.8164	0.0010057	273.2	0.8159
80	353.15	0.0010186	353.5	1.0588	0.0010182	354.3	1.0582	0.0010178	355.1	1.0576	0.0010172	355.9	1.0570
100	373.15	0.0010325	436.5	1.2888	0.0010320	437.2	1.2881	0.0010316	438.0	1.2873	0.0010311	438.8	1.2865
120	393.15	0.0010485	520.1	1.5071	0.0010479	520.8	1.5062	0.0010475	521.5	1.5053	0.0010470	522.3	1.5045
140	413.15	0.0010666	604.2	1.716	0.0010660	604.9	1.715	0.0010654	605.6	1.714	0.0010649	606.3	1.713
150	423.15	0.0010765	646.6	1.817	0.0010759	647.3	1.816	0.0010753	647.9	1.815	0.0010747	648.6	1.814
160	433.15	0.0010871	689.3	1.916	0.0010864	689.9	1.915	0.0010858	690.5	1.914	0.0010851	691.1	1.913
180	453.15	0.0011102	775.2	2.109	0.0011095	775.7	2.108	0.0011087	776.3	2.107	0.0011080	776.9	2.105
200	473.15	0.0011366	862.1	2.296	0.0011357	862.6	2.295	0.0011349	863.0	2.293	0.0011340	863.5	2.292
220	493.15	0.0011668	950.6	2.478	0.0011658	950.9	2.477	0.0011648	951.3	2.475	0.0011638	951.7	2.474
240	513.15	0.0012017	1041.0	2.659	0.0012004	1041.3	2.657	0.0011992	1041.6	2.655	0.0011980	1041.8	2.653
250	523.15	0.0012211	1087.1	2.748	0.0012197	1087.3	2.746	0.0012183	1087.5	2.744	0.0012170	1087.7	2.742
260	533.15	0.0012421	1133.9	2.837	0.0012404	1134.0	2.835	0.0012388	1134.1	2.833	0.0012373	1134.2	2.831
280	553.15	0.0012904	1230.5	3.015	0.0012883	1230.3	3.011	0.0012863	1230.2	3.009	0.0012843	1230.1	3.006
300	573.15	0.0013505	1331.7	3.194	0.0013475	1331.2	3.190	0.0013446	1330.7	3.187	0.0013418	1330.3	3.183
320	593.15	0.001430	1439.4	3.380	0.001425	1438.3	3.375	0.001421	1437.3	3.371	0.001418	1436.4	3.367
340	613.15	0.001542	1559.8	3.581	0.001534	1557.3	3.573	0.001527	1555.3	3.567	0.001522	1553.4	3.561
350	623.15	0.001623	1629	3.694	0.001612	1625	3.684	0.001602	1621	3.675	0.001591	1618	3.667
360	633.15	0.001734	1709	3.821	0.001713	1702	3.807	0.001695	1696	3.794	0.001679	1691	3.782
380	653.15	0.00480	2327	4.802	0.00265	2050	4.348	0.00224	1926	4.149	0.00210	1894	4.096
400	673.15	0.00750	2690	5.324	0.00676	2638	5.236	0.00602	2579	5.137	0.00529	2511	5.028
420	693.15	0.00888	2839	5.546	0.00821	2807	5.484	0.00758	2772	5.420	0.00699	2733	5.355
440	713.15	0.00993	2949	5.704	0.00929	2924	5.655	0.00868	2896	5.604	0.00812	2867	5.553
450	723.15	0.01041	2994	5.769	0.00977	2971	5.723	0.00917	2947	5.677	0.00861	2922	5.631
460	733.15	0.01086	3037	5.828	0.01021	3016	5.785	0.00962	2994	5.743	0.00907	2972	5.701
480	753.15	0.01167	3117	5.936	0.01102	3098	5.898	0.01043	3080	5.860	0.00987	3061	5.823
500	773.15	0.01239	3191	6.034	0.01174	3174	5.999	0.01113	3157	5.965	0.01056	3141	5.932
520	793.15	0.01306	3260	6.125	0.01240	3245	6.092	0.01179	3230	6.059	0.01121	3215	6.028
540	813.15	0.01372	3324	6.207	0.01303	3312	6.176	0.01242	3299	6.145	0.01183	3286	6.116
550	823.15	0.01404	3355	6.246	0.01334	3343	6.216	0.01272	3331	6.186	0.01213	3319	6.157
560	833.15	0.01434	3385	6.283	0.01364	3374	6.254	0.01302	3362	6.225	0.01242	3350	6.196
580	853.15	0.01492	3444	6.353	0.01422	3434	6.325	0.01358	3423	6.298	0.01297	3412	6.270
600	873.15	0.01550	3502	6.421	0.01478	3493	6.394	0.01413	3483	6.367	0.01350	3473	6.340
620	893.15	0.01606	3559	6.485	0.01532	3551	6.459	0.01465	3542	6.433	0.01402	3533	6.408
640	913.15	0.01662	3616	6.549	0.01586	3608	6.523	0.01517	3600	6.498	0.01452	3592	6.473
650	923.15	0.01689	3645	6.580	0.01612	3637	6.554	0.01542	3629	6.529	0.01477	3622	6.505
660	933.15	0.01716	3673	6.609	0.01638	3666	6.584	0.01566	3658	6.560	0.01501	3651	6.536
680	953.15	0.01767	3728	6.666	0.01687	3721	6.641	0.01615	3714	6.618	0.01547	3707	6.595
700	973.15	0.01817	3783	6.721	0.01735	3776	6.697	0.01662	3770	6.674	0.01593	3763	6.651
720	993.15	0.01867	3837	6.776	0.01784	3830	6.752	0.01709	3824	6.729	0.01638	3818	6.707
740	1013.15	0.01917	3890	6.830	0.01832	3884	6.806	0.01756	3878	6.783	0.01684	3872	6.761
750	1023.15	0.01941	3911	6.856	0.01856	3910	6.832	0.01779	3905	6.810	0.01706	3899	6.788
760	1033.15	0.01966	3942	6.882	0.01880	3937	6.858	0.01802	3932	6.836	0.01728	3926	6.814
780	1053.15	0.02014	3995	6.933	0.01927	3990	6.910	0.01847	3985	6.888	0.01772	3979	6.866
800	1073.15	0.02062	4048	6.983	0.01973	4043	6.960	0.01891	4038	6.938	0.01814	4033	6.916
820	1093.15	0.02109	4100	7.031	0.02018	4095	7.008	0.01935	4090	6.986	0.01857	4086	6.964
840	1113.15	0.02155	4152	7.078	0.02063	4148	7.056	0.01979	4143	7.034	0.01899	4139	7.012
850	1123.15	0.02178	4178	7.102	0.02085	4174	7.080	0.02000	4169	7.058	0.01920	4165	7.036
860	1133.15	0.02201	4204	7.125	0.02107	4200	7.103	0.02022	4196	7.081	0.01940	4192	7.069
880	1153.15	0.02248	4256	7.170	0.02151	4252	7.148	0.02064	4248	7.126	0.01981	4244	7.105
900	1173.15	0.02293	4308	7.214	0.02195	4304	7.192	0.02106	4300	7.170	0.02022	4297	7.150
920	1193.15	0.02338	4360	7.258	0.02239	4356	7.236	0.02147	4353	7.214	0.02062	4349	7.194
940	1213.15	0.02382	4412	7.301	0.02282	4408	7.279	0.02188	4405	7.258	0.02102	4401	7.238
950	1223.15	0.02404	4447	7.322	0.02303	4434	7.300	0.02208	4431	7.279	0.02122	4428	7.259
960	1233.15	0.02426	4463	7.343	0.02324	4460	7.321	0.02229	4457	7.300	0.02142	4454	7.280
980	1353.15	0.02470	4515	7.385	0.02366	4512	7.363	0.02270	4509	7.342	0.02181	4506	7.322
1000	1273.15	0.02513	4566	7.424	0.02408	4563	7.403	0.02310	4560	7.383	0.02220	4557	7.363

Pressure p		280 bar			300 bar			350 bar			400 bar		
t	T	v	i	s	v	i	s	v	i	s	v	i	s
°C	K	m³/kg	kJ/kg	kJ/kg K	m³/kg	kJ/kg	kJ/kg K	m³/kg	kJ/kg	kJ/kg K	m³/kg	kJ/kg	kJ/kg K
0	273.15	0.0009866	28.1	0.0013	0.0009857	30.1	0.0013	0.0009833	34.9	0.0008	0.0009810	39.7	0.0004
20	293.15	0.0009894	110.1	0.2905	0.0009886	112.0	0.2902	0.0009864	116.6	0.2885	0.0009843	121.2	0.2873
40	313.15	0.0009957	192.4	0.5610	0.0009949	194.1	0.5603	0.0009928	198.5	0.5577	0.0009908	202.9	0.5557
50	323.15	0.0010000	233.6	0.6898	0.0009992	235.3	0.6889	0.0009971	239.6	0.6863	0.0009951	243.8	0.6840
60	333.15	0.0010049	274.9	0.8150	0.0010041	276.5	0.8140	0.0010020	280.7	0.8114	0.0010000	284.8	0.8090
80	353.15	0.0010164	357.5	1.0558	0.0010156	359.1	1.0545	0.0010133	363.1	1.0518	0.0010113	367.0	1.0489
100	373.15	0.0010302	440.4	1.2850	0.0010293	441.9	1.2834	0.0010269	445.7	1.2796	0.0010248	449.5	1.2757
120	393.15	0.0010459	523.7	1.5030	0.0010450	525.1	1.5014	0.0010425	528.8	1.4965	0.0010401	532.4	1.4919
140	413.15	0.0010638	607.7	1.711	0.0010626	609.0	1.709	0.0010598	612.4	1.704	0.0010572	615.8	1.6994
150	423.15	0.0010735	649.9	1.812	0.0010722	651.2	1.810	0.0010693	654.5	1.805	0.0010665	657.7	1.800
160	433.15	0.0010838	692.4	1.911	0.0010825	693.6	1.908	0.0010794	696.7	1.903	0.0010764	699.8	1.898
180	453.15	0.0011065	778.0	2.103	0.0011050	779.1	2.100	0.0011015	781.8	2.095	0.0010981	784.6	2.089
200	473.15	0.0011323	864.5	2.289	0.0011305	865.4	2.287	0.0011264	867.9	2.280	0.0011225	870.5	2.274
220	493.15	0.0011617	952.5	2.471	0.0011597	953.3	2.468	0.0011548	955.3	2.461	0.0011500	957.5	2.454
240	513.15	0.0011955	1042.4	2.650	0.0011931	1042.9	2.647	0.0011872	1044.4	2.638	0.0011814	1045.9	2.630
250	523.15	0.0012141	1088.1	2.738	0.0012115	1088.5	2.735	0.0012051	1089.7	2.726	0.0011988	1090.9	2.717
260	533.15	0.0012342	1134.5	2.826	0.0012313	1134.7	2.822	0.0012241	1135.6	2.813	0.0012174	1136.6	2.803
280	553.15	0.0012840	1230.0	3.001	0.0012764	1229.9	2.996	0.0012674	1230.1	2.985	0.0012589	1230.2	2.975
300	573.15	0.0013364	1329.6	3.178	0.0013311	1329.0	3.171	0.0013190	1327.8	3.157	0.0013079	1326.7	3.145
320	593.15	0.001410	1434.8	3.359	0.001403	1433.2	3.351	0.001383	1429.6	3.332	0.001367	1426.7	3.315
340	613.15	0.001509	1549.9	3.549	0.001496	1546.8	3.539	0.001466	1539.4	3.513	0.001442	1533.1	3.493
350	623.15	0.001573	1613	3.653	0.001556	1608	3.640	0.001520	1598.1	3.609	0.001489	1589.5	3.585
360	633.15	0.001654	1682	3.763	0.001634	1676	3.747	0.001585	1659.4	3.707	0.001545	1648.1	3.679
380	653.15	0.00196	1860	4.038	0.001887	1836	3.995	0.001766	1797	3.921	0.001691	1773	3.873
400	673.15	0.00387	2348	4.769	0.00283	2155	4.476	0.00213	1991	4.212	0.001922	1934	4.115
420	693.15	0.00589	2650	5.220	0.00493	2559	5.070	0.00310	2309	4.680	0.002371	2154	4.433
440	713.15	0.00710	2806	5.448	0.00621	2743	5.340	0.00440	2573	5.060	0.003211	2401	4.790
450	723.15	0.00760	2870	5.539	0.00672	2816	5.446	0.00494	2670	5.200	0.003691	2518	4.958
460	733.15	0.00807	2926	5.619	0.00719	2880	5.536	0.00542	2753	5.318	0.004145	2613	5.093
480	753.15	0.00887	3024	5.753	0.00800	2986	5.682	0.00625	2884	5.502	0.004961	2770	5.312
500	773.15	0.00956	3107	5.865	0.00869	3073	5.799	0.00694	2986	5.638	0.005627	2898	5.482
520	793.15	0.01019	3186	5.966	0.00932	3155	5.906	0.00754	3081	5.759	0.006209	3006	5.620
540	813.15	0.01080	3258	6.058	0.00989	3232	6.001	0.00808	3166	5.866	0.006737	3101	5.739
550	823.15	0.01108	3293	6.101	0.01016	3268	6.045	0.00834	3206	5.916	0.006983	3144	5.793
560	833.15	0.01135	3327	6.141	0.01043	3303	6.088	0.00858	3244	5.963	0.007221	3185	5.844
580	853.15	0.01187	3391	6.218	0.01094	3370	6.167	0.00906	3317	6.049	0.007673	3264	5.939
600	873.15	0.01239	3453	6.289	0.01144	3434	6.242	0.00951	3386	6.129	0.008094	3338	6.025
620	893.15	0.01289	3515	6.359	0.01191	3497	6.312	0.00993	3453	6.204	0.008493	3409	6.104
640	913.15	0.01337	3576	6.425	0.01237	3559	6.379	0.01035	3518	6.275	0.008873	3477	6.180
650	923.15	0.01360	3606	6.457	0.01259	3590	6.412	0.01056	3549	6.310	0.009057	3510	6.217
660	933.15	0.01383	3635	6.489	0.01281	3620	6.444	0.01076	3580	6.344	0.009238	3542	6.253
680	953.15	0.01427	3693	6.549	0.01323	3678	6.507	0.01116	3641	6.409	0.009589	3606	6.321
700	973.15	0.01471	3749	6.607	0.01365	3736	6.566	0.01153	3701	6.470	0.009937	3668	6.384
720	993.15	0.01514	3805	6.663	0.01405	3792	6.623	0.01189	3760	6.528	0.01026	3728	6.446
740	1013.15	0.01556	3860	6.719	0.01445	3848	6.679	0.01225	3818	6.585	0.01058	3788	6.505
750	1023.15	0.01577	3888	6.746	0.01465	3876	6.706	0.01243	3847	6.613	0.01074	3818	6.534
760	1033.15	0.01598	3915	6.773	0.01485	3904	6.733	0.01260	3875	6.641	0.01090	3848	6.562
780	1053.15	0.01639	3969	6.825	0.01523	3958	6.785	0.01294	3932	6.696	0.01121	3905	6.618
800	1073.15	0.01679	4023	6.875	0.01562	4013	6.837	0.01328	3988	6.750	0.01152	3962	6.673
820	1093.15	0.01720	4076	6.924	0.01600	4067	6.887	0.01360	4043	6.802	0.01182	4019	6.725
840	1113.15	0.01759	4129	6.972	0.01637	4121	6.935	0.01394	4098	6.852	0.01212	4075	6.775
850	1123.15	0.01779	4156	6.996	0.01655	4147	6.959	0.01410	4125	6.876	0.01227	4103	6.800
860	1133.15	0.01798	4183	7.019	0.01674	4174	6.983	0.01426	4153	6.900	0.01241	4131	6.824
880	1153.15	0.01837	4236	7.065	0.01710	4228	7.029	0.01458	4207	6.947	0.01268	4186	6.872
900	1173.15	0.01874	4289	7.111	0.01746	4281	7.075	0.01489	4261	6.993	0.01296	4242	6.918
920	1193.15	0.01911	4342	7.155	0.01782	4334	7.119	0.01520	4315	7.038	0.01325	4296	6.964
940	1213.15	0.01948	4394	7.199	0.01817	4387	7.163	0.01551	4369	7.082	0.01353	4351	7.009
950	1223.15	0.01967	4421	7.221	0.01834	4414	7.185	0.01567	4396	7.104	0.01366	4378	7.031
960	1233.15	0.01986	4447	7.242	0.01851	4440	7.206	0.01581	4422	7.125	0.01380	4405	7.053
980	1253.15	0.02023	4499	7.284	0.01885	4492	7.248	0.01611	4475	7.168	0.01408	4459	7.096
1000	1273.15	0.02059	4551	7.325	0.01919	4544	7.290	0.01641	4529	7.210	0.01434	4513	7.138

Table 46-19 Properties of Superheated Steam (H₂O) in International System of Units—SI (Continued)

Pressure p		450 bar			500 bar			550 bar			600 bar		
t	T	v	i	s	v	i	s	v	i	s	v	i	s
°C	K	m³/kg	kJ/kg	kJ/kg K	m³/kg	kJ/kg	kJ/kg K	m³/kg	kJ/kg	kJ/kg K	m³/kg	kJ/kg	kJ/kg K
0	273.15	0.0009789	44.5	0.0000	0.0009766	49.3	—0.0004	0.0009745	54.0	—0.0008	0.0009722	58.8	—0.0008
20	293.15	0.0009823	125.7	0.2856	0.0009802	130.2	0.2843	0.0009782	134.7	0.2830	0.0009761	139.2	0.2820
40	313.15	0.0009888	207.2	0.5535	0.0009868	211.5	0.5515	0.0009849	215.8	0.5498	0.0009829	220.0	0.5477
50	323.15	0.0009931	248.0	0.6817	0.0009911	252.2	0.6794	0.0009892	256.4	0.6773	0.0009872	260.5	0.6749
60	333.15	0.0009980	288.9	0.8064	0.0009960	293.0	0.8040	0.0009940	297.1	0.8014	0.0009920	301.1	0.7988
80	353.15	0.0010092	371.0	1.0455	0.0010071	374.9	1.0429	0.0010051	378.8	1.0397	0.0010031	382.5	1.0367
100	373.15	0.0010226	453.3	1.2720	0.0010204	457.1	1.2687	0.0010182	460.8	1.2653	0.0010161	464.4	1.2623
120	393.15	0.0010377	536.0	1.4879	0.0010354	539.6	1.4839	0.0010330	543.1	1.4805	0.0010307	546.7	1.4767
140	413.15	0.0010546	619.2	1.6944	0.0010520	622.6	1.6901	0.0010495	626.0	1.6860	0.0010469	629.4	1.6814
150	423.15	0.0010638	661.0	1.795	0.0010610	664.3	1.790	0.0010584	667.6	1.786	0.0010557	670.9	1.781
160	433.15	0.0010735	703.0	1.893	0.0010705	706.2	1.888	0.0010678	709.4	1.883	0.0010650	712.5	1.878
180	453.15	0.0010948	787.5	2.083	0.0010915	790.5	2.078	0.0010882	793.4	2.073	0.0010850	796.3	2.067
200	473.15	0.0011187	873.0	2.268	0.0011149	875.6	2.261	0.0011110	878.2	2.255	0.0011074	880.7	2.249
220	493.15	0.0011455	959.7	2.447	0.0011411	961.8	2.440	0.0011368	963.9	2.433	0.0011326	966.0	2.426
240	513.15	0.0011761	1047.6	2.622	0.0011709	1049.2	2.614	0.0011659	1051.0	2.606	0.0011610	1052.6	2.598
250	523.15	0.0011930	1092.2	2.708	0.0011873	1093.6	2.700	0.0011818	1095.2	2.692	0.0011765	1096.6	2.683
260	533.15	0.0012109	1137.6	2.794	0.0012047	1138.6	2.784	0.0011987	1139.9	2.776	0.0011930	1141.2	2.767
280	553.15	0.0012508	1230.3	2.964	0.0012431	1230.7	2.952	0.0012361	1231.5	2.942	0.0012294	1232.2	2.933
300	573.15	0.0012976	1326.2	3.132	0.0012877	1326.0	3.120	0.0012787	1326.0	3.108	0.0012703	1326.0	3.097
320	593.15	0.001353	1425.1	3.300	0.001340	1424.0	3.286	0.001329	1423.1	3.273	0.001319	1422.3	3.261
340	613.15	0.001422	1528.3	3.472	0.001405	1525.7	3.455	0.001389	1523.5	3.439	0.001376	1521.5	3.425
350	623.15	0.001463	1582.6	3.560	0.001443	1578.4	3.541	0.001425	1575.2	3.523	0.001409	1572.2	3.507
360	633.15	0.001511	1639.1	3.651	0.001486	1632.5	3.628	0.001465	1627.7	3.608	0.001445	1623.8	3.589
380	653.15	0.001635	1758	3.837	0.001595	1747	3.807	0.001560	1739	3.781	0.001530	1732	3.758
400	673.15	0.001816	1902	4.052	0.001741	1878	4.004	0.001682	1860	3.965	0.001638	1847	3.933
420	693.15	0.002087	2078	4.308	0.001939	2030	4.226	0.001848	1996	4.165	0.001778	1973	4.117
440	713.15	0.002595	2277	4.592	0.002272	2199	4.467	0.002090	2148	4.378	0.001966	2111	4.312
450	723.15	0.002920	2383	4.745	0.002492	2290	4.594	0.002247	2228	4.491	0.002085	2183	4.415
460	733.15	0.003274	2485	4.886	0.002742	2381	4.722	0.002430	2310	4.607	0.002226	2258	4.521
480	753.15	0.004004	2660	5.133	0.003319	2558	4.967	0.002870	2480	4.843	0.002564	2418	4.742
500	773.15	0.004637	2809	5.332	0.003892	2722	5.189	0.003350	2646	5.065	0.002955	2579	4.957
520	793.15	0.005191	2931	5.488	0.004413	2857	5.364	0.003816	2785	5.249	0.003362	2720	5.142
540	813.15	0.005693	3035	5.619	0.004891	2969	5.507	0.004257	2906	5.401	0.003759	2844	5.301
550	823.15	0.005929	3082	5.678	0.005114	3020	5.571	0.004464	2961	5.469	0.003951	2902	5.372
560	833.15	0.006157	3127	5.735	0.005327	3069	5.631	0.004664	3013	5.533	0.004137	2957	5.439
580	853.15	0.006585	3210	5.837	0.005731	3160	5.740	0.005047	3110	5.649	0.004493	3060	5.562
600	873.15	0.006982	3289	5.929	0.006109	3243	5.839	0.005405	3198	5.753	0.004828	3152	5.672
620	893.15	0.007355	3364	6.014	0.006462	3322	5.929	0.005741	3280	5.849	0.005148	3238	5.772
640	913.15	0.007712	3436	6.093	0.006797	3397	6.013	0.006058	3358	5.937	0.005449	3319	5.864
650	923.15	0.007886	3471	6.132	0.006959	3433	6.053	0.006211	3396	5.978	0.005593	3359	5.907
660	933.15	0.008057	3505	6.169	0.007118	3468	6.091	0.006360	3432	6.018	0.005733	3397	5.949
680	953.15	0.008389	3571	6.240	0.007426	3537	6.165	0.006647	3503	6.095	0.006007	3470	6.028
700	973.15	0.008707	3636	6.306	0.007721	3603	6.235	0.006926	3571	6.167	0.006271	3540	6.103
720	993.15	0.009012	3699	6.369	0.008010	3669	6.301	0.007197	3639	6.235	0.006525	3608	6.173
740	1013.15	0.009311	3761	6.431	0.008290	3732	6.364	0.007459	3704	6.299	0.006772	3675	6.240
750	1023.15	0.009458	3791	6.461	0.008426	3763	6.395	0.007587	3736	6.330	0.006892	3708	6.272
760	1033.15	0.009602	3821	6.490	0.008560	3794	6.425	0.007713	3768	6.361	0.007009	3741	6.303
780	1053.15	0.009885	3880	6.547	0.008823	3855	6.483	0.007959	3830	6.421	0.007239	3806	6.364
800	1073.15	0.01017	3939	6.602	00.09081	3915	6.539	0.008198	3891	6.479	0.007464	3868	6.423
820	1093.15	0.01045	3997	6.655	0.009333	3975	6.593	0.008432	3952	6.535	0.007684	3929	6.480
840	1113.15	0.01071	4054	6.706	0.009581	4033	6.645	0.008663	4012	6.589	0.007901	3991	6.535
850	1123.15	0.01084	4083	6.731	0.009705	4062	6.671	0.008777	4042	6.615	0.008008	4021	6.562
860	1133.15	0.01097	4111	6.756	0.009827	4091	6.696	0.008891	4071	6.641	0.008113	4051	6.589
880	1153.15	0.01123	4167	6.805	0.010070	4148	6.476	0.009114	4129	6.692	0.008324	4110	6.640
900	1173.15	0.01149	4224	6.853	0.01030	4206	6.794	0.009333	4187	6.740	0.008531	4169	6.690
920	1193.15	0.01174	4279	6.899	0.01053	4262	6.841	0.009551	4244	6.788	0.008735	4227	6.738
940	1213.15	0.01199	4335	6.945	0.01077	4318	6.887	0.009765	4301	6.834	0.008935	4285	6.784
950	1223.15	0.01211	4362	6.967	0.01088	4346	6.910	0.009870	4329	6.857	0.009034	4313	6.807
960	1233.15	0.01223	4390	6.989	0.01099	4374	6.932	0.009975	4357	6.879	0.009133	4342	6.830
980	1253.15	0.01247	4444	7.033	0.01121	4429	6.975	0.01018	4413	6.923	0.009329	4399	6.874
1000	1273.15	0.01271	4498	7.075	0.01143	4483	7.018	0.01038	4468	6.966	0.009522	4455	6.918

Table 46-20 Properties of Superheated Steam (H$_2$O) in International System of Units—SI (*Continued*)

Pressure p		650 bar			700 bar			750 bar			800 bar		
t	T	v	i	s	v	i	s	v	i	s	v	i	s
°C	K	m³/kg	kJ/kg	kJ/kg K	m³/kg	kJ/kg	kJ/kg K	m³/kg	kJ/kg	kJ/kg K	m³/kg	kJ/kg	kJ/kg K
0	273.15	0.0009702	63.5	—0.0008	0.0009681	68.2	—0.0013	0.0009661	72.8	—0.0013	0.0009641	77.5	—0.0013
20	293.15	0.0009741	143.7	0.2807	0.0009722	148.1	0.2794	0.0009703	152.5	0.2780	0.0009684	156.9	0.2766
40	313.15	0.0009810	224.2	0.5459	0.0009791	228.3	0.5438	0.0009773	232.5	0.5415	0.0009755	236.6	0.5397
50	323.15	0.0009853	264.6	0.6729	0.0009834	268.6	0.6706	0.0009816	272.6	0.6682	0.0009798	276.7	0.6660
60	333.15	0.0009901	305.1	0.7963	0.0009882	309.0	0.7938	0.0009864	312.9	0.7914	0.0009846	316.9	0.7888
80	353.15	0.0010012	386.3	1.0334	0.0009991	390.2	1.0300	0.0009972	394.0	1.0271	0.0009954	397.8	1.0245
100	373.15	0.0010140	468.1	1.2588	0.0010119	471.8	1.2554	0.0010099	475.6	1.2519	0.0010080	479.3	1.2486
120	393.15	0.0010284	550.3	1.4725	0.0010262	553.9	1.4690	0.0010242	557.5	1.4651	0.0010221	561.0	1.4616
140	413.15	0.0010445	632.8	1.6772	0.0010421	636.2	1.6727	0.0010398	639.6	1.6685	0.0010375	643.0	1.6646
150	423.15	0.0010532	674.1	1.776	0.0010507	677.4	1.771	0.0010482	680.8	1.767	0.0010458	648.1	1.763
160	433.15	0.0010624	715.6	1.873	0.0010597	718.8	1.868	0.0010571	722.0	1.863	0.0010545	725.2	1.859
180	453.15	0.0010820	799.1	2.061	0.0010791	801.9	2.056	0.0010762	804.8	2.050	0.0010732	807.8	2.045
200	473.15	0.0011039	883.0	2.243	0.0011006	885.5	2.237	0.0010972	888.1	2.230	0.0010941	891.0	2.225
220	493.15	0.0011286	967.8	2.419	0.0011246	970.0	2.412	0.0011209	972.3	2.405	0.0011173	975.0	2.399
240	513.15	0.0011563	1054.1	2.591	0.0011517	1055.9	2.583	0.0011474	1058.0	2.575	0.0011432	1060.1	2.568
250	523.15	0.0011715	1098.1	2.675	0.0011666	1099.7	2.667	0.0011619	1101.5	2.659	0.0011573	1103.2	2.651
260	533.15	0.0011876	1142.6	2.758	0.0011824	1144.0	2.749	0.0011773	1145.4	2.741	0.0011722	1146.7	2.733
280	553.15	0.0012229	1233.0	2.923	0.0012166	1233.9	2.913	0.0012105	1234.6	2.904	0.0012046	1235.2	2.895
300	573.15	0.0012627	1326.0	3.087	0.0012550	1326.0	3.076	0.0012474	1326.0	3.065	0.0012405	1326.0	3.055
320	593.15	0.001308	1421.2	3.249	0.001299	1420.4	3.236	0.001291	1419.7	3.224	0.001283	1419.2	3.213
340	613.15	0.001363	1519.2	3.410	0.001352	1517.5	3.396	0.001341	1515.8	3.382	0.001331	1514.4	3.369
350	623.15	0.001395	1569.5	3.491	0.001382	1567.0	3.476	0.001369	1564.7	3.461	0.001358	1562.7	3.447
360	633.15	0.001429	1620.6	3.572	0.001415	1617.2	3.555	0.001399	1614.2	3.540	0.001386	1611.5	3.524
380	653.15	0.001507	1727	3.737	0.001487	1721	3.716	0.001467	1717	3.697	0.001449	1712	3.679
400	673.15	0.001603	1839	3.906	0.001572	1831	3.880	0.001545	1824	3.857	0.001522	1817	3.835
420	693.15	0.001722	1958	4.080	0.001678	1945	4.048	0.001641	1935	4.019	0.001610	1926	3.992
440	713.15	0.001879	2085	4.262	0.001813	2065	4.219	0.001759	2050	4.183	0.001714	2037	4.153
450	723.15	0.001976	2151	4.356	0.001894	2128	4.308	0.001828	2110	4.267	0.001774	2095	4.234
460	733.15	0.002088	2220	4.453	0.001986	2193	4.400	0.001906	2172	4.355	0.001842	2155	4.317
480	753.15	0.002358	2369	4.661	0.002205	2331	4.594	0.002090	2302	4.538	0.002001	2277	4.489
500	773.15	0.002672	2521	4.865	0.002465	2474	4.785	0.002310	2436	4.718	0.002189	2404	4.660
520	793.15	0.003012	2659	5.044	0.002755	2608	4.960	0.002561	2565	4.888	0.002404	2528	4.824
540	813.15	0.003368	2785	5.205	0.003066	2733	5.120	0.002830	2689	5.046	0.002637	2649	4.978
550	823.15	0.003545	2844	5.279	0.003224	2793	5.195	0.002968	2748	5.121	0.002759	2708	5.052
560	833.15	0.003718	2901	5.349	0.003380	2851	5.267	0.003108	2806	5.193	0.002883	2765	5.124
580	853.15	0.004048	3009	5.479	0.003683	2961	5.401	0.003382	2917	5.329	0.003133	2877	5.262
600	873.15	0.004361	3107	5.594	0.003973	3063	5.522	0.003651	3022	5.453	0.003382	2983	5.388
620	893.15	0.004659	3196	5.700	0.004252	3157	5.631	0.003910	3118	5.566	0.003623	3083	5.504
640	913.15	0.004942	3281	5.796	0.004519	3243	5.730	0.004161	3208	5.668	0.003860	3175	5.609
650	923.15	0.005079	3321	5.841	0.004648	3285	5.777	0.004284	3251	5.716	0.003975	3219	5.658
660	933.15	0.005212	3361	5.884	0.004774	3326	5.821	0.004404	3292	5.762	0.004089	3261	5.705
680	953.15	0.005471	3437	5.965	0.005021	3404	5.905	0.004636	3372	5.848	0.004308	3342	5.793
700	973.15	0.005721	3510	6.042	0.005257	3479	5.984	0.004861	3449	5.928	0.004520	3421	5.876
720	993.15	0.005962	3580	6.114	0.005486	3551	6.058	0.005078	3523	6.004	0.004726	3498	5.953
740	1013.15	0.006195	3648	6.183	0.005706	3622	6.128	0.005286	3596	6.076	0.004924	3572	6.027
750	1023.15	0.006308	3682	6.216	0.005813	3657	6.162	0.005387	3632	6.111	0.005020	3609	6.062
760	1033.15	0.006419	3716	6.248	0.005917	3691	6.195	0.005486	3667	6.145	0.005113	3645	6.097
780	1053.15	0.006637	3781	6.311	0.006122	3758	6.259	0.005682	3737	6.211	0.005299	3715	6.164
800	1073.15	0.006849	3845	6.371	0.006324	3824	6.320	0.005873	3803	6.273	0.005481	3783	6.228
820	1093.15	0.007057	3909	6.429	0.006521	3888	6.380	0.006061	3869	6.334	0.005660	3850	6.290
840	1113.15	0.007261	3971	6.485	0.006715	3952	6.437	0.006244	3934	6.392	0.005835	3915	6.349
850	1123.15	0.007362	4002	6.513	0.006811	3983	6.465	0.006335	3966	6.421	0.005922	3948	6.378
860	1133.15	0.007461	4032	6.540	0.006905	4014	6.492	0.006425	3997	6.449	0.006008	3980	6.407
880	1153.15	0.007660	4092	6.592	0.007091	4076	6.546	0.006603	4060	6.503	0.006177	4044	6.462
900	1173.15	0.007855	4152	6.642	0.007276	4136	6.597	0.006777	4122	6.554	0.006344	4106	6.514
920	1193.15	0.008047	4211	6.691	0.007458	4196	6.647	0.006951	4182	6.605	0.006508	4168	6.566
940	1213.15	0.008235	4270	6.738	0.007637	4256	6.694	0.007120	4243	6.654	0.006669	4230	6.615
950	1223.15	0.008328	4299	6.761	0.007726	4285	6.718	0.007204	4272	6.677	0.006750	4260	6.639
960	1233.15	0.008422	4328	6.784	0.007815	4314	6.741	0.007288	4302	6.700	0.006830	4289	6.662
980	1253.15	0.008607	4385	6.829	0.007990	4372	6.786	0.007454	4361	6.746	0.006987	4349	6.708
1000	1273.15	0.008789	4443	6.873	0.008162	4431	6.831	0.007618	4419	6.791	0.007143	4409	6.753

Pressure p		850 bar			900 bar			950 bar			970 bar		
t	T	v	i	s	v	i	s	v	i	s	v	i	s
°C	K	m³/kg	kJ/kg	kJ/kg K	m³/kg	kJ/kg	kJ/kg K	m³/kg	kJ/kg	kJ/kg K	m³/kg	kJ/kg	kJ/kg K
0	273.15	0.0009620	82.1	−0.0017	0.0009601	86.7	−0.0021	0.0009582	91.3	−0.0025	0.0009574	92.9	−0.0027
20	293.15	0.0009665	161.2	0.2758	0.0009647	165.5	0.2753	0.0009629	169.9	0.2746	0.0009621	171.5	0.2744
40	313.15	0.0009737	240.7	0.5381	0.0009719	244.9	0.5371	0.0009702	249.1	0.5355	0.0009694	250.4	0.5351
50	323.15	0.0009780	280.7	0.6644	0.0009763	284.8	0.6628	0.0009746	289.0	0.6609	0.0009738	290.2	0.6604
60	333.15	0.0009828	320.9	0.7870	0.0009811	324.9	0.7849	0.0009794	329.1	0.7826	0.0009786	330.2	0.7820
80	353.15	0.0009936	401.7	1.0217	0.0009918	405.7	1.0191	0.0009900	409.6	1.0166	0.0009892	410.7	1.0158
100	373.15	0.0010060	483.1	1.2451	0.0010041	486.9	1.2421	0.0010022	490.7	1.2389	0.0010014	491.8	1.2379
120	393.15	0.0010200	564.5	1.4580	0.0010180	568.2	1.4541	0.0010159	571.9	1.4507	0.0010150	573.0	1.4496
140	413.15	0.0010353	646.4	1.6605	0.0010331	649.8	1.6560	0.0010309	653.1	1.6526	0.0010299	654.2	1.6513
150	423.15	0.0010435	687.4	1.758	0.0010411	690.7	1.754	0.0010389	693.8	1.750	0.0010379	694.9	1.748
160	433.15	0.0010520	728.5	1.854	0.0010495	731.7	1.850	0.0010472	734.7	1.845	0.0010462	735.8	1.843
180	453.15	0.0010705	810.9	2.040	0.0010678	814.0	2.035	0.0010651	816.9	2.030	0.0010640	818.0	2.028
200	473.15	0.0010910	894.0	2.220	0.0010879	896.9	2.214	0.0010849	899.5	2.208	0.0010837	900.5	2.206
220	493.15	0.0011137	977.8	2.393	0.0011102	980.4	2.386	0.0011067	982.6	2.380	0.0011054	983.4	2.378
240	513.15	0.0011392	1062.4	2.562	0.0011351	1064.6	2.554	0.0011312	1066.4	2.546	0.0011297	1067.0	2.544
250	523.15	0.0011529	1105.1	2.644	0.0011485	1107.1	2.636	0.0011444	1108.7	2.628	0.0011428	1109.3	2.625
260	533.15	0.0011674	1148.2	2.725	0.0011627	1149.9	2.717	0.0011583	1151.3	2.708	0.0011565	1151.8	2.705
280	553.15	0.0011990	1235.9	2.885	0.0011935	1237.1	2.876	0.0011884	1238.0	2.866	0.0011864	1238.4	2.862
300	573.15	0.0012340	1326.0	3.044	0.0012279	1326.4	3.033	0.001222	1326.8	3.022	0.0012201	1326.9	3.018
320	593.15	0.001275	1418.5	3.200	0.001267	1417.9	3.187	0.001261	1417.4	3.175	0.001258	1417.3	3.171
340	613.15	0.001322	1512.7	3.354	0.001312	1511.3	3.340	0.001305	1509.8	3.326	0.001301	1509.4	3.321
350	623.15	0.001348	1560.5	3.431	0.001337	1558.7	3.416	0.001328	1556.8	3.401	0.001324	1556.2	3.396
360	633.15	0.001374	1608.9	3.508	0.001363	1606.5	3.491	0.001353	1604.2	3.476	0.001349	1603.5	3.470
380	653.15	0.001432	1708	3.660	0.001419	1704	3.641	0.001407	1701	3.624	0.001401	1699	3.618
400	673.15	0.001500	1811	3.813	0.001482	1805	3.791	0.001465	1800	3.772	0.001459	1798	3.764
420	693.15	0.001581	1917	3.967	0.001555	1909	3.943	0.001533	1901	3.919	0.001524	1898	3.910
440	713.15	0.001676	2027	4.124	0.001642	2016	4.096	0.001612	2005	4.067	0.001601	2001	4.056
450	723.15	0.001730	2083	4.203	0.001691	2071	4.172	0.001657	2058	4.141	0.001645	2053	4.129
460	733.15	0.001790	2140	4.283	0.001744	2126	4.249	0.001706	2111	4.215	0.001693	2106	4.202
480	753.15	0.001928	2256	4.444	0.001868	2238	4.403	0.001819	2219	4.363	0.001800	2212	4.348
500	773.15	0.002094	2375	4.606	0.002015	2351	4.557	0.001950	2328	4.513	0.001926	2320	4.496
520	793.15	0.002281	2495	4.765	0.002181	2466	4.712	0.002098	2441	4.665	0.002068	2430	4.647
540	813.15	0.002486	2614	4.918	0.002361	2583	4.864	0.002260	2556	4.815	0.002223	2545	4.796
550	823.15	0.002592	2673	4.992	0.002456	2641	4.938	0.002344	2612	4.888	0.002304	2601	4.869
560	833.15	0.002702	2730	5.064	0.002555	2698	5.009	0.002432	2668	4.959	0.002389	2657	4.940
580	853.15	0.002928	2841	5.201	0.002760	2808	5.145	0.002617	2777	5.093	0.002567	2766	5.073
600	873.15	0.003159	2947	5.327	0.002969	2914	5.271	9.002808	2883	5.219	0.002752	2871	5.198
620	893.15	0.003386	3047	5.444	0.003178	3015	5.389	0.003000	2984	5.337	0.002938	2972	5.316
640	913.15	0.003605	3141	5.552	0.003383	3110	5.497	0.003193	3081	5.447	0.003124	3069	5.426
650	923.15	0.003711	3186	5.602	0.003484	3155	5.549	0.003288	3127	5.499	0.003216	3116	5.478
660	933.15	0.003817	3230	5.650	0.003583	3200	5.598	0.003381	3172	5.549	0.003307	3162	5.529
680	953.15	0.004024	3314	5.741	0.003780	3286	5.691	0.003565	3260	5.644	0.003488	3250	5.625
700	973.15	0.004226	3395	5.826	0.003970	3369	5.777	0.003746	3344	5.731	0.003665	3334	5.714
720	993.10	0.004421	3473	5.904	0.004155	3449	5.857	0.003921	3425	5.813	0.003837	3416	5.797
740	1013.15	0.004608	3549	5.979	0.004333	3525	5.935	0.004091	3503	5.893	0.004003	3495	5.876
750	1023.15	0.004699	3586	6.016	0.004419	3563	5.972	0.004173	3542	5.931	0.004084	3534	5.914
760	1033.15	0.004789	3622	6.051	0.004505	3601	6.009	0.004255	3579	5.968	0.004164	3571	5.952
780	1053.15	0.004967	3694	6.119	0.004674	3673	6.078	0.004415	3653	6.039	0.004321	3646	6.023
800	1073.15	0.005140	3763	6.185	0.004840	3744	6.145	0.004573	3725	6.106	0.004476	3718	6.092
820	1093.15	0.005310	3831	6.249	0.005002	3814	6.209	0.004728	3795	6.171	0.004627	3788	6.157
840	1113.15	0.005478	3898	6.309	0.005161	3881	6.270	0.004880	3864	6.234	0.004776	3857	6.219
850	1123.15	0.005561	3931	6.338	0.005240	3914	6.300	0.004955	3897	6.264	0.004850	3891	6.249
860	1135.15	0.005642	3964	6.367	0.005318	3948	6.329	0.005030	3931	6.293	0.004924	3925	6.279
880	1153.15	0.005802	4028	6.422	0.005473	4013	6.385	0.005177	3998	6.350	0.005069	3992	6.336
900	1173.15	0.005961	4092	6.475	0.005624	4078	6.439	0.005323	4064	6.405	0.005213	4058	6.391
920	1193.15	0.006117	4155	6.527	0.005774	4142	6.491	0.005467	4128	6.458	0.005354	4123	6.444
940	1213.15	0.006273	4217	6.577	0.005923	4204	6.542	0.005608	4192	6.509	0.005493	4187	6.495
950	1223.15	0.006349	4247	6.601	0.005996	4235	6.566	0.005678	4223	6.533	0.005562	4219	6.520
960	1233.15	0.006425	4277	6.625	0.006070	4266	6.590	0.005748	4254	6.557	0.005630	4250	6.545
980	1253.15	0.006576	4337	6.671	0.006213	4326	6.637	0.005887	4316	6.604	0.005766	4312	6.591
1000	1273.15	0.006726	4398	6.717	0.006356	4387	6.682	0.006022	4377	6.650	0.005900	4372	6.638

Table 47-1 Properties of Saturated Diphenyl Vapors (26.5% Diphenyl, $C_{12}H_{10}$; 73.5% Diphenyl Ether, $C_{12}H_{10}O$)

Temperature		Pressure		Density		Specific volume		Specific enthalpy				Heat of vaporization	
				Liquid	Vapor	Liquid	Vapor	Liquid		Vapor		$r=i''-i'$	
t	T	p	p	ρ'	ρ''	v'	v''	i'	i'	i''	i''		
°C	K	kp/cm²	bar	kg/m³	kg/m³	m³/kg	m³/kg	kcal/kg	kJ/kg	kcal/kg	kJ/kg	kcal/kg	kJ/kg
20	293.15	—	—	1062	—	0.0009416	—	0.0	0.0	88.4	370.1	88.4	370.1
30	303.15	—	—	1054	—	0.0009488	—	3.9	16.3	91.5	383.1	87.6	366.8
40	313.15	—	—	1046	—	0.0009560	—	7.9	33.1	94.7	396.5	86.8	363.5
50	323.15	0.00046	0.000451	1037	0.0025	0.0009643	400	11.9	49.8	98.0	410.3	86.1	360.5
60	333.15	0.00085	0.000834	1029	0.0045	0.0009718	222	15.9	66.6	101.3	424.1	85.4	357.6
70	343.15	0.00152	0.001491	1021	0.0079	0.0009794	127	20.1	84.2	104.7	438.4	84.6	354.2
80	353.15	0.00262	0.002569	1013	0.0133	0.0009872	75.2	24.3	101.7	108.1	452.6	83.8	350.9
90	363.15	0.00437	0.004286	1004	0.0282	0.0009960	35.5	28.5	119.3	111.6	467.2	83.1	347.9
100	373.15	0.00709	0.006953	996	0.0347	0.001004	28.8	32.8	137.3	115.1	481.9	82.3	344.6
110	383.15	0.0112	0.01098	987	0.0537	0.001013	18.6	37.2	155.7	118.7	497.0	81.5	341.2
120	393.15	0.0172	0.01687	979	0.0812	0.001021	12.3	41.6	174.2	122.4	512.5	80.8	338.3
130	403.15	0.0259	0.02540	970	0.1195	0.001031	8.368	46.1	193.0	126.1	528.0	80.0	334.9
140	413.15	0.0381	0.03736	962	0.1727	0.001040	5.790	50.7	212.3	129.9	543.9	79.2	331.6
150	423.15	0.0550	0.05394	953	0.2450	0.001049	4.082	55.3	231.5	133.7	559.8	78.4	328.2
160	433.15	0.0778	0.07630	945	0.3406	0.001058	2.936	59.9	250.8	137.6	576.1	77.7	325.3
170	443.15	0.1083	0.10621	936	0.4655	0.001068	2.148	64.7	270.9	141.6	592.9	76.9	322.0
180	453.15	0.1484	0.14553	927	0.6301	0.001079	1.587	69.5	291.0	145.5	609.2	76.0	318.2
190	463.15	0.200	0.1961	918	0.8339	0.001089	1.199	74.3	311.1	149.5	625.9	75.2	314.8
200	473.15	0.266	0.2609	909	1.097	0.001100	0.912	79.2	331.6	153.5	642.7	74.3	311.1
210	483.15	0.350	0.3432	901	1.418	0.001110	0.705	84.1	352.1	157.5	659.4	73.4	307.3
220	493.15	0.453	0.4422	892	1.815	0.001121	0.551	89.2	373.5	161.7	677.0	72.5	303.5
230	503.15	0.581	0.5698	882	2.301	0.001134	0.435	94.2	394.4	165.8	694.2	71.6	299.8
240	513.15	0.737	0.7228	873	2.882	0.001145	0.347	99.4	416.2	170.0	711.8	70.6	295.6
250	523.15	0.925	0.9071	864	3.581	0.001157	0.279	104.6	437.9	174.2	729.3	69.6	291.4
260	533.15	1.15	1.128	855	4.400	0.001170	0.227	109.8	459.7	178.4	746.9	68.6	287.2
270	543.15	1.42	1.393	846	5.381	0.001182	0.186	115.1	481.9	182.6	764.5	67.5	282.6
280	553.15	1.73	1.697	836	6.517	0.001196	0.153	120.5	504.5	187.0	782.9	66.5	278.4
290	563.15	2.10	2.059	827	7.846	0.001209	0.127	125.9	527.1	191.2	800.5	65.3	273.4
300	573.15	2.52	2.471	818	9.339	0.001222	0.107	131.4	550.1	195.6	818.9	64.2	268.8
310	583.15	3.01	2.952	808	11.10	0.001238	0.0901	137.0	573.6	200.0	837.4	63.0	263.8
320	593.15	3.57	3.501	798	13.16	0.001253	0.0760	142.6	597.0	204.3	855.4	61.7	258.3
330	603.15	4.21	4.129	787	15.39	0.001271	0.0650	148.2	620.5	208.6	873.4	60.4	252.9
340	613.15	4.93	4.835	779	17.98	0.001284	0.0556	153.9	644.3	212.9	891.4	59.0	247.0
350	623.15	5.73	5.619	769	20.88	0.001300	0.0479	159.7	668.6	217.2	909.4	57.5	240.7
360	633.15	6.64	6.512	759	24.23	0.001318	0.0413	165.6	693.3	221.5	927.4	55.9	234.0
370	643.15	7.64	7.492	748	27.92	0.001337	0.0358	171.5	718.0	225.8	945.4	54.3	227.3
380	653.15	8.76	8.591	738	32.19	0.001355	0.0311	177.4	742.7	229.9	962.5	52.5	219.8
390	663.15	9.98	9.787	727	36.95	0.001376	0.0271	183.4	767.9	234.0	979.7	50.6	211.9
400	673.15	11.33	11.111	717	42.35	0.001395	0.0236	189.5	793.4	238.1	996.9	48.6	203.5

1 at = 1 kp/cm² = 98 076.5 Pa = 98 066.5 N/m² = 9.806 65 N/cm² = 0.980 665 bar   1 kcal = 4.186 8 k/J

Table 48-1 Properties of Saturated Mercury Vapors (Hg)

Pressure p (kp/cm²)	Pressure p (bar)	Temperature t (°C)	Temperature T (K)	Specific volume Liquid v' (m³/kg)	Specific volume Vapor v'' (m³/kg)	Specific enthalpy Liquid i' (kcal/kg)	Specific enthalpy Liquid i' (kJ/kg)	Specific enthalpy Vapor i'' (kcal/kg)	Specific enthalpy Vapor i'' (kJ/kg)	Heat of vaporization r=i''−i' (kcal/kg)	Heat of vaporization r=i''−i' (kJ/kg)	Specific entropy Liquid s' (kcal/kg K)	Specific entropy Liquid s' (kJ/kg K)	Specific entropy Vapor s'' (kcal/kg K)	Specific entropy Vapor s'' (kJ/kg K)
0.001	0.0009807	118.5	391.65	0.0000752	165.9	3.96	16.580	76.22	319.118	72.26	302.538	0.0119	0.0498	0.1959	0.8202
0.002	0.0019613	134.6	407.75	0.0000754	86.16	4.45	18.631	76.61	320.751	72.16	302.119	0.0132	0.0553	0.1902	0.7963
0.003	0.0029420	144.1	417.25	0.0000755	58.78	4.76	19.929	76.86	321.797	72.10	301.868	0.0139	0.0582	0.1867	0.7817
0.004	0.0039227	151.2	424.35	0.0000756	44.84	4.98	20.850	77.03	322.509	72.05	301.659	0.0145	0.0607	0.1843	0.7716
0.005	0.0049033	161.5	434.65	0.0000758	30.62	5.31	22.232	77.32	323.723	71.98	301.366	0.0152	0.0663	0.1808	0.7570
0.008	0.0078453	168.9	442.05	0.0000759	23.35	5.58	23.362	77.62	324.979	71.94	301.198	0.0158	0.0662	0.1785	0.7473
0.010	0.0098067	175.0	448.15	0.0000760	18.94	5.79	24.242	77.69	325.272	71.90	301.031	0.0162	0.0678	0.1767	0.7398
0.015	0.0147100	186.6	459.75	0.0000761	12.95	6.16	25.791	77.98	326.487	71.82	300.696	0.0171	0.0716	0.1733	0.7256
0.02	0.0196133	195.0	468.15	0.0000762	9.893	6.44	26.963	78.20	327.408	71.76	300.445	0.0178	0.0745	0.1711	0.7164
0.03	0.0294200	207.6	480.75	0.0000764	6.772	6.85	28.680	78.53	328.789	71.68	300.110	0.0186	0.0779	0.1677	0.7021
0.04	0.0392266	216.9	490.05	0.0000765	5.178	7.16	29.977	78.78	329.836	71.62	299.859	0.0193	0.0808	0.1654	0.6925
0.05	0.0490333	221.5	494.65	0.0000766	4.206	7.41	31.024	78.98	330.673	71.57	299.649	0.0198	0.0829	0.1636	0.6850
0.06	0.0588399	230.9	504.05	0.0000767	3.550	7.63	31.945	79.16	331.427	71.53	299.482	0.0202	0.0846	0.1621	0.6787
0.08	0.0784532	241.0	514.15	0.0000769	2.716	7.98	33.411	79.44	332.599	71.46	299.189	0.0208	0.0871	0.1598	0.6691
0.10	0.0980665	249.6	522.75	0.0000770	2.209	8.25	34.541	79.66	333.520	71.41	298.979	0.0213	0.0892	0.1580	0.6615
0.12	0.1176798	256.7	529.85	0.0000771	1.866	8.48	35.504	79.84	334.274	71.36	298.770	0.0218	0.0913	0.1565	0.6552
0.14	0.1372931	262.7	535.85	0.0000772	1.618	8.68	36.341	80.00	334.944	71.32	298.603	0.0222	0.0929	0.1553	0.6502
0.16	0.1569064	268.0	541.15	0.0000772	1.430	8.86	37.095	80.14	335.530	71.28	298.435	0.0225	0.0942	0.1542	0.6456
0.18	0.1765197	272.9	546.15	0.0000769	1.282	9.02	37.765	80.27	336.074	71.25	298.310	0.0228	0.0955	0.1533	0.6418
0.20	0.1961330	277.3	550.45	0.0000774	1.1632	9.16	38.351	80.38	336.535	71.22	298.184	0.0231	0.0967	0.1525	0.6385
0.25	0.2451663	286.7	559.85	0.0000775	0.9464	9.46	39.607	80.62	337.540	71.16	297.933	0.0236	0.0988	0.1507	0.6310
0.3	0.2941995	294.4	567.55	0.0000776	0.7995	9.73	40.738	80.84	338.461	71.11	297.723	0.0241	0.1009	0.1494	0.6255
0.4	0.3922660	308.0	581.15	0.0000779	0.6140	10.18	42.622	81.19	339.926	71.01	297.305	0.0249	0.1043	0.1471	0.6159
0.5	0.4903325	318.8	591.95	0.0000780	0.5003	10.55	44.171	81.49	341.182	70.94	297.012	0.0255	0.1068	0.1458	0.6104
0.6	0.5883990	328.0	601.15	0.0000781	0.4234	10.86	45.469	81.74	342.229	70.88	296.760	0.0260	0.1089	0.1439	0.6025
0.7	0.6864655	335.9	609.05	0.0000783	0.3677	11.12	46.557	81.94	343.066	70.82	296.509	0.0265	0.1110	0.1428	0.5979
0.8	0.7845320	340.7	613.85	0.0000783	0.3253	11.34	47.478	82.01	343.359	70.77	296.300	0.0269	0.1126	0.1418	0.5937
0.9	0.8825985	349.2	622.35	0.0000784	0.2922	11.56	48.399	82.29	344.532	70.73	296.132	0.0272	0.1139	0.1408	0.5895
1.0	0.980665	355.9	629.05	0.0000785	0.2655	11.76	49.237	82.45	345.202	70.68	295.923	0.0275	0.1151	0.1400	0.5862
1.2	1.176798	365.8	638.95	0.0000787	0.2240	12.11	50.702	82.66	346.081	70.62	295.672	0.0280	0.1172	0.1386	0.5803
1.4	1.372931	374.0	647.15	0.0000788	0.1953	12.58	52.670	82.94	347.253	70.56	295.421	0.0285	0.1193	0.1375	0.5757
1.6	1.569064	381.9	655.05	0.0000789	0.1730	12.64	52.921	83.14	348.091	70.50	295.169	0.0290	0.1214	0.1366	0.5719
1.8	1.765197	389.3	662.45	0.0000790	0.1555	12.90	54.010	83.35	348.970	70.45	294.960	0.0294	0.1231	0.1357	0.5681
2.0	1.961330	395.8	668.95	0.0000791	0.1414	13.11	54.889	83.51	349.640	70.40	294.751	0.0297	0.1243	0.1349	0.5648
2.2	2.157463	401.7	674.85	0.0000792	0.1296	13.32	55.768	83.68	350.351	70.36	294.583	0.0300	0.1256	0.1342	0.5619
2.4	2.353596	407.4	680.55	0.0000793	0.1198	13.54	56.689	83.86	351.105	70.32	294.416	0.0303	0.1269	0.1335	0.5589
2.6	2.549729	412.4	685.55	0.0000794	0.1114	13.70	57.359	83.98	351.607	70.28	294.248	0.0305	0.1277	0.1329	0.5564
2.8	2.745862	417.0	690.15	0.0000794	0.1043	13.87	58.071	84.11	352.152	70.24	294.081	0.0307	0.1285	0.1324	0.5543
3.0	2.941995	422.4	695.55	0.0000795	0.09798	14.04	58.783	84.25	352.738	70.21	293.955	0.0309	0.1294	0.1320	0.5527
3.5	3.432328	432.8	705.95	0.0000797	0.08524	14.40	60.290	84.58	354.120	70.13	293.620	0.0315	0.1319	0.1308	0.5476

Table 48-2 Properties of Saturated Mercury Vapors (Hg) (Continued)

Pressure		Temperature		Specific volume		Specific enthalpy				Heat of vaporization		Specific entropy			
				Liquid v'	Vapor v''	Liquid i'		Vapor i''		$r=i''-i'$		Liquid s'		Vapor s''	
p		t	T												
kp/cm²	bar	°C	K	m³/kg	m³/kg	kcal/kg	kJ/kg	kcal/kg	kJ/kg	kcal/kg	kJ/kg	kcal/kg K	kJ/kg K	kcal/kg K	kJ/kg K
4.0	3.922660	442.4	715.55	0.0000798	0.07558	14.74	61.713	84.80	355.041	70.06	293.327	0.0319	0.1336	0.1298	0.5434
4.5	4.412993	451.0	724.15	0.0000799	0.06801	15.03	62.928	85.02	355.962	69.99	293.034	0.0323	0.1352	0.1289	0.5397
5.0	4.903325	458.0	731.15	0.0000801	0.06487	15.30	64.058	85.28	357.050	69.93	292.783	0.0327	0.1369	0.1282	0.5367
5.5	5.393658	466.8	739.95	0.0000802	0.05682	15.56	65.147	85.48	357.888	69.87	292.532	0.0331	0.1386	0.1276	0.5342
6.0	5.883990	472.8	745.95	0.0000803	0.05254	15.78	66.068	85.59	358.348	69.84	292.406	0.0334	0.1398	0.1270	0.5317
7.0	6.864655	485.1	758.25	0.0000805	0.04578	16.20	67.826	85.91	359.688	69.74	291.987	0.0339	0.1419	0.1258	0.5267
8.0	7.845320	496.3	769.45	0.0000806	0.04065	16.59	69.459	86.20	360.902	69.74	291.569	0.0344	0.1440	0.1249	0.5229
9	8.825985	506.3	779.45	0.0000808	0.03660	16.94	70.924	86.47	362.033	69.53	291.108	0.0349	0.1461	0.1241	0.5196
10	9.80665	515.5	788.65	0.0000809	0.03383	17.25	72.222	86.70	362.996	69.45	290.773	0.0356	0.1491	0.1234	0.5167
12	11.76798	532.3	805.45	0.0000812	0.02837	17.85	74.734	87.15	364.880	69.30	290.145	0.0360	0.1507	0.1220	0.5108
14	13.72931	546.7	819.85	0.0000814	0.02476	18.35	76.828	87.54	366.512	69.16	289.559	0.0366	0.1532	0.1210	0.5066
16	15.69064	559.8	832.95	0.0000816	0.02200	18.84	78.879	87.84	367.769	69.03	289.015	0.0372	0.1557	0.1201	0.5028
18	17.65197	571.4	844.55	0.0000818	0.01933	19.28	80.722	88.14	369.025	68.91	288.512	0.0377	0.1578	0.1193	0.4995
20	19.61330	582.4	855.55	0.0000819	0.01806	19.62	82.145	88.42	370.197	68.80	288.052	0.0384	0.1608	0.1185	0.4961
25	24.51663	606.5	879.65	0.0000823	0.01487	20.46	85.662	89.00	372.625	68.54	286.963	0.0391	0.1637	0.1170	0.4899
30	29.41995	627.1	900.25	0.0000827	0.01268	21.18	88.676	89.48	374.635	68.30	285.958	0.0399	0.1671	0.1158	0.4848
35	34.32328	645.0	918.15	0.0000830	0.01109	21.83	91.398	89.91	376.435	68.08	285.037	0.0406	0.1700	0.1147	0.4802
40	39.22660	661.8	934.95	0.0000832	0.009873	22.41	93.826	90.28	377.984	67.87	284.158	0.0412	0.1725	0.1138	0.4765
45	44.12993	677.0	950.15	0.0000835	0.008923	22.90	95.878	90.62	379.408	67.67	283.321	0.0418	0.1750	0.1130	0.4731
50	49.03325	690.9	964.05	0.0000837	0.008148	23.44	98.139	90.91	380.622	67.47	282.483	0.0423	0.1771	0.1123	0.4702

1 at = 1 kp/cm² = 98 066.5 Pa = 98 066.5 N/m² = 9.806 65 N/cm² = 0.980 665 bar   1 kcal = 4.186 8 kJ

Table 49-1 Thermodynamic Properties of Refrigerants

Refrigerant	Chemical formula	Relative molecular mass M (kg/kmol)	Gas constant R (kpm/kg K)	Gas constant R (J/kg K)	Specific heat — Liquid c (kcal/kg K)	Liquid c (kJ/kg K)	Vapor c_p (kcal/kg K)	Vapor c_p (kJ/kg K)	$\kappa = c_p/c_v$ at 0°C	Boiling point t_v (°C)	T_v (°K)	Melting point t_t (°C)	T_t (°K)	Critical temp t_k (°C)	T_k (°K)	Critical pressure p_k (kp/cm²)	p_k (bar)	Critical density ρ_k (kg/m³)
Ammonia	NH_3	17.032	49.789	488.263	1.11	4.647	0.492	2.060	1.312	− 33.35	239.80	− 77.9	195.25	132.4	405.55	115.2	112.9726	235
Carbon dioxide	CO_2	44.01	19.268	188.955	—	—	0.197	0.825	1.30	− 78.48	194.67	− 56.6	216.55	31.0	304.15	75.21	73.7558	460
Chlorodifluoromethane (Freon 22)	CHF_2Cl	86.475	9.806	96.164	0.26	1.089	0.145	0.607	1.19	− 40.80	232.35	−160	113.15	96	369.15	50.33	49.3569	526
Chlorotrifluoromethane (Freon 13)	CF_3Cl	104.47	8.117	79.601	0.203	0.850	0.126	0.528	1.15	− 81.5	191.65	−181	92.15	28.78	301.93	39.46	38.6970	581
Dichlorodifluoromethane (Freon 12)	CF_2Cl_2	120.92	7.0127	68.771	0.204	0.854	0.146	0.611	1.148	− 29.8	243.35	−155	118.15	112.0	385.15	40.87	40.0798	555
Dichlorotetrafluoroethane (Freon 114)	$C_2F_4Cl_2$	170.93	4.961	48.651	0.232	0.971	0.152	0.636	1.106	+ 4.1	277.25	− 94	179.15	146	419.15	34.4	33.7349	–
Ethane	C_2H_6	30.07	28.201	276.557	—	—	0.413	1.729	1.202	− 88.63	184.52	−183.6	89.55	32.1	305.25	50.3	49.3247	210
Ethylene	C_2H_4	28.05	30.25	296.651	—	—	0.385	1.612	1.25	−103.6	169.55	−104	169.15	9.4	282.55	51.4	50.4062	216
Methane	CH_4	16.04	16.03	157.201	—	—	0.520	2.177	1.30	−161.6	111.55	−182.6	90.55	− 82	191.15	47.3	46.3855	162
Methyl chloride	CH_3Cl	50.491	16.80	164.752	0.37	1.549	0.176	0.737	1.27	− 24.0	249.15	− 91.5	181.65	143.1	416.25	68.1	66.7833	370
Propane	C_3H_8	44.09	19.233	188.611	—	—	0.365	1.528	1.15	− 42.5	230.65	−189.9	83.25	96.85	370.00	43.4	42.5609	226
Sulphur dioxide	SO_2	64.06	13.238	129.820	0.324	1.357	0.145	0.607	1.271	− 10.02	263.13	− 75.5	197.65	157.2	430.35	80.3	78.7474	524
Trichlorofluoromethane (Freon 11)	$CFCl_3$	137.38	6.173	60.536	0.208	0.871	0.130	0.544	1.124	+ 23.65	296.80	−111	162.15	198	471.15	44.6	43.7377	555
Trichlorotrifluoroethane (Freon 113)	$C_2F_3Cl_3$	187.39	4.525	44.375	0.226	0.946	0.149	0.624	1.075	+ 47.6	320.75	− 36.5	236.65	214.1	487.25	34.8	34.1271	578
Water (steam)	H_2O	18.02	47.06	461.501	1.008	4.220	0.444	1.859	1.40	+100	373.15	− 0	273.15	374.15	647.30	225.65	221.2871	322

1 at = 1 kp/cm² = 98066.5 Pa = 98066.5 N/m² = 9.80665 N/cm² = 0.980665 bar

1 kcal = 4.186 8 kJ

Table 50-1 Properties of Saturated Ammonia (NH_3)

Temperature t (°C)	Temperature T (K)	Pressure p (kp/cm²)	Pressure p (bar)	Density Liquid ρ' (kg/m³)	Density Vapor ρ'' (kg/m³)	Spec. volume Liquid v' (m³/kg)	Spec. volume Vapor v'' (m³/kg)	Enthalpy Liquid i' (kcal/kg)	Enthalpy Liquid i' (kJ/kg)	Enthalpy Vapor i'' (kcal/kg)	Enthalpy Vapor i'' (kJ/kg)	Heat of vap. r=i''−i' (kcal/kg)	Heat of vap. r=i''−i' (kJ/kg)	Entropy Liquid s' (kcal/kg K)	Entropy Liquid s' (kJ/kg K)	Entropy Vapor s'' (kcal/kg K)	Entropy Vapor s'' (kJ/kg K)
−75	198.15	0.0765	0.07502	731.0	0.0775	0.001368	12.89	20.9	87.504	373.5	1563.770	352.6	1476.266	0.6633	2.7771	2.4431	10.2288
−70	203.15	0.1114	0.10925	725.3	0.1110	0.0013788	9.009	25.9	108.438	375.7	1572.981	349.8	1464.543	0.6878	2.8797	2.4101	10.0906
−68	205.15	0.1287	0.12621	723.0	0.1271	0.0013832	7.870	27.9	116.812	376.6	1576.749	348.7	1459.937	0.6975	2.9203	2.3976	10.0383
−66	207.15	0.1485	0.14563	720.7	0.1453	0.0013876	6.882	29.9	125.185	377.4	1580.098	347.5	1454.913	0.7074	2.9617	2.3853	9.9868
−64	209.15	0.1706	0.16730	718.4	0.1655	0.0013920	6.044	32.9	133.978	378.3	1583.866	346.3	1449.889	0.7173	3.0032	2.3734	9.9370
−62	211.15	0.1954	0.19162	716.1	0.1878	0.0013965	5.324	34.0	142.351	379.1	1587.216	345.1	1444.865	0.7270	3.0438	2.3618	9.8884
−60	213.15	0.2233	0.21898	713.8	0.2128	0.0014010	4.699	36.0	150.725	380.0	1590.984	344.0	1440.259	0.7366	3.0840	2.3507	9.8419
−58	215.15	0.2543	0.24938	711.4	0.2403	0.0014056	4.161	38.1	159.517	380.8	1594.333	342.7	1434.816	0.7461	3.1238	2.3393	9.7942
−56	217.15	0.2889	0.28331	709.1	0.2708	0.0014103	3.693	40.2	168.309	381.7	1598.102	341.5	1429.792	0.7555	3.1631	2.3285	9.7490
−54	219.15	0.3272	0.32087	706.7	0.3041	0.0014150	3.288	42.2	176.683	382.5	1601.451	340.3	1424.768	0.7648	3.2021	2.3180	9.7050
−52	221.15	0.3697	0.36255	704.4	0.3409	0.0014197	2.933	44.2	185.057	383.3	1604.800	339.1	1419.744	0.7741	3.2410	2.3078	9.6623
−50	223.15	0.4168	0.40874	702.0	0.3812	0.0014245	2.623	46.3	193.849	384.1	1608.150	337.8	1414.301	0.7882	3.3000	2.2978	9.6204
−48	225.15	0.4686	0.45904	699.6	0.425	0.0014293	2.351	48.4	202.641	384.9	1611.499	336.6	1409.277	0.7931	3.3206	2.2808	9.5493
−46	227.15	0.5256	0.51544	697.2	0.473	0.0014343	2.112	50.4	211.015	385.7	1614.849	335.3	1403.834	0.8021	3.3582	2.2785	9.5396
−44	229.15	0.5882	0.57683	694.8	0.526	0.0014392	1.901	52.5	219.807	386.5	1618.198	334.0	1398.391	0.8112	3.3963	2.2692	9.5007
−42	231.15	0.6568	0.64410	692.4	0.583	0.0014442	1.715	54.6	228.599	387.3	1621.548	332.7	1392.948	0.8203	3.4344	2.2600	9.4622
−40	233.15	0.7318	0.71765	690.0	0.645	0.0014493	1.550	56.8	237.810	388.1	1624.897	331.3	1387.087	0.8295	3.4730	2.2510	9.4245
−39	234.15	0.7719	0.75698	688.8	0.678	0.0014519	1.4752	57.82	242.081	388.49	1626.530	330.67	1384.449	0.8340	3.4918	2.2465	9.4056
−38	235.15	0.8137	0.79797	687.5	0.712	0.0014545	1.4045	58.88	246.519	388.88	1628.163	329.99	1381.602	0.8385	3.5106	2.2421	9.3872
−37	236.15	0.8573	0.84072	686.3	0.748	0.0014571	1.3377	59.94	250.957	389.27	1629.796	329.31	1378.755	0.8430	3.5295	2.2378	9.3692
−36	237.15	0.9028	0.88534	685.1	0.785	0.0014597	1.2746	61.01	255.437	389.65	1631.387	328.63	1375.908	0.8475	3.5483	2.2336	9.3516
−35	238.15	0.9503	0.93193	683.9	0.823	0.0014623	1.2151	62.08	259.917	390.03	1632.978	327.95	1373.061	0.8520	3.5672	2.2294	9.3341
−34	239.15	0.9999	0.98057	682.6	0.863	0.0014649	1.1589	63.15	264.396	390.41	1634.569	327.26	1370.172	0.8565	3.5860	2.2252	9.3165
−33	240.15	1.0515	1.03117	681.4	0.905	0.0014676	1.1058	64.21	268.834	390.79	1636.160	326.57	1367.283	0.8610	3.6048	2.2211	9.2993
−32	241.15	1.1052	1.08383	680.1	0.948	0.0014703	1.0555	65.28	273.314	391.17	1637.751	325.88	1364.394	0.8654	3.6233	2.2170	9.2821
−31	242.15	1.1610	1.13855	678.9	0.992	0.0014730	1.0080	66.35	277.794	391.54	1639.300	325.19	1361.505	0.8698	3.6417	2.2130	9.2654
−30	243.15	1.2190	1.19543	677.7	1.038	0.0014757	0.9630	67.42	282.274	391.91	1640.849	324.49	1358.575	0.8742	3.6601	2.2090	9.2486
−29	244.15	1.279	1.25427	676.4	1.086	0.0014784	0.9204	68.49	286.754	392.28	1642.398	323.79	1355.644	0.8786	3.6785	2.2050	9.2319
−28	245.15	1.342	1.31605	675.2	1.136	0.0014811	0.8801	69.56	291.234	392.64	1643.905	323.08	1352.671	0.8830	3.6969	2.2011	9.2156
−27	246.15	1.407	1.37980	673.9	1.188	0.0014839	0.8418	70.63	295.714	393.00	1645.412	322.37	1349.699	0.8874	3.7154	2.1972	9.1992
−26	247.15	1.475	1.44648	672.6	1.242	0.0014867	0.8056	71.71	300.235	393.36	1646.920	321.66	1346.726	0.8918	3.7334	2.1934	9.1813
−25	248.15	1.546	1.51611	671.4	1.297	0.0014895	0.7712	72.78	304.715	393.72	1648.427	320.94	1343.712	0.8960	3.7514	2.1896	9.1674
−24	249.15	1.619	1.58770	670.1	1.354	0.0014923	0.7386	73.86	309.237	394.07	1649.892	320.22	1340.697	0.9003	3.7694	2.1858	9.1515
−23	250.15	1.695	1.66223	668.8	1.413	0.0014951	0.7076	74.93	313.717	394.42	1651.358	319.49	1337.641	0.9046	3.7874	2.1821	9.1360
−22	251.15	1.774	1.73970	667.6	1.474	0.0014980	0.6782	76.01	318.239	394.77	1652.823	318.76	1334.584	0.9089	3.8054	2.1784	9.1205
−21	252.15	1.856	1.82011	666.3	1.538	0.0015008	0.6502	77.09	322.760	395.12	1654.288	318.03	1331.528	0.9132	3.8234	2.1747	9.1050
−20	253.15	1.940	1.90249	665.0	1.604	0.0015037	0.6236	78.17	327.282	395.46	1655.712	317.29	1328.430	0.9175	3.8414	2.1710	9.0895
−19	254.15	2.027	1.98781	663.7	1.672	0.0015066	0.5983	79.25	331.804	395.80	1657.135	316.55	1325.332	0.9217	3.8590	2.1674	9.0745
−18	255.15	2.117	2.07607	662.4	1.742	0.0015096	0.5742	80.33	336.326	396.13	1658.517	315.80	1322.191	0.9259	3.8766	2.1638	9.0594
−17	256.15	2.211	2.16825	661.1	1.814	0.0015125	0.5513	81.41	340.847	396.46	1659.899	315.05	1319.051	0.9301	3.8941	2.1602	9.0443

Table 50-2 Properties of Saturated Ammonia (NH$_3$) (Continued)

Temperature		Pressure		Density		Specific volume		Specific enthalpy				Heat of vaporization		Specific entropy			
t	T	p	p	Liquid ρ'	Vapor ρ''	Liquid v'	Vapor v''	Liquid i'		Vapor i''		r = i''−i'		Liquid s'		Vapor s''	
°C	K	kp/cm²	bar	kg/m³	kg/m³	m³/kg	m³/kg	kcal/kg	kJ/kg	kcal/kg	kJ/kg	kcal/kg	kJ/kg	kcal/kg K	kJ/kg K	kcal/kg K	kJ/kg K
−16	257.15	2.309	2.26436	659.8	1.889	0.0015155	0.5295	82.50	345.411	396.79	1661.280	314.29	1315.869	0.9343	3.9117	2.1567	9.0297
−15	258.15	2.410	2.36340	658.5	1.966	0.0015185	0.5087	83.59	349.975	397.12	1662.662	313.53	1312.687	0.9385	3.9293	2.1532	9.0150
−14	259.15	2.514	2.46539	657.2	2.046	0.0015215	0.4889	84.68	354.538	397.44	1664.002	312.76	1309.464	0.9427	3.9469	2.1498	9.0008
−13	260.15	2.621	2.57032	655.9	2.128	0.0015245	0.4700	85.76	359.060	397.75	1665.300	311.99	1306.240	0.9469	3.9645	2.1464	8.9865
−12	261.15	2.732	2.67918	654.6	2.213	0.0015276	0.4520	86.85	363.624	398.06	1666.598	311.21	1302.974	0.9511	3.9821	2.1430	8.9723
−11	262.15	2.847	2.79195	653.3	2.300	0.0015307	0.4348	87.94	368.187	398.37	1667.896	310.43	1299.708	0.9552	3.9992	2.1396	8.9581
−10	263.15	2.966	2.90865	652.0	2.390	0.0015338	0.4184	89.03	372.751	398.67	1669.152	309.64	1296.401	0.9593	4.0164	2.1362	8.9438
−9	264.15	3.089	3.02927	650.7	2.483	0.0015369	0.4028	90.12	377.314	398.97	1670.408	308.85	1293.093	0.9634	4.0336	2.1329	8.9300
−8	265.15	3.216	3.15382	649.3	2.579	0.0015400	0.3878	91.91	381.878	399.27	1671.664	308.06	1289.786	0.9675	4.0507	2.1296	8.9162
−7	266.15	3.347	3.28229	648.0	2.678	0.0015432	0.3735	92.30	386.442	399.56	1672.878	307.25	1286.394	0.9716	4.0679	2.1263	8.9024
−6	267.15	3.481	3.41369	646.7	2.779	0.0015464	0.3599	93.40	391.047	399.85	1674.092	306.45	1283.045	0.9757	4.0851	2.1231	8.8890
−5	268.15	3.619	3.54903	645.3	2.883	0.0015496	0.3469	94.50	395.653	400.14	1675.306	305.64	1279.654	0.9798	4.1022	2.1199	8.8756
−4	269.15	3.761	3.68828	644.0	2.991	0.0015528	0.3344	95.59	400.216	400.42	1676.478	304.83	1276.262	0.9839	4.1194	2.1167	8.8622
−3	270.15	3.908	3.83244	642.6	3.102	0.0015561	0.3225	96.69	404.822	400.70	1677.651	304.01	1272.829	0.9880	4.1366	2.1135	8.8488
−2	271.15	4.060	3.98150	641.3	3.216	0.0015594	0.3111	97.79	409.427	400.98	1678.823	303.19	1269.396	0.9920	4.1533	2.1103	8.8354
−1	272.15	4.217	4.13546	639.9	3.332	0.0015627	0.3002	98.89	414.033	401.25	1679.954	302.36	1265.921	0.9960	4.1701	2.1072	8.8224
0	273.15	4.379	4.29433	638.6	3.452	0.0015660	0.2897	100.00	418.680	401.52	1681.084	301.52	1262.404	1.0000	4.1868	2.1041	8.8094
1	274.15	4.545	4.45712	637.2	3.576	0.0015694	0.2797	101.10	423.285	401.78	1682.173	300.68	1258.887	1.0040	4.2035	2.1010	8.7965
2	275.15	4.716	4.62482	635.8	3.702	0.0015727	0.2700	102.21	427.933	402.04	1683.261	299.84	1255.370	1.0080	4.2203	2.0979	8.7835
3	276.15	4.892	4.79741	634.5	3.834	0.0015761	0.2608	103.32	432.580	402.30	1684.350	298.99	1251.811	1.0120	4.2370	2.0949	8.7709
4	277.15	5.073	4.97491	633.1	3.969	0.0015796	0.2520	104.43	437.228	402.55	1685.396	298.13	1248.211	1.0160	4.2538	2.0919	8.7584
5	278.15	5.259	5.15732	631.7	4.108	0.0015831	0.2435	105.54	441.875	402.80	1686.443	297.26	1244.568	1.0200	4.2705	2.0889	8.7458
6	279.15	5.450	5.34462	630.3	4.250	0.0015866	0.2353	106.65	446.522	403.04	1687.448	296.39	1240.926	1.0240	4.2873	2.0859	8.7332
7	280.15	5.647	5.53682	628.9	4.396	0.0015901	0.2275	107.76	451.170	403.27	1688.411	295.51	1237.241	1.0280	4.3040	2.0829	8.7207
8	281.15	5.849	5.73591	627.5	4.546	0.0015935	0.2200	108.87	455.817	403.50	1689.374	294.63	1233.557	1.0319	4.3204	2.0799	8.7081
9	282.15	6.057	5.93989	626.1	4.700	0.0015972	0.2128	109.99	460.506	403.73	1690.337	293.74	1229.831	1.0358	4.3367	2.0770	8.6960
10	283.15	6.271	6.14975	624.7	4.859	0.0016008	0.2058	111.11	465.195	403.95	1691.258	292.84	1226.063	1.0397	4.3530	2.0741	8.6838
11	284.15	6.490	6.36452	623.3	5.022	0.0016045	0.1992	112.23	469.885	404.17	1692.179	291.94	1222.294	1.0436	4.3693	2.0712	8.6717
12	285.15	6.715	6.58517	621.8	5.189	0.0016081	0.1927	113.35	474.574	404.38	1693.058	291.03	1218.484	1.0475	4.3857	2.0683	8.6596
13	286.15	6.946	6.81170	620.4	5.361	0.0016118	0.1866	114.47	479.263	404.59	1693.937	290.12	1214.674	1.0514	4.4020	2.0654	8.6474
14	287.15	7.183	7.04412	619.0	5.537	0.0016156	0.1806	115.59	483.952	404.79	1694.775	289.20	1210.823	1.0553	4.4183	2.0626	8.6357
15	288.15	7.427	7.28340	617.5	5.718	0.0016193	0.1749	116.72	488.683	404.99	1695.612	288.27	1206.929	1.0592	4.4347	2.0598	8.6240
16	289.15	7.677	7.52857	616.1	5.904	0.0016231	0.1694	117.85	493.414	405.19	1696.449	287.34	1203.035	1.0631	4.4510	2.0570	8.6122
17	290.15	7.933	7.77962	614.6	6.094	0.0016270	0.1642	118.98	498.145	405.38	1697.245	286.40	1199.100	1.0670	4.4673	2.0542	8.6005
18	291.15	8.196	8.03753	613.2	6.289	0.0016308	0.1591	120.11	502.877	405.57	1698.040	285.46	1195.164	1.0709	4.4836	2.0514	8.5888
19	292.15	8.465	8.30133	611.7	6.489	0.0016347	0.1542	121.24	507.608	405.75	1698.794	284.51	1191.186	1.0747	4.4996	2.0486	8.5771
20	293.15	8.741	8.57199	610.3	6.694	0.0016386	0.1494	122.38	512.381	405.93	1699.548	283.55	1187.167	1.0785	4.5155	2.0459	8.5658
21	294.15	9.024	8.84952	608.8	6.904	0.0016426	0.1449	123.52	517.154	406.10	1700.259	282.58	1183.106	1.0824	4.5318	2.0432	8.5545
22	295.15	9.314	9.13391	607.3	7.119	0.0016466	0.1405	124.66	521.926	406.27	1700.971	281.61	1179.045	1.0862	4.5477	2.0405	8.5432
23	296.15	9.611	9.42517	605.8	7.339	0.0016507	0.1363	125.80	526.699	406.43	1701.641	280.63	1174.942	1.0900	4.5636	2.0378	8.5319
24	297.15	9.915	9.72329	604.3	7.564	0.0016546	0.1322	126.94	531.472	406.59	1702.311	279.65	1170.839	1.0938	4.5795	2.0351	8.5206
25	298.15	10.255	10.02730	602.8	7.795	0.0016588	0.1283	128.09	536.287	406.75	1702.981	278.66	1166.694	1.0976	4.5954	2.0324	8.5093
26	299.15	10.544	10.34013	601.3	8.031	0.0016630	0.1245	129.24	541.102	406.89	1703.567	277.66	1162.507	1.1014	4.6113	2.0297	8.4979
27	300.15	10.870	10.65983	599.8	8.273	0.0016672	0.1209	130.39	545.917	407.03	1704.153	276.65	1158.278	1.1052	4.6273	2.0270	8.4866
28	301.15	11.204	10.98737	598.3	8.521	0.0016714	0.1174	131.54	550.732	407.17	1704.739	275.64	1154.050	1.1090	4.6432	2.0243	8.4753

Table 50-3 Properties of Saturated Ammonia (NH₃) *(Continued)*

Temperature t °C	T K	Pressure p kg/cm²	Pressure p bar	Density Liquid ρ' kg/m³	Density Vapor ρ'' kg/m³	Specific volume Liquid v' m³/kg	Specific volume Vapor v'' m³/kg	Specific enthalpy Liquid i' kcal/kg	Liquid i' kJ/kg	Specific enthalpy Vapor i'' kcal/kg	Vapor i'' kJ/kg	Heat of vaporization $r=i''-i'$ kcal/kg	r kJ/kg	Specific entropy Liquid s' kcal/kg K	Liquid s' kJ/kg K	Specific entropy Vapor s'' kcal/kg K	Vapor s'' kJ/kg K
29	302.15	11.546	11.32276	596.8	8.775	0.0016757	0.1140	132.69	555.546	407.30	1705.284	274.62	1149.779	1.1128	4.6591	2.0217	8.4645
30	303.15	11.895	11.66501	595.2	9.034	0.0016800	0.1107	133.84	560.361	407.43	1705.828	273.59	1145.467	1.1165	4.6746	2.0191	8.4536
31	304.15	12.252	12.01511	593.7	9.300	0.0016844	0.1075	135.00	565.218	407.55	1706.330	272.55	1141.112	1.1203	4.6905	2.0165	8.4427
32	305.15	12.617	12.37305	592.1	9.573	0.0016888	0.1045	136.16	570.075	407.67	1706.833	271.50	1136.716	1.1241	4.7064	2.0139	8.4318
33	306.15	12.991	12.73982	590.6	9.852	0.0016932	0.1015	137.32	574.931	407.78	1707.293	270.45	1132.320	1.1278	4.7219	2.0113	8.4209
34	307.15	13.374	13.11541	589.0	10.138	0.0016977	0.0986	138.48	579.788	407.88	1707.712	269.39	1127.882	1.1315	4.7374	2.0087	8.4100
35	308.15	13.765	13.49885	587.5	10.431	0.0017023	0.0959	139.65	584.687	407.97	1708.089	268.32	1123.402	1.1352	4.7529	2.0061	8.3991
36	309.15	14.165	13.89112	585.9	10.731	0.0017069	0.0932	140.82	589.585	408.06	1708.466	267.24	1118.880	1.1390	4.7688	2.0035	8.3883
37	310.15	14.573	14.29123	584.3	11.038	0.0017115	0.0906	141.99	594.484	408.15	1708.842	266.15	1114.317	1.1427	4.7843	2.0009	8.3774
38	311.15	14.990	14.70017	582.7	11.353	0.0017162	0.0881	143.16	599.382	408.23	1709.177	265.06	1109.753	1.1464	4.7997	1.9984	8.3669
39	312.15	15.415	15.11695	581.1	11.675	0.0017209	0.0857	144.34	604.323	408.30	1709.940	263.96	1105.148	1.1501	4.8152	1.9958	8.3560
40	313.15	15.850	15.54354	579.5	12.005	0.0017257	0.0833	145.52	609.263	408.37	1709.764	262.85	1100.500	1.1538	4.8307	1.9933	8.3455
41	314.15	16.294	15.97896	577.9	12.34	0.0017305	0.0810	146.70	614.204	408.43	1710.015	261.73	1095.811	1.1575	4.8462	1.9908	8.3351
42	315.15	16.747	16.42320	576.2	12.69	0.0017354	0.0788	147.88	619.144	408.49	1710.266	260.60	1091.080	1.1612	4.8617	1.9882	8.3242
43	316.15	17.210	16.87724	574.6	13.04	0.0017404	0.0767	149.06	624.084	408.54	1710.475	259.47	1086.349	1.1649	4.8772	1.9857	8.3137
44	317.15	17.682	17.34012	572.9	13.40	0.0017454	0.0746	150.24	629.025	408.58	1710.643	258.33	1081.576	1.1686	4.8927	1.9832	8.3033
45	318.15	18.165	17.81378	571.3	13.77	0.0017504	0.0726	151.43	634.007	408.61	1710.768	257.18	1076.761	1.1722	4.9078	1.9807	8.2928
46	319.15	18.658	18.29725	569.6	14.15	0.0017555	0.0707	152.62	638.989	408.64	1710.894	256.02	1071.905	1.1759	4.9233	1.9781	8.2819
47	320.15	19.161	18.79052	568.0	14.54	0.0017607	0.0688	153.81	643.972	408.66	1710.978	254.85	1067.006	1.1796	4.9387	1.9756	8.2714
48	321.15	19.673	19.29262	566.3	14.94	0.0017659	0.0670	155.00	648.954	408.68	1711.061	253.67	1062.066	1.1832	4.9538	1.9731	8.2610
49	322.15	20.195	19.80453	564.6	15.34	0.0017712	0.0652	156.20	653.978	408.70	1711.145	252.48	1057.083	1.1868	4.9689	1.9706	8.2505
50	323.15	20.727	20.32624	562.8	15.75	0.0017775	0.0635	157.38	658.919	408.72	1711.229	251.34	1052.310	1.1905	4.9844	1.9683	8.2409
52	325.15	21.83	21.40792	559.1	16.59	0.001788	0.0602	159.8	669.051	408.7	1711.145	248.9	1042.095	1.1982	5.0166	1.9638	8.2220
54	327.15	22.97	22.52588	555.4	17.47	0.001800	0.0572	162.2	679.099	408.8	1711.564	246.6	1032.465	1.2056	5.0476	1.9590	8.2019
56	329.15	24.15	23.68306	551.6	18.39	0.001812	0.0543	164.6	689.147	408.8	1711.564	244.2	1022.417	1.2130	5.0786	1.9542	8.1818
58	331.15	25.37	24.87947	547.8	19.35	0.001825	0.0515	167.1	699.614	408.7	1711.145	241.6	1011.531	1.2205	5.1100	1.9494	8.1617
60	333.15	26.66	26.14453	544.0	20.35	0.001838	0.0489	169.6	710.081	408.6	1710.726	238.0	996.458	1.2280	5.1414	1.9445	8.1412
62	335.15	27.98	27.43901	540.2	21.41	0.001851	0.0464	172.2	720.967	408.5	1710.308	236.3	989.341	1.2354	5.1724	1.9396	8.1207
64	337.15	29.36	28.79232	536.4	22.53	0.001864	0.0441	174.8	731.853	408.3	1709.470	233.5	977.618	1.2428	5.2034	1.9347	8.1002
66	339.15	30.77	30.17506	532.6	23.73	0.001877	0.0420	177.4	742.738	408.0	1708.214	230.6	965.476	1.2502	5.2343	1.9297	8.0793
68	341.15	32.25	31.62645	528.8	25.01	0.001891	0.0399	180.0	753.624	407.7	1706.958	227.7	953.334	1.2576	5.2653	1.9247	8.0583
70	343.15	33.77	33.11706	524.8	26.36	0.001905	0.0379	182.7	764.928	407.3	1705.284	224.6	940.355	1.2650	5.2963	1.9196	8.0370

1 at = 1 kp/cm² = 98 066.5 Pa = 98 066.5 N/m² = 9.806 65 N/cm² = 0.980 665 bar

1 kcal = 4.186 8 kJ

Table 51-1 Properties of Saturated Dichlorodifluoromethane "Freon 12" (CF$_2$Cl$_2$)

Temperature t °C	T K	Pressure p kp/cm²	p bar	Density Liquid ρ' kg/m³	Density Vapor ρ" kg/m³	Sp. vol. Liquid v' m³/kg	Sp. vol. Vapor v" m³/kg	Enthalpy Liquid i kcal/kg	Enthalpy Liquid i kJ/kg	Enthalpy Vapor i" kcal/kg	Enthalpy Vapor i" kJ/kg	Heat of vap. r=i"−i' kcal/kg	r kJ/kg	Entropy Liquid s' kcal/kg K	s' kJ/kg K	Entropy Vapor s" kcal/kg K	s" kJ/kg K
−70	203.15	0.1258	0.12337	1604	0.888	0.0006234	1.1259	85.84	359.395	128.88	539.595	42.99	179.991	0.94050	3.9377	1.15219	4.8240
−69	204.15	0.1341	0.13151	1601	0.943	0.0006246	1.0605	86.02	360.149	128.95	539.888	42.93	179.739	0.94139	3.9414	1.15173	4.8221
−68	205.15	0.1429	0.14014	1598	1.000	0.0006258	0.9998	86.20	360.902	129.06	540.348	42.86	179.446	0.94230	3.9452	1.15130	4.8203
−67	206.15	0.1521	0.14916	1595	1.060	0.0006270	0.9437	86.39	361.698	129.19	540.893	42.80	179.195	0.94322	3.9491	1.15087	4.8185
−66	207.15	0.1618	0.15867	1592	1.122	0.0006282	0.8911	86.57	362.451	129.30	541.353	42.73	178.902	0.94411	3.9528	1.15044	4.8167
−65	208.15	0.1721	0.16877	1590	1.189	0.0006289	0.8413	86.75	363.205	129.41	541.814	42.66	178.609	0.94500	3.9565	1.15001	4.8149
−64	209.15	0.1829	0.17936	1587	1.257	0.0006301	0.7954	86.94	364.000	129.54	542.358	42.60	178.358	0.94589	3.9603	1.14961	4.8132
−63	210.15	0.1941	0.19035	1584	1.328	0.0006313	0.7528	87.12	364.754	129.65	542.819	42.53	178.065	0.94678	3.9640	1.14920	4.8115
−62	211.15	0.2059	0.20192	1581	1.403	0.0006325	0.7125	87.31	365.550	129.77	543.321	42.46	177.772	0.94769	3.9678	1.14883	4.8099
−61	212.15	0.2183	0.21408	1578	1.482	0.0006337	0.6749	87.50	366.345	129.89	543.823	42.39	177.478	0.94858	3.9715	1.14844	4.8083
−60	213.15	0.2315	0.22702	1575	1.564	0.0006349	0.6394	87.68	367.099	130.00	544.284	42.32	177.185	0.94946	3.9752	1.14806	4.8067
−59	214.15	0.2451	0.24036	1572	1.649	0.0006361	0.6064	87.87	367.894	130.12	544.786	42.25	176.892	0.95034	3.9789	1.14769	4.8051
−58	215.15	0.2595	0.25448	1569	1.738	0.0006373	0.5752	88.06	368.690	130.24	545.289	42.18	176.599	0.95122	3.9826	1.14731	4.8036
−57	216.15	0.2744	0.26909	1566	1.831	0.0006386	0.5461	88.25	369.485	130.36	545.791	42.11	176.306	0.95211	3.9863	1.14698	4.8022
−56	217.15	0.2900	0.28439	1564	1.927	0.0006394	0.5188	88.44	370.281	130.48	546.294	42.04	176.013	0.95300	3.9900	1.14663	4.8007
−55	218.15	0.3065	0.30057	1561	2.028	0.0006406	0.4930	88.63	371.076	130.59	546.754	41.96	175.678	0.95387	3.9937	1.14627	4.7992
−54	219.15	0.3236	0.31734	1558	2.134	0.0006418	0.4687	88.82	371.872	130.71	547.257	41.89	175.385	0.95474	3.9973	1.14595	4.7979
−53	220.15	0.3414	0.33480	1555	2.242	0.0006431	0.4461	89.01	372.667	130.83	547.759	41.82	175.092	0.95561	4.0009	1.14562	4.7965
−52	221.15	0.3602	0.35324	1552	2.355	0.0006443	0.4246	89.20	373.463	130.95	548.261	41.75	174.799	0.95650	4.0047	1.14531	4.7952
−51	222.15	0.3797	0.37236	1549	2.473	0.0006456	0.4043	89.39	374.258	131.06	548.722	41.67	174.464	0.95737	4.0083	1.14500	4.7939
−50	223.15	0.3999	0.39217	1546	2.595	0.0006468	0.3854	89.59	375.095	131.18	549.224	41.59	174.129	0.95824	4.0120	1.14468	4.7925
−49	224.15	0.4212	0.41306	1543	2.723	0.0006481	0.3673	89.78	375.891	131.30	549.727	41.52	173.836	0.95910	4.0156	1.14438	4.7913
−48	225.15	0.4432	0.43463	1540	2.854	0.0006493	0.3504	89.97	376.686	131.42	550.229	41.45	173.543	0.95997	4.0192	1.14410	4.7901
−47	226.15	0.4662	0.45719	1538	2.990	0.0006502	0.3344	90.17	377.524	131.54	550.732	41.37	173.208	0.96084	4.0228	1.14381	4.7889
−46	227.15	0.4900	0.48053	1535	3.132	0.0006515	0.3193	90.36	378.319	131.65	551.192	41.29	172.873	0.96170	4.0264	1.14352	4.7877
−45	228.15	0.5150	0.50504	1532	3.279	0.0006527	0.3050	90.56	379.157	131.77	551.695	41.21	172.538	0.96256	4.0300	1.14324	4.7865
−44	229.15	0.5409	0.53044	1529	3.432	0.0006540	0.2914	90.76	379.994	131.89	552.197	41.13	172.203	0.96342	4.0336	1.14297	4.7854
−43	230.15	0.5687	0.55682	1526	3.588	0.0006553	0.2787	90.95	380.789	132.01	552.699	41.06	171.910	0.96428	4.0372	1.14271	4.7843
−42	231.15	0.5958	0.58428	1523	3.752	0.0006566	0.2665	91.15	381.627	132.13	553.202	40.98	171.575	0.96515	4.0409	1.14247	4.7833
−41	232.15	0.6247	0.61262	1520	3.920	0.0006579	0.2551	91.35	382.464	132.24	553.662	40.89	171.198	0.96600	4.0444	1.14220	4.7822
−40	233.15	0.6551	0.64243	1517	4.097	0.0006592	0.2441	91.55	383.302	132.36	554.165	40.81	170.863	0.96685	4.0480	1.14193	4.7810
−39	234.15	0.6865	0.67323	1514	4.279	0.0006605	0.2337	91.75	384.139	132.48	554.667	40.73	170.528	0.96770	4.0516	1.14170	4.7801
−38	235.15	0.7189	0.70500	1511	4.466	0.0006618	0.2239	91.95	384.976	132.60	555.170	40.65	170.193	0.96855	4.0551	1.14146	4.7791
−37	236.15	0.7523	0.73775	1508	4.660	0.0006631	0.2146	92.15	385.814	132.72	555.672	40.57	169.858	0.96941	4.0587	1.14124	4.7781
−36	237.15	0.7875	0.77227	1505	4.862	0.0006645	0.2057	92.35	386.651	132.83	556.133	40.48	169.482	0.97026	4.0623	1.14101	4.7772
−35	238.15	0.8238	0.80787	1502	5.069	0.0006658	0.1973	92.55	387.488	132.95	556.635	40.40	169.147	0.97110	4.0658	1.14078	4.7762
−34	239.15	0.8610	0.84435	1499	5.280	0.0006671	0.1894	92.76	388.368	133.07	557.137	40.31	168.770	0.97194	4.0693	1.14055	4.7753
−33	240.15	0.9000	0.88260	1496	5.501	0.0006684	0.1818	92.96	389.205	133.19	557.640	40.23	168.435	0.97278	4.0728	1.14034	4.7744
−32	241.15	0.9400	0.92188	1493	5.724	0.0006698	0.1747	93.16	390.042	133.30	558.100	40.14	168.058	0.97364	4.0764	1.14014	4.7735
−31	242.15	0.9818	0.96282	1490	5.960	0.0006711	0.1678	93.37	390.922	133.43	558.645	40.06	167.723	0.97448	4.0800	1.13993	4.7727

Table 51-2 Properties of Saturated Dichlorodifluoromethane "Freon 12" (CF_2Cl_2) (Continued)

t	T	p	p	ρ'	ρ''	v'	v''	i'	i'	i''	i''	r=i''−i'	r=i''−i'	s'	s'	s''	s''
°C	K	kp/cm²	bar	kg/m³	kg/m³	m³/kg	m³/kg	kcal/kg	kJ/kg	kcal/kg	kJ/kg	kcal/kg	kJ/kg	kcal/kg K	kJ/kg K	kcal/kg K	kJ/kg K
−30	243.15	1.0245	1.00469	1487	6.200	0.0006725	0.1613	93.57	391.759	133.54	559.105	39.97	167.346	0.97532	4.0835	1.13975	4.7719
−29	244.15	1.0688	1.04813	1484	6.447	0.0006739	0.1551	93.78	392.638	133.66	559.608	39.88	166.970	0.97616	4.0870	1.13954	4.7710
−28	245.15	1.1149	1.09334	1481	6.702	0.0006752	0.1492	93.98	393.475	133.77	560.068	39.79	166.593	0.97699	4.0905	1.13934	4.7702
−27	246.15	1.1622	1.13973	1478	6.964	0.0006766	0.1436	94.19	394.355	133.90	560.613	39.71	166.258	0.97783	4.0940	1.13917	4.7695
−26	247.15	1.2109	1.18749	1475	7.236	0.0006780	0.1382	94.40	395.234	134.01	561.073	39.61	165.839	0.97867	4.0975	1.13899	4.7687
−25	248.15	1.2616	1.23721	1472	7.513	0.0006793	0.1331	94.61	396.113	134.13	561.575	39.52	165.462	0.97950	4.1010	1.13879	4.7679
−24	249.15	1.3140	1.28859	1469	7.800	0.0006807	0.1282	94.81	396.951	134.24	562.036	39.43	165.086	0.98033	4.1044	1.13862	4.7672
−23	250.15	1.3678	1.34135	1466	8.097	0.0006821	0.1235	95.02	397.830	134.36	562.538	39.34	164.709	0.98116	4.1079	1.13845	4.7665
−22	251.15	1.4227	1.39519	1463	8.403	0.0006835	0.1190	95.23	398.709	134.47	562.999	39.24	164.290	0.98200	4.1114	1.13829	4.7658
−21	252.15	1.4805	1.45187	1459	8.718	0.0006854	0.1147	95.44	399.588	134.59	563.501	39.15	163.913	0.98283	4.1149	1.13814	4.7652
−20	253.15	1.5396	1.50983	1456	9.034	0.0006868	0.1107	95.65	400.467	134.71	564.003	39.06	163.536	0.98365	4.1183	1.13798	4.7645
−19	254.15	1.6005	1.56955	1453	9.372	0.0006882	0.1067	95.87	401.389	134.83	564.506	38.96	163.118	0.98448	4.1218	1.13783	4.7639
−18	255.15	1.6627	1.63055	1450	9.709	0.0006897	0.1030	96.08	402.268	134.95	565.009	38.87	162.741	0.98531	4.1253	1.13768	4.7632
−17	256.15	1.7275	1.69410	1447	10.06	0.0006911	0.09938	96.29	403.147	135.06	565.469	38.77	162.322	0.98614	4.1288	1.13753	4.7626
−16	257.15	1.7940	1.75931	1444	10.42	0.0006925	0.09597	96.50	404.026	135.17	565.930	38.67	161.904	0.98696	4.1322	1.13738	4.7620
−15	258.15	1.8622	1.82619	1441	10.79	0.0006940	0.09268	96.72	404.947	135.29	566.432	38.57	161.485	0.98778	4.1356	1.13723	4.7614
−14	259.15	1.9321	1.89474	1438	11.17	0.0006954	0.08952	96.93	405.827	135.40	566.893	38.47	161.066	0.98860	4.1391	1.13709	4.7608
−13	260.15	2.0050	1.96623	1434	11.56	0.0006973	0.08650	97.15	406.748	135.52	567.395	38.37	160.648	0.98942	4.1425	1.13695	4.7602
−12	261.15	2.0793	2.03910	1431	11.96	0.0006988	0.08361	97.36	407.627	135.63	567.856	38.27	160.229	0.99025	4.1460	1.13682	4.7596
−11	262.15	2.1555	2.11382	1428	12.37	0.0007003	0.08082	97.58	408.548	135.75	568.358	38.17	159.810	0.99107	4.1494	1.13668	4.7591
−10	263.15	2.2342	2.19100	1425	12.80	0.0007018	0.07813	97.80	409.469	135.87	568.861	38.07	159.391	0.99188	4.1528	1.13657	4.7586
−9	264.15	2.3148	2.27004	1422	13.23	0.0007032	0.07558	98.02	410.390	135.98	569.321	37.96	158.931	0.99270	4.1562	1.13644	4.7580
−8	265.15	2.3984	2.35203	1419	13.68	0.0007047	0.07313	98.23	411.269	136.09	569.782	37.86	158.512	0.99351	4.1596	1.13633	4.7576
−7	266.15	2.4833	2.43529	1416	14.13	0.0007062	0.07078	98.45	412.190	136.20	570.242	37.75	158.052	0.99432	4.1630	1.13620	4.7570
−6	267.15	2.5712	2.52149	1413	14.60	0.0007077	0.06852	98.67	413.112	136.32	570.745	37.65	157.633	0.99514	4.1665	1.13609	4.7566
−5	268.15	2.6602	2.60877	1410	15.08	0.0007092	0.06635	98.89	414.033	136.43	571.205	37.54	157.172	0.99595	4.1698	1.13598	4.7561
−4	269.15	2.7531	2.69987	1407	15.57	0.0007107	0.06427	99.11	414.954	136.54	571.666	37.43	156.712	0.99676	4.1732	1.13586	4.7556
−3	270.15	2.8479	2.79284	1403	16.07	0.0007127	0.06228	99.33	415.875	136.65	572.126	37.32	156.251	0.99757	4.1766	1.13575	4.7552
−2	271.15	2.9439	2.88698	1400	16.59	0.0007143	0.06028	99.56	416.838	136.77	572.629	37.21	155.791	0.99839	4.1801	1.13566	4.7548
−1	272.15	3.0446	2.98573	1397	17.11	0.0007158	0.05844	99.78	417.759	136.88	573.089	37.10	155.330	0.99919	4.1834	1.13555	4.7543
0	273.15	3.1465	3.08566	1394	17.65	0.0007173	0.05667	100.00	418.680	136.99	573.550	36.99	154.870	1.00000	4.1868	1.13546	4.7539
1	274.15	3.2511	3.18824	1391	18.20	0.0007189	0.05496	100.22	419.601	137.10	574.010	36.88	154.409	1.00081	4.1902	1.13535	4.7535
2	275.15	3.3583	3.29337	1388	18.76	0.0007205	0.05330	100.45	420.564	137.21	574.471	36.76	153.907	1.00161	4.1935	1.13524	4.7530
3	276.15	3.4676	3.40055	1385	19.35	0.0007220	0.05168	100.67	421.485	137.32	574.931	36.65	153.446	1.00242	4.1969	1.13515	4.7526
4	277.15	3.5804	3.51117	1381	19.95	0.0007241	0.05012	100.90	422.448	137.43	575.392	36.53	152.944	1.00322	4.2003	1.13506	4.7523
5	278.15	3.6959	3.62444	1378	20.56	0.0007257	0.04863	101.12	423.369	137.54	575.852	36.42	152.483	1.00402	4.2036	1.13497	4.7519
6	279.15	3.8135	3.73977	1375	21.18	0.0007273	0.04721	101.35	424.332	137.65	576.313	36.30	151.981	1.00483	4.2070	1.13488	4.7515
7	280.15	3.9348	3.85872	1372	21.82	0.0007289	0.04583	101.58	425.295	137.76	576.774	36.18	151.478	1.00563	4.2104	1.13480	4.7512
8	281.15	4.0582	3.97973	1368	22.47	0.0007310	0.04450	101.80	426.216	137.86	577.192	36.06	150.976	1.00643	4.2137	1.13471	4.7508
9	282.15	4.1853	4.10438	1365	23.13	0.0007326	0.04323	102.03	427.179	137.97	577.653	35.94	150.474	1.00723	4.2171	1.13462	4.7504

Table 51-3 Properties of Saturated Dichlorodifluoromethane "Freon 12" (CF$_2$Cl$_2$) (Continued)

| Temperature | | Pressure | | Density | | Specific volume | | Specific enthalpy | | | | Heat of vaporization | | Specific entropy | | | |
| t | T | p | p | Liquid ρ' | Vapor ρ'' | Liquid v' | Vapor v'' | Liquid i' | | Vapor i'' | | r = i'' − i' | | Liquid s' | | Vapor s'' | |
°C	K	kp/cm²	bar	kg/m³	kg/m³	m³/kg	m³/kg	kcal/kg	kJ/kg	kcal/kg	kJ/kg	kcal/kg	kJ/kg	kcal/kg K	kJ/kg K	kcal/kg K	kJ/kg K
10	283.15	4.3135	4.23010	1362	23.79	0.0007342	0.04204	102.26	428.142	138.08	578.113	35.82	149.971	1.00803	4.2204	1.13455	4.7501
11	284.15	4.4466	4.36062	1359	24.48	0.0007358	0.04086	102.49	429.105	138.18	578.532	35.69	149.427	1.00883	4.2238	1.13446	4.7498
12	285.15	4.5828	4.49419	1355	25.19	0.0007380	0.03970	102.72	430.068	138.29	578.993	35.57	148.924	1.00963	4.2271	1.13439	4.7495
13	286.15	4.7209	4.62962	1352	25.92	0.0007396	0.03858	102.95	431.031	138.39	579.411	35.44	148.380	1.01042	4.2304	1.13430	4.7491
14	287.15	4.8621	4.76809	1349	26.66	0.0007413	0.03751	103.18	431.994	138.49	579.830	35.31	147.836	1.01122	4.2338	1.13422	4.7488
15	288.15	5.0076	4.91078	1345	27.41	0.0007435	0.03648	103.42	432.999	138.61	580.332	35.19	147.333	1.01201	4.2371	1.13414	4.7484
16	289.15	5.1550	5.05533	1342	28.19	0.0007452	0.03547	103.65	433.962	138.70	580.709	35.05	146.747	1.01281	4.2404	1.13407	4.7481
17	290.15	5.3067	5.20409	1339	28.99	0.0007468	0.03449	103.88	434.925	138.81	581.170	34.93	146.245	1.01361	4.2438	1.13400	4.7478
18	291.15	5.4605	5.35492	1335	29.87	0.0007491	0.03354	104.12	435.930	138.91	581.588	34.79	145.659	1.01440	4.2471	1.13392	4.7475
19	292.15	5.6172	5.50859	1332	30.65	0.0007507	0.03263	104.35	436.893	139.01	582.007	34.66	145.141	1.01519	4.2504	1.13385	4.7472
20	293.15	5.7786	5.66687	1329	31.50	0.0007524	0.03175	104.59	437.897	139.12	582.468	34.53	144.570	1.01598	4.2537	1.13378	4.7469
21	294.15	5.9432	5.82829	1325	32.38	0.0007547	0.03089	104.82	438.860	139.21	582.844	34.39	143.984	1.01678	4.2571	1.13372	4.7467
22	295.15	6.1112	5.99304	1321	33.28	0.0007570	0.03005	105.06	439.865	139.31	583.263	34.25	143.398	1.01757	4.2604	1.13364	4.7463
23	296.15	6.2825	6.16103	1318	34.19	0.0007587	0.02925	105.29	440.828	139.40	583.640	34.11	142.812	1.01835	4.2636	1.13356	4.7460
24	297.15	6.4584	6.33353	1315	35.11	0.0007605	0.02848	105.53	441.833	139.50	584.059	33.97	142.226	1.01914	4.2669	1.13350	4.7457
25	298.15	6.6363	6.50799	1311	36.07	0.0007628	0.02773	105.77	442.838	139.61	584.519	33.84	141.681	1.01993	4.2702	1.13344	4.7455
26	299.15	6.8175	6.68568	1308	37.04	0.0007645	0.02700	106.01	443.843	139.70	584.896	33.69	141.053	1.02072	4.2736	1.13337	4.7452
27	300.15	7.0020	6.86662	1304	38.04	0.0007669	0.02629	106.25	444.848	139.79	585.373	33.54	140.425	1.02151	4.2769	1.13329	4.7449
28	301.15	7.1933	7.05422	1300	39.06	0.0007692	0.02560	106.49	445.852	139.89	585.691	33.40	139.839	1.02229	4.2801	1.13322	4.7446
29	302.15	7.3863	7.24349	1297	40.10	0.0007710	0.02494	106.73	446.857	139.98	586.068	33.25	139.211	1.02307	4.2834	1.13315	4.7443
30	303.15	7.5810	7.43442	1293	41.11	0.0007734	0.02433	106.97	447.862	140.08	586.487	33.11	138.625	1.02387	4.2867	1.13310	4.7441
31	304.15	7.7826	7.63212	1289	42.18	0.0007758	0.02371	107.21	448.867	140.16	586.822	32.95	137.955	1.02465	4.2900	1.13301	4.7437
32	305.15	7.9897	7.83522	1285	43.31	0.0007782	0.02309	107.45	449.872	140.25	587.199	32.80	137.327	1.02543	4.2933	1.13294	4.7434
33	306.15	8.2003	8.04175	1282	44.45	0.0007800	0.02250	107.69	450.876	140.34	587.576	32.65	136.699	1.02620	4.2965	1.13286	4.7431
34	307.15	8.4087	8.24612	1278	45.62	0.0007825	0.02192	107.94	451.923	140.43	587.952	32.49	136.029	1.02699	4.2998	1.13280	4.7428
35	308.15	8.6264	8.45961	1274	46.81	0.0007849	0.02136	108.18	452.928	140.51	588.287	32.33	135.359	1.02778	4.3031	1.13273	4.7425
36	309.15	8.8475	8.67643	1270	48.01	0.0007874	0.02083	108.43	453.975	140.61	588.706	32.18	134.731	1.02856	4.3064	1.13266	4.7422
37	310.15	9.0726	8.89718	1267	49.25	0.0007893	0.02030	108.67	454.980	140.69	589.041	32.02	134.061	1.02934	4.3096	1.13258	4.7419
38	311.15	9.2989	9.11911	1263	50.51	0.0007918	0.01980	108.92	456.026	140.77	589.376	31.85	133.350	1.03011	4.3129	1.13250	4.7416
39	312.15	9.5351	9.35074	1259	51.79	0.0007943	0.01931	109.16	457.031	140.85	589.711	31.69	132.680	1.03089	4.3161	1.13243	4.7413
40	313.15	9.7707	9.58178	1255	53.13	0.0007968	0.01882	109.41	458.078	140.94	590.088	31.53	132.010	1.03167	4.3194	1.13236	4.7410
41	314.15	10.014	9.82038	1251	54.49	0.0007994	0.01835	109.66	459.124	141.02	590.423	31.36	131.298	1.03246	4.3227	1.13229	4.7407
42	315.15	10.257	10.05868	1247	55.90	0.0008019	0.01789	109.91	460.171	141.10	590.757	31.19	130.586	1.03324	4.3260	1.13222	4.7404
43	316.15	10.511	10.30777	1243	57.34	0.0008045	0.01744	110.16	461.218	141.18	591.092	31.02	129.875	1.03400	4.3292	1.13212	4.7400
44	317.15	10.763	10.55490	1239	58.83	0.0008071	0.01700	110.41	462.265	141.25	591.386	30.84	129.121	1.03478	4.3324	1.13204	4.7396
45	318.15	11.023	10.80987	1234	60.38	0.0008104	0.01656	110.66	463.311	141.33	591.720	30.67	128.409	1.03556	4.3357	1.13197	4.7393
46	319.15	11.283	11.06484	1230	61.95	0.0008130	0.01614	110.91	464.358	141.40	592.014	30.49	127.656	1.03634	4.3389	1.13188	4.7390
47	320.15	11.553	11.32962	1226	63.57	0.0008157	0.01573	111.15	465.363	141.48	592.348	30.31	126.902	1.03712	4.3422	1.13180	4.7386
48	321.15	11.828	11.59931	1221	65.24	0.0008190	0.01533	111.42	466.493	141.56	592.683	30.14	126.190	1.03788	4.3454	1.13173	4.7383
49	322.15	12.108	11.87389	1217	66.94	0.0008217	0.01494	111.67	467.540	141.64	593.018	29.97	125.487	1.03865	4.3486	1.13168	4.7381

Table 51-4   Properties of Saturated Dichlorodifluoromethane "Freon 12" (CF_2Cl_2) (Continued)

Temperature t (°C)	Temperature T (K)	Pressure p (kp/cm²)	Pressure p (bar)	Density Liquid ρ' (kg/m³)	Density Vapor ρ" (kg/m³)	Sp. vol. Liquid v' (m³/kg)	Sp. vol. Vapor v" (m³/kg)	Enthalpy Liquid i' (kcal/kg)	Enthalpy Liquid i' (kJ/kg)	Enthalpy Vapor i" (kcal/kg)	Enthalpy Vapor i" (kJ/kg)	Heat of vap. r=i"−i' (kcal/kg)	Heat of vap. r=i"−i' (kJ/kg)	Entropy Liquid s' (kcal/kg K)	Entropy Liquid s' (kJ/kg K)	Entropy Vapor s" (kcal/kg K)	Entropy Vapor s" (kJ/kg K)
50	323.15	12.386	12.14652	1213	68.56	0.0008244	0.01459	111.94	468.670	141.73	593.395	29.79	124.725	1.03945	4.3520	1.13163	4.7379
55	328.15	13.868	13.59986	1189	75.98	0.0008410	0.01316	113.25	474.155	142.13	595.070	28.88	120.915	1.0433	4.3681	1.1314	4.7369
60	333.15	15.481	15.18167	1167	85.69	0.0008568	0.01167	114.57	479.682	142.49	596.577	27.92	116.895	1.0472	4.3844	1.1311	4.7357
65	338.15	17.216	16.88313	1141	96.52	0.0008741	0.01036	115.92	485.334	142.82	597.959	26.90	112.625	1.0511	4.4007	1.1307	4.7340
70	343.15	19.096	18.72678	1119	108.81	0.0008936	0.00919	117.29	491.070	143.09	599.089	25.80	108.019	1.0550	4.4171	1.1302	4.7319
75	348.15	21.125	20.71655	1093	122.85	0.0009149	0.00814	118.69	496.931	143.31	600.010	24.62	103.079	1.0590	4.4338	1.1297	4.7298
80	353.15	23.290	22.83969	1064	138.31	0.0009398	0.00723	120.13	502.960	143.46	600.638	23.33	97.678	1.0629	4.4501	1.1290	4.7269
85	358.15	25.620	25.12464	1033	156.49	0.0009680	0.00639	121.61	509.157	143.51	600.848	21.90	91.691	1.0669	4.4669	1.1281	4.7231
90	363.15	28.107	27.56355	999	177.30	0.0010009	0.00564	123.12	515.479	143.41	600.429	20.29	84.950	1.0700	4.4799	1.1269	4.7181
95	368.15	30.771	30.17605	960	201.20	0.0010416	0.00497	124.69	522.052	143.11	599.173	18.42	77.121	1.0714	4.4857	1.1252	4.7110
100	373.15	33.614	32.96407	913	228.83	0.0010952	0.00437	126.36	529.044	142.51	596.661	16.15	67.617	1.0794	4.5192	1.1227	4.7005
105	378.15	36.654	35.94529	852	278.48	0.0011736	0.00359	128.13	536.455	141.51	592.474	13.38	56.019	1.0841	4.5389	1.1195	4.6871
110	383.15	39.874	39.10304	742	374.93	0.0013513	0.00266	131.44	550.313	138.89	581.505	7.45	31.192	1.0917	4.5707	1.1111	4.6520
115,5 (krit.)	388.65	40.879	40.08860	558	557.59	0.0017934	0.00179	134.75	564.171	134.75	564.171	0	0	1.1016	4.6122	1.1016	4.6122

1 at = 1 kp/cm² = 98 066.5 Pa = 98 066.5 N/m² = 9.806 65 N/cm² = 0.980 665 bar

1 kcal = 4.186 8 kJ

Table 52-1 Properties of Saturated Chlorodifluoromethane "Freon 22" (CHF$_2$Cl)

Temperature t °C	T K	Pressure p kp/cm²	p bar	Density Liquid ρ' kg/m³	Density Vapor ρ'' kg/m³	Spec. vol. Liquid v' m³/kg	Spec. vol. Vapor v'' m³/kg	Enthalpy Liquid i' kcal/kg	i' kJ/kg	Enthalpy Vapor i'' kcal/kg	i'' kJ/kg	Heat of vap. r kcal/kg	r kJ/kg	Entropy Liquid s' kcal/kg K	s' kJ/kg K	Entropy Vapor s'' kcal/kg K	s'' kJ/kg K
−100	173.15	0.0210	0.02059	1560	0.1199	0.0006409	8.340	74.12	310.326	137.92	577.443	63.80	267.118	0.8828	3.6961	1.2512	5.2385
−98	175.15	0.0243	0.02383	1555	0.1433	0.0006429	6.980	74.63	312.461	138.16	578.448	63.55	265.987	0.8858	3.7087	1.2485	5.2272
−96	177.15	0.0292	0.02864	1550	0.1868	0.0006450	5.890	75.14	314.596	138.40	579.453	63.26	264.857	0.8886	3.7204	1.2457	5.2155
−94	179.15	0.0348	0.03413	1545	0.2006	0.0006470	4.985	75.63	316.648	138.62	580.374	62.99	263.727	0.8914	3.7321	1.2430	5.2042
−92	181.15	0.0410	0.04021	1540	0.2353	0.0006490	4.250	76.12	318.699	138.84	581.295	62.72	262.596	0.8942	3.7438	1.2404	5.1933
−90	183.15	0.0489	0.04795	1536	0.2752	0.0006510	3.634	76.63	320.834	139.14	582.551	62.51	261.717	0.8970	3.7556	1.2382	5.1841
−88	185.15	0.0575	0.05639	1531	0.3208	0.0006530	3.117	77.14	322.970	139.34	583.389	62.20	260.419	0.8997	3.7669	1.2356	5.1732
−86	187.15	0.0670	0.06570	1526	0.3691	0.0006550	2.709	77.65	325.105	139.58	584.394	61.93	259.289	0.9024	3.7782	1.2333	5.1636
−84	189.15	0.0781	0.07659	1522	0.4292	0.0006570	2.330	78.15	327.198	139.81	585.357	61.66	258.158	0.9051	3.7895	1.2311	5.1544
−82	191.15	0.0910	0.08924	1517	0.4926	0.0006592	2.030	78.65	329.292	140.05	586.361	61.40	257.070	0.9078	3.8008	1.2290	5.1456
−80	193.15	0.1050	0.10297	1512	0.5634	0.0006612	1.775	79.14	331.343	140.29	587.366	61.15	256.023	0.9104	3.8117	1.2270	5.1372
−78	195.15	0.1213	0.11895	1507	0.6464	0.0006632	1.547	79.65	333.479	140.54	588.413	60.89	254.934	0.9130	3.8225	1.2250	5.1288
−76	197.15	0.1400	0.13729	1503	0.7337	0.0006653	1.363	80.14	335.530	140.77	589.376	60.63	253.846	0.9155	3.8330	1.2230	5.1205
−74	199.15	0.1605	0.15740	1498	0.8292	0.0006675	1.206	80.64	337.624	141.01	590.381	60.37	252.757	0.9180	3.8435	1.2211	5.1125
−72	201.15	0.1832	0.17966	1494	0.9434	0.0006693	1.060	81.15	339.759	141.26	591.427	60.11	251.669	0.9206	3.8544	1.2194	5.1054
−70	203.15	0.2088	0.20476	1489	1.064	0.0006714	0.940	81.64	341.810	141.49	592.390	59.85	250.580	0.9230	3.8644	1.2176	5.0978
−68	205.15	0.2370	0.23242	1484	1.130	0.0006735	0.885	82.15	343.946	141.74	593.437	59.59	249.491	0.9254	3.8745	1.2159	5.0907
−66	207.15	0.267	0.26184	1480	1.341	0.0006756	0.746	82.64	345.997	141.96	594.358	59.32	248.361	0.9278	3.8845	1.2141	5.0832
−64	209.15	0.303	0.29714	1475	1.513	0.0006778	0.661	83.15	348.132	142.21	595.405	59.06	247.272	0.9302	3.8946	1.2126	5.0769
−62	211.15	0.341	0.33441	1470	1.689	0.0006801	0.592	83.65	350.226	142.44	596.368	58.79	246.142	0.9325	3.9042	1.2109	5.0698
−60	213.15	0.382	0.37461	1465	1.869	0.0006824	0.535	84.15	352.319	142.74	597.624	58.59	245.305	0.9350	3.9147	1.2097	5.0648
−58	215.15	0.428	0.41972	1460	2.079	0.0006849	0.481	84.65	354.341	142.91	598.336	58.26	243.923	0.9372	3.9239	1.2080	5.0577
−56	217.15	0.479	0.46964	1455	2.304	0.0006874	0.434	85.16	356.548	143.16	599.382	58.00	242.834	0.9396	3.9339	1.2067	5.0522
−54	219.15	0.534	0.52368	1450	2.545	0.0006897	0.393	85.67	358.683	143.40	600.387	57.73	241.704	0.9419	3.9435	1.2053	5.0464
−52	221.15	0.593	0.58153	1444	2.817	0.0006923	0.355	86.18	360.818	143.65	601.434	57.47	240.615	0.9442	3.9532	1.2041	5.0413
−50	223.15	0.660	0.64724	1439	3.096	0.0006950	0.323	86.70	362.996	143.90	602.481	57.20	239.485	0.9465	3.9628	1.2028	5.0359
−48	225.15	0.730	0.71589	1433	3.413	0.0006977	0.293	87.21	365.131	144.15	603.527	56.94	238.396	0.9488	3.9724	1.2017	5.0313
−46	227.15	0.807	0.79140	1427	3.745	0.0007005	0.267	87.72	367.266	144.39	604.532	56.67	237.266	0.9512	3.9825	1.2007	5.0271
−44	229.15	0.891	0.87377	1422	4.098	0.0007030	0.244	88.25	369.485	144.63	605.537	56.38	236.052	0.9534	3.9917	1.1994	5.0216
−42	231.15	0.979	0.96007	1416	4.484	0.0007058	0.223	88.75	371.579	144.85	606.458	56.10	234.879	0.9557	4.0013	1.1984	5.0175
−40	233.15	1.076	1.05520	1411	4.878	0.0007086	0.205	89.27	373.756	145.12	607.588	55.85	233.833	0.9579	4.0105	1.1974	5.0133
−38	235.15	1.182	1.15915	1405	5.319	0.0007113	0.188	89.77	375.849	145.29	608.300	55.52	232.451	0.9602	4.0202	1.1963	5.0087
−36	237.15	1.295	1.26996	1400	5.780	0.0007142	0.173	90.32	378.152	145.56	609.431	55.24	231.279	0.9624	4.0294	1.1953	5.0045
−34	239.15	1.414	1.38666	1395	6.329	0.0007173	0.158	90.85	380.371	145.79	610.394	54.94	230.023	0.9646	4.0386	1.1943	5.0003
−32	241.15	1.542	1.51219	1388	6.849	0.0007205	0.146	91.37	282.548	146.02	611.357	54.65	228.809	0.9668	4.0478	1.1934	4.9965
−30	243.15	1.679	1.64654	1382	7.407	0.0007235	0.135	91.90	384.767	146.25	612.320	54.35	227.553	0.9690	4.0570	1.1925	4.9928
−28	245.15	1.824	1.78873	1375	8.000	0.0007270	0.125	92.45	387.070	146.48	613.282	54.03	226.313	0.9712	4.0662	1.1916	4.9890
−26	247.15	1.978	1.93976	1369	8.621	0.0007304	0.116	93.00	389.372	146.71	614.245	53.71	224.873	0.9733	4.0750	1.1906	4.9848
−24	249.15	2.14	2.09862	1363	9.259	0.0007337	0.108	93.51	391.508	146.91	615.083	53.40	223.575	0.9754	4.0838	1.1897	4.9810
−22	251.15	2.32	2.27514	1356	10.00	0.0007370	0.100	94.04	393.727	147.12	615.962	53.08	222.235	0.9775	4.0926	1.1888	4.9773
−20	253.15	2.51	2.46147	1350	10.76	0.0007405	0.0929	94.58	395.988	147.35	616.925	52.77	220.937	0.9796	4.1014	1.1880	4.9739
−18	255.15	2.70	2.64780	1344	11.57	0.0007437	0.0864	95.12	398.248	147.58	617.888	52.46	219.640	0.9817	4.1102	1.1873	4.9710
−16	257.15	2.92	2.86354	1338	12.43	0.0007472	0.0805	95.65	400.467	147.80	618.809	52.15	218.342	0.9837	4.1186	1.1865	4.9676
−14	259.15	3.14	3.07929	1331	13.32	0.0007508	0.0751	96.18	402.686	148.02	619.730	51.84	217.044	0.9857	4.1269	1.1857	4.9643
−12	261.15	3.37	3.30484	1325	14.29	0.0007545	0.0700	96.70	404.864	148.23	620.609	51.53	215.746	0.9878	4.1357	1.1851	4.9618

Table 52-2 Properties of Saturated Chlorodifluoromethane "Freon 22" (CHF$_2$Cl) (Continued)

Temperature t (°C)	T (K)	Pressure p (kp/cm²)	p (bar)	Density Liquid ρ' (kg/m³)	Density Vapor ρ'' (kg/m³)	Spec. vol. Liquid v' (m³/kg)	Spec. vol. Vapor v'' (m³/kg)	Enthalpy Liquid i' (kcal/kg)	i' (kJ/kg)	Enthalpy Vapor i'' (kcal/kg)	i'' (kJ/kg)	Heat of vap. r=i''−i' (kcal/kg)	r (kJ/kg)	Entropy Liquid s' (kcal/kg K)	s' (kJ/kg K)	Entropy Vapor s'' (kcal/kg K)	s'' (kJ/kg K)
−10	263.15	3.63	3.55981	1318	15.29	0.0007582	0.0654	97.25	407.166	148.45	621.530	51.20	214.364	0.9898	4.1441	1.1844	4.9588
−8	265.15	3.89	3.81479	1312	16.37	0.0007620	0.0611	97.78	409.385	148.63	622.284	50.85	212.899	0.9918	4.1525	1.1836	4.9555
−6	267.15	4.17	4.08937	1305	17.48	0.0007658	0.0572	98.31	411.604	148.83	623.121	50.52	211.517	0.9938	4.1608	1.1829	4.9526
−4	269.15	4.46	4.37377	1299	18.66	0.0007697	0.0536	98.87	413.949	149.03	623.959	50.16	210.010	0.9959	4.1696	1.1823	4.9501
−2	271.15	4.77	4.67777	1292	19.92	0.0007739	0.0502	99.43	416.294	149.23	624.796	49.80	208.503	0.9979	4.1780	1.1816	4.9471
0	273.15	5.10	5.00139	1285	21.23	0.0007785	0.0471	100.00	418.680	149.43	625.634	49.43	206.954	1.0000	4.1868	1.1810	4.9446
2	275.15	5.44	5.33482	1278	22.57	0.0007823	0.0443	100.58	421.108	149.63	626.471	49.05	205.363	1.0022	4.1960	1.1805	4.9425
4	277.15	5.82	5.70747	1271	24.04	0.0007867	0.0416	101.16	423.537	149.81	627.225	48.65	203.688	1.0043	4.2048	1.1798	4.9396
6	279.15	6.18	6.06051	1264	25.64	0.0007912	0.0390	101.77	426.091	150.01	628.062	48.24	201.971	1.0064	4.2136	1.1792	4.9371
8	281.15	6.57	6.44297	1257	27.25	0.0007957	0.0367	102.40	428.728	150.20	628.857	47.80	200.129	1.0086	4.2228	1.1786	4.9346
10	283.15	6.99	6.85485	1249	28.90	0.0008004	0.0346	103.02	431.324	150.36	629.527	47.40	198.454	1.0107	4.2316	1.1780	4.9321
12	285.15	7.42	7.27653	1242	30.67	0.0008050	0.0326	103.60	433.752	150.52	630.197	46.92	196.445	1.0128	4.2404	1.1773	4.9291
14	287.15	7.87	7.71783	1235	32.57	0.0008096	0.0307	104.25	436.474	150.72	631.034	46.47	194.561	1.0150	4.2496	1.1768	4.9270
16	289.15	8.34	8.17875	1228	34.60	0.0008145	0.0289	104.87	439.210	150.87	631.663	46.00	192.593	1.0172	4.2588	1.1763	4.9249
18	291.15	8.83	8.65927	1220	36.63	0.0008194	0.0273	105.50	441.707	151.00	632.207	45.50	190.499	1.0193	4.2676	1.1756	4.9220
20	293.15	9.35	9.16922	1213	38.76	0.0008244	0.0258	106.13	444.345	151.13	632.751	45.00	188.406	1.0214	4.2764	1.1749	4.9191
22	295.15	9.89	9.69878	1206	41.15	0.0008294	0.0243	106.78	447.067	151.27	633.337	44.49	186.271	1.0236	4.2856	1.1743	4.9166
24	297.15	10.45	10.24795	1198	43.48	0.0008345	0.0230	107.38	449.579	151.38	633.798	43.96	184.052	1.0258	4.2948	1.1737	4.9140
26	299.15	11.03	10.81673	1190	46.08	0.0008398	0.0217	108.10	452.593	151.54	634.468	43.44	181.875	1.0280	4.3040	1.1732	4.9120
28	301.15	11.63	11.40513	1183	48.54	0.0008455	0.0206	108.75	455.315	151.65	634.928	42.90	179.614	1.0302	4.3132	1.1726	4.9094
30	303.15	12.26	12.02295	1176	51.55	0.0008501	0.0194	109.44	458.203	151.78	635.473	42.34	177.269	1.0323	4.3220	1.1720	4.9069
32	305.15	12.92	12.67012	1167	54.34	0.0008570	0.0184	110.10	460.967	151.87	635.849	41.77	174.883	1.0344	4.3308	1.1713	4.9040
34	307.15	13.60	13.33704	1158	57.47	0.0008632	0.0174	110.77	463.772	151.97	636.268	41.20	172.496	1.0365	4.3396	1.1706	4.9011
36	309.15	14.30	14.02351	1150	60.61	0.0008695	0.0165	111.43	466.535	152.03	636.519	40.60	169.984	1.0386	4.3484	1.1699	4.8981
38	311.15	15.02	14.72959	1141	64.10	0.0008760	0.0156	112.10	469.340	152.07	636.687	39.97	167.346	1.0408	4.3576	1.1693	4.8956
40	313.15	15.79	15.48470	1132	67.57	0.0008830	0.0148	112.77	472.145	152.12	636.896	38.35	160.564	1.0429	4.3664	1.1686	4.8927
42	315.15	16.58	16.24531	1123	71.43	0.0008900	0.0140	113.45	474.992	152.19	637.189	38.74	162.197	1.0451	4.3756	1.1680	4.8902
44	317.15	17.39	17.05376	1114	75.19	0.0008972	0.0133	114.13	477.839	152.23	637.357	38.10	159.517	1.0472	4.3844	1.1673	4.8873
46	319.15	18.23	17.87752	1105	79.37	0.0009049	0.0126	114.82	480.728	152.26	637.482	37.44	156.754	1.0493	4.3932	1.1666	4.8843
48	321.15	19.10	18.73070	1095	83.33	0.0009132	0.0120	115.51	483.617	152.29	637.608	36.78	153.991	1.0514	4.4020	1.1659	4.8814
50	323.15	20.00	19.61330	1085	88.50	0.0009214	0.0113	116.27	486.799	152.37	637.943	36.10	151.143	1.0535	4.4108	1.1653	4.8789
52	325.15	20.93	20.52532	1075	93.457	0.0009303	0.0107	116.97	489.730	152.43	638.194	35.46	148.464	1.0556	4.4196	1.1647	4.8764
54	327.15	21.886	21.46283	1064.4	100.000	0.0009395	0.0100	117.65	492.577	152.47	638.361	34.82	145.784	1.0577	4.4284	1.1642	4.8743
56	329.15	22.879	22.43663	1053.6	107.526	0.0009491	0.0093	118.32	495.382	152.50	638.487	34.18	143.105	1.0598	4.4372	1.1637	4.8722
58	331.15	23.905	23.44280	1042.6	116.279	0.0009591	0.0086	118.99	498.187	152.53	638.613	33.54	140.425	1.0619	4.4460	1.1632	4.8701
60	333.15	24.969	24.48622	1031.4	126.582	0.0009696	0.0079	119.66	500.992	152.56	638.738	32.90	137.746	1.0640	4.4548	1.1628	4.8684
70	343.15	30.97	30.37120	971	149.6	0.001029	0.0067	123.95	518.954	152.32	637.733	28.37	118.780	1.074	4.4966	1.158	4.8483

1 at = 1 kp/cm² = 98 066.5 Pa = 98 066.5 N/m² = 9.806 65 N/cm² = 0.980 665 bar   1 kcal = 4.186 8 kJ

Table 53-1 Properties of Saturated Ethane (C₂H₆)

Temperature t °C	T K	Pressure p kp/cm²	p bar	Density Liquid ρ' kg/m³	Vapor ρ'' kg/m³	Sp. vol. Liquid v' m³/kg	Vapor v'' m³/kg	Enthalpy Liquid i' kcal/kg	i' kJ/kg	Vapor i'' kcal/kg	i'' kJ/kg	Heat of vap. r=i''-i' kcal/kg	r kJ/kg	Entropy Liquid s' kcal/kg K	s' kJ/kg K	Vapor s'' kcal/kg K	s'' kJ/kg K
−100	173.15	0.5354	0.52505	558.9	1.125	0.001789	888.8	35.52	148.715	155.07	649.247	119.55	500.532	0.7145	2.9915	1.4049	5.8820
−95	178.15	0.7229	0.70892	553.1	1.486	0.001808	673.1	38.42	160.857	156.39	654.774	117.97	493.917	0.7310	3.0606	1.3932	5.8330
−90	183.15	0.9596	0.94105	547.9	1.932	0.001825	517.7	41.37	173.208	157.69	660.216	116.32	487.009	0.7472	3.1284	1.3823	5.7874
−85	188.15	1.251	1.22681	542.2	2.470	0.001844	404.8	44.33	185.601	158.96	665.534	114.63	479.933	0.7632	3.1954	1.3724	5.7460
−80	193.15	1.606	1.57495	536.7	3.116	0.001863	320.9	47.25	197.826	160.19	670.683	112.94	472.857	0.7785	3.2594	1.3632	5.7074
−75	198.15	2.037	1.99761	530.9	3.819	0.001884	257.0	50.21	210.219	161.39	675.708	111.18	465.488	0.7934	3.3218	1.3545	5.6710
−70	203.15	2.549	2.49972	525.0	4.798	0.001905	208.4	53.17	222.612	162.56	680.606	109.39	457.994	0.8081	3.3834	1.3466	5.6379
−65	208.15	3.154	3.09302	519.0	5.862	0.001927	170.6	56.12	234.963	163.68	685.295	107.56	450.332	0.8223	3.4428	1.3390	5.6061
−60	213.15	3.861	3.78635	512.5	7.097	0.001951	140.9	59.11	247.482	164.76	689.817	105.65	442.335	0.8364	3.5018	1.3320	5.5768
−55	218.15	4.682	4.59147	506.0	8.525	0.001976	117.3	62.12	260.084	165.79	694.130	103.67	434.046	0.8500	3.5588	1.3253	5.5488
−50	223.15	5.626	5.51722	499.3	10.17	0.002003	98.32	65.08	272.477	166.76	698.191	101.68	425.714	0.8634	3.6149	1.3190	5.5224
−45	228.15	6.704	6.57438	492.1	12.05	0.002032	83.01	68.15	285.330	167.69	702.084	99.54	416.754	0.8767	3.6706	1.3130	5.4973
−40	233.15	7.929	7.77569	485.0	14.19	0.002062	70.46	71.30	298.519	168.54	705.643	97.24	407.124	0.8901	3.7267	1.3072	5.4730
−35	238.15	9.309	9.12901	477.8	16.63	0.002093	60.13	74.56	312.168	169.33	708.951	94.77	396.783	0.9037	3.7836	1.3016	5.4495
−30	243.15	10.86	10.65002	470.0	19.41	0.002128	51.53	77.93	326.277	170.05	711.965	92.12	385.688	0.9173	3.8406	1.2962	5.4269
−25	248.15	12.58	12.33677	461.5	22.54	0.002167	44.36	81.32	340.471	170.69	714.645	89.37	374.174	0.9313	3.8992	1.2914	5.4068
−20	253.15	14.51	14.22945	452.6	26.11	0.002209	38.30	84.88	355.376	171.24	716.948	86.36	361.572	0.9446	3.9549	1.2857	5.3830
−15	258.15	16.63	16.30846	443.5	30.16	0.002255	33.16	88.59	370.909	171.70	718.874	83.11	347.965	0.9586	4.0135	1.2805	5.3612
−10	263.15	18.96	18.59341	433.9	34.73	0.002305	28.79	92.27	386.316	172.06	720.381	79.79	334.065	0.9723	4.0708	1.2755	5.3403
−5	268.15	21.52	21.10391	423.0	39.97	0.002364	25.02	96.07	402.226	172.31	721.428	76.24	319.202	0.9861	4.1286	1.2704	5.3189
0	273.15	24.32	23.84977	411.7	45.98	0.002429	21.75	100.00	418.680	172.44	721.972	72.44	303.292	1.0000	4.1868	1.2652	5.2971
5	278.15	27.39	26.86041	399.5	53.19	0.002503	18.80	104.09	435.804	172.17	720.841	68.08	285.037	1.0142	4.2463	1.2590	5.2712
10	283.15	30.75	30.15545	386.5	62.00	0.002587	16.13	108.45	454.058	171.55	718.246	63.10	264.187	1.0290	4.3082	1.2519	5.2415
15	288.15	34.43	33.76430	369.5	73.21	0.002706	13.66	113.11	473.569	170.20	712.593	57.09	239.024	1.0454	4.3769	1.2426	5.2025
20	293.15	38.49	37.74580	350.2	87.49	0.002856	11.43	118.20	494.880	168.41	705.099	50.21	210.219	1.0610	4.4422	1.2323	5.1594
25	298.15	42.98	42.14898	326.0	106.7	0.00307	9.37	123.85	518.535	165.64	693.502	41.79	174.966	1.0791	4.5180	1.2193	5.1050
30	303.15	48.0	47.07192	286.0	142.0	0.00349	7.06	132.09	553.034	159.71	668.674	27.01	113.085	1.1052	4.6273	1.1943	5.0003
31	304.15	49.1	48.15065	271.0	156.0	0.00369	6.43	135.00	565.218	157.30	658.584	21.39	89.556	1.1145	4.6662	1.1848	4.9605
32.1 (crit.)	305.25	50.3	49.32745	213.0	213.0	0.00470	4.70	145.75	610.226	145.75	610.226	0	0	1.1494	4.8123	1.1494	4.8123

1 at = 1 kp/cm² = 98 066.5 Pa = 98 066.5 N/m² = 9.806 65 N/cm² = 0.980 665 bar

1 kcal = 4.186 8 kJ

Table 54-1 Properties of Saturated Methyl Chloride (CH_3Cl)

t °C	T K	p kp/cm²	p bar	ρ' kg/m³	ρ'' kg/m³	v' m³/kg	v'' m³/kg	i' kcal/kg	i' kJ/kg	i'' kcal/kg	i'' kJ/kg	r kcal/kg	r kJ/kg	s' kcal/kg·K	s' kJ/kg·K	s'' kcal/kg·K	s'' kJ/kg·K
−60	213.15	0.159	0.15593	1068	0.448	0.000936	2.235	78.47	328.538	188.46	789.044	109.99	460.506	0.9110	381.4175	1.4271	597.4982
−55	218.15	0.216	0.21182	1059	0.595	0.000944	1.680	80.17	335.656	189.21	792.184	109.04	456.529	0.9191	384.8088	1.4189	594.0651
−50	223.15	0.286	0.28047	1050	0.772	0.000953	1.295	81.94	343.066	189.95	795.283	108.01	452.216	0.9270	388.1164	1.4111	590.7993
−45	228.15	0.375	0.36775	1041	0.992	0.000961	1.008	83.69	350.393	190.67	798.297	106.98	447.904	0.9349	391.4239	1.4037	587.7011
−40	233.15	0.484	0.47464	1031	1.259	0.000970	0.794	85.45	357.762	191.41	801.395	105.96	443.633	0.9425	394.6059	1.3969	584.8541
−37.5	235.65	0.548	0.53740	1027	1.414	0.000974	0.707	86.34	361.488	191.77	802.903	105.43	441.414	0.9463	396.1669	1.3936	583.4724
−35	238.15	0.619	0.60703	1023	1.583	0.000978	0.632	87.23	365.215	192.12	804.368	104.89	439.153	0.9500	397.7460	1.3904	582.1327
−32.5	240.65	0.697	0.68352	1018	1.768	0.000982	0.566	88.13	368.983	192.47	805.833	104.34	436.851	0.9538	399.3370	1.3873	580.8348
−30	243.15	0.783	0.76786	1014	1.969	0.000986	0.508	89.03	372.751	192.83	807.341	103.80	434.590	0.9575	400.8861	1.3843	579.5787
−27.5	245.65	0.877	0.86004	1010	2.188	0.000991	0.457	89.92	376.477	193.17	808.764	103.25	432.287	0.9611	402.3933	1.3814	578.3646
−25	248.15	0.979	0.96007	1005	2.425	0.000995	0.412	90.81	380.203	193.51	810.188	102.70	429.984	0.9648	403.9425	1.3786	577.1922
−22.5	250.65	1.090	1.06892	1001	2.682	0.000999	0.373	91.72	384.013	193.86	811.653	102.14	427.640	0.9684	405.4497	1.3759	576.0618
−20	253.15	1.212	1.18857	997	2.959	0.001003	0.338	92.64	387.865	194.21	813.118	101.57	425.253	0.9720	406.9570	1.3732	574.9314
−17.5	255.65	1.344	1.31801	992	3.260	0.001008	0.307	93.55	391.675	194.55	814.542	101.00	422.867	0.9756	408.4642	1.3707	573.8847
−15	258.15	1.487	1.45825	988	3.582	0.001013	0.279	94.46	395.485	194.89	815.965	100.43	420.480	0.9792	409.9715	1.3682	572.8380
−12.5	260.65	1.641	1.60927	983	3.927	0.001017	0.255	95.37	399.295	195.22	817.347	99.85	418.052	0.9827	411.4368	1.3657	571.7913
−10	263.15	1.808	1.77304	979	4.299	0.001022	0.233	96.29	403.147	195.54	818.687	99.25	415.540	0.9862	412.9022	1.3633	570.7864
−7.5	265.65	1.988	1.94956	974	4.698	0.001027	0.213	97.22	407.041	195.85	819.985	98.63	412.944	0.9897	414.3676	1.3609	569.7816
−5	268.15	2.180	2.13785	970	5.125	0.001032	0.195	98.14	410.893	196.15	821.241	98.01	410.348	0.9931	415.7911	1.3586	568.8186
−2.5	270.65	2.387	2.34085	965	5.582	0.001037	0.179	99.07	414.786	196.45	822.497	97.38	407.711	0.9966	417.2565	1.3564	567.8976
0	273.15	2.609	2.55855	960	6.066	0.001042	0.1648	100.00	418.680	196.75	823.753	96.75	405.073	1.0000	418.6800	1.3542	566.9765
2.5	275.65	2.846	2.79097	955	6.584	0.001047	0.1519	100.94	422.616	197.04	824.967	96.10	402.351	1.0034	420.1035	1.3520	566.0554
5	278.15	3.099	3.03908	950	7.134	0.001053	0.1402	101.88	426.551	197.32	826.139	95.44	399.588	1.0068	421.5270	1.3499	565.1761
7.5	280.65	3.368	3.30288	945	7.719	0.001058	0.1296	102.82	430.487	197.60	827.312	94.78	396.825	1.0102	422.9505	1.3479	564.3388
10	283.15	3.655	3.58433	940	8.342	0.001064	0.1198	103.75	434.381	197.87	828.442	94.12	394.062	1.0135	424.3322	1.3459	563.5014
12.5	285.65	3.961	3.88441	935	9.004	0.001069	0.1111	104.69	438.316	198.13	829.531	93.44	391.215	1.0168	425.7138	1.3439	562.6641
15	288.15	4.284	4.20117	930	9.704	0.001075	0.1031	105.63	442.252	198.39	830.619	92.76	388.368	1.0201	427.0955	1.3420	561.8686
17.5	290.65	4.628	4.53852	925	10.44	0.001081	0.0958	106.58	446.229	198.65	831.708	92.07	385.479	1.0234	428.4771	1.3401	561.0731
20	293.15	4.993	4.89646	921	11.22	0.001086	0.0891	107.54	450.248	198.90	832.755	91.36	382.506	1.0267	429.8588	1.3383	560.3194
22.5	295.65	5.378	5.27402	916	12.06	0.001092	0.0829	108.50	454.268	199.14	833.759	90.64	379.492	1.0299	431.1985	1.3365	559.5658
25	298.15	5.783	5.67119	911	12.93	0.001098	0.0774	109.46	458.287	199.38	834.764	89.92	376.477	1.0331	432.5383	1.3347	558.8122
27.5	300.65	6.209	6.08895	906	13.85	0.001104	0.0722	110.42	462.306	199.60	835.685	89.18	373.379	1.0363	433.8781	1.3329	558.0586
30	303.15	6.658	6.52927	901	14.82	0.001110	0.0675	111.38	466.326	199.82	836.606	88.44	370.281	1.0395	435.2179	1.3312	557.3468
32.5	305.65	7.130	6.99214	896	15.85	0.001116	0.0631	112.35	470.387	200.03	837.486	87.68	367.099	1.0427	436.5576	1.3295	556.6351
35	308.15	7.625	7.47757	891	16.92	0.001123	0.0591	113.32	474.448	200.23	838.323	86.91	363.875	1.0459	437.8974	1.3278	555.9233
37.5	310.65	8.146	7.98850	886	18.05	0.001129	0.0554	114.29	478.509	200.43	839.160	86.14	360.651	1.0490	439.1953	1.3262	555.2534
40	313.15	8.690	8.52198	881	19.22	0.001135	0.0520	115.27	482.612	200.63	839.998	85.36	357.385	1.0521	440.4932	1.3247	554.6254
42.5	315.65	9.262	9.08292	876	20.45	0.001142	0.0489	116.25	486.716	200.82	840.793	84.57	354.078	1.0552	441.7911	1.3231	553.9555
45	318.15	9.861	9.67034	870	21.75	0.001149	0.0460	117.23	490.819	201.00	841.547	83.77	350.728	1.0583	443.0890	1.3215	553.2856
47,5	320.65	10.48	10.27737	865	23.11	0.001156	0.0433	118.21	494.922	201.17	842.259	82.96	347.337	1.0614	444.3870	1.3201	552.6995
50	323.15	11.13	10.91480	859	24.51	0.001164	0.0408	119.20	499.067	201.34	842.970	82.14	343.904	1.0645	445.6849	1.3187	552.1133
52,5	325.65	11.82	11.59146	853	26.00	0.001172	0.0385	120.18	503.170	201.49	843.598	81.31	340.429	1.0676	446.9828	1.3173	551.5272
55	328.15	12.53	12.28773	848	27.55	0.001180	0.0363	121.17	507.315	201.64	844.226	80.47	336.912	1.0706	448.2388	1.3158	550.8991
57,5	330.65	13.26	13.00362	842	29.18	0.001188	0.0343	122.17	511.501	201.79	844.812	79.62	333.311	1.0736	449.4949	1.3144	550.3130
60	333.15	14.03	13.75873	837	30.87	0.001196	0.0324	123.17	515.688	201.93	845.441	78.76	329.752	1.0766	450.7509	1.3130	549.7268

1 at = 1 kp/cm² = 98 066.5 Pa = 98 066.5 N/m² = 9.806 65 N/cm² = 0.980 665 bar   1 kcal = 4.186 8 kJ

Table 55-1 Properties of Saturated Trichlorofluoromethane "Freon 11" (CFCl₃)

Temperature		Pressure		Density		Specific volume		Specific enthalpy				Heat of vaporization		Specific entropy			
				Liquid	Vapor	Liquid	Vapor	Liquid		Vapor		$r = i'' - i'$		Liquid		Vapor	
t	T	p	p	ρ'	ρ''	v'	v''	i'		i''				s'		s''	
°C	K	kp/cm²	bar	kg/m³	kg/m³	m³/kg	m³/kg	kcal/kg	kJ/kg	kcal/kg	kJ/kg	kcal/kg	kJ/kg	kcal/kg K	kJ/kg K	kcal/kg K	kJ/kg K
−40	233.15	0.052	0.05099	1622	0.362	0.0006167	2.760	92.07	385.48	140.67	588.96	48.60	203.48	0.9686	4.0553	1.1770	4.9279
−38	235.15	0.059	0.05786	1617	0.414	0.0006184	2.415	92.46	387.11	140.91	589.96	48.45	202.85	0.9702	4.0620	1.1762	4.9245
−36	237.15	0.066	0.06472	1613	0.471	0.0006201	2.124	92.86	388.79	141.15	590.97	48.29	202.18	0.9719	4.0692	1.1756	4.9220
−34	239.15	0.075	0.07355	1608	0.530	0.0006217	1.888	93.25	390.42	141.38	591.93	48.13	201.51	0.9735	4.0758	1.1748	4.9187
−32	241.15	0.084	0.08238	1604	0.589	0.0006234	1.698	93.64	392.05	141.62	592.94	47.98	200.88	0.9751	4.0825	1.1741	4.9157
−30	243.15	0.094	0.09218	1600	0.652	0.0006250	1.533	94.03	393.68	141.86	593.93	47.83	200.25	0.9767	4.0892	1.1734	4.9128
−28	245.15	0.105	0.10297	1596	0.720	0.0006267	1.389	94.42	395.32	142.09	594.90	47.67	199.58	0.9784	4.0964	1.1729	4.9107
−26	247.15	0.117	0.11474	1591	0.791	0.0006284	1.264	94.82	396.99	142.33	395.91	47.51	198.91	0.9800	4.1031	1.1723	4.9082
−24	249.15	0.130	0.12749	1587	0.865	0.0006300	1.156	95.22	398.67	142.58	596.95	47.36	198.29	0.9816	4.1098	1.1717	4.9057
−22	251.15	0.144	0.14122	1583	0.946	0.0006318	1.057	95.61	400.30	142.81	597.92	47.20	197.62	0.9832	4.1165	1.1712	4.9036
−20	253.15	0.160	0.15691	1579	1.038	0.0006335	0.963	96.01	401.97	143.06	598.96	47.05	196.99	0.9848	4.1232	1.1707	4.9015
−18	255.15	0.177	0.17358	1574	1.138	0.0006352	0.879	96.41	403.65	143.31	600.01	46.90	196.36	0.9863	4.1294	1.1701	4.8990
−16	257.15	0.195	0.19123	1570	1.241	0.0006370	0.806	96.81	405.32	143.55	601.02	46.74	195.69	0.9878	4.1357	1.1696	4.8969
−14	259.15	0.216	0.21182	1565	1.357	0.0006388	0.737	97.20	406.96	143.78	601.98	46.58	195.02	0.9894	4.1424	1.1692	4.8952
−12	261.15	0.238	0.23340	1561	1.486	0.0006406	0.673	97.60	408.63	144.03	603.02	46.43	194.39	0.9809	4.1068	1.1687	4.8931
−10	263.15	0.261	0.25595	1556	1.623	0.0006425	0.616	98.00	410.31	144.27	604.03	46.27	193.72	0.9924	4.1550	1.1682	4.8910
−8	265.15	0.2875	0.28194	1552	1.773	0.0006443	0.564	98.40	411.98	144.52	605.08	46.12	193.10	0.9940	4.1617	1.1679	4.8898
−6	267.15	0.3145	0.30842	1548	1.934	0.0006461	0.517	98.80	413.66	144.76	606.08	45.96	192.43	0.9955	4.1680	1.1675	4.8881
−4	269.15	0.3430	0.33637	1543	2.105	0.0006480	0.475	99.20	415.33	145.00	607.09	45.80	191.76	0.9970	4.1742	1.1671	4.8864
−2	271.15	0.3750	0.36775	1539	2.278	0.0006499	0.439	99.60	417.01	145.24	608.02	45.64	191.09	0.9985	4.1805	1.1668	4.8852
0	273.15	0.4100	0.40207	1534	2.469	0.0006519	0.405	100.00	418.68	145.48	609.10	45.48	190.42	1.0000	4.1868	1.1665	4.8839
2	275.15	0.4460	0.43738	1530	2.674	0.0006538	0.374	100.41	420.40	145.73	610.14	45.32	189.75	1.0014	4.1927	1.1661	4.8822
4	277.15	0.4855	0.47611	1525	2.890	0.0006558	0.346	100.81	422.07	145.97	611.15	45.16	189.08	1.0029	4.1989	1.1659	4.8814
6	279.15	0.5270	0.51681	1520	3.115	0.0006578	0.321	101.21	423.75	146.20	612.11	44.99	188.36	1.0043	4.2048	1.1655	4.8797
8	281.15	0.5715	0.56045	1516	3.356	0.0006598	0.298	101.62	425.46	146.45	613.16	44.83	187.69	1.0058	4.2111	1.1653	4.8789
10	283.15	0.6175	0.60556	1511	3.610	0.0006619	0.277	102.02	427.14	146.69	614.16	44.67	187.02	1.0072	4.2169	1.1650	4.8776
12	285.15	0.6675	0.65459	1506	3.891	0.0006639	0.257	102.43	428.85	146.93	615.17	44.50	186.31	1.0087	4.2232	1.1648	4.8768
14	287.15	0.7210	0.70706	1502	4.184	0.0006660	0.239	102.83	430.53	147.16	616.13	44.33	185.60	1.0101	4.2291	1.1646	4.8759
16	289.15	0.7790	0.76394	1497	4.484	0.0006680	0.223	103.24	432.25	147.41	617.18	44.17	184.93	1.0115	4.2349	1.1643	4.8747
18	291.15	0.8400	0.82376	1492	4.808	0.0006701	0.208	103.66	434.00	147.66	618.22	44.00	184.22	1.0129	4.2408	1.1641	4.8739

Table 55-2 Properties of Saturated Trichlorofluoromethane "Freon 11" (CFCl$_3$) (Continued)

Temperature		Pressure		Density		Specific volume		Specific enthalpy				Heat of vaporization		Specific entropy			
								Liquid		Vapor		$r=i''-i'$		Liquid		Vapor	
t	T	p		Liquid ρ'	Vapor ρ''	Liquid v'	Vapor v''	i'		i''				s'		s''	
°C	K	kp/cm²	bar	kg/m³	kg/m³	m³/kg	m³/kg	kcal/kg	kJ/kg	kcal/kg	kJ/kg	kcal/kg	kJ/kg	kcal/kg K	kJ/kg K	kcal/kg K	kJ/kg K
20	293.15	0.9040	0.88652	1488	5.155	0.0006722	0.194	104.07	435.72	147.90	619.23	43.83	183.51	1.0143	4.2467	1.1638	4.8726
22	295.15	0.9720	0.95321	1483	5.525	0.0006743	0.181	104.48	437.44	148.14	620.23	43.66	182.80	1.0157	4.2525	1.1636	4.8718
24	297.15	1.0445	1.02430	1478	5.882	0.0006765	0.170	104.90	439.20	148.38	621.24	43.48	182.04	1.0171	4.2584	1.1634	4.8709
26	299.15	1.1205	1.09884	1473	6.289	0.0006787	0.159	105.31	440.91	148.61	622.20	43.30	181.29	1.0185	4.2643	1.1632	4.8701
28	301.15	1.2000	1.17680	1469	6.711	0.0006809	0.149	105.73	442.67	148.86	623.25	43.15	180.58	1.0199	4.2701	1.1631	4.8697
30	303.15	1.2855	1.26064	1463	7.143	0.0006833	0.140	106.14	444.39	149.09	624.21	42.95	179.82	1.0213	4.2760	1.1630	4.8692
32	305.15	1.374	1.34743	1459	7.576	0.0006856	0.132	106.56	446.15	149.33	625.21	42.77	179.07	1.0226	4.2814	1.1628	4.8684
34	307.15	1.466	1.43765	1454	8.065	0.0006879	0.124	106.98	447.90	149.56	626.18	42.58	178.27	1.0240	4.2873	1.1627	4.8680
36	309.15	1.565	1.53474	1449	8.621	0.0006903	0.116	107.40	449.66	149.80	627.18	42.40	177.52	1.0254	4.2931	1.1626	4.8676
38	311.15	1.671	1.63869	1444	9.174	0.0006927	0.109	107.82	451.42	150.03	628.15	42.21	176.72	1.0268	4.2990	1.1625	4.8672
40	313.15	1.782	1.74755	1439	9.709	0.0006950	0.103	108.24	453.18	150.27	629.15	42.03	175.97	1.0281	4.3044	1.1623	4.8663
42	315.15	1.899	1.86228	1434	10.204	0.0006975	0.098	108.66	454.94	150.50	630.11	41.84	175.18	1.0295	4.3103	1.1622	4.8659
44	317.15	2.022	1.98290	1429	10.870	0.0007000	0.092	109.09	456.74	150.74	631.12	41.65	174.38	1.0308	4.3158	1.1621	4.8655
46	319.15	2.148	2.10647	1423	11.494	0.0007025	0.087	109.52	458.54	150.97	632.08	41.45	173.54	1.0322	4.3216	1.1621	4.8655
48	321.15	2.275	2.23101	1418	12.195	0.0007050	0.082	109.95	460.34	151.20	633.04	41.25	172.71	1.0335	4.3271	1.1620	4.8651
50	323.15	2.403	2.35654	1413	12.987	0.0007075	0.077	110.38	462.14	151.43	634.01	41.05	171.87	1.0349	4.3329	1.1619	4.8646

1 at = 1 kp/cm² = 98 066.5 Pa = 98 066.5 N/m² = 9.806 65 N/cm² = 0.980 665 bar

1 kcal = 4.186 8 kJ

Table 56-1 Properties of Saturated Propane (C$_3$H$_8$)

Temperature t °C	T K	Pressure kp/cm²	Pressure bar	Density Liquid ρ' kg/m³	Density Vapor ρ'' kg/m³	Spec. vol. Liquid v' m³/kg	Spec. vol. Vapor v'' m³/kg	Enthalpy Liquid i' kcal/kg	Enthalpy Liquid i' kJ/kg	Enthalpy Vapor i'' kcal/kg	Enthalpy Vapor i'' kJ/kg	Heat of vap. r=i''−i' kcal/kg	Heat of vap. r=i''−i' kJ/kg	Entropy Liquid s' kcal/kg K	Entropy Liquid s' kJ/kg K	Entropy Vapor s'' kcal/kg K	Entropy Vapor s'' kJ/kg K
−80	193.15	0.134	0.13141	624.0	0.367	0.001603	2.724	55.81	233.665	164.51	688.770	108.70	455.105	0.8073	3.3800	1.3705	5.7380
−75	198.15	0.184	0.18044	618.6	0.497	0.001616	2.012	58.84	246.351	166.88	698.693	107.94	451.923	0.8221	3.4420	1.3668	5.7225
−70	203.15	0.249	0.24419	613.4	0.648	0.001630	1.544	61.69	258.284	168.74	706.481	107.05	448.197	0.8369	3.5039	1.3637	5.7095
−65	208.15	0.332	0.32558	608.0	0.852	0.001644	1.173	64.33	270.174	170.57	714.142	106.04	443.968	0.8505	3.5609	1.3601	5.6945
−60	213.15	0.435	0.42659	602.5	1.098	0.001659	0.911	67.34	281.939	172.39	721.762	105.05	439.823	0.8637	3.6161	1.3565	5.6794
−55	218.15	0.563	0.55211	597.1	1.389	0.001674	0.720	70.07	293.369	174.12	729.006	104.05	435.637	0.8762	3.6685	1.3531	5.6652
−50	223.15	0.721	0.70706	591.0	1.725	0.001690	0.580	72.72	304.464	175.87	736.333	103.05	431.450	0.8887	3.7208	1.3505	5.6543
−45	228.15	0.908	0.89044	585.3	2.141	0.001707	0.467	75.31	315.308	177.31	742.362	102.00	427.054	0.9006	3.7706	1.3476	5.6421
−40	233.15	1.137	1.11502	579.3	2.630	0.001725	0.380	77.98	326.487	178.88	748.935	100.90	422.448	0.9122	3.8192	1.3450	5.6312
−35	238.15	1.406	1.37881	573.5	3.145	0.001743	0.318	80.71	337.917	180.68	756.471	99.97	418.554	0.9237	3.8673	1.3434	5.6245
−30	243.15	1.705	1.67203	568.0	3.845	0.001761	0.260	83.42	349.263	182.36	763.505	98.94	414.242	0.9351	3.9151	1.3420	5.6187
−25	248.15	2.057	2.01723	561.7	4.651	0.001780	0.215	86.14	360.651	183.86	769.785	97.72	409.134	0.9466	3.9632	1.3403	5.6116
−20	253.15	2.471	2.42322	555.5	5.495	0.001799	0.182	88.84	371.955	185.19	775.353	96.35	403.398	0.9578	4.0101	1.3383	5.6032
−15	258.15	2.946	2.88904	549.3	6.427	0.001820	0.1556	91.54	383.260	186.59	781.215	95.05	397.955	0.9684	4.0545	1.3364	5.5952
−10	263.15	3.472	3.40487	543.0	7.595	0.001842	0.1318	94.29	394.773	187.99	787.077	93.70	392.303	0.9790	4.0989	1.3350	4.5894
− 5	268.15	4.094	4.01484	536.7	8.826	0.001864	0.1133	97.09	406.496	189.19	792.101	92.10	385.604	0.9895	4.1428	1.3330	5.5810
0	273.15	4.776	4.68366	530.0	10.28	0.001887	0.0974	100.00	418.680	190.44	797.334	90.44	378.654	1.0000	4.1868	1.3311	5.5730
5	278.15	5.561	5.45348	523.0	11.82	0.001911	0.0846	102.92	430.905	191.62	802.275	88.70	371.369	1.0102	4.2295	1.3296	5.5668
10	283.15	6.464	6.33902	516.0	13.69	0.001935	0.0731	105.79	442.922	192.77	807.089	86.98	364.168	1.0204	4.2722	1.3276	5.5584
15	288.15	7.442	7.29811	509.0	15.65	0.001963	0.0639	108.74	455.273	193.92	811.904	85.18	356.632	1.0306	4.3149	1.3263	5.5530
20	293.15	8.498	8.33369	501.5	17.80	0.001992	0.0561	111.94	468.670	194.94	816.175	83.25	348.551	1.0408	4.3576	1.3248	5.5467
25	298.15	9.676	9.48891	494.0	20.20	0.002023	0.0495	114.74	480.393	195.97	820.487	81.25	340.178	1.0511	4.4007	1.3236	5.5416
30	303.15	11.02	10.80693	486.0	22.98	0.002055	0.0435	117.84	493.373	196.91	824.423	79.07	331.050	1.0614	4.4439	1.3224	5.5366
35	308.15	12.46	12.21909	477.7	25.97	0.002095	0.0385	121.13	507.147	197.82	828.233	76.69	321.086	1.0716	4.4866	1.3204	5.5283
40	313.15	14.01	13.73912	469.0	29.95	0.002135	0.0339	124.41	520.880	198.65	831.708	74.24	310.828	1.0817	4.5289	1.3187	5.5211
45	318.15	15.76	15.45528	459.5	33.11	0.002178	0.0302	127.88	535.408	199.58	835.602	71.70	300.194	1.0920	4.5720	1.3173	5.5153
50	323.15	17.61	17.26951	450.0	37.33	0.002222	0.0268	131.24	549.476	200.36	838.867	69.12	289.392	1.1023	4.6151	1.3161	5.5102

1 at = 1 kp/cm² = 98 066.5 Pa = 98 066.5 N/m² = 9.806 65 N/cm² = 0.980 665 bar   1 kcal = 4.186 8 kJ

Table 57-1 Properties of Saturated Sulfur Dioxide (SO_2)

Temperature t (°C)	T (K)	Pressure p (kp/cm²)	p (bar)	Density Liquid ρ' (kg/m³)	Density Vapor ρ'' (kg/m³)	Spec. volume Liquid v' (m³/kg)	Spec. volume Vapor v'' (m³/kg)	Enthalpy Liquid i' (kcal/kg)	Enthalpy Liquid i' (kJ/kg)	Enthalpy Vapor i'' (kcal/kg)	Enthalpy Vapor i'' (kJ/kg)	Heat of vap. r=i''−i' (kcal/kg)	Heat of vap. r=i''−i' (kJ/kg)	Entropy Liquid s' (kcal/kg K)	Entropy Liquid s' (kJ/kg K)	Entropy Vapor s'' (kcal/kg K)	Entropy Vapor s'' (kJ/kg K)
−50	223.15	0.118	0.11572	1557	0.4015	0.0006423	2.4907	83.69	350.393	184.91	774.181	101.22	423.788	0.9341	3.9109	1.3877	5.8100
−47.5	225.65	0.139	0.13631	1551	0.4682	0.0006448	2.1359	84.51	353.826	185.24	775.563	100.73	421.736	0.9378	3.9264	1.3842	5.7954
−45	228.15	0.163	0.15985	1545	0.5424	0.0006472	1.8436	85.34	357.302	185.56	776.903	100.22	419.601	0.9412	3.9406	1.3808	5.7811
−42.5	230.65	0.190	0.18633	1539	0.6270	0.0006498	1.5950	86.13	360.609	185.89	778.284	99.76	417.675	0.9449	3.9561	1.3774	5.7669
−40	233.15	0.220	0.21575	1533	0.7209	0.0006523	1.3872	87.00	364.252	186.21	779.624	99.21	415.372	0.9485	3.9712	1.3740	5.7527
−37.5	235.65	0.256	0.25105	1527	0.8275	0.0006549	1.2085	87.76	367.434	186.53	780.964	98.77	413.530	0.9519	3.9854	1.3710	5.7401
−35	238.15	0.294	0.28832	1521	0.9446	0.0006575	1.0586	88.64	371.118	186.85	782.304	98.21	411.186	0.9556	4.0009	1.3680	5.7275
−32.5	240.65	0.339	0.33245	1515	1.0771	0.0006601	0.9284	89.39	374.258	187.16	783.601	97.77	409.343	0.9588	4.0143	1.3651	5.7154
−30	243.15	0.388	0.38050	1509	1.2220	0.0006627	0.8183	90.27	377.942	187.47	784.899	97.20	406.957	0.9624	4.0294	1.3621	5.7028
−27.5	245.65	0.443	0.43443	1503	1.3843	0.0006653	0.7224	91.02	381.083	187.78	786.197	96.75	405.073	0.9655	4.0424	1.3594	5.6915
−25	248.15	0.504	0.49426	1497	1.5610	0.0006680	0.6406	91.90	384.767	188.09	787.495	96.19	402.728	0.9691	4.0574	1.3567	5.6802
−22.5	250.65	0.573	0.56192	1490	1.7578	0.0006710	0.5689	92.65	387.907	188.40	788.793	95.75	400.886	0.9720	4.0696	1.3540	5.6689
−20	253.15	0.648	0.63547	1484	1.9720	0.0006739	0.5071	93.53	391.591	188.70	790.049	95.17	398.458	0.9755	4.0842	1.3514	5.6580
−17.5	255.65	0.732	0.71785	1477	2.2085	0.0006769	0.4528	94.29	394.773	189.00	791.305	94.71	396.532	0.9786	4.0972	1.3490	5.6480
−15	258.15	0.823	0.80709	1471	2.4643	0.0006798	0.4058	95.15	398.374	189.30	792.561	94.15	394.187	0.9819	4.1110	1.3466	5.6379
−12.5	260.65	0.924	0.90613	1464	2.7465	0.0006829	0.3641	95.92	401.598	189.59	793.775	93.67	392.178	0.9848	4.1232	1.3442	5.6279
−10	263.15	1.034	1.01401	1458	3.0488	0.0006859	0.3280	96.76	405.115	189.89	795.031	93.13	389.917	0.9889	4.1361	1.3418	5.6178
−7.5	265.65	1.155	1.13267	1452	3.3829	0.0006888	0.2956	97.55	408.422	190.17	796.204	92.62	387.781	0.9910	4.1491	1.3396	5.6086
−5	268.15	1.286	1.26114	1446	3.7383	0.0006916	0.2675	98.39	411.939	190.46	797.418	92.07	385.479	0.9942	4.1625	1.3375	5.5998
−2.5	270.65	1.430	1.40235	1440	4.1305	0.0006945	0.2421	99.18	415.247	190.74	798.590	91.56	383.343	0.9970	4.1742	1.3353	5.5906
0	273.15	1.585	1.55105	1434	4.5455	0.0006974	0.2200	100.00	418.680	191.02	799.763	91.02	381.083	1.0000	4.1868	1.3332	5.5818
2.5	275.65	1.755	1.72107	1428	5.000	0.0007005	0.2000	100.81	422.071	191.29	800.893	90.48	378.822	1.0030	4.1994	1.3312	5.5735
5	278.15	1.936	1.89857	1422	5.482	0.0007035	0.1824	101.63	425.504	191.57	802.065	89.94	376.561	1.0060	4.2119	1.3293	5.5655
7.5	280.65	2.135	2.09372	1415	6.010	0.0007066	0.1664	102.43	428.854	191.83	803.154	89.40	374.300	1.0088	4.2236	1.3273	5.5571
10	283.15	2.347	2.30162	1409	6.566	0.0007097	0.1523	103.23	432.203	192.09	804.242	88.86	372.039	1.0115	4.2349	1.3253	5.5488
12.5	285.65	2.577	2.52717	1403	7.168	0.0007130	0.1395	104.05	435.637	192.35	805.331	88.30	369.694	1.0144	4.2471	1.3235	5.5412
15	288.15	2.823	2.76842	1396	7.812	0.0007163	0.1280	104.85	438.986	192.61	806.420	87.76	367.434	1.0173	4.2592	1.3218	5.5341
17.5	290.65	3.088	3.02829	1389	8.496	0.0007197	0.1177	105.67	442.419	192.85	807.424	87.19	365.047	1.0200	4.2705	1.3200	5.5266
20	293.15	3.370	3.30484	1383	9.225	0.0007231	0.1084	106.45	445.685	193.10	808.471	86.65	362.786	1.0227	4.2818	1.3183	5.5195
22.5	295.65	3.674	3.60296	1376	10.01	0.0007266	0.0999	107.24	448.992	193.31	809.350	86.07	360.358	1.0255	4.2936	1.3166	5.5123
25	298.15	3.997	3.91972	1370	10.83	0.0007301	0.0923	107.99	452.133	193.52	810.230	85.53	358.097	1.0282	4.3049	1.3150	5.5056
27.5	300.65	4.343	4.25903	1363	11.72	0.0007338	0.0853	108.84	455.691	193.78	811.318	84.94	355.627	1.0308	4.3158	1.3133	5.4985
30	303.15	4.710	4.61893	1356	12.66	0.0007375	0.0790	109.65	459.083	194.04	812.407	84.39	353.324	1.0333	4.3262	1.3117	5.4918
32.5	305.65	5.103	5.00433	1349	13.66	0.0007414	0.0732	110.47	462.516	194.27	813.370	83.80	350.854	1.0360	4.3375	1.3102	5.4855
35	308.15	5.518	5.41131	1342	14.70	0.0007453	0.0680	111.26	465.823	194.49	814.291	83.23	348.467	1.0386	4.3484	1.3087	5.4793
37.5	310.65	5.960	5.84476	1334	15.82	0.0007495	0.0632	112.06	469.173	194.70	815.170	82.65	346.039	1.0412	4.3593	1.3072	5.4730
40	313.15	6.427	6.30273	1327	17.01	0.0007536	0.0588	112.83	472.397	194.92	816.091	82.09	343.694	1.0434	4.3685	1.3057	5.4667
42.5	315.65	6.923	6.78914	1319	18.28	0.0007578	0.0547	113.62	475.704	195.12	816.928	81.50	341.224	1.0461	4.3798	1.3043	5.4608
45	318.15	7.447	7.30301	1311	19.57	0.0007622	0.0511	114.41	479.012	195.32	817.766	80.91	338.754	1.0486	4.3903	1.3029	5.4550
47.5	320.65	8.001	7.84630	1303	20.96	0.0007666	0.0477	115.21	482.361	195.52	818.603	60.31	336.242	1.0511	4.4007	1.3015	5.4491
50	323.15	8.583	8.41705	1295	22.42	0.0007712	0.0446	116.01	485.711	195.72	819.440	79.71	333.730	1.0534	4.4104	1.3001	5.4433
52.5	325.65	9.199	9.02114	1289	23.92	0.0007759	0.0418	116.77	488.893	195.90	820.194	79.13	331.301	1.0558	4.4204	1.2978	5.4336
55	328.15	9.848	9.65759	1281	25.58	0.0007808	0.0391	117.64	492.535	196.09	820.990	78.45	328.454	1.0584	4.4313	1.2974	5.4320
57.5	330.65	10.53	10.32640	1273	27.24	0.0007857	0.0367	118.43	495.843	196.27	821.743	77.84	325.901	1.0607	4.4409	1.2961	5.4265
60	333.15	11.25	11.03248	1264	29.07	0.0007909	0.0344	119.23	499.192	196.44	822.455	77.21	323.263	1.0631	4.4510	1.2949	5.4215

1 at = 1 kp/cm² = 98 066.5 Pa = 98 066.5 N/m² = 9.806 65 N/cm² = 0.980 665 bar   1 kcal = 4.186 8 kJ

Table 58-1 Properties of Saturated Chlorofluoromethane "Freon 13" (CF₃Cl)

Temperature		Pressure		Density		Specific volume		Specific enthalpy				Heat of vaporization		Specific entropy			
				Liquid ρ'	Vapor ρ''	Liquid v'	Vapor v''	Liquid i'		Vapor i''		$r=i''-i'$		Liquid s'		Vapor s''	
t	T	p	p														
°C	K	kp/cm²	bar	kg/m³	kg/m³	m³/kg	m³/kg	kcal/kg	kJ/kg	kcal/kg	kJ/kg	kcal/kg	kJ/kg	kcal/kg K	kJ/kg K	kcal/kg K	kJ/kg K
−140	133.15	0.0087	0.00853	1736	0.0808	0.000576	12.378	68.46	286.628	109.90	460.129	41.44	173.501	0.8441	3.5341	1.1553	4.8370
−135	138.15	0.0157	0.01540	1721	0.1406	0.000581	7.112	69.33	290.271	110.37	462.097	41.04	171.826	0.8505	3.5609	1.1475	4.8044
−130	143.15	0.0271	0.02658	1704	0.2340	0.000587	4.273	70.24	294.081	110.86	464.149	40.62	170.068	0.8570	3.5881	1.1407	4.7759
−125	148.15	0.0448	0.04393	1686	0.3741	0.000593	2.673	71.15	297.891	111.34	466.158	40.19	168.267	0.8632	3.6140	1.1345	4.7499
−120	153.15	0.0714	0.07002	1669	0.5774	0.000599	1.732	72.09	301.826	111.84	468.252	39.75	166.425	0.8694	3.6400	1.1290	4.7269
−115	158.15	0.1100	0.10787	1653	0.8636	0.000605	1.158	73.04	305.804	112.34	470.345	39.30	164.541	0.8755	3.6655	1.1240	4.7060
−110	163.15	0.1643	0.16112	1634	1.2531	0.000612	0.798	74.01	309.865	112.84	472.439	38.83	163.573	0.8816	3.6911	1.1196	4.6875
−105	168.15	0.2391	0.23448	1616	1.7762	0.000619	0.563	74.99	313.968	113.34	474.532	38.35	160.564	0.8876	3.7162	1.1156	4.6708
−100	173.15	0.3392	0.33264	1597	2.4570	0.000626	0.4070	76.00	318.197	113.85	476.667	37.85	158.470	0.8935	3.7409	1.1120	4.6557
−95	178.15	0.4705	0.46140	1577	3.3278	0.000634	0.3005	77.03	322.509	114.36	478.802	37.33	156.293	0.8992	3.7648	1.1087	4.6419
−90	183.15	0.640	0.62763	1558	4.4267	0.000642	0.2259	78.08	326.905	114.86	480.896	36.78	153.991	0.9050	3.7891	1.1058	4.6298
−85	188.15	0.854	0.83749	1541	5.7870	0.000649	0.1728	79.13	331.301	115.36	482.989	36.23	151.688	0.9107	3.8129	1.1032	4.6189
−80	193.15	1.120	1.09834	1520	7.4516	0.000658	0.1342	80.21	335.823	115.86	485.083	35.65	149.259	0.9163	3.8364	1.1009	4.6092
−75	198.15	1.446	1.41804	1502	9.4607	0.000666	0.1057	81.30	340.387	116.35	487.134	35.05	146.747	0.9216	3.8586	1.0987	4.6000
−70	203.15	1.841	1.80540	1481	11.8483	0.000675	0.0844	82.40	344.992	116.84	489.186	34.44	144.193	0.9271	3.8816	1.0968	4.5921
−65	208.15	2.313	2.26828	1460	14.6843	0.000685	0.0681	83.52	349.682	117.32	491.195	33.80	141.514	0.9327	3.9050	1.0951	4.5850
−60	213.15	2.873	2.81745	1439	18.0440	0.000695	0.05542	84.67	354.496	117.78	493.121	33.11	138.625	0.9382	3.9281	1.0935	4.5783
−55	218.15	3.528	3.45979	1416	21.9539	0.000706	0.04555	85.84	359.395	118.23	495.005	32.39	135.610	0.9435	3.9502	1.0920	4.5720
−50	223.15	4.287	4.20411	1395	26.4971	0.000717	0.03774	87.03	364.377	118.66	496.806	31.63	132.428	0.9489	3.9729	1.0906	4.5661
−45	228.15	5.164	5.06415	1374	31.7662	0.000728	0.03148	88.25	369.485	119.08	498.564	30.83	129.079	0.9542	3.9950	1.0893	4.5607
−40	233.15	6.17	6.05070	1350	37.8501	0.000741	0.02642	89.49	374.677	119.48	500.239	29.99	125.562	0.9595	4.0172	1.0881	4.5557
−35	238.15	7.31	7.16866	1326	44.8430	0.000754	0.02230	90.74	379.910	119.85	501.788	29.11	121.878	0.9647	4.0390	1.0869	4.5506
−30	243.15	8.59	8.42391	1300	52.9381	0.000769	0.01889	92.01	385.227	120.19	503.211	28.18	117.984	0.9699	4.0608	1.0858	4.5460
−25	248.15	10.04	9.84608	1274	62.1891	0.000785	0.01608	93.30	390.628	120.50	504.509	27.20	113.881	0.9751	4.0825	1.0847	4.5414
−20	253.15	11.66	11.43455	1247	72.8332	0.000802	0.01373	94.61	396.113	120.77	505.640	26.16	109.527	0.9802	4.1039	1.0835	4.5364
−15	258.15	13.46	13.19975	1218	85.1064	0.000821	0.01175	95.94	401.682	121.02	506.687	25.08	105.005	0.9853	4.1253	1.0824	4.5318
−10	263.15	15.45	15.15127	1188	99.0099	0.000842	0.01010	97.27	407.250	121.22	507.524	23.95	100.274	0.9902	4.1458	1.0812	4.5268
−5	268.15	17.66	17.31854	1155	115.207	0.000866	0.00868	98.61	412.860	121.37	508.152	22.76	95.292	0.9950	4.1659	1.0799	4.5213
0	273.15	20.09	19.70156	1119	133.869	0.000894	0.00747	100.00	418.680	121.48	508.612	21.48	89.932	1.0000	4.1868	1.0786	4.5159
5	278.15	22.76	22.31994	1083	155.763	0.000923	0.00642	101.44	424.709	121.51	508.738	20.07	84.029	1.0050	4.2077	1.0772	4.5100
10	283.15	25.69	25.19328	1040	182.149	0.000962	0.00549	102.99	431.199	121.42	508.361	18.43	77.163	1.0103	4.2299	1.0754	4.5025
15	288.15	28.91	28.35103	989.1	215.983	0.001011	0.00463	104.74	438.525	121.16	507.273	16.42	68.747	1.0162	4.2546	1.0732	4.4933
20	293.15	32.41	31.78335	926.8	261.165	0.001079	0.003829	106.75	446.941	120.59	504.886	13.84	57.945	1.0228	4.2823	1.0700	4.4799
25	298.15	36.24	35.53930	838.2	334.448	0.001193	0.002990	109.29	457.575	119.33	499.611	10.04	42.035	1.0310	4.3166	1.0647	4.4577
28.8	301.95	39.36	38.59897	584.1	581.058	0.001712	0.001721	113.94	477.044	113.94	477.044	0.00	0	1.0462	4.3802	1.0462	4.3802

1 at = 1 kp/cm² = 98 066.5 Pa = 98 066.5 N/m² = 9.806 65 N/cm² = 0.980 665 bar   1 kcal = 4.186 8 kJ

Table 59-1 Properties of Saturated Carbon Dioxide (CO$_2$) Liquid–Vapor

Temperature		Pressure		Density		Specific volume		Specific enthalpy				Heat of vaporization		Specific entropy			
		p		Solid ρ'''	Vapor ρ''	Solid v'''	Vapor v''	Solid i'''		Vapor i''		$r = i'' - i'''$		Solid s'''		Vapor s''	
t	T																
°C	K	kp/cm²	bar	kg/m³	kg/m³	m³/kg	m³/kg	kcal/kg	kJ/kg	kcal/kg	kJ/kg	kcal/kg	kJ/kg	kcal/kg K	kJ/kg K	kcal/kg K	kJ/kg K
−100	173.15	0.142	0.13925	1594	0.428	0.000627	2.336	10.9	45.636	150.7	630.951	139.8	585.315	0.5996	2.5104	1.4070	5.8908
−95	178.15	0.236	0.23144	1590	0.694	0.000629	1.442	12.2	51.079	151.4	633.882	139.2	582.803	0.6074	2.5431	1.3889	5.8150
−90	183.15	0.379	0.37167	1582	1.03	0.000632	0.920	13.6	56.940	152.2	637.231	138.6	580.290	0.6150	2.5749	1.3718	5.7435
−85	188.15	0.596	0.58448	1574	1.67	0.000635	0.598	15.0	62.802	152.9	640.162	137.9	577.360	0.6224	2.6059	1.3554	5.6748
−80	193.15	0.914	0.89633	1566	2.51	0.000639	0.398	16.4	68.664	153.5	642.674	137.1	574.010	0.6299	2.6373	1.3398	5.6095
−78.9	194.25	1.000	0.98067	1566	2.74	0.000639	0.36512	16.73	70.045	153.62	643.176	136.89	573.131	0.6314	2.6435	1.3363	5.5948
−75	198.15	1.37	1.34351	1556	3.72	0.000643	0.2694	17.9	74.944	154.1	645.186	136.2	570.242	0.6376	2.6695	1.3248	5.5467
−70	203.15	2.02	1.98094	1546	5.39	0.000647	0.1854	19.6	82.061	154.5	646.861	134.9	564.799	0.6459	2.7043	1.3103	5.4860
−65	208.15	2.93	2.87335	1534	7.73	0.000652	0.1293	21.5	90.016	154.9	648.535	133.4	558.519	0.6551	2.7428	1.2960	5.4261
−60	213.15	4.18	4.09918	1522	11.0	0.000657	0.0912	23.7	99.227	155.1	649.373	131.4	550.146	0.6655	2.7863	1.2819	5.3671
−56.6	216.55	5.28	51.7791	1513	13.9	0.000661	0.0722	25.2	105.507	155.1	649.373	129.9	543.865	0.6725	2.8156	1.2724	5.3273

1 at = 1 kp/cm² = 98 066.5 Pa = 98 066.5 N/m² = 9.806 65 N/cm² = 0.980 665 bar

1 kcal = 4.186 8 kJ

Table 59-2 Properties of Saturated Carbon Dioxide (CO_2) Solid–Vapor (Continued)

Temperature t °C	Temperature T K	Pressure p kp/cm²	Pressure p bar	Density Liquid ρ' kg/m³	Density Vapor ρ'' kg/m³	Sp. vol. Liquid v' m³/kg	Sp. vol. Vapor v'' m³/kg	Enthalpy Liquid i' kcal/kg	Enthalpy Liquid i' kJ/kg	Enthalpy Vapor i'' kcal/kg	Enthalpy Vapor i'' kJ/kg	Heat of vap. r=i''−i''' kcal/kg	Heat of vap. r kJ/kg	Entropy Liquid s' kcal/kg K	Entropy Liquid s' kJ/kg K	Entropy Vapor s'' kcal/kg K	Entropy Vapor s'' kJ/kg K
−56.6	216.55	5.28	5.17791	1178.0	13.9	0.000849	0.0722	72.00	301.450	155.10	649.373	83.10	347.923	0.8885	3.7200	1.2724	5.3273]
−55	218.15	5.66	5.55056	1172.0	14.8	0.000853	0.0676	72.70	304.380	155.20	649.791	82.50	345.411	0.8917	3.7334	1.2700	5.3172
−50	223.15	6.97	6.83524	1153.5	18.1	0.000867	0.055407	75.01	314.052	155.57	651.340	80.56	337.289	0.9020	3.7765	1.2631	5.2883
−47.5	225.65	7.67	7.52170	1144.4	19.9	0.000873	0.050250	76.18	318.950	155.73	652.010	79.55	333.060	0.9070	3.7974	1.2598	5.2745
−45	228.15	8.49	8.32585	1134.5	21.8	0.000881	0.045809	77.30	323.640	155.89	652.680	78.59	329.041	0.9120	3.8184	1.2565	5.2607
−42.5	230.65	9.33	9.14960	1125.0	23.9	0.000889	0.041780	78.42	328.329	156.03	653.266	77.61	324.938	0.9170	3.8393	1.2534	5.2477
−40	233.15	10.25	10.05182	1115.0	26.2	0.000897	0.038164	79.59	333.227	156.17	653.853	76.58	320.625	0.9218	3.8594	1.2503	5.2348
−37.5	235.65	11.20	10.98345	1105.0	28.7	0.000905	0.034900	80.72	337.958	156.28	654.313	75.56	316.355	0.9266	3.8795	1.2473	5.2222
−35	238.15	12.26	12.02295	1094.9	31.2	0.000913	0.032008	81.80	342.480	156.39	654.774	74.51	311.958	0.9314	3.8996	1.2443	5.2096
−32.5	240.65	13.35	13.09188	1084.5	33.9	0.000922	0.029480	83.01	347.546	156.48	655.150	73.47	307.604	0.9362	3.9197	1.2414	5.1975
−30	243.15	14.55	14.26868	1074.2	37.0	0.000931	0.027001	84.19	352.487	156.56	655.485	72.37	302.999	0.9408	3.9389	1.2385	5.1854
−27.5	245.65	15.76	15.45528	1063.6	40.2	0.000940	0.024850	85.35	357.343	156.62	655.737	71.27	298.393	0.9460	3.9607	1.2355	5.1728
−25	248.15	17.14	16.80860	1052.6	43.8	0.000950	0.022885	86.53	362.284	156.67	655.946	70.14	293.662	0.9501	3.9779	1.2328	5.1615
−22.5	250.65	18.68	18.31882	1041.7	47.5	0.000960	0.021070	87.73	367.308	156.70	656.072	68.97	288.764	0.9550	3.9984	1.2298	5.1489
−20	253.15	20.06	19.67214	1029.9	51.4	0.000971	0.019466	88.93	372.322	156.78	656.407	67.79	283.823	0.9594	4.0168	1.2272	5.1380
−17.5	255.15	21.71	21.29024	1018.5	55.7	0.000982	0.017950	90.18	377.566	156.72	656.155	66.54	278.590	0.9644	4.0377	1.2243	5.1259
−15	258.15	23.34	22.88872	1006.1	60.2	0.000994	0.016609	91.44	382.841	156.70	656.072	65.26	273.231	0.9690	4.0570	1.2218	5.1154
−12.5	260.65	25.10	24.61469	993.8	65.3	0.001006	0.015320	92.75	388.326	156.65	655.862	63.90	267.537	0.9740	4.0779	1.2188	5.1029
−10	263.15	26.99	26.46815	980.8	70.5	0.001019	0.014194	94.09	393.936	156.60	655.653	62.51	261.717	0.9787	4.0976	1.2163	5.0924
−7.5	265.65	29.00	28.43929	968.0	76.2	0.001033	0.013120	95.48	399.756	156.51	655.276	61.03	255.520	0.9835	4.1177	1.2135	5.0807
−5	268.15	31.05	30.44965	953.8	82.4	0.001048	0.012141	96.91	405.743	156.41	654.857	59.50	249.115	0.9890	4.1407	1.2109	5.0698
−2.5	270.65	33.21	32.56788	940.0	89.0	0.001063	0.011230	98.38	411.897	156.27	654.271	57.89	242.374	0.9942	4.1625	1.2082	5.0585
0	273.15	35.54	34.85283	924.8	96.3	0.001081	0.010383	100.00	418.680	156.13	653.685	56.13	235.005	1.0000	4.1868	1.2055	5.0472
2.5	275.65	37.95	37.21624	910.0	104.3	0.001100	0.009584	101.84	426.384	155.82	652.387	53.98	226.003	1.0050	4.2077	1.2022	5.0334
5	278.15	40.50	39.71693	893.1	113.0	0.001120	0.008850	103.10	431.659	155.45	650.838	52.35	219.179	1.0103	4.2299	1.1985	5.0179
7.5	280.65	43.20	42.36473	876.0	122.3	0.001142	0.008175	104.78	438.693	155.08	649.289	50.30	210.596	1.0155	4.2517	1.1952	5.0041
10	283.15	45.95	45.06156	858.0	133.0	0.001166	0.007519	106.50	445.894	154.59	647.237	48.09	201.343	1.0218	4.2781	1.1917	4.9894
12.5	285.65	48.83	47.88587	838.5	144.7	0.001193	0.006910	108.20	453.012	153.95	644.558	45.75	191.546	1.0274	4.3015	1.1875	4.9718
15	288.15	51.93	50.92593	817.9	158.0	0.001223	0.006323	110.10	460.967	153.17	641.292	43.07	180.325	1.0340	4.3292	1.1835	4.9551
17.5	290.65	55.10	54.03464	795.5	173.2	0.001253	0.005774	111.90	468.503	152.27	637.524	40.37	169.021	1.0400	4.3543	1.1790	4.9362
20	293.15	58.46	57.32968	771.1	189.8	0.001297	0.005269	114.00	477.295	151.10	632.625	37.10	155.330	1.0468	4.3827	1.1734	4.9128
22.5	295.65	61.85	60.65413	742.9	210.4	0.001346	0.004753	116.20	486.506	149.50	625.927	33.30	139.420	1.0543	4.4141	1.1666	4.8843
25	298.15	65.59	64.32182	709.5	236.3	0.001409	0.004232	118.80	497.392	147.33	616.841	28.53	119.449	1.0628	4.4497	1.1585	4.8504
27.5	300.65	69.35	68.00912	666.4	271.8	0.001501	0.003679	122.00	510.790	144.55	605.202	22.55	94.412	1.0730	4.4924	1.1487	4.8094
30	303.15	73.34	71.92197	595.1	335.7	0.001680	0.002979	125.90	527.118	140.95	590.129	15.05	63.011	1.0854	4.5444	1.1351	4.7524
31 (crit.)	304.15	74.96	73.51065	463.9	463.9	0.002156	0.002156	133.50	558.938	133.50	558.938	0.00	0	1.1098	4.6465	1.1098	4.6465

1 at = 1 kp/cm² = 98 066.5 Pa = 98 066.5 N/m² = 9.806 65 N/cm² = 0.980 665 bar

1 kcal = 4.186 8 kJ

Table 60-1 Properties of Superheated Dichlorodifluoromethane "Freon 12" (CF_2Cl_2)

Pressure p		0.05 at = 0.049 033 25 bar						0.1 at = 0.098 066 5 bar					
Temperature		$t_s = -82.5\,°C = 190.65\,K$ $\rho'' = 0.3748\,kg/m^3$ $i'' = 127.19\,kcal/kg = 532.52\,kJ/kg$ $v'' = 2.668\,m^3/kg$ $s'' = 1.1592\,kcal/kg\,K = 4.8533\,kJ/kg\,K$						$t_s = -73.3\,°C = 199.85\,K$ $\rho'' = 0.7184\,kg/m^3$ $i'' = 128.46\,kcal/kg = 537.84\,kJ/kg$ $v'' = 1.392\,m^3/kg$ $s'' = 1.1544\,kcal/kg\,K = 4.8332\,kJ/kg\,K$					
t	T	ρ	v	i		s		ρ	v	i		s	
°C	K	kg/m³	m³/kg	kcal/kg	kJ/kg	kcal/kg K	kJ/kg K	kg/m³	m³/kg	kcal/kg	kJ/kg	kcal/kg K	kJ/kg K
−80	193.15	0.3687	2.712	127.67	534.53	1.1608	4.8600	—	—	—	—	—	—
−75	198.15	0.3596	2.781	128.26	537.00	1.1641	4.8739	—	—	—	—	—	—
−70	203.15	0.3509	2.850	128.86	539.51	1.1673	4.8873	0.7047	1.419	128.84	539.43	1.1563	4.8412
−65	208.15	0.3426	2.919	129.47	542.06	1.1704	4.9002	0.6878	1.454	129.45	541.98	1.1593	4.8538
−60	213.15	0.3347	2.988	130.09	544.66	1.1734	4.9128	0.6716	1.489	130.07	544.58	1.1622	4.8659
−55	218.15	0.3271	3.057	130.72	547.30	1.1764	4.9254	0.6562	1.524	130.69	547.17	1.1651	4.8780
−50	223.15	0.3199	3.126	131.35	549.94	1.1793	4.9375	0.6414	1.559	131.32	549.81	1.1679	4.8898
−45	228.15	0.3130	3.195	131.99	552.62	1.1822	4.9496	0.6274	1.594	131.96	552.49	1.1707	4.9015
−40	233.15	0.3063	3.265	132.64	555.34	1.1850	4.9614	0.6139	1.629	132.61	555.21	1.1735	4.9132
−35	238.15	0.2999	3.335	133.29	558.06	1.1878	4.9731	0.6010	1.664	133.26	557.93	1.1762	4.9245
−30	243.15	0.2937	3.405	133.95	560.82	1.1905	4.9844	0.5886	1.699	133.92	560.70	1.1789	4.9358
−25	248.15	0.2878	3.475	134.61	563.59	1.1932	4.9957	0.5767	1.734	134.59	563.50	1.1816	4.9471
−20	253.15	0.2821	3.545	135.28	566.39	1.1959	5.0070	0.5653	1.769	135.26	566.31	1.1843	4.9584
−15	258.15	0.2766	3.615	135.95	569.20	1.1985	5.0179	0.5543	1.804	135.94	569.15	1.1869	4.9693
−10	263.15	0.2713	3.686	136.63	572.04	1.2011	5.0288	0.5438	1.839	136.62	572.00	1.1895	4.9802
− 5	268.15	0.2662	3.756	137.32	574.93	1.2037	5.0397	0.5333	1.875	137.30	574.85	1.1921	4.9911
0	273.15	0.2614	3.826	138.01	577.82	1.2063	5.0505	0.5233	1.911	137.99	577.74	1.1947	5.0020
5	278.15	0.2567	3.895	138.71	580.75	1.2088	5.0610	0.5136	1.947	138.69	580.67	1.1972	5.0124
10	283.15	0.2523	3.964	139.41	583.68	1.2113	5.0715	0.5043	1.983	139.39	583.60	1.1997	5.0229
15	288.15	0.2479	4.034	140.11	586.61	1.2138	5.0819	0.4953	2.019	140.10	586.57	1.2022	5.0334
20	293.15	0.2437	4.104	140.82	589.59	1.2163	5.0924	0.4869	2.054	140.81	589.54	1.2047	5.0438
25	298.15	0.2396	4.174	141.54	592.60	1.2188	5.1029	0.4787	2.089	141.53	592.56	1.2072	5.0543
30	303.15	0.2356	4.244	142.27	595.66	1.2212	5.1129	0.4708	2.124	142.26	595.61	1.2097	5.0648
35	308.15	0.2318	4.314	143.00	598.71	1.2236	5.1230	0.4632	2.159	142.99	589.67	1.2121	5.0748
40	313.15	0.2281	4.385	143.74	601.81	1.2260	5.1330	0.4558	2.194	143.73	601.77	1.2145	5.0849
45	318.15	0.2244	4.456	144.48	604.91	1.2284	5.1431	0.4486	2.229	144.47	604.87	1.2169	5.0949
50	323.15	0.2209	4.527	145.23	608.05	1.2307	5.1527	0.4417	2.264	145.22	608.01	1.2192	5.1045
55	328.15	0.2175	4.598	145.99	611.23	1.2330	5.1623	0.4350	2.299	145.98	611.19	1.2215	5.1142
60	333.15	0.2141	4.670	146.76	614.45	1.2351	5.1711	0.4286	2.333	146.75	614.41	1.2238	5.1238
65	338.15	0.2109	4.742	147.53	617.68	1.2375	5.1812	0.4225	2.367	147.53	617.68	1.2260	5.1330
70	343.15	0.2077	4.815	148.31	620.94	1.2398	5.1908	0.4165	2.401	148.32	620.99	1.2282	5.1422
75	348.15	0.2046	4.888	149.09	624.21	1.2421	5.2004	0.4108	2.434	149.11	624.29	1.2304	5.1514
80	353.15	0.2015	4.962	149.88	627.52	1.2443	5.2096	0.4054	2.467	149.90	627.60	1.2326	5.1606

1 at = 1 kp/cm² = 98 066.5 Pa = 98 066.5 N/m² = 9.806 65 N/cm² = 0.980 665 bar

1 kcal = 4.186 8 kJ

Table 60-2 Properties of Superheated Dichlorodifluoromethane "Freon 12" (CF_2Cl_2) (*Continued*)

Pressure p		0.2 at = 0.196 133 0 bar						0.3 at = 0.294 199 5 bar					
Temperature		$t_s = -62.5\ °C = 210.65\ K$ $i'' = 129.73\ kcal/kg = 426.15\ kJ/kg$ $s'' = 1.1493\ kcal/kg\ K = 4.8119\ kJ/kg\ K$				$\rho'' = 1.368\ kg/m^3$ $v'' = 0.7311\ m^3/kg$		$t_s = -55.4\ °C = 217.75\ K$ $i'' = 130.55\ kcal/kg = 546.59\ kJ/kg$ $s'' = 1.1466\ kcal/kg\ K = 4.8006\ kJ/kg\ K$				$\rho'' = 1.992\ kg/m^3$ $v'' = 0.5021\ m^3/kg$	
t	T	ρ	v	i		s		ρ	v	i		s	
°C	K	kg/m³	m³/kg	kcal/kg	kJ/kg	kcal/kg K	kJ/kg K	kg/m³	m³/kg	kcal/kg	kJ/kg	kcal/kg K	kcal/kg K
−60	213.15	1.349	0.7411	130.02	544.37	1.1506	4.8173	—	—	—	—	—	—
−55	218.15	1.319	0.7581	130.65	547.01	1.1534	4.8291	1.970	0.5077	130.57	546.67	1.1469	4.8018
−50	223.15	1.290	0.7751	131.28	549.64	1.1562	4.8408	1.927	0.5190	131.22	549.39	1.1497	4.8136
−45	228.15	1.262	0.7922	131.92	552.32	1.1590	4.8525	1.885	0.5304	131.87	552.11	1.1524	4.8249
−40	233.15	1.235	0.8095	132.57	555.04	1.1618	4.8642	1.845	0.5420	132.52	554.83	1.1551	4.8362
−35	238.15	1.209	0.8269	133.23	557.81	1.1645	4.8755	1.806	0.5536	133.18	557.60	1.1578	4.8475
−30	243.15	1.184	0.8443	133.89	560.57	1.1673	4.8873	1.769	0.5654	133.85	560.40	1.1605	4.8588
−25	248.15	1.160	0.8618	134.55	563.33	1.1699	4.8981	1.733	0.5772	134.52	563.21	1.1632	4.8701
−20	253.15	1.137	0.8794	135.22	566.14	1.1726	4.9094	1.698	0.5891	135.19	566.01	1.1659	4.8814
−15	258.15	1.115	0.8972	135.89	568.94	1.1752	4.9203	1.664	0.6010	135.86	568.82	1.1686	4.8927
−10	263.15	1.093	0.9151	136.57	571.79	1.1778	4.9312	1.632	0.6129	136.53	571.62	1.1712	4.9036
−5	268.15	1.072	0.9331	137.26	574.68	1.1805	4.9425	1.601	0.6248	137.21	574.47	1.1738	4.9145
0	273.15	1.051	0.9511	137.96	577.61	1.1831	4.9534	1.571	0.6367	137.90	577.36	1.1764	4.9254
5	278.15	1.032	0.9692	138.66	580.54	1.1857	4.9643	1.542	0.6486	138.60	580.29	1.1790	4.9362
10	283.15	1.013	0.9873	139.36	583.47	1.1883	4.9752	1.514	0.6605	139.31	583.26	1.1816	4.9471
15	288.15	0.9950	1.005	140.07	586.45	1.1907	4.9852	1.487	0.6724	140.02	586.24	1.1841	4.9576
20	293.15	0.9775	1.023	140.78	589.42	1.1932	4.9957	1.461	0.6843	140.74	589.25	1.1866	4.9681
25	298.15	0.9606	1.041	141.50	592.43	1.1957	5.0062	1.436	0.6962	141.46	592.26	1.1891	4.9785
30	303.15	0.9443	1.059	142.23	595.49	1.1982	5.0166	1.412	0.7081	142.19	595.32	1.1915	4.9886
35	308.15	0.9285	1.077	142.96	598.54	1.2006	5.0267	1.389	0.7200	142.92	598.38	1.1939	4.9986
40	313.15	0.9132	1.095	143.70	601.64	1.2030	5.0367	1.366	0.7318	143.66	601.48	1.1963	5.0087
45	318.15	0.8985	1.113	144.44	604.74	1.2054	5.0468	1.345	0.7436	144.40	604.57	1.1987	5.0187
50	323.15	0.8842	1.131	145.19	607.88	1.2077	5.0564	1.324	0.7554	145.15	607.71	1.2010	5.0283
55	328.15	0.8703	1.149	145.95	611.06	1.2100	5.0660	1.303	0.7672	145.91	610.90	1.2033	5.0380
60	333.15	0.8569	1.167	146.72	614.29	1.2124	5.0761	1.284	0.7790	146.68	614.12	1.2056	5.0476
65	338.15	0.8439	1.185	147.50	617.55	1.2146	5.0853	1.265	0.7908	147.46	617.39	1.2079	5.0572
70	343.15	0.8313	1.203	148.29	620.86	1.2168	5.0945	1.246	0.8026	148.25	620.69	1.2102	5.0669
75	348.15	0.8190	1.221	149.08	624.17	1.2190	5.1037	1.228	0.8145	149.04	624.00	1.2125	5.0765
80	353.15	0.8078	1.238	149.87	627.48	1.2212	5.1129	1.210	0.8264	149.83	627.31	1.2147	5.0857
85	358.15	0.7968	1.255	150.66	630.78	1.2234	5.1221	1.193	0.8383	150.62	630.62	1.2169	5.0949
90	363.15	0.7862	1.272	151.45	634.09	1.2256	5.1313	1.176	0.8502	151.41	633.92	1.2191	5.1041
95	368.15	0.7758	1.289	152.24	637.40	1.2278	5.1406	1.160	0.8621	152.20	637.23	1.2213	5.1133
100	373.15	—	—	—	—	—	—	1.144	0.8740	153.00	640.58	1.2235	5.1225
105	378.15	—	—	—	—	—	—	1.129	0.8859	153.80	643.93	1.2256	5.1313
110	383.15	—	—	—	—	—	—	1.114	0.8978	154.60	647.28	1.2277	5.1401
115	388.15	—	—	—	—	—	—	1.099	0.9097	155.40	650.63	1.2298	5.1489

1 at = 1 kp/cm² = 98 066.5 Pa = 980 66.5 N/m² = 9.80665 N/cm² = 0.980 665 bar

1 kcal = 4.186 8 kJ

Table 60-3 Properties of Superheated Dichlorodifluoromethane "Freon 12" (CF_2Cl_2) (*Continued*)

| Pressure p | | 0.4 at = 0.392 266 0 bar | | | | | | 0.6 at = 0.588 399 0 bar | | | | | |

| Temperature | | $t_s = -50\ °C = 223.15$ K $i'' = 131.19$ kcal/kg $= 549.27$ kJ/kg $s'' = 1.1446$ kcal/kg K $= 4.7922$ kJ/kg K | | | | $\rho'' = 2.591$ kg/m³ $v'' = 0.3859$ m³/kg | | $t_s = -41.8\ °C = 231.35$ K $i'' = 132.13$ kcal/kg $= 553.20$ kJ/kg $s'' = 1.1423$ kcal/kg K $= 4.7826$ kJ/kg K | | | | $\rho'' = 3.772$ kg/m³ $v'' = 0.2651$ m³/kg |

t	T	ρ	v	i		s		ρ	v	i		s	
°C	K	kg/m³	m³/kg	kcal/kg	kJ/kg	kcal/kg K	kJ/kg K	kg/m³	m³/kg	kcal/kg	kJ/kg	kcal/kg K	kJ/kg K
−45	228.15	2.535	0.3945	131.83	551.95	1.1474	4.8039	—	—	—	—	—	—
−40	233.15	2.476	0.4038	132.47	554.63	1.1502	4.8157	3.697	0.2705	132.38	554.25	1.1432	4.7863
−35	238.15	2.422	0.4128	133.13	557.39	1.1530	4.8274	3.621	0.2762	133.05	557.05	1.1460	4.7981
−30	243.15	2.371	0.4218	133.80	560.19	1.1558	4.8391	3.546	0.2820	133.72	559.86	1.1489	4.8102
−25	248.15	2.321	0.4308	134.47	563.00	1.1585	4.8504	3.475	0.2878	134.40	562.71	1.1516	4.8215
−20	253.15	2.274	0.4398	135.14	565.80	1.1612	4.8617	3.406	0.2936	135.08	565.55	1.1543	4.8328
−15	258.15	2.228	0.4488	135.82	568.65	1.1639	4.8730	3.340	0.2994	135.76	568.40	1.1570	4.8441
−10	263.15	2.184	0.4578	136.50	571.50	1.1666	4.8843	3.275	0.3053	136.44	571.25	1.1596	4.8550
−5	268.15	2.142	0.4668	137.19	574.39	1.1692	4.8952	3.213	0.3112	137.13	574.14	1.1622	4.8659
0	273.15	2.102	0.4758	137.89	577.32	1.1718	4.9061	3.154	0.3171	137.82	577.02	1.1648	4.8768
5	278.15	2.063	0.4848	138.59	580.25	1.1743	4.9166	3.095	0.3231	138.52	579.96	1.1674	4.8877
10	283.15	2.025	0.4938	139.29	583.18	1.1768	4.9270	3.039	0.3291	139.23	582.93	1.1699	4.8981
15	288.15	1.989	0.5028	139.99	586.11	1.1793	4.9375	2.984	0.3351	139.94	585.90	1.1724	4.9086
20	293.15	1.954	0.5118	140.70	589.08	1.1818	4.9480	2.932	0.3411	140.66	588.92	1.1748	4.9187
25	298.15	1.920	0.5208	141.43	592.14	1.1842	4.9580	2.881	0.3471	141.39	591.97	1.1772	4.9287
30	303.15	1.888	0.5298	142.17	595.24	1.1866	4.9681	2.832	0.3531	142.12	595.03	1.1797	4.9392
35	308.15	1.856	0.5388	142.91	598.34	1.1890	4.9781	2.785	0.3591	142.86	598.13	1.1820	4.9488
40	313.15	1.825	0.5478	143.65	601.53	1.1914	4.9882	2.739	0.3651	143.60	601.22	1.1845	4.9593
45	318.15	1.796	0.5568	144.39	604.53	1.1938	4.9982	2.695	0.3711	144.34	604.32	1.1868	4.9689
50	323.15	1.767	0.5658	145.14	607.67	1.1962	5.0083	2.652	0.3771	145.09	607.46	1.1841	4.9576
55	328.15	1.740	0.5748	145.90	610.85	1.1985	5.0179	2.610	0.3831	145.85	610.64	1.1915	4.9886
60	333.15	1.713	0.5838	146.68	614.12	1.2008	5.0275	2.570	0.3891	146.62	613.87	1.1940	4.9990
65	338.15	1.687	0.5928	147.46	617.39	1.2031	5.0371	2.531	0.3951	147.40	617.13	1.1962	5.0083
70	343.15	1.662	0.6018	148.24	620.65	1.2054	5.0468	2.493	0.4011	148.19	620.44	1.1985	5.0179
75	348.15	1.637	0.6108	149.03	623.96	1.2077	5.0564	2.456	0.4071	148.98	623.75	1.2007	5.0271
80	353.15	1.613	0.6198	149.82	627.27	1.2099	5.0656	2.421	0.4131	149.77	627.06	1.2031	5.0371
85	358.15	1.590	0.6288	150.61	630.57	1.2111	5.0706	2.386	0.4191	150.56	630.36	1.2053	5.0464
90	363.15	1.568	0.6378	151.40	633.88	1.2143	5.0840	2.352	0.4251	151.35	633.67	1.2075	5.0556
95	368.15	1.546	0.6468	152.19	637.19	1.2165	5.0932	2.320	0.4311	152.15	637.02	1.2096	5.0644
100	373.15	1.525	0.6558	152.99	640.54	1.2186	5.1020	2.288	0.4371	152.95	640.37	1.2119	5.0740
105	378.15	1.504	0.6648	153.79	643.89	1.2207	5.1108	2.257	0.4431	153.75	643.72	1.2190	5.1037
110	383.15	1.484	0.6738	154.59	647.24	1.2228	5.1196	2.227	0.4491	154.55	647.07	1.2161	5.0916
115	388.15	1.465	0.6828	155.39	650.59	1.2249	5.1284	2.197	0.4551	155.35	650.42	1.2183	5.1008

1 at = 1 kp/cm² = 98 066.5 Pa = 98 066.5 N/m² = 9.806 65 N/cm² = 0.980 665 bar   1 kcal = 4.186 8 kJ

Table 60-4 Properties of Superheated Dichlorodifluoromethane "Freon 12" (CF_2Cl_2) (Continued)

Pressure p		0.8 at = 0.784 532 0 bar						1.0 at = 0.980 665 0 bar					
Temperature		$t_s = -35.6\,°C = 237.55$ K $\quad$ $i'' = 132.86$ kcal/kg $= 556.26$ kJ/kg $\quad$ $s'' = 1.1408$ kcal/kg K $= 4.7763$ kJ/kg K $\quad$ $\rho'' = 4.955$ kg/m³ $\quad$ $v'' = 0.2018$ m³/kg						$t_s = -30.6\,°C = 242.55$ K $\quad$ $i'' = 133.48$ kcal/kg $= 558.85$ kJ/kg $\quad$ $s'' = 1.1398$ kcal/kg K $= 4.7721$ kJ/kg K $\quad$ $\rho'' = 6.075$ kg/m³ $\quad$ $v'' = 0.1646$ m³/kg					
t	T	ρ	v	i		s		ρ	v	i		s	
°C	K	kg/m³	m³/kg	kcal/kg	kJ/kg	kcal/kg K	kJ/kg K	kg/m³	m³/kg	kcal/kg	kJ/kg	kcal/kg K	kJ/kg K
−35	238.15	4.931	0.2028	132.96	556.68	1.1411	4.7776	—	—	—	—	—	—
−30	243.15	4.826	0.2072	133.63	559.48	1.1438	4.7889	6.061	0.1650	133.52	559.02	1.1401	4.7734
−25	248.15	4.726	0.2116	134.31	562.33	1.1466	4.8006	5.938	0.1684	134.21	561.91	1.1428	4.7847
−20	253.15	4.630	0.2160	134.99	565.18	1.1494	4.8123	5.824	0.1717	134.90	564.80	1.1455	4.7960
−15	258.15	4.537	0.2204	135.67	568.02	1.1520	4.8232	5.714	0.1750	135.59	567.69	1.1481	4.8069
−10	263.15	4.448	0.2248	136.35	570.87	1.1547	4.8345	5.609	0.1783	136.29	570.62	1.1507	4.8178
−5	268.15	4.363	0.2292	137.04	573.76	1.1573	4.8454	5.507	0.1816	136.99	573.55	1.1534	4.8291
0	273.15	4.281	0.2336	137.74	576.69	1.1599	4.8563	5.408	0.1849	137.69	576.48	1.1559	4.8395
5	278.15	4.202	0.2380	138.45	579.66	1.1624	4.8667	5.313	0.1882	138.40	579.45	1.1585	4.8504
10	283.15	4.125	0.2424	139.16	582.64	1.1650	4.8776	5.222	0.1915	139.11	582.43	1.1612	4.8617
15	288.15	4.054	0.2467	139.88	585.65	1.1674	4.8877	5.133	0.1948	139.82	585.40	1.1635	4.8713
20	293.15	3.984	0.2510	140.61	588.71	1.1700	4.8986	5.048	0.1981	140.54	588.41	1.1660	4.8818
25	298.15	3.917	0.2553	141.34	591.76	1.1724	4.9086	4.965	0.2014	141.27	591.47	1.1684	4.8919
30	303.15	3.852	0.2596	142.08	594.86	1.1749	4.9191	4.885	0.2047	142.01	594.57	1.1709	4.9023
35	308.15	3.791	0.2638	142.82	597.96	1.1773	4.9291	4.808	0.2080	142.75	597.67	1.1732	4.9120
40	313.15	3.731	0.2680	143.56	601.06	1.1797	4.9392	4.735	0.2112	143.49	600.76	1.1756	4.9220
45	318.15	3.674	0.2722	144.30	604.16	1.1820	4.9488	4.664	0.2144	144.23	603.86	1.1780	4.9321
50	323.15	3.618	0.2764	145.05	607.30	1.1844	4.9588	4.696	0.2176	144.99	607.04	1.1805	4.9425
55	328.15	3.564	0.2806	145.82	610.52	1.1868	4.9689	4.529	0.2208	145.76	610.27	1.1828	4.9521
60	333.15	3.511	0.2848	146.59	613.74	1.1891	4.9785	4.464	0.2240	146.54	613.53	1.1853	4.9626
65	338.15	3.460	0.2890	147.37	617.01	1.1915	4.9886	4.401	0.2272	147.32	616.80	1.1876	4.9722
70	343.15	3.411	0.2932	148.15	620.27	1.1938	4.9982	4.340	0.2304	148.10	620.07	1.1899	4.9819
75	348.15	3.362	0.2974	148.94	623.58	1.1960	5.0074	4.283	0.2335	148.88	623.33	1.1922	4.9915
80	353.15	3.316	0.3016	149.73	626.89	1.1983	5.0170	4.227	0.2366	149.68	626.68	1.1944	5.0007
85	358.15	3.270	0.3058	150.52	630.20	1.2005	5.0263	4.172	0.2397	150.48	630.03	1.1966	5.0099
90	363.15	3.226	0.3100	151.31	633.50	1.2028	5.0359	4.119	0.2428	151.28	633.38	1.1989	5.0196
95	368.15	3.183	0.3142	152.11	636.85	1.2049	5.0447	4.067	0.2459	152.08	636.73	1.2011	5.0288
100	373.15	3.141	0.3184	152.91	640.20	1.2071	5.0539	4.016	0.2490	152.88	640.08	1.2032	5.0376
105	378.15	3.100	0.3226	153.71	643.55	1.2092	5.0627	3.965	0.2522	153.68	643.43	1.2054	5.0468
110	383.15	3.060	0.3268	154.51	646.90	1.2115	5.0723	3.915	0.2554	154.48	646.78	1.2076	5.0560
115	388.15	3.021	0.3310	155.32	650.29	1.2136	5.0811	3.867	0.2586	155.28	650.13	1.2097	5.0648

1 at = 1 kp/cm² = 98 066.5 Pa = 98 066.5 N/m² = 9.806 65 N/cm² = 0.980 665 bar

1 kcal = 4.186 8 kJ

Table 60-5 Properties of Superheated Dichlorodifluoromethane "Freon 12" (CF_2Cl_2) (Continued)

Pressure p		1.5 at = 1.470 997 5 bar						2 at = 1.961 330 0 bar					
Temperature		$t_s = -20.7\,°C = 252.45\,K$ $\quad$ $\rho'' = 8.818\,kg/m^3$						$t_s = -13.1\,°C = 260.05\,K$ $\quad$ $\rho'' = 11.521\,kg/m^3$					
		$i'' = 134.65\,kcal/kg = 563.75\,kJ/kg$ $\quad$ $v'' = 0.1134\,m^3/kg$						$i'' = 135.51\,kcal/kg = 567.35\,kJ/kg$ $\quad$ $v'' = 0.0868\,m^3/kg$					
		$s'' = 1.1381\,kcal/kg\,K = 4.7650\,kJ/kg\,K$						$s'' = 1.1371\,kcal/kg\,K = 4.7608\,kJ/kg\,K$					
t	T	ρ	v	i		s		ρ	v	i		s	
°C	K	kg/m³	m³/kg	kcal/kg	kJ/kg	kcal/kg K	kJ/kg K	kg/m³	m³/kg	kcal/kg	kJ/kg	kcal/kg K	kJ/kg K
−20	253.15	8.795	0.1137	134.83	564.51	1.1385	4.7667	—	—	—	—	—	—
−15	258.15	8.606	0.1162	135.52	567.40	1.1411	4.7776	—	—	—	—	—	—
−10	263.15	8.425	0.1187	136.22	570.33	1.1436	4.7880	11.364	0.0880	135.95	569.20	1.1385	4.7667
− 5	268.15	8.251	0.1212	136.92	573.26	1.1467	4.8010	11.148	0.0897	136.66	572.17	1.1411	4.7776
0	273.15	8.084	0.1237	137.63	576.23	1.1488	4.8098	10.929	0.0915	137.37	575.14	1.1437	4.7884
5	278.15	7.924	0.1262	138.34	579.20	1.1514	4.8207	10.707	0.0934	138.08	578.11	1.1463	4.7993
10	283.15	7.770	0.1287	139.05	582.17	1.1539	4.8311	10.482	0.0954	138.80	581.13	1.1489	4.8102
15	288.15	7.622	0.1312	139.77	585.19	1.1564	4.8416	10.267	0.0974	139.52	584.14	1.1514	4.8207
20	293.15	7.479	0.1337	140.50	588.25	1.1589	4.8521	10.060	0.0994	140.25	587.20	1.1539	4.8311
25	298.15	7.342	0.1362	141.23	591.30	1.1614	4.8625	9.862	0.1014	140.99	590.30	1.1564	4.8416
30	303.15	7.210	0.1387	141.97	594.40	1.1638	4.8725	9.671	0.1034	141.73	593.40	1.1589	4.8521
35	308.15	7.082	0.1412	142.72	597.54	1.1663	4.8831	9.488	0.1054	142.48	596.54	1.1613	4.8621
40	313.15	6.959	0.1437	143.47	600.68	1.1688	4.8935	9.320	0.1073	143.23	599.68	1.1637	4.8722
45	318.15	6.840	0.1462	144.22	603.82	1.1712	4.9036	9.158	0.1092	143.98	602.82	1.1661	4.8822
50	323.15	6.729	0.1486	144.98	607.00	1.1736	4.9136	9.001	0.1111	144.74	606.00	1.1685	4.8923
55	328.15	6.623	0.1510	145.74	610.18	1.1760	4.9237	8.850	0.1130	145.52	609.26	1.1709	4.9023
60	333.15	6.519	0.1534	146.50	613.37	1.1783	4.9333	8.703	0.1149	146.30	612.53	1.1733	4.9124
65	338.15	6.418	0.1558	147.26	616.55	1.1807	4.9434	8.562	0.1168	147.08	615.79	1.1756	4.9220
70	343.15	6.321	0.1582	148.02	619.73	1.1830	4.9530	8.425	0.1187	147.87	619.10	1.1780	4.9321
75	348.15	6.227	0.1606	148.79	622.95	1.1854	4.9630	8.292	0.1206	148.66	622.41	1.1802	4.9413
80	353.15	6.135	0.1630	149.57	626.22	1.1876	4.9722	8.163	0.1225	149.45	625.72	1.1825	4.9509
85	358.15	6.046	0.1654	150.36	629.53	1.1899	4.9819	8.039	0.1244	150.25	629.07	1.1848	4.9605
90	363.15	5.959	0.1678	151.16	632.88	1.1920	4.9907	7.924	0.1262	151.05	632.42	1.1871	4.9702
95	368.15	5.875	0.1702	151.96	636.23	1.1943	5.0003	7.813	0.1280	151.85	635.77	1.1894	4.9798
100	373.15	5.794	0.1726	152.77	639.62	1.1964	5.0091	7.704	0.1298	152.66	639.16	1.1916	4.9890
105	378.15	5.714	0.1750	153.58	643.01	1.1986	5.0183	7.599	0.1316	153.48	642.59	1.1938	4.9982
110	383.15	5.634	0.1775	154.39	646.40	1.2009	5.0279	7.496	0.1334	154.30	646.02	1.1960	5.0074
115	388.15	5.556	0.1800	155.20	649.79	1.2029	5.0363	7.396	0.1352	155.12	649.46	1.1982	5.0166
120	393.15	5.479	0.1825	156.04	653.31	1.2050	5.0451	7.299	0.1370	155.96	652.97	1.2003	5.0254
125	398.15	5.405	0.1850	156.89	656.87	1.2071	5.0539	7.205	0.1388	156.81	156.81	1.2024	5.0342
130	403.15	5.333	0.1875	157.75	660.47	1.2092	5.0627	7.112	0.1406	157.67	660.13	1.2045	5.0430

1 at = 1 kp/cm² = 98 066.5 Pa = 98 066.5 N/m² = 9.806 65 N/cm² = 0.980 665 bar $\qquad$ 1 kcal = 4.186 8 kJ

Table 60-6 Properties of Superheated Dichlorodifluoromethane "Freon 12" (CF_2Cl_2) (Continued)

Pressure p		2.5 at = 2.451 662 5 bar						3.0 at = 2.941 995 0 bar					
Temperature		$t_s = -6.8\,°C = 266.35$ K$\quad$ $\rho'' = 14.184$ kg/m³ $i'' = 136.22$ kcal/kg = 570.33 kJ/kg$\quad$ $v'' = 0.0705$ m³/kg $s'' = 1.1362$ kcal/kg K = 4.7570 kJ/kg K						$t_s = -1.5\,°C = 271.65$ K$\quad$ $\rho'' = 16.892$ kg/m³ $i'' = 136.82$ kcal/kg = 572.84 kJ/kg$\quad$ $v'' = 0.0592$ m³/kg $s'' = 1.1356$ kcal/kg K = 4.7545 kJ/kg K					
t	T	ρ	v	i		s		ρ	v	i		s	
°C	K	kg/m³	m³/kg	kcal/kg	kJ/kg	kcal/kg K	kJ/kg K	kg/m³	m³/kg	kcal/kg	kJ/kg	kcal/kg K	kJ/kg K
− 5	268.15	14.085	0.0710	136.48	571.41	1.1372	4.7612	—	—	—	—	—	—
0	273.15	13.850	0.0722	137.20	574.43	1.1398	4.7721	16.779	0.0596	137.04	573.76	1.1363	4.7575
5	278.15	13.605	0.0735	137.92	577.44	1.1423	4.7826	16.393	0.0610	137.76	576.77	1.1389	4.7683
10	283.15	13.351	0.0749	138.64	580.46	1.1448	4.7930	16.026	0.0624	138.49	579.83	1.1414	4.7788
15	288.15	13.089	0.0764	139.37	583.51	1.1473	4.8035	15.674	0.0638	139.22	582.89	1.1439	4.7893
20	293.15	12.821	0.0780	140.10	586.57	1.1498	4.8140	15.361	0.0651	139.96	585.98	1.1465	4.8002
25	298.15	12.563	0.0796	140.85	589.71	1.1523	4.8244	15.060	0.0664	140.70	589.08	1.1490	4.8106
30	303.15	12.315	0.0812	141.60	592.85	1.1548	4.8349	14.771	0.0677	141.45	592.22	1.1515	4.8211
35	308.15	12.077	0.0828	142.35	595.99	1.1573	4.8454	14.493	0.0690	142.21	595.40	1.1540	4.8316
40	313.15	11.848	0.0844	143.10	599.13	1.1598	4.8559	14.225	0.0703	142.97	598.59	1.1565	4.8420
45	318.15	11.628	0.0860	143.85	602.27	1.1622	4.8659	13.966	0.0716	143.73	601.77	1.1589	4.8521
50	323.15	11.429	0.0875	144.61	605.45	1.1646	4.8759	13.717	0.0729	144.49	604.95	1.1613	4.8621
55	328.15	11.236	0.0890	145.39	608.72	1.1670	4.8860	13.477	0.0742	145.27	608.22	1.1637	4.8722
60	333.15	11.050	0.0905	146.17	611.98	1.1694	4.8960	13.245	0.0755	146.06	611.52	1.1661	4.8822
65	338.15	10.881	0.0919	146.96	615.29	1.1717	4.9057	13.021	0.0768	146.85	614.83	1.1685	4.8923
70	343.15	10.718	0.0933	147.75	618.60	1.1740	4.9153	12.821	0.0780	147.64	618.14	1.1708	4.9019
75	348.15	10.549	0.0948	148.54	621.91	1.1763	4.9249	12.610	0.0793	148.43	621.45	1.1731	4.9115
80	353.15	10.384	0.0963	149.34	625.26	1.1786	4.9346	12.407	0.0806	149.22	624.75	1.1754	4.9212
85	358.15	10.225	0.0978	150.14	628.61	1.1809	4.9442	12.210	0.0819	150.03	628.15	1.1777	4.9308
90	363.15	10.070	0.0939	150.95	632.00	1.1832	4.9538	12.034	0.0831	150.84	631.54	1.1800	4.9404
95	368.15	9.930	0.1007	151.76	635.39	1.1854	4.9630	11.862	0.0843	151.65	634.93	1.1822	4.9496
100	373.15	9.785	0.1022	152.57	638.78	1.1876	4.9722	11.696	0.0855	152.47	638.36	1.1844	4.9588
105	378.15	9.643	0.1037	153.39	642.21	1.1898	4.9815	11.521	0.0868	153.29	641.79	1.1866	4.9681
110	383.15	9.506	0.1052	154.21	645.65	1.1920	4.9907	11.351	0.0881	154.11	645.23	1.1889	4.9777
115	388.15	9.372	0.1067	155.04	649.12	1.1941	4.9995	11.186	0.0894	154.94	648.70	1.1911	4.9869
120	393.15	9.242	0.1082	155.87	652.60	1.1963	5.0087	11.025	0.0907	155.79	652.26	1.1932	4.9957
125	398.15	9.116	0.1097	156.73	656.20	1.1984	5.0175	10.870	0.0920	156.65	655.86	1.1953	5.0045
130	403.15	8.993	0.1112	157.59	659.80	1.2005	5.0263	10.730	0.0932	157.51	659.46	1.1974	5.0133
135	408.15	8.873	0.1127	158.45	663.40	1.2025	5.0346	10.593	0.0944	158.37	663.06	1.1995	5.0221
140	413.15	—	—	—	—	—	—	10.450	0.0956	159.23	666.66	1.2015	5.0304
145	418.15	—	—	—	—	—	—	10.341	0.0967	160.10	670.31	1.2036	5.0392

1 at = 1 kp/cm² = 98 066.5 Pa = 98 066.5 N/m² = 9.806 65 N/cm² = 0.980 665 bar

1 kcal = 4.186 8 kJ

Table 60-7 Properties of Superheated Dichlorodifluoromethane "Freon 12" (CF_2Cl_2) (Continued)

Pressure p		4 at = 3.922 660 bar						5 at = 4.903 325 bar					
Temperature		$t_s = 7.5\,°C = 280.65\ K$ $\qquad$ $\rho'' = 22.173\ kg/m^3$ $i'' = 137.80\ kcal/kg = 576.94\ kJ/kg$ $\qquad$ $v'' = 0.0451\ m^3/kg$ $s'' = 1.1348\ kcal/kg\ K = 4.7512\ kJ/kg\ K$						$t_s = 15\,°C = 288.15\ K$ $\qquad$ $\rho'' = 27.397\ kg/m^3$ $i'' = 138.61\ kcal/kg = 580.33\ kJ/kg$ $\qquad$ $v'' = 0.0365\ m^3/kg$ $s'' = 1.1342\ kcal/kg\ K = 4.7487\ kJ/kg\ K$					
t	T	ρ	v	i		s		ρ	v	i		s	
°C	K	kg/m³	m³/kg	kcal/kg	kJ/kg	kcal/kg K	kJ/kg K	kg/m³	m²/kg	kcal/kg	kJ/kg	kcal/kg K	kJ/kg K
10	283.15	21.786	0.0459	138.18	578.53	1.1361	4.7566	—	—	—	—	—	—
15	288.15	21.277	0.0470	138.91	581.59	1.1387	4.7675	—	—	—	—	—	—
20	293.15	20.833	0.0480	139.66	584.73	1.1413	4.7784	26.525	0.0377	139.36	583.47	1.1367	4.7591
25	298.15	20.408	0.0490	140.42	587.91	1.1438	4.7889	25.974	0.0385	140.12	586.65	1.1392	4.7696
30	303.15	20.000	0.0500	141.18	591.09	1.1463	4.7993	25.445	0.0393	140.88	589.84	1.1417	4.7801
35	308.15	19.608	0.0510	141.94	594.27	1.1488	4.8098	24.938	0.0401	141.64	593.02	1.1442	4.7905
40	313.15	19.231	0.0520	142.70	597.46	1.1513	4.8203	24.450	0.0409	142.42	596.28	1.1466	4.8006
45	318.15	18.868	0.0530	143.47	600.68	1.1537	4.8303	23.981	0.0417	143.20	599.55	1.1491	4.8111
50	323.15	18.519	0.0540	144.24	603.90	1.1561	4.8404	23.529	0.0425	143.98	602.82	1.1516	4.8215
55	338.15	18.182	0.0550	145.01	607.13	1.1585	4.8504	23.095	0.0433	144.78	606.16	1.1540	4.8316
60	333.15	17.857	0.0560	145.81	610.48	1.1609	4.8605	22.676	0.0441	145.58	609.51	1.1565	4.8420
65	338.15	17.544	0.0570	146.61	613.83	1.1633	4.8705	22.272	0.0449	146.38	612.86	1.1590	4.8255
70	343.15	17.241	0.0580	147.41	617.18	1.1657	4.8806	21.882	0.0457	147.18	616.21	1.1614	4.8625
75	348.15	16.949	0.0590	148.21	620.53	1.1680	4.8902	21.505	0.0465	147.98	619.56	1.1637	4.8722
80	353.15	16.667	0.0600	149.01	623.88	1.1703	4.8998	21.142	0.0473	148.78	622.91	1.1660	4.8818
85	358.15	16.393	0.0610	149.81	627.22	1.1726	4.9094	20.790	0.0481	149.60	626.35	1.1684	4.8919
90	363.15	16.129	0.0620	150.63	630.66	1.1749	4.9191	20.450	0.0489	150.42	629.78	1.1706	4.9011
95	368.15	15.873	0.0630	151.45	634.09	1.1771	4.9283	20.121	0.0497	151.24	633.21	1.1729	4.9107
100	373.15	15.625	0.0640	152.27	637.52	1.1793	4.9375	19.802	0.0505	152.06	636.64	1.1751	4.9199
105	378.15	15.408	0.0649	153.10	641.00	1.1815	4.9467	19.569	0.0511	152.88	640.08	1.1774	4.9295
110	383.15	15.198	0.0658	153.94	644.52	1.1837	4.9559	19.231	0.0520	153.72	643.59	1.1796	4.9387
115	388.15	14.993	0.0667	154.79	648.07	1.1859	4.9651	18.939	0.0528	154.57	647.15	1.1818	4.9480
120	393.15	14.793	0.0676	155.64	651.63	1.1880	4.9739	18.657	0.0536	155.44	650.80	1.1840	4.9572
125	398.15	14.599	0.0685	156.50	655.23	1.1901	4.9827	18.416	0.0543	156.31	654.44	1.1861	4.9660
130	403.15	14.409	0.0694	157.36	658.83	1.1923	4.9919	18.149	0.0551	157.18	658.08	1.1881	4.9743
135	408.15	14.205	0.0704	158.22	662.44	1.1944	5.0007	17.889	0.0559	158.05	661.72	1.1903	4.9835
140	413.15	14.006	0.0714	159.09	666.08	1.1965	5.0095	17.637	0.0567	158.92	665.37	1.1924	4.9923
145	418.15	13.812	0.0724	159.97	669.76	1.1986	5.0183	17.391	0.0575	159.80	669.05	1.1945	5.0011
150	423.15	13.624	0.0734	160.86	674.49	1.2006	5.0267	17.153	0.0583	160.68	672.74	1.1966	5.0099
155	428.15	—	—	—	—	—	—	16.920	0.0591	161.56	676.42	1.1986	5.0183
160	433.15	—	—	—	—	—	—	16.667	0.0600	162.44	680.10	1.2007	5.0271
165	438.15	—	—	—	—	—	—	16.447	0.0608	163.32	683.79	1.2027	5.0355

1 at = 1 kp/cm² = 98 066.5 Pa = 98 066.5 N/m² = 9.806 65 N/cm² = 0.980 665 bar $\qquad$ 1 kcal = 4.186 8 kJ

Table 60-8 Properties of Superheated Dichlorodifluoromethane "Freon 12" ($CF_2 Cl_2$) (*Continued*)

Pressure p	6 at = 5.883 990 bar						7 at = 6.864 655 bar					
Temperature	t_s = 21.3 °C = 294.45 K i'' = 139.25 kcal/kg = 583.01 kJ/kg s'' = 1.1337 kcal/kg K = 4.7466 kJ/kg K						t_s = 27 °C = 300.15 K i'' = 139.79 kcal/kg = 585.27 kJ/kg s'' = 1.1333 kcal/kg K = 4.7449 kJ/kg K					
							ρ'' = 32.787 kg/m³ v'' = 0.0305 m³/kg					
							ρ'' = 37.736 kg/m³ v'' = 0.0265 m³/kg					
t	T	ρ	v	i		s		ρ	v	i		s	
°C	K	kg/m³	m³/kg	kcal/kg	kJ/kg	kcal/kg K	kJ/kg K	kg/m³	m³/kg	kcal/kg	kJ/kg	kcal/kg K	kJ/kg K
25	298.15	32.258	0.0310	139.80	585.31	1.1355	4.7541	—	—	—	—	—	—
30	303.15	31.447	0.0318	140.56	588.50	1.1380	4.7646	37.313	0.0268	140.26	587.24	1.1348	4.7512
35	308.15	30.675	0.0326	141.34	591.76	1.1405	4.7750	35.496	0.0274	141.05	590.55	1.1374	4.7621
40	313.15	30.030	0.0333	142.12	595.03	1.1430	4.7855	35.714	0.0280	141.84	593.86	1.1398	4.7721
45	318.15	29.412	0.0340	142.90	598.29	1.1455	4.7690	34.965	0.0286	142.63	597.16	1.1424	4.7830
50	323.15	28.818	0.0347	143.68	601.56	1.1479	4.8060	34.247	0.0292	143.42	600.47	1.1449	4.7935
55	328.15	28.249	0.0354	144.49	604.95	1.1504	4.8165	33.670	0.0297	144.23	603.86	1.1474	4.8039
60	333.15	27.701	0.0361	145.30	608.34	1.1529	4.8270	33.003	0.0303	145.04	607.25	1.1499	4.8144
65	338.15	27.174	0.0368	146.11	611.73	1.1552	4.8366	32.362	0.0309	145.85	610.64	1.1523	4.8244
70	343.15	26.667	0.0375	146.92	615.12	1.1576	6.8466	31.746	0.0315	146.66	614.04	1.1547	4.8345
75	348.15	26.178	0.0382	147.73	618.52	1.1600	4.8567	31.153	0.0321	147.48	617.47	1.1571	4.8445
80	353.15	25.641	0.0390	148.54	621.91	1.1625	4.8672	30.581	0.0327	148.30	620.90	1.1595	4.8546
85	358.15	25.189	0.0397	149.35	625.30	1.1649	4.8772	30.030	0.0333	149.13	624.38	1.1619	4.8646
90	363.15	24.752	0.0404	150.18	628.77	1.1671	4.8864	29.499	0.0339	149.96	627.85	1.1642	4.8743
95	368.15	24.331	0.0411	151.01	632.25	1.1695	4.8965	28.986	0.0345	150.80	631.37	1.1665	4.8839
100	373.15	23.866	0.0419	151.85	635.77	1.1717	4.9057	28.409	0.0352	151.64	634.89	1.1688	4.8935
105	378.15	23.529	0.0425	152.69	639.28	1.1739	4.9149	27.933	0.0358	152.48	638.40	1.1710	4.9027
110	383.15	23.148	0.0432	153.53	642.80	1.1761	4.9241	27.473	0.0364	153.33	641.96	1.1732	4.9120
115	388.15	22.779	0.0439	154.39	646.40	1.1784	4.9337	26.882	0.0372	154.20	645.60	1.1755	4.9216
120	393.15	22.472	0.0445	155.26	650.04	1.1805	4.9425	26.525	0.0377	155.08	649.29	1.1776	4.9304
125	398.15	22.173	0.0451	156.13	653.69	1.1828	4.9521	26.110	0.0383	155.96	652.97	1.1798	4.9396
130	403.15	21.834	0.0458	157.00	657.33	1.1850	4.9614	25.707	0.0389	156.84	656.66	1.1819	4.9484
135	408.15	21.505	0.0465	157.87	660.97	1.1870	4.9697	25.316	0.0395	157.72	660.34	1.1840	4.9572
140	413.15	21.186	0.0472	158.76	664.70	1.1892	4.9789	24.938	0.0401	156.61	664.07	1.1861	4.9660
145	418.15	20.921	0.0478	159.65	668.42	1.1913	4.9877	24.570	0.0407	159.50	667.79	1.1882	4.9748
150	423.15	20.619	0.0485	160.54	672.15	1.1934	4.9965	24.213	0.0413	160.39	671.52	1.1904	4.9840
155	428.15	20.325	0.0492	161.43	675.88	1.1954	5.0049	23.923	0.0418	161.28	675.25	1.1925	4.9928
160	433.15	20.040	0.0499	162.32	679.60	1.1975	5.0137	23.585	0.0424	162.18	679.02	1.1947	5.0020
165	438.15	19.802	0.0505	163.22	683.37	1.1995	5.0221	23.256	0.0430	163.08	682.78	1.1966	5.0099

1 at = 1 kp/cm² = 98 066.5 Pa = 98 066.5 N/m² = 9.806 65 N/cm² = 0.980 665 bar   1 kcal = 4.186 8 kJ

Table 60-9 Properties of Superheated Dichlorodifluoromethane "Freon 12" ($CF_2 Cl_2$) (*Continued*)

Pressure p		8 at = 7.845320 bar						10 at = 98.06650 bar					
Temperature		$t_s = 32\,°C = 305.15\,K$ $i'' = 140.26$ kcal/kg = 587.24 kJ/kg $s'' = 1.1329$ kcal/kg K = 4.7432 kJ/kg K				$\rho'' = 43.290$ kg/m³ $v'' = 0.0231$ m³/kg		$t_s = 41\,°C = 314.15\,K$ $i'' = 141.02$ kcal/kg = 590.42 kJ/kg $s'' = 1.1322$ kcal/kg K = 4.7403 kJ/kg K				$\rho'' = 54.083$ kg/m³ $v'' = 0.01849$ m³/kg	
t	T	ρ	v	i		s		ρ	v	i		s	
°C	K	kg/m³	m³/kg	kcal/kg	kJ/kg	kcal/kg K	kJ/kg K	kg/m³	m³/kg	kcal/kg	kJ/kg	kcal/kg K	kJ/kg K
35	308.15	42.553	0.0235	140.70	589.08	1.1344	4.7495	—	—	—	—	—	—
40	313.15	41.667	0.0240	141.51	592.47	1.1370	4.7604	—	—	—	—	—	—
45	318.15	40.816	0.0245	142.33	595.91	1.1395	4.7709	53.277	0.01877	141.71	593.31	1.1344	4.7495
50	323.15	40.000	0.0250	143.15	599.34	1.1420	4.7813	51.680	0.01935	142.52	596.70	1.1370	4.7604
55	328.15	39.216	0.0255	143.97	602.77	1.1445	4.7918	50.277	0.01989	143.35	600.18	1.1395	4.7709
60	333.15	38.462	0.0260	144.79	606.21	1.1470	4.8023	49.044	0.02039	144.18	603.65	1.1420	4.7813
65	338.15	37.594	0.0266	145.61	609.64	1.1495	4.8127	47.916	0.02087	145.02	607.17	1.1445	4.7918
70	343.15	36.765	0.0272	146.43	613.07	1.1519	4.8228	46.838	0.02135	145.86	610.69	1.1470	4.8023
75	348.15	35.971	0.0278	147.25	616.51	1.1542	4.8324	45.830	0.02182	146.70	614.20	1.1495	4.8127
80	353.15	35.211	0.0284	148.07	619.94	1.1566	4.8425	44.883	0.02228	147.54	617.72	1.1520	4.8232
85	358.15	34.483	0.0290	148.91	623.46	1.1590	4.8525	43.957	0.02274	148.39	621.28	1.1543	4.8328
90	363.15	33.898	0.0295	149.75	626.97	1.1614	4.8625	43.103	0.02320	149.24	624.84	1.1567	4.8429
95	368.15	33.223	0.0301	150.59	630.49	1.1636	4.8718	42.283	0.02365	150.10	628.44	1.1595.	4.8546
100	373.15	32.573	0.0307	151.44	634.05	1.1660	4.8818	41.511	0.02409	150.96	632.04	1.1614	4.8625
105	378.15	31.949	0.0313	152.29	637.61	1.1682	4.8910	40.766	0.02453	151.83	635.68	1.1636	4.8718
110	383.15	31.348	0.0319	153.15	641.21	1.1705	4.9006	40.048	0.02497	152.71	639.37	1.1660	4.8818
115	388.15	30.864	0.0324	154.01	644.81	1.1727	4.9099	39.355	0.02541	153.59	643.05	1.1682	4.8910
120	393.15	30.395	0.0329	154.88	648.45	1.1749	4.9191	38.685	0.02585	154.47	646.73	1.1705	4.9006
125	398.15	29.940	0.0334	155.76	652.14	1.1771	4.9283	38.037	0.02629	155.37	650.50	1.1727	4.9099
130	403.35	29.499	0.0339	156.64	655.82	1.1794	4.9379	37.411	0.02673	156.27	654.27	1.1750	4.9195
135	408.15	28.986	0.0345	157.73	659.55	1.1815	4.9467	36.805	0.02717	157.17	658.04	1.1772	4.9287
140	413.15	28.571	0.0350	158.42	663.27	1.1836	4.9555	36.219	0.02761	158.07	661.81	1.1794	4.9379
145	418.15	28.169	0.0355	159.32	667.04	1.1857	4.9643	35.651	0.02805	158.97	665.58	1.1815	4.9467
150	423.15	27.778	0.0360	160.22	670.81	1.1897	4.9735	35.100	0.02849	159.87	669.34	1.1836	4.9555
155	428.15	27.397	0.0365	161.12	674.58	1.1900	4.9823	34.566	0.02893	160.78	673.15	1.1858	4.9647
160	433.15	27.027	0.0370	162.02	678.35	1.1921	4.9911	34.060	0.02936	161.69	676.96	1.1880	4.9739
165	438.15	26.667	0.0375	162.93	686.16	1.1941	4.9995	33.568	0.02979	162.16	680.82	1.1900	4.9823
170	443.15	26.316	0.0380	163.85	686.01	1.1962	5.0083	33.091	0.03022	163.54	684.71	1.1921	4.9911
175	448.15	25.974	0.0385	164.78	689.90	1.1982	5.0166	32.626	0.03065	164.47	688.60	1.1941	4.9995
180	453.15	25.641	0.0390	165.72	693.84	1.2003	5.0254	32.185	0.03107	165.41	692.54	1.1963	5.0087
185	458.15	—	—	—	—	—	—	31.756	0.03149	166.35	696.47	1.1984	5.0175

1 at = 1 kp/cm² = 98 066.5 Pa = 98 066.5 N/m² = 9.806 65 N/cm² = 0.980 665 bar   1 kcal = 4.186 8 kJ

Table 60-10 Properties of Superheated Dichlorodifluoromethane "Freon 12" (CF$_2$Cl$_2$) (Continued)

Pressure p		12 at = 11.767 980 bar						14 at = 13.729 310 bar					
Temperature		$t_s = 48.6\,°C = 321.75\,K$ $\rho'' = 66.050\,kg/m^3$ $i'' = 141.64\,kcal/kg = 593.02\,kJ/kg$ $v'' = 0.01514\,m^3/kg$ $s'' = 1.1319\,kcal/kg\,K = 4.7390\,kJ/kg\,K$						$t_s = 55.6\,°C = 328.75\,K$ $\rho'' = 77.459\,kg/m^3$ $i'' = 142.16\,kcal/kg = 595.20\,kg$ $v'' = 0.01291\,m^3/kg$ $s'' = 1.1315\,kcal/kg\,K = 4.7374\,kJ/kg\,K$					
t	T	ρ	v	i		s		ρ	v	i		s	
°C	K	kg/m³	m³/kg	kcal/kg	kJ/kg	kcal/kg K	kJ/kg K	kg/m³	m³/kg	kcal/kg	kJ/kg	kcal/kg K	kJ/kg K
50	323.15	65.789	0.01520	141.89	594.07	1.1326	4.7420	—	—	—	—	—	—
55	328.15	63.939	0.01564	142.73	597.58	1.1351	4.7524	—	—	—	—	—	—
60	333.15	62.189	0.01608	148.58	601.14	1.1376	4.7629	76.046	0.01315	142.91	598.34	1.1337	4.7466
65	338.15	60.533	0.01652	144.43	604.70	1.1401	4.7734	73.638	0.01358	143.77	601.94	1.1364	4.7579
70	343.15	58.962	0.01696	145.28	608.26	1.1427	4.7843	71.429	0.01400	144.64	605.58	1.1390	4.7688
75	348.15	57.471	0.01740	146.15	611.90	1.1452	4.7947	69.639	0.01441	145.52	609.26	1.1415	4.7792
80	353.15	56.054	0.01784	147.02	615.54	1.1476	4.8048	67.476	0.01482	146.42	613.03	1.1440	4.7897
85	358.15	54.705	0.01828	147.89	619.19	1.1501	4.8152	65.660	0.01523	147.32	616.80	1.1465	4.8002
90	363.15	53.419	0.01872	148.77	622.87	1.1525	4.8253	63.980	0.01563	148.22	620.57	1.1489	4.8102
95	368.15	52.219	0.01915	149.65	626.55	1.1550	4.8358	62.461	0.01601	149.12	624.34	1.1514	4.8207
100	373.15	51.099	0.01957	150.53	630.24	1.1574	4.8458	61.087	0.01637	150.02	620.10	1.1537	4.8303
105	378.15	50.025	0.01999	151.41	633.92	1.1598	4.8559	59.773	0.01673	150.92	631.87	1.1562	4.8408
110	383.15	49.020	0.02040	152.29	637.61	1.1622	4.8659	58.548	0.01708	151.83	635.68	1.1586	4.8505
115	388.15	48.077	0.02080	153.18	641.33	1.1644	4.8751	57.372	0.01743	152.74	639.49	1.1609	4.8605
120	393.15	47.192	0.02119	154.08	645.10	1.1666	4.8843	56.306	0.01776	153.65	643.30	1.1632	4.8701
125	398.15	46.361	0.02157	154.98	648.87	1.1690	4.8944	55.249	0.01810	154.46	647.11	1.1655	4.8797
130	403.15	45.558	0.02195	155.88	652.64	1.1712	4.9036	54.230	0.01844	155.47	650.92	1.1677	4.8889
135	408.15	44.803	0.02232	156.79	656.45	1.1735	4.9132	53.277	0.01877	156.38	654.73	1.1700	4.8986
140	413.15	44.131	0.02266	157.70	660.26	1.1756	4.9220	52.329	0.01911	157.29	658.54	1.1723	4.9082
145	418.15	43.365	0.02306	158.61	664.07	1.1778	4.9312	41.414	0.01945	158.22	662.44	1.1745	4.9174
150	423.15	42.680	0.02343	159.53	667.92	1.1800	4.9404	50.531	0.01979	159.16	666.37	1.1767	4.9266
155	428.15	41.999	0.02381	160.46	671.81	1.1822	4.9496	49.677	0.02013	160.10	670.31	1.1789	4.9358
160	433.15	41.356	0.02418	161.39	675.71	1.1844	4.9588	48.876	0.02046	161.04	674.24	1.1811	4.9450
165	438.15	40.717	0.02456	162.32	679.60	1.1865	4.9676	48.123	0.02078	161.98	678.18	1.1833	4.9542
170	443.15	40.096	0.02494	163.25	683.50	1.1886	4.9764	47.393	0.02110	162.93	682.16	1.1855	4.9635
175	448.15	39.494	0.02532	164.19	687.43	1.1908	4.9856	46.685	0.02142	163.88	686.13	1.1876	4.9722
180	453.15	38.926	0.02569	165.13	691.37	1.1928	4.9940	45.998	0.02174	164.84	690.15	1.1898	4.9815
185	458.15	38.373	0.02606	166.07	695.30	1.1949	5.0028	45.331	0.02206	165.80	694.17	1.1919	4.9902
190	463.15	—	—	—	—	—	—	44.683	0.02238	166.76	698.19	1.1940	4.9990
195	568.15	—	—	—	—	—	—	44.053	0.02270	167.74	702.29	1.1962	5.0083
200	473.15	—	—	—	—	—	—	43.459	0.02301	168.74	706.48	1.1984	5.0175

1 at = 1 kp/cm² = 98 066.5 Pa = 98 066.5 N/m² = 9.806 65 N/cm² = 0.980 665 bar   1 kcal = 4.186 8 kJ

Table 61-1   Properties of Superheated Chlorodifluoromethane "Freon 22" ($CHClF_2$)

| Pressure p | | 0.02 at = 0.019 613 30 bar | | | | | | 0.03 at = 0.029 419 95 bar | | | | | |

Temperature:

0.02 at = 0.019 613 30 bar
$t_s = -101.2\,°C = 171.95\,K$ $\rho'' = 0.1191\,kg/m^3$
$i'' = 137.94\,kcal/kg = 577.53\,kJ/kg$ $v'' = 8.3935\,m^3/kg$
$s'' = 1.2518\,kcal/kg\,K = 5.2410\,kJ/kg\,K$

0.03 at = 0.029 419 95 bar
$t_s = -95.84\,°C = 177.31\,K$ $\rho'' = 0.1744\,kg/m^3$
$i'' = 138.56\,kcal/kg = 579.70\,kJ/kg$ $v'' = 5.7338\,m^3/kg$
$s'' = 1.2456\,kcal/kg\,K = 5.2151\,kJ/kg\,K$

t	T	ρ	v	i		s		ρ	v	i		s	
°C	K	kg/m³	m³/kg	kcal/kg	kJ/kg	kcal/kg K	kJ/kg K	kg/m³	m³/kg	kcal/kg	kJ/kg	kcal/kg K	kJ/kg K
−100	173.15	0.1190	8.4019	138.05	577.99	1.2519	5.2415	—	—	—	—	—	—
−95	178.15	0.1157	8.6450	138.60	580.29	1.2553	5.2557	0.1736	5.7611	138.56	580.12	1.2462	5.2176
−90	183.15	0.1125	8.8881	139.16	582.64	1.2586	5.2695	0.1688	5.9233	139.14	582.55	1.2495	5.2314
−85	188.15	0.1095	9.1313	139.75	585.11	1.2617	5.2825	0.1643	6.0855	139.74	585.06	1.2527	5.2448
−80	193.15	0.1067	9.3744	140.36	587.66	1.2649	5.2959	0.1601	6.2477	140.36	587.66	1.2558	5.2578
−75	198.15	0.1040	9.6176	140.97	590.21	1.2680	5.3089	0.1560	6.4099	140.97	590.21	1.2589	5.2708
−70	203.15	0.1014	9.8607	141.60	592.85	1.2710	5.3214	0.1522	6.5721	141.59	592.81	1.2620	5.2837
−65	208.15	0.0990	10.1038	142.22	595.45	1.2741	5.3344	0.1485	6.7343	142.22	595.45	1.2650	5.2963
−60	213.15	0.0966	10.3470	142.86	598.13	1.2771	5.3470	0.1450	6.8965	142.85	598.08	1.2680	5.3089
−55	218.15	0.0944	10.5901	143.50	600.81	1.2801	5.3595	0.1417	7.0587	143.50	600.81	1.2710	5.3214
−50	223.15	0.0923	10.8332	144.15	603.53	1.2830	5.3717	0.1385	7.2208	144.14	603.49	1.2739	5.3336
−45	228.15	0.0903	11.0763	144.80	606.25	1.2859	5.3838	0.1354	7.3830	144.80	606.25	1.2768	5.3457
−40	233.15	0.0883	11.3195	145.46	609.01	1.2888	5.3959	0.1325	7.5451	145.46	609.01	1.2797	5.3578
−35	238.15	0.0865	11.5626	146.13	611.82	1.2916	5.4077	0.1297	7.7073	146.13	611.82	1.2826	5.3700
−30	243.15	0.0847	11.8057	146.81	614.66	1.2944	5.4194	0.1271	7.8694	146.80	614.62	1.2854	5.3817
−25	248.15	0.0830	12.0488	147.49	617.51	1.2972	5.4311	0.1245	8.0315	147.48	617.47	1.2882	5.3934
−20	253.15	0.0814	12.2919	148.18	620.40	1.3000	5.4428	0.1220	8.1937	148.17	620.36	1.2909	5.4047
−15	258.15	0.0798	12.5350	148.87	623.29	1.3027	5.4541	0.1197	8.3558	148.87	623.29	1.2938	5.4169
−10	263.15	0.0783	12.7781	149.51	625.97	1.3054	5.4654	0.1174	8.5179	149.57	626.22	1.2963	5.4273
−5	268.15	0.0768	13.0213	150.28	629.19	1.3081	5.4768	0.1152	8.6800	150.27	629.15	1.2990	5.4387
0	273.15	0.0754	13.2644	150.99	632.16	1.3107	5.4867	0.1131	8.8421	150.99	632.16	1.3016	5.4495
5	278.15	0.0740	13.5075	151.71	635.18	1.3133	5.4985	0.1111	9.0042	151.71	635.18	1.3042	5.4604
10	283.15	0.0727	13.7506	152.44	638.24	1.3159	5.5094	0.1091	9.1662	152.44	638.24	1.3068	5.4713
15	288.15	0.0715	13.9937	153.18	641.33	1.3185	5.5203	0.1072	9.3283	153.17	641.29	1.3094	5.4822
20	293.15	0.0702	14.2368	153.92	644.43	1.3210	5.5308	0.1054	9.4904	153.90	644.35	1.3119	5.4927
25	298.15	0.0691	14.4799	154.66	647.53	1.3235	5.5412	0.1036	9.6524	154.66	647.53	1.3145	5.5035
30	303.15	0.0679	14.7229	155.42	650.71	1.3260	5.5517	0.1019	9.8145	155.42	650.71	1.3170	5.5140
35	308.15	0.0668	14.9660	156.18	653.89	1.3285	5.5622	0.1002	9.9765	156.18	653.89	1.3195	5.5245
40	313.15	0.0658	15.2091	156.95	657.12	1.3310	5.5726	0.0986	10.1386	156.94	657.08	1.3219	5.5345
45	318.15	0.0647	15.4522	157.72	660.34	1.3335	5.5831	0.0971	10.3006	157.72	660.34	1.3244	5.5450
50	323.15	0.0637	15.6953	158.50	663.61	1.3359	5.5931	0.0956	10.4626	158.50	663.61	1.3268	5.5550
55	328.15	0.0627	15.9384	159.29	666.92	1.3383	5.6032	0.0941	10.6246	159.29	666.92	1.3293	5.5655
60	333.15	0.0618	16.1815	160.08	670.22	1.3407	5.6132	0.0927	10.7866	160.08	670.22	1.3317	5.5756
65	338.15	0.0609	16.4246	160.88	673.57	1.3432	5.6237	0.0913	10.9486	160.88	673.57	1.3340	5.5852
70	343.15	0.0600	16.6677	161.69	676.96	1.3456	5.6338	0.0900	11.1106	161.69	676.96	1.3365	5.5957

1 at = 1 kp/cm² = 98 066.5 Pa = 98 066.5 N/cm² = 9.806 65 N/cm² = 0.980 665 bar

1 kcal = 4.186 8 kJ

Table 61-2 Properties of Superheated Chlorodifluoromethane "Freon 22" ($CHCIF_2$) (Continued)

Pressure p		0.04 at = 0.039 226 60 bar						0.05 at = 0.049 033 25 bar					
Temperature		$t_s = -92.5\,°C = 180.65\,K$ $i'' = 138.85\,kcal/kg = 581.34\,kJ/kg$ $s'' = 1.2412\,kcal/kg\,K = 5.1967\,kJ/kg\,K$			$\rho'' = 0.2284\,kg/m^3$ $v'' = 4.3791\,m^3/kg$			$t_s = -89.8\,°C = 183.35\,K$ $i'' = 139.17\,kcal/kg = 582.68\,kJ$ $s'' = 1.2379\,kcal/kg\,K = 5.1828\,kJ/kg\,K$			$\rho'' = 0.2817\,kg/m^3$ $v'' = 3.5494\,m^3/kg$		
t	T	ρ	v	i		s		ρ	v	i		s	
°C	K	kg/m³	m³/kg	kcal/kg	kJ/kg	kcal/kg K	kJ/kg K	kg/m³	m³/kg	kcal/kg	kJ/kg	kcal/kg K	kJ/kg K
−90	183.15	0.2252	4.4404	139.14	582.55	1.2429	5.2038	—	—	—	—	—	—
−85	188.15	0.2192	4.5622	139.74	585.06	1.2461	5.2172	0.2745	3.6436	139.74	585.06	1.2410	5.1958
−80	193.15	0.2135	4.6839	140.35	587.62	1.2492	5.2302	0.2673	3.7412	140.35	587.62	1.2422	5.2008
−75	198.15	0.2081	4.8056	140.97	590.21	1.2523	5.2431	0.2605	3.8387	140.96	590.17	1.2473	5.2222
−70	203.15	0.2030	4.9273	141.59	592.81	1.2554	5.2561	0.2541	3.9362	141.58	592.77	1.2503	5.3248
−65	208.15	0.1981	5.0490	142.22	595.45	1.2584	5.2687	0.2479	4.0338	142.21	595.40	1.2534	5.2477
−60	213.15	0.1934	5.1707	142.85	598.08	1.2614	5.2812	0.2421	4.1313	142.85	598.08	1.2564	5.2603
−55	218.15	0.1890	5.2924	143.49	600.76	1.2644	5.2938	0.2365	4.2288	143.49	600.76	1.2593	5.2724
−50	223.15	0.1847	5.4141	144.14	603.49	1.2673	5.3059	0.2311	4.3264	144.14	603.49	1.2623	5.2850
−45	228.15	0.1806	5.5358	144.80	606.25	1.2702	5.3181	0.2260	4.4239	144.79	606.21	1.2652	2.2971
−40	233.15	0.1768	5.6574	145.46	609.01	1.2731	5.3302	0.2212	4.5214	145.45	608.97	1.2681	5.3093
−35	238.15	0.1730	5.7791	146.12	611.78	1.2750	5.3382	0.2165	4.6189	146.12	611.78	1.2709	5.3210
−30	243.15	0.1695	5.9008	146.80	614.62	1.2788	5.3541	0.2120	4.7164	146.80	614.62	1.2737	5.3327
−25	248.15	0.1660	6.0224	147.48	617.47	1.2816	5.3658	0.2077	4.8139	147.48	617.47	1.2765	5.3445
−20	253.15	0.1628	6.1440	148.17	620.36	1.2843	5.3771	0.2036	4.9114	148.16	620.32	1.2793	5.3562
−15	258.15	0.1596	6.2657	148.86	623.25	1.2870	5.3884	0.1996	5.0089	148.86	623.25	1.2820	5.3675
−10	263.15	0.1566	6.3873	149.56	626.18	1.2897	5.3997	0.1958	5.1064	149.56	626.18	1.2847	5.3788
− 5	268.15	0.1536	6.5089	150.27	629.15	1.2924	5.4110	0.1922	5.2039	150.27	629.15	1.2874	5.3901
0	273.15	0.1508	6.6305	150.99	632.16	1.2950	5.4219	0.1886	5.3031	150.98	632.12	1.2899	5.4006
5	278.15	0.1481	6.7521	151.71	635.18	1.2976	5.4328	0.1852	5.3988	151.70	635.14	1.2926	5.4119
10	283.15	0.1455	6.8737	152.43	638.19	1.3002	5.4437	0.1819	5.4963	152.43	638.19	1.2952	5.4227
15	288.15	0.1430	6.9953	153.17	641.29	1.3028	5.4546	0.1788	5.5937	153.17	641.29	1.2977	5.4332
20	293.15	0.1405	7.1169	153.90	644.35	1.3053	5.4650	0.1757	5.6912	153.91	644.39	1.3003	5.4441
25	298.15	0.1382	7.2384	154.66	647.53	1.3078	5.4755	0.1728	5.7887	154.66	647.53	1.3028	5.4546
30	303.15	0.1361	7.3500	155.41	650.67	1.3104	5.4864	0.1699	5.8861	155.41	650.67	1.3053	5.4650
35	308.15	0.1350	7.4816	156.17	653.85	1.3128	5.4964	0.1671	5.9836	156.17	653.85	1.3078	5.4755
40	313.15	0.1315	7.6031	156.94	657.08	1.3153	5.5069	0.1644	6.0810	156.94	657.08	1.3102	5.4860
45	318.15	0.1295	7.7246	157.72	660.34	1.3178	5.5174	0.1619	6.1784	157.71	660.30	1.3127	5.4960
50	323.15	0.1275	7.8462	158.50	663.61	1.3202	5.5274	0.1593	6.2759	158.49	663.57	1.3152	5.5065
55	328.15	0.1255	7.9677	159.28	666.87	1.3227	5.5379	0.1569	6.3733	159.28	666.87	1.3176	5.5165
60	333.15	0.1236	8.0892	160.08	670.22	1.3251	5.5479	0.1545	6.4707	160.08	670.22	1.3200	5.5266
65	338.15	0.1218	8.2107	160.88	673.57	1.3274	5.5576	0.1522	6.5682	160.88	673.57	1.3224	5.5366
70	343.15	0.1200	8.3322	161.68	676.92	1.3298	5.5676	0.1500	6.6656	161.69	676.96	1.3248	5.5467
75	348.15	0.1183	8.4537	162.50	680.36	1.3322	5.5777	0.1479	6.7630	162.50	680.36	1.3272	5.5567
80	353.15	0.1166	8.5752	163.33	683.83	1.3345	5.5873	0.1458	6.8604	163.33	683.83	1.3295	5.5664

1 at = 1 kp/cm² = 98 066.5 Pa = 98 066.5 N/cm² = 9.806 65 N/cm² = 0.980 665 bar

1 kcal = 4.186 8 kJ

Table 61-3 Properties of Superheated Chlorodifluoromethane "Freon 22" ($CHClF_2$) (*Continued*)

Pressure p		0.06 at = 0.058 839 90 bar						0.07 at = 0.068 646 55 bar					
Temperature		$t_s = -87.5\ °C = 185.65\ K$ $i'' = 139.41\ kcal/kg = 583.68\ kJ/kg$ $s'' = 1.2352\ kcal/kg\ K = 5.1715\ kJ/kg\ K$				$\rho'' = 0.3335\ kg/m^3$ $v'' = 2.9981\ m^3/kg$		$t_s = -85.6\ °C = 187.55\ K$ $i'' = 139.67\ kcal/kg = 584.77\ kJ/kg$ $s'' = 1.2329\ kcal/kg\ K = 5.1619\ kJ/kg\ K$				$\rho'' = 0.3852\ kg/m^3$ $v'' = 2.5963\ m^3/kg$	
t	T	ρ	v	i		s		ρ	v	i		s	
°C	K	kg/m³	m³/kg	kcal/kg	kJ/kg	kcal/kg K	kJ/kg K	kg/m³	m³/kg	kcal/kg	kJ/kg	kcal/kg K	kJ/kg K
−85	188.15	0.3290	3.0393	139.74	585.06	1.2367	5.1778	0.3840	2.6044	139.73	585.02	1.2332	5.1632
−80	193.15	0.3205	3.1204	140.34	587.58	1.2398	5.1908	0.3740	2.6740	140.34	587.58	1.2364	5.1766
−75	198.15	0.3124	3.2015	140.96	590.17	1.2429	5.2038	0.3645	2.7435	140.95	590.13	1.2395	5.1895
−70	203.15	0.3046	3.2827	141.58	592.77	1.2460	5.2168	0.3555	2.8131	141.58	592.77	1.2426	5.2025
−65	208.15	0.2973	3.3638	142.21	595.40	1.2490	5.2293	0.3469	2.8827	142.20	595.36	1.2456	5.2151
−60	213.15	0.2903	3.4449	142.84	598.04	1.2520	5.2419	0.3387	2.9522	142.84	598.04	1.2486	5.2276
−55	218.15	0.2836	3.5261	143.49	600.76	1.2550	5.2544	0.3309	3.0218	143.48	600.72	1.2516	5.2402
−50	223.15	0.2772	3.6072	144.13	603.44	1.2579	5.2666	0.3235	3.0913	144.13	603.44	1.2546	5.2528
−45	228.15	0.2711	3.6883	144.79	606.21	1.2608	5.2787	0.3164	3.1609	144.78	606.16	1.2575	5.2649
−40	233.15	0.2653	3.7695	145.45	608.97	1.2637	5.2909	0.3095	3.2305	145.45	608.97	1.2604	5.2770
−35	238.15	0.2597	3.8506	146.12	611.78	1.2666	5.3030	0.3030	3.3000	146.11	611.73	1.2632	5.2888
−30	243.15	0.2543	3.9317	146.79	614.58	1.2694	5.3147	0.2968	3.3696	146.79	614.58	1.2660	5.3005
−25	248.15	0.2492	4.0128	147.47	617.43	1.2722	5.3264	0.2908	3.4391	147.47	617.43	1.2688	5.3122
−20	253.15	0.2443	4.0940	148.16	620.32	1.2749	5.3378	0.2850	3.5086	148.16	620.32	1.2716	5.3239
−15	258.15	0.2385	4.1751	148.86	623.25	1.2776	5.3491	0.2795	3.5782	148.85	623.21	1.2743	5.3352
−10	263.15	0.2350	4.2562	149.46	626.18	1.2803	5.3604	0.2741	3.6477	149.55	626.14	1.2770	5.3465
−5	268.15	0.2306	4.3374	150.26	629.11	1.2830	5.3717	0.2690	3.7173	150.26	629.11	1.2796	5.3574
0	273.15	0.2263	4.4185	150.98	632.12	1.2856	5.3826	0.2641	3.7868	150.98	632.12	1.2823	5.3687
5	278.15	0.2222	4.4996	151.70	635.14	1.2882	5.3934	0.2593	3.8563	151.70	635.14	1.2849	5.3796
10	283.15	0.2183	4.5807	152.43	638.19	1.2908	5.4043	0.2547	3.9259	152.43	638.19	1.2875	5.3905
15	288.15	0.2145	4.6618	153.16	641.25	1.2934	5.4152	0.2503	3.9954	153.16	641.25	1.2901	5.4014
20	293.15	0.2108	4.7430	153.90	644.35	1.2959	5.4257	0.2460	4.0649	153.90	644.35	1.2926	5.4119
25	298.15	0.2073	4.8241	154.65	647.49	1.2984	5.4361	0.2419	4.1345	154.65	647.49	1.2952	5.4227
30	303.15	0.2039	4.9052	155.40	650.63	1.3010	5.4470	0.2379	4.2040	155.40	650.63	1.2977	5.4332
35	308.15	0.2005	4.9863	156.17	653.85	1.3034	5.4571	0.2340	4.2735	156.17	653.85	1.3002	5.4437
40	313.15	0.1973	5.0675	156.94	657.08	1.3059	5.4675	0.2303	4.3430	156.93	657.03	1.3026	5.4537
45	318.15	0.1942	5.1486	157.71	660.30	1.3084	5.4780	0.2266	4.4126	157.71	660.30	1.3051	5.4642
50	323.15	0.1912	5.2297	158.49	663.57	1.3108	4.4881	0.2231	4.4821	158.49	663.57	1.3075	5.4742
55	328.15	0.1883	5.3108	159.28	666.87	1.3133	5.4985	0.2197	4.5516	159.28	666.87	1.3099	5.4843
60	333.15	0.1855	5.3919	160.07	670.18	1.3157	5.5086	0.2164	4.6211	160.07	670.18	1.3123	5.4943
65	338.15	0.1827	5.4730	160.87	673.53	1.3180	5.5182	0.2132	4.6906	160.87	673.53	1.3147	5.5044
70	343.15	0.1800	5.5541	161.68	676.92	1.3204	5.5283	0.2101	4.7601	161.68	676.92	1.3171	5.5144
75	348.15	0.1775	5.6353	162.49	680.41	1.3228	5.5383	0.2071	4.8296	162.49	680.31	1.3194	5.5241
80	353.15	0.1749	5.7164	163.31	683.75	1.3251	5.5479	0.2041	4.8991	163.31	683.75	1.3217	5.5337

1 at = 1 kp/cm² = 98 066.5 Pa = 98 066.5 N/m² = 9.806 65 N/cm² = 0.980 665 bar   1 kcal = 4.186 8 kJ

Table 61-4 Properties of Superheated Chlorodifluoromethane "Freon 22" ($CHClF_2$) (Continued)

Pressure p			0.08 at = 0.078 453 20 bar					0.09 at = 0.088 259 85 bar					
Temperature		$t_s = -83.9\,°C = 189.25\,K$ $\rho'' = 0.4364\,kg/m^3$ $i'' = 139.87\,kcal/kg = 585.61\,kJ/kg$ $v'' = 2.2915\,m^3/kg$ $s'' = 1.2311\,kcal/kg\,K = 5.1544\,kJ/kg\,K$						$t_s = -82.4\,°C = 190.75\,K$ $\rho'' = 0.4873\,kg/m^3$ $i'' = 140.06\,kcal/kg = 586.40\,kJ/kg$ $v'' = 2.0523\,m^3/kg$ $s'' = 1.2294\,kcal/kg\,K = 5.1473\,kJ/kg\,K$					
t	T	ρ	v	i		s		ρ	v	i		s	
°C	K	kg/m³	m³/kg	kcal/kg	kJ/kg	kcal/kg K	kJ/kg K	kg/m³	m³/kg	kcal/kg	kJ/kg	kcal/kg K	kJ/kg K
−80	193.15	0.4275	2.3390	140.34	587.58	1.2334	5.1640	0.4812	2.0783	140.33	587.53	1.2307	5.1527
−75	198.15	0.4167	2.3999	140.95	590.13	1.2365	5.1770	0.4690	2.1324	140.95	590.13	1.2338	5.1657
−70	203.15	0.4064	2.4607	141.57	592.73	1.2396	5.1900	0.4573	2.1866	141.57	592.73	1.2369	5.1787
−65	208.15	0.3966	2.5216	142.20	595.36	1.2427	5.2029	0.4463	2.2407	142.20	595.36	1.2400	5.1916
−60	213.15	0.3872	2.5825	142.84	598.04	1.2457	5.2155	0.4358	2.2948	142.83	598.00	1.2430	5.2042
−55	218.15	0.3783	2.6434	143.48	600.62	1.2487	5.2281	0.4257	2.3490	143.47	600.68	1.2460	5.2168
−50	223.15	0.3698	2.7042	144.13	603.44	1.2516	5.2402	0.4161	2.4031	144.12	603.40	1.2489	5.2289
−45	228.15	0.3617	2.7651	144.78	606.16	1.2545	5.2523	0.4070	2.4572	144.78	606.16	1.2518	5.2410
−40	233.15	0.3539	2.8260	145.44	608.93	1.2574	5.2645	0.3982	2.5114	145.44	608.93	1.2547	5.2532
−35	238.15	0.3464	2.8869	146.11	611.73	1.2603	2.2766	0.3898	2.5655	146.11	611.73	1.2576	5.2653
−30	243.15	0.3392	2.9477	146.79	614.58	1.2631	5.2883	0.3817	2.6196	146.78	614.54	1.2604	5.2770
−25	248.15	0.3324	3.0086	147.47	617.43	1.2659	5.3001	0.3740	2.6737	147.46	617.39	1.2632	5.2888
−20	253.15	0.3258	3.0695	148.15	620.27	1.2686	5.3114	0.3666	2.7279	148.15	620.27	1.2659	5.3001
−15	258.15	0.3195	3.1303	148.85	623.21	1.2713	5.3227	0.3595	2.7820	148.85	623.21	1.2686	5.3114
−10	263.15	0.3134	3.1912	149.55	626.14	1.2740	5.3340	0.3526	2.8361	149.55	626.14	1.2713	5.3227
− 5	268.15	0.3075	3.2520	150.26	629.11	1.2767	5.3453	0.3460	2.8902	150.25	629.07	1.2740	5.3340
0	273.15	0.3019	3.3129	150.97	632.08	1.2793	5.3562	0.3396	2.9443	150.97	632.08	1.2766	5.3449
5	278.15	0.2964	3.3738	151.69	635.10	1.2819	5.3671	0.3335	2.9984	151.69	635.10	1.2792	5.3558
10	283.15	0.2912	3.4346	152.42	638.15	1.2845	5.3779	0.3276	3.0525	152.42	638.15	1.2818	5.3666
15	288.15	0.2861	3.4955	153.16	641.25	1.2871	5.3888	0.3219	3.1066	153.15	641.21	1.2844	5.3775
20	293.15	0.2812	3.5563	153.90	644.35	1.2896	5.3993	0.3164	3.1607	153.90	644.35	1.2869	5.3880
25	298.15	0.2765	3.6172	154.65	647.49	1.2922	5.4102	0.3111	3.2148	154.64	647.45	1.2894	5.3985
30	303.15	0.2719	3.6780	155.40	650.63	1.2947	5.4206	0.3066	3.2619	155.40	650.63	1.2919	5.4089
35	308.15	0.2675	3.7389	156.16	653.81	1.2972	5.4311	0.3009	3.3230	156.16	653.81	1.2944	5.4194
40	313.15	0.2632	3.7997	156.93	657.04	1.2997	5.4416	0.2961	3.3771	156.93	657.03	1.2969	5.4299
45	318.15	0.2590	3.8605	157.71	660.30	1.3021	5.4516	0.2914	3.4312	157.70	660.26	1.2994	5.4403
50	323.15	0.2550	3.9214	158.49	663.57	1.3045	5.4617	0.2869	3.4853	158.48	663.52	1.3018	5.4504
55	328.15	0.2511	3.9822	159.27	666.83	1.3070	5.4721	0.2825	3.5394	159.27	666.83	1.3042	5.4604
60	333.15	0.2473	4.0431	160.07	670.18	1.3094	5.4822	0.2783	3.5935	160.07	670.18	1.3067	5.4709
65	338.15	0.2437	4.1039	160.87	673.53	1.3117	5.4918	0.2742	3.6475	160.87	673.53	1.3090	5.4805
70	343.15	0.2401	4.1647	161.68	676.92	1.3141	5.5019	0.2702	3.7016	161.68	676.92	1.3114	5.4906
75	348.15	0.2367	4.2256	162.49	680.31	1.3164	5.5115	0.2663	3.7557	162.49	680.31	1.3138	5.5006
80	353.15	0.2333	4.2864	163.32	683.79	1.3187	5.5211	0.2625	3.8098	163.31	683.75	1.3161	5.5102
85	358.15	0.2300	4.3472	164.15	687.26	1.3211	5.5312	0.2588	3.8638	164.14	687.22	1.3184	5.5199

1 at = 1 kp/cm² = 98 066.5 Pa = 98 066.5 N/m² = 9.806 65 N/cm² = 0.980 665 bar   1 kcal = 4.186 8 kJ

Table 61-5 Properties of Superheated Chlorodifluoromethane "Freon 22" (CHClF₂) (Continued)

Pressure p		0.1 at = 0.098 066 50 bar						0.12 at = 0.117 679 80 bar					
Temperature		$t_s = -81\ °C = 192.15\ K$ $i'' = 140.22\ kcal/kg = 587.07\ kJ/kg$ $s'' = 1.2280\ kcal/kg\ K = 5.1414\ kJ/kg\ K$ $\rho'' = 0.5376\ kg/m^3$ $v'' = 1.8602\ m^3/kg$						$t_s = -78.4\ °C = 194.75\ K$ $i'' = 140.54\ kcal/kg = 588.41\ kJ/kg$ $s'' = 1.2252\ kcal/kg\ K = 5.1297\ kJ/kg\ K$ $\rho'' = 0.6367\ kg/m^3$ $v'' = 1.5705\ m^3/kg$					
t	T	ρ	v	i		s		ρ	v	i		s	
°C	K	kg/m³	m³/kg	kcal/kg	kJ/kg	kcal/kg K	kJ/kg K	kg/m³	m³/kg	kcal/kg	kJ/kg	kcal/kg K	kJ/kg K
−80	193.15	0.5348	1.8698	140.33	587.53	1.2285	5.1435	—	—	—	—	—	—
−75	198.15	0.5212	1.9185	140.94	590.09	1.2316	5.1565	0.6259	1.5977	140.95	590.13	1.2271	5.1376
−70	203.15	0.5083	1.9673	141.55	592.64	1.2347	5.1694	0.6104	1.6384	141.56	592.68	1.2302	5.1506
−65	208.15	0.4960	2.0160	142.18	595.28	1.2377	5.1820	0.5956	1.6790	142.18	595.28	1.2333	5.1636
−60	213.15	0.4843	2.0648	142.82	597.96	1.2408	5.1950	0.5815	1.7197	142.81	597.92	1.2363	5.1761
−55	218.15	0.4731	2.1135	143.46	600.64	1.2437	5.2071	0.5681	1.7603	143.46	600.64	1.2393	5.1887
−50	223.15	0.4625	2.1622	144.12	603.40	1.2467	5.2197	0.5553	1.8009	144.11	603.36	1.2422	5.2008
−45	228.15	0.4523	2.2110	144.77	606.12	1.2496	5.2318	0.5430	1.8416	144.77	606.12	1.2452	5.2134
−40	233.15	0.4425	2.2597	145.44	608.93	1.2525	5.2440	0.5313	1.8822	145.43	608.89	1.2480	5.2251
−35	238.15	0.4332	2.3084	146.10	611.69	1.2553	5.2557	0.5201	1.9228	146.10	611.69	1.2509	5.2373
−30	243.15	0.4242	2.3572	146.78	614.54	1.2582	5.2678	0.5093	1.9634	146.77	614.50	1.2537	5.2490
−25	248.15	0.4156	2.4059	147.46	617.39	1.2609	5.2791	0.4990	2.0040	147.46	617.39	1.2565	5.2607
−20	253.15	0.4074	2.4546	148.15	620.27	1.2637	5.2909	0.4891	2.0447	148.14	620.23	1.2592	5.2720
−15	258.15	0.3995	2.5033	148.84	623.16	1.2664	5.3022	0.4795	2.0853	148.84	623.16	1.2619	5.2833
−10	263.15	0.3918	2.5521	149.54	626.09	1.2691	5.3135	0.4704	2.1259	149.54	626.09	1.2646	5.2946
− 5	268.15	0.3845	2.6008	150.25	629.07	1.2717	5.3244	0.4616	2.1665	150.25	629.07	1.2673	5.3059
0	273.15	0.3774	2.6495	150.97	632.08	1.2744	5.3357	0.4531	2.2071	150.96	632.04	1.2700	5.3172
5	278.15	0.3706	2.6982	151.69	635.10	1.2770	5.3465	0.4449	2.2477	151.68	635.05	1.2726	5.3281
10	283.15	0.3640	2.7469	152.42	638.15	1.2796	5.3574	0.4370	2.2883	152.41	638.11	1.2751	5.3386
15	288.15	0.3577	2.7956	153.15	641.21	1.2822	5.3683	0.4294	2.3289	153.15	641.21	1.2777	5.3495
20	293.15	0.3516	2.8443	153.89	644.31	1.2847	5.3788	0.4220	2.3695	153.89	644.31	1.2803	5.3604
25	298.15	0.3457	2.8930	154.64	647.45	1.2873	5.3897	0.4149	2.4101	154.64	647.45	1.2828	5.3708
30	303.15	0.3399	2.9417	155.40	650.63	1.2898	5.4001	0.4080	2.4507	155.39	650.59	1.2853	5.3813
35	308.15	0.3344	2.9904	156.16	653.81	1.2923	5.4106	0.4014	2.4913	156.15	653.77	1.2878	5.3918
40	313.15	0.3290	3.0391	156.92	656.99	1.2947	5.4206	0.3950	2.5319	156.92	656.99	1.2903	5.4022
45	318.15	0.3239	3.0877	157.70	660.26	1.2972	5.4311	0.3887	2.5725	157.70	660.26	1.2927	5.4123
50	323.15	0.3188	3.1364	158.48	663.52	1.2996	5.4412	0.3827	2.6131	158.48	663.52	1.2952	5.4227
55	328.15	0.3140	3.1851	159.27	666.83	1.3021	5.4516	0.3768	2.6537	159.27	666.83	1.2976	5.4328
60	333.15	0.3092	3.2338	160.07	670.18	1.3045	5.4617	0.3712	2.6943	160.06	670.14	1.3000	5.4428
65	338.15	0.3046	3.2825	160.87	673.53	1.3069	5.4717	0.3657	2.7348	160.86	673.49	1.3024	5.4529
70	343.15	0.3002	3.3311	161.68	676.92	1.3092	5.4814	0.3603	2.7754	161.67	676.88	1.3048	6.4629
75	348.15	0.2959	3.3798	162.49	680.31	1.3116	5.4914	0.3551	2.8160	162.49	630.31	1.3071	5.4726
80	353.15	0.2917	3.4285	163.34	683.87	1.3139	5.5010	0.3501	2.8566	163.31	683.75	1.3095	5.4826
85	358.15	0.2876	3.4771	164.14	687.22	1.3163	5.5111	0.3452	2.8971	164.14	687.22	1.3118	5.4922
90	363.15	0.2869	3.4858	164.97	690.70	1.3186	5.5207	0.3404	2.9377	164.97	690.70	1.3142	5.5023

1 at = 1 kp/cm² = 98 066.5 Pa = 98 066.5 N/m² = 9.806 65 N/cm² = 0.980 66 bar   1 kcal = 4.186 8 kJ

Table 61-6 Properties of Superheated Chlorodifluoromethane "Freon 22" ($CHClF_2$) (Continued)

Pressure p		0.14 at = 0.137 293 10 bar						0.16 at = 0.156 906 40 bar					
Temperature		$t_s = -76.1\ °C = 197.05\ K$ $i'' = 140.80\ kcal/kg = 589.50\ kJ/kg$ $s'' = 1.2231\ kcal/kg\ K = 5.1209\ kJ/kg\ K$			$\rho'' = 0.7348\ kg/m^3$ $v'' = 1.3609\ m^3/kg$			$t_s = -74.2\ °C = 198.95\ K$ $i'' = 141.04\ kcal/kg = 590.51\ kJ/kg$ $s'' = 1.2212\ kcal/kg\ K = 5.1129\ kJ/kg\ K$			$\rho'' = 0.8322\ kg/m^3$ $v'' = 1.2017\ m^3/kg$		
t	T	ρ	v	i		s		ρ	v	i		s	
°C	K	kg/m³	m³/kg	kcal/kg	kJ/kg	kcal/kg K	kJ/kg K	kg/m³	m³/kg	kcal/kg	kJ/kg	kcal/kg K	kJ/kg K
−75	198.15	0.7307	1.3685	140.94	590.09	1.2236	5.1230	—	—	—	—	—	—
−70	203.15	0.7126	1.4034	141.55	592.64	1.2267	5.1349	0.8149	1.2272	141.55	592.64	1.2237	5.1234
−65	208.15	0.6953	1.4382	142.17	595.24	1.2297	5.1485	0.7951	1.2577	142.17	595.24	1.2267	5.1359
−60	213.15	0.6788	1.4731	142.81	597.92	1.2328	5.1615	0.7763	1.2882	142.80	597.88	1.2297	5.1485
−55	218.15	0.6631	1.5080	143.45	600.60	1.2357	5.1736	0.7583	1.3187	143.44	600.55	1.2327	5.1611
−50	223.15	0.6482	1.5428	144.11	603.36	1.2387	5.1862	0.7411	1.3493	144.10	603.32	1.2357	5.1736
−45	228.15	0.6338	1.5777	144.76	606.08	1.2416	5.1983	0.7247	1.3798	144.76	606.08	1.2386	5.1858
−40	233.15	0.6202	1.6125	145.42	608.84	1.2445	5.2105	0.7091	1.4103	145.42	608.84	1.2414	5.1975
−35	238.15	0.6071	1.6473	146.09	611.65	1.2473	5.2222	0.6941	1.4408	146.09	611.65	1.2443	5.2096
−30	243.15	0.5945	1.6822	146.77	614.50	1.2502	5.2343	0.6797	1.4713	146.76	614.45	1.2471	5.2214
−25	248.15	0.5824	1.7170	147.45	617.34	1.2529	5.2456	0.6659	1.5018	147.44	617.30	1.2499	5.2331
−20	253.15	0.5708	1.7519	148.14	620.23	1.2557	5.2574	0.6526	1.5323	148.13	620.19	1.2527	5.2448
−15	258.15	0.5597	1.7867	148.83	623.12	1.2584	5.2687	0.6399	1.5628	148.83	623.12	1.2554	5.2561
−10	263.15	0.5490	1.8215	149.53	626.05	1.2611	5.2800	0.6276	1.5933	149.53	626.05	1.2581	5.2674
−5	268.15	0.5387	1.8563	150.24	629.02	1.2638	5.2913	0.6159	1.6237	150.24	629.02	1.2608	5.2787
0	273.15	0.5288	1.8912	150.96	632.04	1.2664	5.3022	0.6045	1.6542	150.95	632.00	1.2634	5.2896
5	278.15	0.5192	1.9260	151.68	635.05	1.2690	5.3130	0.5936	1.6867	151.67	635.01	1.2660	5.3005
10	283.15	0.5100	1.9608	152.41	638.11	1.2716	5.3239	0.5830	1.7152	152.40	638.07	1.2686	5.3114
15	288.15	0.5011	1.9956	153.14	641.17	1.2742	5.3348	0.5728	1.7457	153.14	641.17	1.2712	5.3223
20	293.15	0.4925	2.0305	153.88	644.26	1.2767	5.3453	0.5630	1.7761	153.88	644.26	1.2737	5.3327
25	298.15	0.4842	2.0653	154.63	647.40	1.2793	5.3562	0.5535	1.8066	154.63	647.40	1.2763	5.3436
30	303.15	0.4762	2.1001	155.39	650.59	1.2818	5.3666	0.5443	1.8371	155.38	650.54	1.2788	5.3541
35	308.15	0.4684	2.1349	156.15	653.77	1.2843	5.3771	0.5354	1.8676	156.14	653.73	1.2812	5.3641
40	313.15	0.4609	2.1679	156.92	656.99	1.2868	5.3876	0.5269	1.8980	156.91	656.95	1.2837	5.3746
45	318.15	0.4536	2.2045	157.69	660.22	1.2892	5.3976	0.5185	1.9285	157.69	660.22	1.2862	5.3851
50	323.15	0.4466	2.2393	158.47	663.48	1.2917	5.4081	0.5105	1.9589	158.47	663.48	1.2886	5.3591
55	328.15	0.4397	2.2741	159.26	666.79	1.2941	5.4181	0.5027	1.9894	159.26	666.79	1.2910	5.4052
60	333.15	0.4331	2.3089	160.06	670.14	1.2965	5.4282	0.4941	2.0199	160.05	670.10	1.2935	5.4156
65	338.15	0.4267	2.3437	160.86	673.49	1.2989	5.4382	0.4877	2.0503	160.85	673.45	1.2958	5.4253
70	343.15	0.4204	2.3785	161.67	676.88	1.3012	5.4479	0.4806	2.0808	161.66	676.84	1.2982	5.4353
75	348.15	0.4144	2.4133	162.48	680.27	1.3036	5.4579	0.4737	2.1112	162.48	680.27	1.3006	5.4454
80	353.15	0.4085	2.4481	163.30	683.70	1.3059	5.4675	0.4669	2.1417	163.30	683.70	1.3029	5.4550
85	358.15	0.4028	2.4829	164.15	687.26	1.3083	5.4776	0.4604	2.1721	164.13	687.18	1.3052	5.4646
90	363.15	0.3972	2.5176	164.98	690.74	1.3107	5.4876	0.4540	2.2025	164.97	690.70	1.3077	5.4751
95	368.15	0.3918	2.5524	165.81	694.21	1.3130	5.4973	0.4478	2.2330	165.81	694.21	1.3100	5.4847

1 at = 1 kp/cm² = 98 066.5 Pa = 98 066.5 N/m² = 9.806 65 N/cm² = 0.980 665 bar

1 kcal = 4.186 8 kJ

Table 61-7 Properties of Superheated Chlorodifluoromethane "Freon 22" (CHClF$_2$) (Continued)

Pressure p		0.18 at = 0.176 519 70 bar						0.2 at = 0.196 133 00 bar					
Temperature		t_s = −72.4 °C = 200.75 K i'' = 141.25 kcal/kg = 591.39 kJ/kg s'' = 1.2197 kcal/kg K = 5.1066 kJ/kg K				ρ'' = 0.9277 kg/m³ v'' = 1.0779 m³/kg		t_s = −70.7 °C = 202.45 K i'' = 141.44 kcal/kg = 592.18 kJ/kg s'' = 1.2183 kcal/kg K = 5.1008 kJ/kg K				ρ'' = 1.0238 kg/m³ v'' = 0.9768 m³/kg	
t	T	ρ	v	i		s		ρ	v	i		s	
°C	K	kg/m³	m³/kg	kcal/kg	kJ/kg	kcal/kg K	kJ/kg K	kg/m³	m³/kg	kcal/kg	kJ/kg	kcal/kg K	kJ/kg K
−70	203.15	0.9173	1.0901	141.54	592.60	1.2210	5.1121	1.0200	0.9804	141.52	592.52	1.2186	5.1020
−65	208.15	0.8950	1.1173	142.16	595.20	1.2241	5.1251	0.9951	1.0049	142.15	595.15	1.2217	5.1150
−60	213.15	0.8738	1.1444	142.79	597.83	1.2271	5.1376	0.9715	1.0293	142.78	597.79	1.2247	5.1276
−55	218.15	0.8536	1.1715	143.44	600.55	1.2300	5.1498	0.9489	1.0538	143.43	600.51	1.2277	5.1401
−50	223.15	0.8342	1.1987	144.09	603.28	1.2330	5.1623	0.9275	1.0782	144.08	603.23	1.2306	5.1523
−45	228.15	0.8158	1.2258	144.75	606.04	1.2359	5.1745	0.9069	1.1027	144.74	606.00	1.2335	5.1644
−40	233.15	0.7981	1.2530	145.41	608.80	1.2388	5.1866	0.8872	1.1271	145.40	608.76	1.2364	5.1766
−35	238.15	0.7812	1.2801	146.08	611.61	1.2416	5.1983	0.8684	1.1515	146.07	611.57	1.2392	5.1883
−30	243.15	0.7650	1.3072	146.75	614.41	1.2445	5.2105	0.8503	1.1760	146.75	614.41	1.2421	5.2004
−25	248.15	0.7495	1.3343	147.44	617.30	1.2472	5.2218	0.8331	1.2004	147.43	617.26	1.2449	5.2121
−20	253.15	0.7345	1.3615	148.13	620.19	1.2500	5.2335	0.8165	1.2248	148.12	620.15	1.2476	5.2235
−15	258.15	0.7201	1.3886	148.82	623.08	1.2527	5.2448	0.8005	1.2492	148.82	623.08	1.2503	5.2348
−10	263.15	0.7065	1.4157	149.52	626.01	1.2554	5.2561	0.7851	1.2737	149.52	626.01	1.2530	5.2461
− 5	268.15	0.6931	1.4428	150.23	628.98	1.2581	5.2674	0.7704	1.2981	150.23	628.98	1.2557	5.2574
0	273.15	0.6803	1.4699	150.95	632.00	1.2607	5.2783	0.7561	1.3225	150.94	631.96	1.2583	5.2683
5	278.15	0.6680	1.4970	151.67	635.01	1.2633	5.2892	0.7424	1.3469	151.66	634.97	1.2609	5.2791
10	283.14	0.6561	1.5241	152.40	638.07	1.2659	5.3001	0.7292	1.3713	152.39	638.03	1.2635	5.2900
15	288.15	0.6447	1.5513	153.13	641.12	1.2685	5.3110	0.7165	1.3957	153.13	641.12	1.2661	5.3009
20	293.15	0.6336	1.5784	153.87	644.22	1.2710	5.3214	0.7042	1.4201	153.87	644.22	1.2687	5.3118
25	298.15	0.6229	1.6054	154.62	647.36	1.2736	5.3323	0.6923	1.4445	154.62	647.36	1.2712	5.3223
30	303.15	0.6126	1.6325	155.38	650.54	1.2761	5.3428	0.6808	1.4689	155.37	650.50	1.2737	5.3327
35	308.15	0.6026	1.6596	156.14	653.73	1.2786	5.3532	0.6697	1.4933	156.13	653.69	1.2762	5.3432
40	313.15	0.5929	1.6867	156.91	656.95	1.2811	5.3637	0.6589	1.5157	156.90	656.91	1.2787	5.3537
45	318.15	0.5835	1.7138	157.68	660.17	1.2835	5.3738	0.6485	1.5421	157.68	660.17	1.2812	5.3641
50	323.15	0.5744	1.7409	158.46	663.44	1.2859	5.3838	0.6384	1.5665	158.46	663.44	1.2836	5.3742
55	328.15	0.5656	1.7680	159.25	666.75	1.2884	5.3934	0.6286	1.5909	159.25	666.75	1.2860	5.3842
60	333.15	0.5571	1.7951	160.05	670.10	1.2908	5.4043	0.6191	1.6152	160.04	670.06	1.2884	5.3943
65	338.15	0.5488	1.8222	160.85	673.45	1.2932	5.4144	0.6099	1.6396	160.85	673.45	1.2908	5.4043
70	343.15	0.5408	1.8492	161.66	676.84	1.2956	5.4244	0.6010	1.6640	161.66	676.84	1.2932	5.4144
75	348.15	0.5330	1.8763	162.47	680.23	1.2979	5.4340	0.5923	1.6884	162.47	680.23	1.2955	5.4240
80	353.15	0.5254	1.9034	163.30	683.70	1.3003	5.4449	0.5839	1.7127	163.29	683.66	1.2979	5.4340
85	358.15	0.5180	1.9304	164.13	687.18	1.3026	5.4537	0.5757	1.7371	164.12	687.14	1.3002	5.4437
90	363.15	0.5109	1.9575	164.96	690.65	1.3050	5.4638	0.5677	1.7615	164.96	690.65	1.3026	5.4537
95	368.15	0.5039	1.9846	165.80	694.17	1.3073	5.4734	0.5600	1.7858	165.80	694.17	1.3049	5.4634

1 at = 1 kp/cm² = 98 066.5 Pa = 98 066.5 N/m² = 9.806 65 N/cm² = 0.980 665 bar   1 kcal = 4.186 8 kJ

Table 61-8 Properties of Superheated Chlorodifluoromethane "Freon 22" (CHClF$_2$) (*Continued*)

Pressure p		0.25 at = 0.245 166 25 bar						0.3 at = 0.294 199 50 bar					
Temperature		$t_s = -67.3\,°C = 205.85\,K$ $i'' = 141.87$ kcal/kg = 593.98 kJ/kg $s'' = 1.2153$ kcal/kg K = 5.0882 kJ/kg K				$\rho'' = 1.2601$ kg/m³ $v'' = 0.7936$ m³/kg		$t_s = -64.4\,°C = 208.75\,K$ $i'' = 142.23$ kcal/kg = 595.49 kJ/kg $s'' = 1.2129$ kcal/kg K = 5.0782 kJ/kg K				$\rho'' = 1.4932$ kg/m³ $v'' = 0.6697$ m³/kg	
t	T	ρ	v		i		s	ρ	v		i		s
°C	K	kg/m³	m³/kg	kcal/kg	kJ/kg	kcal/kg K	kJ/kg K	kg/m³	m³/kg	kcal/kg	kJ/kg	kcal/kg K	kJ/kg K
−65	208.15	1.2460	0.8026	142.15	595.15	1.2165	5.0932	—	—	—	—	—	—
−60	213.15	1.2162	0.8222	142.77	597.75	1.2196	5.1062	1.4618	0.6841	142.77	597.75	1.2155	5.0891
−55	218.15	1.1879	0.8418	143.41	600.43	1.2225	5.1184	1.4276	0.7005	143.40	600.39	1.2184	5.1012
−50	223.15	1.1609	0.8614	144.07	603.19	1.2255	5.1309	1.3951	0.7168	144.05	603.11	1.2214	5.1138
−45	228.15	1.1351	0.8810	144.73	605.96	1.2284	5.1431	1.3639	0.7332	144.71	605.87	1.2243	5.1259
−40	233.15	1.1104	0.9006	145.39	608.72	1.2313	5.1552	1.3342	0.7495	145.37	608.64	1.2272	5.1380
−35	238.15	1.0868	0.9201	146.06	611.52	1.2341	5.1669	1.3057	0.7659	146.04	611.44	1.2300	5.1498
−30	243.15	1.0642	0.9397	146.73	614.33	1.2369	5.1787	1.2784	0.7822	146.72	614.29	1.2328	5.1615
−25	248.15	1.0424	0.9593	147.42	617.22	1.2397	5.1904	1.2666	0.7895	147.40	617.13	1.2356	5.1732
−20	253.15	1.0216	0.9789	148.11	620.11	1.2425	5.2021	1.2271	0.8149	148.09	620.02	1.2384	5.1849
−15	258.15	1.0016	0.9984	148.81	623.04	1.2452	5.2134	1.2031	0.8312	148.79	622.95	1.2411	5.1962
−10	263.15	0.9823	1.0180	149.50	625.93	1.2479	5.2247	1.1799	0.8475	149.49	625.88	1.2438	5.2075
− 5	268.15	0.9639	1.0375	150.21	628.90	1.2506	5.2360	1.1577	0.8638	150.20	628.86	1.2465	5.2188
0	273.15	0.9460	1.0571	150.93	631.91	1.2532	5.2469	1.1361	0.8802	150.92	631.87	1.2491	5.2572
5	278.15	0.9289	1.0766	151.65	634.93	1.2558	5.2578	1.1154	0.8965	151.64	634.89	1.2517	5.2406
10	283.15	0.9192	1.0962	152.38	637.98	1.2584	5.2687	1.0955	0.9128	152.37	637.94	1.2543	5.2515
15	288.15	0.8963	1.1157	153.11	641.04	1.2610	5.2796	1.0763	0.9291	153.10	641.00	1.2569	5.2624
20	293.15	0.8808	1.1353	153.86	644.18	1.2635	5.2900	1.0578	0.9454	153.84	644.10	1.2594	5.2729
25	298.15	0.8660	1.1548	154.60	647.28	1.2661	5.3009	1.0398	0.9617	154.59	647.24	1.2620	5.2837
30	303.15	0.8515	1.1744	155.36	650.46	1.2686	5.3114	1.0225	0.9780	155.35	650.42	1.2645	5.2942
35	308.15	0.8376	1.1939	156.12	653.64	1.2711	5.3218	1.0057	0.9943	156.11	653.60	1.2670	5.3047
40	313.15	0.8241	1.2134	156.89	656.87	1.2736	5.3323	0.9895	1.0106	156.88	656.83	1.2695	5.3151
45	318.15	0.8110	1.2330	157.67	660.13	1.2760	5.3424	0.9738	1.0269	157.66	660.09	1.2719	5.3252
50	323.15	0.7984	1.2525	158.45	663.40	1.2784	5.3524	0.9586	1.0432	158.44	663.36	1.2743	5.3352
55	328.15	0.7862	1.2720	159.24	666.71	1.2809	5.3629	0.9439	1.0594	159.23	666.66	1.2768	5.3457
60	333.15	0.7743	1.2915	160.03	670.01	1.2833	5.3729	0.9296	1.0757	160.02	669.97	1.2792	5.3558
65	338.15	0.7628	1.3110	160.84	673.40	1.2857	5.3830	0.9158	1.0920	160.83	673.36	1.2816	5.3658
70	343.15	0.7515	1.3306	161.64	676.75	1.2881	5.3930	0.9023	1.1083	161.63	676.71	1.2840	5.3759
75	348.15	0.7407	1.3501	162.46	680.19	1.2904	5.4026	0.8893	1.1245	162.45	680.15	1.2863	5.3855
80	353.15	0.7301	1.3696	163.28	683.62	1.2928	5.4127	0.8766	1.1408	163.27	683.58	1.2887	5.3955
85	358.15	0.7199	1.3891	164.11	687.10	1.2951	5.4223	0.8642	1.1571	164.10	687.05	1.2910	5.4052
90	363.15	0.7099	1.4086	164.95	690.91	1.2975	5.4324	0.8523	1.1733	164.94	690.57	1.2934	5.4152
95	368.15	0.7002	1.4281	165.79	694.13	1.2998	5.4420	0.8406	1.1896	165.78	694.09	1.2957	5.4248
100	373.15	0.6908	1.4476	166.64	697.69	1.3021	5.4516	0.8293	1.2058	166.63	697.65	1.2980	5.4345
105	378.15	0.6816	1.4671	167.50	701.29	1.3045	5.4617	0.8183	1.2221	167.49	701.25	1.3001	5.4433

1 at = 1 kp/cm² = 98 066.5 Pa = 98 066.5 N/m² = 9.806 65 N/cm² = 0.980 665 bar   1 kcal = 4.186 8 kJ

Table 61-9 Properties of Superheated Chlorodifluoromethane "Freon 22" ($CHClF_2$) (*Continued*)

Pressure p		0.35 at = 0.343 232 75 bar						0.4 at = 0.392 266 00 bar					
Temperature		$t_s = -61.6\,°C = 211.55\,K$ $i'' = 142.54\,kcal/kg = 596.79\,kJ/kg$ $s'' = 1.2109\,kcal/kg\,K = 5.0698\,kJ/kg\,K$				$\rho'' = 1.7215\,kg/m^3$ $v'' = 0.5809\,m^3/kg$		$t_s = -59.3\,°C = 213.85\,K$ $i'' = 142.81\,kcal/kg = 597.92\,kJ/kg$ $s'' = 1.2092\,kcal/kg\,K = 5.0627\,kJ/kg\,K$				$\rho'' = 1.9486\,kg/m^3$ $v'' = 0.5132\,m^3/kg$	
t	T	ρ	v	i		s		ρ	v	i		s	
°C	K	kg/m³	m³/kg	kcal/kg	kJ/kg	kcal/kg K	kJ/kg K	kg/m³	m³/kg	kcal/kg	kJ/kg	kcal/kg K	kJ/kg K
—60	213.15	1.7082	0.5854	142.74	597.62	1.2128	5.0778	—	—	—	—	—	—
—55	218.15	1.6683	0.5994	143.38	600.30	1.2148	5.0861	1.9091	0.5238	143.36	600.22	1.2117	5.0731
—50	223.15	1.6300	0.6135	144.03	603.02	1.2177	5.0983	1.8653	0.5361	144.01	602.94	1.2146	5.0853
—45	228.15	1.5936	0.6275	144.69	605.79	1.2206	5.1104	1.8235	0.5484	144.68	605.75	1.2176	5.0978
—40	233.15	1.5584	0.6417	145.36	608.59	1.2235	5.1225	1.7835	0.5607	145.34	608.51	1.2205	5.1069
—35	238.15	1.5251	0.6557	146.03	611.40	1.2264	5.1347	1.7452	0.5730	146.01	611.31	1.2233	5.1217
—30	243.15	1.4932	0.6697	146.70	614.20	1.2292	5.1464	1.7085	0.5853	146.69	614.16	1.2261	5.1334
—25	248.15	1.4626	0.6837	147.39	617.09	1.2320	5.1581	1.6734	0.5976	147.37	617.01	1.2289	5.1452
—20	253.15	1.4333	0.6977	148.08	619.98	1.2347	5.1694	1.6396	0.6099	148.06	619.90	1.2316	5.1565
—15	258.15	1.4049	0.7118	148.77	622.87	1.2375	5.1812	1.6072	0.6222	148.76	622.83	1.2344	5.1682
—10	263.15	1.3778	0.7258	149.48	625.84	1.2402	5.1925	1.5760	0.6435	149.46	625.76	1.2371	5.1795
— 5	268.15	1.3517	0.7398	150.19	628.82	1.2429	5.2038	1.5463	0.6467	150.17	628.73	1.2398	5.1908
0	273.15	1.3266	0.7538	150.90	631.79	1.2455	5.2147	1.5175	0.6590	150.89	631.75	1.2424	5.2017
5	278.15	1.3024	0.7678	151.63	634.84	1.2481	5.2255	1.4896	0.6713	151.62	634.80	1.2450	5.2126
10	283.15	1.2791	0.7818	152.35	637.86	1.2507	5.2364	1.4631	0.6835	152.34	637.82	1.2476	5.2235
15	288.15	1.2566	0.7958	153.09	640.96	1.2533	5.2473	1.4372	0.6958	153.08	640.92	1.2502	5.2343
20	293.15	1.2349	0.8098	153.83	644.06	1.2558	5.2578	1.4124	0.7080	153.82	644.01	1.2527	5.2448
25	298.15	1.2139	0.8238	154.58	647.20	1.2583	5.2683	1.3883	0.7203	154.57	647.15	1.2552	6.2553
30	303.15	1.1937	0.8377	155.33	650.34	1.2608	5.2787	1.3652	0.7325	155.33	650.34	1.2578	5.2662
35	308.15	1.1741	0.8517	156.10	653.56	1.2633	5.2892	1.2436	0.7448	156.09	653.52	1.2603	5.2766
40	313.15	1.1551	0.8657	156.87	656.78	1.2658	5.2997	1.3210	0.7570	156.86	656.74	1.2627	5.2867
45	318.18	1.1368	0.8797	157.64	660.01	1.2683	5.3101	1.2999	0.7693	157.63	659.97	1.2652	5.2971
50	323.15	1.1191	0.8936	158.43	663.31	1.2707	5.3202	1.2796	0.7815	158.42	663.27	1.2676	5.3072
55	328.15	1.1018	0.9076	159.22	666.62	1.2731	5.3302	1.2599	0.7937	159.21	666.58	1.2700	5.3172
60	333.15	1.0851	0.9216	160.01	669.93	1.2756	5.3407	1.2407	0.8060	160.00	669.89	1.2725	5.3277
65	338.15	1.0689	0.9355	160.82	673.32	1.2779	5.3503	1.2222	0.8182	160.81	673.28	1.2749	5.3378
70	343.15	1.0537	0.9495	161.62	676.67	1.2803	5.3604	1.2042	0.8304	161.61	676.63	1.2772	5.3474
75	348.15	1.0380	0.9634	162.44	680.10	1.2827	5.3704	1.1868	0.8426	162.43	680.06	1.2796	5.3574
80	353.15	1.0231	0.9774	163.26	683.54	1.2850	5.3800	1.1699	0.8548	163.25	683.50	1.2819	5.3671
85	358.15	1.0088	0.9913	164.09	687.01	1.2873	5.3897	1.1534	0.8670	164.08	686.97	1.2843	5.3771
90	363.15	0.9947	1.0053	164.93	690.53	1.2898	5.4001	1.1374	0.8792	164.92	690.49	1.2867	5.3872
95	368.15	0.9812	1.0192	165.77	694.05	1.2921	5.4098	1.1218	0.8914	165.76	694.00	1.2980	5.3968
100	373.15	0.9679	1.0332	166.62	697.60	1.2943	5.4190	1.1067	0.9036	166.61	697.56	1.2913	5.4064
105	378.15	0.9549	1.0472	167.48	701.21	1.2966	5.4286	1.0919	0.9158	167.47	701.16	1.2935	5.4156

1 at = 1 kp/cm² = 98 066.5 Pa = 98 066.5 N/m² = 9.806 65 N/cm² = 0.980 665 bar   1 kcal = 4.186 8 kJ

Table 61-10 Properties of Superheated Chlorodifluoromethane "Freon 22" ($CHClF_2$) (*Continued*)

Pressure p		0.45 at = 0.441 299 25 bar						0.5 at = 0.490 332 50 bar					
Temperature		$t_s = -57.2\,°C = 215.95\,°C$ $\quad \rho'' = 2.1730\ kg/m^3$ $i'' = 143.07\ kcal/kg = 599.01\ kJ/kg \quad v'' = 0.4602\ m^3/kg$ $s'' = 1.2077\ kcal/kg\,K = 5.0564\ kJ/kg\,K$						$t_s = -55.2\,°C = 217.95\,K$ $\quad \rho'' = 2.3958\ kg/m^3$ $i'' = 143.30\ kcal/kg = 599.97\ kJ/kg \quad v'' = 0.4174\ m^3/kg$ $s'' = 1.2064\ kcal/kg\,K = 5.0510\ kJ/kg\,K$					
t	T	ρ	v	i		s		ρ	v	i		s	
°C	K	kg/m³	m³/kg	kcal/kg	kJ/kg	kcal/kg K	kJ/kg K	kg/m³	m³/kg	kcal/kg	kJ/kg	kcal/kg K	kJ/kg K
−55	218.15	2.1510	0.4649	143.34	600.14	1.2089	5.0614	2.3935	0.4178	143.32	600.05	1.2066	5.0518
−50	223.15	2.1013	0.4759	144.00	602.90	1.2119	5.0740	2.3381	0.4277	143.98	602.82	1.2095	5.0639
−45	228.15	2.0542	0.4868	144.66	605.66	1.2148	5.0861	2.2852	0.4376	144.64	605.58	1.2124	5.0761
−40	233.15	2.0088	0.4978	145.33	608.47	1.2177	5.0983	2.2351	0.4474	145.31	608.38	1.2152	5.0878
−35	238.15	1.9658	0.5087	145.99	611.23	1.2205	5.1100	2.1867	0.4573	145.98	611.19	1.2181	5.0999
−30	243.15	1.9242	0.5197	146.67	614.08	1.2233	5.1217	2.1404	0.4672	146.66	614.04	1.2209	5.1117
−25	248.15	1.8847	0.5306	147.36	616.97	1.2261	5.1334	2.0964	0.4770	147.34	616.88	1.2237	5.1234
−20	253.15	1.8464	0.5416	148.05	619.86	1.2289	5.1452	2.0538	0.4869	148.03	619.77	1.2265	5.1351
−15	258.15	1.8100	0.5525	148.75	622.79	1.2317	5.1569	2.0133	0.4967	148.73	622.70	1.2292	5.1464
−10	263.15	1.7749	0.5634	149.45	625.72	1.2344	5.1682	1.9739	0.5066	149.43	625.63	1.2319	5.1577
− 5	268.15	1.7413	0.5743	150.16	628.69	1.2371	5.1795	1.9365	0.5164	150.15	628.65	1.2346	5.1690
0	273.15	1.7085	0.5853	150.88	631.70	1.2397	5.1904	1.9001	0.5263	150.86	631.62	1.2373	5.1803
5	278.15	1.6773	0.5962	151.61	634.76	1.2423	5.2013	1.8653	0.5361	151.59	634.68	1.2399	5.1912
10	283.15	1.6472	0.6071	152.33	637.78	1.2449	5.2121	1.8315	0.5460	152.32	637.73	1.2425	5.2021
15	288.15	1.6181	0.6180	153.06	640.83	1.2475	5.2230	1.7992	0.5558	153.05	640.79	1.2451	5.2130
20	293.15	1.5901	0.6289	153.81	643.97	1.2500	5.2335	1.7680	0.5656	153.80	643.93	1.2476	5.2235
25	298.15	1.5630	0.6398	154.56	647.11	1.2526	5.2444	1.7379	0.5754	154.54	647.03	1.2501	5.2339
30	303.15	1.5368	0.6507	155.31	650.25	1.2551	5.2549	1.7085	0.5853	155.30	650.21	1.2527	5.2448
35	308.15	1.5115	0.6616	156.08	653.48	1.2576	5.2653	1.6804	0.5951	156.06	653.39	1.2551	5.2549
40	313.15	1.4870	0.6725	156.85	656.70	1.2600	5.2754	1.6532	0.6049	156.83	656.62	1.2576	5.2653
45	318.15	1.4633	0.6834	157.62	659.92	1.2625	5.2858	1.6268	0.6147	157.61	659.88	1.2601	5.2758
50	323.15	1.4403	0.6943	158.41	663.23	1.2650	5.2963	1.6013	0.6245	158.39	663.15	1.2625	5.2858
55	328.15	1.4180	0.7052	159.20	666.54	1.2674	5.3064	1.5768	0.6342	159.18	666.45	1.2650	5.2963
60	333.15	1.3965	0.7161	159.99	669.85	1.2698	5.3164	1.5526	0.6441	159.98	669.80	1.2674	5.3064
65	338.15	1.3557	0.7269	160.79	673.20	1.2722	5.3264	1.5293	0.6539	160.78	673.15	1.2698	5.3164
70	343.15	1.3554	0.7378	161.60	676.59	1.2746	5.3365	1.5067	0.6637	161.59	676.55	1.2722	5.3264
75	348.15	1.3356	0.7487	162.42	680.02	1.2769	5.3461	1.4848	0.6735	162.41	679.98	1.2745	5.3361
80	353.15	1.3167	0.7595	163.24	683.45	1.2793	5.3562	1.4635	0.6833	163.23	683.41	1.2769	5.3461
85	358.15	1.2980	0.7704	164.07	686.93	1.2815	5.3654	1.4428	0.6931	164.06	686.89	1.2792	5.3558
90	363.15	1.2799	0.7813	164.91	690.45	1.2839	5.3754	1.4229	0.7028	164.90	690.40	1.2815	5.3654
95	368.15	1.2625	0.7921	165.75	693.96	1.2862	5.3851	1.4033	0.7126	165.74	693.92	1.2838	5.3750
100	373.15	1.2453	0.8030	166.60	697.52	1.2885	5.3947	1.3843	0.7224	166.59	697.48	1.2861	5.3846
105	378.15	1.2288	0.8138	167.46	701.12	1.2908	5.4043	1.3659	0.7321	167.45	701.08	1.2884	5.3943
110	383.15	1.2127	0.8246	168.32	704.72	1.2930	5.4135	1.3479	0.7419	168.31	704.68	1.2906	5.4035

1 at = 1 kp/cm² = 98 066.5 Pa = 98 066.5 N/m² = 9.806 65 N/cm² = 0.980 665 bar

1 kcal = 4.186 8 kJ

Table 61-11 Properties of Superheated Chlorodifluoromethane "Freon 22" ($CHClF_2$) (Continued)

Pressure p		0.6 at = 0.588 399 0 bar						0.7 at = 0.686 465 5 bar					
Temperature		$t_s = -51.8\,°C = 221.35\,K$ $i'' = 143.71$ kcal/kg = 601.69 kJ/kg $s'' = 1.2042$ kcal/kg K = 5.0417 kJ/kg K				$\rho'' = 2.8377$ kg/m³ $v'' = 0.3524$ m³/kg		$t_s = -48.9\,°C = 224.25\,K$ $i'' = 144.06$ kcal/kg = 603.15 kJ/kg $s'' = 1.2023$ kcal/kg K = 5.0338 kJ/kg K				$\rho'' = 3.2755$ kg/m³ $v'' = 0.3053$ m³/kg	
t	T	ρ	v	i		s		ρ	v	i		s	
°C	K	kg/m³	m³/kg	kcal/kg	kJ/kg	kcal/kg K	kJ/kg K	kg/m³	m³/kg	kcal/kg	kJ/kg	kcal/kg K	kJ/kg K
−50	223.15	2.8137	0.3554	143.94	602.65	1.2052	5.0459	—	—	—	—	—	—
−45	228.15	2.7503	0.3636	144.61	605.45	1.2081	5.0581	3.2175	0.3108	144.58	605.33	1.2046	5.0434
−40	233.15	2.6889	0.3719	145.28	608.26	1.2110	5.0702	3.1456	0.3179	145.25	608.13	1.2075	5.0556
−35	238.15	2.6309	0.3801	145.95	611.06	1.2139	5.0824	3.0769	0.3250	145.01	610.90	1.2104	5.0677
−30	243.15	2.5747	0.3884	146.63	613.91	1.2167	5.0941	3.0111	0.3321	146.59	613.74	1.2132	5.0794
−25	248.15	2.5214	0.3966	147.31	616.76	1.2195	5.1058	2.9481	0.3392	147.28	616.63	1.2160	5.0911
−20	253.15	2.4697	0.4049	148.00	619.65	1.2223	5.1175	2.8877	0.3463	147.98	619.56	1.2188	5.1029
−15	258.15	2.4207	0.4131	148.70	622.58	1.2250	5.1288	2.8297	0.3534	148.68	622.49	1.2215	5.1142
−10	263.15	2.3730	0.4214	149.41	625.55	1.2277	5.1401	2.7739	0.3605	149.38	625.42	1.2244	5.1263
−5	268.15	2.3277	0.4296	150.12	628.52	1.2305	5.1519	2.7211	0.3675	150.09	628.40	1.2269	5.1368
0	273.15	2.2841	0.4378	150.84	631.54	1.2331	5.1627	2.6695	0.3746	150.81	631.41	1.2296	5.1481
5	278.15	2.2422	0.4460	151.57	634.59	1.2357	5.1736	2.6199	0.3817	151.55	634.51	1.2322	5.1590
10	283.15	2.2017	0.4542	152.29	637.61	1.2383	5.1845	2.5727	0.3887	152.27	637.52	1.2348	5.1699
15	288.15	2.1622	0.4625	153.03	640.71	1.2409	5.1954	2.5265	0.3958	153.00	640.58	1.2374	5.1807
20	293.15	2.1245	0.4707	153.77	643.80	1.2434	5.2059	2.4820	0.4029	153.75	643.72	1.2399	5.1912
25	298.15	2.0881	0.4789	154.52	646.94	1.2460	5.2168	2.4396	0.4099	154.50	646.86	1.2425	5.2021
30	303.15	2.0530	0.4871	155.28	650.13	1.2485	5.2272	2.3981	0.4170	155.26	650.04	1.2450	5.2126
35	308.15	2.0190	0.4953	156.04	653.31	1.2510	5.2377	2.3585	0.4240	156.02	653.22	1.2475	5.2230
40	313.15	1.9861	0.5035	156.81	656.53	1.2534	5.2477	2.3202	0.4310	156.79	656.45	1.2499	5.2331
45	318.15	1.9543	0.5117	157.59	659.80	1.2559	5.2582	2.2826	0.4381	157.57	659.71	1.2524	5.2435
50	323.15	1.9234	0.5199	158.37	663.06	1.2583	5.2683	2.2467	0.4451	158.35	662.98	1.2548	5.2536
55	328.15	1.8939	0.5280	159.16	666.37	1.2608	5.2787	2.2119	0.4521	159.14	666.29	1.2573	5.2641
60	333.15	1.8650	0.5362	159.96	669.72	1.2632	5.2888	2.1777	0.4592	159.94	669.64	1.2597	5.2741
65	338.15	1.8369	0.5444	160.76	673.07	1.2656	5.2988	2.1450	0.4662	160.74	672.99	1.2621	5.2842
70	343.15	1.8096	0.5526	161.57	676.46	1.2680	5.3089	2.1133	0.4732	161.55	676.38	1.2645	5.2942
75	348.15	1.7835	0.5607	162.39	679.89	1.2703	5.3185	2.0825	0.4802	162.37	679.81	1.2669	5.3043
80	353.15	1.7578	0.5689	163.21	683.33	1.2727	5.3285	2.0525	0.4872	163.19	683.24	1.2692	5.3139
85	358.15	1.7328	0.5771	164.04	686.80	1.2750	5.3382	2.0235	0.4942	164.02	686.72	1.2715	5.3235
90	363.15	1.7088	0.5852	164.88	690.32	1.2773	5.3478	1.9952	0.5012	164.86	690.24	1.2738	5.3331
95	368.15	1.6852	0.5934	165.72	693.84	1.2796	5.3574	1.9681	0.5081	165.70	693.75	1.2761	5.3428
100	373.15	1.6625	0.6015	166.57	697.40	1.2819	5.3671	1.9410	0.5152	166.55	697.31	1.2784	5.3524
105	378.15	1.6402	0.6097	167.43	701.00	1.2842	5.3767	1.9150	0.5222	167.41	700.91	1.2807	5.3620
110	383.15	1.6186	0.6178	168.29	704.60	1.2864	5.3859	1.8900	0.5291	168.27	704.51	1.2829	5.3712
115	388.15	1.5977	0.6259	169.16	708.24	1.2887	5.3955	1.8653	0.5361	169.14	708.16	1.2852	5.3809
120	393.15	1.5773	0.6340	170.04	711.92	1.2910	5.4052	1.8413	0.5431	170.02	711.84	1.2874	5.3901

1 at = 1 kp/cm² = 98 066.5 Pa = 98 066.5 N/m² = 9.806 65 N/cm² = 0.980 665 bar

1 kcal = 4.186 8 kJ

Table 61-12 Properties of Superheated Chlorodifluoromethane "Freon 22" (CHClF$_2$) (*Continued*)

Pressure p		0.8 at = 0.784 532 0 bar						0.9 at = 0.882 598 5 bar					
Temperature		$t_s = -46.3$ °C = 226.85 K $i'' = 144.38$ kcal/kg = 604.49 kJ/kg $s'' = 1.2007$ kcal/kg K = 5.0271 kJ/kg K				$\rho'' = 3.7092$ kg/m³ $v'' = 0.2696$ m³/kg		$t_s = -43.8$ °C = 229.35 K $i'' = 144.67$ kcal/kg = 605.70 kJ/kg $s'' = 1.1994$ kcal/kg K = 5.0216 kJ/kg K				$\rho'' = 4.1356$ kg/m³ $v'' = 0.2418$ m³/kg	
t	T	ρ	v	i		s		ρ	v	i		s	
°C	K	kg/m³	m³/kg	kcal/kg	kJ/kg	kcal/kg K	kJ/kg K	kg/m³	m³/kg	kcal/kg	kJ/kg	kcal/kg K	kJ/kg K
−45	228.15	3.6873	0.2712	144.54	605.16	1.2015	5.0304	—	—	—	—	—	—
−40	233.15	3.6036	0.2775	145.21	607.97	1.2044	5.0426	4.0650	0.2460	145.17	607.80	1.2015	5.0304
−35	238.15	3.5249	0.2837	145.88	610.77	1.2072	5.0543	3.9746	0.2516	145.85	610.64	1.2044	5.0426
−30	243.15	3.4495	0.2899	146.56	613.62	1.2101	5.0664	3.8895	0.2571	146.53	613.49	1.2073	5.0547
−25	248.15	3.3772	0.2961	147.25	616.51	1.2129	5.0782	3.8081	0.2626	147.22	616.38	1.2101	5.0664
−20	253.15	3.3069	0.3024	147.95	619.44	1.2157	5.0899	3.7286	0.2682	147.92	619.31	1.2128	5.0778
−15	258.15	3.2404	0.3086	148.65	622.37	1.2184	5.1012	3.6536	0.2737	148.62	622.24	1.2156	5.0892
−10	263.15	3.1766	0.3148	149.35	625.30	1.2212	5.1129	3.5804	0.2793	149.32	625.17	1.2184	5.1012
−5	268.15	3.1162	0.3209	150.06	628.27	1.2238	5.1238	3.5112	0.2848	150.04	628.19	1.2210	5.1121
0	273.15	3.0562	0.3272	150.79	631.33	1.2265	5.1351	3.4447	0.2903	150.75	631.16	1.2237	5.1234
5	278.15	2.9994	0.3334	151.52	634.38	1.2282	5.1422	3.3807	0.2958	151.48	634.22	1.2254	5.1305
10	283.15	2.9446	0.3396	152.24	637.40	1.2317	5.1569	3.3179	0.3014	152.21	637.27	1.2289	5.1452
15	288.15	2.8918	0.3458	152.98	640.50	1.2343	5.1678	3.2584	0.3069	152.95	640.37	1.2315	5.1560
20	293.15	2.8409	0.3520	153.72	643.59	1.2369	5.1787	3.2010	0.3124	153.70	643.51	1.2341	5.1669
25	298.15	2.7917	0.3582	154.47	646.73	1.2394	5.1891	3.1456	0.3179	154.45	646.65	1.2366	5.1774
30	303.15	2.7450	0.3643	155.23	649.92	1.2419	5.1996	3.0921	0.3234	155.21	649.83	1.2391	5.1879
35	308.15	2.6991	0.3705	156.00	653.14	1.2444	5.2101	3.0404	0.3289	155.97	653.02	1.2416	5.1983
40	313.15	2.6546	0.3767	156.77	656.36	1.2469	5.2205	2.9904	0.3344	156.74	656.24	1.2441	5.2088
45	318.15	2.6116	0.3829	157.55	659.63	1.2493	5.2306	2.9420	0.3399	157.52	659.50	1.2466	5.2193
50	323.15	2.5707	0.3890	158.33	662.90	1.2518	5.2410	2.8952	0.3454	158.21	662.39	1.2490	5.2293
55	328.15	2.5304	0.3952	159.12	666.20	1.2542	5.2511	2.8498	0.3509	159.10	666.12	1.2514	5.2394
60	333.15	2.4913	0.4014	159.92	669.55	1.2566	5.2611	2.8058	0.3564	159.90	669.47	1.2539	5.2498
65	338.15	2.4540	0.4075	160.73	672.94	1.2590	5.2712	2.7632	0.3619	160.70	672.82	1.2563	5.2599
70	343.15	2.4172	0.4137	161.55	676.38	1.2614	5.2812	2.7218	0.3674	161.51	676.21	1.2586	5.2695
75	348.15	2.3821	0.4198	162.35	679.73	1.2638	5.2913	2.6824	0.3728	162.33	679.64	1.2610	5.2796
80	353.15	2.3474	0.4260	163.17	683.16	1.2661	5.3009	2.6434	0.3783	163.15	683.08	1.2634	5.2896
85	358.15	2.3143	0.4321	164.00	686.64	1.2684	5.3105	2.6055	0.3838	163.98	686.55	1.2657	5.2992
90	363.15	2.2815	0.4383	164.84	690.15	1.2708	5.3206	2.5687	0.3893	164.82	690.07	1.2680	5.3089
95	368.15	2.2502	0.4444	165.68	693.67	1.2731	5.3302	2.5336	0.3947	165.67	693.63	1.2703	5.3185
100	373.15	2.2198	0.4505	166.53	697.23	1.2754	5.3398	2.4988	0.4002	166.52	697.19	1.2726	5.3281
105	378.15	2.1896	0.4567	167.39	700.83	1.2776	5.3491	2.4655	0.4056	167.37	700.74	1.2749	5.3378
110	383.15	2.1608	0.4628	168.25	704.43	1.2799	5.3587	2.4325	0.4111	168.24	704.39	1.2772	5.3474
115	388.15	2.1327	0.4689	169.12	708.07	1.2821	5.3679	2.4010	0.4165	169.11	708.03	1.2794	5.3566
120	393.15	2.1053	0.4750	170.00	711.76	1.2844	5.3775	2.3697	0.4220	169.98	711.67	1.2817	5.3662
125	398.15	2.0786	0.4811	170.89	715.48	1.2866	5.3867	2.3397	0.4274	170.87	715.40	1.2839	5.3754

1 at = 1 kp/cm² = 98 066.5 Pa = 98 066.5 N/m² = 9.806 65 N/cm² = 0.980 665 bar   1 kcal = 4.186 8 kJ

Table 61-13 Properties of Superheated Chlorodifluoromethane "Freon 22" (CHClF$_2$) (Continued)

Pressure p		1.0 at = 0.980 665 bar						1.2 at = 1.176 798 0 bar					
Temperature		$t_s = -41.7\,°C = 231.45\,K$ $\rho'' = 4.5620\,kg/m^3$ $i'' = 144.93\,kcal/kg = 606.79\,kJ/kg$ $v'' = 0.2192\,m^3/kg$ $s'' = 1.1981\,kcal/kg\,K = 5.0162\,kJ/kg\,K$						$t_s = -37.7\,°C = 235.45\,K$ $\rho'' = 5.4054\,kg/m^3$ $i'' = 145.38\,kcal/kg = 608.68\,kJ/kg$ $v'' = 0.1850\,m^3/kg$ $s'' = 1.1961\,kcal/kg\,K = 5,0078\,kJ/kg\,K$					
t	T	ρ	v	i		s		ρ	v	i		s	
°C	K	kg/m³	m³/kg	kcal/kg	kJ/kg	kcal/kg K	kJ/kg K	kg/m³	m³/kg	kcal/kg	kJ/kg	kcal/kg K	kJ/kg K
−40	233.15	4.5290	0.2208	145.15	607.71	1.1991	5.0204	—	—	—	—	—	—
−35	238.15	4.4287	0.2258	145.82	610.52	1.2020	5.0325	5.3419	0.1872	145.76	610.27	1.1977	5.0145
−30	243.15	4.3328	0.2308	146.50	613.37	1.2048	5.0443	5.2247	0.1914	146.44	613.11	1.2005	5.0263
−25	248.15	4.2409	0.2358	147.19	616.26	1.2076	5.0560	5.1125	0.1956	147.13	616.00	1.2033	5.0380
−20	253.15	4.1528	0.2408	147.89	619.19	1.2104	5.0677	5.0050	0.1998	147.83	618.93	1.2061	5.0497
−15	258.15	4.0683	0.2458	148.59	622.12	1.2132	5.0794	4.9020	0.2040	148.53	621.87	1.2089	5.0614
−10	263.15	3.9872	0.2508	149.29	625.05	1.2160	5.0911	4.8031	0.2082	149.24	624.84	1.2117	5.0731
− 5	268.15	3.9093	0.2558	150.01	628.06	1.2186	5.1020	4.7081	0.2124	149.95	627.81	1.2143	5.0840
0	273.15	3.8344	0.2608	150.73	631.08	1.2212	5.1129	4.6168	0.2166	150.67	630.83	1.2170	5.0953
5	278.15	3.7622	0.2658	151.45	634.09	1.2239	5.1242	4.5290	0.2208	151.40	633.88	1.2187	5.1025
10	283.15	3.6928	0.2708	152.19	637.19	1.2265	5.1351	4.4464	0.2249	152.13	636.94	1.2222	5.1171
15	288.15	3.6258	0.2758	152.93	640.29	1.2291	5.1460	4.3649	0.2291	152.87	640.04	1.2248	5.1280
20	293.15	3.5625	0.2807	153.67	643.39	1.2316	5.1565	4.2863	0.2333	153.62	643.18	1.2274	5.1389
25	298.15	3.5002	0.2857	154.43	646.57	1.2342	5.1673	4.2123	0.2374	154.38	646.36	1.2299	5.1493
30	303.15	3.4400	0.2907	155.18	649.71	1.2367	5.1778	4.1391	0.2416	155.14	649.54	1.2325	5.1602
35	308.15	3.3818	0.2957	155.95	652.93	1.2392	5.1883	4.0700	0.2457	155.90	652.72	1.2350	5.1707
40	313.15	3.3267	0.3006	156.72	656.16	1.2417	5.1987	4.0016	0.2499	156.68	655.99	1.2375	5.1812
45	318.15	3.2723	0.3056	157.50	659.42	1.2441	5.2088	3.9370	0.2540	157.46	659.25	1.2399	5.1912
50	323.15	3.2206	0.3105	158.19	662.31	1.2466	5.2193	3.8730	0.2582	158.15	662.14	1.2424	5.2017
55	328.15	3.1696	0.3155	159.08	666.04	1.2490	5.2293	3.8124	0.2623	159.03	665.83	1.2448	5.2117
60	333.15	3.1211	0.3204	159.88	669.39	1.2514	5.2394	3.7523	0.2665	159.83	669.18	1.2472	5.2218
65	338.15	3.0731	0.3254	160.68	672.74	1.2538	5.2494	3.6955	0.2706	160.64	672.57	1.2496	5.2318
70	343.15	3.0276	0.3303	161.49	676.13	1.2562	5.2595	3.6403	0.2747	161.45	675.96	1.2520	5.2419
75	348.15	2.9824	0.3353	162.31	679.56	1.2586	5.2695	3.5855	0.2789	162.27	679.39	1.2544	5.2519
80	353.15	2.9394	0.3402	163.13	682.99	1.2609	5.2791	3.5336	0.2830	163.10	682.87	1.2567	5.2616
85	358.15	2.8977	0.3451	163.97	686.51	1.2633	5.2892	3.4831	0.2871	163.93	686.34	1.2590	5.2712
90	363.15	2.8563	0.3501	164.80	689.98	1.2656	5.2988	3.4341	0.2912	164.77	689.86	1.2614	5.2812
95	368.15	2.8169	0.3550	165.65	693.54	1.2679	5.3084	3.3864	0.2953	165.61	693.38	1.2637	5.2909
100	373.15	2.7785	0.3599	166.50	697.10	1.2702	5.3181	3.3389	0.2995	166.46	696.93	1.2660	5.3005
105	378.15	2.7412	0.3648	167.36	700.70	1.2725	5.3277	3.2938	0.3036	167.32	700.54	1.2683	5.3101
110	383.15	2.7049	0.3697	168.22	704.30	1.2747	5.3369	3.2499	0.3077	168.18	704.14	1.2706	5.3197
115	388.15	2.6695	0.3746	169.09	707.95	1.2770	5.3465	3.2072	0.3118	169.06	707.82	1.2728	5.3290
120	393.15	2.6350	0.3795	169.97	711.63	1.2792	5.3558	3.1656	0.3159	169.93	711.46	1.2751	5.3386
125	398.15	2.6008	0.3845	170.85	715.31	1.2815	5.3654	3.1250	0.3200	170.82	715.19	1.2773	5.3478
130	403.15	2.5681	0.3894	171.73	719.00	1.2838	5.3750	3.0855	0.3241	171.71	718.92	1.2795	5.3570

1 at = 1 kp/cm² = 98 066.5 Pa = 98 066.5 N/m² = 9.806 65 N/cm² = 0.980 665 bar   1 kcal = 4.186 8 kJ

Table 61-14 Properties of Superheated Chlorodifluoromethane "Freon 22" (CHClF$_2$) (*Continued*)

Pressure p		1.4 at = 1.372 931 0 bar						1.6 at = 1.569 064 0 bar					
Temperature		$t_s = -34.2\,°C = 238.95\,K$ $i'' = 145.79\,kcal/kg = 610.39\,kJ/kg$ $s'' = 1.1944\,kcal/kg\,K = 5.0007\,kJ/kg\,K$				$\rho'' = 6.2422\,kg/m^3$ $v'' = 0.1602\,m^3/kg$		$t_s = -31.1\,°C = 242.05\,K$ $i'' = 146.15\,kcal/kg = 611.90\,kJ/kg$ $s'' = 1.1930\,kcal/kg\,K = 4.9949\,kJ/kg\,K$				$\rho'' = 7.0671\,kg/m^3$ $v'' = 0.1415\,m^3/kg$	
t	T	ρ	v	i		s		ρ	v	i		s	
°C	K	kg/m³	m³/kg	kcal/kg	kJ/kg	kcal/kg K	kJ/kg K	kg/m³	m³/kg	kcal/kg	kJ/kg	kcal/kg K	kJ/kg K
−30	243.15	6.1237	0.1633	146.38	612.86	1.1968	5.0108	7.0323	0.1422	146.32	612.61	1.1937	4.9978
−25	248.15	5.9916	0.1669	147.07	615.75	1.1996	5.0225	6.8776	0.1454	147.01	615.50	1.1965	5.0095
−20	253.15	5.8651	0.1705	147.77	618.68	1.2024	5.0342	6.7295	0.1486	147.71	618.43	1.1993	5.0212
−15	258.15	5.7438	0.1741	148.47	621.61	1.2052	5.0459	6.5920	0.1517	148.41	621.36	1.2020	5.0325
−10	263.15	5.6275	0.1777	149.18	624.59	1.2080	5.0577	6.4558	0.1549	149.12	624.34	1.2048	5.0443
− 5	268.15	5.5127	0.1814	149.89	627.56	1.2106	5.0685	6.3052	0.1586	149.84	627.35	1.2075	5.0556
0	273.15	5.4054	0.1850	150.62	630.62	1.2133	5.0798	6.2035	0.1612	150.56	630.36	1.2102	5.0669
5	278.15	5.3022	0.1886	151.34	633.63	1.2151	5.0874	6.0827	0.1644	151.29	633.42	1.2128	5.0778
10	283.15	5.2056	0.1921	152.08	636.73	1.2186	5.1020	5.9666	0.1676	152.03	636.52	1.2155	5.0891
15	288.15	5.1099	0.1957	152.82	639.83	1.2212	5.1129	5.8582	0.1707	152.77	639.62	1.2181	5.0999
20	293.15	5.0176	0.1993	153.57	642.97	1.2238	5.1238	5.7504	0.1739	153.52	642.76	1.2206	5.1104
25	298.15	4.9285	0.2029	154.33	646.15	1.2263	5.1343	5.6497	0.1770	154.28	645.94	1.2232	5.1213
30	303.15	4.8426	0.2065	155.09	649.33	1.2289	5.1452	5.5494	0.1802	155.04	649.12	1.2257	5.1318
35	308.15	4.7596	0.2101	155.86	652.55	1.2314	5.1556	5.4555	0.1833	155.81	652.35	1.2283	5.1426
40	313.15	4.6816	0.2136	156.63	655.78	1.2338	5.1657	5.3619	0.1865	156.59	655.61	1.2307	5.1527
45	318.15	4.6041	0.2172	157.41	659.04	1.2363	5.1761	5.2743	0.1896	157.37	658.88	1.2332	5.1632
50	323.15	4.5290	0.2208	158.10	661.93	1.2387	5.1862	5.1894	0.1927	158.06	661.77	1.2357	5.1736
55	328.15	4.4563	0.2244	158.99	665.66	1.2412	5.1967	5.1046	0.1959	158.95	665.49	1.2381	5.1837
60	333.15	4.3879	0.2279	159.79	669.01	1.2436	5.2067	5.0251	0.1990	159.75	668.84	1.2405	5.1937
65	338.15	4.3197	0.2315	160.60	672.40	1.2460	5.2168	4.9480	0.2021	160.56	672.23	1.2429	5.2038
70	343.15	4.2553	0.2350	161.41	675.79	1.2484	5.2268	4.8733	0.2052	161.37	675.62	1.2453	5.2138
75	348.15	4.1911	0.2386	162.23	679.22	1.2508	5.2368	4.8008	0.2083	162.19	679.06	1.2477	5.2239
80	353.15	4.1305	0.2421	163.06	682.70	1.2531	5.2465	4.7281	0.2115	163.02	682.53	1.2500	5.2335
85	358.15	4.0700	0.2457	163.89	686.17	1.2555	5.2565	4.6598	0.2146	163.85	686.01	1.2524	5.2435
90	363.15	4.0128	0.2492	164.73	689.69	1.2578	5.2662	4.5935	0.2177	164.69	689.52	1.2547	5.2532
95	368.15	3.9573	0.2527	165.57	693.21	1.2601	5.2758	4.5290	0.2208	165.54	693.08	1.2570	5.2628
100	373.15	3.9017	0.2563	166.43	696.81	1.2623	5.2850	4.4663	0.2239	166.39	696.64	1.2593	5.2724
105	378.15	3.8491	0.2598	167.28	700.37	1.2647	5.2950	4.4053	0.2270	167.25	700.24	1.2616	5.2821
110	383.15	3.7979	0.2633	168.15	704.01	1.2670	5.3047	4.3459	0.2301	168.12	703.88	1.2639	5.2917
115	388.15	3.7467	0.2669	169.02	707.65	1.2693	5.3143	4.2882	0.2332	168.99	707.53	1.2662	5.3013
120	393.15	3.6982	0.2704	169.90	711.34	1.2715	5.3235	4.2319	0.2363	169.87	711.21	1.2684	5.3105
125	398.15	3.6510	0.2739	170.78	715.02	1.2738	5.3331	4.1771	0.2394	170.75	714.90	1.2707	5.3202
130	403.15	3.6049	0.2774	171.68	718.79	1.2760	5.3424	4.1237	0.2425	171.64	718.62	1.2729	5.3294
135	408.15	3.5600	0.2809	172.57	722.52	1.2782	5.3516	4.0717	0.2456	172.54	722.39	1.2751	5.3386

1 at = 1 kp/cm² = 98 066.5 Pa = 98 066.5 N/m² = 9.806 65 N/cm² = 0.980 666 bar

1 kcal = 4.186 8 kJ

Table 61-15 Properties of Superheated Chlorodifluoromethane "Freon 22" (CHClF$_2$) (*Continued*)

Pressure p		1.8 at = 1.765 197 0 bar						2.0 at = 1.961 330 0 bar					
Temperature		$t_s = -28.4\,°C = 244.75\,K$		$\rho'' = 7.8864\,kg/m^3$				$t_s = -25.8\,°C = 247.35\,K$		$\rho'' = 8.7032\,kg/m^3$			
		$i'' = 146.48\,kcal/kg = 613.28\,kJ/kg$		$v'' = 0.1268\,m^3/kg$				$i'' = 146.77\,kcal/kg = 614.50\,kJ/kg$		$v'' = 0.1149\,m^3/kg$			
		$s'' = 1.1918\,kcal/kg\,K = 4.9898\,kJ/kg\,K$						$s'' = 1.1906\,kcal/kg\,K = 4.9848\,kJ/kg\,K$					
t	T	ρ	v	i		s		ρ	v	i		s	
°C	K	kg/m³	m³/kg	kcal/kg	kJ/kg	kcal/kg K	kJ/kg K	kg/m³	m³/kg	kcal/kg	kJ/kg	kcal/kg K	kJ/kg K
−25	248.15	7.7700	0.1287	146.94	615.21	1.1973	4.9978	8.6730	0.1153	146.88	614.96	1.1911	4.9869
−20	253.15	7.5988	0.1316	147.64	618.14	1.1965	5.0095	8.4818	0.1179	147.58	617.89	1.1939	4.9986
−15	258.15	7.4405	0.1344	148.35	621.11	1.1993	5.0212	8.3056	0.1204	148.29	620.86	1.1966	5.0099
−10	263.15	7.2886	0.1372	149.06	624.08	1.2020	5.0325	8.1301	0.1230	149.00	623.83	1.1994	5.0216
− 5	268.15	7.1429	0.1400	149.78	627.10	1.2047	5.0438	7.9681	0.1255	149.72	626.85	1.2021	5.0330
0	273.15	7.0028	0.1428	150.51	630.16	1.2074	5.0551	7.8064	0.1281	150.46	629.95	1.2048	5.0443
5	278.15	6.8634	0.1457	151.23	633.17	1.2100	5.0660	7.6570	0.1306	151.18	632.96	1.2074	5.0551
10	283.15	6.7340	0.1485	151.97	636.27	1.2127	5.0773	7.5075	0.1332	151.92	636.06	1.2101	5.0664
15	288.15	6.6094	0.1513	152.72	639.41	1.2153	5.0882	7.3692	0.1357	152.66	639.16	1.2127	5.0773
20	293.15	6.4893	0.1541	153.47	642.55	1.2179	5.0991	7.2307	0.1383	153.42	642.34	1.2153	5.0882
25	298.15	6.3735	0.1569	154.23	645.73	1.2204	5.1096	7.1023	0.1408	154.18	645.52	1.2179	5.0091
30	303.15	6.2617	0.1597	154.99	648.91	1.2230	5.1205	6.9784	0.1433	154.94	648.70	1.2204	5.1096
35	308.15	6.1538	0.1625	155.76	652.14	1.2255	5.1309	6.8540	0.1459	155.71	651.93	1.2230	5.1205
40	313.15	6.0496	0.1653	156.44	654.98	1.2280	5.1414	6.7385	0.1484	156.39	654.77	1.2254	5.1305
45	318.15	5.9488	0.1681	157.32	658.67	1.2306	5.1519	6.6269	0.1509	157.28	658.50	1.2279	5.1410
50	323.15	5.8514	0.1709	158.11	661.97	1.2329	5.1619	6.5147	0.1535	158.07	661.81	1.2304	5.1514
55	328.15	5.7571	0.1737	158.91	665.32	1.2353	5.1720	6.4103	0.1560	158.86	665.12	1.2328	5.1615
60	333.15	5.6657	0.1765	159.71	668.67	1.2377	5.1820	6.3091	0.1585	159.66	668.46	1.2353	5.1720
65	338.15	5.5772	0.1793	160.51	672.02	1.2401	5.1921	6.2112	0.1610	160.47	671.86	1.2377	5.1820
70	343.15	5.4915	0.1821	161.33	675.46	1.2425	5.2021	6.1162	0.1635	161.29	675.29	1.2400	5.1916
75	348.15	5.4083	0.1849	162.15	678.89	1.2449	5.2121	6.0205	0.1661	162.11	678.22	1.2424	5.2017
80	353.15	5.3305	0.1876	162.98	682.36	1.2473	5.2222	5.9312	0.1686	162.94	682.20	1.2448	5.2117
85	358.15	5.2521	0.1904	163.81	685.84	1.2496	5.2318	5.8445	0.1711	163.77	685.67	1.2471	5.2214
90	363.15	5.1760	0.1932	164.65	689.36	1.2519	5.2415	5.7604	0.1736	164.61	689.19	1.2495	5.2314
95	368.15	5.1020	0.1960	165.50	692.92	1.2543	5.2515	5.6786	0.1761	165.46	692.75	1.2518	5.2410
100	373.15	5.0327	0.1987	166.35	696.47	1.2566	5.2611	5.5991	0.1786	166.32	696.45	1.2541	5.2507
105	378.15	4.9628	0.2015	167.21	700.07	1.2589	5.2708	5.5218	0.1811	167.18	699.95	1.2564	5.2603
110	383.15	4.8948	0.2043	168.08	703.72	1.2611	5.2800	5.4466	0.1836	168.04	703.55	1.2587	5.2699
115	388.15	4.8309	0.2070	168.95	707.36	1.2634	5.2896	5.3735	0.1861	168.92	707.23	1.2609	5.2791
120	393.15	4.7664	0.2098	169.83	711.04	1.2657	5.2992	5.3022	0.1886	169.80	710.92	1.2632	5.2888
125	398.15	4.7059	0.2125	170.72	714.77	1.2679	5.3084	5.2356	0.1910	170.68	714.60	1.2654	5.2980
130	403.15	4.6447	0.2153	171.61	718.50	1.2701	5.3177	5.1680	0.1935	171.58	718.37	1.2676	5.3072
135	408.15	4.5872	0.2180	172.51	722.26	1.2723	5.3269	5.1020	0.1960	172.48	722.14	1.2699	5.3168
140	413.15	4.5290	0.2208	173.41	726.03	1.2745	5.3361	5.0378	0.1985	173.38	725.91	1.2720	5.3256

1 at = 1 kp/cm² = 98 066.5 Pa = 98 066.5 N/m² = 9.806 65 N/cm² = 0.980 665 bar   1 kcal = 4.186 8 kJ

Table 61-16 Properties of Superheated Chlorodifluoromethane "Freon 22" (CHClF$_2$) (Continued)

Pressure p		2.5 at = 2.451 662 5 bar						3.0 at = 2.941 995 0 bar					
Temperature		$t_s = -20.1$ °C = 253.05 K $i'' = 147.40$ kcal/kg = 617.13 kJ/kg $s'' = 1.1884$ kcal/kg K = 4.9756 kJ/kg K				$\rho'' = 10.730$ kg/m³ $v'' = 0.0932$ m³/kg		$t_s = -15.1$ °C = 258.05 K $i'' = 147.93$ kcal/kg = 619.35 kJ/kg $s'' = 1.1865$ kcal/kg K = 4.9676 kJ/kg K				$\rho'' = 12.739$ kg/m³ $v'' = 0.0785$ m³/kg	
t	T	ρ	v	i		s		ρ	v	i		s	
°C	K	kg/m³	m³/kg	kcal/kg	kJ/kg	kcal/kg K	kJ/kg K	kg/m³	m³/kg	kcal/kg	kJ/kg	kcal/kg K	kJ/kg K
−20	253.15	10.730	0.0932	147.42	617.22	1.1884	4.9756	—	—	—	—	—	—
−15	258.15	10.493	0.0953	148.13	620.19	1.1912	4.9873	12.739	0.0785	147.94	619.40	1.1872	4.9706
−10	263.15	10.267	0.0974	148.85	623.21	1.1939	4.9986	12.453	0.0803	148.66	622.41	1.1899	4.9819
− 5	268.15	10.060	0.0994	149.58	626.26	1.1967	5.0103	12.195	0.0820	149.39	624.47	1.1927	4.9936
0	273.15	9.8522	0.1015	150.32	629.36	1.1994	5.0216	11.933	0.0838	150.13	628.56	1.1954	5.0049
5	278.15	9.6525	0.1036	151.04	632.37	1.2020	5.0325	11.696	0.0855	150.87	631.66	1.1980	5.0158
10	283.15	9.4697	0.1056	151.79	635.51	1.2047	5.0438	11.455	0.0873	151.62	634.80	1.2007	5.0271
15	288.15	9.2851	0.1077	152.53	638.61	1.2073	5.0547	11.236	0.0890	153.28	637.98	1.2033	5.0380
20	293.15	9.1075	0.1098	153.29	641.79	1.2099	5.0656	11.025	0.0907	153.15	641.21	1.2059	5.0489
25	298.15	8.9445	0.1118	154.05	644.98	1.2125	5.0765	10.811	0.0925	153.92	644.43	1.2084	5.0593
30	303.15	8.7796	0.1139	154.82	648.20	1.2151	5.0874	10.616	0.0942	154.70	647.70	1.2110	5.0702
35	308.15	8.6281	0.1159	155.59	651.42	1.2176	5.0978	10.428	0.0959	155.47	650.92	1.2135	5.0807
40	313.15	8.4746	0.1180	156.27	654.27	1.2202	5.1087	10.235	0.0977	156.26	654.23	1.2160	5.0911
45	318.15	8.3333	0.1200	157.16	658.00	1.2227	5.1192	10.060	0.0994	157.05	657.54	1.2185	5.1016
50	323.15	8.1900	0.1221	157.95	661.31	1.2251	5.1292	9.8912	0.1011	157.84	660.84	1.2209	5.1117
55	328.15	8.0580	0.1241	158.75	664.65	1.2276	5.1397	9.7276	0.1028	158.64	664.19	1.2218	5.1154
60	333.15	7.9302	0.1261	159.56	668.05	1.2300	5.1498	9.5694	0.1045	159.45	667.59	1.2258	5.1322
65	338.15	7.8003	0.1282	160.37	671.44	1.2324	5.1598	9.4073	0.1063	160.26	670.98	1.2282	5.1422
70	343.15	7.6805	0.1302	161.19	674.87	1.2348	5.1699	9.2593	0.1080	161.09	647.45	1.2306	5.1423
75	348.15	7.5643	0.1322	162.01	678.30	1.2372	5.1799	9.1158	0.1097	161.91	677.88	1.2329	5.1619
80	353.15	7.4460	0.1343	162.48	681.78	1.2396	5.1900	8.9767	0.1114	162.74	681.36	1.2353	5.1720
85	358.15	7.3368	0.1363	163.68	685.30	1.2419	5.1996	8.8417	0.1131	163.57	684.83	1.2377	5.1820
90	363.15	7.2307	0.1383	164.52	688.81	1.2442	5.2092	8.7108	0.1148	164.42	688.39	1.2400	5.1916
95	368.15	7.1276	0.1403	165.37	692.37	1.2466	5.2193	8.5837	0.1165	165.28	691.99	1.2423	5.2013
100	373.15	7.0274	0.1423	166.22	695.93	1.2489	5.2289	8.4602	0.1182	166.13	695.55	1.2446	5.2109
105	378.15	6.9300	0.1443	167.09	699.57	1.2512	5.2385	8.3403	0.1199	167.00	699.20	1.2469	5.2205
110	383.15	6.8306	0.1464	167.95	703.17	1.2534	5.2477	8.2237	0.1216	167.87	702.84	1.2492	5.2302
115	388.15	6.7385	0.1484	168.83	706.86	1.2557	5.2574	8.1169	0.1232	168.74	706.48	1.2515	5.2398
120	393.15	6.6489	0.1504	169.71	710.54	1.2580	5.2670	8.0064	0.1249	169.63	710.21	1.2537	5.2490
125	398.15	6.5617	0.1524	170.60	714.27	1.2602	5.2762	7.8989	0.1266	170.52	713.93	1.2559	5.2582
130	403.15	6.4767	0.1544	171.49	717.99	1.2624	5.2854	7.7942	0.1283	171.41	717.66	1.2582	5.2678
135	408.15	6.3939	0.1564	172.40	721.80	1.2647	5.2950	7.6923	0.1300	172.32	721.47	1.2604	5.2770
140	413.15	6.3131	0.1584	173.30	725.57	1.2668	5.3038	7.5988	0.1316	173.22	725.24	1.2626	5.2863
145	418.15	6.2344	0.1604	174.22	729.42	1.2690	5.3130	7.5019	0.1333	175.14	729.09	1.2648	5.2955
150	423.15	6.1576	0.1624	175.14	733.28	1.2711	5.3218	7.4074	0.1350	175.06	732.94	1.2670	5.3047

1 at = 1 kp/cm² = 98 066.5 Pa = 98 066.5 N/m² = 9.806 65 N/cm² = 0.980 665 bar

1 kcal = 4.186 8 kJ

Pressure p		3.5 at = 3.432 327 5 bar						4.0 at = 3.922 660 0 bar					
Temperature		$t_s = -10.95\ °C = 262.20\ K$ $i'' = 148.38\ kcal/kg = 621.24\ kJ/kg$ $s'' = 1.1851\ kcal/kg\ K = 4.9618\ kJ/kg\ K$			$\rho'' = 14.771\ kg/m^3$ $v'' = 0.06770\ m^3/kg$			$t_s = -7.3\ °C = 265.85\ K$ $i'' = 148.77\ kcal/kg = 622.87\ kJ/kg$ $s'' = 1.1837\ kcal/kg\ K = 4.9559\ kJ/kg\ K$			$\rho'' = 16.798\ kg/m^3$ $v'' = 0.05953\ m^3/kg$		
t	T	ρ	v	i		s		ρ	v	i		s	
°C	K	kg/m³	m²/kg	kcal/kg	kJ/kg	kcal/kg K	kJ/kg K	kg/m³	m³/kg	kcal/kg	kJ/kg	kcal/kg K	kJ/kg K
−10	263.15	14.706	0.06800	148.53	621.87	1.1856	4.9639	—	—	—	—	—	—
−5	268.15	14.372	0.06958	149.65	626.55	1.1883	4.9752	16.620	0.06017	149.10	624.25	1.1848	4.9605
0	273.15	14.065	0.07110	149.99	627.98	1.1910	4.9865	16.250	0.06154	149.84	627.35	1.1876	4.9722
5	278.15	13.770	0.07262	150.73	631.08	1.1937	4.9978	15.901	0.06289	150.58	630.45	1.1903	4.9835
10	283.15	13.490	0.07413	151.49	634.26	1.1964	5.0091	15.567	0.06424	151.34	633.63	1.1930	4.9949
15	288.15	13.221	0.07564	152.25	637.44	1.1991	5.0204	15.246	0.06559	152.10	636.81	1.1957	5.0062
20	293.15	12.962	0.07715	153.01	640.62	1.2017	5.0313	14.941	0.06693	152.87	640.04	1.1983	5.0170
25	298.15	12.713	0.07866	153.79	643.89	1.2043	5.0422	14.650	0.06826	153.65	643.30	1.2009	5.0279
30	303.15	12.475	0.08016	154.57	647.15	1.2069	5.0530	14.368	0.06960	154.43	646.57	1.2035	5.0388
35	308.15	12.246	0.08166	155.35	650.42	1.2095	5.0639	14.098	0.07093	155.22	649.88	1.2061	5.0497
40	313.15	12.026	0.08315	156.14	653.73	1.2120	5.0744	13.839	0.07226	156.01	653.18	1.2087	5.0606
45	318.15	11.813	0.08465	156.93	657.03	1.2145	5.0849	13.589	0.07359	156.81	656.53	1.2112	5.0711
50	323.15	11.609	0.08614	157.73	660.38	1.2170	5.0953	13.349	0.07491	157.61	659.88	1.2137	5.0815
55	328.15	11.411	0.08763	158.53	663.73	1.2195	5.1058	13.118	0.07623	158.42	663.27	1.2162	5.0920
60	333.15	11.221	0.08912	159.34	667.12	1.2219	5.1159	12.895	0.07755	159.23	666.66	1.2186	5.1020
65	338.15	11.038	0.09060	160.16	670.56	1.2243	5.1259	12.681	0.07886	160.05	670.10	1.2211	5.1125
70	343.15	10.860	0.09208	160.98	673.99	1.2268	5.1364	12.473	0.08017	160.88	673.57	1.2235	5.1225
75	348.15	10.688	0.09356	161.81	677.47	1.2291	5.1460	12.273	0.08148	161.71	677.05	1.2259	5.1326
80	353.15	10.523	0.09503	162.64	680.94	1.2315	5.1560	12.080	0.08278	162.54	680.52	1.2283	5.1426
85	358.15	10.363	0.09650	163.48	686.46	1.2339	5.1661	11.892	0.08409	163.38	684.40	1.2307	5.1527
90	363.15	10.207	0.09797	164.33	688.22	1.2362	5.1757	11.711	0.08539	164.23	687.60	1.2330	5.1623
95	368.15	10.057	0.09943	165.18	691.58	1.2386	5.1858	11.537	0.08668	165.09	691.20	1.2354	5.1724
100	373.15	9.9108	0.10090	166.04	695.18	1.2409	5.1954	11.368	0.08797	165.95	694.80	1.2377	5.1820
105	378.15	9.7694	0.10236	166.91	698.82	1.2432	5.2050	11.203	0.08926	166.81	698.40	1.2400	5.1916
110	383.15	9.6321	0.10382	167.80	702.55	1.2455	5.2147	11.044	0.09055	167.71	702.17	1.2423	5.2013
115	388.15	9.4985	0.10528	168.66	706.15	1.2478	5.2243	10.889	0.09184	168.57	705.77	1.2446	5.2109
120	393.15	9.3694	0.10673	169.54	709.83	1.2500	5.2335	10.739	0.09312	169.45	709.45	1.2469	5.2205
125	398.15	9.2439	0.10818	170.43	713.56	1.2523	5.2431	10.594	0.09439	170.35	713.22	1.2492	5.2302
130	403.15	9.1224	0.10962	171.33	717.32	1.2545	5.2523	10.453	0.09567	171.25	716.99	1.2514	5.2394
135	408.15	9.0033	0.11107	172.23	721.09	1.2568	5.2620	10.316	0.09694	172.15	720.76	1.2536	5.2486
140	413.15	8.8881	0.11251	173.14	724.90	1.2590	5.2712	10.182	0.09821	173.06	724.57	1.2559	5.2582
145	418.15	8.7758	0.11395	174.06	728.75	1.2612	5.2804	10.052	0.09948	173.98	728.42	1.2581	5.2674
150	423.15	8.6670	0.11538	174.98	732.61	1.2634	5.2896	9.9265	0.10074	174.90	732.27	1.2603	5.2766
155	428.15	8.5602	0.11682	175.87	736.33	1.2656	5.2988	9.8039	0.10200	175.83	736.17	1.2625	5.2858
160	433.15	8.4567	0.11825	176.75	740.02	1.2678	5.3080	9.6843	0.10326	176.77	740.10	1.2647	5.2950

1 at = 1 kp/cm² = 98 066.5 Pa = 98 066.5 N/m² = 9.806 65 N/cm² = 0.980 665 bar   1 kcal = 4.186 8 kJ

Table 61-18 Properties of Superheated Chlorodifluoromethane "Freon 22" (CHClF$_2$) (Continued)

| Pressure p | | \multicolumn{6}{c}{4.5 at = 4.412 992 5 bar} | | | | | | \multicolumn{6}{c}{5.0 at = 4.903 325 bar} | | | | | |

| Temperature | | \multicolumn{6}{l}{$t_s = -3.9\,°C = 269.25$ K $\quad \rho'' = 18.793$ kg/m³} | | | | | | \multicolumn{6}{l}{$t_s = -0.67\,°C = 272.48$ K $\quad \rho'' = 20.816$ kg/m³} | | | | | |

$i'' = 149.12$ kcal/kg $= 624.34$ kJ/kg $\quad v'' = 0.05321$ m³/kg
$s'' = 1.1824$ kcal/kg K $= 4.9505$ kJ/kg K

$i'' = 149.43$ kcal/kg $= 625.63$ kJ/kg $\quad v'' = 0.04804$ m³/kg
$s'' = 1.1814$ kcal/kg K $= 4.9463$ kJ/kg K

t	T	ρ	v	\multicolumn{2}{c}{i}		\multicolumn{2}{c}{s}		ρ	v	\multicolumn{2}{c}{i}		\multicolumn{2}{c}{s}	
°C	K	kg/m³	m³/kg	kcal/kg	kJ/kg	kcal/kg K	kJ/kg K	kg/m³	m³/kg	kcal/kg	kJ/kg	kcal/kg K	kJ/kg K
0	273.15	18.464	0.05416	149.69	626.72	1.1845	4.9593	20.747	0.04820	149.54	626.09	1.1815	4.9467
5	278.15	18.060	0.05537	150.44	629.86	1.1873	4.9710	20.268	0.04934	150.28	629.19	1.1844	4.9588
10	283.15	17.668	0.05660	151.19	633.00	1.1900	4.9823	19.822	0.05045	151.04	632.37	1.1871	4.9702
15	288.15	17.298	0.05781	151.96	636.23	1.1927	4.9936	19.395	0.05156	151.81	635.60	1.1898	4.9815
20	293.15	16.943	0.05902	152.73	639.45	1.1953	5.0045	18.990	0.05266	152.59	638.86	1.1915	4.9928
25	298.15	16.603	0.06023	153.52	642.76	1.1980	5.0158	18.601	0.05376	153.37	642.13	1.1952	5.0041
30	303.15	16.279	0.06143	154.30	646.02	1.2006	5.0267	18.228	0.05486	154.16	645.44	1.1978	5.0149
35	308.15	15.969	0.06262	155.09	649.33	1.2032	5.0476	17.873	0.05595	154.96	648.79	1.2004	5.0258
40	313.15	15.672	0.06381	155.89	652.68	1.2057	5.0480	17.535	0.05703	155.76	652.14	1.2030	5.0367
45	318.15	15.382	0.06501	156.69	656.03	1.2083	5.0589	17.206	0.05812	156.56	655.49	1.2055	5.0472
50	323.15	15.108	0.06619	157.49	659.38	1.2108	5.0694	16.889	0.05921	157.37	658.88	1.2081	5.0581
55	328.15	14.843	0.06737	158.30	662.77	1.2133	5.0798	16.589	0.06028	158.19	662.31	1.2106	5.0685
60	333.15	14.588	0.06855	159.12	666.20	1.2157	5.0899	16.300	0.06135	159.01	665.74	1.2131	5.0790
65	338.15	14.343	0.06972	159.94	669.64	1.2182	5.1004	16.021	0.06242	159.83	669.18	1.2155	5.0891
70	343.15	14.106	0.07089	160.77	673.11	1.2206	5.1104	15.753	0.06348	160.66	672.65	1.2180	5.0995
75	348.15	13.877	0.07206	161.60	676.59	1.2230	5.1205	15.494	0.06454	161.50	676.17	1.2204	5.1096
80	353.15	13.657	0.07322	162.44	680.10	1.2254	5.1305	15.244	0.06560	162.34	679.69	1.2228	5.1196
85	358.15	13.443	0.07439	163.28	683.62	1.2278	5.1406	15.004	0.06665	163.18	683.20	1.2252	5.1297
90	363.15	13.236	0.07555	164.13	687.18	1.2302	5.1506	14.769	0.06771	164.03	686.76	1.2275	5.1393
95	368.15	13.036	0.07671	164.99	690.78	1.2325	5.1602	14.543	0.06876	164.89	690.36	1.2299	5.1493
100	373.15	12.844	0.07786	165.85	694.38	1.2348	5.1699	14.325	0.06981	165.76	694.00	1.2322	5.1590
105	378.15	12.655	0.07902	166.72	698.02	1.2372	5.1799	14.112	0.07086	166.63	697.65	1.2346	5.1690
110	383.15	12.473	0.08017	167.62	701.79	1.2395	5.1895	13.908	0.07190	167.52	701.37	1.2369	5.1787
115	388.15	12.297	0.08132	168.48	705.39	1.2418	5.1992	13.710	0.07294	168.39	705.02	1.2392	5.1883
120	393.15	12.127	0.08246	169.37	709.12	1.2440	5.2084	13.517	0.07398	169.28	708.74	1.2415	5.1979
125	398.15	11.960	0.08361	170.26	712.84	1.2463	5.2180	13.330	0.07502	170.17	712.47	1.2437	5.2071
130	403.15	11.799	0.08475	171.16	716.61	1.2486	5.2276	13.149	0.07605	171.08	716.28	1.2460	5.2168
135	408.15	11.643	0.08589	172.07	720.42	1.2508	5.2368	12.972	0.07709	171.98	720.05	1.2482	5.2260
140	413.15	11.492	0.08702	172.98	724.23	1.2530	5.2461	12.801	0.07812	172.90	723.90	1.2505	5.2356
145	418.15	11.343	0.08816	173.90	728.08	1.2552	5.2553	12.636	0.07914	173.82	727.75	1.2527	5.2448
150	423.15	11.199	0.08929	174.83	731.98	1.2575	5.2649	12.473	0.08017	175.74	731.64	1.2549	5.2540
155	428.15	11.060	0.09042	175.76	735.87	1.2597	5.2741	12.317	0.08119	175.68	735.54	1.2571	5.2632
160	433.15	10.923	0.09155	176.70	739.81	1.2619	5.2833	12.164	0.08221	176.62	739.47	1.2593	5.2724
165	438.15	10.790	0.09268	177.63	743.70	1.2641	5.2925	12.015	0.08323	177.57	743.45	1.2615	5.2816

1 at = 1 kp/cm² = 98 066.5 Pa = 98 066.5 N/m² = 9.806 65 N/cm² = 0.980 665 bar

1 kcal = 4.186 8 kJ

Pressure p		6.0 at = 5.883 990 0 bar						7.0 at = 6.864 655 0 bar					
Temperature		t_s = +5.1 °C = 278.25 K i'' = 149.97 kcal/kg = 627.89 kJ/kg s'' = 1.1796 kcal/kg K = 4.9387 kJ/kg K			ρ'' = 24.888 kg/m³ v'' = 0.04018 m³/kg			t_s = +10.6 °C = 283.75 K i'' = 150.42 kcal/kg = 629.78 kJ/kg s'' = 1.1781 kcal/kg K = 4.9325 kJ/kg K			ρ'' = 28.977 kg/m³ v'' = 0.03451 m³/kg		
t	T	ρ	v	i		s		ρ	v	i		s	
°C	K	kg/m³	m³/kg	kcal/kg	kJ/kg	kcal/kg K	kJ/kg K	kg/m³	m³/kg	kcal/kg	kJ/kg	kcal/kg K	kJ/kg K
10	283.15	24.295	0.04116	150.72	631.03	1.1823	4.9501	—	—	—	—	—	—
15	288.15	23.736	0.04213	151.50	634.30	1.1850	4.9614	28.313	0.03532	151.15	632.83	1.1806	4.9429
20	293.15	23.213	0.04308	152.29	637.61	1.1877	4.9727	27.617	0.03621	151.97	636.27	1.1833	4.9542
25	298.15	22.712	0.04403	153.09	640.96	1.1904	4.9840	26.983	0.03706	152.78	639.66	1.1860	4.9655
30	303.15	22.232	0.04498	153.88	644.26	1.1930	4.9949	26.385	0.03790	153.69	643.05	1.1887	4.9768
35	308.15	21.782	0.04591	154.69	647.66	1.1957	5.0062	25.820	0.03873	154.40	646.44	1.1914	4.9882
40	313.15	21.345	0.04685	155.49	651.01	1.1983	5.0170	25.278	0.03956	155.22	649.88	1.1940	4.9990
45	318.15	20.929	0.04778	156.30	654.40	1.2009	5.0279	24.765	0.04038	156.04	653.31	1.1966	5.0099
50	323.15	20.538	0.04869	157.12	657.83	1.2034	5.0384	24.272	0.04120	156.86	656.74	1.1992	5.0208
55	328.15	20.157	0.04961	157.94	661.26	1.2060	5.0493	23.804	0.04201	157.69	660.22	1.2018	5.0317
60	333.15	19.794	0.05052	158.77	664.74	1.2085	5.0597	23.364	0.04280	158.53	663.73	1.2043	5.0422
65	338.15	19.444	0.05143	159.60	668.21	1.2110	5.0702	22.936	0.04360	159.37	667.25	1.2068	5.0526
70	343.15	19.109	0.05223	160.44	671.73	1.2134	5.0803	22.528	0.04439	160.21	670.77	1.2093	5.0631
75	348.15	18.786	0.05323	161.28	675.25	1.2159	5.0907	22.134	0.04518	161.06	674.33	1.2118	5.0736
80	353.15	18.474	0.05413	162.12	678.76	1.2183	5.1008	21.758	0.04596	161.91	677.88	1.2142	5.0836
85	358.15	18.172	0.05503	162.97	682.32	1.2207	5.1108	21.390	0.04675	162.76	681.44	1.2167	5.0941
90	363.15	17.879	0.05593	163.83	685.92	1.2231	5.1209	21.039	0.04753	163.62	685.04	1.2191	5.1041
95	368.15	17.599	0.05682	164.69	689.52	1.2254	5.1305	20.704	0.04830	164.49	688.69	1.2214	5.1138
100	373.15	17.328	0.05771	165.56	693.17	1.2278	5.1406	20.375	0.04908	165.36	692.33	1.2238	5.1238
105	378.15	17.065	0.05860	166.44	696.85	1.2301	5.1502	20.060	0.04985	166.24	696.01	1.2262	5.1339
110	383.15	16.812	0.05948	167.34	700.62	1.2324	5.1598	19.755	0.05062	167.15	699.82	1.2285	5.1435
115	388.15	16.567	0.06036	168.21	704.26	1.2347	5.1694	19.459	0.05139	168.02	703.47	1.2308	5.1531
120	393.15	16.329	0.06124	169.10	707.99	1.2370	5.1791	19.175	0.05215	168.92	707.23	1.2331	5.1627
125	398.15	16.098	0.06212	170.00	711.76	1.2393	5.1887	18.900	0.05291	169.82	711.00	1.2354	5.1724
130	403.15	15.876	0.06299	170.90	715.52	1.2416	5.1983	18.632	0.05367	170.73	714.81	1.2377	5.1820
135	408.15	15.659	0.06386	171.82	719.38	1.2438	5.2075	18.372	0.05443	171.65	718.66	1.2400	5.1916
140	413.15	15.449	0.06473	172.73	723.19	1.2461	5.2172	18.123	0.05518	172.57	722.52	1.2422	5.2008
145	418.15	15.244	0.06560	173.66	727.08	1.2483	5.2264	17.879	0.05593	173.50	726.41	1.2445	5.2105
150	423.15	15.047	0.06646	174.59	730.97	1.2505	5.2356	17.643	0.05668	174.43	730.30	1.2467	5.2197
155	428.15	14.852	0.06733	175.53	734.91	1.2527	5.2448	17.413	0.05743	175.37	734.24	1.2489	5.2289
160	433.15	14.667	0.06818	176.47	738.84	1.2549	5.2540	17.191	0.05817	176.32	738.22	1.2511	5.2381
165	438.15	14.484	0.06904	177.42	742.82	1.2571	5.2632	16.975	0.05891	177.27	742.19	1.2533	5.2473
170	443.15	14.308	0.06989	178.36	746.46	1.2583	5.2724	16.764	0.05965	178.23	746.21	1.2555	5.2565
175	448.15	14.134	0.07075	179.30	750.69	1.2615	5.2816	16.562	0.06038	719.19	750.23	1.2577	5.2657

1 at = 1 kp/cm² = 98 066.5 Pa = 98 066.5 N/m² = 9.806 65 N/cm² = 0.980 665 bar   1 kcal = 4.186 8 kJ

Table 61-20 Properties of Superheated Chlorodifluoromethane "Freon 22" ($CHClF_2$) (Continued)

Pressure p		8.0 at = 7.845 320 0 bar						9.0 at = 8.825 985 0 bar					
Temperature		$t_s = +14.6\,°C = 287.75\,K$ $\rho'' = 33.102\,kg/m^3$ $i'' = 150.78\,kcal/kg = 631.29\,kJ/kg$ $v'' = 0.03021\,m^3/kg$ $s'' = 1.1768\,kcal/kg\,K = 4.9270\,kJ/kg\,K$						$t_s = +18.7\,°C = 291.85\,K$ $\rho'' = 37.300\,kg/m^3$ $i'' = 151.09\,kcal/kg = 632.58\,kJ/kg$ $v'' = 0.02681\,m^3/kg$ $s'' = 1.1755\,kcal/kg\,K = 4.9216\,kJ/kg\,K$					
t	T	ρ	v	i		s		ρ	v	i		s	
°C	K	kg/m³	m³/kg	kcal/kg	kJ/kg	kcal/kg K	kJ/kg K	kg/m³	m³/kg	kcal/kg	kJ/kg	kcal/kg K	kJ/kg K
15	288.15	33.025	0.03028	150.85	631.58	1.1770	4.9279	—	—	—	—	—	—
20	293.15	32.196	0.03106	151.65	634.93	1.1798	4.9396	37.037	0.02700	151.33	633.59	1.1763	4.9249
25	298.15	31.407	0.03184	152.47	638.36	1.1825	4.9509	36.075	0.02772	152.15	637.02	1.1790	4.9362
30	303.15	30.684	0.03259	153.29	641.79	1.1852	4.9622	35.125	0.02847	152.98	640.50	1.1828	4.9521
35	308.15	29.994	0.03334	154.11	645.23	1.1879	4.9735	34.223	0.02922	153.81	643.97	1.1845	4.9593
40	313.15	29.343	0.03408	154.93	648.66	1.1905	4.9844	33.434	0.02991	154.64	647.45	1.1872	4.9706
45	318.15	28.719	0.03482	155.76	652.14	1.1931	4.9953	32.690	0.03059	155.48	650.96	1.1898	4.9815
50	323.15	28.129	0.03555	156.60	655.65	1.1948	5.0024	31.990	0.03126	156.32	654.48	1.1913	4.9877
55	328.15	27.563	0.03628	157.43	659.13	1.1983	4.0170	31.348	0.03190	157.17	658.04	1.1950	5.0032
60	333.15	27.027	0.03700	158.28	662.69	1.2009	5.0279	30.703	0.03257	158.02	661.60	1.1976	5.0141
65	338.15	26.518	0.03771	159.12	666.20	1.2034	5.0384	30.120	0.03320	158.87	665.16	1.2002	5.0250
70	343.15	26.021	0.03843	159.97	669.76	1.2059	5.0489	29.525	0.03387	159.73	668.76	1.2027	5.0355
75	348.15	25.556	0.03913	160.83	673.36	1.2084	5.0593	28.986	0.03450	160.59	672.36	1.2052	5.0459
80	353.15	25.107	0.03983	161.68	676.92	1.2108	5.0694	28.458	0.03514	161.45	675.96	1.2077	5.0564
85	358.15	24.673	0.04053	162.54	680.52	1.2133	5.0798	27.949	0.03578	162.32	679.60	1.2101	5.0664
90	363.15	24.260	0.04122	163.41	686.16	1.2157	5.0899	27.465	0.03641	163.20	683.29	1.2126	5.0769
95	368.15	23.861	0.04191	164.29	687.85	1.2181	5.0999	27.005	0.03703	164.08	686.97	1.2130	5.0786
100	373.15	23.474	0.04260	165.16	691.49	1.2204	5.1096	26.539	0.03768	164.96	690.65	1.2184	5.1012
105	378.15	23.105	0.04328	166.05	695.22	1.2228	5.1196	26.110	0.03830	165.85	694.38	1.2198	5.1071
110	383.15	22.748	0.04396	166.96	699.03	1.2252	5.1297	25.641	0.03900	166.76	698.19	1.2221	5.1167
115	388.15	22.396	0.04465	167.83	702.67	1.2275	5.1393	25.265	0.03958	167.64	701.88	1.2214	5.1263
120	393.15	22.065	0.04532	168.73	706.44	1.2298	5.1489	24.876	0.04020	168.55	705.69	1.2268	5.1364
125	398.15	21.739	0.04600	169.64	710.25	1.2321	5.1586	24.510	0.04080	169.46	709.50	1.2291	5.1460
130	403.15	21.427	0.04667	170.55	714.06	1.2344	5.1682	24.160	0.04139	170.38	713.35	1.2314	5.1556
135	408.15	21.119	0.04735	171.47	717.91	1.2366	5.1774	23.827	0.04197	171.30	717.20	1.2336	5.1648
140	413.15	20.825	0.04802	172.40	721.80	1.2389	5.1870	23.485	0.04258	172.23	721.09	1.2359	5.1745
145	418.15	20.538	0.04869	173.33	725.70	1.2412	5.1967	23.159	0.04318	173.17	725.03	1.2382	5.1841
150	423.15	20.263	0.04935	174.27	729.63	1.2434	5.2059	22.841	0.04378	174.11	728.96	1.2404	5.1933
155	428.15	19.992	0.05002	175.22	733.61	1.2456	5.2151	22.538	0.04437	175.06	732.94	1.2427	5.2029
160	433.15	19.732	0.05068	176.17	737.59	1.2478	5.2243	22.237	0.04497	176.01	736.92	1.2449	5.2121
165	438.15	19.478	0.05134	177.13	741.61	1.2500	5.2335	21.949	0.04556	176.97	740.94	1.2471	5.2214
170	443.15	19.231	0.05200	178.10	745.67	1.2522	5.2427	21.664	0.04616	177.93	744.96	1.2493	5.2306
175	448.15	18.990	0.05266	179.07	749.73	1.2544	5.2519	21.395	0.04674	178.90	749.02	1.2515	5.2398
180	453.15	18.755	0.05332	180.05	753.83	1.2566	5.2611	21.133	0.04732	179.88	753.12	1.2537	5.2490
185	458.15	18.525	0.05398	181.02	757.89	1.2588	5.2703	20.859	0.04794	180.86	757.22	1.2558	5.2578

1 at = 1 kp/cm² = 98 066.5 Pa = 98 066.5 N/m² = 9.806 65 N/cm² = 0.980 665 bar   1 kcal = 4.186 8 kJ

Table 61-21 Properties of Superheated Chlorodifluoromethane "Freon 22" (CHClF$_2$) (Continued)

Pressure p		10.0 at = 9.806 650 bar						12.0 at = 11.767 980 bar					
Temperature		t_s = +22.4 °C = 295.55 K ρ'' = 41.545 kg/m³						t_s = +29.4 °C = 302.55 K ρ'' = 50.277 kg/m³					
		i'' = 151.34 kcal/kg = 633.63 kJ/kg v'' = 0.02407 m³/kg						i'' = 151.73 kcal/kg = 635.26 kJ/kg v'' = 0.01989 m³/kg					
		s'' = 1.1744 kcal/kg K = 4.9170 kJ/kg K						s'' = 1.1724 kcal/kg K = 4.9086 kJ/kg K					
t	T	ρ	v	i		s		ρ	v	i		s	
°C	K	kg/m³	m³/kg	kcal/kg	kJ/kg	kcal/kg K	kJ/kg K	kg/m³	m³/kg	kcal/kg	kJ/kg	kcal/kg K	kJ/kg K
25	298.15	40.950	0.02442	151.82	635.64	1.1753	4.9207	—	—	—	—	—	—
30	303.15	39.888	0.02507	152.66	639.16	1.1781	4.9325	50.025	0.01999	151.84	635.72	1.1728	4.9103
35	308.15	38.865	0.02573	153.49	642.63	1.1809	4.9442	48.567	0.02059	152.77	639.62	1.1756	4.9220
40	313.15	37.908	0.02638	154.34	646.19	1.1837	4.9559	47.214	0.02118	153.68	643.43	1.1785	4.9341
45	318.15	37.037	0.02700	155.18	649.71	1.1864	4.9672	45.977	0.02175	154.56	647.11	1.1812	4.9454
50	323.15	36.193	0.02763	156.04	653.31	1.1891	4.9785	44.823	0.02231	155.43	650.75	1.1840	4.9572
55	328.15	35.411	0.02824	156.89	656.87	1.1918	4.9898	43.764	0.02285	156.31	654.44	1.1867	4.9685
60	333.15	34.662	0.02885	157.75	660.47	1.1944	5.0007	42.753	0.02339	157.19	658.12	1.1894	4.9798
65	338.15	33.956	0.02945	158.61	664.07	1.1970	5.0116	41.806	0.02392	158.08	661.85	1.1920	4.9907
70	343.15	33.278	0.03005	159.47	667.67	1.1996	5.0225	40.900	0.02445	158.96	665.53	1.1946	5.0016
75	348.15	32.637	0.03064	160.35	671.35	1.2021	5.0330	40.048	0.02497	159.85	669.26	1.1972	5.0124
80	353.15	32.031	0.03122	161.22	675.00	1.2047	5.0438	39.231	0.02549	160.74	672.99	1.1998	5.0233
85	358.15	31.447	0.03180	162.10	678.68	1.2072	5.0543	38.462	0.02600	161.63	676.71	1.2023	5.0338
90	363.15	30.893	0.03237	162.98	682.36	1.2096	5.0644	37.736	0.02650	162.52	680.44	1.2048	5.0443
95	368.15	30.340	0.03296	163.86	686.05	1.2121	5.0748	37.037	0.02700	163.42	684.21	1.2072	5.0543
100	373.15	29.824	0.03353	164.75	689.78	1.2145	5.0849	36.364	0.02750	164.79	689.94	1.2096	5.0644
105	378.15	29.326	0.03410	165.65	693.54	1.2169	5.0949	35.740	0.02798	165.23	691.78	1.2121	5.0748
110	383.15	28.852	0.03466	166.56	697.35	1.2193	5.1050	35.125	0.02847	166.15	695.64	1.2145	5.0849
115	388.15	28.393	0.03522	167.45	701.08	1.2216	5.1146	34.542	0.02895	167.06	699.45	1.2168	5.0945
120	393.15	27.949	0.03578	168.36	704.89	1.2239	5.1242	33.979	0.02943	167.98	703.30	1.2192	5.1045
125	398.15	27.518	0.03634	169.27	708.70	1.2263	5.1343	33.434	0.02991	168.91	707.19	1.2215	5.1142
130	403.15	27.100	0.03690	170.20	712.59	1.2286	5.1439	32.916	0.03038	169.84	711.09	1.2239	5.1242
135	408.15	26.702	0.03745	171.12	716.45	1.2309	5.1535	32.415	0.03085	170.77	714.98	1.2262	5.1339
140	413.15	26.316	0.03800	172.06	720.38	1.2332	5.1632	31.939	0.03131	171.71	718.92	1.2285	5.1435
145	418.15	25.940	0.03855	173.00	724.32	1.2355	5.1728	31.466	0.03178	172.66	722.89	1.2307	5.1527
150	423.15	25.575	0.03910	173.94	728.25	1.2377	5.1820	31.027	0.03223	173.61	726.87	1.2330	5.1623
155	428.15	25.227	0.03964	174.90	732.27	1.2400	5.1916	30.590	0.03269	174.52	730.68	1.2353	5.1720
160	433.15	24.888	0.04018	175.85	736.25	1.2422	5.2008	30.175	0.03314	175.53	734.91	1.2375	5.1812
165	438.15	24.558	0.04072	176.81	740.27	1.2444	5.2101	29.771	0.03359	176.51	739.01	1.2397	5.1904
170	443.15	24.237	0.04126	177.78	744.33	1.2466	5.2193	29.377	0.03404	177.48	743.07	1.2420	5.2000
175	448.15	23.929	0.04179	178.76	748.43	1.2488	5.2285	29.002	0.03448	178.46	747.18	1.2442	5.2092
180	453.15	23.624	0.04233	179.74	752.54	1.2510	5.2377	28.637	0.03492	179.45	751.32	1.2464	5.2184
185	458.15	23.332	0.04286	180.72	756.64	1.2532	5.2469	28.289	0.03535	180.43	755.42	1.2485	5.2272
190	463.15	23.052	0.04338	181.71	760.78	1.2553	5.2557	27.941	0.03579	181.43	759.61	1.2507	5.2364
195	468.15	22.826	0.04381	182.69	764.89	1.2575	5.2649	27.609	0.03622	182.42	763.76	1.2529	5.2456

1 at = 1 kp/cm² = 98 066.5 Pa = 98 066.5 N/m² = 9.806 65 N/cm² = 0.980 665 bar   1 kcal = 4.186 8 kJ

Table 61-22 Properties of Superheated Chlorodifluoromethane "Freon 22" (CHClF$_2$) (Continued)

Pressure p		14.0 at = 13.729 310 bar						16.0 at = 15.690 640 bar					
Temperature		$t_s = +35.1\,°C = 308.25\,K$ $\rho'' = 59.277\,kg/m^3$ $i'' = 152.04\,kcal/kg = 636.56\,kJ/kg$ $v'' = 0.01687\,m^3/kg$ $s'' = 1.1704\,kcal/kg\,K = 4.9002\,kJ/kg\,K$						$t_s = +40.5\,°C = 313.65\,K$ $\rho'' = 68.540\,kg/m^3$ $i'' = 152.21\,kcal/kg = 637.27\,kJ/kg$ $v'' = 0.01459\,m^3/kg$ $s'' = 1.1686\,kcal/kg\,K = 4.8927\,kJ/kg\,K$					
t	T	ρ	v	i		s		ρ	v	i		s	
°C	K	m³/kg	kg/m³	kcal/kg	kJ/kg	kcal/kg K	kJ/kg K	m³/kg	kg/m³	kcal/kg	kJ/kg	kcal/kg K	kJ/kg K
40	313.15	57.504	0.01739	152.96	640.41	1.1733	4.9124	—	—	—	—	—	—
45	318.15	55.772	0.01793	153.86	644.18	1.1762	4.9245	66.445	0.01505	153.08	640.92	1.1712	4.9036
50	323.15	54.142	0.01847	154.76	647.95	1.1790	4.9362	64.309	0.01555	154.03	644.89	1.1742	4.9161
55	328.15	52.715	0.01897	155.68	651.80	1.1818	4.9480	62.383	0.01603	154.98	648.87	1.1771	4.9283
60	333.15	51.387	0.01946	156.58	655.57	1.1846	4.9597	60.643	0.01649	155.92	652.81	1.1800	4.9404
65	338.15	50.125	0.01995	157.50	659.42	1.1873	4.9710	58.997	0.01695	156.87	656.78	1.1828	4.9521
70	343.15	48.948	0.02043	158.41	663.23	1.1900	4.9823	57.471	0.01740	157.45	659.21	1.1856	4.9639
75	348.15	47.847	0.02090	159.31	667.00	1.1926	4.9932	56.054	0.01784	158.73	664.57	1.1883	4.9752
80	353.15	46.795	0.02137	160.22	670.81	1.1952	5.0041	54.705	0.01828	159.68	668.55	1.1910	4.9865
85	358.15	45.809	0.02183	161.13	674.62	1.1978	5.0149	53.447	0.01871	160.61	672.44	1.1937	4.9978
90	363.15	44.863	0.02229	162.04	678.43	1.2003	5.0254	52.274	0.01913	161.54	676.34	1.1963	5.0087
95	368.15	43.975	0.02274	162.96	682.28	1.2028	5.0359	51.177	0.01954	162.47	680.23	1.1988	5.0191
100	373.15	43.141	0.02318	163.88	686.13	1.2053	5.0464	50.150	0.01994	163.41	684.16	1.2013	5.0296
105	378.15	42.337	0.02362	164.80	689.98	1.2077	5.0564	49.164	0.02034	164.35	688.10	1.2038	5.0401
110	383.15	41.580	0.02405	165.73	693.88	1.2102	5.0669	48.239	0.02073	165.30	692.08	1.2063	5.0505
115	388.15	40.850	0.02448	166.66	697.77	1.2125	5.0765	47.348	0.02112	166.24	696.01	1.2087	5.0606
120	393.15	40.161	0.02490	167.59	701.67	1.2149	5.0865	46.512	0.02150	167.19	699.99	1.2112	5.0711
125	398.15	39.494	0.02532	168.52	705.56	1.2173	5.0966	45.704	0.02188	168.13	703.93	1.2136	5.0811
130	403.15	38.850	0.02574	169.47	709.54	1.2197	5.1066	44.924	0.02226	169.09	707.95	1.2160	5.0911
135	408.15	38.241	0.02615	170.41	713.47	1.2220	5.1163	44.189	0.02263	170.04	711.92	1.2183	5.1008
140	413.15	37.651	0.02656	171.36	717.45	1.2243	5.1259	43.478	0.02300	171.00	715.94	1.2206	5.1104
145	418.15	37.092	0.02696	172.31	721.43	1.2266	5.1355	42.808	0.02336	171.96	719.96	1.2230	5.1205
150	423.15	36.536	0.02737	173.28	725.49	1.2289	5.1452	42.159	0.02372	172.94	724.07	1.2253	5.1301
155	428.15	36.023	0.02776	174.14	729.09	1.2312	5.1548	41.528	0.02408	173.76	727.50	1.2276	5.1397
160	433.15	35.511	0.02816	175.21	733.57	1.2334	5.1640	40.933	0.02443	174.89	732.23	1.2299	5.1493
165	438.15	35.026	0.02855	176.19	737.67	1.2357	5.1736	40.355	0.02478	175.87	736.33	1.2321	5.1586
170	443.15	34.554	0.02894	177.17	741.78	1.2379	5.1828	39.809	0.02512	176.86	740.48	1.2344	5.1682
175	448.15	34.095	0.02933	178.16	745.92	1.2401	5.1921	39.277	0.02546	177.86	744.66	1.2366	5.1774
180	453.15	33.659	0.02971	179.15	750.07	1.2423	5.2013	38.760	0.02580	178.85	748.81	1.2388	5.1866
185	458.15	33.234	0.03009	180.15	754.25	1.2445	5.2105	38.270	0.02613	179.85	753.00	1.2411	5.1962
190	463.15	32.830	0.03046	181.15	758.44	1.2467	5.2197	37.793	0.02646	180.87	757.27	1.2433	5.2054
195	468.15	32.436	0.03083	182.16	762.67	1.2488	5.2285	37.327	0.02679	181.88	761.50	1.2455	5.2147
200	473.15	32.051	0.03120	183.17	766.90	1.2509	5.2373	36.887	0.02711	182.89	765.72	1.2477	5.2239
205	478.15	31.686	0.03156	184.19	771.17	1.2530	5.2461	36.456	0.02743	183.91	769.99	1.2498	5.2327
210	483.15	31.319	0.03193	185.20	775.40	1.2551	5.2549	36.049	0.02774	184.92	774.22	1.2520	5.2419

1 at = 1 kp/cm² = 98 066.5 Pa = 98 066.5 N/m² = 9.806 65 N/cm² = 0.980 665 bar

1 kcal = 4.186 8 kJ

Pressure p			18 at = 17.651 970 bar					
Temperature			$t_s = +45.5\ °C = 318.65$ K $\quad\rho'' = 78.186$ kg/m^3 $i'' = 152.28$ kcal/kg = 637.57 kJ/kg $\quad v'' = 0.01279$ m^3/kg $s'' = 1.1666$ kcal/kg K = 4.8843 kJ/kg K					
t	T	ρ	v	i		s		
°C	K	kg/m³	m³/kg	kcal/kg	kJ/kg	kcal/kg K	kJ/kg K	
50	323.15	75.301	0.01328	153.22	641.50	1.1696	4.8969	
55	328.15	72.834	0.01373	154.23	645.73	1.1727	4.9099	
60	333.15	70.522	0.01418	155.22	649.88	1.1757	4.9224	
65	338.15	68.399	0.01462	156.20	653.98	1.1787	4.9350	
70	343.15	66.445	0.01505	157.17	658.04	1.1816	4.9471	
75	348.15	64.641	0.01547	158.13	662.06	1.1844	4.9588	
80	353.15	62.972	0.01588	159.09	666.08	1.1871	4.9702	
85	358.15	61.425	0.01628	160.05	670.10	1.1899	4.9819	
90	363.15	59.988	0.01667	161.01	674.12	1.1925	4.9928	
95	368.15	58.651	0.01705	161.97	678.14	1.1951	5.0036	
100	373.15	57.372	0.01743	162.92	682.11	1.1977	5.0145	
105	378.15	56.211	0.01779	163.88	686.13	1.2013	5.0296	
110	383.15	55.127	0.01814	164.85	690.19	1.2028	5.0359	
115	388.15	54.083	0.01849	165.80	694.17	1.2052	5.0459	
120	393.15	53.135	0.01882	166.77	698.23	1.2077	5.0564	
125	398.15	52.056	0.01921	167.74	702.29	1.2102	5.0669	
130	403.15	51.125	0.01956	168.70	706.31	1.2125	5.0769	
135	408.15	50.251	0.01990	169.66	710.33	1.2150	5.0870	
140	413.15	49.432	0.02023	170.63	714.39	1.2173	5.0966	
145	418.15	48.614	0.02057	171.60	718.45	1.2197	5.1066	
150	423.15	47.870	0.02089	172.60	722.64	1.2220	5.1163	
155	428.15	47.125	0.02122	173.58	726.74	1.2243	5.1259	
160	433.15	46.447	0.02153	174.56	730.85	1.2266	5.1355	
165	438.15	45.767	0.02185	175.55	734.99	1.2289	5.1452	
170	443.15	45.126	0.02216	176.54	739.14	1.2312	5.1548	
175	448.15	44.524	0.02246	177.55	743.37	1.2335	5.1644	
180	453.15	43.937	0.02276	178.55	747.55	1.2357	5.1736	
185	458.15	43.365	0.02306	179.56	751.78	1.2379	5.1828	
190	463.15	42.827	0.02335	180.59	756.09	1.2402	5.1925	
195	468.15	42.301	0.02364	181.61	760.36	1.2423	5.2013	
200	473.15	41.806	0.02392	182.62	764.59	1.2445	5.2105	
205	478.15	41.322	0.02420	183.63	768.82	1.2467	5.2197	
210	483.15	40.866	0.02447	184.65	773.09	1.2489	5.2289	

1 at = 1 kp/cm² = 98 066.5 Pa = 98 066.5 N/m² = 9.806 65 N/cm² = 0.980 665 bar   1 kcal = 4.186 8 kJ

Table 62-1 Viscosities of Refrigerants at Saturation

Substance	Chemical formula	Saturation temperature		Viscosity			
				Liquid		Vapor	
		t_s	T_s	$\eta \times 10^6$		$\eta \times 10^6$	
		°C	°K	kp s/m²	Pa s	kp s/m²	Pa s
Ammonia	NH_3	− 20	253.15	25.799	253	1.111	10.90
		− 10	263.15	25.085	246	1.152	11.30
		0	273.15	24.371	239	1.203	11.80
		10	283.15	23.453	230	1.264	12.40
		20	293.15	22.332	219	1.315	12.90
Carbon dioxide	CO_2	− 20	253.15	12.237	120.0	—	—
		− 15	258.15	11.778	115.5	1.683	16.50
		− 10	263.15	11.339	111.2	1.703	16.70
		0	273.15	10.268	100.7	1.774	17.40
		10	283.15	8.861	86.9	1.866	18.30
		20	293.15	7.148	70.1	2.070	20.30
		30	303.15	4.844	47.5	2.396	23.50
		31	304.15	3.222	31.6	3.222	31.60
Dichlorodifluoromethane (Freon 12)	CF_2Cl_2	−150	123.15	99.32	973.996	0.910	8.924
		−140	133.15	92.27	904.860	0.920	9.022
		−130	143.15	85.68	840.234	0.931	9.130
		−120	153.15	79.47	779.334	0.943	9.248
		−110	163.15	73.63	722.064	0.956	9.375
		−100	173.15	68.15	668.323	0.970	9.512
		− 90	183.15	62.99	617.721	0.985	9.660
		− 80	193.15	58.15	570.257	1.003	9.836
		− 70	203.15	53.63	525.931	1.021	10.013
		− 60	213.15	49.37	484.154	1.042	10.219
		− 50	223.15	45.37	444.928	1.065	10.444
		− 40	233.15	41.80	409.918	1.086	10.650
		− 30	243.15	38.12	373.829	1.118	10.964
		− 20	253.15	34.82	341.468	1.150	11.278
		− 10	263.15	31.74	311.263	1.185	11.621
		0	273.15	28.84	282.824	1.225	12.013
		10	283.15	26.13	256.248	1.270	12.454
		20	293.15	23.59	231.339	1.321	12.955
		30	303.15	21.20	207.901	1.381	13.543
		40	313.15	18.96	185.934	1.450	14.220
		50	323.15	16.85	165.242	1.533	15.034
		60	333.15	14.85	145.639	1.634	16.024
		70	343.15	12.96	127.094	1.759	17.250
		80	353.15	11.14	109.246	1.922	18.848
		90	363.15	9.357	91.761	2.148	21.065
		100	373.15	7.535	73.893	2.507	24.585
		110	383.15	5.216	51.151	3.401	33.352
Methyl chloride	CH_3Cl	− 20	253.15	31.509	309	1.050	10.30
		− 10	263.15	30.693	301	1.101	10.80
		0	273.15	29.878	293	1.162	11.40
		10	283.15	28.654	281	1.234	12.10
		20	293.15	27.430	269	1.326	13.00
		30	303.15	26.920	264	—	—
Sulphur dioxide	SO_2	− 20	253.15	49.456	485	1.081	10.60
		− 10	263.15	44.561	437	1.152	11.30
		0	273.15	39.259	385	1.254	12.30
		10	283.15	33.752	331	1.377	13.50
		20	293.15	27.736	272	1.540	15.10
		30	303.15	—	—	1.723	16.90
		40	313.15	—	—	1.866	18.30

1 kp s/m² = 9.80665 Pa s = 9.80665 N s/m²

Table 63-1 Viscosities η of Ammonia (NH_3) at Various Pressures and Temperatures

| Pressure p | | −20 °C = 253.15 K | | −10 °C = 263.15 K | | 0 °C = 273.15 K | | 10 °C = 283.15 K | | 20 °C = 293.15 K | | 45 °C = 318.15 K | | 80 °C = 353.15 K | |
| | | $\eta \times 10^6$ | | $\eta \times 10^6$ | | $\eta \times 10^6$ | | $\eta \times 10^6$ | | $\eta \times 10^6$ | | $\eta \times 10^6$ | | $\eta \times 10^6$ | |
kp/cm²	bar	kp s/m²	Pa s	kp s/m²	Pa s	kp s/m²	Pa s	kp s/m²	Pa s	kp s/m²	Pa s	kp s/m²	Pa s	kp s/m²	Pa s
1	0.9807	0.88	8.630	0.92	9.022	0.95	9.316	0.99	9.709	1.02	10.003	1.12	10.983	1.24	12.160
2	1.9613	25.78	252.815	0.98	9.611	0.98	9.611	1.01	9.905	1.04	10.199	1.12	10.983	1.24	12.160
4	3.9227	26.47	259.582	25.52	250.266	1.09	10.689	1.06	10.395	1.09	10.689	1.13	11.082	1.25	12.258
6	5.8840	27.04	265.172	26.16	256.542	24.98	244.970	1.20	11.768	1.16	11.376	1.15	11.278	1.26	12.356
8	7.8453	27.52	269.879	26.67	261.543	25.58	250.854	24.10	236.340	1.27	12.454	1.19	11.670	1.29	12.651
10	9.8067	27.89	273.507	27.10	265.760	26.07	255.659	24.64	241.636	22.81	223.690	1.23	12.062	1.31	12.847
12	11.7680	28.20	276.548	27.44	269.094	26.45	259.386	25.08	245.951	23.33	228.789	1.30	12.749	1.36	13.337
14	13.7293	28.45	278.999	27.75	272.135	26.77	262.524	25.43	249.383	23.72	232.614	1.39	13.631	1.41	13.827
16	15.6906	28.66	281.059	27.98	274.390	27.02	264.976	25.70	252.031	24.03	235.654	1.49	14.612	1.49	14.612
18	17.6520	28.85	282.922	28.18	276.351	27.24	267.133	25.93	254.286	24.28	238.105	—	—	1.60	15.691
20	19.6133	29.02	284.589	28.36	278.117	27.42	268.898	26.13	256.248	24.49	240.165	—	—	1.73	16.966
22	21.5746	29.16	285.962	28.52	279.686	27.57	270.369	26.30	257.915	24.66	241.832	—	—	1.89	18.535
24	23.5360	29.28	287.139	28.64	280.862	27.70	271.644	26.45	259.386	24.82	243.401	—	—	—	—
26	25.4973	29.40	288.316	28.76	282.039	27.82	272.821	26.57	260.563	24.97	244.872	—	—	—	—

1 kp s/m² = 9.806 65 Pa s = 9.806 65 N s/m²

Table 64-1 Viscosities η of Methyl Chloride (CH_3Cl) at Various Pressures and Temperatures

| Pressure p | | −20 °C = 253.15 K | | −10 °C = 263.15 K | | 0 °C = 273.15 K | | 10 °C = 283.15 K | | 20 °C = 293.15 K | | 30 °C = 303.15 K | |
| | | $\eta \times 10^6$ | | $\eta \times 10^6$ | | $\eta \times 10^6$ | | $\eta \times 10^6$ | | $\eta \times 10^6$ | | $\eta \times 10^6$ | |
kp/cm²	bar	kp s/m²	Pa s	kp s/m²	Pa s	kp s/m²	Pa s	kp s/m²	Pa s	kp s/m²	Pa s	kp s/m²	Pa s
0.5	0.4903	0.89	8.728	0.96	9.414	0.99	9.709	1.04	10.199	1.09	10.689	1.13	11.082
1.0	0.9807	0.95	9.316	0.99	9.709	1.02	10.003	1.05	10.297	1.10	10.787	1.13	11.082
1.5	1.4710	31.77	311.557	1.05	10.297	1.05	10.297	1.06	10.395	1.10	10.787	1.14	11.180
2.0	1.9613	32.15	315.284	30.91	303.124	1.09	10.689	1.08	10.591	1.11	10.885	1.14	11.180
2.5	2.4517	32.48	318.520	31.24	306.360	1.14	11.180	1.11	10.885	1.13	11.082	1.15	11.278
3.0	2.9420	32.74	321.070	31.55	309.400	30.15	295.670	1.15	11.278	1.15	11.278	1.16	11.376
3.5	3.4323	32.97	323.325	31.80	311.851	30.42	298.318	1.21	11.866	1.18	11.572	1.18	11.572
4.0	3.9227	33.18	325.385	32.03	314.107	30.65	300.574	28.90	283.412	1.22	11.964	1.21	11.866
4.5	4.4130	33.36	327.150	32.23	316.068	30.87	302.731	29.15	285.864	1.27	12.454	1.24	12.160
5.0	4.9033	33.53	328.317	32.40	317.735	31.05	304.496	29.36	287.923	27.53	269.977	1.29	12.651
5.5	5.3937	33.66	330.092	32.54	319.108	31.18	305.771	29.55	289.787	27.72	271.840	1.36	13.337
6.0	5.8840	33.77	331.171	32.67	320.383	31.32	307.144	29.70	291.258	27.87	273.311	1.43	14.024
6.5	6.3743	33.85	331.955	32.78	321.462	31.42	308.125	29.82	292.434	27.98	274.390	1.54	15.102
7.0	6.8647	33.90	332.445	32.84	322.050	31.50	308.909	29.90	293.219	28.06	275.175	25.90	253.992

1 kp s/m² = 9.806 65 Pa s = 9.806 65 N s/m²

1 at = 1 kp/cm² = 98 066.5 Pa = 98 066.5 N/m² = 9.806 65 N/cm² = 0.980 665 bar
Values listed above the lines refer to the vapor and those under the line to the liquid

Table 65-1 Viscosities η of Sulfur Dioxide (SO₂) at Various Pressures and Temperatures

Pressure p		−20 °C = 253.15 K $\eta \times 10^6$		−10 °C = 263.15 K $\eta \times 10^6$		0 °C = 273.15 K $\eta \times 10^6$		10 °C = 283.15 K $\eta \times 10^6$		20 °C = 293.15 K $\eta \times 10^6$		30 °C = 303.15 K $\eta \times 10^6$		40 °C = 313.15 K $\eta \times 10^6$	
kp/cm²	bar	kp s/m²	Pa s	kp s/m²	Pa s	kp s/m²	Pa s	kp s/m²	Pa s	kp s/m²	Pa s	kp s/m²	Pa s	kp s/m²	Pa s
0.5	0.4903	1.07	10.394	1.11	10.885	1.15	11.278	1.21	11.866	1.26	12.356	1.32	12.945	1.39	13.631
1.0	0.9807	50.00	490.333	1.14	11.180	1.19	11.670	1.24	12.160	1.28	12.553	1.33	13.043	1.40	13.729
1.5	1.4710	50.60	496.216	45.20	443.160	1.24	12.160	1.28	12.553	1.31	12.847	1.34	13.141	1.41	13.827
2.0	1.9613	51.20	502.100	45.85	449.635	40.00	392.266	1.33	13.043	1.36	13.337	1.36	13.337	1.43	14.024
2.5	2.4517	51.63	506.317	46.33	454.342	40.57	397.856	33.95	332.936	1.41	13.827	1.38	13.533	1.45	14.220
3.0	2.9420	52.00	509.946	46.80	458.951	41.05	402.563	34.48	338.133	1.48	14.514	1.42	13.925	1.47	14.416
3.5	3.4323	52.35	512.888	47.20	462.874	41.45	406.486	34.95	342.742	27.94	273.998	1.47	14.416	1.50	14.710
4.0	3.9227	52.55	515.339	47.55	466.306	41.83	410.212	35.35	346.665	28.40	278.509	1.55	15.200	1.54	15.102
4.5	4.4130	52.82	517.987	47.86	469.346	42.13	413.154	35.72	350.294	28.85	282.922	1.66	16.279	1.59	15.593
5.0	4.9033	53.02	519.949	48.15	472.190	42.43	416.096	36.00	353.039	29.24	286.746	—	—	1.65	16.181
5.5	5.3937	53.20	521.714	48.41	474.740	42.70	418.744	36.28	355.785	29.55	289.787	—	—	1.72	16.867
6.0	5.8840	53.35	523.185	48.65	477.094	42.92	420.901	36.50	357.943	29.88	293.023	—	—	1.80	17.652
6.5	6.3743	53.48	524.460	48.85	479.055	43.13	422.961	36.72	360.100	30.12	295.376	—	—	—	—
7.0	6.8647	53.62	525.833	49.05	481.016	43.32	424.824	36.90	361.865	30.37	297.828	—	—	—	—
7.5	7.3550	53.74	527.009	49.25	482.978	43.50	426.589	37.06	363.434	30.60	300.083	—	—	—	—
8.0	7.8453	53.85	528.088	49.44	484.841	43.65	428.060	37.25	365.298	30.80	302.045	—	—	—	—

1 at = 1 kp/cm² = 98 066.5 Pa = 98 066.5 N/m² = 9.806 65 N/cm² = 0.980 665 bar

1 kp s/m² = 9.806 65 Pa s = 9.806 65 N s/m²

Values listed above the lines refer to the vapor and those under the line to the liquid

Table 66-1 Viscosities η of Carbon Dioxide (CO₂) at Various Pressures and Temperatures

Pressure p		−15 °C = 258.15 K $\eta \times 10^6$		−10 °C = 263.15 K $\eta \times 10^6$		0 °C = 273.15 K $\eta \times 10^6$		10 °C = 283.15 K $\eta \times 10^6$		20 °C = 293.15 K $\eta \times 10^6$		30 °C = 303.15 K $\eta \times 10^6$		40 °C = 313.15 K $\eta \times 10^6$	
kp/cm²	bar	kp s/m²	Pa s	kp s/m²	Pa s	kp s/m²	Pa s	kp s/m²	Pa s	kp s/m²	Pa s	kp s/m²	Pa s	kp s/m²	Pa s
5	4.9033	1.38	13.533	1.40	13.729	1.42	13.925	1.45	14.220	1.49	14.612	1.52	14.906	1.60	15.691
10	9.8067	1.43	14.024	1.45	14.220	1.45	14.220	1.47	14.416	1.51	14.808	1.54	15.102	1.62	15.887
15	14.7100	1.50	14.710	1.50	14.710	1.49	14.612	1.48	14.514	1.53	15.004	1.56	15.298	1.64	16.083
20	19.6133	1.58	15.495	1.56	15.298	1.53	15.004	1.51	14.808	1.55	15.200	1.58	15.495	1.67	16.377
25	24.5166	11.85	116.209	1.64	16.083	1.57	15.396	1.55	15.200	1.58	15.495	1.62	15.887	1.69	16.573
30	29.4200	12.04	118.072	11.44	112.188	1.63	15.985	1.59	15.593	1.61	15.789	1.65	16.181	1.72	16.867
35	34.3233	12.24	120.033	11.66	114.346	1.76	17.260	1.63	15.945	1.65	16.181	1.69	16.573	1.76	17.260
40	39.2266	12.33	120.916	11.85	116.209	10.56	103.558	1.71	16.769	1.70	16.671	1.73	16.966	1.80	17.652
45	44.1299	12.47	122.289	12.02	117.876	10.82	106.108	1.83	17.946	1.75	17.162	1.77	17.358	1.85	18.142
50	49.0333	12.60	123.564	12.16	119.249	11.06	108.462	9.14	89.633	1.82	17.848	1.82	17.848	1.90	18.633
55	53.9366	12.72	124.741	12.30	120.622	11.27	110.521	9.45	92.673	1.92	18.829	1.88	18.437	1.95	19.123
60	58.8399	12.82	125.721	12.43	121.897	11.47	112.482	9.73	95.419	7.27	71.294	1.96	19.221	2.01	19.711
65	63.7432	12.93	126.800	12.54	122.975	11.66	114.346	10.00	98.067	7.66	75.119	2.07	20.300	2.09	20.496
70	68.6466	13.04	127.879	12.66	124.152	11.83	116.013	10.22	100.224	8.01	78.551	2.24	21.967	2.18	21.378
75	73.5499	13.13	128.761	12.78	125.329	12.01	117.778	10.45	102.479	8.32	81.591	5.06	49.622	2.30	22.555
80	78.4532	13.23	129.742	12.90	126.506	12.16	119.249	10.66	104.539	8.60	84.337	5.61	55.015	2.47	24.222
85	83.3565	13.34	130.821	13.01	127.585	12.28	120.426	10.86	106.500	8.87	86.985	6.08	59.624	2.80	27.459
90	88.2599	13.44	131.801	13.12	128.663	12.43	121.897	11.06	108.462	9.10	89.241	6.51	63.841	3.32	32.558
95	93.1632	13.53	132.684	13.23	129.742	12.55	123.073	11.23	110.129	9.33	91.496	6.87	67.372	4.02	39.423
100	98.0665	13.62	133.567	13.33	130.723	12.67	124.250	11.41	111.894	9.54	93.555	7.21	70.706	4.70	46.091
105	102.9698	13.72	134.547	13.43	131.703	12.78	125.329	11.57	113.463	9.74	95.517	7.52	73.746	5.27	51.681
110	107.8732	13.81	135.430	13.53	132.684	12.90	126.506	11.72	114.934	9.93	97.380	7.82	76.688	5.72	56.094
115	112.7765	13.89	136.214	13.63	133.665	13.01	127.585	11.87	116.405	10.11	99.145	8.08	79.238	6.10	59.821
120	117.6798	13.97	136.999	13.72	134.547	13.11	128.565	12.01	117.778	10.28	100.812	8.35	81.886	6.44	63.155

1 at = 1 kp/cm² = 98 066.5 Pa = 98 066.5 N/m² = 9.806 65 N/cm² = 0.980 665 bar

1 kp s/m² = 9.806 65 Pa s = 9.806 65 N s/m²

Values listed above the lines refer to the vapor and those under the line to the liquid

239

Table 67-1 Properties of Saturated Moist Air

Temperature		Pressure		Density	at the pressure of 760 mm Hg				at the pressure of 735.5 mm Hg			
					Density	Moisture	Specific enthalpy		Density	Moisture	Specific enthalpy	
t	T	p	p	ρ	ρ'	x'	$i_{1+x'}$	$i_{1+x'}$	ρ'	x'	$i_{1+x'}$	$i_{1+x'}$
°C	K	kp/m²	Pa	kg/m³	kg/m³	kg/kg	kcal/kg	kJ/kg	kg/m³	kg/kg	kcal/kg	kJ/kg
−20	253.15	10.50	102.970	1.396	1.395	0.00063	−4.43	−18.548	1.3490	0.000654	−4.415	−18.485
−19	254.15	11.56	113.365	1.394	1.393	0.00070	−4.15	−17.375	1.3436	0.000720	−4.136	−17.317
−18	255.15	12.71	124.643	1.385	1.384	0.00077	−3.87	−16.203	1.3383	0.000792	−3.854	−16.136
−17	256.15	13.96	136.091	1.379	1.378	0.00085	−3.58	−14.989	1.3330	0.000870	−3.567	−14.934
−16	257.15	15.33	150.336	1.374	1.373	0.00093	−3.29	−13.775	1.3278	0.000955	−3.276	−13.716
−15	258.15	16.82	164.948	1.368	1.367	0.00101	−3.01	−12.602	1.3225	0.001048	−2.981	−12.481
−14	259.15	18.44	180.835	1.363	1.362	0.00111	−2.71	−11.346	1.3174	0.001149	−2.681	−11.225
−13	260.15	20.19	197.996	1.358	1.357	0.00122	−2.40	−10.048	1.3132	0.001258	−2.376	−9.948
−12	261.15	22.12	216.923	1.353	1.352	0.00134	−2.09	−8.750	1.3071	0.001379	−2.064	−8.642
−11	262.15	24.20	237.321	1.348	1.347	0.00146	−1.78	−7.453	1.3020	0.001509	−1.746	−7.310
−10	263.15	26.46	259.484	1.342	1.341	0.00160	−1.45	−6.071	1.2969	0.001650	−1.422	−5.954
−9	264.15	28.89	283.314	1.337	1.336	0.00175	−1.13	−4.731	1.2919	0.001802	−1.091	−4.568
−8	265.15	31.56	309.498	1.332	1.331	0.00191	−0.79	−3.008	1.2869	0.001969	−0.751	−3.144
−7	266.15	34.33	337.643	1.327	1.325	0.00208	−0.45	−1.884	1.2819	0.002149	−0.403	−1.687
−6	267.15	37.54	368.142	1.322	1.320	0.00227	−0.10	−0.419	1.2770	0.002344	−0.046	−0.193
−5	268.15	40.90	401.092	1.317	1.315	0.00247	0.26	1.089	1.2721	0.002554	0.320	1.340
−4	269.15	44.54	436.788	1.312	1.310	0.00269	0.64	2.680	1.2672	0.002783	0.697	2.918
−3	270.15	48.48	475.426	1.308	1.306	0.00294	1.08	4.522	1.2623	0.003030	1.086	4.547
−2	271.15	52.74	517.203	1.303	1.301	0.00319	1.41	5.903	1.2574	0.003298	1.487	6.226
−1	272.15	57.32	562.117	1.298	1.295	0.00347	1.82	7.620	1.2526	0.003586	1.900	7.955
0	273.15	62.28	610.758	1.293	1.290	0.00378	2.25	9.420	1.2478	0.003898	2.328	9.747
1	274.15	66.94	656.457	1.288	1.285	0.00407	2.66	11.137	1.2430	0.004192	2.745	11.493
2	275.15	71.93	705.392	1.284	1.281	0.00437	3.08	12.895	1.2382	0.004506	3.175	13.293
3	276.15	77.23	757.368	1.279	1.275	0.00470	3.52	14.738	1.2335	0.004841	3.618	15.148
4	277.15	82.89	812.873	1.275	1.271	0.00503	3.96	16.580	1.2288	0.005199	4.074	17.057
5	278.15	88.90	871.811	1.270	1.266	0.00540	4.42	18.605	1.2241	0.005579	4.544	19.025
6	279.15	95.30	934.574	1.265	1.261	0.00579	4.90	20.515	1.2194	0.005985	5.030	21.060
7	280.15	102.10	1001.259	1.261	1.256	0.00621	5.40	22.609	1.2148	0.006416	5.532	23.161
8	281.15	108.32	1072.063	1.256	1.251	0.00665	5.90	24.702	1.2101	0.006875	6.050	25.330
9	282.15	116.99	1147.280	1.252	1.247	0.00713	6.43	26.921	1.2054	0.007363	6.587	27.578
10	283.15	125.13	1227.106	1.248	1.242	0.00763	6.97	29.182	1.2008	0.007882	7.142	29.902
11	284.15	133.76	1311.738	1.243	1.237	0.00805	7.53	31.527	1.1962	0.008433	7.717	32.310
12	285.15	142.91	1401.468	1.239	1.232	0.00875	8.14	34.081	1.1916	0.009018	8.314	34.809
13	286.15	152.81	1496.593	1.235	1.228	0.00935	8.74	36.593	1.1871	0.009634	8.932	37.396
14	287.15	162.89	1597.405	1.230	1.223	0.00997	9.36	39.188	1.1824	0.010300	9.575	40.089
15	288.15	173.76	1704.004	1.226	1.218	0.0106	9.98	41.784	1.1778	0.010999	10.242	42.881
16	289.15	185.27	1816.878	1.222	1.214	0.0114	10.7	44.799	1.1732	0.011741	10.935	45.783
17	290.15	197.45	1936.323	1.217	1.208	0.0121	11.4	47.730	1.1686	0.012529	11.657	48.806
18	291.15	210.3	2012.238	1.213	1.204	0.0129	12.1	50.660	1.1641	0.013362	12.406	51.941
19	292.15	223.9	2195.709	1.209	1.200	0.0138	12.9	54.010	1.1595	0.014246	13.188	55.216

1 kp/m² = 9.806 65 Pa = 9.806 65 N/m²

Table 67-2 Properties of Saturated Moist Air (Continued)

Temperature t (°C)	T (K)	Pressure p (kp/m²)	p (Pa)	Density ρ (kg/m³)	at the pressure of 760 mm Hg Density ρ' (kg/m³)	Moisture x' (kg/kg)	Specific enthalpy i₁₊ₓ' (kcal/kg)	(kJ/kg)	at the pressure of 735.5 mm Hg Density ρ' (kg/m³)	Moisture x' (kg/kg)	Specific enthalpy i₁₊ₓ' (kcal/kg)	(kJ/kg)
20	293.15	238.3	2336.925	1.205	1.195	0.0147	13.8	57.778	1.1549	0.015184	14.003	58.628
21	294.15	253.4	2485.005	1.201	1.190	0.0156	14.6	61.127	1.1503	0.016171	14.848	62.166
22	295.15	269.4	2641.912	1.197	1.185	0.0166	15.3	64.058	1.1457	0.017221	15.726	65.842
23	296.15	286.3	2807.644	1.193	1.181	0.0177	16.2	67.826	1.1411	0.018333	16.656	69.735
24	297.15	304.1	2982.202	1.189	1.176	0.0188	17.2	72.013	1.1365	0.019508	17.618	73.763
25	298.15	322.9	3166.567	1.185	1.171	0.0200	18.1	75.781	1.1319	0.020755	18.625	77.979
26	299.15	342.6	3359.758	1.181	1.166	0.0214	19.2	80.387	1.1272	0.022066	19.673	82.367
27	300.15	363.4	3563.737	1.177	1.161	0.0226	20.2	84.573	1.1226	0.023456	20.769	87.069
28	301.15	385.3	3778.502	1.173	1.156	0.0240	21.3	89.179	1.1179	0.024926	21.916	91.758
29	302.15	408.3	4004.055	1.169	1.151	0.0256	22.5	94.203	1.1132	0.026477	23.114	96.774
30	303.15	432.5	4241.376	1.165	1.146	0.0272	23.8	99.646	1.1085	0.028118	24.367	102.020
31	304.15	458.0	4491.446	1.161	1.141	0.0288	25.0	104.670	1.1038	0.029855	25.681	107.521
32	305.15	484.7	4753.283	1.157	1.136	0.0306	26.3	110.113	1.0990	0.031684	27.053	113.266
33	305.15	512.8	5028.850	1.154	1.131	0.0325	27.7	115.974	1.0943	0.033620	28.491	119.286
34	307.15	542.3	5318.146	1.150	1.126	0.0344	29.2	122.255	1.0895	0.035665	29.999	125.600
35	308.15	573.3	5622.152	1.146	1.121	0.0366	30.8	128.953	1.0846	0.037828	31.580	132.219
36	309.15	605.7	5939.888	1.142	1.116	0.0388	32.4	135.652	1.0798	0.040104	33.232	139.136
37	310.15	639.8	6274.295	1.139	1.111	0.0411	34.0	142.351	1.0749	0.042516	34.970	146.412
38	311.15	675.5	6624.392	1.135	1.107	0.0435	35.7	149.469	1.0699	0.045060	36.791	154.037
39	312.15	712.9	6991.161	1.132	1.102	0.0460	37.6	157.424	1.0650	0.047746	38.702	162.038
40	313.15	752.0	7374.601	1.128	1.097	0.0488	39.6	165.797	1.0599	0.050578	40.713	170.457
41	314.15	793.0	7776.673	1.124	1.091	0.0517	41.6	174.171	1.0549	0.053573	42.819	179.275
42	315.15	836.0	8198.359	1.121	1.086	0.0548	43.7	182.963	1.0498	0.056743	45.036	188.557
43	316.15	880.9	8638.678	1.117	1.081	0.0580	45.9	192.174	1.0446	0.060085	47.361	198.291
44	317.15	927.9	9099.591	1.114	1.076	0.0613	48.3	202.222	1.0394	0.063619	49.807	208.532
45	318.15	977.1	9582.078	1.110	1.070	0.0650	50.8	212.689	1.0341	0.067357	52.384	219.321
46	319.15	1028.4	10085.16	1.107	1.065	0.0689	53.4	223.575	1.0288	0.071299	55.089	230.647
47	320.15	1082.1	10611.78	1.103	1.059	0.0728	56.2	235.298	1.0235	0.075474	57.941	242.587
48	321.15	1138.2	11161.93	1.100	1.054	0.0770	59.0	247.021	1.0180	0.079889	60.940	255.144
49	322.15	1196.7	11735.62	1.096	1.048	0.0815	62.1	260.000	1.0125	0.084553	64.109	268.412
50	323.15	1257.8	12334.80	1.093	1.043	0.0862	65.3	273.398	1.0069	0.089491	67.446	282.383
51	324.15	1321.6	12960.47	1.090	1.037	0.0913	68.6	287.214	1.0013	0.094722	70.969	297.133
52	325.15	1388.1	13612.61	1.086	1.031	0.0966	72.3	302.706	0.9956	0.100256	74.685	312.691
53	326.15	1457.5	14293.19	1.083	1.025	0.102	75.9	317.778	0.9898	0.106124	78.612	329.133
54	327.15	1529.8	15002.21	1.080	1.019	0.108	80.0	334.944	0.9839	0.112339	82.761	346.504
55	328.15	1605.1	15740.65	1.076	1.013	0.114	84.1	352.110	0.9779	0.118926	87.147	364.867
56	329.15	1683.5	16509.50	1.073	1.007	0.121	88.6	370.950	0.9719	0.125910	91.785	384.285
57	330.15	1765.3	17311.68	1.070	1.001	0.128	93.2	390.210	0.9657	0.133340	96.711	404.910
58	331.15	1850.4	18146.23	1.067	0.995	0.136	98.5	412.400	0.9595	0.141221	101.921	426.723
59	332.15	1939.0	19015.09	1.063	0.987	0.144	104	435.427	0.9532	0.149616	107.455	449.893

1 kp/m² = 9.806 65 Pa = 9.806 65 N/m²

Table 67-3 Properties of Saturated Moist Air (Continued)

Temperature t (°C)	T (K)	Pressure kp/m²	Pressure Pa	Density ρ (kg/m³)	760 mm Hg — Density ρ' (kg/m³)	760 mm Hg — Moisture x' (kg/kg)	760 mm Hg — $i_{1+x'}$ (kcal/kg)	760 mm Hg — $i_{1+x'}$ (kJ/kg)	735.5 mm Hg — Density ρ' (kg/m³)	735.5 mm Hg — Moisture x' (kg/kg)	735.5 mm Hg — $i_{1+x'}$ (kcal/kg)	735.5 mm Hg — $i_{1+x'}$ (kJ/kg)
60	333.15	2031	19917.31	1.060	0.981	0.152	109	456.361	0.9467	0.158524	113.320	474.448
61	334.15	2127	20858.74	1.057	0.974	0.161	115	481.482	0.9402	0.16804	119.573	500.628
62	335.15	2227	21839.41	1.054	0.968	0.171	121	506.603	0.9335	0.17821	126.240	528.542
63	336.15	2330	22849.49	1.051	0.961	0.181	128	535.910	0.9268	0.18895	133.279	558.013
64	337.15	2438	23908.61	1.048	0.954	0.192	135	565.218	0.9199	0.20053	140.851	589.715
65	338.15	2550	25006.96	1.044	0.946	0.204	143	598.712	0.9129	0.21296	148.924	623.515
66	339.15	2666	26144.53	1.041	0.939	0.216	151	632.207	0.9058	0.22610	157.534	659.563
67	340.15	2787	27331.13	1.038	0.932	0.230	160	669.888	0.8985	0.24033	166.797	698.346
68	341.15	2912	28556.96	1.035	0.924	0.244	169	707.569	0.8912	0.25554	176.688	739.757
69	342.15	3042	29831.83	1.032	0.917	0.259	179	749.437	0.8837	0.27194	187.339	784.351
70	343.15	3177	31155.73	1.029	0.909	0.276	190	795.492	0.8760	0.28962	198.823	832.432
71	344.15	3317	32528.66	1.026	0.901	0.294	202	845.734	0.8682	0.30872	211.202	884.261
72	345.15	3463	33960.43	1.023	0.893	0.314	214	895.975	0.8602	0.32951	224.662	940.615
73	346.15	3613	35431.43	1.020	0.885	0.335	227	950.404	0.8522	0.35185	239.122	1001.156
74	347.15	3769	36961.26	1.017	0.877	0.357	242	1013.206	0.8439	0.37623	254.887	1067.161
75	348.15	3931	38549.94	1.014	0.868	0.382	258	1080.194	0.8355	0.40288	272.100	1139.228
76	349.15	4098	40187.65	1.011	0.859	0.408	275	1151.370	0.8269	0.43188	290.824	1217.622
77	350.15	4272	41894.01	1.009	0.851	0.437	293	1226.732	0.8181	0.46389	311.475	1304.084
78	351.15	4451	43649.40	1.006	0.842	0.470	315	1318.842	0.8092	0.49892	334.062	1398.651
79	352.15	4637	45473.44	1.003	0.833	0.506	338	1415.138	0.8001	0.53780	359.114	1503.538
80	353.15	4829	47356.31	1.000	0.823	0.545	363	1519.808	0.7900	0.58086	386.859	1619.701
81	354.15	5028	49307.84	0.997	0.813	0.589	391	1637.039	0.7813	0.62901	417.852	1749.463
82	355.15	5234	51328.01	0.994	0.803	0.639	425	1779.390	0.7716	0.68308	452.645	1895.134
83	356.15	5447	53416.82	0.992	0.794	0.695	460	1925.928	0.7617	0.74413	491.917	2059.558
84	357.15	5667	55574.29	0.989	0.783	0.756	500	2093.400	0.7516	0.81350	536.517	2246.289
85	358.15	5894	57800.40	0.986	0.773	0.828	545	2281.806	0.7414	0.89286	587.528	2459.862
86	359.15	6129	60104.96	0.983	0.762	0.908	597	2499.520	0.7308	0.98482	646.623	2707.281
87	360.15	6372	62487.97	0.981	0.751	1.000	657	2750.728	0.7201	1.09244	715.757	2996.731
88	361.15	6623	64949.44	0.978	0.740	1.110	725	3035.430	0.7091	1.21987	797.598	3339.383
89	362.15	6882	67489.37	0.975	0.729	1.240	810	3391.308	0.6979	1.37287	895.835	3750.682
90	363.15	7149	70107.74	0.973	0.718	1.400	912	3818.362	0.6865	1.55969	1015.78	4252.868
91	364.15	7425	72814.38	0.970	0.706	1.590	1035	4333.338	0.6749	1.7935	1165.87	4881.265
92	365.15	7710	75609.27	0.967	0.694	1.830	1185	4961.358	0.6629	2.0942	1358.01	5685.716
93	366.15	8004	78492.43	0.965	0.681	2.135	1380	5777.784	0.6507	2.4942	1615.55	6763.985
94	367.15	8307	81463.84	0.962	0.669	2.546	1645	6887.286	0.6383	3.0520	1973.40	8262.231
95	368.15	8619	84523.52	0.959	0.656	3.120	2015	8436.402	0.6256	3.8820	2505.95	10491.911
96	369.15	8942	87691.06	0.957	0.643	3.990	2575	10781.010	0.6126	5.2570	3388.10	14185.297
97	370.15	9274	90946.87	0.954	0.630	5.450	3510	14695.668	0.5994	7.9455	5112.83	21406.397
98	371.15	9616	94300.75	0.951	0.616	8.350	5360	22441.248	0.5859	15.576	10007.8	41900.657
99	372.15	9969	97762.49	0.949	0.602	17.000	10910	45677.988	0.5721	200.023	12854.4	538188.019
100	373.15	10332	101322.31	0.947	0.589	—	—	—	—	—	—	—

1 kp/m² = 9.806 65 Pa = 9.806 65 N/m²

1 kcal = 4.186 8 kJ

Table 68-1 Properties of Superheated Steam (H₂O)

Pressure		Temperature		Density	Specific heat		Thermal conductivity		Viscosity		Kinematic viscosity
p		t	T	ρ	c_p		λ		$\eta \times 10^6$		$v \times 10^6$
kp/cm²	bar	°C	K	kg/m³	kcal/kg K	kJ/kg K̃	kcal/h m K	W/m K	kp s/m²	Pa s	m/s²
1	0.980 7	100	373.15	0.577	0.485	2.030	0.0204	0.0237	1.28	12.553	21.8
		120	393.15	0.547	0.477	1.997	0.0216	0.0251	1.36	13.337	24.4
		140	413.15	0.520	0.473	1.980	0.0228	0.0265	1.43	14.024	27.0
		160	433.15	0.494	0.471	1.972	0.0241	0.0280	1.51	14.808	30.0
		180	453.15	0.473	0.469	1.963	0.0253	0.0294	1.58	15.495	32.8
		200	473.15	0.452	0.469	1.963	0.0266	0.0309	1.66	16.279	36.0
		220	493.15	0.433	0.470	1.968	0.0278	0.0323	1.73	16.966	38.2
		240	513.15	0.416	0.471	1.972	0.0291	0.0338	1.81	17.750	42.7
		260	533.15	0.400	0.472	1.976	0.0305	0.0354	1.89	18.535	46.4
		280	553.15	0.386	0.474	1.985	0.0317	0.0369	1.96	19.221	49.8
		300	573.15	0.372	0.477	1.997	0.0331	0.0385	2.04	20.006	54.8
		320	593.15	0.359	0.480	2.010	0.0345	0.0401	2.11	20.692	58.7
		340	613.15	0.348	0.483	2.022	0.0358	0.0416	2.19	21.477	61.2
2	1.961 3	120	393.15	1.108	0.498	2.085	0.0228	0.0265	1.38	13.533	12.2
		140	413.15	1.048	0.489	2.047	0.0238	0.0277	1.45	14.220	13.6
		160	433.15	0.995	0.483	2.022	0.0249	0.0290	1.53	15.004	15.1
		180	453.15	0.950	0.479	2.005	0.0260	0.0302	1.60	15.691	16.5
		200	473.15	0.908	0.477	1.997	0.0272	0.0316	1.68	16.475	18.1
		220	493.15	0.870	0.477	1.997	0.0284	0.0330	1.75	17.162	19.7
		240	513.15	0.835	0.477	1.997	0.0297	0.0345	1.83	17.946	21.5
		260	533.15	0.803	0.478	2.001	0.0310	0.0361	1.90	18.633	23.2
		280	553.15	0.773	0.479	2.005	0.0322	0.0374	1.98	19.417	25.2
		300	573.15	0.745	0.481	2.014	0.0336	0.0391	2.05	20.104	27.0
		320	593.15	0.720	0.483	2.022	0.0350	0.0407	2.13	20.888	29.0
		340	613.15	0.696	0.485	2.031	0.0363	0.0422	2.20	21.575	31.0
4	3.992 1	160	433.15	2.02	0.512	2.143	0.0265	0.0308	1.56	15.298	7.57
		180	453.15	1.93	0.502	2.101	0.0273	0.0317	1.63	15.985	8.29
		200	473.15	1.84	0.495	2.072	0.0284	0.0330	1.71	16.769	9.11
		220	493.15	1.76	0.491	2.055	0.0294	0.0342	1.78	17.456	9.91
		240	513.15	1.68	0.488	2.043	0.0306	0.0356	1.86	18.240	10.9
		260	533.15	1.62	0.487	2.039	0.0318	0.0370	1.93	18.927	11.7
		280	553.15	1.55	0.487	2.039	0.0330	0.0384	2.01	19.711	12.7
		300	573.15	1.50	0.488	2.043	0.0343	0.0399	2.08	20.398	13.5
		320	593.15	1.45	0.489	2.047	0.0356	0.0414	2.15	21.084	14.5
		340	613.15	1.40	0.491	2.056	0.0369	0.0429	2.23	21.869	15.6
6	5.884 0	160	433.15	3.09	0.549	2.298	0.0283	0.0329	1.60	15.691	5.08
		180	453.15	2.93	0.528	2.218	0.0289	0.0336	1.67	16.377	5.59
		200	473.15	2.78	0.515	2.156	0.0296	0.0344	1.74	17.064	6.14
		220	493.15	2.66	0.506	2.118	0.0305	0.0355	1.82	17.848	6.71
		240	513.15	2.54	0.501	2.097	0.0315	0.0366	1.89	18.535	7.30
		260	533.15	2.44	0.498	2.085	0.0326	0.0379	1.96	19.221	7.87
		280	553.15	2.35	0.496	2.076	0.0338	0.0393	2.05	20.104	8.55
		300	573.15	2.26	0.495	2.072	0.0351	0.0408	2.11	20.692	9.06
		320	593.15	2.18	0.496	2.077	0.0364	0.0423	2.19	21.477	9.86
		340	613.15	2.10	0.496	2.077	0.0378	0.0440	2.26	22.163	10.05
8	7.845 3	180	453.15	3.96	0.561	2.348	0.0308	0.0358	1.72	16.867	4.26
		200	473.15	3.75	0.539	2.256	0.0314	0.0365	1.79	17.554	4.68
		220	493.15	3.58	0.524	2.193	0.0321	0.0373	1.87	18.338	5.12
		240	513.15	3.42	0.515	2.156	0.0329	0.0383	1.94	19.025	5.56
		260	533.15	3.28	0.509	2.131	0.0339	0.0394	2.01	19.711	6.01
		280	553.15	3.14	0.505	2.114	0.0349	0.0406	2.08	20.398	6.49
		300	573.15	3.03	0.503	2.106	0.0361	0.0420	2.16	21.182	6.99
		320	593.15	2.92	0.502	2.101	0.0374	0.0435	2.23	21.869	7.49
		340	613.15	2.82	0.502	2.101	0.0387	0.0450	2.30	22.555	8.00
10	9.806 7	180	453.15	5.04	0.606	2.537	0.0350	0.0407	1.80	17.652	3.50
		200	473.15	4.76	0.563	2.357	0.0343	0.0399	1.87	18.338	3.85
		220	493.15	4.52	0.540	2.260	0.0345	0.0401	1.94	19.025	4.21
		240	513.15	4.31	0.528	2.210	0.0351	0.0408	2.01	19.711	4.57
		260	533.15	4.12	0.521	2.181	0.0359	0.0418	2.09	20.496	4.97
		280	553.15	3.95	0.516	2.160	0.0369	0.0429	2.16	21.182	5.36
		300	573.15	3.80	0.514	2.152	0.0382	0.0444	2.23	21.869	5.76
		320	593.15	3.67	0.512	2.143	0.0394	0.0458	2.30	22.555	6.15
		340	613.15	3.54	0.511	2.139	0.0407	0.0473	2.37	23.242	6.56

1 at = 1 kp/cm² = 98 066.5 Pa = 98 066.5 N/m² = 9.806 65 N/cm² = 0.980 665 bar

1 kcal = 4.186 8 kJ
1 kcal/h = 1.163 0 W

Table 69-1 Coefficient λ of Thermal Conductivity of Steam (H₂O) at Various Pressures

Pressure p		1 at = 0.980 665 bar				20 at = 19.613 3 bar		40 at = 39.226 6 bar		60 at = 58.839 9 bar	
Temperature		λzas		λ		λ		λ		λ	
t °C	T K	kcal/h m K	W/m K	kcal/h m K	W/m K	kcal/h m K	W/m K	kcal/h m K	W/m K	kcal/h m K	W/m K
100	373.15	21.3	24.772	21.3	24.772	—	—	—	—	—	—
150	423.15	25.4	29.540	25.1	29.191	—	—	—	—	—	—
200	473.15	30.5	35.472	29.0	33.727	—	—	—	—	—	—
250	523.15	38.9	45.241	34.2	39.775	35.9	41.752	38.7	45.008	—	—
300	573.15	53.0	61.639	39.3	45.706	41.0	47.683	43.2	50.242	46.2	53.731
350	623.15	96.4	112.113	44.1	51.288	45.8	53.265	47.8	55.591	50.0	58.150
400	673.15	—	—	49.1	57.103	50.7	58.964	52.5	61.058	54.6	63.500
450	723.15	—	—	54.4	63.267	55.9	65.012	57.7	67.105	59.7	69.431
500	773.15	—	—	60.2	70.013	61.7	71.757	63.4	73.734	65.2	75.828
550	823.15	—	—	66.0	76.758	67.4	78.386	69.1	80.363	70.9	82.457
600	873.15	—	—	72.2	83.969	73.5	85.481	75.0	87.225	76.8	89.318

Table 69-2 Coefficient λ of Thermal Conductivity of Steam (H₂O) at Various Pressures (Continued)

Pressure p		80 at = 78.453 2 bar		100 at = 98.066 5 bar		150 at = 147.099 75 bar		200 at = 196.133 0 bar		250 at = 245.166 25 bar		300 at = 294.199 5 bar	
Temperature		λ		λ		λ		λ		λ		λ	
t °C	T K	kcal/h m K	W/m K	kcal/h m K	W/m K	kcal/h m K	W/m K	kcal/h m K	W/m K	kcal/h m K	W/m K	kcal/h m K	W/m K
350	623.15	52.9	61.523	56.7	65.942	75.3	87.574	—	—	133.0	154.679	—	—
400	673.15	56.9	66.175	59.7	69.431	69.2	80.480	82.5	95.948	96.3	111.997	122.2	142.119
450	723.15	61.5	71.525	64.0	74.432	71.2	82.806	81.3	94.552	93.3	108.508	106.7	124.092
500	773.15	67.2	78.154	69.3	80.596	75.4	87.690	83.1	96.645				
550	823.15	72.7	84.550	74.9	87.109	82.2	95.599	86.8	100.948	95.4	110.950	103.8	120.719
600	873.15	78.7	91.528	80.7	93.854	85.6	99.553	91.8	106.763	99.3	115.486	106.3	123.627

1 at = 1 kp/cm² = 98 066.5 Pa = 98 066.5 N/m² = 9.806 65 N/m² = 0.980 665 bar

1 kcal/h = 1.163 0 W

GASES

IMPORTANT NOTE: Users of the Gases tables are alerted to the fact that the symbol for pressure, *at*, refers to the **technical atmosphere** and not to the **standard atmosphere**, which is designated by *atm*.

The relationship of conversion factors between the two atmospheres is clearly explained in Table 119 on page 337.

Table 70-1 Thermal Properties of Gases (Density, Gas Constant, Specific Heat, Melting Point and Heat of Fusion)

Gas	Chemical formula	Relative molecular mass M (kg/kmol)	Density at 0°C 760 mmHg ρ (kg/m³)	Gas constant R (kp m/kg K)	Gas constant R (J/kg K)	Specific heat at 0°C c_p (kcal/kg K)	Specific heat at 0°C c_p (kJ/kg K)	$\kappa = \dfrac{c_p}{c_v}$	Melting point t_t (°C)	Melting point T_t (°K)	Heat of fusion (kcal/kg)	Heat of fusion (kg/kJ)
Acetylene	C_2H_2	26.04	1.1709	32.29	319.599	0.392	1.641	1.23	− 81	192.15	−	−
Air	—	28.96	1.2928	29.27	287.041	0.239	1.227	1.40	−	−	−	−
Ammonia	NH_3	17.031	0.7714	49.78	488.175	0.492	2.060	1.32	− 77.7	195.15	81	339.131
Argon	Ar	39.944	1.7839	21.23	208.195	0.125	0.523	1.67	− 189.3	83.15	7.0	29.308
Arsine (hydrogen arsenide)	H_3As	77.93	3.48	10.9	106.892	−	−	−	− 113.5	159.65	−	−
n−Butane	C_4H_{10}	58.12	2.703	14.60	143.177	0.458*	1.918*	1.11	− 135	138.15	18.0	75.362
Carbon dioxide	CO_2	44.01	1.9768	19.25	188.778	0.197	0.825	1.31	− 56	217.15	44	184.219
Carbon monoxide	CO	28.01	1.2500	30.28	296.945	0.251	1.051	1.40	− 205	68.15	7.2	30.145
Carbonyl sulfide	COS	60.07	2.72	14.2	139.254	0.160	0.670	−	− 138.2	134.95	−	−
Chlorine	Cl_2	70.914	3.22	11.96	117.288	0.120	0.502	1.34	− 103	170.15	45	188.406
Cyanogen	C_2N_2	52.04	2.32	16.60	162.790	−	−	1.26	− 34.4	238.75	−	−
Dichlorodifluoromethane	CF_2Cl_2	120.92	5.083	7.0127	68.771	0.141	0.590	1.14	− 155	118.15	−	−
Ethane	C_2H_6	30.07	1.356	28.22	276.744	0.398	1.666	1.22	− 183.6	89.55	22.2	92.947
Ethylene	C_2H_4	28.05	1.2605	30.25	296.651	0.350	1.465	1.24	− 169.4	103.75	25.0	104.670
Fluorine	F_2	38.000	1.695	22.30	218.688	−	−	−	− 220	53.15	9.0	37.681
Helium	He	4.002	0.1785	212.00	2079.010	1.250	5.234	1.66	−	−	1.365	5.715
Hydrogen	H_2	2.0156	0.08987	420.3	4121.735	3.400	14.235	1.41	− 259.20	13.95	14	58.615
Hydrogen bromide	HBr	80.924	3.644	10.49	102.872	0.082	0.343	1.36	− 87	186.15	7.4	30.982
Hydrogen chloride	HCl	36.465	1.6391	23.25	228.005	0.194	0.812	1.42	− 112	161.15	13.4	56.103
Hydrogen iodide	HJ	127.93	5.789	6.64	65.116	0.055	0.230	1.40	− 51	222.15	5.5	23.027
Hydrogen sulfide	H_2S	34.08	1.5392	24.90	244.186	0.264	1.105	1.30	− 85.6	187.55	16.6	69.501
Isobutane	C_4H_{10}	58.12	2.668	14.60	143.177	0.390*	1.633*	−	− 145	128.15	18.7	78.293
Krypton	Kr	83.7	3.74	10.23	100.322	0.060*	0.251*	1.68	− 157.2	115.95	4.7	19.678
Methane	CH_4	16.04	0.7168	52.90	518.772	0.520	2.177	1.30	− 182.5	90.65	14	58.615
Methylamine	CH_5N	31.06	1.39	27.30	267.722	−	−	−	− 92.5	180.65	−	−
Methyl chloride	CH_3Cl	50.49	2.307	16.80	164.752	0.176	0.737	1.20	− 91.5	181.65	−	−
Methyl ether	C_2H_6O	46.07	2.1097	18.40	180.442	0.3633	1.521	1.11	− 138.5	134.65	−	−
Methyl fluoride	CH_3F	34.03	1.545	24.91	244.284	−	−	−	−	−	−	−
Neon	Ne	20.183	0.8999	41.98	411.683	0.246	1.030	1.67	− 248.60	24.55	4.0	16.747
Nitric oxide	NO	30.008	1.3402	28.26	277.136	0.241	1.009	1.40	− 163.5	109.65	18.4	77.037
Nitrogen	N_2	28.016	1.2505	30.26	296.749	0.249	1.043	1.40	− 210.10	63.13	6.15	25.749
Nitrous oxide	N_2O	44.016	1.9780	19.26	188.876	0.205	0.858	1.31	− 90.8	182.35	35.5	148.631
Nitrosyl chloride	$NOCl$	65.465	2.9919	12.93	126.800	−	−	−	− 61.5	211.65	−	−
Oxygen	O_2	32.000	1.42895	26.49	259.778	0.218	0.913	1.40	− 218.83	54.32	3.3	13.816
Ozone	O_3	48.000	2.22	17.68	173.382	−	−	1.29	− 252	21.5	−	−
Phosphine (hydrogen phosphide)	PH_3	34.04	1.530	24.90	244.186	0.3701	1.550	1.14	− 133.5	139.65	7.85	32.866
Propane	C_3H_8	44.09	2.019	19.25	188.778	0.3406	1.425	−	− 189.9	83.25	19.2	80.387
Propene	C_3H_6	42.08	1.915	20.19	197.996	−	−	−	− 185.2	87.95	16.7	69.920
Sulfur dioxide	SO_2	64.06	2.9263	13.24	129.840	0.151	0.632	1.40	− 75.3	197.85	27.9	116.812
Xenon	X	131.3	5.89	6.51	63.841	0.038*	0.159*	1.66	− 111.9	161.25	4.2	17.585

*Specific heat at 20°C

Table 70-2 Thermal Properties of Gases (Boiling Point, Density, and Heat of Vaporization at the Boiling Point and Critical Constants)

Gas	Chemical formula	Number of atoms in a molecule	Boiling point at 760 mmHg t_v °C	T_v °K	Density ρ kg/m³	Heat of vaporization r kcal/kg	r kJ/kg	Temperature t_k °C	T_k °K	Pressure p_k kp/cm²	p_k bar	Density ρ_k kg/m³
Acetylene	C_2H_2	4	−83.6	189.55	613	198	828.986	35.7	308.85	64.7	63.44903	231
Air	—	—	−194	79.15	875	47	196.780	−140.7	132.45	38.4	37.65754	310
Ammonia	NH_3	4	−33.4	239.75	680	327	1369.084	132.4	405.55	115.2	112.97261	235
Argon	Ar	1	−185.9	87.25	1404	37.6	157.424	−122.4	150.75	49.6	48.64098	531
Arsine (hydrogen arsenide)	AsH_3	4	−55	218.15	—	—	—	—	—	—	—	—
n−Butane	C_4H_{10}	14	+ 0.5	273.65	600	96.4	403.608	153.2	426.35	37.2	36.48074	—
Carbon dioxide	CO_2	3	−78.48	194.67	—	137	573.592	31.0	304.15	75	73.54988	460
Carbon monoxide	CO	2	−191.5	81.65	801	51.6	216.039	−140.2	132.95	35.6	34.91167	301
Carbonyl sulfide	COS	3	−48	225.51	1200	—	—	105	378.15	67.3	65.99875	—
Chlorine	Cl_2	2	−35.0	238.15	1558	62	259.582	144	417.15	78.5	76.98220	573
Cyanogen	C_2N_2	4	−21	252.15	—	—	—	128.3	401.45	62	60.80123	—
Dichlorodifluoromethane	CF_2Cl_2	5	−30.0	243.15	1486	40	167.472	111.5	384.65	4C.9	40.10920	555
Ethane	C_2H_6	8	−88.6	184.55	546	129	540.097	35	308.15	50.6	49.62165	210
Ethylene	C_2H_4	6	−103.5	169.65	568	125	523.350	9.5	282.65	52.4	51.38685	216
Fluorine	F_2	2	−188	85.15	—	38	159.098	−101	172.15	—	—	—
Helium	He	1	−268.9	4.25	125	5	20.934	−267.9	5.25	2.33	2.28495	69
Hydrogen	H_2	2	−252.78	20.37	70.8	110	460.548	−239.9	33.25	13.2	12.94478	31
Hydrogen bromide	HBr	2	−67	206.15	—	52	217.714	90	363.15	87	85.31786	807
Hydrogen chloride	HCl	2	−85	188.15	—	106	443.801	51.4	324.55	86	84.33719	610
Hydrogen iodide	HJ	2	−36	237.15	—	37	154.912	150.8	423.95	—	—	—
Hydrogen sulfide	H_2S	3	−60.4	212.75	920	131	548.471	100.4	373.55	92	90.22118	—
Isobutane	C_4H_{10}	14	−10.2	262.95	595	94.4	395.234	133.7	406.85	37.7	36.97107	—
Krypton	Kr	1	−153.2	119.95	2160	28	117.230	−63.8	209.35	56.0	54.91724	909
Methane	CH_4	5	−161.7	111.45	415	131	548.471	−82.5	190.65	47.2	46.28739	162
Methylamine	CH_5N	7	6.5	266.65	—	206	862.481	157	430.15	76	74.53054	—
Methyl chloride	CH_3Cl	5	−24	249.15	997	100	418.680	143.1	416.25	68.1	66.78329	370
Methyl ether	C_2H_6O	9	−24	249.15	720	112	468.922	127	400.15	55	53.93658	—
Methyl fluoride	CH_3F	5	−78	195.15	—	—	—	44.9	318.05	64.1	62.86063	—
Neon	Ne	1	−246.1	27.05	1207	25	104.670	−228.7	44.45	27.8	27.26249	484
Nitric oxide	NO	2	−152.00	121.15	—	110	460.548	−94	179.15	66	64.72389	520
Nitrogen	N_2	2	−195.81	77.34	810	47.6	199.292	−147.1	126.05	34.6	33.93101	311
Nitrous oxide	N_2O	3	−88.7	184.45	—	90	376.812	36.5	309.65	74	72.56921	460
Nitrosyl chloride	$NOCl$	3	5.5	267.65	—	—	—	165	438.15	95.5	93.65351	—
Oxygen	O_2	2	−182.97	90.18	1131	51	213.527	−118.8	154.35	51.4	50.40618	530
Ozone	O_3	3	−112	161.15	—	—	—	5	268.15	95.4	93.55544	540
Phosphine (hydrogen phosphide)	PH_3	4	−87.5	185.65	—	—	—	52	325.15	66.7	65.41036	—
Propane	C_3H_8	11	−42.6	230.55	585	107	447.988	96.8	369.95	43.3	42.46279	226
Propene	C_3H_6	9	−47.0	226.15	609	109	456.361	92.0	365.15	46.8	45.89512	460
Sulfur dioxide	SO_2	3	−10.0	263.15	1460	96	401.933	157.3	430.45	80.4	78.84547	524
Xenon	X	1	−108.0	165.15	3060	23	96.296	16.6	289.75	60.1	58.93797	1150

Table 71-1 Specific Heat c_p of Gases at Constant Pressure p

Temperature t °C	T K	O₂ kcal/kg K	O₂ kJ/kg K	H₂ kcal/kg K	H₂ kJ/kg K	NO kcal/kg K	NO kJ/kg K	OH kcal/kg K	OH kJ/kg K	H₂O kcal/kg K	H₂O kJ/kg K	N₂ kcal/kg K	N₂ kJ/kg K	Air kcal/kg K	Air kJ/kg K
0	273.15	0.2185	0.9148	3.3904	14.1949	0.2386	0.9990	0.4212	1.7635	0.4441	1.8594	0.2482	1.0392	0.2397	1.0036
100	373.15	0.2230	0.9337	3.4509	14.4482	0.2381	0.9969	0.4163	1.7430	0.4515	1.8903	0.2489	1.0421	0.2413	1.0103
200	473.15	0.2300	0.9630	3.4643	14.5043	0.2414	1.0107	0.4143	1.7346	0.4635	1.9406	0.2512	1.0517	0.2447	1.0245
300	573.15	0.2376	0.9948	3.4712	14.5332	0.2472	1.0350	0.4142	1.7342	0.4778	2.0005	0.2554	1.0693	0.2495	1.0446
400	673.15	0.2445	1.0237	3.4826	14.5809	0.2534	1.0609	0.4157	1.7405	0.4931	2.0645	0.2607	1.0915	0.2552	1.0685
500	773.15	0.2504	1.0484	3.5020	14.6622	0.2594	1.0861	0.4191	1.7547	0.5092	2.1319	0.2664	1.1154	0.2609	1.0923
600	873.15	0.2553	1.0689	3.5298	14.7786	0.2648	1.1087	0.4237	1.7739	0.5258	2.2014	0.2721	1.1392	0.2663	1.1149
700	973.15	0.2593	1.0856	3.5660	14.9301	0.2695	1.1283	0.4294	1.7978	0.5429	2.2730	0.2774	1.1614	0.2712	1.1355
800	1073.15	0.2627	1.0999	3.6101	15.1148	0.2736	1.1455	0.4358	1.8246	0.5601	2.3450	0.2822	1.1815	0.2756	1.1539
900	1173.15	0.2656	1.1120	3.6572	15.3120	0.2770	1.1597	0.4424	1.8522	0.5769	2.4154	0.2864	1.1991	0.2795	1.1702
1000	1273.15	0.2682	1.1229	3.7063	15.5175	0.2799	1.1719	0.4489	1.8795	0.5929	2.4824	0.2902	1.2150	0.2829	1.1844
1100	1373.15	0.2703	1.1317	3.7584	15.7357	0.2824	1.1824	0.4553	1.9063	0.6080	2.5456	0.2935	1.2288	0.2859	1.1970
1200	1473.15	0.2723	1.1401	3.8095	15.9496	0.2845	1.1911	0.4615	1.9322	0.6220	2.6042	0.2964	1.2410	0.2886	1.2083
1300	1573.15	0.2743	1.1484	3.8611	16.1657	0.2864	1.1991	0.4674	1.9569	0.6350	2.6586	0.2989	1.2514	0.2909	1.2179
1400	1673.15	0.2762	1.1564	3.9097	16.3691	0.2881	1.2062	0.4728	1.9795	0.6470	2.7089	0.3011	1.2606	0.2930	1.2267
1500	1773.15	0.2780	1.1639	3.9563	16.5642	0.2895	1.2121	0.4779	2.0009	0.6581	2.7553	0.3030	1.2686	0.2949	1.2347
1600	1873.15	0.2797	1.1710	4.0000	16.7472	0.2907	1.2171	0.4827	2.0210	0.6683	2.7980	0.3048	1.2761	0.2966	1.2418
1700	1973.15	0.2815	1.1786	4.0417	16.9218	0.2918	1.2217	0.4872	2.0398	0.6770	2.8382	0.3063	1.2824	0.2982	1.2485
1800	2073.15	0.2832	1.1857	4.0808	17.0855	0.2928	1.2259	0.4913	2.0570	0.6865	2.8742	0.3077	1.2883	0.2996	1.2544
1900	2173.15	0.2849	1.1928	4.1185	17.2433	0.2938	1.2301	0.4951	2.0729	0.6944	2.9073	0.3089	1.2933	0.3010	1.2602
2000	2273.15	0.2867	1.2004	4.1533	17.3890	0.2947	1.2338	0.4986	2.0875	0.7017	2.9379	0.3100	1.2979	0.3022	1.2653
2100	2373.15	0.2884	1.2075	4.1860	17.5259	0.2954	1.2368	0.5020	2.1018	0.7086	2.9668	0.3110	1.3021	0.3034	1.2703
2200	2473.15	0.2900	1.2142	4.2182	17.6608	0.2960	1.2393	0.5051	2.1148	0.7150	2.9936	0.3120	1.3063	0.3045	1.2749
2300	2573.15	0.2917	1.2213	4.2475	17.7834	0.02967	1.2422	0.5081	2.1273	0.7208	3.0178	0.3128	1.3096	0.3055	1.2791
2400	2673.15	0.2933	1.2280	4.2758	17.9019	0.2973	1.2447	0.5109	2.1390	0.7263	3.0409	0.3136	1.3130	0.3065	1.2833
2500	2773.15	0.2948	1.2343	4.3026	18.0141	0.2978	1.2468	0.5137	2.1508	0.7313	3.0618	0.3143	1.3159	0.3074	1.2870
2600	2873.15	0.2964	1.2410	4.3279	18.1201	0.2984	1.2493	0.5194	2.1746	0.7361	3.0819	0.3155	1.3209	0.3087	1.2925
2700	2973.15	0.2979	1.2472	4.3517	18.2197	0.2989	1.2514	0.5190	2.1729	0.7406	3.1007	0.3162	1.3239	0.3087	1.2925
2800	3073.15	0.2984	1.2493	4.3808	18.3415	0.2994	1.2535	0.5215	2.1834	0.7449	3.1187	0.3170	1.3272	0.3100	1.2979
2900	3173.15	0.2997	1.2548	4.4056	18.4454	0.2999	1.2556	0.5239	2.1935	0.7489	3.1355	0.3173	1.3285	0.3107	1.3008
3000	3273.15	0.3010	1.2602	4.4300	18.5475	0.3004	1.2577	0.5263	2.2035	0.7489	3.1355	0.3180	1.3314	0.3110	1.3021

1 kcal = 4.186 8 kJ

Table 71-2 Specific Heat c_p of Gases at Constant Pressure p (Continued)

t °C	T K	CO c_p kcal/kg K	CO c_p kJ/kg K	CO$_2$ c_p kcal/kg K	CO$_2$ c_p kJ/kg K	N$_2$O c_p kcal/kg K	N$_2$O c_p kJ/kg K	SO$_2$ c_p kcal/kg K	SO$_2$ c_p kJ/kg K	H$_2$S c_p kcal/kg K	H$_2$S c_p kJ/kg K	NH$_3$ c_p kcal/kg K	NH$_3$ c_p kJ/kg K
0	273.15	0.2483	1.0396	0.1946	0.8148	0.2032	0.8508	0.145	0.607	0.237	0.992	0.481	2.056
100	373.15	0.2495	1.0446	0.2182	0.9136	0.2269	0.9500	0.158	0.662	0.245	1.026	0.527	2.206
200	473.15	0.2528	1.0584	0.2371	0.9927	0.2456	1.0283	0.170	0.712	0.255	1.068	0.570	2.386
300	573.15	0.2580	1.0802	0.2524	1.0567	0.2611	1.0932	0.180	0.754	0.268	1.122	0.615	2.575
400	673.15	0.2641	1.1057	0.2652	1.1103	0.2740	1.1472	0.187	0.783	0.280	1.172	0.654	2.738
500	773.15	0.2704	1.1321	0.2758	1.1547	0.2849	1.1928	0.193	0.808	0.293	1.227	0.700	2.931
600	873.15	0.2763	1.1568	0.2847	1.1920	0.2941	1.2313	0.197	0.825	0.304	1.273	0.740	3.098
700	973.15	0.2816	1.1790	0.2921	1.2230	0.3017	1.2632	0.200	0.837	0.315	1.319	0.778	3.257
800	1073.15	0.2863	1.1987	0.2984	1.2493	0.3084	1.2912	0.203	0.850	0.325	1.361	0.812	3.400
900	1173.15	0.2904	1.2158	0.3037	1.2715	0.3141	1.3151	0.205	0.858	0.334	1.398	0.844	3.534
1000	1273.15	0.2939	1.2305	0.3081	1.2900	0.3189	1.3352	0.207	0.867	0.342	1.432	0.873	3.655
1100	1373.15	0.2970	1.2435	0.3119	1.3059	0.3232	1.3532	0.208	0.871	0.349	1.461	—	—
1200	1473.15	0.2996	1.2544	0.3152	1.3197	—	—	0.209	0.875	0.354	1.482	—	—
1300	1573.15	0.3020	1.2644	0.3180	1.3314	—	—	0.210	0.879	—	—	—	—
1400	1673.15	0.3040	1.2728	0.3204	1.3415	—	—	0.211	0.883	—	—	—	—
1500	1773.15	0.3057	1.2799	0.3224	1.3498	0.3254	1.3624	0.212	0.888	0.379	1.587	0.979	4.099
1600	1873.15	0.3073	1.2866	0.3242	1.3574	—	—	0.212	0.888	—	—	—	—
1700	1973.15	0.3087	1.2925	0.3257	1.3636	—	—	0.213	0.892	—	—	—	—
1800	2073.15	0.3100	1.2979	0.3271	1.3695	—	—	0.213	0.892	—	—	—	—
1900	2173.15	0.3111	1.3025	0.3282	1.3741	—	—	0.213	0.892	—	—	—	—
2000	2273.15	0.3121	1.3067	0.3292	1.3783	0.3281	1.3737	0.214	0.896	0.381	1.595	1.040	4.354
2100	2373.15	0.3130	1.3105	0.3300	1.3816	—	—	0.214	0.896	—	—	—	—
2200	2473.15	0.3138	1.3138	0.3306	1.3842	—	—	0.214	0.896	—	—	—	—
2300	2573.15	0.3146	1.3172	0.3311	1.3862	—	—	0.214	0.896	—	—	—	—
2400	2673.15	0.3153	1.3201	0.3314	1.3875	0.3315	1.3879	0.214	0.896	0.401	1.679	1.076	4.505
2500	2773.15	0.3160	1.3230	0.3315	1.3879	—	—	0.215	0.900	—	—	—	—
3000	3273.15	0.3190	1.3356	0.3330	1.3942	0.3330	1.3942	0,215	0.900	0.407	1.704	1.099	4.601

1 kcal = 4.1868 kJ

Table 71-3 Specific Heat c_p of Gases at Constant Pressure p (Continued)

Temperature		CS$_2$		COS		CH$_4$		C$_2$H$_6$		C$_3$H$_8$	
t	T	c_p		c_p		c_p		c_p		c_p	
°C	K	kcal/kg K	kJ/kg K	kcal/kg K	kJ/kg K	kcal/kg K	kJ/kg K	kcal/kg K	kJ/kg K	kcal/kg K	kJ/kg K
0	273.15	0.140	0.585	0.160	0.670	0.5172	2.1654	0.3934	1.6471	0.3701	1.5495
100	373.15	0.153	0.641	0.178	0.745	0.5848	2.4484	0.4938	2.0674	0.4817	2.0168
200	473.15	0.162	0.678	0.191	0.800	0.6704	2.8068	0.5947	2.4899	0.5871	2.4581
300	573.15	0.169	0.708	0.201	0.842	0.7584	3.1753	0.6854	2.8696	0.6770	2.8345
400	673.15	0.174	0.729	0.209	0.875	0.8430	3.5295	0.7676	3.2138	0.7550	3.1610
500	773.15	0.178	0.745	0.214	0.896	0.9210	3.8560	0.8405	3.5190	0.8237	3.4487
600	873.15	0.181	0.758	0.219	0.917	0.9919	4.1529	0.9045	3.7870	0.8831	3.6974
700	973.15	0.183	0.766	0.223	0.934	1.0560	4.4213	0.9607	4.0223	0.9353	3.9159
800	1073.15	0.185	0.775	0.227	0.950	1.1129	4.6595	1.0069	4.2157	0.9775	4.0926
900	1173.15	0.187	0.783	0.229	0.959	1.1638	4.8726	1.0483	4.3890	1.0151	4.2500
1000	1273.15	0.188	0.787	0.232	0.971	1.2089	5.0614	1.0863	4.5481	1.0496	4.3945
1100	1373.15	0.189	0.791	0.234	0.980	1.2483	5.2264	1.1209	4.6930	1.0811	4.5263
1200	1473.15	0.190	0.795	0.235	0.984	1.2820	5.3675	1.1521	4.8236	1.1094	4.6448

Table 71-4 Specific Heat c_p of Gases at Constant Pressure p (Continued)

Temperature		C$_2$H$_4$		C$_2$H$_2$		C$_6$H$_6$		C$_3$H$_6$		C$_2$H$_5$OH	
t	T	c_p		c_p		c_p		c_p		c_p	
°C	K	kcal/kg K	kJ/kg K	kcal/kg K	kJ/kg K	kcal/kg K	kJ/kg K	kcal/kg K	kJ/kg K	kcal/kg K	kJ/kg K
0	273.15	0.3486	1.4595	0.38447	1.6097	0.2253	0.9433	0.3406	1.4260	0.3633	1.5211
100	373.15	0.4363	1.8267	0.44669	1.8702	0.3189	1.3352	0.4299	1.7999	0.4363	1.8267
200	473.15	0.5197	2.1759	0.48817	2.0439	0.4003	1.6760	0.5159	2.1600	0.5053	2.1156
300	573.15	0.5918	2.4777	0.51928	2.1741	0.4673	1.9565	0.5915	2.4765	0.5664	2.3714
400	673.15	0.6534	2.7357	0.54501	2.2818	0.5213	2.1826	0.6576	2.7532	0.6198	2.5950
500	773.15	0.7065	2.9580	0.56756	2.3763	0.5659	2.3693	0.7146	2.9919	0.6662	2.7892
600	873.15	0.7532	3.1535	0.58784	2.4612	0.6029	2.5242	0.7643	3.2000	0.7071	2.9605
700	973.15	0.7942	3.3252	0.60643	2.5390	0.6342	2.6553	0.8078	3.3821	0.7427	3.1095
800	1073.15	0.8295	3.4730	0.62337	2.6099	0.6609	2.7571	0.8456	3.5404	0.7742	3.2414
900	1173.15	0.8609	3.6044	0.63881	2.6746	0.6834	2.8613	0.8784	3.6777	0.8017	3.3566
1000	1273.15	0.8887	3.7208	0.65279	2.7331	0.7029	2.9429	0.9071	3.7978	0.8256	3.4566
1100	1373.15	0.9126	3.8209	0.66542	2.7860	0.7196	3.0128	0.9321	3.9025	0.8463	3.5433
1200	1473.15	0.9336	3.9088	0.67672	2.8333	0.7340	3.0731	0.9535	3.9921	0.8644	3.6191

1 kcal = 4.1868 kJ

Table 72-1 Specific Heat c_v of Gases at Constant Volume v

Temperature t °C	T K	O₂ c_v kcal/kg K	O₂ c_v kJ/kg K	H₂ c_v kcal/kg K	H₂ c_v kJ/kg K	NO c_v kcal/kg K	NO c_v kJ/kg K	OH c_v kcal/kg K	OH c_v kJ/kg K	H₂O c_v kcal/kg K	H₂O c_v kJ/kg K	N₂ c_v kcal/kg K	N₂ c_v kJ/kg K	Air c_v kcal/kg K	Air c_v kJ/kg K
0	273.15	0.1564	0.6548	2.4053	10.0705	0.1724	0.7218	0.3044	1.2745	0.3339	1.3980	0.1773	0.7423	0.1711	0.7164
100	373.15	0.1609	0.6737	2.4658	10.3238	0.1719	0.7197	0.2995	1.2539	0.3413	1.4290	0.1780	0.7453	0.1727	0.7231
200	473.15	0.1679	0.7030	2.4792	10.3799	0.1752	0.7335	0.2975	1.2456	0.3533	1.4792	0.1804	0.7553	0.1761	0.7373
300	573.15	0.1755	0.7348	2.4861	10.4088	0.1810	0.7578	0.2974	1.2452	0.3675	1.5386	0.1845	0.7725	0.1810	0.7578
400	673.15	0.1824	0.7637	2.4975	10.4565	0.1872	0.7838	0.2989	1.2514	0.3828	1.6027	0.1898	0.7947	0.1866	0.7813
500	773.15	0.1833	0.7674	2.5169	10.5378	0.1932	0.8089	0.3023	1.2657	0.3989	1.6701	0.1955	0.8185	0.1923	0.8051
600	873.15	0.1932	0.8089	2.5446	10.6537	0.1986	0.8315	0.3069	1.2849	0.4156	1.7400	0.2012	0.8424	0.1978	0.8281
700	973.15	0.1973	0.8261	2.5808	10.8053	0.2033	0.8512	0.3126	1.3088	0.4327	1.8116	0.2065	0.8646	0.2027	0.8487
800	1073.15	0.2007	0.8403	2.6250	10.9904	0.2074	0.8683	0.3190	1.3356	0.4499	1.8836	0.2113	0.8847	0.2071	0.8671
900	1173.15	0.2035	0.8520	2.6721	11.1875	0.2108	0.8826	0.3256	1.3632	0.4666	1.9536	0.2156	0.9027	0.2110	0.8834
1000	1273.15	0.2060	0.8625	2.7212	11.3931	0.2137	0.8947	0.3321	1.3904	0.4827	2.0210	0.2193	0.9182	0.2144	0.8976
1100	1373.15	0.2082	0.8717	2.7733	11.6113	0.2165	0.9064	0.3385	1.4172	0.4977	2.0838	0.2226	0.9320	0.2174	0.9102
1200	1473.15	0.2103	0.8805	2.8244	11.8252	0.2183	0.9140	0.3447	1.4432	0.5113	2.1407	0.2255	0.9441	0.2200	0.9211
1300	1573.15	0.2122	0.8884	2.8760	12.0412	0.2202	0.9219	0.3506	1.4679	0.5248	2.1972	0.2280	0.9546	0.2224	0.9311
1400	1673.15	0.2141	0.8964	2.9246	12.2447	0.2219	0.9291	0.3560	1.4905	0.5368	2.2475	0.2302	0.9638	0.2245	0.9399
1500	1773.15	0.2159	0.9039	2.9712	12.4398	0.2233	0.9349	0.3611	1.5119	0.5479	2.2939	0.2322	0.9722	0.2264	0.9479
1600	1873.15	0.2177	0.9115	3.0149	12.6228	0.2245	0.9399	0.3659	1.5320	0.5581	2.3367	0.2339	0.9793	0.2281	0.9550
1700	1973.15	0.2194	0.9186	3.0565	12.7970	0.2256	0.9445	0.3704	1.5508	0.5677	2.3768	0.2354	0.9856	0.2296	0.9613
1800	2073.15	0.2212	0.9261	3.0957	12.9611	0.2267	0.9491	0.3745	1.5680	0.5763	2.4129	0.2368	0.9914	0.2311	0.9676
1900	2173.15	0.2229	0.9332	3.1334	13.1189	0.2276	0.9529	0.3783	1.5839	0.5842	2.4459	0.2380	0.9965	0.2324	0.9730
2000	2273.15	0.2246	0.9404	3.1681	13.2642	0.2285	0.9567	0.3818	1.5985	0.5915	2.4765	0.2391	1.0011	0.2337	0.9785
2100	2373.15	0.2253	0.9475	3.2009	13.4015	0.2292	0.9596	0.3852	1.6128	0.5984	2.5054	0.2401	1.0053	0.2348	0.9831
2200	2473.15	0.2280	0.9546	3.2331	13.5363	0.2299	0.9625	0.3883	1.6257	0.6047	2.5318	0.2411	1.0094	0.2359	0.9877
2300	2573.15	0.2296	0.9613	3.2624	13.6590	0.2305	0.9651	0.3913	1.6383	0.6106	2.5565	0.2419	1.0128	0.2369	0.9919
2400	2673.15	0.2312	0.9680	3.2907	13.7775	0.2311	0.9676	0.3941	1.6500	0.6160	2.5791	0.2427	1.0161	0.2379	0.9960
2500	2773.15	0.2328	0.9747	3.3175	13.8897	0.2317	0.9701	0.3969	1.6617	0.6211	2.6004	0.2434	1.0191	0.2388	0.9998
2600	2873.15	0.2343	0.9810	3.3428	13.9956	0.2322	0.9722	0.3996	1.6730	0.6259	2.6205				
2700	2973.15	0.2358	0.9872	3.3666	14.0953	0.2323	0.9726	0.4022	1.6839	0.6304	2.6394				
2800	3073.15	—	—	—	—	0.2333	0.9768	0.4047	1.6944	0.6347	2.6574				
2900	3173.15	—	—	—	—	0.2338	0.9789	0.4071	1.7044	0.6387	2.6741				
3000	3273.15	—	—	—	—	0.2342	0.9805	0.4095	1.7145	—	—				

1 kcal = 4.186 8 kJ

Table 72-2 Specific Heat c_v of Gases at Constant Volume v (Continued)

| Temperature | | CO c_v | | CO_2 c_v | | N_2O c_v | | SO_2 c_v | | H_2S c_v | | CS_2 c_v | | COS c_v | |
t °C	T K	kcal/kg K	kJ/kg K	kcal/kg K	kJ/kg K	kcal/kg K	kJ/kg K	kcal/kg K	kJ/kg K	kcal/kg K	kJ/kg K	kcal/kg K	kJ/kg K	kcal/kg K	kJ/kg K
0	273.15	0.1774	0.7427	0.1495	0.6259	0.1581	0.6619	0.114	0.477	0.178	0.745	0.113	0.473	0.127	0.532
100	373.15	0.1786	0.7478	0.1731	0.7247	0.1817	0.7607	0.127	0.532	0.186	0.779	0.126	0.528	0.145	0.607
200	473.15	0.1819	0.7616	0.1920	0.8039	0.2005	0.8395	0.139	0.582	0.197	0.825	0.136	0.569	0.158	0.662
300	573.15	0.1871	0.7834	0.2073	0.8679	0.2160	0.9043	0.149	0.624	0.209	0.875	0.143	0.599	0.168	0.703
400	673.15	0.1932	0.8089	0.2200	0.9211	0.2289	0.9584	0.156	0.653	0.222	0.929	0.148	0.620	0.175	0.733
500	773.15	0.1995	0.8353	0.2307	0.9659	0.2398	1.0040	0.162	0.678	0.234	0.980	0.152	0.636	0.181	0.758
600	873.15	0.2054	0.8600	0.2395	1.0027	0.2490	1.0425	0.166	0.695	0.246	1.030	0.155	0.649	0.186	0.779
700	973.15	0.2107	0.8822	0.2470	1.0341	0.2566	1.0743	0.169	0.708	0.257	1.076	0.157	0.657	0.190	0.795
800	1073.15	0.2154	0.9018	0.2532	1.0601	0.2632	1.1020	0.172	0.720	0.267	1.118	0.159	0.666	0.194	0.812
900	1173.15	0.2195	0.9190	0.2585	1.0823	0.2689	1.1258	0.174	0.729	0.276	1.156	0.161	0.674	0.196	0.821
1000	1273.15	0.2230	0.9337	0.2630	1.1011	0.2738	1.1463	0.176	0.737	0.283	1.185	0.162	0.678	0.199	0.833
1100	1373.15	0.2261	0.9466	0.2668	1.1170	0.2789	1.1677	0.177	0.741	0.290	1.214	0.163	0.682	0.201	0.842
1200	1473.15	0.2287	0.9575	0.2700	1.1304	—	—	0.178	0.745	0.296	1.239	0.164	0.687	0.202	0.846
1300	1573.15	0.2311	0.9676	0.2728	1.1422										
1400	1673.15	0.2330	0.9755	0.2752	1.1522										
1500	1773.15	0.2348	0.9831	0.2773	1.1610										
1600	1873.15	0.2364	0.9898	0.2791	1.1685										
1700	1973.15	0.2378	0.9956	0.2806	1.1748										
1800	2073.15	0.2390	1.0006	0.2820	1.1807										
1900	2173.15	0.2402	1.0057	0.2831	1.1853										
2000	2273.15	0.2412	1.0099	0.2840	1.1891										
2100	2373.15	0.2421	1.0136	0.2848	1.1924										
2200	2473.15	0.2429	1.0170	0.2855	1.1953										
2300	2573.15	0.2437	1.0203	0.2860	1.1974										
2400	2673.15	0.2444	1.0233	0.2863	1.1987										
2500	2773.15	0.2450	1.0258	0.2864	1.1991										

1 kcal = 4.186 8 kJ

Table 72-3 Specific Heat c_v of Gases at Constant Volume v (Continued)

Temperature		CH₄		C₂H₆		C₃H₈		C₂H₄	
		c_v		c_v		c_v		c_v	
t	T								
°C	K	kcal/kg K	kJ/kg K	kcal/kg K	kJ/kg K	kcal/kg K	kJ/kg K	kcal/kg K	kJ/kg K
0	273.15	0.3934	1.6471	0.3274	1.3708	0.3250	1.3607	0.2777	1.1627
100	373.15	0.4610	1.9301	0.4278	1.7911	0.4366	1.8280	0.3654	1.5299
200	473.15	0.5466	2.2885	0.5287	2.2136	0.5420	2.2692	0.4488	1.8790
300	573.15	0.6346	2.6569	0.6194	2.5933	0.6318	2.6452	0.5208	2.1805
400	673.15	0.7192	3.0111	0.7016	2.9375	0.7098	2.9718	0.5825	2.4388
500	773.15	0.7972	3.3377	0.7744	3.2423	0.7786	3.2598	0.6356	2.6611
600	873.15	0.8681	3.6346	0.8385	3.5106	0.8380	3.5085	0.6823	2.8567
700	973.15	0.9322	3.9029	0.8946	3.7455	0.8901	3.7267	0.7233	3.0283
800	1073.15	0.9891	4.1412	0.9408	3.9389	0.9323	3.9034	0.7586	3.1761
900	1173.15	1.0400	4.3543	0.9822	4.1123	0.9700	4.0612	0.7900	3.3076
1000	1273.15	1.0851	4.5431	1.0202	4.2714	1.0044	4.2052	0.8178	3.4240
1100	1373.15	1.1245	4.7081	1.0548	4.4162	1.0360	4.3375	0.8416	3.5236
1200	1473.15	1.1582	4.8492	1.0861	4.5473	1.0643	4.4560	0.8627	3.6120

Table 72-4 Specific Heat c_v of Gases at Constant Volume v (Continued)

Temperature		C₂H₂		C₆H₆		C₃H₆		C₂H₅OH	
		c_v		c_v		c_v		c_v	
t	T								
°C	K	kcal/kg K	kJ/kg K	kcal/kg K	kJ/kg K	kcal/lgk K	kJ/kg K	kcal/kg K	kJ/kg K
0	273.15	0.3080	1.2895	0.1998	0.8365	0.2933	1.2280	0.3202	1.3406
100	373.15	0.3703	1.5504	0.2934	1.2284	0.3826	1.6019	0.3932	1.6462
200	473.15	0.4117	1.7237	0.3749	1.5696	0.4686	1.9619	0.4622	1.9351
300	573.15	0.4428	1.8539	0.4418	1.8497	0.5442	2.2785	0.5233	2.1910
400	673.15	0.4686	1.9619	0.4958	2.0758	0.6103	2.5552	0.5767	2.4145
500	773.15	0.49128	2.0569	0.5404	2.2625	0.6673	2.7939	0.6231	2.6088
600	873.15	0.51156	2.1418	0.5774	2.4175	0.7170	3.0019	0.6640	2.7800
700	973.15	0.53015	2.2196	0.6088	2.5489	0.7605	3.1841	0.6996	2.9291
800	1073.15	0.54709	2.2906	0.6354	2.6603	0.7983	3.3423	0.7311	3.0610
900	1173.15	0.56253	2.3552	0.6579	2.7545	0.8311	3.4796	0.7586	3.1761
1000	1273.15	0.57651	2.4137	0.6774	2.8361	0.8598	3.5998	0.7825	3.2762
1100	1373.15	0.58915	2.4667	0.6942	2.9065	0.8848	3.7045	0.8032	3.3628
1200	1473.15	0.60044	2.5139	0.7085	2.9663	0.9062	3.7941	0.8213	3.4386

1 kcal = 4.186 8 kJ

Table 73-1 Specific Heat C_p of Gases at Constant Pressure p

Temperature t °C	T K	O₂ C_p kcal/kmol K	O₂ C_p kJ/kmol K	H₂ C_p kcal/kmol K	H₂ C_p kJ/kmol K	NO C_p kcal/kmol K	NO C_p kJ/kmol K	OH C_p kcal/kmol K	OH C_p kJ/kmol K	H₂O C_p kcal/kmol K	H₂O C_p kJ/kmol K	N₂ C_p kcal/kmol K	N₂ C_p kJ/kmol K	Air C_p kcal/kmol K	Air C_p kJ/kmol K
0	273.15	6.992	29.274	6.835	28.617	7.160	29.977	7.163	29.990	8.001	33.499	6.954	29.115	6.944	29.073
100	373.15	7.136	29.877	6.957	29.128	7.146	29.919	7.081	29.647	8.134	34.055	6.974	29.199	6.990	29.266
200	473.15	7.360	30.815	6.984	29.241	7.245	30.333	7.046	29.500	8.351	34.964	7.039	29.471	7.088	29.676
300	573.15	7.603	31.832	6.998	29.299	7.418	31.058	7.045	29.496	8.607	36.036	7.154	29.952	7.229	30.266
400	673.15	7.824	32.758	7.021	29.396	7.603	31.832	7.071	29.605	8.883	37.191	7.303	30.576	7.392	30.949
500	773.15	8.013	33.549	7.060	29.559	7.784	32.590	7.128	29.844	9.173	38.406	7.464	31.250	7.557	31.640
600	873.15	8.169	34.202	7.116	29.793	7.946	33.268	7.206	30.170	9.473	39.662	7.624	31.920	7.715	32.301
700	973.15	8.299	34.746	7.189	30.099	8.087	33.859	7.304	30.580	9.781	40.951	7.772	32.540	7.858	32.900
800	1073.15	8.408	35.203	7.278	30.472	8.209	34.369	7.412	31.033	10.091	42.249	7.906	33.101	7.985	33.432
900	1173.15	8.499	35.584	7.373	30.869	8.311	34.796	7.524	31.501	10.393	43.513	8.025	33.599	8.098	33.905
1000	1273.15	8.578	35.914	7.472	31.284	8.399	35.165	7.635	31.966	10.682	44.723	8.130	34.043	8.196	34.315
1100	1373.15	8.650	36.216	7.577	31.723	8.473	35.475	7.744	32.423	10.953	45.858	8.222	34.424	8.283	34.679
1200	1473.15	8.715	36.488	7.680	32.155	8.537	35.743	7.849	32.862	11.205	46.913	8.303	34.763	8.360	35.002
1300	1573.15	8.778	36.752	7.784	32.590	8.594	35.981	7.949	33.281	11.440	47.897	8.374	35.060	8.429	35.291
1400	1673.15	8.837	36.999	7.882	33.000	8.644	36.191	8.041	33.666	11.656	48.801	8.436	35.320	8.490	35.546
1500	1773.15	8.895	37.242	7.976	33.394	8.687	36.371	8.128	34.030	11.856	49.639	8.490	35.546	8.544	35.772
1600	1873.15	8.952	37.480	8.064	33.762	8.724	36.526	8.210	34.374	12.040	50.409	8.538	35.747	8.593	35.977
1700	1973.15	9.008	37.715	8.148	34.114	8.757	36.664	8.286	34.692	12.213	51.133	8.581	35.927	8.639	36.170
1800	2073.15	9.063	37.945	8.227	34.445	8.788	36.794	8.356	34.985	12.368	51.782	8.620	36.090	8.681	36.346
1900	2173.15	9.118	38.175	8.303	34.763	8.816	36.911	8.421	35.257	12.510	52.377	8.655	36.237	8.720	36.509
2000	2273.15	9.173	38.406	8.373	35.056	8.843	37.024	8.481	35.508	12.642	52.930	8.686	36.367	8.755	36.655
2100	2373.15	9.228	38.636	8.439	35.332	8.865	37.116	8.538	35.747	12.766	53.449	8.714	36.484	8.789	36.798
2200	2473.15	9.281	38.858	8.504	35.605	8.884	37.196	8.591	35.969	12.881	53.930	8.740	36.593	8.820	36.928
2300	2573.15	9.334	39.080	8.563	35.852	8.903	37.275	8.641	36.178	12.986	54.370	8.764	36.693	8.850	37.053
2400	2673.15	9.385	39.293	8.620	36.090	8.921	37.350	8.689	36.379	13.084	54.780	3.786	36.785	8.878	37.170
2500	2773.15	9.435	39.502	8.674	36.316	8.938	37.422	8.737	36.580	13.175	55.161	8.806	36.869	8.904	37.279
2600	2873.15	9.484	39.708	8.725	36.530	8.955	37.493	8.783	36.773	13.262	55.525	8.840	37.022	8.940	37.430
2700	2973.15	9.532	39.909	8.773	36.731	8.971	37.560	8.827	36.957	13.343	55.864	8.860	37.106	8.960	37.514
2800	3073.15	9.550	39.984	8.830	36.969	8.986	37.623	8.870	37.137	13.420	56.187	8.880	37.189	8.980	37.597
2900	3173.15	9.590	40.152	8.880	37.199	9.001	37.685	8.911	37.309	13.492	56.488	8.890	37.231	9.000	37.681
3000	3273.15	9.620	40.277	8.930	37.388	9.015	37.744	8.951	37.476	13.500	56.522	8.900	37.263	9.020	37.765
M*		32		2.0156		30.008		17.008		18.020		28.016		28.964	

*M — relative molecular mass in kg/mol

1 kcal = 4.186 8 kJ

Table 73-2 Specific Heat C_p of Gases at Constant Pressure p (Continued)

t °C	T K	CO $\bar{C}_p$ kcal/kmol K	CO kJ/kmol K	CO₂ C_p kcal/kmol K	CO₂ kJ/kmol K	N₂O C_p kcal/kmol K	N₂O kJ/kmol K	SO₂ C_p kcal/kmol K	SO₂ kJ/kmol K	H₂S C_p kcal/kmol K	H₂S kJ/kmol K	NH₃ C_p kcal/kmol K	NH₃ kJ/kmol K
0	273.15	6.956	29.123	8.565	35.860	8.945	37.451	9.28	38.854	8.07	33.787	8.36	35.002
100	373.15	6.989	29.262	9.603	40.206	9.986	41.809	10.13	42.412	8.34	34.918	8.98	37.597
200	473.15	7.081	29.647	10.435	43.689	10.811	45.263	10.88	45.552	8.70	36.425	9.72	40.696
300	573.15	7.226	30.254	11.110	46.515	11.493	48.119	11.52	48.232	9.12	38.184	10.47	43.836
400	673.15	7.398	30.974	11.670	48.860	12.062	50.501	12.00	50.242	9.55	39.984	11.13	46.599
500	773.15	7.573	31.707	12.137	50.815	12.542	52.511	12.35	51.707	9.97	41.742	11.93	49.949
600	873.15	7.739	32.402	12.528	52.452	12.946	54.202	12.63	52.879	10.37	43.417	12.61	52.796
700	973.15	7.888	33.025	12.856	53.826	13.282	55.609	12.84	53.759	10.75	45.008	13.24	55.433
800	1073.15	8.019	33.574	13.131	54.977	13.573	56.827	13.00	54.428	11.08	46.390	13.83	57.903
900	1173.15	8.134	34.055	13.364	55.952	13.824	57.878	13.14	55.015	11.38	47.646	14.38	60.206
1000	1273.15	8.233	34.470	13.560	56.773	14.038	58.774	13.24	55.433	11.65	48.776	14.88	62.300
1100	1373.15	8.318	34.826	13.727	57.472	14.226	59.561	13.32	55.768	11.88	49.739	—	—
1200	1473.15	8.393	35.140	13.870	58.071	—	—	13.39	56.061	12.08	50.577	—	—
1300	1573.15	8.458	35.412	13.993	58.586	—	—	13.46	56.354	—	—	—	—
1400	1673.15	8.514	35.646	14.099	59.030	—	—	13.51	56.564	—	—	—	—
1500	1773.15	8.564	35.856	14.190	59.411	14.29	59.829	13.56	56.773	12.54	52.502	16.67	69.794
1600	1873.15	8.608	36.040	14.268	59.737	—	—	13.59	56.899	—	—	—	—
1700	1973.15	8.647	36.203	14.336	60.022	—	—	13.62	57.024	—	—	—	—
1800	2073.15	8.682	36.350	14.395	60.269	—	—	13.65	57.150	—	—	—	—
1900	2173.15	8.713	36.480	14.445	60.478	—	—	13.67	57.234	—	—	—	—
2000	2273.15	8.741	36.597	14.487	60.654	14.44	60.457	13.69	57.317	13.00	54.428	17.71	74.148
2100	2373.15	8.767	36.706	14.522	60.801	—	—	13.70	57.359	—	—	—	—
2200	2473.15	8.790	36.802	14.550	60.918	—	—	13.72	57.443	—	—	—	—
2300	2573.15	8.812	36.894	14.571	61.006	—	—	13.73	57.485	—	—	—	—
2400	2673.15	8.832	36.978	14.584	61.060	—	—	13.74	57.527	—	—	—	—
2500	2773.15	8.850	37.053	14.590	61.085	14.59	61.085	13.76	57.610	13.28	55.601	18.33	76.744
3000	3273.15	8.930	37.388	14.660	61.378	14.67	61.420	13.79	57.736	13.44	56.271	18.72	78.379
M*		28.010		44.010		44.016		64.04		34.08		17.031	

*M — relative molecular mass in kg/mol

1 kcal = 4.186 8 kJ

Table 73-3 Specific Heat C_p of Gases at Constant Pressure p (Continued)

Temperature		CS₂		COS		CH₄		C₂H₆		C₃H₈	
t	T	C_p		C_p		C_p		C_p		C_p	
°C	K	$\dfrac{kcal}{kmol\ K}$	$\dfrac{kJ}{kmol\ K}$	$\dfrac{kcal}{kmol\ K}$	$\dfrac{kJ}{kmol\ K}$	$\dfrac{kcal}{kmol\ K}$	$\dfrac{kJ}{kmol\ K}$	$\dfrac{kcal}{kmol\ K}$	$\dfrac{kJ}{kmol\ K}$	$\dfrac{kcal}{kmol\ K}$	$\dfrac{kJ}{kmol\ K}$
0	273.15	10.63	44.506	9.62	40.277	8.297	34.738	11.830	49.530	16.32	68.329
100	373.15	11.62	48.651	10.70	44.799	9.382	39.281	14.849	62.170	21.24	88.928
200	473.15	12.35	51.707	11.49	48.106	10.755	45.029	17.883	74.873	25.89	108.396
300	573.15	12.87	53.884	12.06	50.493	12.167	50.941	20.610	86.290	29.85	124.976
400	673.15	13.27	55.559	12.53	52.461	13.524	56.622	23.081	96.636	33.29	139.379
500	773.15	13.57	56.815	12.87	53.884	14.774	61.856	25.271	105.805	36.32	152.065
600	873.15	13.80	57.778	13.17	55.140	15.912	66.620	27.197	113.868	38.94	163.034
700	973.15	13.97	58.490	13.42	56.187	16.941	70.929	28.885	120.936	41.24	172.664
800	1073.15	14.11	59.076	13.62	57.024	17.853	74.747	30.275	126.755	43.10	180.451
900	1173.15	14.22	59.536	13.79	57.736	18.670	78.168	31.519	131.964	44.76	187.401
1000	1273.15	14.32	59.955	13.93	58.322	19.393	81.195	32.663	136.753	46.28	193.765
1100	1373.15	14.40	60.290	14.05	58.825	20.026	83.845	33.703	141.108	47.67	199.585
1200	1473.15	14.46	60.541	14.14	59.201	20.566	86.106	34.642	145.039	48.92	204.818
$M*$		76.13		60.07		16.031		30.07		44.06	

*M — relative molecular mass in kg/mol

Table 73-4 Specific Heat C_p of Gases at Constant Pressure p (Continued)

Temperature		C₂H₄		C₂H₂		C₆H₆		C₃H₆		C₂H₅OH	
t	T	C_p		$\bar{C}_p$		C_p		C_p		C_p	
°C	K	$\dfrac{kcal}{kmol\ K}$	$\dfrac{kJ}{kmol\ K}$	$\dfrac{kcal}{kmol\ K}$	$\dfrac{kJ}{kmol\ K}$	$\dfrac{kcal}{kmol\ K}$	$\dfrac{kJ}{kmol\ K}$	$\dfrac{kcal}{kmol\ K}$	$\dfrac{kJ}{kmol\ K}$	$\dfrac{kcal}{kmol\ K}$	$\dfrac{kJ}{kmol\ K}$
0	273.15	9.78	40.947	10.01	41.910	17.60	73.688	14.33	59.997	16.74	70.087
100	373.15	12.24	51.246	11.63	48.692	24.91	104.293	18.09	75.739	20.11	84.197
200	473.15	14.58	61.044	12.71	53.214	31.27	130.921	21.71	90.895	23.28	97.469
300	573.15	16.60	69.501	13.52	56.606	36.50	152.818	24.89	104.209	26.10	109.275
400	673.15	18.33	76.744	14.19	59.411	40.72	170.486	27.67	115.849	28.56	119.575
500	773.15	19.82	82.982	14.777	61.868	44.20	185.057	30.07	125.897	30.70	128.535
600	873.15	21.13	88.467	15.305	64.079	47.09	197.156	32.16	134.647	32.59	136.448
700	973.15	22.28	93.282	15.789	66.105	49.54	207.414	33.99	142.309	34.23	143.314
800	1073.15	23.27	97.427	16.230	67.952	51.62	216.123	35.58	148.966	35.68	149.385
900	1173.15	24.15	101.111	16.632	69.635	53.38	223.491	36.96	154.744	36.95	154.702
1000	1273.15	24.93	104.377	16.996	71.159	54.90	229.855	38.17	159.810	38.05	159.308
1100	1373.15	25.60	107.182	17.325	72.536	56.21	235.340	39.22	164.206	39.00	163.285
1200	1473.15	26.19	109.652	17.619	73.767	57.33	240.029	40.12	167.974	39.83	166.760
$M*$		28.031		26.040		78.108		42.08		46.07	

*M — relative molecular mass in kg/mol

1 kcal = 4.186 8 kJ

Table 74-1 Specific Heat C_v of Gases at Constant Volume v

Temperature			O_2		H_2		NO		OH		H_2O		N_2		Air	
t	T		C_v		C_v		C_v		C_v		C_v		C_v		C_v	
°C	K		$\frac{kcal}{kmol\,K}$	$\frac{kJ}{kmol\,K}$	$\frac{kcal}{kmol\,K}$	$\frac{kJ}{kmol\,K}$	$\frac{kcal}{kmol\,K}$	$\frac{kJ}{kmol\,K}$	$\frac{kcal}{kmol\,K}$	$\frac{kJ}{kmol\,K}$	$\frac{kcal}{kmol\,K}$	$\frac{kJ}{kmol\,K}$	$\frac{kcal}{kmol\,K}$	$\frac{kJ}{kmol\,K}$	$\frac{kcal}{kmol\,K}$	$\frac{kJ}{kmol\,K}$
0	273.15		5.006	20.959	4.849	20.302	5.174	21.663	5.177	21.675	6.015	25.184	4.968	20.800	4.958	20.758
100	373.15		5.150	21.562	4.971	20.813	5.160	21.604	5.095	21.332	6.148	25.740	4.988	20.884	5.004	20.951
200	473.15		5.374	22.500	4.998	20.926	5.259	22.019	5.060	21.185	6.365	26.649	5.053	21.156	5.102	21.361
300	573.15		5.617	23.517	5.012	20.984	5.432	22.743	5.059	21.181	6.621	27.721	5.168	21.637	5.243	21.951
400	673.15		5.838	24.443	5.035	21.081	5.617	23.517	5.085	21.290	6.897	28.876	5.317	22.261	5.406	22.634
500	773.15		6.027	25.234	5.074	21.244	5.798	24.275	5.142	21.529	7.187	30.091	5.478	22.935	5.571	23.325
600	873.15		6.183	25.887	5.130	21.478	5.960	24.953	5.220	21.855	7.487	31.347	5.638	23.605	5.729	23.986
700	973.15		6.313	26.431	5.203	21.784	6.101	25.544	5.318	22.265	7.795	32.636	5.786	24.225	5.872	24.585
800	1073.15		6.422	26.888	5.292	22.157	6.223	26.054	5.426	22.718	8.105	33.934	5.920	24.786	5.999	25.117
900	1173.15		6.513	27.269	5.387	22.554	6.325	26.482	5.538	23.186	8.407	35.198	6.039	25.284	6.112	25.590
1000	1273.15		6.592	27.599	5.486	22.969	6.413	26.850	5.649	23.651	8.696	36.408	6.144	25.724	6.210	26.000
1100	1373.15		6.664	27.901	5.591	23.408	6.497	27.208	5.758	24.108	8.967	37.543	6.236	26.109	6.297	26.364
1200	1473.15		6.729	28.173	5.694	23.840	6.551	27.428	5.863	24.547	9.219	38.598	6.317	26.448	6.374	26.687
1300	1573.15		6.792	28.437	5.798	24.275	6.608	27.666	5.963	24.966	9.454	39.582	6.388	26.745	6.443	26.976
1400	1673.15		6.851	28.684	5.896	24.685	6.658	27.876	6.055	25.351	9.670	40.486	6.450	27.005	6.504	27.231
1500	1773.15		6.909	28.927	5.990	25.079	6.701	28.056	6.142	25.715	9.870	41.324	6.504	27.231	6.558	27.457
1600	1873.15		6.966	29.165	6.078	25.447	6.738	28.211	6.224	26.059	10.054	42.094	6.552	27.432	6.607	27.662
1700	1973.15		7.022	29.400	6.162	25.799	6.771	28.349	6.300	26.377	10.227	42.818	6.595	27.612	6.653	27.855
1800	2073.15		7.077	29.630	6.241	26.130	6.802	28.479	6.370	26.670	10.382	43.467	6.634	27.775	6.695	28.031
1900	2173.15		7.132	29.860	6.317	26.448	6.830	28.596	6.435	26.942	10.524	44.062	6.669	27.922	6.734	28.194
2000	2273.15		7.187	30.091	6.387	26.741	6.857	28.709	6.495	27.193	10.656	44.615	6.700	28.052	6.769	28.340
2100	2373.15		7.242	30.321	6.453	27.017	6.879	28.801	6.552	27.432	10.780	45.134	6.728	28.169	6.808	28.504
2200	2473.15		7.295	30.543	6.518	27.290	6.898	28.881	6.605	27.654	10.895	45.615	6.754	28.278	6.834	28.613
2300	2573.15		7.348	30.765	6.577	27.537	6.917	28.960	6.655	27.863	11.000	46.055	6.778	28.378	6.864	28.738
2400	2673.15		7.399	30.978	6.634	27.775	6.935	29.035	6.703	28.064	11.098	46.465	6.800	28.470	6.892	28.855
2500	2773.15		7.449	31.187	6.688	28.001	6.952	29.107	6.751	28.265	11.189	46.846	6.820	28.554	6.918	28.964
2600	2873.15		7.498	31.393	6.739	28.215	6.969	29.178	6.797	28.458	11.276	47.210				
2700	2973.15		7.546	31.594	6.787	28.416	6.985	29.245	6.841	28.642	11.357	47.549				
2800	3073.15		—	—	—	—	7.000	29.308	6.884	28.822	11.434	47.872				
2900	3173.15		—	—	—	—	7.015	29.370	6.925	28.994	11.506	48.173				
3000	3273.15		—	—	—	—	7.029	29.429	6.965	29.161	—	—				
M^*			32		2.0156		30.008		17.008		18.020		28.016		28.964	

$*M$ — relative molecular mass in kg/mol

1 kcal = 4.186 8 kJ

Table 74-2 Specific Heat C_v of Gases at Constant Volume v (Continued)

Temperature		CO C_v		CO$_2$ C_v		N$_2$O $\bar{C}_v$		SO$_2$ C_v		H$_2$S C_v	
t °C	T K	kcal/kmol K	kJ/kmol K	kcal/kmol K	kJ/kmol K	kcal/kmol K	kJ/kmol K	kcal/kmol K	kJ/kmol K	kcal/kmol K	kJ/kmol K
0	273.15	4.970	20.808	6.579	27.545	6.959	29.136	7.29	30.522	6.08	25.456
100	373.15	5.003	20.947	7.617	31.891	8.000	33.494	8.14	34.081	6.35	26.586
200	473.15	5.095	21.332	8.449	35.374	8.825	36.949	8.89	37.221	6.71	28.093
300	573.15	5.240	21.939	9.124	38.200	9.507	39.804	9.53	39.900	7.13	29.852
400	673.15	5.412	22.659	9.684	40.545	10.076	42.186	10.01	41.910	7.56	31.652
500	773.15	5.587	23.392	10.151	42.500	10.556	44.196	10.36	43.375	7.98	33.441
600	873.15	5.753	24.087	10.542	44.137	10.960	45.887	10.64	44.548	8.38	35.085
700	973.15	5.902	24.710	10.870	45.511	11.296	47.294	10.85	45.427	8.76	36.676
800	1073.15	6.033	25.259	11.145	46.662	11.587	48.512	11.01	46.097	9.09	38.058
900	1173.15	6.148	25.740	11.378	47.637	11.838	49.563	11.15	46.683	9.39	39.314
1000	1273.15	6.247	26.155	11.574	48.458	12.052	50.459	11.25	47.102	9.66	40.444
1100	1373.15	6.332	26.511	11.741	49.157	12.240	51.246	11.33	47.436	9.89	41.407
1200	1473.15	6.407	26.825	11.884	49.756	—	—	11.40	47.730	10.09	42.245
1300	1573.15	6.472	27.097	12.007	50.271						
1400	1673.15	6.528	27.331	12.113	50.715						
1500	1773.15	6.578	27.541	12.204	51.096						
1600	1873.15	6.622	27.725	12.282	51.422						
1700	1973.15	6.661	27.888	12.350	51.707						
1800	2073.15	6.696	28.035	12.409	51.954						
1900	2173.15	6.727	28.165	12.459	52.163						
2000	2273.15	6.755	28.282	12.501	52.339						
2100	2373.15	6.781	28.391	12.536	52.486						
2200	2473.15	6.804	28.487	12.564	52.603						
2300	2573.15	6.826	28.579	12.585	52.691						
2400	2673.15	6.846	28.663	12.598	52.745						
2500	2773.15	6.864	28.738	12.604	52.770						
M*		28.010		44.010		44.016		64.06		34.08	

*M — relative molecular mass in kg/mol

1 kcal = 4.186 8 kJ

Table 74-3 Specific Heat C_v of Gases at Constant Volume v (Continued)

Temperature		CS₂		COS		CH₄		C₂H₆		C₃H₈	
t	$\overline{T}$	C_v		C_v		C_v		C_v		C_v	
°C	K	kcal / kmol K	kJ / kmol K	kcal / kmol K	kJ / kmol K	kcal / kmol K	kJ / kmol K	kcal / kmol K	kJ / kmol K	kcal / kmol K	kJ / kmol K
0	273.15	8.64	36.174	7.63	31.945	6.311	26.423	9.844	41.215	14.33	59.997
100	373.15	9.63	40.319	8.71	36.467	7.396	30.966	12.863	53.855	19.25	80.596
200	473.15	10.36	43.375	9.50	39.755	8.769	36.714	15.897	66.558	23.90	100.065
300	573.15	10.88	45.552	10.07	42.161	10.181	42.626	18.624	77.975	27.86	116.644
400	673.15	11.28	47.227	10.54	44.129	11.538	48.307	21.095	88.321	31.30	131.047
500	773.15	11.58	48.483	10.88	45.552	12.788	53.541	23.285	97.490	34.33	143.733
600	873.15	11.81	49.446	11.18	46.808	13.926	58.305	25.211	105.553	36.95	154.702
700	973.15	11.98	50.158	11.43	47.855	14.955	62.614	26.899	112.621	39.25	164.332
800	1073.15	12.12	50.744	11.63	48.692	15.867	66.432	28.289	118.440	41.11	172.119
900	1173.15	12.23	51.205	11.80	49.404	16.684	69.853	29.533	123.649	42.77	179.069
1000	1273.15	12.33	51.623	11.94	49.990	17.407	72.880	30.677	128.438	44.29	185.433
1100	1373.15	12.41	51.958	12.06	50.493	18.040	75.530	31.717	132.793	45.68	191.253
1200	1473.15	12.47	52.209	12.15	50.870	18.580	77.791	32.656	136.724	46.93	196.487
M*		76.13		60.07		16.031		30.07		44.06	

*M — relative molecular mass in kg/mol

Table 74-4 Specific Heat C_v of Gases at Constant Volume v (Continued)

Temperature		C₂H₄		C₂H₂		C₆H₆		C₃H₆		C₂H₅OH	
t	T	C_v		C_v		C_v		C_v		C_v	
°C	K	kcal / kmol K	kJ / kmol K	kcal / kmol K	kJ / kmol K	kcal / kmol K	kJ / kmol K	kcal / kmol K	kJ / kmol K	kcal / kmol K	kJ / kmol K
0	273.15	7.79	32.615	8.02	33.578	15.61	65.356	12.34	51.665	14.75	61.755
100	373.15	10.25	42.915	9.64	40.361	22.92	95.961	16.10	67.407	18.12	75.865
200	473.15	12.59	52.712	10.72	44.882	29.28	122.590	19.72	82.564	21.29	89.137
300	573.15	14.61	61.169	11.53	48.274	34.51	144.486	22.90	95.878	24.11	100.944
400	673.15	16.34	68.412	12.20	51.079	38.73	162.155	25.68	107.517	26.57	111.243
500	773.15	17.83	74.651	12.791	53.553	42.21	176.725	28.08	117.565	28.71	120.203
600	873.15	19.14	80.135	13.319	55.764	45.10	188.825	30.17	126.316	30.60	128.116
700	973.15	20.29	84.950	13.803	57.790	47.45	199.082	32.00	133.978	32.24	134.982
800	1073.15	21.28	89.095	14.244	59.637	49.63	207.791	33.59	140.635	33.69	141.053
900	1173.15	22.16	92.779	14.646	61.320	51.39	215.160	34.97	146.412	34.96	146.371
1000	1273.15	22.94	96.045	15.010	62.844	52.91	221.524	36.18	151.478	36.06	150.976
1100	1373.15	23.61	98.850	15.339	64.221	54.22	227.008	37.23	155.875	37.01	154.953
1200	1473.15	24.20	101.321	15.633	65.452	55.34	231.698	38.13	159.643	37.84	158.429
M*		28.031		26.04		78.108		42.08		46.07	

*M — relative molecular mass in kg/mol

1 kcal = 4.186 8 kJ

Table 75-1 Specific Heat c_p' of Gases at Constant Pressure p

| Temperature | | O₂ | | H₂ | | NO | | OH | | H₂O | | N₂ | | Air | |
°C	K	$\frac{kcal}{m_n^3 K}$	$\frac{kJ}{m_n^3 K}$	$\frac{kcal}{m_n^3 K}$	$\frac{kJ}{m_n^3 K}$	$\frac{kcal}{m_n^3 K}$	$\frac{kJ}{m_n^3 K}$	$\frac{kcal}{m_n^3 K}$	$\frac{kJ}{m_n^3 K}$	$\frac{kcal}{m_n^3 K}$	$\frac{kJ}{m_n^3 K}$	$\frac{kcal}{m_n^3 K}$	$\frac{kJ}{m_n^3 K}$	$\frac{kcal}{m_n^3 K}$	$\frac{kJ}{m_n^3 K}$
0	273.15	0.3119	1.3059	0.3049	1.2766	0.3194	1.3373	0.3196	1.3381	0.3569	1.4943	0.3102	1.2987	0.3098	1.2971
100	373.15	0.3184	1.3331	0.3104	1.2996	0.3188	1.3348	0.3159	1.3226	0.3629	1.5194	0.3111	1.3025	0.3119	1.3059
200	473.15	0.3284	1.3749	0.3116	1.3046	0.3232	1.3532	0.3143	1.3159	0.3726	1.5600	0.3140	1.3147	0.3162	1.3239
300	573.15	0.3392	1.4202	0.3122	1.3071	0.3309	1.3854	0.3143	1.3159	0.3840	1.6077	0.3192	1.3364	0.3225	1.3502
400	673.15	0.3491	1.4616	0.3132	1.3113	0.3392	1.4202	0.3155	1.3209	0.3963	1.6592	0.3258	1.3641	0.3298	1.3808
500	773.15	0.3575	1.4968	0.3150	1.3188	0.3473	1.4541	0.3180	1.3314	0.4092	1.7132	0.3330	1.3942	0.3372	1.4118
600	873.15	0.3644	1.5257	0.3175	1.3293	0.3545	1.4842	0.3215	1.3461	0.4226	1.7693	0.3401	1.4239	0.3442	1.4411
700	973.15	0.3702	1.5500	0.3207	1.3427	0.3608	1.5106	0.3259	1.3645	0.4364	1.8271	0.3467	1.4516	0.3506	1.4679
800	1073.15	0.3751	1.5705	0.3247	1.3595	0.3662	1.5332	0.3307	1.3846	0.4502	1.8849	0.3527	1.4767	0.3563	1.4918
900	1173.15	0.3792	1.5876	0.3289	1.3770	0.3708	1.5525	0.3357	1.4055	0.4637	1.9414	0.3580	1.4989	0.3613	1.5127
1000	1273.15	0.3827	1.6023	0.3333	1.3955	0.3747	1.5688	0.3406	1.4260	0.4766	1.9954	0.3627	1.5186	0.3657	1.5311
1100	1373.15	0.3859	1.6157	0.3380	1.4151	0.3780	1.5826	0.3455	1.4465	0.4886	2.0457	0.3668	1.5357	0.3695	1.5470
1200	1473.15	0.3888	1.6278	0.3426	1.4344	0.3809	1.5984	0.3502	1.4662	0.4999	2.0930	0.3704	1.5508	0.3730	1.5617
1300	1573.15	0.3916	1.6396	0.3473	1.4541	0.3834	1.6052	0.3546	1.4846	0.5104	2.1369	0.3736	1.5642	0.3760	1.5742
1400	1673.15	0.3942	1.6504	0.3516	1.4721	0.3856	1.6144	0.3587	1.5018	0.5200	2.1771	0.3764	1.5759	0.3788	1.5860
1500	1773.15	0.3968	1.6613	0.3558	1.4897	0.3876	1.6228	0.3626	1.5181	0.5289	2.2144	0.3788	1.5860	0.3812	1.5960
1600	1873.15	0.3994	1.6722	0.3598	1.5064	0.3892	1.6295	0.3663	1.5336	0.5371	2.2487	0.3809	1.5948	0.3834	1.6052
1700	1973.15	0.4019	1.6827	0.3635	1.5219	0.3907	1.6358	0.3697	1.5479	0.5449	2.2814	0.3828	1.6027	0.3854	1.6136
1800	2073.15	0.4043	1.6927	0.3670	1.5366	0.3921	1.6416	0.3728	1.5608	0.5518	2.3103	0.3846	1.6102	0.3873	1.6215
1900	2173.15	0.4068	1.7032	0.3704	1.5508	0.3933	1.6467	0.3757	1.5730	0.5581	2.3367	0.3861	1.6165	0.3890	1.6287
2000	2273.15	0.4092	1.7132	0.3735	1.5638	0.3945	1.6517	0.3784	1.5843	0.5640	2.3614	0.3875	1.6224	0.3906	1.6354
2100	2373.15	0.4117	1.7273	0.3765	1.5763	0.3955	1.6559	0.3809	1.5948	0.5695	2.3844	0.3888	1.6278	0.3921	1.6416
2200	2473.15	0.4141	1.7338	0.3794	1.5885	0.3963	1.6592	0.3833	1.6048	0.5747	2.4062	0.3899	1.6324	0.3935	1.6475
2300	2573.15	0.4164	1.7434	0.3820	1.5994	0.3972	1.6630	0.3855	1.6140	0.5793	2.4254	0.3910	1.6370	0.3948	1.6529
2400	2673.15	0.4197	1.7572	0.3846	1.6102	0.3980	1.6663	0.3876	1.6228	0.5837	2.4438	0.3920	1.6412	0.3961	1.6584
2500	2773.15	0.4209	1.7622	0.3870	1.6203	0.3988	1.6697	0.3898	1.6320	0.5878	2.4610	0.3929	1.6450	0.3972	1.6630
2600	2873.15	0.4231	1.7714	0.3893	1.6299	0.3995	1.6726	0.3918	1.6404	0.5917	2.4773	—	—	—	—
2700	2973.15	0.4253	1.7806	0.3914	1.6387	0.4002	1.6756	0.3938	1.6488	0.5953	2.4924	—	—	—	—
2800	3073.15	—	—	—	—	0.4009	1.6785	0.3957	1.6567	0.5987	2.5066	—	—	—	—
2900	3173.15	—	—	—	—	0.4016	1.6814	0.3976	1.6647	0.6019	2.5200	—	—	—	—
3000	3273.15	0.4290	1.7961	0.3980	1.6663	0.4022	1.6839	0.3993	1.6718	0.6019	2.5200	0.3970	1.6622	0.4020	1.6831

1 kcal = 4.186 8 kJ

Table 75-2 Specific Heat c_p' of Gases at Constant Pressure p (Continued)

Temperature t (°C)	T (K)	CO c_p' kcal/m_n^3 K	CO c_p' kJ/m_n^3 K	CO_2 c_p' kcal/m_n^3 K	CO_2 c_p' kJ/m_n^3 K	N_2O c_p' kcal/m_n^3 K	N_2O c_p' kJ/m_n^3 K	SO_2 c_p' kcal/m_n^3 K	SO_2 c_p' kJ/m_n^3 K	H_2S c_p' kcal/m_n^3 K	H_2S c_p' kJ/m_n^3 K	NH_3 c_p' kcal/m_n^3 K	NH_3 c_p' kJ/m_n^3 K
0	273.15	0.3103	1.2992	0.3821	1.5998	0.3991	1.6710	0.414	1.733	0.360	1.507	0.373	1.562
100	373.15	0.3118	1.3054	0.4284	1.7936	0.4455	1.8652	0.452	1.892	0.372	1.557	0.400	1.675
200	473.15	0.3159	1.3226	0.4655	1.9490	0.4823	2.0193	0.485	2.031	0.388	1.624	0.434	1.817
300	573.15	0.3224	1.3498	0.4957	2.0754	0.5127	2.1466	0.514	2.152	0.407	1.704	0.467	1.955
400	673.15	0.3300	1.3816	0.5206	2.1796	0.5381	2.2529	0.535	2.240	0.426	1.784	0.497	2.081
500	773.15	0.3379	1.4147	0.5415	2.2672	0.5595	2.3425	0.551	2.307	0.445	1.863	0.532	2.227
600	873.15	0.3453	1.4457	0.5589	2.3400	0.5776	2.4183	0.563	2.357	0.463	1.938	0.562	2.353
700	973.15	0.3519	1.4733	0.5736	2.4015	0.5926	2.4811	0.573	2.399	0.480	2.010	0.591	2.474
800	1073.15	0.3577	1.4976	0.5858	2.4526	0.6055	2.5351	0.580	2.428	0.494	2.068	0.617	2.583
900	1173.15	0.3629	1.5194	0.5962	2.4926	0.6167	2.5820	0.586	2.453	0.508	2.127	0.642	2.688
1000	1273.15	0.3673	1.5378	0.6050	2.5330	0.5263	2.6222	0.591	2.474	0.520	2.177	0.664	2.780
1100	1373.15	0.3711	1.5537	0.6124	2.5640	0.6347	4.6574	0.594	2.487	0.530	2.219	—	—
1200	1473.15	0.3744	1.5675	0.6188	2.5908	—	—	0.597	2.500	0.531	2.223	—	—
1300	1573.15	0.3773	1.5797	0.6243	2.6138	—	—	—	—	—	—	—	—
1400	1673.15	0.3798	1.5901	0.6290	2.6335	—	—	—	—	—	—	—	—
1500	1773.15	0.3821	1.5998	0.6331	2.6507	0.6330	2.6502	0.605	2.533	0.560	2.345	0.744	3.115
1600	1873.15	0.3840	1.6077	0.6365	2.6649	—	—	—	—	—	—	—	—
1700	1973.15	0.3858	1.6153	0.6396	2.6779	—	—	—	—	—	—	—	—
1800	2073.15	0.3873	1.6215	0.6422	2.6888	—	—	—	—	—	—	—	—
1900	2173.15	0.3887	1.6274	0.6444	2.6980	—	—	—	—	—	—	—	—
2000	2273.15	0.3900	1.6329	0.6463	2.7059	0.6440	2.6963	0.610	2.554	0.580	2.428	0.790	3.308
2100	2373.15	0.3911	1.6375	0.6479	2.7126	—	—	—	—	—	—	—	—
2200	2473.15	0.3921	1.6416	0.6491	2.7177	—	—	—	—	—	—	—	—
2300	2573.15	0.3931	1.6458	0.6501	2.7218	—	—	—	—	—	—	—	—
2400	2673.15	0.3940	1.6496	0.6506	2.7239	—	—	—	—	—	—	—	—
2500	2773.15	0.3948	1.6529	0.6509	2.7252	0.6510	2.7256	0.614	2.571	0.592	2.479	0.818	3.425
3000	3273.15	0.398	1.6663	0.6540	2.7382	0.6540	2.7382	0.616	2.579	0.600	2.512	0.835	3.496

1 kcal = 4.186 8 kJ

Table 75-3 Specific Heat c_p' of Gases at Constant Pressure p (Continued)

Temperature		CS$_2$		COS		CH$_4$		C$_2$H$_6$		C$_3$H$_8$	
		c_p'		c_p'		c_p'		c_p'		c_p'	
t	T										
°C	K	$\frac{kcal}{m_n^3 K}$	$\frac{kJ}{m_n^3 K}$	$\frac{kcal}{m_n^3 K}$	$\frac{kJ}{m_n^3 K}$	$\frac{kcal}{m_n^3 K}$	$\frac{kJ}{m_n^3 K}$	$\frac{kcal}{m_n^3 K}$	$\frac{kJ}{m_n^3 K}$	$\frac{kcal}{m_n^3 K}$	$\frac{kJ}{m_n^3 K}$
0	273.15	0.474	1.985	0.429	1.796	0.3702	1.5500	0.5278	2.2098	0.7281	3.0484
100	373.15	0.518	2.169	0.477	1.997	0.4186	1.7526	0.6625	2.7738	0.9476	3.9674
200	473.15	0.551	2.307	0.513	2.148	0.4798	2.0088	0.7978	3.3402	1.1550	4.8358
300	573.15	0.574	2.403	0.538	2.252	0.5428	2.2726	0.9195	3.8498	1.3317	5.5756
400	673.15	0.592	2.479	0.559	2.340	0.6034	2.5263	1.0297	4.3111	1.4852	6.2182
500	773.15	0.605	2.533	0.574	2.403	0.6591	2.7595	1.1274	4.7202	1.6204	6.7843
600	874.15	0.616	2.579	0.588	2.462	0.7099	2.9722	1.2134	5.0803	1.7373	7.2737
700	973.15	0.623	2.608	0.599	2.508	0.7558	3.1644	1.2887	5.3955	1.8399	7.7033
800	1073.15	0.629	2.633	0.608	2.546	0.7965	3.3348	1.3507	5.6551	1.9228	8.0504
900	1173.15	0.634	2.654	0.615	2.575	0.8329	3.4872	1.4062	5.8875	1.9969	8.3606
1000	1273.15	0.639	2.675	0.621	2.600	0.8652	3.6224	1.4572	5.1010	2.0647	8.6445
1100	1373.15	0.642	2.688	0.627	2.625	0.8934	3.7405	1.5036	6.2953	2.1267	8.9041
1200	1473.15	0.645	2.700	0.631	2.642	0.9175	3.8414	1.5455	6.4707	2.1825	9.1377

Table 75-4 Specific Heat c_p' of Gases at Constant Pressure p (Continued)

Temperature		C$_2$H$_4$		C$_2$H$_2$		C$_6$H$_6$		C$_3$H$_6$		C$_2$H$_5$OH	
		c_p'		c_p'		c_p'		c_p'		c_p'	
t	T										
°C	K	$\frac{kcal}{m_n^3 K}$	$\frac{kJ}{m_n^3 K}$	$\frac{kcal}{m_n^3 K}$	$\frac{kJ}{m_n^3 K}$	$\frac{kcal}{m_n^3 K}$	$\frac{kJ}{m_n^3 K}$	$\frac{kcal}{m_n^3 K}$	$\frac{kJ}{m_n^3 K}$	$\frac{kcal}{m_n^3 K}$	$\frac{kJ}{m_n^3 K}$
0	273.15	0.4363	1.8267	0.4466	1.8698	0.7852	3.2875	0.6393	2.6766	0.7468	3.1267
100	373.15	0.5461	2.2864	0.5189	2.1725	1.1113	4.6528	0.8071	3.3792	0.8972	3.7564
200	473.15	0.6505	2.7235	0.5670	2.3729	1.3951	5.8410	0.9686	4.0553	1.0386	4.3484
300	573.15	0.7406	3.1007	0.6032	2.5255	1.6284	6.8178	1.1104	4.6490	1.1644	4.8751
400	673.15	0.8178	3.4240	0.6331	2.6507	1.8167	7.6062	1.2345	5.1686	1.2742	5.3348
500	773.15	0.8842	3.7020	0.65926	2.7602	1.9719	8.2560	1.3415	5.6166	1.3696	5.7342
600	873.15	0.9427	3.9469	0.68281	2.8588	2.1009	8.7960	1.4368	6.0072	1.4540	6.0876
700	973.15	0.9940	4.1617	0.70441	2.9492	2.2102	9.2537	1.5164	6.3489	1.5271	6.3937
800	1073.15	1.0382	4.3467	0.72408	3.0316	2.3030	9.6422	1.5874	6.6461	1.5918	6.6645
900	1173.15	1.0774	4.5109	0.74202	3.1067	2.3815	9.9709	1.6489	6.9036	1.6485	6.9019
1000	1273.15	1.1122	4.6566	0.75826	3.1747	2.4493	10.2547	1.7029	7.1297	1.6976	7.1075
1100	1373.15	1.1421	4.7817	0.77293	3.2361	2.5077	10.4992	1.7497	7.3256	1.7399	7.2846
1200	1473.15	1.1684	4.8919	0.78605	3.2910	2.5577	10.7086	1.7899	7.4940	1.7770	7.4399

1 kcal = 4.186 8 kJ

Table 76-1 Specific Heat c_v' of Gases at Constant Volume v

Temperature t (°C)	T (K)	O₂ $\frac{kcal}{m_n^3 K}$	O₂ $\frac{kJ}{m_n^3 K}$	H₂ $\frac{kcal}{m_n^3 K}$	H₂ $\frac{kJ}{m_n^3 K}$	NO $\frac{kcal}{m_n^3 K}$	NO $\frac{kJ}{m_n^3 K}$	OH $\frac{kcal}{m_n^3 K}$	OH $\frac{kJ}{m_n^3 K}$	H₂O $\frac{kcal}{m_n^3 K}$	H₂O $\frac{kJ}{m_n^3 K}$	N₂ $\frac{kcal}{m_n^3 K}$	N₂ $\frac{kJ}{m_n^3 K}$	Air $\frac{kcal}{m_n^3 K}$	Air $\frac{kJ}{m_n^3 K}$
0	273.15	0.2233	0.9349	0.2163	0.9056	0.2308	0.9663	0.2310	0.9672	0.2684	1.1237	0.2216	0.9278	0.2212	0.9261
100	373.15	0.2298	0.9621	0.2218	0.9286	0.2302	0.9638	0.2273	0.9517	0.2743	1.1484	0.2225	0.9316	0.2233	0.9349
200	473.15	0.2397	1.0036	0.2230	0.9337	0.2346	0.9822	0.2257	0.9450	0.2840	1.1891	0.2254	0.9437	0.2276	0.9529
300	573.15	0.2506	1.0492	0.2236	0.9362	0.2423	1.0145	0.2257	0.9450	0.2954	1.2368	0.2306	0.9655	0.2339	0.9793
400	673.15	0.2605	1.0907	0.2246	0.9404	0.2506	1.0492	0.2269	0.9500	0.3077	1.2883	0.2372	0.9931	0.2412	1.0099
500	773.15	0.2689	1.1258	0.2264	0.9479	0.2587	1.0831	0.2294	0.9605	0.3206	1.3423	0.2444	1.0233	0.2486	1.0408
600	873.15	0.2758	1.1547	0.2289	0.9584	0.2659	1.1133	0.2329	0.9751	0.3340	1.3984	0.2515	1.0530	0.2556	1.0701
700	973.15	0.2816	1.1790	0.2321	0.9718	0.2722	1.1396	0.2373	0.9935	0.3478	1.4562	0.2581	1.0806	0.2620	1.0969
800	1073.15	0.2865	1.1995	0.2361	0.9885	0.2776	1.1623	0.2421	1.0136	0.3616	1.5139	0.2641	1.1057	0.2677	1.1208
900	1173.15	0.2906	1.2167	0.2403	1.0061	0.2822	1.1815	0.2471	1.0346	0.3751	1.5704	0.2694	1.1279	0.2727	1.1417
1000	1273.15	0.2941	1.2313	0.2447	1.0245	0.2861	1.1978	0.2520	1.0551	0.3880	1.6245	0.2741	1.1476	0.2771	1.1602
1100	1373.15	0.2973	1.2447	0.2494	1.0442	0.2899	1.2138	0.2569	1.0756	0.4000	1.6747	0.2782	1.1648	0.2809	1.1761
1200	1473.15	0.3002	1.2569	0.2540	1.0634	0.2923	1.2238	0.2616	1.0953	0.4113	1.7220	0.2818	1.1798	0.2844	1.1907
1300	1573.15	0.3030	1.2686	0.2587	1.0831	0.2948	1.2343	0.2660	1.1137	0.4218	1.7660	0.2850	1.1932	0.2874	1.2033
1400	1673.15	0.3056	1.2795	0.2630	1.1011	0.2970	1.2435	0.2701	1.1309	0.4314	1.8062	0.2878	1.2050	0.2901	1.2146
1500	1773.15	0.3082	1.2904	0.2672	1.1187	0.2990	1.2519	0.2740	1.1472	0.4403	1.8434	0.2902	1.2150	0.2926	1.2251
1600	1873.15	0.3108	1.3013	0.2712	1.1355	0.3006	1.2586	0.2777	1.1627	0.4485	1.8778	0.2923	1.2238	0.2948	1.2343
1700	1973.15	0.3133	1.3117	0.2749	1.1510	0.3021	1.2648	0.2811	1.1769	0.4563	1.9104	0.2942	1.2318	0.2968	1.2426
1800	2073.15	0.3157	1.3218	0.2784	1.1656	0.3035	1.2707	0.2842	1.1899	0.4632	1.9393	0.2960	1.2393	0.2987	1.2506
1900	2173.15	0.3182	1.3322	0.2818	1.1798	0.3047	1.2757	0.2871	1.2020	0.4695	1.9657	0.2975	1.2456	0.3004	1.2577
2000	2273.15	0.3206	1.3423	0.2849	1.1928	0.3059	1.2807	0.2898	1.2133	0.4754	1.9904	0.2989	1.2514	0.3020	1.2644
2100	2373.15	0.3231	1.3528	0.2879	1.2054	0.3069	1.2849	0.2923	1.2238	0.4809	2.0134	0.3002	1.2569	0.3035	1.2707
2200	2473.15	0.3255	1.3628	0.2908	1.2175	0.3077	1.2883	0.2947	1.2338	0.4861	2.0352	0.3013	1.2615	0.3049	1.2766
2300	2573.15	0.3287	1.3724	0.2934	1.2284	0.3086	1.2920	0.2969	1.2431	0.4907	2.0545	0.3024	1.2661	0.3062	1.2820
2400	2673.15	0.3301	1.3821	0.2960	1.2383	0.3094	1.2954	0.2990	1.2519	0.4951	2.0729	0.3034	1.2703	0.3075	1.2874
2500	2773.15	0.3328	1.3934	0.2984	1.2493	0.3101	1.2983	0.3012	1.2611	0.4992	2.0901	0.3043	1.2740	0.3086	1.2920
2600	2873.15	0.3345	1.4005	0.3006	1.2586	0.3109	1.3017	0.3032	1.2694	0.5031	2.1064				
2700	2973.15	0.3367	1.4097	0.3028	1.2678	0.3116	1.3046	0.3052	1.2778	0.5067	2.1215				
2800	3073.15	—	—	—	—	0.3123	1.3075	0.3071	1.2858	0.5101	2.1357				
2900	3173.15	—	—	—	—	0.3130	1.3105	0.3090	1.2937	0.5133	2.1491				
3000	3273.15	—	—	—	—	0.3136	1.3130	0.3107	1.3008	—	—				

1 kcal = 4.186 8 kJ

Table 76-2 Specific Heat c_v' of Gases at Constant Volume v (Continued)

Temperature t °C	T K	CO c_v' $\frac{kcal}{m_n^3 K}$	CO c_v' $\frac{kJ}{m_n^3 K}$	CO_2 c_v' $\frac{kcal}{m_n^3 K}$	CO_2 c_v' $\frac{kJ}{m_n^3 K}$	N_2O c_v' $\frac{kcal}{m_n^3 K}$	N_2O c_v' $\frac{kJ}{m_n^3 K}$	SO_2 c_v' $\frac{kcal}{m_n^3 K}$	SO_2 c_v' $\frac{kJ}{m_n^3 K}$	H_2S c_v' $\frac{kcal}{m_n^3 K}$	H_2S c_v' $\frac{kJ}{m_n^3 K}$	CS_2 c_v' $\frac{kcal}{m_n^3 K}$	CS_2 c_v' $\frac{kJ}{m_n^3 K}$	COS c_v' $\frac{kcal}{m_n^3 K}$	COS c_v' $\frac{kJ}{m_n^3 K}$
0	273.15	0.2217	0.9282	0.2935	1.2288	0.3105	1.3000	0.325	1.361	0.271	1.135	0.385	1.612	0.340	1.424
100	373.15	0.2232	0.9345	0.3398	1.4227	0.3596	1.4943	0.363	1.520	0.283	1.185	0.430	1.800	0.389	1.629
200	473.15	0.2273	0.9517	0.3769	1.5780	0.3937	1.6483	0.397	1.662	0.299	1.252	0.462	1.934	0.424	1.775
300	573.15	0.2338	0.9789	0.4071	1.7044	0.4241	1.7756	0.425	1.779	0.318	1.331	0.485	2.031	0.449	1.880
400	673.15	0.2414	1.0107	0.4320	1.8087	0.4495	1.8820	0.447	1.871	0.337	1.411	0.503	2.106	0.470	1.968
500	773.15	0.2492	1.0434	0.4529	1.8962	0.4709	1.9716	0.462	1.934	0.356	1.491	0.517	2.615	0.485	2.031
600	873.15	0.2567	1.0748	0.4703	1.9691	0.4890	2.0473	0.475	1.989	0.374	1.566	0.527	2.206	0.499	2.089
700	973.15	0.2633	1.1024	0.4850	2.0306	0.5040	2.1101	0.484	2.026	0.391	1.637	0.534	2.236	0.510	2.135
800	1073.15	0.2691	1.1267	0.4972	2.0817	0.5169	2.1642	0.491	2.056	0.406	1.700	0.541	2.265	0.519	2.173
900	1173.15	0.2743	1.1484	0.5076	2.1252	0.5281	2.2110	0.497	2.081	0.419	1.754	0.546	2.286	0.526	2.202
1000	1273.15	0.2787	1.1669	0.5164	2.1621	0.5377	2.2512	0.502	2.102	0.431	1.805	0.550	2.303	0.533	2.232
1100	1373.15	0.2825	1.1828	0.5238	2.1930	0.5461	2.2864	0.505	2.114	0.441	1.846	0.554	2.319	0.538	2.252
1200	1473.15	0.2858	1.1966	0.5302	2.2198	—	—	0.509	2.131	0.450	1.884	0.556	2.328	0.542	2.269
1300	1573.15	0.2887	1.2087	0.5357	2.2429										
1400	1673.15	0.2912	1.2192	0.5404	2.2625										
1500	1773.15	0.2935	1.2288	0.5445	2.2797										
1600	1873.15	0.2954	1.2368	0.5479	2.2939										
1700	1973.15	0.2972	1.2443	0.5510	2.3069										
1800	2073.15	0.2987	1.2506	0.5536	2.3178										
1900	2173.15	0.3001	1.2565	0.5558	2.3270										
2000	2273.15	0.3014	1.2619	0.5577	2.3350										
2100	2373.15	0.3025	1.2665	0.5592	2.3413										
2200	2473.15	0.3035	1.2707	0.5605	2.3467										
2300	2573.15	0.3045	1.2749	0.5615	2.3509										
2400	2673.15	0.3054	1.2786	0.5620	2.3530										
2500	2773.15	0.3062	1.2820	0.5623	2.3542										

1 kcal = 4.1868 kJ

Table 76-3 Specific Heat c_v' of Gases at Constant Volume v (Continued)

Temperature		CH₄		C₂H₆		C₃H₈		C₂H₄	
t	T	c_v'		c_v'		c_v'		c_v'	
°C	K	$\dfrac{kcal}{m_n^3\,K}$	$\dfrac{kJ}{m_n^3\,K}$	$\dfrac{kcal}{m_n^3\,K}$	$\dfrac{kJ}{m_n^3\,K}$	$\dfrac{kcal}{m_n^3\,K}$	$\dfrac{kJ}{m_n^3\,K}$	$\dfrac{kcal}{m_n^3\,K}$	$\dfrac{kJ}{m_n^3\,K}$
0	273.15	0.2816	1.1790	0.4392	1.8388	0.6393	2.6766	0.3475	1.4549
100	373.15	0.3300	1.3816	0.5739	2.4028	0.8588	3.5956	0.4573	1.9146
200	473.15	0.3912	1.6379	0.7092	2.9693	1.0663	4.4644	0.5617	2.3517
390	573.15	0.4542	1.9016	0.8309	3.4788	1.2429	5.2038	0.6518	2.7290
400	673.15	0.5147	2.1549	0.9411	3.9402	1.3964	5.8464	0.7290	3.0522
500	773.15	0.5705	2.3886	1.0388	4.3492	1.5316	6.4125	0.7955	3.3306
600	873.15	0.6213	2.6013	1.1248	4.7093	1.6485	6.9019	0.8539	3.5751
700	973.15	0.6672	2.7934	1.2001	5.0246	1.7511	7.3315	0.9052	3.7899
800	1073.15	0.7079	2.9638	1.2621	5.2842	1.8341	7.6790	0.9494	3.9749
900	1173.15	0.7443	3.1162	1.3176	5.5165	1.9081	7.9888	0.9886	4.1391
1000	1273.15	0.7766	3.2515	1.3686	5.7301	1.9759	8.2727	1.0234	4.2848
1100	1373.15	0.8048	3.3695	1.4150	5.9243	2.0380	8.5327	1.0533	4.4100
1200	1473.15	0.8289	3.4704	1.4569	6.0997	2.0937	8.7659	1.0796	4.5201

Table 76-4 Specific Heat c_v' of Gases at Constant Volume v (Continued)

Temperature		C₂H₂		C₆H₆		C₃H₆		C₂H₅OH	
t	T	c_v'		c_v'		c_v'		c_v'	
°C	K	$\dfrac{kcal}{m_n^3\,K}$	$\dfrac{kJ}{m_n^3\,K}$	$\dfrac{kcal}{m_n^3\,K}$	$\dfrac{kJ}{m_n^3\,K}$	$\dfrac{kcal}{m_n^3\,K}$	$\dfrac{kJ}{m_n^3\,K}$	$\dfrac{kcal}{m_n^3\,K}$	$\dfrac{kJ}{m_n^3\,K}$
0	273.15	0.3578	1.4980	0.6964	2.9157	0.5505	2.3048	0.6582	2.7558
100	373.15	0.4301	1.8007	1.0225	4.2810	0.7183	3.0074	0.8086	3.3854
200	473.15	0.4783	2.0025	1.3063	5.4692	0.8798	3.6835	0.9500	3.9775
300	573.15	0.5144	2.1537	1.5396	6.4460	1.0217	4.2777	1.0758	4.5042
400	673.15	0.5443	2.2789	1.7279	7.2344	1.1457	4.7968	1.1856	4.9639
500	773.15	0.57065	2.3892	1.8831	7.8842	1.2527	5.2448	1.2810	5.3633
600	873.15	0.59421	2.4878	2.0121	8.4243	1.3460	5.6354	1.3654	5.7167
700	973.15	0.61580	2.5782	2.1214	8.8819	1.4276	5.9771	1.4386	6.0231
800	1073.15	0.63548	2.6606	2.2142	9.2704	1.4986	6.2743	1.5032	6.2936
900	1173.15	0.65341	2.7357	2.2927	9.5991	1.5601	6.5318	1.5599	6.5310
1000	1273.15	0.66965	2.8037	2.3605	9.8829	1.6141	6.7579	1.6090	6.7366
1100	1373.15	0.68433	2.8652	2.4190	10.1279	1.6610	6.9543	1.6513	6.9137
1200	1473.15	0.69745	2.9201	2.4689	10.3368	1.7011	7.1222	1.6884	7.0690

1 kcal = 4.186 8 kJ

Table 77-1 Mean Specific Heat c_{pm} of Gases at Constant Pressure p Between $0°C$ (273.15 K) and $t°C$ (TK)

Temperature		H₂		N₂		O₂		CO		CO₂	
t	T	c_{pm}		c_{pm}		c_{pm}		c_{pm}		c_{pm}	
°C	K	$\frac{kcal}{kg\,K}$	$\frac{kJ}{kg\,K}$	$\frac{kcal}{kg\,K}$	$\frac{kJ}{kg\,K}$	$\frac{kcal}{kg\,K}$	$\frac{kJ}{kg\,K}$	$\frac{kcal}{kg\,K}$	$\frac{kJ}{kg\,K}$	$\frac{kcal}{kg\,K}$	$\frac{kJ}{kg\,K}$
0	273.15	3,3904	14.195	0.2482	1.039	0.2185	0.915	0.2483	1.040	0.1946	0.815
100	373.15	3.4281	14.353	0.2485	1.040	0.2205	0.923	0.2488	1.042	0.2068	0.866
200	473.15	3.4444	14.421	0.2492	1.043	0.2234	0.935	0.2499	1.046	0.2174	0.910
300	573.15	3.4504	14.446	0.2505	1.049	0.2269	0.950	0.2517	1.054	0.2266	0.949
400	673.15	3.4578	14.477	0.2524	1.057	0.2305	0.965	0.2540	1.063	0.2347	0.983
500	773.15	3.4653	14.509	0.2546	1.066	0.2339	0.979	0.2567	1.075	0.2419	1.013
600	873.15	3.4732	14.542	0.2570	1.076	0.2371	0.993	0.2594	1.086	0.2483	1.040
700	973.15	3.4841	14.587	0.2596	1.087	0.2400	1.005	0.2622	1.098	0.2541	1.064
800	1073.15	3.4970	14.641	0.2621	1.097	0.2426	1.016	0.2649	1.109	0.2592	1.085
900	1173.15	3.5124	14.706	0.2646	1.108	0.2450	1.026	0.2675	1.120	0.2638	1.104
1000	1273.15	3.5293	14.776	0.2670	1.118	0.2472	1.035	0.2700	1.130	0.2681	1.122
1100	1373.15	3.5476	14.853	0.2692	1.127	0.2492	1.043	0.2723	1.140	0.2719	1.138
1200	1473.15	3.5670	14.934	0.2713	1.136	0.2510	1.051	0.2745	1.149	0.2754	1.153
1300	1573.15	3.5883	15.023	0.2734	1.145	0.2527	1.058	0.2765	1.158	0.2785	1.166
1400	1673.15	3.6096	15.113	0.2753	1.153	0.2543	1.065	0.2784	1.166	0.2814	1.178
1500	1773.15	3.6309	15.202	0.2771	1.160	0.2559	1.071	0.2802	1.173	0.2841	1.189
1600	1873.15	3.6528	15.294	0.2788	1.167	0.2573	1.077	0.2818	1.180	0.2865	1.200
1700	1973.15	3.6741	15.383	0.2803	1.174	0.2587	1.083	0.2834	1.187	0.2888	1.209
1800	2073.15	3.6954	15.472	0.2818	1.180	0.2600	1.089	0.2848	1.192	0.2909	1.218
1900	2173.15	3.7168	15.561	0.2832	1.186	0.2613	1.094	0.2862	1.198	0.2928	1.226
2000	2273.15	3.7376	15.649	0.2845	1.191	0.2625	1.099	0.2874	1.203	0.2946	1.233
2100	2373.15	3.7584	15.736	0.2858	1.197	0.2637	1.104	0.2886	1.208	0.2963	1.241
2200	2473.15	3.7783	15.819	0.2869	1.201	0.2648	1.109	0.2897	1.213	0.2978	1.247
2300	2573.15	3.7981	15.902	0.2880	1.206	0.2660	1.114	0.2908	1.218	0.2993	1.253
2400	2673.15	3.8175	15.983	0.2891	1.210	0.2671	1.118	0.2918	1.222	0.3006	1.259
2500	2773.15	3.8368	16.064	0.2900	1.214	0.2682	1.123	0.2928	1.226	0.3018	1.264
2600	2873.15	3.8552	16.141	—	—	0.2692	1.127				
2700	2973.15	3.8730	16.215	—	—	0.2702	1.131				

Table 77-2 Mean Specific Heat c_{pm} of Gases at Constant Pressure p Between $0°C$ (273.15 K) and $t°C$ (TK) (Continued)

Temperature		COS		C₂H₆		C₃H₈		C₆H₆		C₃H₆		C₂H₅OH	
t	T	c_{pm}		c_{pm}		c_{pm}		c_{pm}		c_{pm}		c_{pm}	
°C	K	$\frac{kcal}{kg\,K}$	$\frac{kJ}{kg\,K}$	$\frac{kcal}{kg\,K}$	$\frac{kJ}{kg\,K}$	$\frac{kcal}{kg\,K}$	$\frac{kJ}{kg\,K}$	$\frac{kcal}{kg\,K}$	$\frac{kJ}{kg\,K}$	$\frac{kcal}{kg\,K}$	$\frac{kJ}{kg\,K}$	$\frac{kcal}{kg\,K}$	$\frac{kJ}{kg\,K}$
0	273.15	0.160	0.670	0.3934	1.647	0.3701	1.550	0.2253	0.943	0.3406	1.426	0.3633	1.521
100	373.15	0.169	0.708	0.4442	1.860	0.4261	1.784	0.2736	1.146	0.3878	1.624	0.4005	1.677
200	473.15	0.177	0.741	0.4940	2.068	0.4815	2.016	0.3166	1.326	0.4299	1.800	0.4348	1.820
300	573.15	0.184	0.770	0.5420	2.269	0.5305	2.221	0.3564	1.492	0.4715	1.974	0.4688	1.963
400	673.15	0.189	0.791	0.5891	2.466	0.5779	2.420	0.3911	1.637	0.5095	2.133	0.5001	2.094
500	773.15	0.193	0.808	0.6325	2.648	0.6184	2.589	0.4216	1.765	0.5449	2.281	0.5287	2.214
600	873.15	0.197	0.825	0.6726	2.816	0.6595	2.761	0.4490	1.880	0.5775	2.418	0.5553	2.325
700	973.15	0.201	0.842	0.7098	2.972	0.6949	2.909	0.4732	1.981	0.6077	2.544	0.5794	2.426
800	1073.15	0.204	0.854	0.7444	3.117	0.7271	3.044	0.4950	2.072	0.6350	2.659	0.6018	2.420
900	1173.15	0.206	0.862	0.7765	3.251	0.7566	3.168	0.5147	2.155	0.6604	2.765	0.6224	2.606
1000	1273.15	0.209	0.875	0.8076	3.377	0.7845	3.285	0.5326	2.230	0.6835	2.862	0.6410	2.684
1100	1373.15	0.211	0.883	0.8339	3.491	0.8108	3.935	0.5489	2.298	0.7049	2.951	0.6588	2.758
1200	1473.15	0.213	0.892	0.8592	3.597	0.8355	3.498	0.5634	2.359	0.7248	3.035	0.6749	2.826

1 kcal = 4.186 8 kJ

Table 77-3 Mean Specific Heat c_{pm} of Gases at Constant Pressure p Between $0°C$ (273.15 K) and $t°C$ (TK) (Continued)

Temperature		Air		H_2O		NO		OH		SO_2	
t	T	c_{pm}		c_{pm}		c_{pm}		c_{pm}		c_{pm}	
°C	K	$\dfrac{kcal}{kg\,K}$	$\dfrac{kJ}{kg\,K}$	$\dfrac{kcal}{kg\,K}$	$\dfrac{kJ}{kg\,K}$	$\dfrac{kcal}{kg\,K}$	$\dfrac{kJ}{kg\,K}$	$\dfrac{kcal}{kg\,K}$	$\dfrac{kJ}{kg\,K}$	$\dfrac{kcal}{kg\,K}$	$\dfrac{kJ}{kg\,K}$
0	273.15	0.2397	1.004	0.4441	1.859	0.2386	0.999	0.4212	1.763	0.145	0.607
100	373.15	0.2403	1.006	0.4473	1.873	0.2380	0.996	0.4179	1.750	0.152	0.636
200	473.15	0.2416	1.012	0.4523	1.894	0.2388	1.000	0.4163	1.743	0.158	0.662
300	573.15	0.2434	1.019	0.4584	1.919	0.2406	1.007	0.4157	1.740	0.164	0.687
400	673.15	0.2456	1.028	0.4652	1.948	0.2430	1.017	0.4155	1.740	0.169	0.708
500	773.15	0.2481	1.039	0.4724	1.978	0.2457	1.029	0.4157	1.740	0.173	0.724
600	873.15	0.2507	1.050	0.4799	2.009	0.2484	1.040	0.4167	1.745	0.176	0.737
700	973.15	0.2533	1.061	0.4877	2.042	0.2511	1.051	0.4181	1.751	0.180	0.754
800	1073.15	0.2558	1.071	0.4957	2.075	0.2537	1.062	0.4199	1.758	0.182	0.762
900	1173.15	0.2583	1.081	0.5039	2.110	0.2561	1.072	0.4220	1.767	0.185	0.775
1000	1273.15	0.2605	1.091	0.5120	2.144	0.2583	1.081	0.4244	1.777	0.187	0.783
1100	1373.15	0.2627	1.100	0.5200	2.177	0.2604	1.090	0.4270	1.788	0.189	0.791
1200	1473.15	0.2647	1.108	0.5280	2.211	0.2623	1.098	0.4297	1.799	0.190	0.795
1300	1573.15	0.2667	1.117	0.5357	2.243	0.2641	1.106	0.4324	1.810		
1400	1673.15	0.2685	1.124	0.5432	2.274	0.2658	1.113	0.4350	1.821		
1500	1773.15	0.2702	1.131	0.5505	2.305	0.2673	1.119	0.4376	1.832		
1600	1873.15	0.2718	1.138	0.5576	2.335	0.2688	1.125	0.4403	1.843		
1700	1973.15	0.2733	1.144	0.5644	2.363	0.2701	1.131	0.4430	1.855		
1800	2073.15	0.2747	1.150	0.5710	2.391	0.2713	1.136	0.4455	1.865		
1900	2173.15	0.2761	1.156	0.5772	2.417	0.2725	1.141	0.4480	1.876		
2000	2273.15	0.2773	1.161	0.5833	2.442	0.2736	1.146	0.4504	1.886		
2100	2373.15	0.2786	1.166	0.5891	2.466	0.2746	1.150	0.4528	1.896		
2200	2473.15	0.2797	1.171	0.5946	2.489	0.2756	1.154	0.4552	1.906		
2300	2573.15	0.2808	1.176	0.6000	2.512	0.2765	1.158	0.4575	1.915		
2400	2673.15	0.2819	1.180	0.6051	2.533	0.2773	1.161	0.4597	1.925		
2500	2773.15	0.2828	1.184	0.6101	2.554	0.2781	1.164	0.4618	1.933		
2600	2873.15	—	—	0.6149	2.574	0.2789	1.168	0.4639	1.942		
2700	2973.15	—	—	0.6195	2.594	0.2796	1.171	0.4659	1.951		
2800	3073.15	—	—	0.6239	2.612	0.2803	1.174	0.4678	1.959		
2900	3173.15	—	—	0.6281	2.630	0.2810	1.176	0.4697	1.967		
3000	3273.15	—	—	—	—	0.2816	1.179	0.4715	1.974		

Table 77-4 Mean Specific Heat c_{pm} of Gases at Constant Pressure p Between $0°C$ (273.15 K) and $t°C$ (TK) (Continued)

Temperature		CH_4		C_2H_4		C_2H_2		N_2O		H_2S		CS_2	
t	T	c_{pm}		c_{pm}		c_{pm}		c_{pm}		c_{pm}		c_{pm}	
°C	K	$\dfrac{kcal}{kg\,K}$	$\dfrac{kJ}{kg\,K}$	$\dfrac{kcal}{kg\,K}$	$\dfrac{kJ}{kg\,K}$	$\dfrac{kcal}{kg\,K}$	$\dfrac{kJ}{kg\,K}$	$\dfrac{kcal}{kg\,K}$	$\dfrac{kJ}{kg\,K}$	$\dfrac{kcal}{kh\,K}$	$\dfrac{kJ}{kg\,K}$	$\dfrac{kcal}{kg\,K}$	$\dfrac{kJ}{kg\,K}$
0	273.15	0.5172	2.165	0.3486	1.460	0.38447	1.610	0.2032	0.851	0.237	0.992	0.140	0.586
100	373.15	0.5480	2.294	0.3936	1.648	0.42080	1.762	0.2155	0.886	0.241	1.009	0.146	0.611
200	473.15	0.5870	2.458	0.4356	1.824	0.44508	1.863	0.2261	0.947	0.245	1.026	0.152	0.636
300	573.15	0.6294	2.635	0.4763	1.994	0.46489	1.946	0.2353	0.985	0.251	1.051	0.157	0.657
400	673.15	0.6727	2.816	0.5126	2.146	0.48195	2.018	0.2434	1.019	0.258	1.080	0.160	0.670
500	773.15	0.7143	2.991	0.5465	2.288	0.49685	2.080	0.2506	1.049	0.262	1.097	0.163	0.682
600	873.15	0.7545	3.159	0.5775	2.418	0.51041	2.137	0.2572	1.077	0.268	1.122	0.166	0.695
700	973.15	0.7932	3.321	0.6049	2.533	0.52289	2.189	0.2630	1.101	0.274	1.147	0.169	0.708
800	1073.15	0.8323	3.485	0.6313	2.643	0.53453	2.238	0.2684	1.124	0.280	1.172	0.171	0.716
900	1173.15	0.8685	3.636	0.6549	2.742	0.54532	2.283	0.2731	1.143	0.285	1.193	0.172	0.720
1000	1273.15	0.9008	3.771	0.6770	2.834	0.55535	2.325	0.2775	1.162	0.291	1.218	0.174	0.729
1100	1373.15	0.9299	3.893	0.6976	2.921	0.56483	2.365	0.2814	1.178	0.296	1.239	0.175	0.743
1200	1473.15	0.9555	4.000	0.7162	2.999	0.57374	2.402	—	—	0.300	1.256	0.176	0.737

1 kcal = 4.186 8 kJ

Table 78-1 Mean Specific Heat c_{vm} of Gases at Constant Volume v Between $0°$C (273.15 K) and $t°$C (TK)

Temperature		H$_2$		N$_2$		O$_2$		CO		CO$_2$	
t	T	c_{vm}		c_{vm}		c_{vm}		c_{vm}		c_{vm}	
°C	K	$\frac{kcal}{kg\,K}$	$\frac{kJ}{kg\,K}$	$\frac{kcal}{kg\,K}$	$\frac{kJ}{kg\,K}$	$\frac{kcal}{kg\,K}$	$\frac{kJ}{kg\,K}$	$\frac{kcal}{kg\,K}$	$\frac{kJ}{kg\,K}$	$\frac{kcal}{kg\,K}$	$\frac{kJ}{kg\,K}$
0	273.15	2.4053	10.071	0.1773	0.742	0.1564	0.655	0.1774	0.743	0.1495	0.626
100	373.15	2.4430	10.228	0.1776	0.744	0.1584	0.663	0.1779	0.745	0.1617	0.677
200	473.15	2.4593	10.297	0.1783	0.747	0.1613	0.675	0.1790	0.749	0.1723	0.721
300	573.15	2.4653	10.322	0.1796	0.752	0.1648	0.690	0.1808	0.757	0.1815	0.760
400	673.15	2.4727	10.353	0.1815	0.760	0.1684	0.705	0.1831	0.767	0.1896	0.794
500	773.15	2.4802	10.384	0.1837	0.769	0.1718	0.719	0.1857	0.777	0.1968	0.824
600	873.15	2.4881	10.417	0.1861	0.779	0.1750	0.733	0.1885	0.789	0.2032	0.851
700	973.15	2.4990	10.463	0.1887	0.790	0.1779	0.745	0.1913	0.801	0.2089	0.875
800	1073.15	2.5119	10.517	0.1912	0.801	0.1805	0.756	0.1940	0.812	0.2141	0.896
900	1173.15	2.5273	10.581	0.1937	0.811	0.1829	0.766	0.1966	0.823	0.2187	0.916
1000	1273.15	2.5441	10.652	0.1961	0.821	0.1851	0.775	0.1991	0.834	0.2229	0.933
1100	1373.15	2.5625	10.729	0.1983	0.830	0.1871	0.783	0.2014	0.843	0.2268	0.950
1200	1473.15	2.5818	10.809	0.2005	0.839	0.1890	0.791	0.2046	0.857	0.2302	0.964
1300	1573.15	2.6032	10.899	0.2025	0.848	0.1907	0.798	0.2056	0.861	0.2334	0.977
1400	1673.15	2.6245	10.988	0.2044	0.856	0.1923	0.805	0.2075	0.869	0.2363	0.989
1500	1773.15	2.6458	11.077	0.2062	0.863	0.1938	0.811	0.2093	0.876	0.2390	1.001
1600	1873.15	2.6677	11.169	0.2079	0.870	0.1952	0.817	0.2109	0.883	0.2414	1.011
1700	1973.15	2.6890	11.258	0.2095	0.877	0.1966	0.823	0.2124	0.889	0.2437	1.020
1800	2073.15	2.7103	11.347	0.2109	0.883	0.1979	0.829	0.2139	0.896	0.2458	1.029
1900	2173.15	2.7316	11.437	0.2123	0.889	0.1992	0.834	0.2153	0.901	0.2477	1.037
2000	2273.15	2.7525	11.524	0.2136	0.894	0.2004	0.839	0.2165	0.906	0.2495	1.045
2100	2373.15	2.7733	11.611	0.2149	0.900	0.2016	0.844	0.2177	0.911	0.2512	1.052
2200	2473.15	2.7931	11.694	0.2161	0.905	0.2028	0.849	0.2188	0.916	0.2527	1.058
2300	2573.15	2.8180	11.798	0.2172	0.909	0.2039	0.854	0.2199	0.921	0.2541	1.064
2400	2673.15	2.8323	11.858	0.2182	0.914	0.2050	0.858	0.2209	0.925	0.2555	1.070
2500	2773.15	2.8517	11.939	0.2192	0.918	0.2061	0.863	0.2219	0.929	0.2567	1.075
2600	2873.15	2.8700	12.016	—	—	0.2072	0.868				
2700	2973.15	2.8879	12.091	—	—	0.2082	0.872				

Table 78-2 Mean Specific Heat c_{vm} of Gases at Constant Volume v Between $0°$C (273.15 K) and $t°$C (TK)

Temperature		COS		C$_2$H$_6$		C$_3$H$_8$		C$_6$H$_6$		C$_3$H$_6$		C$_2$H$_5$OH	
t	T	c_{vm}		c_{vm}		c_{vm}		c_{vm}		c_{vm}		c_{vm}	
°C	K	$\frac{kcal}{kg\,K}$	$\frac{kJ}{kg\,K}$	$\frac{kcal}{kg\,K}$	$\frac{kJ}{kg\,K}$	$\frac{kcal}{kg\,K}$	$\frac{kJ}{kg\,K}$	$\frac{kcal}{kg\,K}$	$\frac{kJ}{kg\,K}$	$\frac{kcal}{kg\,K}$	$\frac{kJ}{kg\,K}$	$\frac{kcal}{kg\,K}$	$\frac{kJ}{kg\,K}$
0	273.15	0.127	0.532	0.3274	1.371	0.3250	1.361	0.1998	0.837	0.2933	1.228	0.3202	1.341
100	373.15	0.136	0.569	0.3781	1.583	0.3810	1.595	0.2481	1.039	0.3406	1.426	0.3574	1.496
200	473.15	0.144	0.603	0.4280	1.792	0.4363	1.827	0.2911	1.219	0.3826	1.602	0.3917	1.640
300	573.15	0.150	0.628	0.4759	1.992	0.4853	2.032	0.3309	1.385	0.4242	1.776	0.4257	1.782
400	673.15	0.156	0.653	0.5230	2.190	0.5327	2.230	0.3656	1.531	0.4622	1.935	0.4570	1.913
500	773.15	0.160	0.670	0.5665	2.372	0.5733	2.400	0.3961	1.658	0.4976	2.083	0.4856	2.033
600	873.15	0.164	0.687	0.6066	2.440	0.6144	2.572	0.4235	1.773	0.5302	2.220	0.5122	2.144
700	973.15	0.168	0.703	0.6437	2.695	0.6497	2.720	0.4477	1.874	0.5604	2.346	0.5363	2.245
800	1073.15	0.171	0.716	0.6784	2.840	0.6819	2.855	0.4695	1.966	0.5877	2.461	0.5587	2.339
900	1173.15	0.173	0.724	0.7106	2.975	0.7114	2.978	0.4892	2.048	0.6131	2.567	0.5793	2.425
1000	1273.15	0.176	0.737	0.7404	3.100	0.7393	3.095	0.5071	2.123	0.6263	2.664	0.5979	2.503
1100	1373.15	0.178	0.745	0.7679	3.215	0.7656	3.205	0.5234	2.191	0.6576	2.753	0.6157	2.578
1200	1473.15	0.180	0.754	0.7931	3.321	0.7904	3.309	0.5380	2.252	0.6775	2.837	0.6318	2.645

1 kcal = 4.186 8 kJ

Table 78-3 Mean Specific Heat c_{vm} of Gases at Constant Volume v Between $0°C$ (273.15 K) and $t°C$ (TK) (Continued)

Temperature		Air		H_2O		NO		OH		SO_2	
		c_{vm}		c_{vm}		c_{vm}		c_{vm}		c_{vm}	
t	T	$\dfrac{kcal}{kg\,K}$	$\dfrac{kJ}{kg\,K}$	$\dfrac{kcal}{kg\,K}$	$\dfrac{kJ}{kg\,K}$	$\dfrac{kcal}{kg\,K}$	$\dfrac{kJ}{kg\,K}$	$\dfrac{kcal}{kg\,K}$	$\dfrac{kJ}{kg\,K}$	$\dfrac{kcal}{kg\,K}$	$\dfrac{kJ}{kg\,K}$
°C	K										
0	273.15	0.1711	0.716	0.3339	1.398	0.1724	0.722	0.3044	1.274	0.114	0.477
100	373.15	0.1718	0.719	0.3371	1.411	0.1718	0.719	0.3011	1.261	0.121	0.507
200	473.15	0.1730	0.724	0.3421	1.432	0.1726	0.723	0.2995	1.254	0.127	0.532
300	573.15	0.1748	0.732	0.3481	1.457	0.1744	0.730	0.2989	1.251	0.133	0.557
400	673.15	0.1771	0.741	0.3550	1.486	0.1768	0.740	0.2987	1.251	0.138	0.578
500	773.15	0.1796	0.752	0.3621	1.516	0.1795	0.752	0.2989	1.251	0.142	0.595
600	873.15	0.1821	0.762	0.3696	1.547	0.1822	0.763	0.2999	1.256	0.145	0.607
700	973.15	0.1847	0.773	0.3775	1.581	0.1849	0.774	0.3013	1.261	0.149	0.624
800	1073.15	0.1873	0.784	0.3855	1.614	0.1875	0.785	0.3031	1.269	0.151	0.632
900	1173.15	0.1897	0.794	0.3937	1.648	0.1899	0.795	0.3052	1.278	0.154	0.645
1000	1273.15	0.1920	0.804	0.4018	1.682	0.1921	0.804	0.3076	1.288	0.156	0.653
1100	1373.15	0.1941	0.813	0.4098	1.716	0.1942	0.813	0.3102	1.299	0.158	0.662
1200	1473.15	0.1962	0.821	0.4177	1.749	0.1961	0.821	0.3129	1.310	0.159	0.666
1300	1573.15	0.1981	0.829	0.4255	1.781	0.1979	0.829	0.3156	1.321		
1400	1673.15	0.1999	0.837	0.4330	1.813	0.1996	0.836	0.3182	1.332		
1500	1773.15	0.2016	0.844	0.4403	1.843	0.2011	0.842	0.3208	1.343		
1600	1873.15	0.2032	0.851	0.4473	1.873	0.2026	0.848	0.3235	1.354		
1700	1973.15	0.2047	0.857	0.4542	1.902	0.2039	0.854	0.3262	1.366		
1800	2073.15	0.2062	0.863	0.4608	1.929	0.2051	0.859	0.3287	1.376		
1900	2173.15	0.2075	0.869	0.4670	1.955	0.2063	0.864	0.3312	1.387		
2000	2273.15	0.2088	0.874	0.4730	1.980	0.2074	0.868	0.3336	1.397		
2100	2373.15	0.2100	0.879	0.4789	2.005	0.2084	0.873	0.3360	1.408		
2200	2473.15	0.2112	0.884	0.4844	2.028	0.2094	0.877	0.3384	1.417		
2300	2573.15	0.2123	0.889	0.4897	2.050	0.2103	0.880	0.3407	1.426		
2400	2673.15	0.2133	0.893	0.4949	2.072	0.2111	0.884	0.3429	1.436		
2500	2773.15	0.2143	0.897	0.4998	2.093	0.2119	0.887	0.3450	1.444		
2600	2873.15	—	—	0.5047	2.113	0.2127	0.891	0.3471	1.453		
2700	2973.15	—	—	0.5093	2.132	0.2134	0.893	0.3491	1.462		
2800	3073.15	—	—	0.5137	2.151	0.2141	0.896	0.3510	1.470		
2900	3173.15	—	—	0.5179	2.168	0.2148	0.899	0.3529	1.478		
3000	3273.15	—	—	—	—	0.2154	0.902	0.3547	1.485		

Table 78-4 Mean Specific Heat c_{vm} of Gases at Constant Volume v Between $0°C$ (273.15 K) and $t°C$ (TK) (Continued)

Temperature		CH_4		C_2H_4		C_2H_2		N_2O		H_2S		CS_2	
		c_{vm}		c_{vm}		c_{vm}		c_{vm}		c_{vm}		c_{vm}	
t	T	$\dfrac{kcal}{kg\,K}$	$\dfrac{kJ}{kg\,K}$	$\dfrac{kcal}{kg\,K}$	$\dfrac{kJ}{kg\,K}$	$\dfrac{kcal}{kg\,K}$	$\dfrac{kJ}{kg\,K}$	$\dfrac{kcal}{kg\,K}$	$\dfrac{kJ}{kg\,K}$	$\dfrac{kcal}{kg\,K}$	$\dfrac{kJ}{kg\,K}$	$\dfrac{kcal}{kg\,K}$	$\dfrac{kJ}{kg\,K}$
°C	K												
0	273.15	0.3934	1.647	0.2777	1.163	0.3080	1.290	0.1581	0.662	0.178	0.745	0.113	0.473
100	373.15	0.4242	1.776	0.3226	1.351	0.3445	1.442	0.1704	0.713	0.182	0.762	0.120	0.502
200	473.15	0.4632	1.939	0.3647	1.527	0.3688	1.544	0.1810	0.758	0.187	0.783	0.126	0.528
300	573.15	0.5056	2.117	0.4053	1.697	0.38862	1.627	0.1902	0.796	0.192	0.804	0.130	0.544
400	673.15	0.5489	2.298	0.4417	1.849	0.40567	1.698	0.1983	0.830	0.198	0.829	0.134	0.561
500	773.15	0.5905	2.472	0.4755	1.991	0.42057	1.761	0.2055	0.860	0.204	0.854	0.137	0.574
600	873.15	0.6307	2.641	0.5066	2.121	0.43413	1.818	0.2120	0.888	0.210	0.879	0.140	0.586
700	973.15	0.6694	2.803	0.5340	2.236	0.44661	1.870	0.2179	0.912	0.216	0.904	0.142	0.595
800	1073.15	0.7085	2.966	0.5604	2.346	0.45825	1.919	0.2232	0.934	0.222	0.929	0.144	0.603
900	1173.15	0.7447	3.118	0.5839	2.445	0.46904	1.964	0.2280	0.955	0.227	0.950	0.146	0.611
1000	1273.15	0.7770	3.253	0.6060	2.537	0.47907	2.006	0.2323	0.973	0.232	0.971	0.148	0.620
1100	1373.15	0.8061	3.375	0.6267	2.624	0.48855	2.045	0.2363	0.989	0.237	0.992	0.149	0.624
1200	1473.15	0.8317	3.482	0.6452	2.701	0.49746	2.083	—	—	0.242	1.013	0.150	0.628

1 kcal = 4.186 8 kJ

Table 79-1 Mean Molar Heat Capacity C_{pm} of Gases at Constant Pressure p Between $0°C$ (273.15 K) and $t°C$ (TK)

| Temperature | | H$_2$ | | N$_2$ | | O$_2$ | | CO | | CO$_2$ | |
| t | T | C_{pm} | | C_{pm} | | C_{pm} | | C_{pm} | | C_{pm} | |
°C	K	$\frac{kcal}{kmol\,K}$	$\frac{kJ}{kmol\,K}$	$\frac{kcal}{kmol\,K}$	$\frac{kJ}{kmol\,K}$	$\frac{kcal}{kmol\,K}$	$\frac{kJ}{kmol\,K}$	$\frac{kcal}{kmol\,K}$	$\frac{kJ}{kmol\,K}$	$\frac{kcal}{kmol\,K}$	$\frac{kJ}{kmol\,K}$
0	273.15	6.835	28.617	6.954	29.115	6.992	29.274	6.956	29.123	8.565	35.860
100	373.15	6.911	28.935	6.961	29.144	7.055	29.538	6.969	29.178	9.103	38.112
200	473.15	6.944	29.073	6.981	29.228	7.149	29.931	6.999	29.303	9.568	40.059
300	573.15	6.956	29.123	7.018	29.383	7.261	30.400	7.050	29.517	9.973	41.755
400	673.15	6.971	29.186	7.070	29.601	7.375	30.878	7.115	29.789	10.330	43.250
500	773.15	6.986	29.249	7.133	29.864	7.484	31.334	7.189	30.099	10.646	44.573
600	873.15	7.002	29.316	7.201	30.149	7.586	31.761	7.267	30.425	10.928	45.753
700	973.15	7.024	29.408	7.273	30.451	7.679	32.150	7.345	30.752	11.181	46.813
800	1073.15	7.050	29.517	7.344	30.748	7.763	32.502	7.421	31.070	11.408	47.763
900	1173.15	7.081	29.647	7.413	31.037	7.840	32.825	7.494	31.376	11.612	48.617
1000	1273.15	7.115	29.789	7.479	31.313	7.910	33.118	7.563	31.665	11.797	49.392
1100	1373.15	7.152	29.944	7.542	31.577	7.974	33.386	7.628	31.937	11.966	50.099
1200	1473.15	7.191	30.107	7.602	31.828	8.033	33.633	7.689	32.192	12.119	50.740
1300	1573.15	7.234	30.287	7.659	32.067	8.088	33.863	7.745	32.427	12.258	51.322
1400	1673.15	7.277	30.467	7.713	32.293	8.139	34.076	7.799	32.653	12.386	51.858
1500	1773.15	7.320	30.647	7.763	32.502	8.188	34.282	7.848	32.858	12.503	52.348
1600	1873.15	7.364	30.832	7.810	32.699	8.234	34.474	7.894	33.051	12.611	52.800
1700	1973.15	7.407	31.012	7.854	32.883	8.278	34.658	7.937	33.231	12.711	53.218
1800	2073.15	7.450	31.192	7.895	33.055	8.320	34.834	7.978	33.402	12.803	53.604
1900	2173.15	7.493	31.372	7.934	33.218	8.361	35.006	8.016	33.561	12.888	53.959
2000	2273.15	7.535	31.548	7.971	33.373	8.400	35.169	8.051	33.708	12.967	54.290
2100	2373.15	7.577	31.723	8.006	33.520	8.438	35.328	8.085	33.850	13.040	54.596
2200	2473.15	7.617	31.891	8.039	33.658	8.475	35.483	8.116	33.980	13.108	54.881
2300	2573.15	7.657	32.058	8.070	33.787	8.511	35.634	8.146	34.106	13.171	55.144
2400	2673.15	7.696	32.222	8.099	33.909	8.547	35.785	8.174	34.223	13.230	55.391
2500	2773.15	7.735	32.385	8.126	34.022	8.581	35.927	8.201	34.336	13.284	55.617
2600	2873.15	7.772	32.540	8.170	34.206	8.615	36.069	8.240	34.499	13.340	55.852
2700	2973.15	7.808	32.691	8.190	34.290	8.648	36.207	8.260	34.583	13.390	56.061
2800	3073.15	7.850	32.866	8.220	34.415	8.680	36.341	8.280	34.667	13.430	56.229
2900	3173.15	7.890	33.034	8.240	34.499	8.720	36.509	8.300	34.750	13.480	56.438
3000	3273.15	7.920	33.159	8.260	34.583	8.760	36.676	8.320	34.834	13.520	56.606
$M*$		2.0156		28.016		32.000		28.010		44.010	

*M — relative molecular mass in kg/mol

Table 79-2 Mean Molar Heat Capacity C_{pm} of Gases at Constant Pressure p Between $0°C$ (273.15 K) and $t°C$ (TK) (*Continued*)

| Temperature | | COS | | C$_2$H$_6$ | | C$_3$H$_8$ | | C$_6$H$_6$ | | C$_3$H$_6$ | | C$_2$H$_5$OH | |
| t | T | C_{pm} | | C_{pm} | | C_{pm} | | C_{pm} | | C_{pm} | | C_{pm} | |
°C	K	$\frac{kcal}{kmol\,K}$	$\frac{kJ}{kmol\,K}$	$\frac{kcal}{kmol\,K}$	$\frac{kJ}{kmol\,K}$	$\frac{kcal}{kmol\,K}$	$\frac{kJ}{kmol\,K}$	$\frac{kcal}{kmol\,K}$	$\frac{kJ}{kmol\,K}$	$\frac{kcal}{kmol\,K}$	$\frac{kJ}{kmol\,K}$	$\frac{kcal}{kmol\,K}$	$\frac{kJ}{kmol\,K}$
0	273.15	9.62	40.277	11.830	49.530	16.32	68.329	17.60	73.688	14.33	59.997	16.74	70.087
100	373.15	10.18	42.622	13.356	55.919	18.79	78.670	21.37	89.472	16.32	68.329	18.46	77.288
200	473.15	10.65	44.589	14.855	62.195	21.23	88.886	24.73	103.540	18.09	75.739	20.04	83.903
300	573.15	11.03	46.180	16.297	68.232	23.39	97.929	27.84	116.561	19.84	83.066	21.61	90.477
400	673.15	11.35	47.520	17.713	74.161	25.48	106.680	30.55	127.907	21.44	89.765	23.04	96.464
500	773.15	11.62	48.651	19.019	79.629	27.27	114.174	32.93	137.871	22.93	96.003	24.36	101.990
600	873.15	11.85	49.614	20.224	84.674	29.08	121.752	35.07	146.831	24.30	101.739	25.59	107.140
700	973.15	12.06	50.493	21.342	89.355	30.64	128.284	36.96	154.744	25.57	107.056	26.70	111.788
800	1073.15	12.24	51.246	22.383	93.713	32.06	134.229	38.66	161.862	26.72	111.871	27.73	116.100
900	1173.15	12.40	51.916	23.351	97.766	33.86	141.765	40.20	168.309	27.79	116.351	28.68	120.077
1000	1273.15	12.55	52.544	24.249	101.526	34.59	144.821	41.60	174.171	28.76	120.412	29.54	123.678
1100	1373.15	12.68	53.089	25.075	104.984	35.75	149.678	42.87	179.488	29.66	124.170	30.36	127.111
1200	1473.15	12.80	53.591	25.834	108.162	36.84	154.242	44.01	184.261	30.50	127.697	31.10	130.209
$M*$		60.07		30.047		44.06		78.108		42.08		46.07	

*M — relative molecular mass in kg/mol

Table 79-3 Mean Molar Heat Capacity C_{pm} of Gases at Constant Pressure p Between $0°C$ (273.15 K) and $t°C$ (TK) (*Continued*)

Temperature		Air		H₂O		NO		OH		SO₂	
		C_{pm}		C_{pm}		C_{pm}		C_{pm}		C_{pm}	
t	T	$\dfrac{kcal}{kmol\,K}$	$\dfrac{kJ}{kmol\,K}$	$\dfrac{kcal}{kmol\,K}$	$\dfrac{kJ}{kmol\,K}$	$\dfrac{kcal}{kmol\,K}$	$\dfrac{kJ}{kmol\,K}$	$\dfrac{kcal}{kmol\,K}$	$\dfrac{kJ}{kmol\,K}$	$\dfrac{kcal}{kmol\,K}$	$\dfrac{kJ}{kmol\,K}$
°C	K										
0	273.15	6.944	29.073	8.001	33.499	7.160	29.977	7.163	29.990	9.28	38.854
100	373.15	6.963	29.153	8.059	33.741	7.143	29.906	7.108	29.760	9.71	40.654
200	473.15	6.998	29.299	8.149	34.118	7.165	29.998	7.080	29.643	10.11	42.329
300	573.15	7.051	29.521	8.258	34.575	7.220	30.229	7.070	29.601	10.48	43.878
400	673.15	7.115	29.789	8.381	35.090	7.292	30.530	7.066	29.584	10.80	45.217
500	773.15	7.188	30.095	8.510	35.630	7.373	30.869	7.071	29.605	11.08	46.390
600	873.15	7.262	30.405	8.645	36.195	7.455	31.213	7.087	29.672	11.31	47.353
700	973.15	7.338	30.723	8.787	36.789	7.536	31.552	7.111	29.772	11.52	48.232
800	1073.15	7.411	31.028	8.931	37.392	7.613	31.874	7.141	29.898	11.69	48.944
900	1173.15	7.481	31.321	9.078	38.008	7.684	32.171	7.177	30.049	11.85	49.614
1000	1273.15	7.547	31.598	9.224	38.619	7.752	32.456	7.218	30.220	11.98	50.158
1100	1373.15	7.610	31.862	9.369	39.226	7.814	32.716	7.263	30.409	12.10	50.660
1200	1473.15	7.669	32.109	9.512	39.825	7.872	32.958	7.309	30.601	12.20	51.079
1300	1573.15	7.725	32.343	9.651	40.407	7.925	33.180	7.354	30.790	12.33	51.623
1400	1673.15	7.778	32.565	9.787	40.976	7.975	33.390	7.398	30.974	12.41	51.958
1500	1773.15	7.828	32.774	9.918	41.525	8.021	33.582	7.447	31.179	12.48	52.251
1600	1873.15	7.874	32.967	10.045	42.056	8.065	33.767	7.489	31.355	12.55	52.544
1700	1973.15	7.918	33.151	10.169	42.576	8.105	33.934	7.534	31.543	12.61	52.796
1800	2073.15	7.958	33.319	10.287	43.070	8.142	34.089	7.577	31.723	12.67	53.047
1900	2173.15	7.997	33.482	10.399	43.539	8.176	34.231	7.619	31.899	12.71	53.214
2000	2273.15	8.035	33.641	10.508	43.995	8.209	34.369	7.661	32.075	12.77	53.465
2100	2373.15	8.070	33.787	10.613	44.435	8.240	34.499	7.702	32.247	12.81	53.633
2200	2473.15	8.103	33.926	10.713	44.853	8.269	34.621	7.742	32.414	12.85	53.800
2300	2573.15	8.135	34.060	10.809	45.255	8.296	34.734	7.781	32.577	12.89	53.968
2400	2673.15	8.165	34.185	10.902	45.644	8.321	34.838	7.819	32.737	12.93	54.135
2500	2773.15	8.194	34.307	10.991	46.017	8.346	34.943	7.855	32.887	12.96	54.261
2600	2873.15	8.200	34.332	11.078	46.381	8.369	35.039	7.890	33.034	12.99	54.387
2700	2973.15	8.230	34.457	11.161	46.729	8.391	35.131	7.924	33.176	13.02	54.512
2800	3073.15	8.250	34.541	11.240	47.060	8.412	35.219	7.957	33.314	13.04	54.596
2900	3173.15	8.270	34.625	11.316	47.378	8.432	35.303	7.989	33.448	13.07	54.721
3000	3273.15	8.290	34.709	—	—	8.451	35.383	8.020	33.578	13.10	54.847
$M*$		28.964		18.020		30.008		17.008		64.06	

*M — relative molecular mass in kg/mol

Table 79-4 Mean Molar Heat Capacity C_{pm} of Gases at Constant Pressure p Between $0°C$ (273.15 K) and $t°C$ (TK) (*Continued*)

Temperature		CH₄		C₂H₄		C₂H₂		N₂O		H₂S		CS₂	
		C_{pm}		C_{pm}		C_{pm}		C_{pm}		C_{pm}		C_{pm}	
t	T	$\dfrac{kcal}{kmol\,K}$	$\dfrac{kJ}{kmol\,K}$	$\dfrac{kcal}{kmol\,K}$	$\dfrac{kJ}{kmol\,K}$	$\dfrac{kcal}{kmol\,K}$	$\dfrac{kJ}{kmol\,K}$	$\dfrac{kcal}{kmol\,K}$	$\dfrac{kJ}{kmol\,K}$	$\dfrac{kcal}{kmol\,K}$	$\dfrac{kJ}{kmol\,K}$	$\dfrac{kcal}{kmol\,K}$	$\dfrac{kJ}{kmol\,K}$
°C	K												
0	273.15	8.297	34.738	9.78	40.947	10.01	41.910	8.945	37.451	8.07	33.787	10.63	44.506
100	373.15	8.791	36.806	11.04	46.222	10.956	45.871	9.485	39.712	8.20	34.332	11.15	46.683
200	473.15	9.417	39.427	12.22	51.163	11.588	48.517	9.954	41.675	8.36	35.002	11.58	48.483
300	573.15	10.097	42.274	13.36	55.936	12.104	50.677	10.359	43.371	8.54	35.755	11.92	49.907
400	673.15	10.791	45.180	14.38	60.206	12.548	52.536	10.715	44.862	8.74	36.593	12.21	51.121
500	773.15	11.459	47.977	15.33	64.184	12.936	54.160	11.031	46.185	8.94	37.430	12.45	52.126
600	873.15	12.103	50.673	16.20	67.826	13.289	55.638	11.319	47.390	9.15	37.309	12.66	53.005
700	973.15	12.725	53.277	16.97	71.050	13.614	56.999	11.578	11.475	9.35	39.147	12.84	53.759
800	1073.15	13.352	55.902	17.71	74.148	13.917	58.268	11.812	49.454	9.55	39.984	12.99	54.387
900	1173.15	13.932	58.330	18.37	76.912	14.198	59.444	12.022	50.334	9.73	40.738	13.12	54.931
1000	1273.15	14.451	60.503	18.99	79.507	14.459	60.537	12.213	51.133	9.91	41.491	13.23	55.391
1100	1373.15	14.917	62.454	19.57	81.936	14.706	61.571	12.388	51.866	10.08	42.203	13.34	55.852
1200	1473.15	15.328	64.175	20.09	84.113	14.938	62.542	—	—	10.23	42.831	13.43	56.229
$M*$		16.031		28.031		26.04		44.016		34.08		76.13	

*M — relative molecular mass in kg/mol

1 kcal = 4.186 8 kJ

Table 80-1 Mean Molar Heat Capacity C_{vm} of Gases at Constant Volume v Between 0°C (273.15 K) and t°C (TK)

Temperature		H₂		N₂		O₂		CO		CO₂	
t	T	C_{vm}		C_{vm}		C_{vm}		C_{vm}		C_{vm}	
°C	K	$\frac{kcal}{kmol\ K}$	$\frac{kJ}{kmol\ K}$	$\frac{kcal}{kmol\ K}$	$\frac{kJ}{kmol\ K}$	$\frac{kcal}{kmol\ K}$	$\frac{kJ}{kmol\ K}$	$\frac{kcal}{kmol\ K}$	$\frac{kJ}{kmol\ K}$	$\frac{kcal}{kmol\ K}$	$\frac{kJ}{kmol\ K}$
0	273.15	4.849	20.302	4.968	20.800	5.006	20.959	4.970	20.808	6.579	27.545
100	373.15	4.925	20.620	4.975	20.829	5.069	21.223	4.983	20.863	7.117	29.797
200	473.15	4.958	20.758	4.995	20.913	5.163	21.616	5.013	20.988	7.582	31.744
300	573.15	4.970	20.808	5.032	21.068	5.275	22.085	5.064	21.202	7.987	33.440
400	673.15	4.985	20.871	5.084	21.286	5.389	22.563	5.129	21.474	8.344	34.935
500	773.15	5.000	20.934	5.147	21.549	5.498	23.019	5.203	21.784	8.660	36.258
600	873.15	5.016	21.001	5.215	21.834	5.600	23.446	5.281	22.100	8.942	37.438
700	973.15	5.038	21.093	5.287	22.136	5.693	23.835	5.359	22.437	9.195	38.498
800	1073.15	5.064	21.202	5.358	22.433	5.777	24.187	5.435	22.755	9.322	39.448
900	1173.15	5.095	21.332	5.427	22.722	5.854	24.510	5.508	23.061	9.626	40.302
1000	1273.15	5.129	21.474	5.493	22.998	5.924	24.803	5.577	23.350	9.811	41.077
1100	1373.15	5.166	21.629	5.556	23.262	5.988	25.071	5.642	23.622	9.980	41.784
1200	1473.15	5.205	21.792	5.616	23.513	6.047	25.318	5.703	23.877	10.133	42.425
1300	1573.15	5.248	21.972	5.673	23.752	6.102	25.548	5.759	24.112	10.272	43.007
1400	1673.15	5.291	22.152	5.727	23.978	6.153	25.761	5.813	24.338	10.400	43.543
1500	1773.15	5.334	22.332	5.777	24.187	6.202	25.967	5.862	24.543	10.517	44.033
1600	1873.15	5.378	22.517	5.824	24.384	6.248	26.159	5.908	24.736	10.625	44.485
1700	1973.15	5.421	22.697	5.868	24.568	6.292	26.343	5.951	24.916	10.725	44.903
1800	2073.15	5.464	22.877	5.909	24.740	6.334	26.519	5.992	25.087	10.817	45.289
1900	2173.15	5.507	23.057	5.948	24.903	6.375	26.691	6.030	25.246	10.902	45.644
2000	2273.15	5.549	23.233	5.985	25.058	6.414	26.854	6.065	25.393	10.981	45.975
2100	2373.15	5.591	23.408	6.020	25.205	6.452	27.013	6.099	25.535	11.054	46.281
2200	2473.15	5.631	23.576	6.053	25.343	6.489	27.168	6.130	25.665	11.122	46.566
2300	2573.15	5.671	23.743	6.084	25.472	6.525	27.319	6.160	25.791	11.185	46.829
2400	2673.15	5.710	23.907	6.113	25.594	6.561	27.470	6.188	25.908	11.244	47.076
2500	2773.15	5.749	24.070	6.140	25.707	6.595	27.612	6.215	26.021	11.298	47.302
2600	2873.15	5.786	24.225	—	—	6.629	27.754				
2700	2973.15	5.822	24.376	—	—	6.662	27.892				
M*		2.0156		28.016		32.000		28.010		44.010	

*M — relative molecular mass in kg/mol

Table 80-2 Mean Molar Heat Capacity C_{vm} of Gases at Constant Volume v Between 0°C (273.15 K) and t°C (TK) (Continued)

Temperature		COS		C₂H₆		C₃H₈		C₆H₆		C₃H₆		C₂H₅OH	
t	T	C_{vm}		C_{vm}		C_{vm}		C_{vm}		C_{vm}		C_{vm}	
°C	K	$\frac{kcal}{kmol\ K}$	$\frac{kJ}{kmol\ K}$	$\frac{kcal}{kmol\ K}$	$\frac{kJ}{kmol\ K}$	$\frac{kcal}{kmol\ K}$	$\frac{kJ}{kmol\ K}$	$\frac{kcal}{kmol\ K}$	$\frac{kJ}{kmol\ K}$	$\frac{kcal}{kmol\ K}$	$\frac{kJ}{kmol\ K}$	$\frac{kcal}{kmol\ K}$	$\frac{kJ}{kmol\ K}$
0	273.15	7.63	31.945	9.844	41.215	14.33	59.997	15.61	65.356	12.34	51.665	14.75	61.755
100	373.15	8.19	34.290	11.370	47.604	16.80	70.338	19.38	81.140	14.33	59.997	16.47	68.957
200	473.15	8.66	36.258	12.869	53.880	19.24	80.554	22.74	95.208	16.10	67.407	18.05	75.572
300	573.15	9.04	37.849	14.311	59.917	21.40	89.598	25.85	108.229	17.85	74.734	19.62	82.145
400	673.15	9.36	39.188	15.727	65.846	23.49	98.348	28.56	119.575	19.45	81.433	21.05	88.132
500	773.15	9.63	40.319	17.033	71.314	25.28	105.842	30.94	129.540	20.94	87.672	22.37	93.659
600	873.15	9.86	41.282	18.238	76.359	27.09	113.420	33.08	138.499	22.31	93.408	23.60	98.808
700	973.15	10.07	42.161	19.356	81.040	28.65	119.952	34.97	146.412	23.58	98.725	24.71	103.456
800	1073.15	10.25	42.916	20.397	85.398	30.07	125.897	36.67	153.530	24.73	103.540	25.74	107.768
900	1173.15	10.41	43.585	21.365	89.451	31.37	131.340	38.24	160.103	25.80	108.019	26.79	112.164
1000	1273.15	10.56	44.213	22.263	93.211	32.60	136.490	39.61	165.839	26.77	112.081	27.55	115.346
1100	1373.15	10.69	44.757	23.089	96.669	33.76	141.346	40.88	171.156	27.67	115.849	28.37	118.780
1200	1473.15	10.81	45.259	23.848	99.847	34.85	145.910	42.02	175.929	28.51	119.366	29.11	121.878
M*		60.07		30.47		44.06		78.108		42.08		46.07	

*M — relative molecular mass in kg/mol

1 kcal = 4.186 8 kJ

Table 80-3 Mean Molar Heat Capacity C_{vm} of Gases at Constant Volume v Between $0°C$ (273.15 K) and $t°C$ (TK) (Continued)

Temperature		Air		H₂O		NO		OH		SO₂	
		C_{vm}		C_{vm}		C_{vm}		C_{vm}		C_{vm}	
t	T	$\dfrac{kcal}{kmol\,K}$	$\dfrac{kJ}{kmol\,K}$	$\dfrac{kcal}{kmol\,K}$	$\dfrac{kJ}{kmol\,K}$	$\dfrac{kcal}{kmol\,K}$	$\dfrac{kJ}{kmol\,K}$	$\dfrac{kcal}{kmol\,K}$	$\dfrac{kJ}{kmol\,K}$	$\dfrac{kcal}{kmol\,K}$	$\dfrac{kJ}{kmol\,K}$
°C	K										
0	273.15	4.958	20.758	6.015	25.184	5.174	21.663	5.177	21.675	7.29	30.522
100	373.15	4.977	20.838	6.073	25.426	5.157	21.591	5.122	21.445	7.72	32.322
200	473.15	5.012	20.984	6.163	25.803	5.179	21.683	5.094	21.328	8.12	33.997
300	573.15	5.065	21.206	6.272	26.260	5.234	21.914	5.084	21.286	8.49	35.546
400	673.15	5.129	21.474	6.395	26.775	5.306	22.215	5.080	21.269	8.81	36.886
500	773.15	5.202	21.780	6.524	27.315	5.387	22.554	5.085	21.290	9.09	38.058
600	873.15	5.276	22.090	6.659	27.880	5.469	22.898	5.101	21.357	9.32	39.021
700	973.15	5.352	22.408	6.801	28.474	5.550	23.237	5.125	21.457	9.53	39.900
800	1073.15	5.425	22.713	6.945	29.077	5.627	23.559	5.155	21.583	9.70	40.612
900	1173.15	5.495	23.006	7.092	29.693	5.698	23.856	5.191	21.734	9.86	41.282
1000	1273.15	5.561	23.283	7.238	30.304	5.766	24.141	5.232	21.905	9.99	41.826
1100	1373.15	5.624	23.547	7.383	30.911	5.828	24.401	5.277	22.094	10.11	42.329
1200	1473.15	5.683	23.794	7.526	31.510	5.886	24.644	5.323	22.286	10.21	42.747
1300	1573.15	5.739	24.028	7.665	32.092	5.939	24.865	5.368	22.475		
1400	1673.15	5.792	24.250	7.801	32.661	5.989	25.075	5.412	22.659		
1500	1773.15	5.842	24.459	7.932	33.210	6.035	25.267	5.457	22.847		
1600	1873.15	5.888	24.652	8.059	33.741	6.079	25.452	5.503	23.040		
1700	1973.15	5.932	24.836	8.183	34.261	6.119	25.619	5.548	23.228		
1800	2073.15	5.972	25.004	8.301	34.755	6.156	25.774	5.591	23.408		
1900	2173.15	6.011	25.167	8.413	35.224	6.190	25.916	5.633	23.584		
2000	2273.15	6.049	25.326	8.522	35.680	6.223	26.054	5.675	23.760		
2100	2373.15	6.084	25.472	8.627	36.120	6.254	26.184	5.716	23.932		
2200	2473.15	6.117	25.611	8.727	36.538	6.283	26.306	5.756	24.099		
2300	2573.15	6.149	25.745	8.823	36.940	6.310	26.419	5.795	24.263		
2400	2673.15	6.179	25.870	8.916	37.330	6.335	26.523	5.833	24.422		
2500	2773.15	6.208	25.992	9.005	37.702	6.360	26.628	5.869	24.572		
2600	2873.15	—	—	9.092	38.066	6.383	26.724	5.904	24.719		
2700	2973.15	—	—	9.175	38.414	6.405	26.816	5.938	24.861		
2800	3073.15	—	—	9.254	38.745	6.426	26.904	5.971	24.999		
2900	3173.15	—	—	9.330	39.063	6.446	26.988	6.003	25.133		
3000	3273.15	—	—	—	—	6.465	27.068	6.034	25.263		
$M*$		28.964		18.020		30.008		17.008		64.06	

*M — relative molecular mass in kg/mol

Table 80-4 Mean Molar Heat Capacity C_{vm} of Gases at Constant Volume v Between $0°C$ (273.15 K) and $t°C$ (TK) (Continued)

Temperature		CH₄		C₂H₄		C₂H₂		N₂O		H₂S		CS₂	
		C_{vm}		C_{vm}		C_{vm}		C_{vm}		C_{vm}		C_{vm}	
t	T	$\dfrac{kcal}{kmol\,K}$	$\dfrac{kJ}{kmol\,K}$	$\dfrac{kcal}{kmol\,K}$	$\dfrac{kJ}{kmol\,K}$	$\dfrac{kcal}{kmol\,K}$	$\dfrac{kJ}{kmol\,K}$	$\dfrac{kcal}{kmol\,K}$	$\dfrac{kJ}{kmol\,K}$	$\dfrac{kcal}{kmol\,K}$	$\dfrac{kJ}{kmol\,K}$	$\dfrac{kcal}{kmol\,K}$	$\dfrac{kJ}{kmol\,K}$
°C	K												
0	273.15	6.311	26.423	7.79	32.615	8.02	33.578	6.959	29.136	6.08	25.456	8.64	36.174
100	373.15	6.805	28.491	9.05	37.891	8.970	37.556	7.499	31.397	6.21	26.000	9.16	38.351
200	473.15	7.431	31.112	10.23	42.831	9.602	40.202	7.968	33.360	6.37	26.670	9.59	40.151
300	573.15	8.111	33.959	11.37	47.604	10.118	42.362	8.373	35.056	6.55	27.424	9.93	41.575
400	673.15	8.805	36.865	12.38	51.833	10.562	44.221	8.729	36.547	6.75	28.261	10.22	42.789
500	773.15	9.473	39.662	13.34	55.852	10.950	45.845	9.045	37.870	6.95	29.098	10.46	43.794
600	873.15	10.117	42.358	14.21	59.494	11.303	47.323	9.333	39.075	7.16	29.977	10.67	44.673
700	973.15	10.739	44.962	14.98	62.718	11.628	48.684	9.592	40.160	7.86	30.908	10.85	45.427
800	1073.15	11.366	47.587	15.72	65.816	11.931	49.953	9.826	41.139	7.56	31.652	11.00	46.055
900	1173.15	11.946	50.016	16.38	68.580	12.212	51.129	10.036	42.019	7.74	32.406	11.13	46.599
1000	1273.15	12.465	52.188	17.00	71.176	12.473	52.222	10.227	42.818	7.92	33.159	11.24	47.060
1100	1373.15	12.931	54.140	17.58	73.604	12.720	53.256	10.402	43.551	8.09	33.871	11.35	47.520
1200	1473.15	13.342	55.860	18.10	75.781	12.952	54.227	—	—	8.24	34.499	11.44	47.897
$M*$		16.031		28.031		26.040		44.016		34.08		76.13	

*M — relative molecular mass in kg/mol

1 kcal = 4.186 8 kJ

Table 81-1 Mean Volume Heat Capacity c'_{pm} of Gases at Constant Pressure p Between 0°C (273.15 K) and t°C (TK)

| Temperature | | H₂ | | N₂ | | O₂ | | CO | | CO₂ | |
| t | T | c'_{pm} | | c'_{pm} | | c'_{pm} | | c'_{pm} | | c'_{pm} | |
°C	K	$\frac{kcal}{m_n^3\,K}$	$\frac{kJ}{m_n^3\,K}$	$\frac{kcal}{m_n^3\,K}$	$\frac{kJ}{m_n^3\,K}$	$\frac{kcal}{m_n^3\,K}$	$\frac{kJ}{m_n^3\,K}$	$\frac{kcal}{m_n^3\,K}$	$\frac{kJ}{m_n^3\,K}$	$\frac{kcal}{m_n^3\,K}$	$\frac{kJ}{m_n^3\,K}$
0	273.15	0.3049	1.277	0.3102	1.299	0.3119	1.306	0.3103	1.299	0.3821	1.600
100	373.15	0.3083	1.291	0.3106	1.300	0.3147	1.318	0.3109	1.302	0.4061	1.700
200	473.15	0.3098	1.297	0.3114	1.304	0.3189	1.335	1.3122	1.307	0.4269	1.787
300	573.15	0.3103	1.299	0.3131	1.311	0.3239	1.356	0.3145	1.317	0.4449	1.863
400	673.15	0.3110	1.302	0.3154	1.321	0.3290	1.377	0.3174	1.329	0.4609	1.930
500	773.15	0.3117	1.305	0.3182	1.332	0.3339	1.398	0.3207	1.343	0.4750	1.989
600	873.15	0.3124	1.308	0.3213	1.345	0.3384	1.417	0.3242	1.357	0.4875	2.041
700	973.15	0.3134	1.312	0.3245	1.359	0.3426	1.434	0.3277	1.372	0.4988	2.088
800	1073.15	0.3145	1.317	0.3276	1.372	0.3468	1.450	0.3311	1.386	0.5090	2.131
900	1173.15	0.3159	1.323	0.3307	1.385	0.3498	1.465	0.3343	1.400	0.5181	2.169
1000	1273.15	0.3174	1.329	0.3337	1.397	0.3529	1.478	0.3374	1.413	0.5263	2.204
1100	1373.15	0.3191	1.336	0.3365	1.409	0.3557	1.489	0.3403	1.425	0.5338	2.235
1200	1473.15	0.3208	1.343	0.3392	1.420	0.3584	1.501	0.3430	1.436	0.5407	2.264
1300	1573.15	0.3227	1.351	0.3417	1.431	0.3608	1.511	0.3455	1.447	0.5469	2.290
1400	1673.15	0.3246	1.359	0.3441	1.441	0.3631	1.520	0.3479	1.457	0.5526	2.314
1500	1773.15	0.3266	1.367	0.3463	1.450	0.3653	1.529	0.3501	1.466	0.5578	2.335
1600	1873.15	0.3285	1.375	0.3484	1.459	0.3673	1.538	0.3522	1.475	0.5626	2.355
1700	1973.15	0.3304	1.383	0.3504	1.467	0.3693	1.546	0.3541	1.483	0.5671	2.374
1800	2073.15	0.3324	1.392	0.3522	1.475	0.3712	1.554	0.3559	1.490	0.5712	2.392
1900	2173.15	0.3343	1.400	0.3540	1.482	0.3730	1.562	0.3576	1.497	0.5750	2.407
2000	2273.15	0.3362	1.408	0.3556	1.489	0.3748	1.569	0.3592	1.504	0.5785	2.422
2100	2373.15	0.3380	1.415	0.3572	1.496	0.3764	1.576	0.3607	1.510	0.5818	2.436
2200	2473.15	0.3398	1.423	0.3587	1.502	0.3781	1.583	0.3621	1.516	0.5848	2.448
2300	2573.15	0.3416	1.430	0.3600	1.507	0.3797	1.590	0.3634	1.521	0.5876	2.460
2400	2673.15	0.3433	1.437	0.3613	1.513	0.3813	1.596	0.3647	1.527	0.5902	2.471
2500	2773.15	0.3451	1.445	0.3625	1.518	0.3828	1.603	0.3659	1.532	0.5926	2.481
2600	2873.15	0.3467	1.452	—	—	0.3843	1.609				
2700	2973.15	0.3483	1.458	—	—	0.3858	1.615				

Table 81-2 Mean Volume Heat Capacity c'_{pm} of Gases at Constant Pressure p Between 0°C (273.15 K) and t°C (TK) (Continued)

| Temperature | | COS | | C₂H₆ | | C₃H₈ | | C₆H₆ | | C₃H₆ | | C₂H₅OH | |
| t | T | c'_{pm} | | c'_{pm} | | c'_{pm} | | c'_{pm} | | c'_{pm} | | c'_{pm} | |
°C	K	$\frac{kcal}{m_n^3\,K}$	$\frac{kJ}{m_n^3\,K}$	$\frac{kcal}{m_n^3\,K}$	$\frac{kJ}{m_n^3\,K}$	$\frac{kcal}{m_n^3\,K}$	$\frac{kJ}{m_n^3\,K}$	$\frac{kcal}{m_n^3\,K}$	$\frac{kJ}{m_n^3\,K}$	$\frac{kcal}{m_n^3\,K}$	$\frac{kJ}{m_n^3\,K}$	$\frac{kcal}{m_n^3\,K}$	$\frac{kJ}{m_n^3\,K}$
0	273.15	0.429	1.796	0.5278	2.210	0.7281	3.048	0.7856	3.287	0.6393	2.677	0.7468	3.127
100	373.15	0.454	1.901	0.5959	2.495	0.8383	3.510	0.9434	3.950	0.7281	3.048	0.8236	3.448
200	473.15	0.475	1.989	0.6627	2.775	0.9471	3.965	1.1033	4.619	0.8071	3.379	0.8941	3.743
300	573.15	0.492	2.060	0.7271	3.044	1.0435	4.369	1.2420	5.200	0.8851	3.706	0.9641	4.036
400	673.15	0.506	2.119	0.7902	3.308	1.1368	4.760	1.3629	5.706	0.9565	4.005	1.0279	4.304
500	773.15	0.518	2.169	0.8485	3.552	1.2166	5.094	1.4691	6.151	1.0230	4.283	1.0868	4.550
600	873.15	0.529	2.215	0.9023	3.778	1.2974	5.432	1.5646	6.551	1.0841	4.539	1.1417	4.780
700	973.15	0.538	2.252	0.9521	3.986	1.3670	5.723	1.6489	6.904	1.1408	4.776	1.1912	4.987
800	1073.15	0.546	2.286	0.9986	4.181	1.4803	6.198	1.7248	7.221	1.1921	4.991	1.2371	5.179
900	1173.15	0.553	2.315	1.0418	4.362	1.4883	6.231	1.7935	7.509	1.2398	5.191	1.2795	5.357
1000	1273.15	0.560	2.345	1.0818	4.529	1.5432	6.461	1.8559	7.770	1.2831	5.372	1.3179	5.518
1100	1373.15	0.566	2.370	1.1187	4.684	1.5949	6.678	1.9126	8.008	1.3232	5.540	1.3545	5.671
1200	1473.15	0.571	2.391	1.1525	4.825	1.6436	6.881	1.9634	8.220	1.3607	5.697	1.3875	5.809

1 kcal = 4.1868 kJ

Table 81-3 Mean Volume Heat Capacity c'_{pm} of Gases at Constant Pressure p Between $0°C$ (273.15 K) and $t°C$ (TK) (Continued)

| Temperature | | Air | | H_2O | | NO | | OH | | SO_2 | |
| t | T | c'_{pm} | | c'_{pm} | | c'_{pm} | | c'_{pm} | | c'_{pm} | |
°C	K	$\frac{kcal}{m_n^3 K}$	$\frac{kJ}{m_n^3 K}$	$\frac{kcal}{m_n^3 K}$	$\frac{kJ}{m_n^3 K}$	$\frac{kcal}{m_n^3 K}$	$\frac{kJ}{m_n^3 K}$	$\frac{kcal}{m_n^3 K}$	$\frac{kJ}{m_n^3 K}$	$\frac{kcal}{m_n^3 K}$	$\frac{kJ}{m_n^3 K}$
0	273.15	0.3098	1.297	0.3569	1.494	0.3194	1.337	0.3196	1.338	0.414	1.733
100	373.15	0.3106	1.300	0.3595	1.505	0.3187	1.334	0.3171	1.328	0.433	1.813
200	473.15	0.3122	1.307	0.3636	1.522	0.3197	1.339	0.3159	1.323	0.451	1.888
300	573.15	0.3146	1.317	0.3684	1.542	0.3221	1.349	0.3154	1.321	0.467	1.955
400	673.15	0.3174	1.329	0.3739	1.565	0.3253	1.362	0.3152	1.320	0.482	2.018
500	773.15	0.3207	1.343	0.3797	1.590	0.3289	1.377	0.3155	1.321	0.494	2.068
600	873.15	0.3240	1.357	0.3857	1.615	0.3326	1.393	0.3162	1.324	0.505	2.114
700	973.15	0.3274	1.371	0.3920	1.641	0.3362	1.408	0.3172	1.328	0.514	2.152
800	1073.15	0.3306	1.384	0.3984	1.668	0.3396	1.422	0.3186	1.334	0.521	2.181
900	1173.15	0.3338	1.398	0.4050	1.696	0.3428	1.435	0.3202	1.341	0.529	2.215
1000	1273.15	0.3367	1.410	0.4115	1.723	0.3458	1.448	0.3220	1.348	0.534	2.236
1100	1373.15	0.3395	1.421	0.4180	1.750	0.3486	1.460	0.3240	1.357	0.540	2.261
1200	1473.15	0.3422	1.433	0.4244	1.777	0.3512	1.470	0.3261	1.365	0.544	2.278
1300	1573.15	0.3447	1.443	0.4306	1.803	0.3536	1.480	0.3281	1.374		
1400	1673.15	0.3470	1.453	0.4366	1.828	0.3558	1.490	0.3301	1.382		
1500	1773.15	0.3492	1.462	0.4425	1.853	0.3578	1.498	0.3321	1.390		
1600	1873.15	0.3513	1.471	0.4481	1.876	0.3598	1.506	0.3341	1.399		
1700	1973.15	0.3532	1.479	0.4537	1.900	0.3616	1.514	0.3361	1.407		
1800	2073.15	0.3551	1.487	0.4589	1.921	0.3632	1.521	0.3380	1.415		
1900	2173.15	0.3568	1.494	0.4639	1.942	0.3648	1.527	0.3399	1.423		
2000	2273.15	0.3585	1.501	0.4688	1.963	0.3662	1.533	0.3418	1.431		
2100	2373.15	0.3600	1.507	0.4735	1.982	0.3676	1.539	0.3436	1.439		
2200	2473.15	0.3615	1.514	0.4779	2.001	0.3689	1.545	0.3454	1.446		
2300	2573.15	0.3629	1.519	0.4822	2.019	0.3701	1.550	0.3471	1.453		
2400	2673.15	0.3643	1.525	0.4864	2.036	0.3712	1.554	0.3488	1.460		
2500	2773.15	0.3655	1.530	0.4903	2.053	0.3723	1.559	0.3504	1.467		
2600	2873.15	—	—	0.4942	2.069	0.3734	1.563	0.3520	1.474		
2700	2973.15	—	—	0.4949	2.085	0.3743	1.567	0.3535	1.480		
2800	3073.15	—	—	0.5015	2.100	0.3753	1.571	0.3550	1.486		
2900	3173.15	—	—	0.5048	2.113	0.3762	1.575	0.3564	1.492		
3000	3273.15	—	—	—	—	0.3770	1.578	0.3578	1.498		

Table 81-4 Mean Volume Heat Capacity c'_{pm} of Gases at Constant Pressure p Between $0°C$ (273.15 K) and $t°C$ (TK) (Continued)

| Temperature | | CH_4 | | C_2H_4 | | C_2H_2 | | N_2O | | H_2S | | CS_2 | |
| t | T | c'_{pm} | | c'_{pm} | | c'_{pm} | | c'_{pm} | | c'_{pm} | | c'_{pm} | |
°C	K	$\frac{kcal}{m_n^3 K}$	$\frac{kJ}{m_n^3 K}$	$\frac{kcal}{m_n^3 K}$	$\frac{kJ}{m_n^3 K}$	$\frac{kcal}{m_n^3 K}$	$\frac{kJ}{m_n^3 K}$	$\frac{kcal}{m_n^3 K}$	$\frac{kJ}{m_n^3 K}$	$\frac{kcal}{m_n^3 K}$	$\frac{kJ}{m_n^3 K}$	$\frac{kcal}{m_n^3 K}$	$\frac{kJ}{m_n^3 K}$
0	273.15	0.3702	1.550	0.4363	1.827	0.4466	1.870	0.3991	1.671	0.360	1.507	0.474	1.985
100	373.15	0.3922	1.642	0.4925	2.062	0.48879	2.046	0.4232	1.772	0.366	1.532	0.497	2.081
200	473.15	0.4201	1.759	0.5452	2.283	0.51698	2.164	0.4441	1.859	0.373	1.562	0.517	2.165
300	573.15	0.4505	1.886	0.5960	2.495	0.54000	2.261	0.4621	1.935	0.381	1.595	0.532	2.227
400	673.15	0.4814	2.016	0.6415	2.686	0.55981	2.344	0.4780	2.001	0.390	1.633	0.545	2.282
500	773.15	0.5112	2.140	0.6839	2.863	0.57712	2.416	0.4921	2.060	0.399	1.671	0.555	2.324
600	873.15	0.5400	2.261	0.7227	3.026	0.59287	2.482	0.5050	2.114	0.408	1.708	0.565	2.366
700	973.15	0.5677	2.377	0.7571	3.170	0.60737	2.543	0.5165	2.162	0.417	1.746	0.573	2.399
800	1073.15	0.5957	2.494	0.7901	3.308	0.62089	2.600	0.5270	2.206	0.426	1.784	0.579	2.424
900	1173.15	0.6216	2.603	0.8196	3.432	0.63343	2.652	0.5363	2.245	0.434	1.817	0.585	2.449
1000	1273.15	0.6447	2.699	0.8472	3.547	0.64507	2.701	0.5449	2.281	0.442	1.851	0.590	2.470
1100	1373.15	0.6655	2.786	0.8731	3.655	0.65609	2.747	0.5527	2.314	0.450	1.884	0.595	2.491
1200	1473.15	0.6838	2.863	0.8963	3.753	0.66644	2.790	—	—	0.456	1.909	0.599	2.508

1 kcal = 4.186 8 kJ

Table 82-1 Mean Volume Heat Capacity c'_{vm} of Gases at Constant Volume v Between 0°C (273.15 K) and t°C (TK)

Temperature		H₂		N₂		O₂		CO		CO₂	
		c'_{vm}		c'_{vm}		c'_{vm}		c'_{vm}		c'_{vm}	
t	T	kcal	kJ	kcal	kJ	kcal	kJ	kcal	kJ	kcal	kJ
°C	K	$\frac{kcal}{m_n^3 K}$	$\frac{kJ}{m_n^3 K}$	$\frac{kcal}{m_n^3 K}$	$\frac{kJ}{m_n^3 K}$	$\frac{kcal}{m_n^3 K}$	$\frac{kJ}{m_n^3 K}$	$\frac{kcal}{m_n^3 K}$	$\frac{kJ}{m_n^3 K}$	$\frac{kcal}{m_n^3 K}$	$\frac{kJ}{m_n^3 K}$
0	273.15	0.2163	0.906	0.2216	0.928	0.2233	0.935	0.2217	0.928	0.2935	1.229
100	373.15	0.2197	0.920	0.2220	0.929	0.2261	0.947	0.2223	0.931	0.3175	1.329
200	473.15	0.2212	0.926	0.2228	0.933	0.2303	0.964	0.2236	0.936	0.3383	1.416
300	573.15	0.2217	0.928	0.2245	0.940	0.2353	0.985	0.2259	0.946	0.3563	1.492
400	673.15	0.2224	0.931	0.2268	0.950	0.2404	1.007	0.2288	0.958	0.3723	1.559
500	773.15	0.2231	0.934	0.2296	0.961	0.2453	1.027	0.2321	0.972	0.3864	1.618
600	873.15	0.2238	0.937	0.2327	0.974	0.2498	1.046	0.2356	0.986	0.3989	1.670
700	973.15	0.2248	0.941	0.2359	0.988	0.2540	1.063	0.2391	1.001	0.4102	1.717
800	1073.15	0.2259	0.946	0.2390	1.001	0.2577	1.079	0.2425	1.015	0.4204	1.760
900	1173.15	0.2273	0.952	0.2421	1.014	0.2612	1.094	0.2457	1.029	0.4295	1.798
1000	1273.15	0.2288	0.958	0.2451	1.026	0.2643	1.107	0.2488	1.042	0.4377	1.833
1100	1373.15	0.2305	0.965	0.2479	1.038	0.2671	1.118	0.2517	1.054	0.4452	1.864
1200	1473.15	0.2322	0.972	0.2506	1.049	0.2698	1.130	0.2544	1.065	0.4521	1.893
1300	1573.15	0.2341	0.980	0.2531	1.060	0.2722	1.140	0.2569	1.076	0.4583	1.919
1400	1673.15	0.2360	0.988	0.2555	1.070	0.2745	1.149	0.2593	1.086	0.4640	1.943
1500	1773.15	0.2380	0.996	0.2577	1.079	0.2767	1.158	0.2615	1.095	0.4629	1.964
1600	1873.15	0.2399	1.004	0.2598	1.088	0.2787	1.167	0.2636	1.104	0.4740	1.985
1700	1973.15	0.2418	1.012	0.2618	1.096	0.2807	1.175	0.2655	1.112	0.4785	2.003
1800	2073.15	0.2438	1.021	0.2636	1.104	0.2826	1.183	0.2673	1.119	0.4826	2.021
1900	2173.15	0.2457	1.029	0.2654	1.111	0.2844	1.191	0.2690	1.126	0.4864	2.036
2000	2273.15	0.2476	1.037	0.2670	1.118	0.2861	1.198	0.2706	1.133	0.4899	2.051
2100	2373.15	0.2494	1.044	0.2686	1.125	0.2878	1.205	0.2721	1.139	0.4932	2.065
2200	2473.15	0.2512	1.052	0.2700	1.130	0.2895	1.212	0.2735	1.145	0.4962	2.077
2300	2573.15	0.2530	1.059	0.2714	1.136	0.2911	1.219	0.2748	1.151	0.4990	2.089
2400	2673.15	0.2547	1.066	0.2727	1.142	0.2927	1.225	0.2761	1.156	0.5016	2.100
2500	2773.15	0.2565	1.074	0.2739	1.147	0.2942	1.232	0.2773	1.161	0.5040	2.110
2600	2873.15	0.2581	1.081	—	—	0.2956	1.238	—	—	—	—
2700	2973.15	0.2597	1.087	—	—	0.2972	1.244	—	—	—	—

Table 82-2 Mean Volume Heat Capacity c'_{vm} of Gases at Constant Volume v Between 0°C (273.15 K) and t°C (TK) (Continued)

Temperature		COS		C₂H₆		C₃H₈		C₆H₆		C₃H₆		C₂H₅OH	
		c'_{vm}		c'_{vm}		c'_{vm}		c'_{vm}		c'_{vm}		c'_{vm}	
t	T	kcal	kJ	kcal	kJ	kcal	kJ	kcal	kJ	kcal	kJ	kcal	kJ
°C	K	$\frac{kcal}{m_n^3 K}$	$\frac{kJ}{m_n^3 K}$	$\frac{kcal}{m_n^3 K}$	$\frac{kJ}{m_n^3 K}$	$\frac{kcal}{m_n^3 K}$	$\frac{kJ}{m_n^3 K}$	$\frac{kcal}{m_n^3 K}$	$\frac{kJ}{m_n^3 K}$	$\frac{kcal}{m_n^3 K}$	$\frac{kJ}{m_n^3 K}$	$\frac{kcal}{m_n^3 K}$	$\frac{kJ}{m_n^3 K}$
0	273.15	0.340	1.424	0.4392	1.839	0.6393	2.677	0.6964	2.916	0.5505	2.305	0.6582	2.756
100	373.15	0.365	1.528	0.5073	2.124	0.7495	3.138	0.8646	3.620	0.6393	2.677	0.7350	3.077
200	473.15	0.386	1.616	0.5741	2.404	0.8584	3.594	1.0145	4.248	0.7183	3.007	0.8055	3.372
300	573.15	0.403	1.687	0.6385	2.673	0.9547	3.997	1.1533	4.829	0.7964	3.334	0.8755	3.666
400	673.15	0.418	1.750	0.7016	2.937	1.0480	4.388	1.2742	5.335	0.8677	3.633	0.9393	3.933
500	773.15	0.430	1.800	0.7599	3.182	1.1278	4.722	1.3803	5.779	0.9342	3.911	0.9982	4.179
600	873.15	0.440	1.842	0.8137	3.407	1.2086	5.060	1.4758	6.179	0.9953	4.167	1.0531	4.409
700	973.15	0.449	1.880	0.8635	3.615	1.2782	5.352	1.5601	6.532	1.0520	4.405	1.1026	4.616
800	1073.15	0.457	1.913	0.9100	3.810	1.3415	5.617	1.6360	6.850	1.1033	4.619	1.1485	4.809
900	1173.15	0.464	1.943	0.9532	3.991	1.3995	5.859	1.7047	7.137	1.1510	4.819	1.1909	4.986
1000	1273.15	0.471	1.972	0.9932	4.158	1.4544	6.089	1.7671	7.398	1.1943	5.000	1.2293	5.147
1100	1373.15	0.477	1.997	1.0301	4.313	1.5062	6.306	1.8238	7.636	1.2345	5.169	1.2659	5.300
1200	1473.15	0.482	2.018	1.0639	4.454	1.5548	6.510	1.8747	7.849	1.2719	5.325	1.2989	5.438

1 kcal = 4.186 8 kJ

Table 82-3 Mean Volume Heat Capacity c'_{vm} of Gases at Constant Volume v Between $0°C$ (273.15 K) and $t°C$ (TK) (Continued)

| Temperature | | Air | | H_2O | | NO | | OH | | SO_2 | |
| t | T | c'_{vm} | | c'_{vm} | | c'_{vm} | | c'_{vm} | | c'_{vm} | |
°C	K	$\frac{kcal}{m_n^3 K}$	$\frac{kJ}{m_n^3 K}$	$\frac{kcal}{m_n^3 K}$	$\frac{kJ}{m_n^3 K}$	$\frac{kcal}{m_n^3 K}$	$\frac{kJ}{m_n^3 K}$	$\frac{kcal}{m_n^3 K}$	$\frac{kJ}{m_n^3 K}$	$\frac{kcal}{m_n^3 K}$	$\frac{kJ}{m_n^3 K}$
0	273.15	0.2212	0.926	0.2684	1.124	0.2308	0.966	0.2310	0.967	0.325	1.361
100	373.15	0.2220	0.929	0.2709	1.134	0.2301	0.963	0.2287	0.958	0.344	1.440
200	473.15	0.2236	0.936	0.2750	1.151	0.2310	0.967	0.2273	0.952	0.362	1.516
300	573.15	0.2260	0.946	0.2798	1.171	0.2335	0.978	0.2268	0.950	0.379	1.587
400	673.15	0.2288	0.958	0.2853	1.194	0.2367	0.991	0.2266	0.949	0.393	1.645
500	773.15	0.2321	0.972	0.2911	1.219	0.2403	1.006	0.2269	0.950	0.406	1.700
600	873.15	0.2354	0.986	0.2971	1.244	0.2440	1.022	0.2276	0.953	0.416	1.742
700	973.15	0.2388	1.000	0.3034	1.270	0.2476	1.037	0.2286	0.957	0.425	1.779
800	1073.15	0.2420	1.013	0.3098	1.297	0.2510	1.051	0.2300	0.963	0.433	1.813
900	1173.15	0.2451	1.026	0.3164	1.325	0.2542	1.064	0.2316	0.970	0.440	1.842
1000	1273.15	0.2481	1.039	0.3229	1.352	0.2572	1.077	0.2334	0.977	0.446	1.867
1100	1373.15	0.2509	1.050	0.3294	1.379	0.2600	1.089	0.2354	0.986	0.451	1.888
1200	1473.15	0.2536	1.062	0.3358	1.406	0.2626	1.099	0.2375	0.994	0.455	1.905
1300	1573.15	0.2561	1.072	0.3420	1.432	0.2650	1.110	0.2395	1.003		
1400	1673.15	0.2584	1.082	0.3480	1.457	0.2672	1.119	0.2415	1.011		
1500	1773.15	0.2606	1.091	0.3539	1.482	0.2692	1.127	0.2435	1.019		
1600	1873.15	0.2627	1.100	0.3595	1.505	0.2712	1.135	0.2455	1.028		
1700	1973.15	0.2646	1.108	0.3651	1.529	0.2730	1.143	0.2475	1.036		
1800	2072.15	0.2665	1.116	0.3703	1.550	0.2746	1.150	0.2494	1.044		
1900	2173.15	0.2682	1.123	0.3753	1.571	0.2762	1.156	0.2513	1.052		
2000	2273.15	0.2698	1.130	0.3802	1.592	0.2776	1.162	0.2532	1.060		
2100	2373.15	0.2714	1.136	0.3849	1.611	0.2790	1.168	0.2550	1.068		
2200	2473.15	0.2729	1.143	0.3893	1.630	0.2803	1.174	0.2568	1.075		
2300	2573.15	0.2743	1.148	0.3936	1.648	0.2815	1.179	0.2585	1.082		
2400	2673.15	0.2757	1.154	0.3978	1.666	0.2826	1.183	0.2602	1.089		
2500	2773.15	0.2769	1.159	0.4017	1.682	0.2873	1.188	0.2618	1.096		
2600	2873.15	—	—	0.4056	1.698	0.2848	1.192	0.2634	1.103		
2700	2973.15	—	—	0.4093	1.714	0.2857	1.196	0.2649	1.109		
2800	3073.15	—	—	0.4129	1.729	0.2867	1.200	0.2664	1.115		
2900	3173.15	—	—	0.4162	1.743	0.2876	1.204	0.2678	1.121		
3000	32731.5	—	—	—	—	0.2884	1.207	0.2692	1.127		

Table 82-4 Mean Volume Heat Capacity c'_{vm} of Gases at Constant Volume v Between $0°C$ (273.15 K) and $t°C$ (TK) (Continued)

| Temperature | | CH_4 | | C_2H_4 | | C_2H_2 | | N_2O | | H_2S | | CS_2 | |
| t | T | c_{vm} | | c'_{vm} | | c'_{vm} | | c'_{vm} | | c'_{vm} | | c'_{vm} | |
°C	K	$\frac{kcal}{m_n^3 K}$	$\frac{kJ}{m_n^3 K}$	$\frac{kcal}{m_n^3 K}$	$\frac{kJ}{m_n^3 K}$	$\frac{kcal}{m_m^3 K}$	$\frac{kJ}{m_n^3 K}$	$\frac{kcal}{m_n^3 K}$	$\frac{kJ}{m_n^3 K}$	$\frac{kcal}{m_n^3 K}$	$\frac{kJ}{m_n^3 K}$	$\frac{kcal}{m_n^3 K}$	$\frac{kJ}{m_n^3 K}$
0	273.15	0.2816	1.179	0.3475	1.455	0.3578	1.498	0.3105	1.300	0.271	1.135	0.385	1.612
100	373.15	0.3036	1.271	0.4037	1.690	0.4002	1.676	0.3346	1.401	0.277	1.160	0.409	1.712
200	473.15	0.3315	1.388	0.4564	1.911	0.4284	1.794	0.3555	1.488	0.283	1.185	0.428	1.792
300	573.15	0.3619	1.515	0.5073	2.124	0.45140	1.890	0.3735	1.564	0.292	1.223	0.443	1.855
400	673.15	0.3928	1.645	0.5528	2.314	0.47121	1.973	0.3894	1.630	0.301	1.260	0.456	1.909
500	773.15	0.4226	1.769	0.5951	2.492	0.48852	2.045	0.4035	1.689	0.310	1.298	0.467	1.955
600	873.15	0.4514	1.890	0.6340	2.654	0.50427	2.111	0.4164	1.743	0.319	1.336	0.476	1.993
700	973.15	0.4791	2.006	0.6683	2.798	0.51877	2.172	0.4279	1.792	0.328	1.373	0.484	2.026
800	1073.15	0.5071	2.123	0.7013	2.936	0.53229	2.229	0.4384	1.835	0.337	1.411	0.491	2.056
900	1173.15	0.5330	2.232	0.7308	3.060	0.54482	2.281	0.4477	1.874	0.345	1.444	0.496	2.077
1000	1273.15	0.5561	2.328	0.7584	3.175	0.55647	2.330	0.4563	1.910	0.353	1.478	0.501	2.098
1100	1373.15	0.5769	2.415	0.7843	3.284	0.56749	2.376	0.4641	1.943	0.361	1.511	0.506	2.119
1200	1473.15	0.5952	2.492	0.8075	3.381	0.57784	2.419	—	—	0.368	1.541	0.510	2.135

1 kcal = 4.186 8 kJ

Table 83-1　Specific Enthalpy i of Gases

Temperature		H₂		N₂		O₂		CO		CO₂		Air	
t	T	i		i		i		i		i		i	
°C	K	kcal/kg	kJ/kg	kcal/kg	kJ/kg	kcal/kg	kJ/kg	kcal/kg	kJ/kg	kcal/kg	kJ/kg	kcal/kg	kJ/kg
0	273.15	0.0	0.0	0.00	0.00	0.00	0.00	0.00	0.00	0.00	0.00	0.00	0.00
100	373.15	342.8	1435.2	24.85	104.04	22.05	92.32	24.88	104.17	20.68	86.58	24.03	100.61
200	473.15	688.9	2884.3	49.84	208.67	44.68	187.07	49.98	209.28	43.48	182.04	48.32	202.31
300	573.15	1035	4333.3	75.15	314.64	68.07	284.90	75.51	316.15	67.98	284.62	73.02	305.72
400	673.15	1383	5790.3	100.1	419.10	92.20	386.02	101.6	425.38	93.88	393.06	98.24	411.31
500	773.15	1733	7255.7	127.3	532.98	116.9	489.44	128.4	537.59	121.0	506.60	124.1	519.58
600	873.15	2084	8725.3	154.2	645.61	142.3	595.78	155.6	651.47	149.0	623.8	150.4	629.70
700	973.15	2439	10211.7	181.7	760.74	168.0	703.38	183.5	768.28	177.9	744.8	177.3	742.32
800	1073.15	2798	11714.5	209.7	877.97	194.1	812.66	211.9	887.18	207.4	868.3	204.6	856.61
900	1173.15	3161	13234	238.1	996.88	220.5	923.19	240.8	1008.12	237.4	993.9	232.5	973.43
1000	1273.15	3529	14775	267.0	1117.89	247.2	1034.90	270.0	1130.44	268.1	1122.5	260.5	1090.70
1100	1373.15	3902	16337	296.1	1239.7	274.1	1147.6	299.5	1253.9	299.1	1252.3	289.0	1210.00
1200	1473.15	3280	17920	325.6	1363.2	301.2	1261.1	329.4	1379.1	330.5	1383.7	317.6	1329.76
1300	1573.15	4665	19531	355.4	1487.0	328.5	1375.4	359.5	1505.2	362.1	1516.0	346.7	1451.60
1400	1673.15	5053	21156	385.4	1613.6	356.0	1490.5	389.8	1632.0	394.0	1649.6	375.9	1573.81
1500	1773.15	5446	22801	415.7	1740.5	383.8	1606.9	420.3	1759.7	426.2	1784.4	405.3	1696.91
1600	1873.15	5844	24468	446.1	1867.7	411.7	1723.7	450.9	1887.8	458.4	1919.2	434.9	1820.8
1700	1973.15	6246	26151	476.5	1995.0	439.8	1841.4	481.8	2017.2	491.0	2055.7	464.6	1945.1
1800	2073.15	6652	27851	507.2	2123.5	468.0	1959.4	512.6	2146.2	523.6	2192.2	494.5	2070.3
1900	2173.15	7062	29567	538.1	2252.9	496.5	2078.7	543.8	2276.8	556.3	2329.1	524.6	2196.3
2000	2273.15	7475	31296	569.0	2382.3	525.0	2198.1	574.8	2406.6	589.2	2466.9	554.6	2322.0
2100	2373.15	7893	33046	600.2	2512.9	553.8	2318.7	606.1	2537.6	622.2	2605.0	585.1	2449.7
2200	2473.15	8312	34801	631.2	2642.7	582.6	2439.2	637.3	2668.2	655.2	2743.2	615.3	2576.1
2300	2573.15	8736	36576	662.4	2773.3	611.8	2561.5	668.8	2800.1	688.4	2882.2	645.8	2703.8
2400	2673.15	9162	38359	693.8	2904.8	641.0	2683.7	700.3	2932.0	721.4	3020.4	676.6	2832.8
2500	2773.15	9592	40160	725.0	3035.4	670.5	2807.3	732.0	3064.7	754.5	3158.9	707.0	2960.1
2600	2873.15	10020	41952	—	—	699.9	2930.3	—	—	—	—		
2700	2973.15	10460	43794	—	—	729.5	3054.3	—	—	—	—		
3000	3273.15	11790	49362	884.0	3701.1	820.3	3434.4	891.4	3732.1	921.6	3858.6		

1 kcal = 4.186 8 kJ

Table 83-2 Specific Enthalpy *i* of Gases (*Continued*)

Temperature		NO		OH		N₂O		H₂S		NH₃	
t	*T*	*i*		*i*		*i*		*i*		*i*	
°C	K	kcal/kg	kJ/kg	kcal/kg	kJ/kg	kcal/kg	kJ/kg	kcal/kg	kJ/kg	kcal/kg	kJ/kg
0	273.15	0.00	0.000	0.00	0.00	0.00	0.000	0.0	0.00	0.00	0.00
100	373.15	23.80	99.646	41.79	174.97	21.55	90.226	24.1	100.90	50.9	213.11
200	473.15	47.76	199.96	83.26	348.59	45.22	189.33	49.0	205.15	105.6	442.13
300	573.15	72.18	302.20	124.7	522.09	70.61	295.63	75.3	315.27	167.2	700.03
400	673.15	97.20	406.96	166.2	695.85	97.37	407.67	103.2	432.08	230.8	966.31
500	773.15	122.8	514.14	207.9	870.44	125.3	524.61	131.0	548.47	298.6	1250.2
600	873.15	149.0	623.83	250.0	1046.7	154.3	646.02	160.8	673.24	370.4	1550.8
700	973.15	175.8	736.04	292.7	1225.5	184.1	770.79	191.8	803.03	446.3	1868 6
800	1073.15	203.0	849.92	335.9	1406.3	214.7	898.91	224.0	937.84	525.7	2201 0
900	1173.15	230.5	965.06	379.8	1590.1	245.8	1029.1	256.5	1073.9	608.5	2547 7
1000	1273.15	258.3	1081.45	424.4	1776.9	277.5	1161.8	291.0	1218.4	697.2	2919.0
1100	1373.15	286.4	1199.1	469.7	1966.5	309.6	1296.2	325.6	1363.2	—	—
1200	1473.15	314.8	1318.0	515.6	2058.7	—	—	360.0	1507.2	—	—
1300	1573.15	343.3	1437.3	562.1	2353.4	—	—	—	—	—	—
1400	1673.15	372.1	1557.9	609.0	2549.8	—	—	—	—	—	—
1500	1773.15	401.0	1678.9	656.4	2748.2	432.5	1810.8	484.3	2027.7	1172	4906.9
1600	1873.15	430.1	1800.7	704.5	2949.6	—	—	—	—	—	—
1700	1973.15	459.2	1922.6	753.1	3153.1	—	—	—	—	—	—
1800	2073.15	488.3	2044.4	801.9	3357.4	—	—	—	—	—	—
1900	2173.15	517.8	2167.9	851.2	3563.8	—	—	—	—	—	—
2000	2273.15	547.2	2291.0	900.8	3771.5	594.8	2490.3	677.8	2837.8	1375	5756.9
2100	2373.15	576.7	2414.5	950	3981.2	—	—	—	—	—	—
2200	2473.15	606.3	2538.5	1001	4191.0	—	—	—	—	—	—
2300	2573.15	636.0	2662.8	1052	4404.5	—	—	—	—	—	—
2400	2673.15	665.5	2786.2	1103	4618.0	—	—	—	—	—	—
2500	2773.15	695.3	2911.1	1155	4835.8	760.0	3182.0	877	3671.8	1597	6686.3
2600	2873.15	725.1	3035.8	1206	5049.3	—	—	—	—	—	—
2700	2973.15	754.9	3160.6	1258	5267.0	—	—	—	—	—	—
2800	3073.15	784.8	3285.8	1310	5484.7	—	—	—	—	—	—
2900	3173.15	814.9	3411.8	1362	5702.4	—	—	—	—	—	—
3000	3273.15	844.8	3537.0	1415	5924.3	926.3	3878.2	1079	4517.6	1805	7557.2

1 kcal = 4.186 8 kJ

Table 83-3 Specific Enthalpy *i* of Gases (*Continued*)

Temperature		H₂O		SO₂		CH₄		C₂H₄		C₂H₂		CS₂	
		i		*i*		*i*		*i*		*i*		*i*	
°C	K	kcal/kg	kJ/kg	kcal/kg	kJ/kg	kcal/kg	kJ/kg	kcal/kg	kJ/kg	kcal/kg	kJ/kg	kcal/kg	kJ/kg
0	273.15	0.00	0.00	0.00	0.000	0.0	0.00	0.00	0.00	0.000	0.00	0.00	0.000
100	373.15	44.73	187.28	15.16	63.47	54.8	229.44	39.36	164.79	42.080	176.18	14.64	61.295
200	473.15	90.46	378.74	31.56	132.14	117.4	491.53	87.12	364.75	89.016	372.69	30.42	127.36
300	573.15	137.5	575.69	49.07	205.45	188.8	790.47	142.9	598.29	139.47	583.93	46.98	196.79
400	673.15	186.1	779.16	67.43	282.32	269.1	1126.67	205.0	858.29	192.78	807.13	64.16	268.62
500	773.15	236.2	988.92	86.47	362.09	357.1	1495.1	273.2	1143.8	248.42	1040.1	81.75	342.27
600	873.15	287.9	1205.4	105.9	443.38	452.7	1895.4	346.5	1450.7	306.25	1282.2	99.78	417.76
700	973.15	341.4	1429.4	125.9	527.12	555.2	2324.5	423.4	1772.7	366.02	1532.4	118.0	494.04
800	1073.15	396.6	1660.5	146.0	611.27	665.8	2787.6	505.0	2114.3	427.62	1790.4	136.5	571.50
900	1173.15	453.5	1898.7	166.5	697.10	781.6	3272.4	539.4	2258.4	490.79	2054.8	155.1	649.37
1000	1273.15	512.0	2143.6	187.0	782.93	900.8	3771.4	677.0	2834.5	555.35	2325.1	173.8	727.67
1100	1373.15	572.0	2394.9	207.8	870.02	1023	4283.1	767.4	3213.0	621.31	2601.3	192.7	806.80
1200	1473.15	633.6	2652.8	228.5	956.69	1147	4802.3	859.4	3598.1	688.49	2882.6	211.7	886.35
1300	1573.15	696.4	2915.7	—	—								
1400	1673.15	760.5	3184.1	—	—								
1500	1773.15	825.8	3457.5	292.2	1223.38								
1600	1873.15	892.1	3735.0	—	—								
1700	1973.15	959.6	4017.7	—	—								
1800	2073.15	1027.8	4303.2	—	—								
1900	2173.15	1096.7	4591.7	—	—								
2000	2273.15	1168.5	4892.3	398.6	1668.86								
2100	2373.15	1237.1	5179.5										
2200	2473.15	1308.2	5477.2	—	—								
2300	2573.15	1380.0	5777.8	—	—								
2400	2673.15	1452.3	6080.5	—	—								
2500	2773.15	1525.2	6385.7	518.2	2169.60								
2600	2873.15	1598.8	6693.9	—	—								
2700	2973.15	1672.7	7003.3	—	—								
3000	3273.15	1870.0	7829.3	613.4	2568.18								

Table 83-4 Specific Enthalpy *i* of Gases (*Continued*)

Temperature		COS		C₂H₆		C₃H₈		C₆H₆		C₃H₆		C₂H₅OH	
t	*T*	*i*		*i*		*i*		*i*		*i*		*i*	
°C	K	kcal/kg	kJ/kg	kcal/kg	kJ/kg	kcal/kg	kJ/kg	kcal/kg	kJ/kg	kcal/kg	kJ/kg	kcal/kg	kJ/kg
0	273.15	0.00	0.000	0.00	0.00	0.00	0.00	0.00	0.00	0.00	0.00	0.00	0.00
100	373.15	16.95	70.966	44.42	185.98	42.61	178.40	27.36	114.55	38.78	162.36	40.05	167.68
200	473.15	35.46	148.45	98.80	413.66	96.30	403.19	63.32	265.12	85.98	359.98	86.96	364.08
300	573.15	55.08	230.61	162.6	680.77	159.1	666.12	106.9	447.57	141.4	592.01	140.6	588.66
400	673.15	75.57	316.40	235.6	986.41	231.2	967.99	156.4	654.82	203.8	853.27	200.0	837.36
500	773.15	96.71	404.91	316.2	1323.9	309.2	1294.6	210.8	882.58	272.4	1140.5	264.4	1107.0
600	873.15	118.4	495.72	403.6	1689.8	395.7	1656.7	269.4	1127.9	346.5	1450.7	333.2	1395.0
700	973.15	140.5	588.25	496.9	2080.4	486.4	2036.5	331.2	1386.7	425.4	1781.1	405.6	1698.2
800	1073.15	163.0	682.45	595.5	2493.2	581.7	2435.5	396.0	1658.0	508.0	2126.9	481.4	2015.5
900	1173.15	185.8	777.91	698.8	2925.7	680.9	2850.8	463.2	1939.3	594.4	2488.6	560.2	2345.4
1000	1273.15	208.9	874.62	806.5	3376.7	784.5	3284.5	532.6	2229.9	683.5	2861.7	641.0	2683.7
1100	1373.15	232.2	972.18	917.3	3840.6	891.9	3734.2	603.8	2528.0	775.4	3246.4	724.7	3034.2
1200	1473.15	255.7	1070.56	103.1	4316.6	1003	4199.4	676.1	2830.7	869.8	3641.7	809.9	3390.9

1 kcal = 4.186 8 kJ

Table 84-1 Molar Specific Enthalpy i_{mol} of Gases

Temperature		H₂		N₂		O₂		CO		CO₂		Air	
t	T	i_{mol}		i_{mol}		i_{mol}		i_{mol}		i_{mol}		i_{mol}	
°C	K	$\frac{kcal}{kmol}$	$\frac{kJ}{kmol}$	$\frac{kcal}{kmol}$	$\frac{kJ}{kmol}$	$\frac{kcal}{kmol}$	$\frac{kJ}{kmol}$	$\frac{kcal}{kmol}$	$\frac{kJ}{kmol}$	$\frac{kcal}{kmol}$	$\frac{kJ}{kmol}$	$\frac{kcal}{kmol}$	$\frac{kJ}{kmol}$
0	273.15	0.0	0.0	0.0	0.0	0.0	0.0	0.0	0.0	0.0	0.0	0.0	0.0
100	373.15	691.1	2893.5	696.1	2914.4	705.5	2953.8	696.9	2917.8	910.3	3811.2	696.3	2915.3
200	473.15	1389	5815.5	1396	5844.8	1430	5987.1	1400	5861.5	1914	8013.5	1400	5861.5
300	573.15	2087	8737.8	2105	8813	2178	9118.9	2115	8855.1	2992	12527	2115	8855.1
400	673.15	2788	11673	2828	11840	2950	12351	2846	11915.6	4132	17300	2846	11915.6
500	773.15	3493	14624	3567	14934	3742	15667	3595	15051.2	5323	22286	3594	15047.7
600	873.15	4201	17589	4321	18091	4552	19058	4360	18254	6557	27453	4357	18242
700	973.15	4917	20586	5091	21315	5375	22504	5142	21529	7827	32770	5137	21508
800	1073.15	5640	23614	5875	24597	6210	26000	5937	24857	9126	38209	5929	24824
900	1173.15	6373	26682	6672	27934	7056	29542	6745	28240	10450	43752	6733	28190
1000	1273.15	7115	29789	7479	31313	7910	33118	7563	31665	11800	49404	7547	31598
1100	1373.15	7867	32938	8296	34734	8771	36722	8391	35131	13160	55098	8371	35048
1200	1473.15	8629	36128	9122	38192	9640	40361	9227	38632	14540	60876	9203	38531
1300	1573.15	9404	39373	9957	41688	10510	44003	10070	42161	15940	66738	10040	42035
1400	1673.15	10190	42663	10800	45217	11390	47688	10920	45720	17340	72599	10890	45594
1500	1773.15	10980	45971	11640	48734	12280	51414	11770	49279	18750	78503	11740	49153
1600	1873.15	11780	49321	12500	52335	13170	55140	12630	52879	20180	84490	12600	52754
1700	1973.15	12590	52712	13350	55894	14070	58908	13490	56480	21610	90477	13460	56354
1800	2073.15	13410	56145	14210	59494	14980	62718	14360	60122	23050	96506	14320	59955
1900	2173.15	14240	59620	15070	63095	15890	66528	15230	63765	24490	102535	15190	63597
2000	2273.15	15070	63095	15940	66738	16800	70338	16100	67407	25930	108564	16070	67282
2100	2373.15	15910	66612	16810	70380	17720	74190	16980	71092	27380	114635	16950	70961
2200	2473.15	16760	70171	17690	74064	18640	78042	17860	74776	28840	120747	17830	74656
2300	2573.15	17610	73730	18560	77707	19570	81936	18740	78461	30290	126818	18710	78335
2400	2673.15	18470	77330	19440	81391	20510	85871	19620	82145	31750	132931	19600	82061
2500	2773.15	19340	80973	20320	85076	21450	89807	20500	85829	33210	139044	20490	85786
2600	2873.15	20210	84615	—	—	22400	93784	—	—	—	—		
2700	2973.15	21080	88257	—	—	23350	97762	—	—	—	—		
3000	3273.15	23770	99520	24780	103749	26250	109904	24960	104503	40550	169775		
M^*		2.0156		28.016		32.000		28.010		44.010		28.964	

*M — relative molecular mass in kg/mol

1 kcal = 4.186 8 kJ

Table 84-2 Molar Specific Enthalpy i_{mol} of Gases (Continued)

Temperature		NO		OH		N₂O		H₂S		NH₃	
t	T	i_{mol}		i_{mol}		i_{mol}		i_{mol}		i_{mol}	
°C	K	$\frac{kcal}{kmol}$	$\frac{kJ}{kmol}$	$\frac{kcal}{kmol}$	$\frac{kJ}{kmol}$	$\frac{kcal}{kmol}$	$\frac{kJ}{kmol}$	$\frac{kcal}{kmol}$	$\frac{kJ}{kmol}$	$\frac{kcal}{kmol}$	$\frac{kJ}{kmol}$
0	273.15	0.0	0.000	0.0	0.0	0	0.0	0	0.0	0.0	0.0
100	373.15	714.3	2990.6	710.8	2976.0	948.5	3971.2	820	3433.2	870	3642.5
200	473.15	1433	5999.7	1416	5928.5	1991	8335.9	1672	7000.3	1800	7536.2
300	573.15	2166	9068	2121	8880.2	3108	13013	2562	10727	2850	11933
400	673.15	2917	12212	2826	11832	4286	17945	3496	14637	3930	16454
500	773.15	3686	15432	3536	14805	5516	23094	4470	18715	5090	21311
600	873.15	4473	18728	4252	17802	6791	28433	5490	22986	6310	26419
700	973.15	5275	22085	4978	20842	8111	33959	6545	27403	7600	31820
800	1073.15	6090	25498	5713	23919	9450	39565	7640	31987	8950	37472
900	1173.15	6916	28956	6459	27043	10820	45301	8757	36664	10360	43375
1000	1273.15	7752	32456	7218	30220	12210	51121	9910	41491	11870	49697
1100	1373.15	8595	35986	7989	33448	13630	57066	11088	46423	—	—
1200	1473.15	9446	39549	8771	36722	—	—	12276	51397	—	—
1300	1573.15	10300	43124	9560	40026	—	—	—	—	—	—
1400	1673.15	11160	46725	10360	43375	—	—	—	—	—	—
1500	1773.15	12030	50367	11160	46725	19030	79674	16020	67073	19970	83610
1600	1873.15	12900	54010	11980	50158	—	—	—	—	—	—
1700	1973.15	13780	57694	12810	53633	—	—	—	—	—	—
1800	2073.15	14660	61378	13640	57108	—	—	—	—	—	—
1900	2173.15	15530	65021	14480	60625	—	—	—	—	—	—
2000	2273.15	16420	68747	15320	64142	26180	109610	22420	93868	23420	98055
2100	2373.15	17300	72432	16170	67701	—	—	—	—	—	—
2200	2473.15	18190	76158	17030	71301	—	—	—	—	—	—
2300	2573.15	19080	79884	17900	74944	—	—	—	—	—	—
2400	2673.15	19970	83610	18770	78586	—	—	—	—	—	—
2500	2773.15	20860	87337	19640	82229	33450	140048	29000	121417	27030	113169
2600	2873.15	21760	91105	20510	85871	—	—	—	—	—	—
2700	2973.15	22660	94873	21390	89556	—	—	—	—	—	—
2800	3073.15	23550	98599	22280	93282	—	—	—	—	—	—
2900	3173.15	24450	102367	23170	97008	—	—	—	—	—	—
3000	3273.15	25350	106135	24060	100734	40770	170696	35700	149469	30740	128702
M^*		30.008		17.008		44.016		34.08		17.031	

$^*M$ — relative molecular mass in kg/mol

1 kcal = 4.186 8 kJ

Table 84-3 Molar Specific Enthalpy i_{mol} of Gases (*Continued*)

Temperature		H₂O		SO₂		CH₄		C₂H₄		C₂H₂		CS₂	
t	T	i_{mol}		i_{mol}		i_{mol}		i_{mol}		i_{mol}		i_{mol}	
°C	K	$\frac{kcal}{kmol}$	$\frac{kJ}{kmol}$	$\frac{kcal}{kmol}$	$\frac{kJ}{kmol}$	$\frac{kcal}{kmol}$	$\frac{kJ}{kmol}$	$\frac{kcal}{kmol}$	$\frac{kJ}{kmol}$	$\frac{kcal}{kmol}$	$\frac{kJ}{kmol}$	$\frac{kcal}{kmol}$	$\frac{kJ}{kmol}$
0	273.15	0.0	0.0	0	0.0	0.0	0.0	0	0.0	0.0	0.0	0	0.0
100	373.15	805.9	3374.1	971	4065.4	879.1	3680.6	1104	4622.2	1095.6	4587.1	1115	4668.3
200	473.15	1630	6824.4	2022	8465.7	1883	7883.7	2440	10233	2317.6	9703.3	2316	9696.6
300	573.15	2477	10371	3144	131634	3029	12681	4008	16781	3631.2	15203	3576	14972
400	673.15	3352	14034	4320	180867	4316	18070	5752	24082	5019.2	21014	4884	20448
500	773.15	4255	17815	5540	231945	5729	23986	7665	32092	6468.0	27080	6225	26063
600	873.15	5187	21717	6786	28412	7262	30405	9720	40696	7973.4	33383	7596	31803
700	973.15	6151	25753	8064	33762	8907	37292	11880	49739	9529.8	39899	8988	37671
800	1073.15	7145	29915	9352	39154	10680	44715	14170	59327	11134	46616	10390	43501
900	1173.15	8170	34206	10665	44652	12540	52502	16530	69207	12778	53499	11810	49446
1000	1273.15	9224	38619	11980	50158	14450	60499	18990	79508	14459	60537	13230	55391
1100	1373.15	10310	43166	13310	55726	16410	68705	21530	90142	16177	67730	14670	61420
1200	1473.15	11410	47771	14640	61295	18390	76995	24110	100944	17926	75053	16120	67491
1300	1573.15	12550	52544	—	—								
1400	1673.15	13700	57359	—	—								
1500	1773.15	14880	62300	18720	78377								
1600	1873.15	16070	67282	--	—								
1700	1973.15	17290	72390	—	—								
1800	2073.15	18520	77540	—	—								
1900	2173.15	19760	82731	—	—								
2000	2273.15	21020	88007	25540	106931								
2100	2373.15	22290	93324	—	—								
2200	2473.15	23570	98683	—	—								
2300	2573.15	24860	104084	—	—								
2400	2673.15	26165	109548	—	—								
2500	2773.15	27480	115053	33200	139002								
2600	2873.15	28810	120622	—	—								
2700	2973.15	30140	126190	—	—								
3000	3273.15	33690	141053	39300	164541								
M*		18.020		64.060		16.031		28.031		26.040		76.13	

*M — relative molecular mass in kg/mol

Table 84-4 Molar Specific Enthalpy i_{mol} of Gases (*Continued*)

Temperature		COS		C₂H₆		C₃H₈		C₆H₆		C₃H₆		C₂H₅OH	
t	T	i_{mol}		i_{mol}		i_{mol}		i_{mol}		i_{mol}		i_{mol}	
°C	K	$\frac{kcal}{kmol}$	$\frac{kJ}{kmol}$	$\frac{kcal}{kmol}$	$\frac{kJ}{kmol}$	$\frac{kcal}{kmol}$	$\frac{kJ}{kmol}$	$\frac{kcal}{kmol}$	$\frac{kJ}{kmol}$	$\frac{kcal}{kmol}$	$\frac{kJ}{kmol}$	$\frac{kcal}{kmol}$	$\frac{kJ}{kmol}$
0	273.15	0	0.0	0	0.0	0	0.0	0	0.0	0	0.0	0	0.0
100	373.15	1018	4262.2	1336	5593.6	1879	7867.0	2137	8947.2	1632	6832.8	1846	7728.8
200	473.15	2130	8917.9	2971	12439	4246	17777	4946	20708	3618	15148	4008	16781
300	573.15	3309	13854	4889	20469	7017	29379	8352	34968	5952	24920	6483	27143
400	673.15	4540	19008	7085	29663	10190	42663	12220	51163	8576	35906	9216	38586
500	773.15	5810	24325	9509	39812	13630	57066	16460	68915	11460	47981	12180	50995
600	873.15	7110	29768	12130	50786	17450	73060	21040	88090	14580	61044	15350	64267
700	973.15	8442	35345	14940	62551	21450	89807	25870	108313	17900	74944	18690	78251
800	1073.15	9792	40997	17910	74986	25650	107391	30930	129498	21380	89514	22180	92863
900	1173.15	11160	46725	21020	88007	30020	125688	36180	151478	25010	104712	25810	108061
1000	1273.15	12550	52544	24250	101530	34590	144821	41600	174171	28760	120412	29540	123678
1100	1373.15	13950	58406	27580	115472	39320	164625	47160	197449	32630	136615	33400	139839
1200	1473.15	15360	64309	31000	129791	44210	185098	52810	221105	36600	153237	37320	156251
M*		60.07		30.07		44.06		78.108		42.08		46.07	

*M — relative molecular mass in kg/mol

1 kcal = 4.186 8 kJ

Table 85-1 Volume Specific Enthalpy i' of Gases

Temperature		H₂		N₂		O₂		CO		CO₂		Air	
t	T	i'		i'		i'		i'		i'		i'	
°C	K	kcal/m³ₙ	kJ/m³ₙ	kcal/m³ₙ	kJ/m³ₙ	kcal/m³ₙ	kJ/m³ₙ	kcal/m³ₙ	kJ/m³ₙ	kcal/m³ₙ	kJ/m³ₙ	kcal/m³ₙ	kJ/m³ₙ
0	273.15	0.00	0.00	0.00	0.00	0.0	0.0	0.00	0.00	0.00	0.00	0.00	0.00
100	373.15	30.83	129.08	31.06	130.04	31.47	131.76	31.09	130.17	40.61	170.03	31.06	130.04
200	473.15	61.96	259.41	62.28	260.75	63.78	267.03	62.44	261.42	85.38	357.47	62.44	261.42
300	573.15	93.09	389.75	93.93	393.27	97.17	406.83	94.35	395.03	133.5	558.94	94.38	395.15
400	673.15	124.4	520.84	126.2	528.37	131.6	550.98	127.0	531.72	184.4	772.05	127.0	531.72
500	773.15	155.8	652.30	159.1	666.12	166.9	698.78	160.4	671.56	237.5	994.37	160.4	671.56
600	873.15	187.4	784.61	192.8	807.22	203.0	849.92	194.5	814.33	292.5	1224.6	194.4	813.91
700	973.15	219.4	918.58	227.2	951.24	239.8	1004.0	229.4	960.45	349.2	1462.0	229.2	959.62
800	1073.15	251.6	1053.4	262.1	1097.34	277.0	1159.7	264.9	1109.1	407.2	1704.9	264.5	1107.4
900	1173.15	284.3	1190.3	297.6	1246.0	314.8	1318.0	300.9	1259.8	466.3	1952.3	300.4	1257.7
1000	1273.15	317.4	1328.9	333.7	1397.1	352.9	1477.5	362.0	1412.6	526.3	2203.5	336.7	1409.7
1100	1373.15	351.0	1469.6	370.2	1550.0	391.3	1638.3	374.3	1567.1	587.2	2458.5	373.5	1563.8
1200	1473.15	385.0	1611.9	407.0	1704.0	430.1	1800.7	411.6	1723.3	648.8	2716.4	410.6	1719.1
1300	1573.15	419.5	1756.4	444.2	1859.8	469.0	1963.6	449.3	1880.7	711.0	2976.8	448.1	1876.1
1400	1673.15	454.4	1902.5	481.7	2016.8	508.3	2128.2	487.1	2039.4	773.6	3238.9	485.8	2033.9
1500	1773.15	489.9	2051.1	519.5	2175.0	547.9	2293.9	525.2	2198.9	836.7	3503.1	523.8	2193.0
1600	1873.15	525.6	2200.6	557.4	2333.7	587.7	2460.6	563.5	2359.3	900.1	3768.5	562.1	2353.4
1700	1973.15	561.7	2351.7	595.7	2494.1	627.8	2628.5	602.0	2520.5	964.1	4036.5	600.4	2513.6
1800	2073.15	598.3	2505.0	634.0	2654.4	668.2	2797.6	640.6	2682.1	1028	4304.0	639.2	2676.2
1900	2173.15	635.2	2659.5	672.6	2816.0	708.7	2967.2	679.4	2844.5	1093	4576.2	677.9	2838.2
2000	2273.15	672.4	2815.2	711.2	2977.7	749.6	3138.4	718.4	3007.8	1157	4844.1	717.0	3001.9
2100	2373.15	709.8	2971.8	750.1	3140.5	790.4	3309.2	757.5	3171.5	1222	5116.3	756.0	3165.2
2200	2473.15	747.6	3130.1	789.1	3303.8	831.8	3482.6	796.6	3335.2	1287	5388.4	795.3	3329.8
2300	2573.15	785.7	3289.6	828.0	3466.7	873.3	3556.3	835.8	3499.3	1351	5656.4	834.7	3494.7
2400	2673.15	823.9	3449.5	867.1	3630.4	915.1	3831.3	875.8	3666.8	1416	5928.5	874.3	3660.5
2500	2773.15	862.7	3612.0	906.3	3794.5	957.0	4006.8	914.8	3830.1	1482	6204.8	913.8	3825.9
2600	2873.15	901.4	3774.0	—	—	999.2	4183.4	—	—	—	—		
2700	2973.15	904.4	3937.3	—	—	1042	4362.6	—	—	—	—		
3000	3273.15	1060	4438,0	1106	4630.6	1171	4902.7	1114	4664.1	1809	7573.9		

1 kcal = 4.186 8 kJ

Table 85-2   Volume Specific Enthalpy i' of Gases (Continued)

Temperature		NO		OH		N₂O		H₂S		NH₃	
t	T	i'		i'		i'		i'		i'	
°C	K	kcal/m_n^3	kJ/m_n^3	kcal/m_n^3	kJ/m_n^3	kcal/m_n^3	kJ/m_n^3	kcal/m_n^3	kJ/m_n^3	kcal/m_n^3	kJ/m_n^3
0	273.15	0.00	0.00	0.00	0.00	0.00	0.00	0.00	0.00	0.00	0.00
100	373.15	31.87	133.43	31.71	132.76	42.32	177.19	36.6	153.24	38.7	162.03
200	473.15	63.94	267.70	63.18	264.52	88.82	371.87	74.6	312.34	80.2	335.78
300	573.15	96.63	404.57	94.62	396.16	138.7	580.71	114.3	578.55	127.1	532.14
400	673.15	130.1	544.70	126.1	527.96	191.2	800.52	156.0	653.14	175.3	733.95
500	773.15	164.4	688.31	157.8	660.67	246.1	1030.4	199.5	835.27	226.9	949.99
600	873.15	199.6	835.68	189.7	794.24	303.0	1268.6	244.8	1024.9	281.5	1178.6
700	973.15	235.3	985.15	222.0	929.47	361.6	1513.9	291.9	1222.1	339.1	1419.7
800	1073.15	271.7	1137.6	254.9	1067.22	421.6	1765.2	340.8	1426.9	399.4	1672.2
900	1173.15	308.5	1291.6	288.2	1206.6	482.7	2021.0	390.6	1635.4	462.4	1936.0
1000	1273.15	345.8	1447.8	322.0	1348.2	544.9	2281.4	442.0	1850.6	529.7	2217.7
1100	1373.15	383.5	1605.6	356.4	1492.2	607.9	2545.2	494.7	2071.2	—	—
1200	1473.15	421.4	1764.3	391.3	1638.3	—	—	547.7	2293.1	—	—
1300	1573.15	459.7	1924.7	426.5	1785.7	—	—	—	—	—	—
1400	1673.15	498.1	2085.4	462.1	1934.7	—	—	—	—	—	—
1500	1773.15	536.7	2247.1	498.2	2085.9	849.2	3555.4	714.7	2992.3	891.1	3730.9
1600	1873.15	575.7	2410.3	534.6	2238.3	—	—	—	—	—	—
1700	1973.15	614.7	2573.6	571.4	2392.3	—	—	—	—	—	—
1800	2073.15	653.8	2737.3	608.4	2547.2	—	—	—	—	—	—
1900	2173.15	693.1	2901.9	645.8	2703.8	—	—	—	—	—	—
2000	2273.15	732.4	3066.4	683.6	2862.1	1169	4894.4	1000	4186.8	1045	4375.2
2100	2373.15	772.0	3232.2	721.6	3021.2	—	—	—	—	—	—
2200	2473.15	811.6	3398.0	759.9	3181.5	—	—	—	—	—	—
2300	2573.15	851.2	3563.8	798.3	3342.3	—	—	—	—	—	—
2400	2673.15	890.9	3730.0	837.1	3504.8	—	—	—	—	—	—
2500	2773.15	930.8	3897.1	876.0	3667.6	1492	6246.7	1294	5417.7	1206	5049.3
2600	2873.15	970.8	4064.5	915.2	3831.8	—	—	—	—	—	—
2700	2973.15	1011	4232.9	964.5	3996.3	—	—	—	—	—	—
2800	3073.15	1051	4400.3	994.0	4161.7	—	—	—	—	—	—
2900	3173.15	1091	4567.8	1034	4329.2	—	—	—	—	—	—
3000	3273.15	1131	4735.3	1073	4492.4	1819	7615.8	1593	6669.6	1372	5744.3

1 kcal = 4.186 8 kJ

Table 85-3 Volume Specific Enthalpy i' of Gases (*Continued*)

Temperature		H_2O		SO_2		CH_4		C_2H_4		C_2H_2		CS_2	
t	T	i'		i'		i'		i'		i'		i'	
°C	K	kcal/m$_n^3$	kJ/m$_n^3$	kcal/m$_n^3$	kJ/m$_n^3$	kcal/m$_n^3$	kJ/m$_n^3$	kcal/m$_n^3$	kJ/m$_n^3$	kcal/m$_n^3$	kJ/m$_n^3$	kcal/m$_n^3$	kJ/m$_n^3$
0	273.15	0.00	0.00	0.00	0.00	0.00	0.00	0.00	0.00	0.00	0.00	0.00	0.00
100	373.15	35.95	150.52	43.32	181.37	39.22	164.21	49.25	206.20	48.88	204.65	49.74	208.25
200	473.15	72.72	304.46	90.21	377.69	84.02	351.76	109.0	456.36	103.40	432.9	103.3	432.50
300	573.15	110.5	462.64	140.3	587.41	135.1	565.64	178.0	745.25	162.00	678.26	159.5	667.80
400	673.15	149.6	626.35	192.7	806.80	192.6	806.39	256.6	1074.3	223.92	937.51	217.9	912.30
500	773.15	189.9	795.07	247.2	1035.0	255.6	1070.1	341.9	1431.5	288.56	1208.1	277.7	1162.70
600	873.15	231.4	968.3	302.7	1267.3	324.0	1356.5	433.6	1815.4	355.72	1489.3	338.9	1418.9
700	973.15	274.4	1148.9	359.8	1506.4	397.4	1663.8	530.0	2219.0	425.16	1780.1	401.0	1678.9
800	1073.15	318.7	1334.3	417.2	1746.7	476.6	1995.4	632.1	2646.5	496.71	2079.6	463.6	1941.0
900	1173.15	364.5	1526.1	475.8	1992.1	559.4	2342.1	737.6	3088.1	570.09	2386.9	526.8	2205.6
1000	1273.15	411.5	1722.9	534.5	2237.8	644.7	2699.2	847.2	2547.2	645.07	2700.8	590.2	2471.0
1100	1373.15	459.8	1925.1	594	2487.0	732.0	3064.7	960.4	4021.0	721.70	3021.6	655	2742.3
1200	1473.15	509.3	2132.3	653	2734.0	820.6	3435.7	1075.6	4503.3	799.73	3348.3	719	3010.3
1300	1573.15	559.8	2343.8	—	—								
1400	1673.15	611.2	2559.0	—	—								
1500	1773.15	663.8	2779.2	835.2	3496.8								
1600	1873.15	717.0	3001.9	—	—								
1700	1973.15	771.3	3229.3	—	—								
1800	2073.15	826.1	3458.7	—	—								
1900	2173.15	881.5	3690.7	—	—								
2000	2273.15	937.6	3925.5	1139	4768.8								
2100	2373.15	994.3	4162.9	—	—								
2200	2473.15	1051.5	4402.4	—	—								
2300	2573.15	1109.1	4643.6	—	—								
2400	2673.15	1167.3	4887.3	—	—								
2500	2773.15	1225.9	5132.6	1473	6167.2								
2600	2873.15	1285.0	5380.0	—	—								
2700	2973.15	1344.4	5628.7	—	—								
3000	3273.15	1503	6292.7	1753	7339.5								

Table 85-4 Volume Specific Enthalpy i' of Gases (*Continued*)

Temperature		COS		C_2H_6		C_3H_8		C_6H_6		C_3H_6		C_2H_5OH	
t	T	i'		i'		i'		i'		i'		i'	
°C	K	kcal/m$_n^3$	kJ/m$_n^3$	kcal/m$_n^3$	kJ/m$_n^3$	kcal/m$_n^3$	kJ/m$_n^3$	kcal/m$_n^3$	kJ/m$_n^3$	kcal/m$_n^3$	kJ/m$_n^3$	kcal/m$_n^3$	kJ/m$_n^3$
0	273,15	0.00	0.00	0.00	0.00	0.00	0.00	0.00	0.00	0.00	0.00	0.00	0.00
100	373.15	45.42	190.16	59.59	249.49	83.83	350.98	95.34	399.17	72.81	304.84	82.36	344.82
200	473.15	95.03	397.87	135.2	554.75	189.4	792.98	220.7	924.03	161.4	675.75	178.8	748.60
300	573.15	147.6	617.97	218.1	913.14	313.0	1310.5	372.6	1560.0	265.5	1111.6	289.2	1210.8
400	673.15	202.5	847.8	316.1	1323.4	454.7	1903.7	545.2	2282.6	382.6	1601.9	411.2	1721.6
500	773.15	259.2	1085.2	424.2	1776.0	608.3	2546.8	734.5	3075.2	511.5	2141.5	543.4	2275.1
600	873.15	317.2	1328.1	541.4	2266.7	778.4	3259.0	938.8	3930.6	650.5	2723.50	685.0	2868.0
700	973.15	376.6	1576.7	666.5	2790.5	956.9	4006.3	1154	4831.6	798.6	3343.58	833.8	3491.0
800	1073.15	436.9	1829.2	798.9	3344.8	1144	4789.7	1380	5777.8	953.7	3993.0	989.7	4143.2
900	1173.15	497.9	2084.6	937.6	3925.5	1339	5606.1	1614	6757.5	1116	4672.5	1152	4823.2
1000	1273.15	569.9	2386.1	1082	4530.1	1543	6460.2	1856	7770.7	1283	5371.7	1318	5518.2
1100	1373.15	622	2604.2	1231	5154.0	1754	7343.6	2104	8809.0	1455	6091.8	1490	6238.3
1200	1473.15	685	2868.0	1383	5790.3	1972	8256.4	2356	9864.1	1633	6837.0	1665	6971.0

1 kcal = 4.186 8 kJ

Table 86-1 Specific Entropy s of Gases

Temperature		H₂		N₂		O₂		CO		CO₂		Air	
t	T	s		s		s		s		s		s	
°C	K	$\frac{kcal}{kg\,K}$	$\frac{kJ}{kg\,K}$	$\frac{kcal}{kg\,K}$	$\frac{kJ}{kg\,K}$	$\frac{kcal}{kg\,K}$	$\frac{kJ}{kg\,K}$	$\frac{kcal}{kg\,K}$	$\frac{kJ}{kg\,K}$	$\frac{kcal}{kg\,K}$	$\frac{kJ}{kg\,K}$	$\frac{kcal}{kg\,K}$	$\frac{kJ}{kg\,K}$
0	273.15	0.000	0.000	0.0000	0.0000	0.0000	0.0000	0.0000	0.0000	0.0000	0.0000	0.0000	0.0000
100	373.15	1.048	4.388	0.0775	0.3245	0.0675	0.2826	0.0776	0.3249	0.0643	0.2692	0.0747	0.3127
200	473.15	1.872	7.838	0.1368	0.5728	0.1214	0.5083	0.1372	0.5744	0.1185	0.4957	0.1324	0.5543
300	573.15	2.537	10.622	0.1854	0.7762	0.1662	0.6958	0.1861	0.7792	0.1653	0.6921	0.1797	0.7524
400	673.15	3.096	12.962	0.2268	0.9496	0.2050	0.8583	0.2281	0.9550	0.2069	0.8662	0.2203	0.9224
500	773.15	3.579	14.985	0.2633	1.1024	0.2393	1.0019	0.2651	1.1099	0.2444	1.0233	0.2560	1.0718
600	873.15	4.007	16.777	0.2961	1.2397	0.2700	1.1304	0.2983	1.2489	0.2785	1.1660	0.2880	1.2058
700	973.15	4.392	18.388	0.3259	1.3645	0.2979	1.2472	0.3286	1.3758	0.3098	1.2371	0.3172	1.3281
800	1073.15	4.743	19.858	0.3532	1.4788	0.3234	1.3540	0.3563	1.4918	0.3386	1.4177	0.3439	1.4398
900	1173.15	5.066	21.210	0.3786	1.5851	0.3470	1.4528	0.3820	1.5994	0.3655	1.5303	0.3687	1.5437
1000	1273.15	5.368	22.475	0.4021	1.6835	0.3688	1.5441	0.4059	1.6994	0.3905	1.6349	0.3917	1.6400
1100	1373.15	5.649	23.651	0.4242	1.7760	0.3892	1.6295	0.4283	1.7932	0.4139	1.7329	0.4132	1.7300
1200	1473.15	5.915	24.765	0.4449	1.8627	0.4082	1.7091	0.4492	1.8807	0.4360	1.8254	0.4334	1.8146
1300	1573.15	6.168	25.824	0.4645	1.9448	0.4262	1.7844	0.4690	1.9636	0.4568	1.9125	0.4522	1.8933
1400	1673.15	6.407	26.825	0.4830	2.0222	0.4432	1.8556	0.4877	2.0419	0.4764	1.9946	0.4704	1.9695
1500	1773.15	6.635	27.779	0.5005	2.0955	0.4592	1.9226	0.5054	2.1160	0.4951	2.0729	0.4875	2.0411
1600	1873.15	6.854	28.696	0.5172	2.1654	0.4745	1.9866	0.5222	2.1863	0.5128	2.1470	0.5037	2.1089
1700	1973.15	7.063	29.571	0.5331	2.2320	0.4891	2.0478	0.5382	2.2533	0.5297	2.2177	0.5224	2.1872
1800	2073.15	7.263	30.409	0.5483	2.2956	0.5031	2.1064	0.5535	2.3174	0.5459	2.2856	0.5339	2.2353
1900	2173.15	7.456	21.217	0.5628	2.3563	0.5165	2.1625	0.5681	2.3785	0.5613	2.3501	0.5481	2.2948
2000	2273.15	7.642	31.996	0.5767	2.4145	0.5293	2.2161	0.5821	2.4371	0.5761	2.4120	0.5617	2.3517
2100	2373.15	7.822	32.749	0.5901	2.4706	0.5417	2.2680	0.5956	2.4937	0.5903	2.4715	0.5747	2.4062
2200	2473.15	7.995	33.473	0.6029	2.5242	0.5536	2.3178	0.6085	2.5477	0.6039	2.5284	0.5872	2.4585
2300	2573.15	8.163	34.177	0.6153	2.5761	0.5652	2.3664	0.6210	2.6000	0.6170	2.5833	0.6031	2.5251
2400	2673.15	8.326	34.859	0.6273	2.6264	0.5763	2.4192	0.6330	2.6502	0.6297	2.6364	0.6110	2.5581
2500	2773.15	8.484	35.521	0.6388	2.6745	0.5871	2.4581	0.6446	2.6988	0.6418	2.6871	0.6223	2.6054
2600	2873.15	8.636	36.157	—	—	0.5976	2.5020	—	—	—	—		
2700	2973.15	8.785	36.781	—	—	0.6078	2.5447	—	—	—	—		
3000	3273.15	9.19	38.477	0.687	2.8763	0.650	2.7214	0.695	2.9098	0.699	2.9266		

1 kcal = 4.186 8 kJ

288

Table 86-2 Specific Entropy s of Gases (Continued)

Temperature		NO		OH		N_2O		H_2S		NH_3	
t	T	s		s		s		s		s	
°C	K	$\frac{kcal}{kg\,K}$	$\frac{kJ}{kg\,K}$	$\frac{kcal}{kg\,K}$	$\frac{kJ}{kg\,K}$	$\frac{kcal}{kg\,K}$	$\frac{kJ}{kg\,K}$	$\frac{kcal}{kg\,K}$	$\frac{kJ}{kg\,K}$	$\frac{kcal}{kg\,K}$	$\frac{kJ}{kg\,K}$
0	273.15	0.0000	0.0000	0.000	0.0000	0.0000	0.0000	0.000	0.0000	0.000	0.0000
100	373.15	0.0742	0.3107	0.130	0.5443	0.0681	0.2851	0.074	0.3098	0.158	0.6615
200	473.15	0.1311	0.5489	0.226	0.9462	0.1243	0.5204	0.133	0.5568	0.288	1.2058
300	573.15	0.1780	0.7453	0.307	1.2853	0.1730	0.7243	0.183	0.7662	0.401	1.6789
400	673.15	0.2182	0.9136	0.374	1.5659	0.2161	0.9048	0.227	0.9504	0.504	2.1101
500	773.15	0.2537	1.0622	0.433	1.8129	0.2548	1.0668	0.267	1.1179	0.598	2.5037
600	873.15	0.2856	1.1958	0.485	2.0306	0.2918	1.2217	0.303	1.2686	0.686	2.8721
700	973.15	0.3146	1.3172	0.532	2.2274	0.3224	1.3498	0.336	1.4068	0.768	3.2155
800	1073.15	0.3413	1.4290	0.575	2.4074	0.3523	1.4750	0.368	1.5407	0.843	3.5295
900	1173.15	0.3658	1.5315	0.614	2.5707	0.3801	1.5914	0.397	1.6622	0.920	3.8519
1000	1273.15	0.3885	1.6266	0.650	2.7214	0.4060	1.6998	0.425	1.7794	0.990	4.1449
1100	1373.15	0.4098	1.7158	0.684	2.8638	0.4303	1.8016	0.451	1.8882	—	—
1200	1473.15	0.4298	1.7995	0.716	2.9977	—	—	0.475	1.9887	—	—
1300	1573.15	0.4486	1.8782	0.746	3.1234	—	—	—	—	—	—
1400	1673.15	0.4663	1.9523	0.774	3.2406	—	—	—	—	—	—
1500	1773.15	0.4831	2.0226	0.801	3.3526	0.528	2.2106	0.546	2.2860	1.297	5.4303
1600	1873.15	0.4990	2.0892	0.827	3.4625	—	—	—	—	—	—
1700	1973.15	0.5141	2.1524	0.852	3.5672	—	—	—	—	—	—
1800	2073.15	0.5285	2.2127	0.876	3.6676	—	—	—	—	—	—
1900	2173.15	0.5423	2.2705	0.900	3.7681	—	—	—	—	—	—
2000	2273.15	0.5555	2.3258	0.923	3.8644	0.608	2.5456	0.639	2.6754	1.547	6.4770
2100	2373.15	0.5682	2.3789	0.945	3.9565	—	—	—	—	—	—
2200	2473.15	0.5805	2.4304	0.966	4.0444	—	—	—	—	—	—
2300	2573.15	0.5923	2.4798	0.986	4.1282	—	—	—	—	—	—
2400	2673.15	0.6036	2.5272	1.005	4.2077	—	—	—	—	—	—
2500	2773.15	0.6146	2.5732	1.024	4.2873	0.676	2.8303	0.716	2.9977	1.759	7.3646
2600	2873.15	0.6251	2.6172	1.042	4.3626	—	—	—	—	—	—
2700	2973.15	0.6353	2.6599	1.059	4.4338	—	—	—	—	—	—
2800	3073.15	0.6452	2.7013	1.075	4.5008	—	—	—	—	—	—
2900	3173.15	0.6548	2.7415	1.091	4.5678	—	—	—	—	—	—
3000	3273.15	0.6641	2.7805	1.107	4.6348	0.729	3.0522	0.782	3.2741	1.939	8.1182

1 kcal = 4.186 8 kJ

Table 86-3 Specific Entropy *s* of Gases (*Continued*)

Temperature		H₂O		SO₂		CH₄		C₂H₄		C₂H₂		CS₂	
		s		s		s		s		s		s	
t	T	$\dfrac{kcal}{kg\,K}$	$\dfrac{kJ}{kg\,K}$	$\dfrac{kcal}{kg\,K}$	$\dfrac{kJ}{kg\,K}$	$\dfrac{kcal}{kg\,K}$	$\dfrac{kJ}{kg\,K}$	$\dfrac{kcal}{kg\,K}$	$\dfrac{kJ}{kg\,K}$	$\dfrac{kcal}{kg\,K}$	$\dfrac{kJ}{kg\,K}$	$\dfrac{kcal}{kg\,K}$	$\dfrac{kJ}{kg\,K}$
°C	K												
0	273.15	0.0000	0.0000	0.0000	0.0000	0.000	0.000	0.000	0.000	0.0000	0.0000	0.0000	0.0000
100	373.15	0.1377	0.5765	0.0468	0.1959	0.168	0.703	0.123	0.515	0.1300	0.5443	0.0453	0.1897
200	473.15	0.2468	1.0333	0.0860	0.3601	0.317	1.327	0.235	0.984	0.2412	1.0099	0.0829	0.3471
300	573.15	0.3367	1.4097	0.1194	0.4999	0.454	1.901	0.341	1.428	0.3379	1.4147	0.1147	0.4802
400	673.15	0.415	1.7375	0.1489	0.6234	0.583	2.441	0.441	1.846	0.4236	1.7735	0.1421	0.5949
500	773.15	0.484	2.0264	0.1753	0.7339	0.704	2.948	0.536	2.244	0.5007	2.0963	0.1667	0.6979
600	873.15	0.547	2.2902	0.1990	0.8332	0.821	3.437	0.624	2.613	0.5711	2.3911	0.1885	0.7892
700	973.15	0.605	2.5330	0.2206	0.9236	0.932	3.902	0.708	2.964	0.6359	2.6624	0.2082	0.8717
800	1073.15	0.659	2.7591	0.2404	1.0065	1.036	4.338	0.788	3.299	0.6961	2.9144	0.2262	0.9471
900	1173.15	0.709	2.9684	0.2585	1.0823	1.137	4.760	0.863	3.613	0.7524	3.1501	0.2429	1.0170
1000	1273.15	0.757	3.1694	0.2754	1.1530	1.233	5.162	0.935	3.915	0.8053	3.3716	0.2582	1.0810
1100	1373.15	0.803	3.3620	0.2911	1.2188	1.328	5.560	1.003	4.199	0.8551	3.5801	0.2725	1.1409
1200	1473.15	0.846	3.5420	0.3058	1.2803	1.418	5.937	1.068	4.472	0.9024	3.7782	0.2859	1.1970
1300	1573.15	0.887	3.7137	—	—								
1400	1673.15	0.926	3.8770	—	—								
1500	1773.15	0.964	4.0361	0.350	1.4654								
1600	1873.15	1.001	4.1910	—	—								
1700	1973.15	1.036	4.3375	—	—								
1800	2073.15	1.069	4.4757	—	—								
1900	2173.15	1.102	4.6139	—	—								
2000	2273.15	1.134	4.7478	0.402	1.6831								
2100	2373.15	1.164	4.8734	—	—								
2200	2473.15	1.193	4.9949	—	—								
2300	2573.15	1.221	5.1121	—	—								
2400	2673.15	1.249	5.2293	—	—								
2500	2773.15	1.276	5.3424	0.445	1.8631								
2600	2873.15	1.301	5.4470	—	—								
2700	2973.15	1.326	5.5517	—	—								
3000	3273.15	1.39	5.8197	0.481	2.0139								

Table 86-4 Specific Entropy *s* of Gases (*Continued*)

Temperature		COS		C₂H₆		C₃H₈		C₆H₆		C₃H₆		C₂H₅OH	
		s		s		s		s		s		s	
t	T	$\dfrac{kcal}{kg\,K}$	$\dfrac{kJ}{kg\,K}$	$\dfrac{kcal}{kg\,K}$	$\dfrac{kJ}{kg\,K}$	$\dfrac{kcal}{kg\,K}$	$\dfrac{kJ}{kg\,K}$	$\dfrac{kcal}{kg\,K}$	$\dfrac{kJ}{kg\,K}$	$\dfrac{kcal}{kg\,K}$	$\dfrac{kJ}{kg\,K}$	$\dfrac{kcal}{kg\,K}$	$\dfrac{kJ}{kg\,K}$
°C	K												
0	273.15	0.0000	0.0000	0.000	0.000	0.000	0.000	0.000	0.0000	0.000	0.0000	0.000	0.0000
100	373.15	0.0528	0.2211	0.138	0.5778	0.131	0.5485	0.085	0.3559	0.121	0.5066	0.123	0.5150
200	473.15	0.0966	0.4044	0.267	1.1179	0.258	1.0802	0.170	0.7118	0.232	0.9713	0.235	0.9839
300	573.15	0.1342	0.5619	0.389	1.6287	0.378	1.5826	0.253	1.0593	0.338	1.4151	0.338	1.4151
400	673.15	0.1673	0.7005	0.506	2.1185	0.494	2.0683	0.333	1.3942	0.438	1.8338	0.433	1.8129
500	773.15	0.1964	0.8223	0.618	2.5874	0.603	2.5246	0.408	1.7082	0.533	2.2316	0.522	2.1855
600	873.15	0.2227	0.9324	0.724	3.0312	0.685	2.8680	0.479	2.0055	0.624	2.6126	0.606	2.5372
700	973.15	0.2467	1.0329	0.825	3.4541	0.807	3.3787	0.546	2.2860	0.709	2.9684	0.685	2.8680
800	1073.15	0.2687	1.1250	0.921	3.8560	0.899	3.7639	0.609	2.5498	0.790	3.3076	0.759	3.1778
900	1173.15	0.2892	1.2108	1.012	4.2370	0.987	4.1324	0.669	2.8010	0.866	3.6258	0.829	3.4709
1000	1273.15	0.3081	1.2900	1.100	4.6055	1.072	4.4882	0.726	3.0396	0.929	3.8895	0.895	3.7472
1100	1373.15	0.3258	1.3641	1.185	4.9614	1.154	4.8316	0.780	3.2657	1.009	4.2245	0.959	4.0151
1200	1473.15	0.3423	1.4331	1.266	5.3005	1.231	5.1540	0.831	3.4792	1.075	4.5008	1.019	4.2663

1 kcal = 4.186 8 kJ

Table 87-1 Molar Specific Entropy s_{mol} of Gases

| Temperature | | H_2 | | N_2 | | O_2 | | CO | | CO_2 | | Air | |
| t | T | s_{mol} | | s_{mol} | | s_{mol} | | s_{mol} | | s_{mol} | | s_{mol} | |
°C	K	$\dfrac{kcal}{kmol\ K}$	$\dfrac{kJ}{kmol\ K}$	$\dfrac{kcal}{kmol\ K}$	$\dfrac{kJ}{kmol\ K}$	$\dfrac{kcal}{kmol\ K}$	$\dfrac{kJ}{kmol\ K}$	$\dfrac{kcal}{kmol\ K}$	$\dfrac{kJ}{kmol\ K}$	$\dfrac{kcal}{kmol\ K}$	$\dfrac{kJ}{kmol\ K}$	$\dfrac{kcal}{kmol\ K}$	$\dfrac{kJ}{kmol\ K}$
0	273.15	0.000	0.000	0.000	0.000	0.000	0.000	0.000	0.000	0.000	0.000	0.000	0.000
100	373.15	2.112	8.843	2.171	9.090	2.161	9.084	2.174	9.102	2.831	11.853	2.164	9.060
200	473.15	3.775	15.805	3.833	16.048	3.884	16.262	3.842	16.086	5.210	21.813	3.834	16.052
300	573.15	5.114	21.411	5.193	21.742	5.318	22.265	5.213	21.826	7.252	30.363	5.206	21.796
400	673.15	6.242	26.134	6.355	26.607	6.560	27.465	6.388	26.745	9.107	38.129	6.381	26.716
500	773.15	7.216	30.212	7.377	30.886	7.656	32.054	7.425	31.087	10.756	45.033	7.416	31.049
600	873.15	8.078	33.821	8.294	34.725	8.640	36.174	8.356	34.985	12.256	51.313	8.345	34.939
700	973.15	8.855	37.074	9.129	38.221	9.534	39.917	9.203	38.531	13.632	57.074	9.189	38.473
800	1073.15	9.562	40.034	9.896	41.433	10.350	43.333	9.981	41.788	14.903	62.396	9.964	41.717
900	1173.15	10.214	42.764	10.606	44.405	11.103	46.486	10.701	44.803	16.084	67.340	10.680	44.715
1000	1273.15	10.822	45.310	11.266	47.168	11.802	49.413	11.370	47.604	17.185	71.950	11.347	47.508
1100	1373.15	11.389	47.683	11.885	49.760	12.453	52.128	11.996	50.225	18.217	76.271	11.970	50.116
1200	1473.15	11.925	49.928	12.466	52.193	13.063	54.692	12.583	52.683	19.187	80.332	12.555	52.565
1300	1573.15	12.434	52.059	13.013	54.483	13.638	57.100	13.137	55.002	20.102	84.163	13.101	54.851
1400	1673.15	12.917	54.081	13.531	56.652	14.181	59.373	13.660	57.192	20.968	87.789	13.628	57.058
1500	1773.15	13.377	56.007	14.022	58.707	14.695	61.525	14.155	59.264	21.789	91.226	14.122	59.126
1600	1873.15	13.817	57.849	14.489	60.663	15.185	63.577	14.626	61.236	22.569	94.492	14.592	61.094
1700	1973.15	14.239	59.616	14.935	62.530	15.652	65.532	15.075	63.116	23.313	97.607	15.040	62.969
1800	2073.15	14.643	61.307	15.360	64.309	16.098	67.399	15.503	64.908	24.026	100.592	15.468	64.761
1900	2173.15	15.032	62.936	15.767	66.013	16.527	69.195	15.913	66.625	24.703	103.427	15.878	66.478
2000	2273.15	15.407	64.506	16.157	67.646	16.939	70.920	16.306	68.270	25.354	106.152	16.271	68.123
2100	2373.15	15.769	66.022	16.531	69.212	17.335	72.578	16.683	69.848	25.978	108.765	16.649	69.706
2200	2473.15	16.118	67.483	16.892	70.723	17.716	74.173	17.045	71.364	26.578	111.277	17.012	71.226
2300	2573.15	16.457	68.902	17.239	72.176	18.085	75.718	17.394	72.825	27.155	113.693	17.363	72.695
2400	2673.15	16.785	70.275	17.573	73.575	18.442	77.213	17.730	74.232	27.711	116.020	17.700	74.106
2500	2773.15	17.103	71.607	17.896	74.927	18.788	78.662	18.055	75.593	28.247	118.265	18.027	75.475
2600	2873.15	17.411	72.896	—	—	19.122	80.060	—	—	—	—		
2700	2973.15	17.710	74.148	—	—	19.448	81.425	—	—	—	—		
3000	3273.15	18.52	77.540	19.24	80.554	20.79	87.044	19.45	81.433	30.76	128.786		
M*		2.0156		28.016		32.000		28.010		44.010		28.964	

*M — relative molecular mass in kg/mol

1 kcal = 4.186 8 kJ

Table 87-2 Molar Specific Entropy s_{mol} of Gases (Continued)

Temperature		NO		OH		N$_2$O		H$_2$S		NH$_3$	
t	T	s_{mol}		s_{mol}		s_{mol}		s_{mol}		s_{mol}	
°C	K	$\frac{kcal}{kmol\,K}$	$\frac{kJ}{kmol\,K}$	$\frac{kcal}{kmol\,K}$	$\frac{kJ}{kmol\,K}$	$\frac{kcal}{kmol\,K}$	$\frac{kJ}{kmol\,K}$	$\frac{kcal}{kmol\,K}$	$\frac{kJ}{kmol\,K}$	$\frac{kcal}{kmol\,K}$	$\frac{kJ}{kmol\,K}$
0	273.15	0.000	0.000	0.00	0.000	0.000	0.000	0.00	0.000	0.00	0.000
100	373.15	2.229	9.332	2.21	9.253	2.999	12.556	2.51	10.509	2.69	11.263
200	473.15	3.936	16.479	3.85	16.119	5.472	22.910	4.53	18.966	4.90	20.515
300	573.15	5.343	22.370	5.22	21.855	7.614	31.878	6.23	26.084	6.83	28.596
400	673.15	6.549	27.419	6.36	26.628	9.510	39.816	7.73	32.364	8.58	35.923
500	773.15	7.614	31.878	7.36	30.815	11.22	46.976	9.09	38.058	10.19	42.664
600	873.15	8.571	35.885	8.25	34.541	12.85	53.800	10.33	43.250	11.68	48.902
700	973.15	9.443	39.536	9.05	37.891	14.19	59.411	11.47	48.023	13.07	54.722
800	1073.15	10.243	42.885	9.78	40.947	15.51	64.937	12.54	52.503	14.36	60.122
900	1173.15	10.979	45.967	10.45	43.752	16.73	70.045	13.54	56.689	15.66	65.565
1000	1273.15	11.662	48.827	11.06	46.306	17.87	74.818	14.48	60.625	16.86	70.589
1100	1373.15	12.298	51.489	11.63	48.693	18.94	79.298	15.37	64.351	—	—
1200	1473.15	12.897	53.997	12.17	50.953	—	—	16.21	67.868	—	—
1300	1573.15	13.460	56.354	12.68	53.089	—	—	—	—	—	—
1400	1673.15	13.991	58.578	13.16	55.098	—	—	—	—	—	—
1500	1773.15	14.495	60.688	13.62	57.024	23.25	97.343	18.61	77.916	22.10	92.528
1600	1873.15	14.973	62.689	14.06	58.866	—	—	—	—	—	—
1700	1973.15	15.427	64.590	14.49	60.667	—	—	—	—	—	—
1800	2073.15	15.890	66.528	14.90	62.383	—	—	—	—	—	—
1900	2173.15	16.273	68.132	15.30	64.058	—	—	—	—	—	—
2000	2273.15	16.669	69.790	15.69	65.691	26.78	112.123	21.78	91.189	26.36	110.364
2100	2373.15	17.051	71.389	16.07	67.282	—	—	—	—	—	—
2200	2473.15	17.419	72.930	16.43	68.789	—	—	—	—	—	—
2300	2573.15	17.772	74.408	16.77	70.213	—	—	—	—	—	—
2400	2673.15	18.113	75.836	17.10	71.594	—	—	—	—	—	—
2500	2773.15	18.442	77.213	17.42	72.934	29.73	124.474	24.40	102.158	29.97	125.478
2600	2873.15	18.759	78.540	17.72	74.190	—	—	—	—	—	—
2700	2973.15	19.065	79.821	18.01	75.404	—	—	—	—	—	—
2800	3073.15	19.361	81.061	18.29	76.577	—	—	—	—	—	—
2900	3173.15	19.648	82.262	18.56	77.707	—	—	—	—	—	—
3000	3273.15	19.928	83.435	18.82	78.796	32.09	134.354	26.63	111.495	33.02	138.248
M*		30.008		17.008		44.016		34.08		17.031	

*M — relative molecular mass in kg/mol

1 kcal = 4.186 8 kJ

Table 87-3 Molar Specific Entropy s_{mol} of Gases (Continued)

Temperature		H₂O		SO₂		CH₄		C₂H₄		C₂H₂		CS₂	
t	T	s_{mol}		s_{mol}		s_{mol}		s_{mol}		s_{mol}		s_{mol}	
°C	K	$\frac{kcal}{kmol\,K}$	$\frac{kJ}{kmol\,K}$	$\frac{kcal}{kmol\,K}$	$\frac{kJ}{kmol\,K}$	$\frac{kcal}{kmol\,K}$	$\frac{kJ}{kmol\,K}$	$\frac{kcal}{kmol\,K}$	$\frac{kJ}{kmol\,K}$	$\frac{kcal}{kmol\,K}$	$\frac{kJ}{kmol\,K}$	$\frac{kcal}{kmol\,K}$	$\frac{kJ}{kmol\,K}$
0	273.15	0.000	0.000	0.00	0.000	0.00	0.000	0.00	0.000	0.000	0.000	0.00	0.000
100	373.15	2.481	10.387	3.00	12.560	2.70	11.304	3.45	14.444	3.384	14.168	3.45	14.444
200	473.15	4.446	18.615	5.51	23.069	5.08	21.269	6.61	27.675	6.280	26.293	6.31	26.419
300	573.15	6.066	25.397	7.65	32.029	7.28	30.480	9.58	40.110	8.798	36.835	8.73	36.551
400	673.15	7.471	31.280	9.54	39.942	9.35	39.147	12.39	51.874	11.029	46.176	10.82	45.301
500	773.15	8.720	36.509	11.23	47.018	11.30	47.311	15.04	62.969	13.036	54.579	12.69	53.130
600	873.15	9.854	41.257	12.75	53.382	13.17	55.140	17.52	73.353	14.868	62.249	14.35	60.081
700	973.15	10.899	45.632	14.14	59.201	14.95	62.593	19.88	83.234	16.555	69.312	15.85	66.361
800	1073.15	11.870	49.697	15.40	64.477	16.62	69.585	22.11	92.570	18.123	75.877	17.22	72.097
900	1173.15	12.781	53.511	16.56	69.333	18.24	76.367	24.22	101.404	19.589	82.015	18.49	77.414
1000	1273.15	13.642	57.116	17.64	73.855	19.79	82.857	26.23	109.820	20.965	87.776	19.66	82.313
1100	1373.15	14.459	60.537	18.65	78.084	21.30	89.179	28.14	117.817	22.263	93.211	20.75	86.876
1200	1473.15	15.237	63.794	19.59	82.019	22.75	95.250	29.97	125.478	23.493	98.360	21.77	91.147
1300	1573.15	15.98	66.905	—	—								
1400	1673.15	16.69	69.878	—	—								
1500	1773.15	17.37	72.725	22.41	93.826								
1600	1873.15	18.03	75.488	—	—								
1700	1973.15	18.66	78.126	—	—								
1800	2073.15	19.27	80.680	—	—								
1900	2173.15	19.86	83.150	—	—								
2000	2273.15	20.43	85.536	25.79	107.978								
2100	2373.15	20.97	87.797	—	—								
2200	2473.15	21.50	90.016	—	—								
2300	2573.15	22.01	92.151	—	—								
2400	2673.15	22.51	94.245	—	—								
2500	2773.15	22.99	96.255	28.52	119.408								
2600	2873.15	23.45	98.180	—	—								
2700	2973.15	23.89	100.023	—	—								
3000	3273.15	24.98	104.586	30.83	129.079								
M*		18.020		64.06		16.031		28.031		26.040		76.13	

*M — relative molecular mass in kg/mol

Table 87-4 Molar Specific Entropy s_{mol} of Gases (Continued)

Temperature		COS		C₂H₆		C₃H₈		C₆H₆		C₃H₆		C₂H₅OH	
t	T	s_{mol}		s_{mol}		s_{mol}		s_{mol}		s_{mol}		s_{mol}	
°C	K	$\frac{kcal}{kmol\,K}$	$\frac{kJ}{kmol\,K}$	$\frac{kcal}{kmol\,K}$	$\frac{kJ}{kmol\,K}$	$\frac{kcal}{kmol\,K}$	$\frac{kJ}{kmol\,K}$	$\frac{kcal}{kmol\,K}$	$\frac{kJ}{kmol\,K}$	$\frac{kcal}{kmol\,K}$	$\frac{kJ}{kmol\,K}$	$\frac{kcal}{kmol\,K}$	$\frac{kJ}{kmol\,K}$
0	273.15	0.00	0.000	0.00	0.000	0.00	0.000	0.00	0.000	0.00	0.000	0.00	0.000
100	373.15	3.17	13.272	4.15	17.375	5.80	24.283	6.60	27.633	5.04	21.102	5.68	23.781
200	473.15	5.80	24.283	8.02	33.578	11.39	47.688	13.26	55.517	9.75	40.821	10.83	45.343
300	573.15	8.06	33.746	11.69	48.943	16.69	69.878	19.76	82.731	14.20	59.453	15.56	65.147
400	673.15	10.05	42.077	15.21	63.681	21.79	91.230	25.97	108.731	18.43	77.163	19.95	83.527
500	773.15	11.81	49.446	18.58	77.790	26.61	111.411	31.86	133.391	22.42	93.868	24.06	100.734
600	873.15	13.39	56.061	21.77	91.146	30.22	126.525	37.40	156.586	26.22	109.778	27.92	116.896
700	973.15	14.82	62.048	24.82	103.916	35.58	148.966	42.65	178.567	29.81	124.809	31.54	132.052
800	1073.15	16.14	67.575	27.69	115.932	39.65	166.007	47.60	199.292	33.21	139.044	34.95	146.329
900	1173.15	17.37	72.725	30.44	127.446	43.55	182.335	52.27	218.844	36.43	152.525	38.19	159.894
1000	1273.15	18.51	77.498	33.09	138.541	47.29	197.994	56.71	237.433	39.51	165.421	41.25	172.706
1100	1373.15	19.57	81.936	35.63	149.175	50.88	213.024	60.91	255.018	42.44	177.688	44.17	184.931
1200	1473.15	20.56	86.081	38.07	159.391	54.31	227.385	64.90	271.723	45.23	189.369	46.94	196.528
M*		60.07		30.07		44.06		78.108		42.08		46.07	

*M — relative molecular mass in kg/mol

1 kcal = 4.186 8 kJ

Table 88-1 Volume Specific Entropy s' of Gases

Temperature		H_2		N_2		O_2		CO		CO_2		Air	
t	T	s'		s'		s'		s'		s'		s'	
°C	K	$\dfrac{kcal}{m_n^3\,K}$	$\dfrac{kJ}{m_n^3\,K}$	$\dfrac{kcal}{m_n^3\,K}$	$\dfrac{kJ}{m_n^3\,K}$	$\dfrac{kcal}{m_n^3\,K}$	$\dfrac{kJ}{m_n^3\,K}$	$\dfrac{kcal}{m_n^3\,K}$	$\dfrac{kJ}{m_n^3\,K}$	$\dfrac{kcal}{m_n^3\,K}$	$\dfrac{kJ}{m_n^3\,K}$	$\dfrac{kcal}{m_n^3\,K}$	$\dfrac{kJ}{m_n^3\,K}$
0	273.15	0.0000	0.0000	0.0000	0.0000	0.0000	0.0000	0.0000	0.0000	0.0000	0.0000	0.0000	0.0000
100	373.15	0.0942	0.3944	0.0969	0.4057	0.0964	0.4036	0.0970	0.4061	0.1263	0.5288	0.0965	0.4040
200	473.15	0.1684	0.7051	0.1710	0.7159	0.1732	0.7252	0.1714	0.7176	0.2324	0.9730	0.1711	0.7164
300	573.15	0.2281	0.9550	0.2317	0.9701	0.2372	0.9931	0.2326	0.9738	0.3246	1.3590	0.2323	0.9726
400	673.15	0.2785	1.1660	0.2835	1.1870	0.2926	1.2251	0.2850	1.1932	0.4063	1.7011	0.2847	1.1920
500	773.15	0.3219	1.3477	0.3291	1.3779	0.3415	1.4298	0.3312	1.3867	0.4799	2.0092	0.3309	1.3854
600	873.15	0.3604	1.5089	0.3700	1.5491	0.3854	1.6136	0.3728	1.5608	0.5468	2.2893	0.3723	1.5587
700	973.15	0.3950	1.6538	0.4073	1.7053	0.4253	1.7806	0.4108	1.7199	0.6082	2.5464	0.4100	1.7166
800	1073.15	0.4266	1.7861	0.4415	1.8485	0.4617	1.9330	0.4453	1.8644	0.6649	2.7838	0.4445	1.8610
900	1173.15	0.4557	1.9079	0.4732	1.9812	0.4953	2.0737	0.4774	1.9988	0.7176	3.0044	0.4765	1.9950
1000	1273.15	0.4828	2.0214	0.5026	2.1043	0.5265	2.2044	0.5073	2.1240	0.7667	3.2100	0.5062	2.1194
1100	1373.15	0.5081	2.1273	0.5302	2.2198	0.5555	2.3258	0.5352	2.2408	0.8127	3.4026	0.5340	2.2358
1200	1473.15	0.5320	2.2274	0.5561	2.3283	0.5827	2.4396	0.5614	2.3505	0.8560	3.5839	0.5601	2.3450
1300	1573.15	0.5547	2.3224	0.5806	2.4309	0.6084	2.5472	0.5861	2.4539	0.8968	3.7547	0.5845	2.4472
1400	1673.15	0.5763	2.4129	0.6037	2.5276	0.6326	2.6486	0.6094	2.5514	0.9354	3.9163	0.6080	2.5456
1500	1773.15	0.5968	2.4987	0.6256	2.6193	0.6555	2.7444	0.6315	2.6440	0.9721	4.0700	0.6300	2.6377
1600	1873.15	0.6164	2.5807	0.6464	2.7063	0.6774	2.8361	0.6525	2.7319	1.0069	4.2157	0.6510	2.7256
1700	1973.15	0.6352	2.6595	0.6663	2.7897	0.6982	2.9232	0.6726	2.8160	1.0401	4.3547	0.6710	2.8093
1800	2073.15	0.6533	2.7352	0.6853	2.8692	0.7181	3.0065	0.6917	2.8960	1.0719	4.4878	0.6901	2.8893
1900	2173.15	0.6706	2.8077	0.7034	2.9450	0.7373	3.0869	0.7099	2.9722	1.1021	4.6143	0.7084	2.9659
2000	2273.15	0.6873	2.8776	0.7208	3.0178	0.7557	3.1640	0.7275	3.0459	1.1311	4.7357	0.7259	3.0392
2100	2373.15	0.7035	2.9454	0.7375	3.0878	0.7733	3.2377	0.7443	3.1162	1.1590	4.8525	0.7428	3.1100
2200	2473.15	0.7191	3.0107	0.7536	3.1552	0.7903	3.3088	0.7604	3.1836	1.1858	4.9647	0.7590	3.1778
2300	2573.15	0.7342	3.0739	0.7691	3.2201	0.8068	3.3779	0.7760	3.2490	1.2115	5.0723	0.7746	3.2431
2400	2673.15	0.7488	3.1351	0.7840	3.2825	0.8227	3.4445	0.7910	3.3118	1.2363	5.1761	0.7879	3.3063
2500	2773.15	0.7630	3.1945	0.7984	3.3427	0.8382	3.5094	0.8055	3.3725	1.2602	5.2762	0.8043	3.3674
2600	2873.15	0.7768	3.2523	—	—	0.8531	3.5718	—	—	—	—		
2700	2973.15	0.7901	3.3080	—	—	0.8676	3.6325	—	—	—	—		
3000	3273.15	0.826	3.4583	0.858	3.5923	0.929	3.8895	0.868	3.6341	1.373	5.7485		

1 kcal = 4.186 8 kJ

Table 88-2 Volume Specific Entropy s' of Gases (*Continued*)

| Temperature | | NO | | OH | | N_2O | | H_2S | | NH_3 | |
| t | T | s' | | s' | | s' | | s' | | s' | |
°C	K	$\dfrac{kcal}{m_n^3\,K}$	$\dfrac{kJ}{m_n^3\,K}$	$\dfrac{kcal}{m_n^3\,K}$	$\dfrac{kJ}{m_n^3\,K}$	$\dfrac{kcal}{m_n^3\,K}$	$\dfrac{kJ}{m_n^3\,K}$	$\dfrac{kcal}{m_n^3\,K}$	$\dfrac{kJ}{m_n^3\,K}$	$\dfrac{kcal}{m_n^3\,K}$	$\dfrac{kJ}{m_n^3\,K}$
0	273.15	0.0000	0.0000	0.000	0.0000	0.0000	0.0000	0.000	0.0000	0.000	0.0000
100	373.15	0.0994	0.4162	0.099	0.4145	0.1338	0.5602	0.112	0.4689	0.120	0.5024
200	473.15	0.1756	0.7352	0.172	0.7201	0.2441	1.0220	0.202	0.8457	0.218	0.9127
300	573.15	0.2384	0.9981	0.233	0.9755	0.3397	1.4223	0.278	1.1639	0.304	1.2728
400	673.15	0.2922	1.2234	0.284	1.1891	0.4243	1.7765	0.345	1.4444	0.383	1.6035
500	773.15	0.3397	1.4223	0.328	1.3733	0.5003	2.0947	0.406	1.6998	0.454	1.9008
600	873.15	0.3824	1.6010	0.368	1.5407	0.5731	2.3995	0.461	1.9301	0.521	2.1813
700	973.15	0.4213	1.7639	0.404	1.6915	0.6331	2.6507	0.512	2.1436	0.580	2.4283
800	1073.15	0.4569	1.9129	0.436	1.8254	0.6918	2.8964	0.559	2.3404	0.641	2.6837
900	1173.15	0.4898	2.0507	0.466	1.9510	0.7464	3.1250	0.604	2.5288	0.699	2.9266
1000	1273.15	0.5203	2.1784	0.493	2.0641	0.7973	3.3381	0.646	2.7047	0.752	3.1485
1100	1373.15	0.5486	2.2969	0.519	2.1729	0.8449	3.5374	0.686	2.8721	—	—
1200	1473.15	0.5753	2.4087	0.543	2.2734	—	—	0.723	3.0271	—	—
1300	1573.15	0.6004	2.5138	0.566	2.3697	—	—	—	—	—	—
1400	1673.15	0.6242	2.6134	0.587	2.4577	—	—	—	—	—	—
1500	1773.15	0.6466	2.7022	0.608	2.5456	1.037	4.3417	0.830	3.4750	0.985	4.1240
1600	1873.15	0.6680	2.7968	0.627	2.6251	—	—	—	—	—	—
1700	1973.15	0.6883	2.8818	0.646	2.7047	—	—	—	—	—	—
1800	2073.15	0.7076	2.9626	0.665	2.7842	—	—	—	—	—	—
1900	2173.15	0.7261	3.0400	0.683	2.8596	—	—	—	—	—	—
2000	2273.15	0.7437	3.1137	0.700	2.9308	1.195	5.0032	0.972	4.0696	1.176	4.9237
2100	2373.15	0.7607	3.1849	0.717	3.0019	—	—	—	—	—	—
2200	2473.15	0.7771	3.2536	0.733	3.0689	—	—	—	—	—	—
2300	2573.15	0.7929	3.3197	0.748	3.1317	—	—	—	—	—	—
2400	2673.15	0.8081	3.3834	0.763	3.1945	—	—	—	—	—	—
2500	2773.15	0.8228	3.4449	0.777	4.2531	1.326	5.5517	1.088	4.5552	1.336	5.5936
2600	2873.15	0.8369	3.5039	0.791	3.3118	—	—	—	—	—	—
2700	2973.15	0.8506	3.5613	0.803	3.3620	—	—	—	—	—	—
2800	3073.15	0.8638	3.6166	0.816	3.4164	—	—	—	—	—	—
2900	3173.15	0.8766	3.6701	0.828	3.4667	—	—	—	—	—	—
3000	3273.15	0.8891	3.7225	0.840	3.5169	1.432	5.9955	1.187	4.9697	1.472	6.1630

1 kcal = 4.186 8 kJ

Table 88-3 Volume Specific Entropy s' of Gases (*Continued*)

Temperature		H$_2$O		SO$_2$		CH$_4$		C$_2$H$_4$		C$_2$H$_2$		CS$_2$	
		s'		s'		s'		s'		s'		s'	
t	T	kcal	kJ	kcal	kJ	kcal	kJ	kcal	kJ	kcal	kJ	kcal	kJ
°C	K	$\frac{\text{kcal}}{\text{m}_n^3\text{K}}$	$\frac{\text{kJ}}{\text{m}_n^3\text{K}}$	$\frac{\text{kcal}}{\text{m}_n^3\text{K}}$	$\frac{\text{kJ}}{\text{m}_n^3\text{K}}$	$\frac{\text{kcal}}{\text{m}_n^3\text{K}}$	$\frac{\text{kJ}}{\text{m}_n^3\text{K}}$	$\frac{\text{kcal}}{\text{m}_n^3\text{K}}$	$\frac{\text{kJ}}{\text{m}_n^3\text{K}}$	$\frac{\text{kcal}}{\text{m}_n^3\text{K}}$	$\frac{\text{kJ}}{\text{m}_n^3\text{K}}$	$\frac{\text{kcal}}{\text{m}_n^3\text{K}}$	$\frac{\text{kJ}}{\text{m}_n^3\text{K}}$
0	273.15	0.0000	0.0000	0.000	0.000	0.000	0.000	0.000	0.000	0.0000	0.0000	0.000	0.0000
100	373.15	0.1107	0.4635	0.134	0.561	0.121	0.507	0.153	0.641	0.1510	0.6322	0.154	0.6448
200	473.15	0.1984	0.8307	0.246	1.030	0.227	0.950	0.294	1.231	0.2802	1.1731	0.282	1.1807
300	573.15	0.2707	1.1334	0.341	1.428	0.325	1.361	0.427	1.788	0.3925	1.6433	0.389	1.6287
400	673.15	0.333	1.394	0.426	1.784	0.417	1.746	0.552	2.311	0.4920	2.0599	0.483	2.0222
500	773.15	0.389	1.629	0.501	2.098	0.504	2.110	0.670	2.805	0.5816	2.4350	0.567	2.3739
600	873.15	0.440	1.842	0.569	2.382	0.588	2.462	0.781	3.270	0.6633	2.7771	0.641	2.6837
700	973.15	0.486	2.035	0.630	2.638	0.667	2.793	0.886	3.710	0.7386	3.0924	0.708	2.9643
800	1073.15	0.530	2.219	0.687	2.876	0.742	3.107	0.986	4.128	0.8085	3.3850	0.769	3.2196
900	1173.15	0.570	2.386	0.739	3.094	0.814	3.408	1.080	4.522	0.8739	3.6588	0.825	3.4541
1000	1273.15	0.609	2.550	0.787	3.295	0.883	3.697	1.170	4.899	0.9353	3.9159	0.878	3.6760
1100	1373.15	0.645	2.700	0.832	3.483	0.951	3.982	1.255	5.254	0.9932	4.1583	0.926	3.8770
1200	1473.15	0.680	2.847	0.874	3.659	1.015	4.250	1.337	5.598	1.0481	4.3882	0.972	4.0696
1300	1573.15	0.712	2.981	—	—								
1400	1673.15	0.744	3.115	—	—								
1500	1773.15	0.774	3.241	1.000	4.187								
1600	1873.15	0.804	3.366	—	—								
1700	1973.15	0.832	3.483	—	—								
1800	2073.15	0.859	3.596	—	—								
1900	2173.15	0.885	3.705	—	—								
2000	2273.15	0.911	3.814	1.150	4.815								
2100	2373.15	0.935	3.915	—	—								
2200	2473.15	0.958	4.011	—	—								
2300	2573.15	0.981	4.107	—	—								
2400	2673.15	1.003	4.199	—	—								
2500	2773.15	1.025	4.291	1.272	5.326								
2600	2873.15	1.045	4.375	—	—								
2700	2973.15	1.065	4.459	—	—								
3000	3273.15	1.114	4.664	1.376	5.761								

Table 88-4 Volume Specific Entropy s' of Gases (*Continued*)

Temperature		COS		C$_2$H$_6$		C$_3$H$_8$		C$_6$H$_6$		C$_3$H$_6$		C$_2$H$_5$OH	
		s'		s'		s'		s'		s'		s'	
t	T	kcal	kJ	kcal	kJ	kcal	kJ	kcal	kJ	kcal	kJ	kcal	kJ
°C	K	$\frac{\text{kcal}}{\text{m}_n^3\text{K}}$	$\frac{\text{kJ}}{\text{m}_n^3\text{K}}$	$\frac{\text{kcal}}{\text{m}_n^3\text{K}}$	$\frac{\text{kJ}}{\text{m}_n^3\text{K}}$	$\frac{\text{kcal}}{\text{m}_n^3\text{K}}$	$\frac{\text{kJ}}{\text{m}_n^3\text{K}}$	$\frac{\text{kcal}}{\text{m}_n^3\text{K}}$	$\frac{\text{kJ}}{\text{m}_n^3\text{K}}$	$\frac{\text{kcal}}{\text{m}_n^3\text{K}}$	$\frac{\text{kJ}}{\text{m}_n^3\text{K}}$	$\frac{\text{kcal}}{\text{m}_n^3\text{K}}$	$\frac{\text{kJ}}{\text{m}_n^3\text{K}}$
0	273.15	0.000	0.0000	0.000	0.0000	0.000	0.0000	0.000	0.0000	0.000	0.0000	0.000	0.0000
100	373.15	0.141	0.5903	0.185	0.7746	0.258	1.0802	0.294	1.2309	0.255	0.9420	0.253	1.0593
200	473.15	0.258	1.0802	0.358	1.4989	0.508	2.1269	0.591	2.4744	0.435	1.8213	0.483	2.0222
300	573.15	0.359	1.5031	0.522	2.1855	0.744	3.1150	0.881	3.6886	0.633	2.6502	0.694	2.9056
400	673.15	0.448	1.8757	0.679	2.8428	0.972	4.0696	1.158	4.8483	0.822	3.4415	0.890	3.7263
500	773.15	0.526	2.2023	0.829	3.4709	1.287	5.3884	1.421	5.9494	1.000	4.1868	1.073	4.4924
600	873.15	0.597	2.4995	0.971	4.0654	1.448	6.0625	1.668	6.9836	1.169	4.8944	1.246	5.2168
700	973.15	0.661	2.7675	1.107	4.6348	1.687	7.0631	1.902	7.9633	1.329	5.5643	1.408	5.8950
500	1073.15	0.720	3.0145	1.235	5.1707	1.869	7.8251	2.123	8.8886	1.481	6.2007	1.559	6.5272
900	1173.15	0.775	3.2448	1.358	5.6857	2.043	8.5536	2.331	9.7594	1.625	6.8036	1.704	7.1343
1000	1273.15	0.825	3.4541	1.476	6.1797	2.209	9.2486	2.530	10.5926	1.762	7.3771	1.840	7.7037
1100	1373.15	0.873	3.6551	1.590	6.6570	2.370	9.9227	2.717	11.3755	1.893	7.9256	1.971	8.2522
1200	1473.15	0.917	3.8393	1.699	7.1134	2.523	10.5633	2.895	12.1208	2.017	8.4448	2.094	8.7672

1 kcal = 4.186 8 kJ

Table 89-1 Viscosities η of Gases

Gas	Chemical formula	t (°C)	T (°K)	$\eta \times 10^6$ (kp s/m²)	$\eta \times 10^6$ (Pa s)
Acetone	C_3H_6O	0	273.15	0.739	7.25
		18	291.15	0.795	7.80
		100	373.15	0.962	9.43
		119	392.15	1.011	9.91
		160	433.15	1.123	11.01
		190	463.15	1.209	11.86
		217	490.15	1.278	12.53
		248	521.15	1.360	13.34
		279	552.15	1.444	14.16
		306	579.15	1.510	14.81
Acetylene	C_2H_2	0	273.15	0.962	9.43
		20	293.15	1.040	10.20
		40	313.15	1.100	10.79
		50	323.15	1.135	11.13
		60	333.15	1.153	11.31
		80	353.15	1.222	11.98
		100	373.15	1.279	12.54
		120	393.15	1.344	13.18
Air		−194	79.15	0.562	5.51
		−183	90.15	0.639	6.27
		−150	123.15	0.887	8.70
		−100	173.15	1.203	11.80
		−50	223.15	1.489	14.60
		0	273.15	1.753	17.19
		50	323.15	1.964	19.26
		100	373.15	2.166	21.24
		150	423.15	2.365	23.19
		200	473.15	2.562	25.12
		250	523.15	2.757	27.04
		300	573.15	2.943	28.86
		350	623.15	3.128	30.68
		400	673.15	3.309	32.45
		450	723.15	3.480	34.13
		500	773.15	3.640	35.70
		550	823.15	3.793	37.20
		600	873.15	3.944	38.68
		650	923.15	4.096	40.17
		700	973.15	4.244	41.62
		750	1023.15	4.386	43.01
		800	1073.15	4.519	44.32

Gas	Chemical formula	t (°C)	T (°K)	$\eta \times 10^6$ (kp s/m²)	$\eta \times 10^6$ (Pa s)
Air		850	1123.15	4.650	45.60
		900	1173.15	4.780	46.88
		950	1223.15	4.907	48.12
		1000	1273.15	5.030	49.33
		1100	1373.15	5.252	51.50
		1200	1473.15	5.48	53.74
		1400	1673.15	5.90	57.86
		1600	1873.15	6.28	61.59
Ammonia	NH_3	−60	213.15	0.724	7.1
		−40	233.15	0.795	7.8
		−20	253.15	0.877	8.6
		0	273.15	0.948	9.3
		20	293.15	1.020	10.0
		50	323.15	1.132	11.1
		100	373.15	1.326	13.0
		150	423.15	1.509	14.8
		200	473.15	1.693	16.6
		250	523.15	1.876	18.4
		300	573.15	2.060	20.2
Argon	Ar	−200	73.15	0.530	5.2
		−150	123.15	1.020	10.0
		−100	173.15	1.448	14.2
		−50	223.15	1.825	17.9
		0	273.15	2.162	21.2
		20	293.15	2.264	22.2
		50	323.15	2.468	24.2
		100	373.15	2.763	27.1
		200	473.15	2.373	32.1
		300	573.15	3.742	36.7
		440	673.15	4.181	41.0
		493	766.15	4.568	44.8
		600	873.15	4.966	48.7
		714	987.15	5.364	52.6
		880	1153.15	5.649	55.4
Benzene	C_6H_6	0	273.15	0.714	7.0
		20	293.15	0.765	7.5
		50	323.15	0.836	8.2
		100	373.15	0.969	9.5
		150	423.15	1.101	10.8

1 kp s/m² = 9.806 65 Pa s = 9.806 65 N s/m²

Table 89-2 Viscosities η of Gases (Continued)

Gas	Chemical formula	Temperature t °C	Temperature T °K	Viscosity $\eta \times 10^6$ kp s/m²	Viscosity $\eta \times 10^6$ Pa s	Gas	Chemical formula	Temperature t °C	Temperature T °K	Viscosity $\eta \times 10^6$ kp s/m²	Viscosity $\eta \times 10^6$ Pa s
Benzene	C_6H_6	200	473.15	1.224	12.0	Carbon dioxide	CO_2	850	1123.15	4.444	43.58
		250	523.15	1.346	13.2			1008	1281.15	4.872	47.78
		300	573.15	1.479	14.5			1052	1325.15	4.880	47.86
Bromine	Br_2	0	273.15	1.489	14.60	Carbon disulfide	CS_2	0	273.15	0.942	9.24
		20	293.15	1.572	15.42			114	387.15	1.329	13.03
		25	298.15	1.558	15.28			153	426.15	1.462	14.34
		138	411.15	2.138	20.97			190	463.15	1.592	15.61
		190	463.15	2.416	23.69			228	501.15	1.725	16.92
		242	515.15	2.678	26.26			263	536.15	1.866	18.30
		316	589.15	3.058	29.99			310	583.15	2.005	19.66
		349	622.15	3.225	31.63						
		410	683.15	3.545	34.76	Carbon monoxide	CO	$-$ 80	193.15	1.295	12.7
		535	808.15	4.187	41.06			$-$ 60	213.15	1.397	13.7
		588	861.15	4.385	43.00			$-$ 40	233.15	1.499	14.7
		594	867.15	4.377	42.92			$-$ 20	253.15	1.601	15.7
n—Butane	C_4H_{10}	0	273.15	0.693	6.80			0	273.15	1.693	16.6
		20	293.15	0.754	7.39			20	293.15	1.795	17.6
		40	313.15	0.803	7.87			50	323.15	1.927	18.9
		60	333.15	0.856	8.39			100	373.15	2.141	21.0
		80	353.15	0.903	8.85			150	423.15	2.335	22.9
		100	373.15	0.966	9.47			200	473.15	2.519	24.7
		120	393.15	1.018	9.98			250	523.15	2.692	26.4
Carbon dioxide	CO_2	$-$ 98	175.15	0.914	8.96			300	573.15	2.845	27.9
		$-$ 78	195.15	0.991	9.72	Carbon tetrachloride (Tetrachloromethane)	CCl_4	0	273.15	0.948	9.3
		$-$ 60	213.15	1.082	10.61			20	293.15	1.010	9.9
		$-$ 40	233.15	1.178	11.55			50	323.15	1.101	10.8
		$-$ 19	254.15	1.285	12.60			100	373.15	1.254	12.3
		0	273.15	1.409	13.82			150	423.15	1.407	13.8
		22	295.15	1.500	14.71			200	473.15	1.560	15.3
		50	323.15	1.652	16.20			250	523.15	1.713	16.8
		100	373.15	1.881	18.45			300	573.15	1.866	18.3
		145	418.15	2.081	20.41	Chlorine	Cl_2	0	273.15	1.254	12.30
		235	508.15	2.463	24.15			20	293.15	1.353	13.27
		300	573.15	2.733	26.80			50	323.15	1.498	14.96
		417	690.15	3.167	31.06			100	373.15	1.712	16.79
		490	763.15	3.365	33.00			150	423.15	1.912	18.75
		574	847.15	3.745	37.63			200	473.15	2.126	20.85
		685	958.15	3.875	38.00			250	523.15	2.321	22.76
		764	1037.15	4.164	40.84			300	573.15	2.549	25.00

1 kp s/m² = 9.806 65 Pa s = 9.806 65 N s/m²

Table 89-3 Viscosities η of Gases (*Continued*)

Gas	Chemical formula	Temperature t °C	Temperature T °K	Viscosity η × 10⁶ kp s/m²	Viscosity η × 10⁶ Pa s	Gas	Chemical formula	Temperature t °C	Temperature T °K	Viscosity η × 10⁶ kp s/m²	Viscosity η × 10⁶ Pa s
Chloroform	CHCl₃	0	273.15	0.978	9.59	Ethyl ether	C₄H₁₀O	0	273.15	0.704	6.9
		20	293.15	1.021	10.01			20	293.15	0.755	7.4
		100	373.15	1.333	13.07			50	323.15	0.836	8.2
		161	434.15	1.520	14.91			100	373.15	0.959	9.4
		189	462.15	1.610	15.79			150	423.15	1.081	10.6
		250	523.15	1.811	17.76			200	473.15	1.203	11.8
		308	581.15	1.985	19.47			250	523.15	1.315	12.9
Cyanogen	C₂N₂	0	273.15	0.967	9.48			300	573.15	1.438	14.1
		20	293.15	1.091	10.70	Ethylene	C₂H₄	− 80	193.15	0.678	6.65
		100	373.15	1.295	12.70			− 60	213.15	0.749	7.35
Ethane	C₂H₆	− 78	195.15	0.657	6.44			− 40	233.15	0.821	8.05
		0	273.15	0.872	8.55			− 20	253.15	0.892	8.75
		20	293.15	0.947	9.29			0	273.15	0.959	9.40
		40	313.15	1.005	9.86			20	293.15	1.028	10.08
		60	333.15	1.071	10.50			50	323.15	1.125	11.03
		80	353.15	1.133	11.11			100	373.15	1.282	12.57
		100	373.15	1.190	11.67			150	323.15	1.431	14.03
		120	393.15	1.254	12.30			200	473.15	1.571	15.41
		150	423.15	1.303	12.78			250	523.15	1.699	16.66
		200	473.15	1.437	14.09	Helium	He	− 250	23.15	0.377	3.7
		250	523.15	1.556	15.26			− 200	73.15	0.816	8.0
Ethyl acetate	C₄H₈O₂	0	273.15	0.704	6.90			− 100	173.15	1.417	13.9
		100	373.15	0.974	9.55			0	273.15	1.897	18.6
		128	401.15	1.038	10.18			50	323.15	2.121	20.8
		159	432.15	1.120	10.98			100	373.15	2.335	22.9
		193	466.15	1.219	11.95			200	473.15	2.753	27.0
		218	491.15	1.275	12.50			300	573.15	3.131	30.7
		249	522.15	1.358	13.32			400	673.15	3.487	34.2
		280	553.15	1.437	14.09			600	873.15	4.150	40.7
		314	587.15	1.527	14.97			800	1073.15	4.742	46.5
Ethyl alcohol	C₂H₆O	0	273.15	0.846	8.3	n−Heptane	C₇H₁₆	100	373.15	0.731	7.17
		100	373.15	1.111	10.9			150	423.15	0.827	8.11
		150	423.15	1.264	12.4			202	475.15	0.940	9.22
		200	473.15	1.407	13.8			252	525.15	1.101	10.80
		250	523.15	1.550	15.2	n−Hexane	C₆H₁₄	0	273.15	0.602	5.90
		300	573.15	1.683	16.5			121	394.15	0.883	8.66
Ethyl chloride	C₂H₅Cl	0	273.15	0.959	9.4			161	434.15	0.977	9.58
								189	462.15	1.041	10.21

1 kp s/m² = 9.806 65 Pa s = 9.806 65 N s/m²

Table 89-4 Viscosities η of Gases (*Continued*)

Gas	Chemical formula	Temperature		Viscosity		Gas	Chemical formula	Temperature		Viscosity	
		t	T	$\eta \times 10^6$				t	T	$\eta \times 10^6$	
		°C	°K	kp s/m²	Pa s			°C	°K	kp s/m²	Pa s
n—Hexane	C_6H_{14}	220	493.15	1.109	10.88	Hydrogen sulfide	H_2S	0	273.15	1.198	11.75
		248	521.15	1.167	11.44			20	293.15	1.326	13.00
		280	553.15	1.237	12.13			100	373.15	1.642	16.10
		307	580.15	1.290	12.65	Iodine	J_2	0	273.15	1.254	12.30
Hydrogen	H_2	— 250	23.15	0.112	1.1			106	379.15	1.820	17.85
		— 200	73.15	0.337	3.3			232	505.15	2.365	23.19
		— 150	123.15	0.489	4.8			279	552.15	2.611	25.61
		— 100	173.15	0.622	6.1			329	602.15	2.802	27.48
		— 50	223.15	0.744	7.3			397	670.15	3.125	30.65
		0	273.15	0.857	8.4			438	711.15	3.314	32.50
		50	323.15	0.959	9.4			523	796.15	3.675	36.04
		100	373.15	1.050	10.3	Isoamylene	C_5H_{10}	22	295.15	0.728	7.14
		200	473.15	1.234	12.1			40	313.15	0.786	7.71
		300	573.15	1.417	13.9			50	323.15	0.809	7.93
		400	673.15	1.570	15.4			60	333.15	0.845	8.29
		500	773.15	1.723	16.9			80	353.15	0.888	8.71
		600	873.15	1.866	18.3			100	373.15	0.933	9.15
		700	973.15	1.999	19.6			120	393.15	0.986	9.67
		800	1073.15	2.141	21.0	Isobutane	C_4H_{10}	0	273.15	0.704	6.90
Hydrogen arsenide	H_3As	15	288.15	1.583	15.52			20	293.15	0.759	7.44
Hydrogen bromide	HBr	0	273.15	1.744	17.10			40	313.15	0.808	7.92
		100	373.15	2.412	23.65			60	333.15	0.862	8.45
Hydrogen chloride	HCl	0	273.15	1.358	13.32			80	353.15	0.906	8.88
		23	296.15	1.473	14.45			100	373.15	0.966	9.47
		53	326.15	1.626	15.95			120	393.15	1.015	9.95
		100	373.15	1.873	18.37	Isopropyl alcohol	C_3H_8O	0	273.15	0.714	7.0
		151	424.15	2.058	20.18			120	393.15	1.050	10.3
		202	475.15	2.358	23.12			150	423.15	1.132	11.1
		251	524.15	2.584	25.34			200	473.15	1.275	12.5
Hydrogen iodide	HJ	0	273.15	1.764	17.30			250	523.15	1.407	13.8
		20	293.15	1.892	18.55	Krypton	Kr	0	273.15	2.376	23.3
		50	323.15	2.058	20.18			20	293.15	2.508	24.6
		100	373.15	2.362	23.16			100	373.15	3.120	30.6
		150	423.15	2.679	26.27	Mercury	Hg	218	491.15	4.802	47.09
		200	473.15	2.982	29.24			281	554.15	5.415	53.10
		250	523.15	3.252	31.89			300	573.15	5.609	55.01
Hydrogen phosphide	PH_3	0	273.15	1.091	10.7			330	603.15	5.946	58.31
		100	373.15	1.479	14.5			421	694.15	6.991	68.56

1 kp s/m² = 9.806 65 Pa s = 9.806 65 N s/m²

Table 89-5   Viscosities η of Gases (Continued)

Gas	Chemical formula	Temperature t °C	Temperature T °K	Viscosity $\eta \times 10^6$ kp s/m²	Viscosity $\eta \times 10^6$ Pa s	Gas	Chemical formula	Temperature t °C	Temperature T °K	Viscosity $\eta \times 10^6$ kp s/m²	Viscosity $\eta \times 10^6$ Pa s
Mercury	Hg	496	769.15	7.760	76.10	Methyl chloride	CH_3Cl	0	273.15	0.999	9.80
		565	838.15	8.507	83.43			20	293.15	1.082	10.61
		610	883.15	8.975	88.02			30	303.15	1.123	11.01
Methane	CH_4	− 80	193.15	0.755	7.4			40	313.15	1.162	11.40
		− 60	213.15	0.826	8.1			50	323.15	1.198	11.75
		− 40	233.15	0.897	8.8			60	333.15	1.233	12.09
		− 20	253.15	0.969	9.5			70	343.15	1.275	12.50
		0	273.15	1.055	10.35			80	353.15	1.312	12.87
		20	293.15	1.108	10.87			90	363.15	1.349	13.23
		50	323.15	1.203	11.80			100	373.15	1.384	13.57
		100	373.15	1.357	13.31			110	383.15	1.428	14.00
		150	423.15	1.500	14.71			120	393.15	1.468	14.40
		200	473.15	1.637	16.05			130	403.15	1.500	14.71
		250	523.15	1.759	17.25			219	492.15	1.804	17.69
		300	573.15	1.897	18.60			257	530.15	1.932	18.95
		380	653.15	2.066	20.26			300	573.15	2.084	20.44
		499	772.15	2.309	22.64	Methylene chloride	CH_2Cl_2	0	273.15	0.928	9.1
Methyl acetate	$C_3H_6O_2$	100	373.15	1.035	10.15			22	295.15	1.011	9.91
		143	416.15	1.161	11.39			100	373.15	1.292	12.67
		178	451.15	1.260	12.36			219	492.15	1.700	16.67
		219	492.15	1.375	13.48			259	532.15	1.833	17.98
		248	521.15	1.457	14.29			309	582.15	1.995	19.56
		278	551.15	1.538	15.08	Methyl ether	C_2H_6O	0	273.15	0.867	8.50
		307	580.15	1.614	15.83			20	293.15	0.927	9.09
Methyl alcohol	CH_4O	0	273.15	0.887	8.7			40	313.15	1.003	9.84
		100	373.15	1.244	12.2			60	333.15	1.065	10.44
		150	423.15	1.428	14.0			80	353.15	1.131	11.09
		200	473.15	1.591	15.6			100	373.15	1.190	11.67
		250	523.15	1.764	17.3			120	393.15	1.252	12.28
		300	573.15	1.927	18.9	Neon	Ne	− 78	195.15	2.414	23.67
Methyl bromide	OH_3Br	0	273.15	1.056	10.36			0	273.15	3.040	29.81
		10	283.15	1.302	12.77			20	293.15	3.172	31.11
		20	293.15	1.353	13.27			100	373.15	3.718	36.46
		30	303.15	1.405	13.78			200	473.15	4.332	42.48
		40	313.15	1.444	14.16			250	523.15	4.621	45.32
		50	323.15	1.486	14.57			285	558.15	4.801	47.08
		60	333.15	1.547	15.17			429	702.15	5.561	54.54
		120	393.15	1.832	17.97			502	775.15	5.916	58.02

1 kp s/m² = 9.806 65 Pa s = 9.806 65 N s/m²

Table 89-6 Viscosities η of Gases (*Continued*)

Gas	Chemical formula	Temperature t °C	Temperature T °K	Viscosity η × 10⁶ kp s/m²	Viscosity η × 10⁶ Pa s	Gas	Chemical formula	Temperature t °C	Temperature T °K	Viscosity η × 10⁶ kp s/m²	Viscosity η × 10⁶ Pa s
		594	867.15	6.353	62.30	Propane		60	333.15	0.940	9.22
		686	959.15	6.757	66.26			80	353.15	0.997	9.78
		827	1100.15	7.352	72.10			100	373.15	1.049	10.29
Nitric oxide	NO	0	273.15	1.832	17.97			120	393.15	1.103	10.82
		20	293.15	1.913	18.76			150	423.15	1.152	11.30
		50	323.15	2.076	20.36			200	473.15	1.275	12.50
		100	373.15	2.317	22.72			250	523.15	1.387	13.60
		150	323.15	2.523	24.74	Propene	C_3H_6	0	273.15	0.795	7.80
		200	473.15	2.735	26.82			20	293.15	0.851	8.35
		250	523.15	2.927	28.70			40	313.15	0.911	8.93
Nitrogen	N_2	— 150	123.15	0.857	8.4			60	333.15	0.978	9.59
		— 100	173.15	1.162	11.4			80	353.15	1.043	10.23
		— 50	223.15	1.438	14.1			100	373.15	1.092	10.71
		0	273.15	1.693	16.6			120	393.15	1.144	11.22
		50	323.15	1.917	18.8	n—Propyl alcohol	C_3H_8O	0	273.15	0.693	6.8
		100	373.15	2.121	20.8			120	393.15	1.050	10.3
		200	473.15	2.508	24.6			150	423.15	1.122	11.0
		400	673.15	3.171	31.1	Sulfur dioxide	SO_2	— 75	198.15	0.875	8.58
		600	873.15	3.732	36.6			— 36	237.15	1.032	10.12
		800	1073.15	4.211	41.3			— 20	253.15	1.099	10.78
Nitrous oxide	N_2O	0	273.15	1.397	13.7			— 6	267.15	1.153	11.31
		20	293.15	1.489	14.6			0	273.15	1.181	11.58
		50	323.15	1.632	16.0			20	293.15	1.279	12.54
		100	373.15	1.866	18.3			40	313.15	1.379	13.52
		150	423.15	2.080	20.4			60	333.15	1.484	14.55
		200	473.15	2.294	22.5			80	353.15	1.570	15.40
		250	523.15	2.508	24.6			100	373.15	1.644	16.12
		300	573.15	2.702	26.5			120	393.15	1.750	17.16
Oxygen	O_2	— 200	73.15	0.581	5.7			150	423.15	1.867	18.31
		— 150	123.15	0.989	9.7			200	473.15	2.078	20.38
		— 100	173.15	1.346	13.2			293	566.15	2.495	24.47
		— 50	223.15	1.662	16.3			421	694.15	2.946	28.89
		0	273.15	1.958	19.2			490	763.15	3.176	31.15
		50	323.15	2.223	21.8			595	868.15	3.489	34.22
		100	373.15	2.488	24.4			679	952.15	3.774	37.01
		200	473.15	2.957	29.0			823	1096.15	4.181	41.00
		400	673.15	3.763	36.9	Water vapor (steam)	H_2O	100	373.15	1.280	12.55
		600	873.15	4.436	43.5			150	423.15	1.473	14.45
		800	1073.15	5.027	49.3			200	473.15	1.667	16.35
Pentane	C_5H_{12}	0	273.15	0.632	6.2			250	523.15	1.863	18.27
		120	393.15	0.928	9.1			300	573.15	2.064	20.24
		160	333.15	1.020	10.0			350	623.15	2.262	22.18
		219	492.15	1.152	11.3			400	673.15	2.460	24.12
		250	523.15	1.213	11.9			500	773.15	2.73	26.77
Propane	C_3H_8	0	273.15	0.765	7.50	Xenon	X	0	273.15	2.152	21.1
		20	293.15	0.822	8.06			20	293.15	2.305	22.6
		40	313.15	0.890	8.73			100	373.15	2.927	28.7

1 kp s/m² = 9.806 65 Pa s = 9.806 65 N s/m²

Table 90-1 Thermal Conductivities λ of Gases

Gas	Chemical formula	t (°C)	T (°K)	λ (kcal/h m K)	λ (W/m K)
Acetone	C_3H_6O	0	273.15	0.0084	0.00977
		20	293.15	0.0094	0.01093
		50	323.15	0.0112	0.01303
		100	373.15	0.0145	0.01686
		150	423.15	0.0285	0.02152
		200	473.15	0.0233	0.02710
Acetylene	C_2H_2	−75	198.15	0.0101	0.01175
		0	273.15	0.0161	0.01872
		100	373.15	0.0256	0.02977
Air	—	−180	93.15	0.00756	0.00879
		−150	123.15	0.0133	0.01549
		−100	173.15	0.0140	0.01633
		−50	223.15	0.0176	0.02052
		−20	253.15	0.0194	0.02256
		0	273.15	0.0204	0.02373
		20	293.15	0.0216	0.02512
		40	313.15	0.0228	0.02652
		50	323.15	0.0230	0.02680
		60	333.15	0.0240	0.02791
		80	353.15	0.0252	0.02931
		100	373.15	0.0264	0.03070
		120	393.15	0.0275	0.03198
		140	413.15	0.0286	0.03326
		160	433.15	0.0296	0.03442
		180	453.15	0.0307	0.03570
		200	473.15	0.0318	0.03698
		250	523.15	0.0344	0.04001
		300	573.15	0.0369	0.04291
		350	623.15	0.0393	0.04571
		400	673.15	0.0417	0.04850
		500	773.15	0.0464	0.05396
		600	873.15	0.0500	0.05815
		800	1073.15	0.0575	0.06687
		1000	1273.15	0.0655	0.07618
		1200	1473.15	0.0727	0.08455
		1400	1673.15	0.080	0.09304
		1600	1873.15	0.087	0.10118
Ammonia	NH_3	−50	223.15	0.0148	0.01721
		0	273.15	0.016	0.01861
		20	293.15	0.021	0.02442
		40	313.15	0.022	0.02559
		60	333.15	0.024	0.02791
		80	353.15	0.026	0.03024
		100	373.15	0.028	0.03256
		200	473.15	0.040	0.04652
		300	573.15	0.050	0.05815
Amylamine	$C_5H_{13}N$	6.5	279.65	0.0101	0.01175
Argon	Ar	−180	93.15	0.00504	0.00586
		−150	123.15	0.00684	0.00795
		−100	173.15	0.00936	0.01089
		−50	223.15	0.01152	0.01340
		0	273.15	0.01404	0.01633
		20	293.15	0.01512	0.01758
		50	323.15	0.01620	0.01884
		100	373.15	0.01872	0.02177
Benzene	C_6H_6	0	273.15	0.0076	0.00884
		20	293.15	0.0090	0.01047
		50	323.15	0.0111	0.01291
		100	373.15	0.0151	0.01756
		150	423.15	0.0194	0.02256
		200	473.15	0.0244	0.02838
n−Butane	C_4H_{10}	0	273.15	0.0116	0.01349
		20	293.15	0.0133	0.01547
		100	373.15	0.0201	0.02338
Butyl alcohol	$C_4H_{10}O$	100	373.15	0.017	0.01977
Butylamine	$C_4H_{11}N$	6.5	279.65	0.0108	0.01256
Carbon dioxide	CO_2	−150	123.15	0.004	0.00465
		−100	173.15	0.007	0.00814
		−50	223.15	0.010	0.01163
		0	273.15	0.0122	0.01424
		20	293.15	0.0137	0.01591
		50	323.15	0.0153	0.01779
		100	373.15	0.0180	0.02093
		200	473.15	0.0245	0.02847
		300	573.15	0.0302	0.03517

1 kcal/h = 1.163 W

Table 90-2 Thermal Conductivities λ of Gases (Continued)

Gas	Chemical formula	Temperature t °C	T °K	Thermal conductivity kcal/h m K	λ W/m K
Carbon dioxide	CO₂	496	769.15	0.0425	0.04943
		546	819.15	0.0511	0.05943
Carbon disulfide	CS₂	0	273.15	0.0058	0.00675
Carbon monoxide	CO	— 180	93.15	0.00684	0.00795
		— 150	123.15	0.00936	0.01089
		— 100	173.15	0.01296	0.01507
		— 75	198.15	0.01470	0.01710
		— 50	223.15	0.01656	0.01926
		— 25	248.15	0.01820	0.02117
		0	273.15	0.01908	0.02219
Carbon tetrachloride	CCl₄	0	273.15	0.0050	0.00582
		20	293.15	0.0055	0.00640
		50	323.15	0.0062	0.00721
		100	373.15	0.0075	0.00872
		150	423.15	0.0087	0.01012
		200	473.15	0.0099	0.01151
Chlorine	Cl₂	0	273.15	0.00684	0.00795
Chloroform	CHCl₃	0	273.15	0.0056	0.00651
		20	293.15	0.00605	0.00704
		50	323.15	0.0069	0.00802
		100	373.15	0.0086	0.01000
		150	423.15	0.0102	0.01186
		200	473.15	0.0120	0.01396
Cyclohexane	C₆H₁₂	100	373.15	0.0116	0.01349
Deuterium	D₂	0	273.15	0.1102	0.12812
		100	373.15	0.1357	0.15784
Dichlorodifluoromethane	CF₂Cl₂	0	273.15	0.00714	0.00830
		20	293.15	0.00809	0.00941
		50	323.15	0.00951	0.01106
		100	373.15	0.0119	0.01384
		150	423.15	0.0144	0.06175
Diethylamine	C₄H₁₁N	6.5	279.65	0.0108	0.01256
Dimethylamine	C₂H₇N	6.5	279.65	0.0126	0.01465
Dipropylamine	C₆H₁₅N	6.5	279.65	0.0094	0.01093

Gas	Chemical formula	Temperature t °C	T °K	Thermal conductivity kcal/h m K	λ W/m K
Ethane	C₂H₆	— 75	198.15	0.0098	0.01140
		— 50	223.15	0.0114	0.01326
		— 25	248.15	0.0133	0.01547
		0	273.15	0.0157	0.01826
		20	293.15	0.0178	0.02070
		50	323.15	0.0214	0.02489
		100	373.15	0.0282	0.03280
Ethyl acetate	C₄H₈O₂	0	273.15	0.0078	0.00907
		20	293.15	0.0090	0.01047
		50	323.15	0.0108	0.01256
		100	373.15	0.0142	0.01651
		150	423.15	0.0180	0.02093
		200	473.15	0.0224	0.02605
Ethyl alcohol	C₂H₆O	0	273.15	0.0119	0.01384
		20	293.15	0.0131	0.01524
		50	323.15	0.0150	0.01745
		100	373.15	0.0183	0.02128
Ethylamine	C₂H₇N	6.5	279.65	0.0115	0.01337
Ethyl bromide	C₂H₅Br	0	273.15	0.0062	0.00721
Ethyl chloride	C₂H₅Cl	0	273.15	0.00815	0.00948
		20	293.15	0.00925	0.01076
		50	323.15	0.0110	0.01279
		100	373.15	0.0141	0.01640
		150	423.15	0.0175	0.02035
		200	473.15	0.0213	0.02477
Ethylene	C₂H₄	— 75	198.15	0.0092	0.01070
		— 50	223.15	0.0110	0.01279
		— 25	248.15	0.0129	0.01500
		0	273.15	0.0150	0.02745
Ethyl ether	C₄H₁₀O	0	273.15	0.0114	0.01326
		20	293.15	0.0127	0.01477
		50	323.15	0.0150	0.01745
		100	373.15	0.0194	0.02256
		150	423.15	0.0242	0.02814
		200	473.15	0.0296	0.03442
Ethyl iodide	C₂H₅J	0	273.15	0.0051	0.00593

1 kcal/h = 1.163 W

Table 90-3 Thermal Conductivities λ of Gases (*Continued*)

Gas	Chemical formula	Temperature t (°C)	Temperature T (°K)	Thermal conductivity λ (kcal/h m K)	Thermal conductivity λ (W/m K)
Helium	He	−200	73.15	0.0508	0.05908
		−180	93.15	0.0587	0.06827
		−150	123.15	0.0702	0.08164
		−100	173.15	0.0887	0.10316
		−50	223.15	0.1065	0.12386
		0	273.15	0.1235	0.14363
		20	293.15	0.1300	0.15119
		50	323.15	0.1380	0.16049
		100	373.15	0.1465	0.17038
n−Heptane	C₇H₁₆	100	373.15	0.0152	0.01768
		200	473.15	0.0167	0.01942
n−Hexane	C₆H₁₄	0	273.15	0.0107	0.01244
		20	293.15	0.0119	0.01384
n−Hexane	C₆H₁₂	0	273.15	0.00895	0.01041
		20	293.15	0.0104	0.01210
		50	323.15	0.0125	0.01454
		100	373.15	0.0161	0.01872
Hydrogen	H₂	−200	73.15	0.0443	0.05152
		−150	123.15	0.0792	0.09211
		−100	173.15	0.1001	0.11639
		−50	223.15	0.1260	0.14654
		0	273.15	0.1508	0.17543
		20	293.15	0.1602	0.18631
		40	313.15	0.169	0.19655
		60	333.15	0.179	0.20818
		80	353.15	0.188	0.21864
		100	373.15	0.197	0.22911
		120	393.15	0.206	0.23958
		140	413.15	0.215	0.25005
		160	433.15	0.223	0.25935
		180	453.15	0.230	0.26749
		200	473.15	0.237	0.27563
		220	493.15	0.243	0.28261
		240	513.15	0.248	0.28842
		260	533.15	0.251	0.29911
		280	553.15	0.253	0.29424
		300	573.15	0.266	0.30940
		500	773.15	0.330	0.38379
		1000	1273.15	0.510	0.59313

Gas	Chemical formula	Temperature t (°C)	Temperature T (°K)	Thermal conductivity λ (kcal/h m K)	Thermal conductivity λ (W/m K)
Hydrogen sulfide	H₂S	0	273.15	0.0108	0.01256
Isobutane	C₄H₁₀	0	273.15	0.0119	0.01384
		100	373.15	0.0207	0.02407
Isopentane	C₅H₁₂	0	273.15	0.0107	0.01244
		20	293.15	0.0121	0.01407
		50	323.15	0.0144	0.01675
		100	373.15	0.0188	0.02186
		150	423.15	0.0238	0.02768
		200	473.15	0.0295	0.03431
Krypton	Kr	0	273.15	0.00756	0.00879
Mercury	Hg	200	473.15	0.0065	0.00754
Methane	CH₄	−150	123.15	0.0111	0.01291
		−100	173.15	0.0159	0.01838
		−75	198.15	0.0183	0.02128
		−50	223.15	0.0208	0.02419
		−25	248.15	0.0234	0.02721
		0	273.15	0.0260	0.03024
		20	293.15	0.0285	0.03315
		50	323.15	0.0320	0.03722
Methyl acetate	C₂H₆O₂	0	273.15	0.0087	0.01012
		20	293.15	0.010	0.01163
Methyl alcohol	CH₄O	0	273.15	0.0123	0.01430
		20	293.15	0.0136	0.01582
		50	323.15	0.0156	0.01814
		100	373.15	0.0187	0.02177
Methylamine	CH₅N	6.5	279.65	0.0137	0.01591
Methyl bromide	CH₃Br	0	273.15	0.0054	0.00628
		20	293.15	0.0061	0.00709
		50	323.15	0.0072	0.00837
		100	373.15	0.0091	0.01058
Methyl chloride	CH₃Cl	0	273.15	0.0078	0.00907
		20	293.15	0.0090	0.01047
		50	323.15	0.0108	0.01256
		100	373.15	0.0139	0.01617
		150	423.15	0.0171	0.01989
		200	473.15	0.0207	0.02407

1 kcal/h = 1.163 W

Table 90-4   Thermal Conductivities λ of Gases (Continued)

Gas	Chemical formula	Temperature t °C	Temperature T °K	Thermal conductivity λ kcal/h m K	Thermal conductivity λ W/m K
Methylene chloride	CH_2Cl_2	0	273.15	0.0057	0.00663
		20	293.15	0.0063	0.00733
		50	323.15	0.0073	0.00849
		100	373.15	0.00925	0.01076
		150	423.15	0.0113	0.01314
		200	473.15	0.0135	0.01570
Methyl iodide	CH_3J	0	273.15	0.0040	0.00465
		20	293.15	0.0045	0.00523
		50	323.15	0.0053	0.00616
		100	373.15	0.0065	0.00756
Neon	Ne	−180	93.15	0.0176	0.02052
		−100	173.15	0.0295	0.03433
		0	273.15	0.0392	0.04564
		100	373.15	0.0479	0.05568
Nitric oxide	NO	−75	198.15	0.0150	0.01745
		−50	223.15	0.0168	0.01954
		−25	248.15	0.0186	0.02163
		0	273.15	0.0204	0.02373
		50	323.15	0.0194	0.02261
Nitrogen	N_2	−180	93.15	0.00756	0.00879
		−150	123.15	0.01044	0.01214
		−100	173.15	0.01404	0.01633
		−50	223.15	0.01764	0.02052
		0	273.15	0.02052	0.02386
		20	293.15	0.02196	0.02554
		50	323.15	0.02376	0.02763
		100	373.15	0.02628	0.03056
		150	423.15	0.0285	0.03315
		200	473.15	0.0306	0.03559
		250	523.15	0.0325	0.03780
		300	573.15	0.0342	0.03977
		500	773.15	0.0403	0.04689
Nitrous oxide	N_2O	−75	198.15	0.0098	0.01140
		−50	223.15	0.0109	0.01268

Gas	Chemical formula	Temperature t °C	Temperature T °K	Thermal conductivity λ kcal/h m K	Thermal conductivity λ W/m K
		−25	248.15	0.0120	0.01396
		0	273.15	0.0130	0.01512
		100	373.15	0.0180	0.02093
Oxygen	O_2	−180	93.15	0.00720	0.00837
		−150	123.15	0.01008	0.01172
		−100	173.15	0.01404	0.01633
		50	223.15	0.01764	0.02052
		0	273.15	0.02088	0.02428
		20	293.15	0.02232	0.02596
		50	323.15	0.02448	0.02847
		100	373.15	0.02736	0.03182
		150	423.15	0.03000	0.03489
n−Pentane	C_5H_{12}	0	273.15	0.0110	0.01279
		20	293.15	0.01225	0.01435
Propane	C_3H_8	0	273.15	0.0130	0.01512
		20	293.15	0.0149	0.01733
		100	373.15	0.0225	0.02617
Propylamine	C_3H_9N	6.5	279.65	0.0108	0.01256
Sulfur dioxide	SO_2	0	273.15	0.0072	0.00837
		100	373.15	0.0103	0.01198
Triethylamine	$C_6H_{15}N$	6.5	279.65	0.0097	0.01130
Trimethylamine	C_3H_9N	6.5	279.65	0.0119	0.01382
Water vapor (steam)	H_2O	100	373.15	0.0208	0.02419
		200	473.15	0.0282	0.03280
		300	573.15	0.0367	0.04268
		400	673.15	0.0474	0.05513
		500	773.15	0.0647	0.07525
Xenon	X	0	273.15	0.00432	0.00502

1 kcal/h = 1.163 W

Table 91-1 Thermal Conductivities λ of Diatomic and Triatomic Gases at Various Temperatures

| Temperature | | N$_2$ | | O$_2$ | | CO$_2$ | | H$_2$O | | H$_2$ | |
| t | T | $\lambda \times 10^3$ | | $\lambda \times 10^3$ | | $\lambda \times 10^3$ | | $\lambda \times 10^3$ | | $\lambda \times 10^3$ | |
°C	K	kcal/h m K	W/m K	kcal/h m K	W/m K	kcal/h m K	W/m K	kcal/h m K	W/m K	kcal/h m K	W/m K
0	273.15	21.38	24.865	21.55	24.714	12.42	14.444	13.89	16.154	150	174.450
100	373.15	27.09	31.506	27.99	32.552	19.52	22.702	21.19	24.644	186	216.318
200	473.15	32.30	37.565	34.37	39.972	26.70	31.052	28.94	33.657	222	258.186
300	573.15	37.31	43.392	40.64	47.264	33.86	39.379	39.24	45.636	258	300.054
400	673.15	42.44	49.358	46.65	54.254	40.84	47.497	49.06	57.057	294	341.922
500	773.15	47.47	55.208	52.40	60.941	47.60	55.359	60.16	69.966	330	383.790
600	873.15	52.35	60.883	57.72	67.128	54.07	62.883	72.10	83.852	366	425.658
700	973.15	57.08	66.384	62.82	73.060	60.27	70.094	84.68	98.483	402	467.526
800	1073.15	61.63	71.676	67.69	78.723	66.12	76.898	98.10	114.090	438	509.394
900	1173.15	66.03	76.793	72.00	83.736	71.74	83.434	111.90	130.140	474	551.262
1000	1273.15	70.27	81.724	86.36	88.807	77.10	89.667	126.10	146.654	510	593.130
1100	1373.15	74.29	86.399	80.60	93.738	82.26	95.668	140.50	163.402	546	634.998
1200	1473.15	78.17	90.912	84.60	98.390	87.11	101.309	155.00	180.265	582	676.866

Table 92-1 Thermal Conductivities λ of Flue Gases Containing 13% CO$_2$ at Various Temperatures

Temperature		Water content, per cent (H$_2$O)									
		5		10		15		20		25	
t	T	$\lambda \times 10^3$		$\lambda \times 10^3$		$\lambda \times 10^3$		$\lambda \times 10^3$		$\lambda \times 10^3$	
°C	K	kcal/h m K	W/m K	kcal/h m K	W/m K	kcal/h m K	W/m K	kcal/h m K	W/m K	kcal/h m K	W/m K
0	273.15	19.3	22.446	19.8	23.027	19.9	23.144	19.9	23.144	20.0	23.260
100	373.15	26.6	30.936	27.2	31.634	27.6	32.099	27.8	32.331	28.0	32.564
200	473.15	33.7	39.193	34.5	40.123	35.2	40.938	35.8	41.635	36.1	41.984
300	573.15	42.2	49.079	43.7	50.823	44.9	52.219	45.8	53.265	46.6	54.196
400	673.15	47.0	54.661	48.9	56.871	50.7	58.964	52.1	60.592	53.1	61.755
500	773.15	53.2	61.872	55.9	65.012	58.2	67.687	60.1	69.896	61.6	71.641
600	873.15	59.4	69.082	62.8	73.036	65.7	76.409	68.2	79.317	70.5	81.992
700	973.15	65.0	75.595	69.3	80.596	72.9	84.783	76.1	88.504	78.8	91.644
800	1073.15	70.4	81.875	75.5	87.807	80.1	93.156	84.1	97.808	87.2	101.414
900	1173.15	75.5	87.807	81.7	95.017	87.1	101.297	92.1	107.112	96.3	111.997
1000	1273.15	80.8	93.970	87.8	102.111	94.4	109.787	100	116.300	105	122.115
1100	1373.15	85.8	99.785	93.7	108.973	101	117.463	108	125.604	114	132.582

1 kcal/h = 1.16300 W

Table 93-1 Specific Heat c_p of Dry Air at Various Pressures

Pressure p		1 at = 0.980665 bar		10 at = 9.806650 bar		20 at = 19.613300 bar		40 at = 39.226600 bar		60 at = 58.839900 bar	
Temperature		Specific heat									
		c_p		c_p		c_p		c_p		c_p	
t	T	kcal/kg K	kJ/kg K	kcal/kg K	kJ/kg K	kcal/kg K	kJ/kg K	kcal/kg K	kJ/kg K	kcal/kg K	kJ/kg K
°C	K										
−140	133.15	0.242	1.0132	0.408	1.7082	0.638	2.6712	—	—	—	—
−100	173.15	0.241	1.0090	0.258	1.0802	0.283	1.1849	—	—	—	—
−50	223.15	0.240	1.0048	0.244	1.0216	0.252	1.0551	0.333	1.3942	—	—
0	273.15	0.240	1.0048	—	—	0.249	1.0425	0.274	1.1472	0.266	1.1137
50	323.15	0.240	1.0048	—	—	0.248	1.0383	—	—	0.260	1.0886
100	373.15	0.241	1.0090	—	—	0.247	1.0341	—	—	0.256	1.0718
150	423.15	0.243	1.0174	—	—	0.247	1.0341	—	—	0.253	1.0593
200	473.15	0.245	1.0258	—	—	0.247	1.0341	—	—	0.251	1.0509
280	553.15	0.249	1.0425	—	—	0.247	1.0341	—	—	0.249	1.0425

Table 93-2 Specific Heat c_p of Dry Air at Various Pressures (*Continued*)

Pressure p		70 at = 68.646550 bar		100 at = 98.066500 bar		140 at = 137.293100 bar		180 at = 176.519700 bar		220 at = 215.746300 bar	
Temperature		Specific heat									
		c_p		c_p		c_p		c_p		c_p	
t	T	kcal/kg K	kJ/kg K	kcal/kg K	kJ/kg K	kcal/kg K	kJ/kg K	kcal/kg K	kJ/kg K	kcal/kg K	kJ/kg K
°C	K										
−100	173.15	0.459	1.9217	—	—	—	—	—	—	—	—
−50	223.15	0.313	1.3015	—	—	—	—	—	—	—	—
0	273.15	—	—	0.280	1.1723	—	—	—	—	—	—
50	323.15	—	—	0.272	1.1388	0.282	1.1807	0.290	1.2142	0.296	1.2393
100	373.15	—	—	0.264	1.1053	0.272	1.1388	0.279	1.1681	0.284	1.1891
150	423.15	—	—	0.260	1.0886	0.266	1.1137	0.271	1.1346	0.275	1.1514
200	473.15	—	—	0.257	1.0760	0.260	1.0886	0.265	1.1095	0.269	1.1262
280	553.15	—	—	0.252	1.0551	0.254	1.0634	0.257	1.0760	0.259	1.0844

1 at = 1 kp/cm² = 98066.5 N/m² = 9.80665 N/cm² = 0.980665 bar   1 kcal = 4.1868 kJ

Table 94-1 Properties of Dry Air at Atmospheric Pressure 1

Temperature t (°C)	Temperature T (K)	Density ρ (kg/m³)	Specific heat c_p (kcal/kg K)	Specific heat c_p (kJ/kg K)	Thermal conductivity λ (kcal/hm K)	Thermal conductivity λ (W/m K)	Thermal diffusivity α × 10⁶ (m²/s)	Viscosity η × 10⁶ (kps/m²)	Viscosity η × 10⁶ (Ns/m²)	Kinematic viscosity ν × 10⁶ (m²/s)	Prandtl number Pr
−180	93.15	3.72	0.250	1.047	0.0650	0.0076	1.94	0.66	6.472	1.75	—
−150	123.15	2.78	0.248	1.038	0.0100	0.0116	4.03	0.876	8.591	3.14	—
−100	173.15	1.948	0.244	1.022	0.014	0.0163	8.0	1.21	11.856	5.96	—
−50	223.15	1.534	0.242	1.013	0.017	0.0198	13.1	1.51	14.808	9.65	0.71
−20	253.15	1.365	0.240	1.005	0.0194	0.0226	16.8	1.66	16.279	12.0	0.71
0	273.15	1.252	0.2414	1.011	0.0204	0.0237	19.2	1.78	17.456	13.9	0.71
10	283.15	1.206	0.2413	1.010	0.0210	0.0244	20.7	1.82	17.848	14.66	0.71
20	293.15	1.164	0.2416	1.012	0.0216	0.0251	22.0	1.86	18.240	15.7	0.71
30	303.15	1.127	0.2419	1.013	0.0222	0.0258	23.4	1.905	18.682	16.58	0.71
40	313.15	1.092	0.2422	1.014	0.0228	0.0265	24.8	1.95	19.123	17.6	0.71
50	323.15	1.057	0.2426	1.016	0.0234	0.0272	26.2	1.99	19.515	18.58	0.71
60	333.15	1.025	0.2429	1.017	0.0240	0.0279	27.6	2.03	19.907	19.4	0.71
70	343.15	0.996	0.2432	1.018	0.0246	0.0286	29.2	2.08	20.398	20.65	0.71
80	353.15	0.968	0.2435	1.019	0.0252	0.0293	30.6	2.12	20.790	21.5	0.71
90	3631.5	0.942	0.2438	1.021	0.0258	0.0300	32.2	2.165	21.231	22.82	0.71
100	373.15	0.916	0.2441	1.022	0.0264	0.0307	33.6	2.21	21.673	23.6	0.71
120	393.15	0.870	0.2447	1.025	0.0275	0.0320	37.0	2.30	22.555	25.9	0.71
140	413.15	0.827	0.2453	1.027	0.0286	0.0333	40.0	2.38	23.340	28.2	0.71
150	423.15	0.810	0.2456	1.028	0.0289	0.0336	41.2	2.42	23.732	29.4	0.71
160	433.15	0.789	0.2460	1.030	0.0296	0.0344	43.3	2.46	24.124	30.6	0.71
180	453.15	0.755	0.2466	1.032	0.0307	0.0357	47.0	2.54	24.909	33.00	0.71
200	473.15	0.723	0.2472	1.035	0.0318	0.0370	49.7	2.62	25.693	35.5	0.71
250	523.15	0.653	0.249	1.043	0.0344	0.0400	60.0	2.81	27.557	42.2	0.71
300	573.15	0.596	0.250	1.047	0.0369	0.0429	68.9	2.99	39.322	49.2	0.71
350	623.15	0.549	0.252	1.055	0.0393	0.0457	80.0	3.16	30.989	56.5	0.72
400	673.15	0.508	0.253	1.059	0.0417	0.0485	89.4	3.34	32.754	64.6	0.72
500	773.15	0.442	0.257	1.076	0.0464	0.0540	113.2	3.65	35.794	81.0	0.72
600	873.15	0.391	0.260	1.089	0.0500	0.0581	133.6	3.94	38.638	98.8	0.73
700	973.15	0.351	0.263	1.101	0.0515	0.0599	162.0	4.24	41.580	118.95	0.73
800	1073.15	0.318	0.266	1.114	0.0575	0.0669	182	4.45	43.640	137	0.73
900	1173.15	0.291	0.269	1.126	0.0579	0.0673	216	4.78	46.876	160	0.74
1000	1273.15	0.268	0.722	1.139	0.0655	0.0762	240	4.94	48.445	181	0.74
1100	1373.15	0.248	0.276	1.156	0.0710	0.0826	277	5.22	51.191	206	0.74
1200	1473.15	0.232	0.278	1.164	0.0727	0.0845	301	5.37	52.662	227	0.74
1400	1673.15	0.204	0.284	1.189	0.080	0.0930	370	5.79	56.781	278	0.76
1600	1873.15	0.182	0.291	1.218	0.087	0.1012	447	6.16	60.409	332	0.76
1800	2073.15	0.165	0.297	1.243	0.094	0.1093	—	6.51	63.841	387	—

1 at = 1 kp/cm² = 98076.5 N/m² = 9.80665 N/cm² = 0.980665 bar   1 kcal/h = 1.163 W   1 kcal = 4.1868 kJ

309

Table 95-1 Adiabatic and Polytropic Changes of Condition of Gases for $p_1/p_2 < 1$

$\dfrac{p_1}{p_2}$	for $n =$				for $n =$			
	1.4 (Adiabate)	1.3	1.2	1.1	1.4 (Adiabate)	1.3	1.2	1.1
	eat: $(p_1/p_2)^{1/n} = V_1/V_2 =$				eat: $(p_1/p_2)^{(n-1/n)} = T_1/T_2 =$			
1.1	1.070	1.076	1.083	1.090	1.028	1.022	1.016	1.009
1.2	1.139	1.151	1.164	1.80	1.053	1.043	1.031	1.017
1.3	1.206	1.224	1.244	1.078	1.078	1.062	1.045	1.024
1.4	1.271	1.295	1.323	1.358	1.101	1.081	1.058	1.031
1.5	1.336	1.366	1.401	1.445	1.123	1.098	1.070	1.038
1.6	1.399	1.436	1.479	1.533	1.144	1.115	1.081	1.044
1.7	1.461	1.504	1.557	1.620	1.164	1.130	1.092	1.050
1.8	1.522	1.571	1.633	1.706	1.183	1.145	1.103	1.055
1.9	1.581	1.638	1.706	1.791	1.201	1.160	1.113	1.060
2.0	1.631	1.705	1.782	1.879	1.219	1.174	1.123	1.065
2.5	1.924	2.023	2.145	2.300	1.299	1.235	1.165	1.087
3.0	2.193	2.330	2.498	2.715	1.369	1.289	1.201	1.105
3.5	2.449	2.624	2.842	3.126	1.431	1.336	1.232	1.121
4.0	2.692	2.907	3.177	3.505	1.487	1.378	1.260	1.134
4.5	2.926	3.178	3.500	3.925	1.537	1.415	1.285	1.147
5.0	3.156	3.449	3.824	4.320	1.583	1.449	1.307	1.157
5.5	3.378	3.712	4.142	4.710	1.627	1.482	1.328	1.167
6.0	3.598	3.970	4.447	5.100	1.668	1.512	1.348	1.177
6.5	3.809	4.218	4.760	5.483	1.707	1.540	1.366	1.186
7.0	4.012	4.467	5.058	5.861	1.742	1.566	1.383	1.194
7.5	4.217	4.710	5.360	6.250	1.778	1.591	1.399	1.201
8.0	4.415	4.950	5.650	6.620	1.811	1.616	1.414	1.208
8.5	4.612	5.187	5.950	6.997	1.843	1.639	1.429	1.215
9.0	4.800	5.420	6.240	7.370	1.873	1.660	1.442	1.221
9.5	4.993	5.651	6.528	7.742	1.903	1.681	1.455	1.227
10	5.188	5.885	6.820	8.120	1.931	1.701	1.468	1.233
11	5.544	6.325	7.376	8.845	1.984	1.739	1.491	1.244
12	5.900	6.763	7.931	9.574	2.034	1.774	1.513	1.253
13	6.247	7.193	8.478	10.30	2.081	1.807	1.533	1.263
14	6.587	7.614	9.018	11.01	2.126	1.839	1.549	1.271
15	6.919	8.030	9.551	11.73	2.168	1.868	1.570	1.279
16	7.246	8.438	10.08	12.44	2.208	1.896	1.587	1.287
17	7.566	8.841	10.60	13.14	2.247	1.923	1.604	1.294
18	7.882	9.238	11.12	13.84	2.284	1.948	1.619	1.301
19	8.192	9.631	11.63	14.54	2.319	1.973	1.633	1.307
20	8.498	10.02	12.14	15.23	2.354	1.996	1.648	1.313
21	8.803	10.40	12.64	15.93	2.387	2.019	1.661	1.319
22	9.097	10.78	13.14	16.61	2.418	2.041	1.674	1.324
23	9.390	11.15	13.64	17.30	2.449	2.062	1.688	1.330
24	9.680	11.53	14.13	17.97	2.479	2.082	1.698	1.335
25	9.967	11.89	14.62	18.65	2.508	2.102	1.710	1.340
26	10.25	12.26	15.10	19.34	2.537	2.121	1.721	1.345
27	10.53	12.62	15.58	20.01	2.564	2.140	1.732	1.349
28	10.81	12.98	16.07	20.68	2.591	2.158	1.743	1.354
29	11.08	13.33	16.54	21.36	2.617	2.175	1.753	1.358
30	11.35	13.68	17.02	22.02	2.643	2.192	1.763	1.362
31	11.62	14.03	17.49	22.69	2.667	2.209	1.773	1.366
32	11.89	14.38	17.96	23.35	2.692	2.225	1.782	1.370
33	12.15	14.69	18.43	24.01	2.715	2.241	1.792	1.374
34	12.42	15.06	18.89	24.58	2.739	2.256	1.800	1.378
35	12.67	15.41	19.35	25.34	2.761	2.272	1.809	1.382
36	12.93	15.74	19.81	25.99	2.784	2.287	1.817	1.385
37	13.19	16.07	20.26	26.65	2.806	2.301	1.826	1.389
38	13.44	16.41	20.72	27.30	2.827	2.315	1.834	1.392
39	13.69	16.74	21.18	27.95	2.848	2.329	1.842	1.395
40	13.94	17.07	21.63	28.60	2.869	2.343	1.850	1.398

Table 96-1 Heating Values H_s and H_i of Gases

Gas	Chemical formula	Relative molecular mass M	Density ρ (kg/m³n)	Characteristic σ	Oxygen O_{min} (m³n)	Air L_{min} (m³n)	H_s (kcal/kmol)	H_s (kJ/kmol)	H_i (kcal/kmol)	H_i (kJ/kmol)	H_s (kcal/kg)	H_s (kJ/kg)	H_i (kcal/kg)	H_i (kJ/kg)	H_s (kcal/m³n)	H_s (kJ/m³n)	H_i (kcal/m³n)	H_i (kJ/m³n)
Acetylene	C_2H_2	26.00	1.171	1.25	2.5	11.9	313000	1310468	302240	1265418	12030	50367	11620	48651	14090	58992	13600	56940
Ammonia	NH_3	27.03	0.7714	—	—	—	91000	380999	74870	313466	5340	22358	4400	18422	4120	17205	3390	14193
Benzene	C_6H_6	78.05	3.490	—	6.5	31.0	783000	3278264	750730	3143156	10030	41994	9620	40277	34960	146371	33520	140342
n—Butane	C_4H_{10}	58.08	2.703	1.625	6.5	31.0	687900	2880100	634980	2654677	11840	49572	10920	45720	32010	134019	29510	123552
Butene	C_4H_8	56.06	2.50	1.50	6	28.6	652000	2729794	608980	2549677	11600	48692	10860	45469	29110	121878	27190	113839
Carbon monoxide	CO	28.00	1.250	0.50	0.5	2.38	67700	283446	67700	283446	2420	10132	2420	10132	3020	12644	3020	12644
Ethane	C_2H_6	30.05	1.356	1.75	3.5	16.7	372800	1560839	340530	1425731	12410	51958	11330	47436	16820	70422	15370	64351
Ethylene	C_2H_4	28.00	1.260	1.50	3	14.3	340000	1423512	318490	1333454	12130	50786	11360	47562	15290	64016	14320	59955
Hydrogen	H_2	2.016	0.08987	—	0.5	2.38	68350	286168	57590	241118	33910	141974	28570	119617	3050	12770	2570	10760
Hydrogen sulfide	$H_2S—SO_2$	34.08	1.539	—	—	—	136000	569405	125240	524355	3990	16705	3680	15407	6140	25707	5660	23697
Hydrogen sulfide	$H_2S—SO_3$	34.08	1.539	—	—	—	159600	667795	148740	622745	4680	19594	4360	18254	7200	30145	6720	28135
Isobutane	C_4H_{10}	58.08	2.668	—	—	—	686300	2873401	632520	2648235	11820	49488	10890	45594	31530	132010	29050	121627
Methane	CH_4	16.00	0.7168	2.00	2	9.52	212800	890951	191290	800893	13280	55601	11930	49949	9520	39858	8550	35797
Methyl chloride	CH_3Cl	50.48	2.307	1.666	—	—	170000	711756	153870	644223	3370	14110	3050	12770	7770	32531	7030	29433
Propane	C_3H_8	44.06	2.019	—	5	23.8	530600	2221516	487580	2041400	12040	50409	11070	46348	24320	101823	22350	93575
Propene	C_3H_6	42.95	1.915	1.50	4.5	21.4	495000	2072466	462730	1937358	11770	49279	11000	46055	22540	94370	21070	88216

Table 96-2 Heating Values H_s and H_i of Some Technical Gaseous Heating Fuels

Gas	Relative molecular mass M	Density ρ (kg/m³n)	Characteristic σ	Characteristic v	H_2	CO	CH_4	C_2H_4	CO_2	N_2	H_s (kcal/m³n)	H_s (kJ/m³n)	H_i (kcal/m³n)	H_i (kJ/m³n)
Blast furnace gas	28.2	1.25	0.45	1.67	4	28	—	—	8	60	970	4061	950	3977
Blue water gas	26.6	1.19	0.67	1.97	6	23	3	0.2	5	62	1200	5024	1150	4815
Carbonization gas, stone coal	15.7	0.70	1.81	0.024	27	7	48	13	3	2	7630	31945	6920	28973
Coke oven gas	11.85	0.53	2.11	0.149	50	8	29	4	2	7	5150	21562	4600	19259
Dowson gas	25.1	1.12	0.77	1.57	12	28	3	0.2	2	54	1540	6448	1440	6029
Illuminating gas	11.2	0.50	2.11	0.060	51	8	32	4	2	3	5480	22944	4890	20473
Mond's gas	23.7	1.06	0.84	1.32	25	12	4	0.3	16	43	1550	6490	1390	5820
Water gas	15.9	0.71	0.98	0.063	49	42	0.5	—	5	3	2810	11765	2580	10802

1 kcal = 4.1868 kJ

UNITS AND MEASURES

IMPORTANT NOTE: Users of Liquids, Vapors, and Gases tables are alerted to the fact that the symbol for pressure, *at*, refers to the **technical atmosphere** and not to the **standard atmosphere**, which is designated by *atm*.

The relationship of conversion factors between the two atmospheres is clearly explained in Table 119 on page 337.

Table 97-1   Prefixes Denoting Decimal Multiples or Submultiples

Multiples unit			Example for use	
Prefix	Abbreviation	Significance	Examples	Name
atto	a	$0.000\,000\,000\,000\,000\,001 = 10^{-18}$	$aJ = 10^{-18}$ J	atojoule
femto	f	$0.000\,000\,000\,000\,001 = 10^{-15}$	$fm = 10^{-15}$ m	femtometer
pico	p	$0.000\,000\,000\,001 = 10^{-12}$	$pF = 10^{-12}$ F	picofarad
nano	n	$0.000\,000\,001 = 10^{-9}$	$nA = 10^{-9}$ A	nanoampere
micro	μ	$0.000\,001 = 10^{-6}$	$\mu W = 10^{-6}$ W	microwat
mili	m	$0.001 = 10^{-3}$	$mg = 10^{-3}$ g	miligram
centi	c	$0.01 = 10^{-2}$	$cm = 10^{-2}$ m	centimeter
deci	d	$0.1 = 10^{-1}$	$dl = 10^{-1}$ l	deciliter
deca	da	$10 = 10^{1}$	$dalm = 10^{1}$ lm	decalumen
hecto	h	$100 = 10^{2}$	$hl = 10^{2}$ l	hectoliter
kilo	k	$1\,000 = 10^{3}$	$kN = 10^{3}$ N	kilonewton
mega	M	$1\,000\,000 = 10^{6}$	$M\Omega = 10^{6}$ Ω	megaohm
giga	G	$1\,000\,000\,000 = 10^{9}$	$GJ = 10^{9}$ J	gigajoule
tera	T	$1\,000\,000\,000\,000 = 10^{12}$	$TWh = 10^{12}$ Wh	teraliter

Table 98-1   Powers of Ten (10)

$10^{25} =$	$10\,000\,000\,000\,000\,000\,000\,000\,000$	10^{-1}	$= 0.1$
$10^{24} =$	$1\,000\,000\,000\,000\,000\,000\,000\,000$	10^{-2}	$= 0.01$
$10^{23} =$	$100\,000\,000\,000\,000\,000\,000\,000$	10^{-3}	$= 0.001$
$10^{22} =$	$10\,000\,000\,000\,000\,000\,000\,000$	10^{-4}	$= 0.000\,1$
$10^{21} =$	$1\,000\,000\,000\,000\,000\,000\,000$	10^{-5}	$= 0.000\,01$
$10^{20} =$	$100\,000\,000\,000\,000\,000\,000$	10^{-6}	$= 0.000\,001$
$10^{19} =$	$10\,000\,000\,000\,000\,000\,000$	10^{-7}	$= 0.000\,000\,1$
$10^{18} =$	$1\,000\,000\,000\,000\,000\,000$	10^{-8}	$= 0.000\,000\,01$
$10^{17} =$	$100\,000\,000\,000\,000\,000$	10^{-9}	$= 0.000\,000\,001$
$10^{16} =$	$10\,000\,000\,000\,000\,000$	10^{-10}	$= 0.000\,000\,000\,1$
$10^{15} =$	$1\,000\,000\,000\,000\,000$	10^{-11}	$= 0.000\,000\,000\,01$
$10^{14} =$	$100\,000\,000\,000\,000$	10^{-12}	$= 0.000\,000\,000\,001$
$10^{13} =$	$10\,000\,000\,000\,000$	10^{-13}	$= 0.000\,000\,000\,000\,1$
$10^{12} =$	$1\,000\,000\,000\,000$	10^{-14}	$= 0.000\,000\,000\,000\,01$
$10^{11} =$	$100\,000\,000\,000$	10^{-15}	$= 0.000\,000\,000\,000\,001$
$10^{10} =$	$10\,000\,000\,000$	10^{-16}	$= 0.000\,000\,000\,000\,000\,1$
$10^{9} =$	$1\,000\,000\,000$	10^{-17}	$= 0.000\,000\,000\,000\,000\,01$
$10^{8} =$	$100\,000\,000$	10^{-18}	$= 0.000\,000\,000\,000\,000\,001$
$10^{7} =$	$10\,000\,000$	10^{-19}	$= 0.000\,000\,000\,000\,000\,000\,1$
$10^{6} =$	$1\,000\,000$	10^{-20}	$= 0.000\,000\,000\,000\,000\,000\,01$
$10^{5} =$	$100\,000$	10^{-21}	$= 0.000\,000\,000\,000\,000\,000\,001$
$10^{4} =$	$10\,000$	10^{-22}	$= 0.000\,000\,000\,000\,000\,000\,000\,1$
$10^{3} =$	$1\,000$	10^{-23}	$= 0.000\,000\,000\,000\,000\,000\,000\,01$
$10^{2} =$	100	10^{-24}	$= 0.000\,000\,000\,000\,000\,000\,000\,001$
$10^{1} =$	10	10^{-25}	$= 0.000\,000\,000\,000\,000\,000\,000\,000\,1$
$10^{0} =$	1		

Table 99-1 Basic Units of the International System of Units—SI

No	Quantity		Unit		Definition
	Term	Symbol	Name	Abbreviation	
1	Length	l, s, a	meter	m	The meter is the length equal to 1,650,763.73 wave lengths in vacuum of the radiation corresponding to the transition between the levels $2_{p_{10}}$ and 5_{d_5} of the krypton-86 atom.
2	Mass	m, M	kilogram	kg	The kilogram is the fundamental unit of mass in the International System.
3	Time	t, τ	second	s	The second is the duration of 9,192,631,770 periods of the radiation corresponding to the transition between the two hyperfine levels of the ground state of the caesium-133 atom.
4	Current	I	ampere	A	The ampere is a constant current which, if maintained in two straight parallel conductors of infinite length, of negligible circular cross-section, and placed 1 meter apart in vacuum, produces a force equal to 2×10^{-7} newton per meter of length between these conductors.
5	Thermodynamic temperature	T	kelvin	K	The kelvin, a unit of thermodynamic temperature, is the fraction 1/273.16 of the thermodynamic temperature of the triple point of water.
6	Luminous intensity	I, I_v	candela	cd	The candela is the luminous intensity, in the perpendicular direction, of a surface of 1/600,000 square meter of a black body at the temperature of freezing platinum under a pressure of 101,325 newtons per square meter.

Table 100-1 Supplementary Units of the International System of Units—SI

No	Quantity		Unit		Definition
	Term	Symbol	Name	Abbreviation	
1	Plane angle	$\alpha, \beta, \gamma, \ldots$	radian	rad	The radian is a unit of plane angular measurement equal to the angle at the center of a circle subtended by an arc equal in length to the radius.
2	Solid angle	Ω, ω	steradian	sr	The steradian is a unit of measure of solid angle which, having its vertex in the center of a sphere, cuts off an area of the surface of the sphere equal to that of a square having sides equal in length to the radius of the sphere.

Table 101-1 Derived Units of the International System of Units—SI

No	Quantity		Unit		Definition
	Term	Symbol	Name	Abbreviation	
1	Area	A, S	square meter	m^2	The square meter is a unit of area equivalent to the area of a square having sides equal to one meter.
2	Volume	V, τ	cubic meter	m^3	The cubic meter is a unit of volume equivalent to that of a cube having edges equal to one meter.
3	Linear density	—	kilogram per meter	kg/m	The kilogram per meter is the linear density of a homogeneous body which, having a constant cross sectional area over the whole of its length, has a mass of 1 kg for each meter of length.
4	Surface density	—	kilogram per square meter	kg/m^2	The kilogram per square meter is the surface density of a homogeneous body which, having a constant thickness, has a mass of 1 kg for each square meter of area.
5	Density	ρ	kilogram per cubic meter	kg/m^3	The kilogram per cubic meter is the density of a homogeneous material containing, in 1 cubic meter, a mass of 1 kilogram.
6	Frequency	f, υ	hertz (cycle per second)	Hz	The hertz (cycle per second) is the frequency of a periodic phenomenon, equal to one second.
7	Velocity (linear)	v, u, c, w	meter per second	m/s	The meter per second is the velocity (speed) of a body which in uniform motion traverses 1 meter in 1 second.
8	Acceleration	a	meter per second squared	m/s^2	The meter per second squared is the acceleration of a moving body which undergoes a velocity change of 1 meter per second.
9	Angular velocity	ω	radian per second	rad/s	The radian per second is the angular velocity of a body in uniform rotation which in 1 s turns by an angle of 1 radian around the axis of revolution.
10	Angular acceleration	ϵ	radian by second squared	rd/s^2	The radian by second squared is the angular acceleration of a body whose angular velocity undergoes the uniform change of 1 m/s.
11	Volume discharge	Q	cubic meter per second	m^3/s	The cubic meter per second is the volume discharge of a homogeneous fluid having the volume of 1 m^3 and flowing in uniform flow through a certain cross section.
12	Mass discharge	q	kilogram per second	kg/s	The kilogram per second is the mass discharge of a homogeneous fluid having the mass of 1 kg and flowing in uniform flow through a given cross section.
13	Force	F	newton	N	The newton is the force that imparts acceleration of 1 m/s^2 to a body having a mass of 1 kg.

Table 101-2 Derived Units of the International System of Units—SI (Continued)

No	Quantity		Unit		Definition
	Term	Symbol	Name	Abbreviation	
14	Pressure	p	pascal	Pa	The pascal is the uniformly exerted pressure on a surface when a force of 1 N is acting at right angles to the area of 1 m².
15	Dynamic (absolute) viscosity	η, μ	pascal second	Pa s	The pascal second is the dynamic (absolute) viscosity of a fluid in (laminar) flow in which a tangential stress of 1 Pa is set up when a velocity gradient of 1 s⁻¹, perpendicular to the direction of flow, obtains.
16	Kinematic viscosity	υ	meter squared per second	m²/s	The meter squared per second is the kinematic viscosity of a fluid having a density of 1 kg/m³ and a dynamic viscosity of 1 Pa s.
17	Energy, work heat	A, W	joule	J	The joule is the work performed when the point of application of the force is moved 1 m in the direction of the force.
18	Power	P	watt	W	The watt is the power dissipated when the energy of 1 J is expended in 1 s.
19	Potential difference	U	volt	V	The volt is a unit of electric potential and electromotive force, equal to the difference in electrical potential between two points on a conducting wire carrying a constant current of 1 A when the power dissipated between these points is 1 W.
20	Resistance	R	ohm	Ω	The ohm is a unit of electric resistance equal to that of a conductor in which a current of one A is produced by a potential of 1 V across its terminals.
21	Conductivity	G	siemens	S	The siemens is the conductivity of a conductor that has the resistance of 1 ohm.
22	Electrical charge, quantity of electricity	Q, q	coulomb	C	The coulomb is the quantity of electricity transported in 1 s through a given cross section of a conductor that carries a constant current of 1 A.
23	Capacitance	C	farad	F	The farad is the capacitance of a capacitor when a charge of 1 C raises the potential difference between its plates to 1 V.
24	Electric flux density, electric displacement	D	coulomb per square meter	C/m²	The coulomb per square meter is the electric flux density or displacement in a plate capacitor whose two parallel plates of infinite area, arranged in a vacuum, are uniformly charged per square meter of area with 1 C of electricity.

No	Quantity		Unit		Definition
	Term	Symbol	Name	Abbreviation	
25	Electric field strength, electric (field) intensity	E	volt per meter	V/m	The volt per meter is the field strength of a homogeneous electric field in which the potential difference between two points 1 m apart in the direction of the field is equal to 1 V.
26	Magnetic flux	Φ	weber	Wb	The weber is the magnetic flux which, uniformly changed and reduced to zero in the course of 1 s, induces, in a single turn of conductor linked by it, the potential difference of 1 V.
27	Magnetic flux density, magnetic induction	B	tesla	T	The tesla is the surface density of a homogeneous magnetic flux of 1 Wb which traverses the area of 1 m² at right angles.
28	Inductance, self-inductance	L	henry	H	The henry is the inductance of a closed single turn of conductor which carries a current of 1 A and is linked in a vacuum by a magnetic flux of 1 Wb.
29	Magnetic field strength, magnetic (field) intensity	H	ampere per meter	A/m	The ampere per meter is the magnetic field strength that a current of 1 A, carried by a straight conductor of infinite length and a circular cross section, is established in a vacuum on the border of a circular plane, concentric to the cross section of the conductor and has a perimeter of 1 m.
30	Luminance	B	candela per square meter	cd/m²	The candela per square meter is the 600000th part of the luminance of a full radiator at the temperature of solidification of platinum under a pressure of 101.325 Pa.
31	Luminous flux	Φ	lumen	lm	The lumen is the luminous flux emitted from a point source of light of uniform intensity of 1 cd into unit solid angle of 1 sr.
32	Illumination	E	lux	lx	The lux is the illumination obtaining on a surface when the luminous flux of 1 lm falls evenly distributed onto each square meter of area of the surface.

Length equivalents

The unit of length in the International System (SI) is the meter, abbreviation m; 1 m = 1.09361 yd.
The Anglo-American unit of length is the yard, abbreviation yd; 1 yd = 0.9144 m.

Table 102-1 Conversion Factors for Units of Length

Unit	Abbreviation	m	dm	cm	mm	μm	km	in.	ft
1 meter =	m	**1**	**10^1**	**10^2**	**10^3**	**10^6**	**10^{-3}**	39.370 1	3.280 84
1 decimeter =	dm	**10^{-1}**	**1**	**10^1**	**10^2**	**10^5**	**10^{-4}**	39.370 1 · 10^{-1}	3.280 84 · 10^{-1}
1 centimeter =	cm	**10^{-2}**	**10^{-1}**	**1**	**10^1**	**10^4**	**10^{-5}**	39.370 1 · 10^{-2}	3.280 84 · 10^{-2}
1 millimeter =	mm	**10^{-3}**	**10^{-2}**	**10^{-1}**	**1**	**10^3**	**10^{-6}**	39.370 1 · 10^{-3}	3.280 84 · 10^{-3}
1 micrometer =	μm	**10^{-6}**	**10^{-5}**	**10^{-4}**	**10^{-3}**	**1**	**10^{-9}**	39.370 1 · 10^{-6}	3.280 84 · 10^{-6}
1 kilometer =	km	**10^3**	**10^4**	**10^5**	**10^6**	**10^9**	**1**	39.370 1 · 10^3	3.280 84 · 10^3
1 inch =	in.	**2.54 · 10^{-2}**	**2.54 · 10^{-1}**	**2.54**	**2.54 · 10^1**	**2.54 · 10^4**	**2.54 · 10^{-5}**	1	0.083 333 3
1 foot =	ft	**3.048 · 10^{-1}**	**3.048**	**3.048 · 10^1**	**3.048 · 10^2**	**3.048 · 10^5**	**3.048 · 10^{-4}**	12	1
1 yard =	yd	**9.144 · 10^{-1}**	**9.144**	**9.144 · 10^1**	**9.144 · 10^2**	**9.144 · 10^5**	**9.144 · 10^{-4}**	36	3
1 fathom =	fathom	**1.828 8**	**1.828 8 · 10^1**	**1.828 8 · 10^2**	**1.828 8 · 10^3**	**1.828 8 · 10^6**	**1.828 8 · 10^{-3}**	72	6
1 rod =	rod	**5.029 2**	**5.029 2 · 10^1**	**5.029 2 · 10^2**	**5.029 2 · 10^3**	**5.029 2 · 10^6**	**5.029 2 · 10^{-3}**	198	16.5
1 chain =	chain	**20.116 8**	**20.116 8 · 10^1**	**20.116 8 · 10^2**	**20.116 8 · 10^3**	**20.116 8 · 10^6**	**20.116 8 · 10^{-3}**	792	66
1 furlong =	furlong	**201.168**	**201.168 · 10^1**	**201.168 · 10^2**	**201.168 · 10^3**	**201.168 · 10^6**	**201.168 · 10^{-3}**	7 920	660
1 mile =	mile	**1 609.344**	**1 609.344 · 10^1**	**1 609.344 · 10^2**	**1 609.344 · 10^3**	**1 609.344 · 10^6**	**1 609.344 · 10^{-3}**	63 360	5 280
1 UK nautical mile =	UK mile	**1 853.18**	**1 853.18 · 10^1**	**1 853.18 · 10^2**	**1 853.18 · 10^3**	**1 853.18 · 10^6**	**1 853.18 · 10^{-3}**	72 960	6 080
1 international nautic. mile* =	n mile*	**1852**	**1852 · 10^1**	**1852 · 10^2**	**1852 · 10^3**	**1852 · 10^6**	**1852 · 10^{-3}**	729 13.39	6 076.12

Unit	Abbreviation	yd	fathom	rod	chain	furlong	mile	UK mile	n mile*
1 meter =	m	1.093 61	5.468 07 · 10^{-1}	1.988 39 · 10^{-1}	4.970 97 · 10^{-2}	4.970 97 · 10^{-3}	6.213 71 · 10^{-4}	5.396 12 · 10^{-4}	5.399 57 · 10^{-4}
1 decimeter =	dm	1.093 61 · 10^{-1}	5.468 07 · 10^{-2}	1.988 39 · 10^{-2}	4.970 97 · 10^{-3}	4.970 97 · 10^{-4}	6.213 71 · 10^{-5}	5.396 12 · 10^{-5}	5.399 57 · 10^{-5}
1 centimeter =	cm	1.093 61 · 10^{-2}	5.468 07 · 10^{-3}	1.988 39 · 10^{-3}	4.970 97 · 10^{-4}	4.970 97 · 10^{-5}	6.213 71 · 10^{-6}	5.396 12 · 10^{-6}	5.399 57 · 10^{-6}
1 millimeter =	mm	1.093 61 · 10^{-3}	5.468 07 · 10^{-4}	1.988 39 · 10^{-4}	4.970 97 · 10^{-5}	4.970 97 · 10^{-6}	6.213 71 · 10^{-7}	5.396 12 · 10^{-7}	5.399 57 · 10^{-7}
1 micrometer =	μm	1.093 61 · 10^{-6}	5.468 07 · 10^{-7}	1.988 39 · 10^{-7}	4.970 97 · 10^{-8}	4.970 97 · 10^{-9}	6.213 71 · 10^{-10}	5.396 12 · 10^{-10}	5.399 57 · 10^{-10}
1 kilometer =	km	1.093 61 · 10^3	5.468 07 · 10^2	1.988 39 · 10^2	4.970 97 · 10^1	4.970 97	6.213 71 · 10^{-1}	5.396 12 · 10^{-1}	5.399 57 · 10^{-1}
1 inch =	in.	0.027 777 8	0.013 888 9	5.050 51 · 10^{-3}	1.262 63 · 10^{-3}	1.262 63 · 10^{-4}	1.578 28 · 10^{-5}	1.370 62 · 10^{-5}	1.371 49 · 10^{-5}
1 foot =	ft	0.333 333	0.166 667	0.060 606 1	0.015 151 5	1.515 15 · 10^{-3}	1.893 94 · 10^{-4}	1.644 74 · 10^{-4}	1.645 79 · 10^{-4}
1 yard =	yd	1	0.5	0.181 818	0.045 454 5	4.545 45 · 10^3	5.681 82 · 10^{-4}	4.934 21 · 10^{-4}	4.937 37 · 10^{-4}
1 fathom =	fathom	2	1	0.363 636	0.090 909 1	9.090 91 · 10^{-3}	1.136 36 · 10^{-3}	9.868 42 · 10^{-4}	9.874 73 · 10^{-4}
1 rod =	rod	5.5	2.75	1	0.25	0.025	0.003 125	2.713 82 · 10^{-3}	2.715 55 · 10^{-3}
1 chain =	chain	22	11	4	1	0.1	0.012 5	0.010 855 3	0.010 862 2
1 furlong =	furlong	220	110	40	10	1	0.125	0.108 553	0.108 622
1 mile =	mile	1 760	880	320	80	8	1	0.868 421	0.868 976
1 UK nautical mile =	UK mile	2 026.67	1 013.33	368.484 8	92.121 2	9.212 12	1.151 52	1	1.000 64
1 international nautic. mile* =	n mile*	2 025.37	1 012.69	368.249	92.062 4	9.206 24	1.150 78	0.999 361	1

Example: 1 m = 10^3 mm = 1000 mm; 1 yd = 9.144 · 10^{-1} m = 0.914 4 m;
1 rod = 2.713 82 · 10^{-3} UK mile = 0.002 713 82 UK mile;
1 fathom = 9.874 73 · 10^{-4} n mile = 0.000 987 473 n mile

Exact values are printed in bold type.
The United Kingdom units are denoted by the prefix UK
The American units are denoted by the prefix US
*International nautical mile = US nautical mile (1 n mile = 1 US mile = 1852 m)

Some other Anglo-American units of length are:

1 mil = 0.001 inch
1 point (printers) = 1/2592 yard
1 line (button) = 1/1440 yard
1 hand = 1/9 yard
1 link = 22/100 yard
1 span = 9 inch

Table 102-2 The Metric Units of Length

Unit		Abbreviation	Tm	Gm	Mm	km	hm	dam	m	dm	cm	mm	µm	nm	pm	Å*
1 terameter	=	Tm	1	10^3	10^6	10^9	10^{10}	10^{11}	10^{12}	10^{13}	10^{14}	10^{15}	10^{18}	10^{21}	10^{24}	10^{22}
1 gigameter	=	Gm	10^{-3}	1	10^3	10^6	10^7	10^8	10^9	10^{10}	10^{11}	10^{12}	10^{15}	10^{18}	10^{21}	10^{19}
1 megameter	=	Mm	10^{-6}	10^{-3}	1	10^3	10^4	10^5	10^6	10^7	10^8	10^9	10^{12}	10^{15}	10^{18}	10^{16}
1 kilometer	=	km	10^{-9}	10^{-6}	10^{-3}	1	10^1	10^2	10^3	10^4	10^5	10^6	10^9	10^{12}	10^{15}	10^{13}
1 hectometer	=	hm	10^{-10}	10^{-7}	10^{-4}	10^{-1}	1	10^1	10^2	10^3	10^4	10^5	10^8	10^{11}	10^{14}	10^{12}
1 decameter	=	dam	10^{-11}	10^{-8}	10^{-5}	10^{-2}	10^{-1}	1	10^1	10^2	10^3	10^4	10^7	10^{10}	10^{13}	10^{11}
1 meter	=	m	10^{-12}	10^{-9}	10^{-6}	10^{-3}	10^{-2}	10^{-1}	1	10^1	10^2	10^3	10^6	10^9	10^{12}	10^{10}
1 decimeter	=	dm	10^{-13}	10^{-10}	10^{-7}	10^{-4}	10^{-3}	10^{-2}	10^{-1}	1	10^1	10^2	10^5	10^8	10^{11}	10^9
1 centimeter	=	cm	10^{-14}	10^{-11}	10^{-8}	10^{-5}	10^{-4}	10^{-3}	10^{-2}	10^{-1}	1	10^1	10^4	10^7	10^{10}	10^8
1 millimeter	=	mm	10^{-15}	10^{-12}	10^{-9}	10^{-6}	10^{-5}	10^{-4}	10^{-3}	10^{-2}	10^{-1}	1	10^3	10^6	10^9	10^7
1 micrometer	=	µm	10^{-18}	10^{-15}	10^{-12}	10^{-9}	10^{-8}	10^{-7}	10^{-6}	10^{-5}	10^{-4}	10^{-3}	1	10^3	10^6	10^4
1 nanometer	=	nm	10^{-21}	10^{-18}	10^{-15}	10^{-12}	10^{-11}	10^{-10}	10^{-9}	10^{-8}	10^{-7}	10^{-6}	10^{-3}	1	10^3	10^1
1 picometer	=	pm	10^{-24}	10^{-21}	10^{-18}	10^{-15}	10^{-14}	10^{-13}	10^{-12}	10^{-11}	10^{-10}	10^{-9}	10^{-6}	10^{-3}	1	10^{-2}
1 angstrom	=	Å*	10^{-22}	10^{-19}	10^{-16}	10^{-13}	10^{-12}	10^{-11}	10^{-10}	10^{-9}	10^{-8}	10^{-7}	10^{-4}	10^{-1}	10^{-2}	1

Example: 1 Tm = 10^{12} m = 1 000 000 000 000 m; 1 km = 10^5 cm = 100 000 cm;
1 m = 10^{-6} Mm = 0.000 001 Mm; 1 nm = 10^{-3} µm = 0.001 µm.

Table 103-1 Angular Measure Equivalents: Conversion Factors for Angular Units

Unit		Abbreviation	radiant* rad*	right angle ⌐	degree °	minute '	second "	gon g	new minute c	new second cc
1 radian*	=	rad*	1	0.636 619 8	57.295 78	3437.747	206 264.8	63.661 98	$63.661\ 98 \cdot 10^2$	$63.661\ 98 \cdot 10^4$
1 right angle = $\frac{\pi}{2}$ radian	=	⌐	1.570 796	1	90	5 400	324 000	10^2	10^4	10^6
1 degree = $\frac{\pi}{180}$ radian	=	°	0.017 453 29	0.011 111 1	1	60	3 600	1.111 111	$1.111\ 111 \cdot 10^2$	$1.111\ 111 \cdot 10^4$
1 minute = $\frac{1}{60}$ degree	=	'	$2.908\ 882 \cdot 10^{-4}$	$1.851\ 852 \cdot 10^{-4}$	0.016 666 67	1	60	$1.851\ 852 \cdot 10^{-2}$	1.851 852	$1.851\ 850 \cdot 10^2$
1 second = $\frac{1}{60}$ minute	=	"	$4.848\ 137 \cdot 10^{-6}$	$3.086\ 420 \cdot 10^{-6}$	$2.777\ 778 \cdot 10^{-4}$	0.016 666 67	1	$3.086\ 420 \cdot 10^{-4}$	$3.086\ 20 \cdot 10^{-2}$	3.086 420
1 gon grade = $\frac{\pi}{200}$ radian	=	g	$1.570\ 796 \cdot 10^{-2}$	10^{-2}	0.9	54	3 240	1	10^2	10^4
1 new minute = $\frac{1}{100}$ gon	=	c	$1.570\ 796 \cdot 10^{-4}$	10^{-4}	$0.9 \cdot 10^{-2}$	$54 \cdot 10^{-2}$	$3\ 240 \cdot 10^{-2}$	10^{-2}	1	10^2
1 new second = $\frac{1}{10000}$ gon	=	cc	$1.570\ 796 \cdot 10^{-6}$	10^{-6}	$0.9 \cdot 10^{-4}$	$54 \cdot 10^{-4}$	$3\ 200 \cdot 10^{-4}$	10^{-4}	10^{-2}	1

Example: 1° = 3 600"; 1 g = $1.570\ 796 \cdot 10^{-2}$ rad = 0.015 707 96 rad.
Exact values are printed in bold type.
*1 radian (rad) = 360°/2π = 57° 17'45".

The unit of area in the International System (SI) is the square meter, abbreviation m²; 1 m² = 1.19599 yd².
The Anglo-American unit of area is the square yard, abbreviation yd² or sq yd; 1 yd² = 0.836 127 36 m².

Table 104-1 Conversion Factors for Units of Area

Unit		Abbreviation	m^2	dm^2	cm^2	mm^2	a	ha	km^2	$in.^2$
1 square meter	=	m^2	1	10^2	10^4	10^6	10^{-2}	10^{-4}	10^{-6}	1 550
1 square decimeter	=	dm^2	10^{-2}	1	10^2	10^4	10^{-4}	10^{-6}	10^{-8}	$1\,550 \cdot 10^{-2}$
1 square centimeter	=	cm^2	10^{-4}	10^{-2}	1	10^2	10^{-6}	10^{-8}	10^{-10}	$1\,550 \cdot 10^{-4}$
1 square millimeter	=	mm^2	10^{-6}	10^{-4}	10^{-2}	1	10^{-8}	10^{-10}	10^{-12}	$1\,550 \cdot 10^{-6}$
1 are	=	a	10^2	10^4	10^6	10^8	1	10^{-2}	10^{-4}	$1\,550 \cdot 10^2$
1 square hectare	=	ha	10^4	10^6	10^8	10^{10}	10^2	1	10^{-2}	$1\,550 \cdot 10^4$
1 square kilometer	=	km^2	10^6	10^8	10^{10}	10^{12}	10^4	10^2	1	$1\,550 \cdot 10^6$
1 square inch	=	$in.^2$	$6.451\,6 \cdot 10^{-4}$	$6.451\,6 \cdot 10^{-2}$	$6.451\,6$	$6.451\,6 \cdot 10^2$	$6.451\,6 \cdot 10^{-6}$	$6.451\,6 \cdot 10^{-8}$	$6.451\,6 \cdot 10^{-10}$	1
1 square foot	=	ft^2	$9.290\,30 \cdot 10^{-2}$	$9.290\,30$	$9.290\,30 \cdot 10^2$	$9.290\,30 \cdot 10^4$	$9.290\,30 \cdot 10^{-4}$	$9.290\,30 \cdot 10^{-6}$	$9.290\,30 \cdot 10^{-8}$	144
1 square yard	=	yd^2	$8.361\,27 \cdot 10^{-1}$	$8.361\,27 \cdot 10^1$	$8.361\,27 \cdot 10^3$	$8.361\,27 \cdot 10^5$	$8.361\,27 \cdot 10^{-3}$	$8.361\,27 \cdot 10^{-5}$	$8.361\,27 \cdot 10^{-7}$	1 296
1 circular mil	=	circ. mil	$5.067\,07 \cdot 10^{-10}$	$5.067\,07 \cdot 10^{-8}$	$5.067\,07 \cdot 10^{-6}$	$5.067\,07 \cdot 10^{-4}$	$5.067\,07 \cdot 10^{-12}$	$5.067\,07 \cdot 10^{-14}$	$5.067\,07 \cdot 10^{-16}$	$7.853\,98 \cdot 10^{-7}$
1 circular inch	=	circ. inch	$5.067\,07 \cdot 10^{-4}$	$5.067\,07 \cdot 10^{-2}$	$5.067\,07$	$5.067\,07 \cdot 10^2$	$5.067\,07 \cdot 10^{-6}$	$5.067\,07 \cdot 10^{-8}$	$5.067\,07 \cdot 10^{-10}$	$7.853\,98 \cdot 10^{-1}$
1 square rod	=	sq. rod	$25.292\,9$	$25.292\,9 \cdot 10^2$	$25.292\,9 \cdot 10^4$	$25.292\,9 \cdot 10^6$	$25.292\,9 \cdot 10^{-2}$	$25.292\,9 \cdot 10^{-4}$	$25.292\,9 \cdot 10^{-6}$	39 204
1 rood	=	rood	$1\,011.71$	$1\,011.71 \cdot 10^2$	$1\,011.71 \cdot 10^4$	$1\,011.71 \cdot 10^6$	$1\,011.71 \cdot 10^{-2}$	$1\,011.71 \cdot 10^{-4}$	$1\,011.71 \cdot 10^{-6}$	1 568 160
1 acre	=	acre	$4\,046.86$	$4\,046.86 \cdot 10^2$	$4\,046.86 \cdot 10^4$	$4\,046.86 \cdot 10^6$	$4\,046.86 \cdot 10^{-2}$	$4\,046.86 \cdot 10^{-4}$	$4\,046.86 \cdot 10^{-6}$	6 272 640
1 square mile	=	sq. mile	$25\,899.9 \cdot 10^2$	$25\,899.9 \cdot 10^4$	$25\,899.9 \cdot 10^6$	$25\,899.9 \cdot 10^8$	$25\,899.9$	$25\,899.9 \cdot 10^{-2}$	$25\,899.9 \cdot 10^{-4}$	4 014 489 600

Unit		Abbreviation	ft^2	yd^2	circ. mil	circ. inch	sq. rod	rood	acre	sq. mile
1 square meter	=	m^2	$10.763\,9$	$1.195\,99$	$19.735\,3 \cdot 10^8$	$19.735\,3 \cdot 10^2$	$3.953\,69 \cdot 10^{-2}$	$9.884\,22 \cdot 10^{-4}$	$2.471\,05 \cdot 10^{-4}$	$3.861\,02 \cdot 10^{-7}$
1 square decimeter	=	dm^2	$10.763\,9 \cdot 10^{-2}$	$1.195\,99 \cdot 10^{-2}$	$19.735\,3 \cdot 10^6$	$19.735\,3$	$3.953\,69 \cdot 10^{-4}$	$9.884\,22 \cdot 10^{-6}$	$2.471\,05 \cdot 10^{-6}$	$3.861\,02 \cdot 10^{-9}$
1 square centimeter	=	cm^2	$10.763\,9 \cdot 10^{-4}$	$1.195\,99 \cdot 10^{-4}$	$19.735\,3 \cdot 10^4$	$19.735\,3 \cdot 10^{-2}$	$3.953\,69 \cdot 10^{-6}$	$9.884\,22 \cdot 10^{-8}$	$2.471\,05 \cdot 10^{-8}$	$3.861\,02 \cdot 10^{-11}$
1 square millimeter	=	mm^2	$10.763\,9 \cdot 10^{-6}$	$1.195\,99 \cdot 10^{-6}$	$19.735\,3 \cdot 10^2$	$19.735\,3 \cdot 10^{-4}$	$3.953\,69 \cdot 10^{-8}$	$9.884\,22 \cdot 10^{-10}$	$2.471\,05 \cdot 10^{-10}$	$3.861\,02 \cdot 10^{-13}$
1 are	=	a	$10.763\,9 \cdot 10^2$	$1.195\,99 \cdot 10^2$	$19.753\,3 \cdot 10^{10}$	$19.735\,3 \cdot 10^4$	$3.953\,69$	$9.884\,22 \cdot 10^{-2}$	$2.471\,05 \cdot 10^{-2}$	$3.861\,02 \cdot 10^{-5}$
1 square hectare	=	ha	$10.763\,9 \cdot 10^4$	$1.195\,99 \cdot 10^4$	$19.735\,3 \cdot 10^{12}$	$19.735\,3 \cdot 10^6$	$3.953\,69 \cdot 10^2$	$9.884\,22$	$2.471\,05$	$3.861\,02 \cdot 10^{-3}$
1 square kilometer	=	km^2	$10.763\,9 \cdot 10^6$	$1.195\,99 \cdot 10^6$	$19.735\,3 \cdot 10^{14}$	$19.735\,3 \cdot 10^8$	$3.953\,69 \cdot 10^4$	$9.884\,22 \cdot 10^2$	$2.471\,05 \cdot 10^2$	$3.861\,02 \cdot 10^{-1}$
1 square inch	=	$in.^2$	$6.944\,44 \cdot 10^{-3}$	$7.716\,05 \cdot 10^{-4}$	$1.273\,24 \cdot 10^6$	$1.273\,24$	$3.935\,69 \cdot 10^{-5}$	$6.376\,90 \cdot 10^{-7}$	$1.594\,23 \cdot 10^{-7}$	$2.490\,98 \cdot 10^{-10}$
1 square foot	=	ft^2	1	$0.111\,111$	$183.347 \cdot 10^6$	183.347	$3.673\,09 \cdot 10^{-3}$	$9.182\,74 \cdot 10^{-5}$	$2.295\,66 \cdot 10^{-5}$	$3.587\,01 \cdot 10^{-8}$
1 square yard	=	yd^2	9	1	$1\,650.12 \cdot 10^6$	$1\,650.12$	$0.033\,057\,8$	$8.264\,46 \cdot 10^{-4}$	$2.066\,12 \cdot 10^{-4}$	$3.228\,31 \cdot 10^{-7}$
1 circular mil	=	circ. mil	$5.454\,15 \cdot 10^{-9}$	$6.060\,17 \cdot 10^{-10}$	1	10^{-6}	$2.003\,36 \cdot 10^{-11}$	$5.008\,40 \cdot 10^{-13}$	$1.252\,10 \cdot 10^{-13}$	$1.956\,41 \cdot 10^{-16}$
1 circular inch	=	circ. inch	$5.454\,15 \cdot 10^{-3}$	$6.060\,17 \cdot 10^{-4}$	10^6	1	$2.003\,36 \cdot 10^{-5}$	$5.008\,40 \cdot 10^{-7}$	$1.252\,10 \cdot 10^{-7}$	$1.956\,41 \cdot 10^{-10}$
1 square rod	=	sq. rod	272.25	30.25	$49\,916.1 \cdot 10^6$	$49\,916.1$	1	0.025	$0.006\,25$	$9.765\,62 \cdot 10^{-6}$
1 rood	=	rood	$108\,90$	$1\,210$	$1\,996\,644 \cdot 10^6$	$1\,996\,644$	40	1	0.25	$3.906\,25 \cdot 10^{-4}$
1 acre	=	acre	$43\,560$	$4\,840$	$7\,986\,576 \cdot 10^6$	$7\,986\,576$	160	4	1	$1.562\,5 \cdot 10^{-3}$
1 square mile	=	sq. mile	$27\,878\,400$	$3\,097\,600$	$511\,141 \cdot 10^{10}$	$511\,141 \cdot 10^4$	$102\,400$	$2\,560$	640	1

Example: 1 m² = 10⁶ mm² = 1 000 000 mm²; 1 cm² = 10.763 9 · 10⁻⁴ ft² = 0.001 076 39 ft²;
1 yd² = 8.361 27 · 10⁻¹ m² = 0.836 127 m²; 1 circ. inch = 1.252 10 · 10⁻⁷ acre = 0.000 000 125 210 acre.

Exact values are printed in bold type.

Table 104-2   The Metric Units of Area

Unit	Abbreviation	km^2	ha	a	m^2	dm^2	cm^2	mm^2	μm^2	nm^2	pm^2
1 square kilometer =	km^2	1	10^2	10^4	10^6	10^8	10^{10}	10^{12}	10^{18}	10^{24}	10^{30}
1 hectare =	ha	10^{-2}	1	10^2	10^4	10^6	10^8	10^{10}	10^{16}	10^{22}	10^{28}
1 are =	a	10^{-4}	10^{-2}	1	10^2	10^4	10^6	10^8	10^{14}	10^{20}	10^{26}
1 square meter =	m^2	10^{-6}	10^{-4}	10^{-2}	1	10^2	10^4	10^6	10^{12}	10^{18}	10^{24}
1 square decimeter =	dm^2	10^{-8}	10^{-6}	10^{-4}	10^{-2}	1	10^2	10^4	10^{10}	10^{16}	10^{22}
1 square centimeter =	cm^2	10^{-10}	10^{-8}	10^{-6}	10^{-4}	10^{-2}	1	10^2	10^8	10^{14}	10^{20}
1 square millimeter =	mm^2	10^{-12}	10^{-10}	10^{-8}	10^{-6}	10^{-4}	10^{-2}	1	10^6	10^{12}	10^{18}
1 square micrometer =	μm^2	10^{-18}	10^{-16}	10^{-14}	10^{-12}	10^{-10}	10^{-8}	10^{-6}	1	10^6	10^{12}
1 square nanometer =	nm^2	10^{-24}	10^{-22}	10^{-20}	10^{-18}	10^{-16}	10^{-14}	10^{-12}	10^{-6}	1	10^6
1 square picometer =	pm^2	10^{-30}	10^{-28}	10^{-26}	10^{-24}	10^{-22}	10^{-20}	10^{-18}	10^{-12}	10^{-6}	1

Example: $1\ km^2 = 10^6\ m^2 = 1\ 000\ 000\ m^2$; $1\ cm^2 = 10^2\ mm^2 = 100\ mm^2$; $1\ nm^2 = 10^{-6}\ \mu m^2 = 0.000\ 001\ \mu m^2$.

Solid angle equivalents

The unit of solid angle in the International System (SI) is the steradian, abbreviation sr.

Table 105-1   Conversion Factors for Units of Solid Angles

Unit	Abbreviation	sr	$(°)^2$	$(g)^2$
1 steradian $\doteq$	sr	1	3282.806	4052.847
1 old square degree =	$(°)^2$	$3.046\ 174 \cdot 10^{-4}$	1	1.234 567
1 square gon grade =	$(g)^2$	$2.467\ 401 \cdot 10^{-4}$	0.81	1

Example: How many steradians the solid angle... $\Omega = 500 \cdot (g)^2 = 500 \cdot 2.467\ 401 \cdot 10^{-4}\ sr = 0.123\ 370\ sr.$

Volume and capacity equivalents

The unit of volume and capacity in the International System (SI) is the cubic meter, abbreviation m³ ; 1 m³ = 1.307 95 yd³ .
The Anglo-American unit of volume and capacity is the cubic yard, abbreviation yd³ or cu yd; 1 yd³ = 0.764 555 m³ .

Table 106-1 Conversion Factors for Units of Volume and Capacity

Unit		Abbreviation	m³	dm³ = l*	cm³ = ml	mm³ = μl	in.³	ft³	yd³	UK bushel	US dry pt
1 cubic meter	=	m³	1	10^3	10^6	10^9	61 023.7	35.314 7	1.307 95	27.496 2	1 816.17
1 cubic decimeter = 1 liter*	=	dm³ = l*	10^{-3}	1	10^3	10^6	61 023.7 · 10^{-3}	35.314 7 · 10^{-3}	1.307 95 · 10^{-3}	27.496 2 · 10^{-3}	1 816.17 · 10^{-3}
1 cubic centimeter = 1 milliliter	=	cm³ = ml	10^{-6}	10^{-3}	1	10^3	61 023.7 · 10^{-6}	35.314 7 · 10^{-6}	1.307 95 · 10^{-6}	27.496 2 · 10^{-6}	1 816.17 · 10^{-6}
1 cubic millimeter = 1 microliter	=	mm³ = μl	10^{-9}	10^{-6}	10^{-3}	1	61 023.7 · 10^{-9}	35.314 7 · 10^{-9}	1.307 95 · 10^{-9}	27.496 2 · 10^{-9}	1 816.17 · 10^{-9}
1 cubic inch	=	in.³	16.387 1 · 10^{-6}	16.387 1 · 10^{-3}	16.387 1	16.387 1 · 10^3	1	5.787 04 · 10^{-4}	2.143 35 · 10^{-5}	4.505 82 · 10^{-4}	0.029 761 6
1 cubic foot	=	ft³	28.316 8 · 10^{-3}	28.316 8	28.316 8 · 10^3	28.316 8 · 10^6	1 728	1	0.037 037 0	0.778 605	51.428 1
1 cubic yard	=	yd³	764.555 · 10^{-3}	764.555	764.555 · 10^3	764.555 · 10^6	46 656	27	1	21.022 3	1 388.56
1 UK bushel	=	UK bushel	36.368 7 · 10^{-3}	36.368 7	36.368 7 · 10^3	36.368 7 · 10^6	2 219.35	1.284 35	0.047 568 5	1	66.051 6
1 US dry pint	=	US dry pt	5.506 10 · 10^{-4}	5.506 10 · 10^{-1}	5.506 10 · 10^2	5.506 10 · 10^5	33.600 3	0.019 444 6	7.201 71 · 10^{-4}	0.015 139 7	1
1 US bushel	=	US bushel	35.239 1 · 10^{-3}	35.239 1	35.239 1 · 10^3	35.239 1 · 10^6	2 150.42	1.244 46	0.046 091 0	0.968 940	64
1 UK pint	=	UK pt	568.261 · 10^{-6}	568.261 · 10^{-3}	568.261	568.261 · 10^3	34.677 4	0.020 067 9	7.432 57 · 10^{-4}	0.015 625 0	1.032 05
1 UK gallon	=	UK gal	4.546 09 · 10^{-3}	4.546 09	4.546 09 · 10^3	4.546 09 · 10^6	277.419	0.016 544	5.946 07 · 10^{-3}	0.125	8.256 43
1 US liquid pint	=	US liq pt	473.176 · 10^{-6}	473.176 · 10^{-3}	473.176	473.176 · 10^3	28.875	0.016 710 1	6.188 92 · 10^{-4}	0.013 010 6	0.859 366
1 US gallon	=	US gal	3.785 41 · 10^{-3}	3.785 41	3.785 41 · 10^3	3.785 41 · 10^6	231	0.133 681	4.951 14 · 10^{-3}	0.104 084	6.874 93
1 UK minim	=	UK min	59.193 8 · 10^{-9}	59.193 8 · 10^{-6}	59.193 8 · 10^{-3}	59.193 8	3.612 23 · 10^{-3}	2.090 41 · 10^{-6}	7.742 27 · 10^{-8}	1.627 61 · 10^{-6}	1.075 06 · 10^{-4}
1 UK fluid drachm	=	UK fl dr	3.551 63 · 10^{-6}	3.551 63 · 10^{-3}	3.551 63	3.551 63 · 10^3	0.216 734	1.254 25 · 10^{-4}	4.645 37 · 10^{-6}	9.765 64 · 10^{-5}	6.450 35 · 10^{-3}
1 UK fluid ounce	=	UK fl oz	28.413 0 · 10^{-6}	28.413 0 · 10^{-3}	28.413 0	28.413 0 · 10^3	1.733 87	1.003 40 · 10^{-3}	3.716 29 · 10^{-5}	7.812 51 · 10^{-4}	0.051 602 7
1 US fluid ounce	=	US fl oz	29.573 5 · 10^{-6}	29.573 5 · 10^{-3}	29.573 5	29.573 5 · 10^3	1.804 69	1.044 38 · 10^{-3}	3.868 08 · 10^{-5}	8.131 61 · 10^{-4}	0.053 710 5

Unit		Abbreviation	US bushel	UK pt	UK gal	US liq pt	US gal	UK min	UK fl dr	UK fl oz	US fl oz
1 cubic meter	=	m³	28.377 6	1.759 76 · 10^3	219.969	2.113 38 · 10^3	264.172	16.893 6 · 10^6	281.561 · 10^3	35.195 1 · 10^3	33.814 0 · 10^3
1 cubic decimeter = 1 liter*	=	dm³ = l*	28.377 6 · 10^{-3}	1.759 76	219.969 · 10^{-3}	2.113 38	264.172 · 10^{-3}	16.893 6 · 10^3	281.561	35.195 1	33.814 0
1 cubic centimeter = 1 milliliter	=	cm³ = ml	28.377 6 · 10^{-6}	1.759 76 · 10^{-3}	219.969 · 10^{-6}	2.113 38 · 10^{-3}	264.172 · 10^{-6}	16.893 6	281.561 · 10^{-3}	35.195 1 · 10^{-3}	33.814 0 · 10^{-3}
1 cubic millimeter = 1 microliter	=	mm³ = μl	28.377 6 · 10^{-9}	1.759 76 · 10^{-6}	219.969 · 10^{-9}	2.113 38 · 10^{-6}	264.172 · 10^{-9}	16.893 6 · 10^{-3}	281.561 · 10^{-6}	35.195 1 · 10^{-6}	33.814 0 · 10^{-6}
1 cubic inch	=	in.³	4.650 25 · 10^{-4}	0.028 837 2	3.604 65 · 10^{-3}	0.034 632 0	4.329 00 · 10^{-3}	276.837	4.613 95	0.576 744	0.554 113
1 cubic foot	=	ft³	0.803 564	49.830 7	6.228 84	59.844 2	7.480 52	478 374	7 972.91	996.614	957.507
1 cubic yard	=	yd³	21.696 2	1 345.43	168.178	1 615.79	201.974	12 916 107	215 268	26 908.6	25 852.7
1 UK bushel	=	UK bushel	1.032 06	63.999 8	8	76.860 5	9.607 57	614 398	10 240.0	1 280.00	1 229.77
1 US dry pint	=	US dry pt	0.015 625	0.968 939	0.121 117	1.163 65	0.145 456	9 301.81	155.030	19.378 8	18.618 4
1 US bushel	=	US bushel	1	62.012 1	7.751 51	74.473 3	9.309 17	595 316	9 921.93	1 240.24	1 191.58
1 UK pint	=	UK pt	0.016 125 9	1	0.125	1.200 95	0.150 119	9 600	160	20	19.215 2
1 UK gallon	=	UK gal	0.129 007	8	1	9.607 59	1.200 59	76 800	1 280	160	153.721
1 US liquid pint	=	US liq pt	0.013 427 6	0.832 675	0.104 084	1	0.125	7 993.67	133.228	16.653 5	16
1 US gallon	=	US gal	0.107 421	6.661 40	0.832 675	8	1	63 949.3	1 065.82	133.228	128
1 UK minim	=	UK min	1.679 78 · 10^{-6}	1.041 67 · 10^{-4}	1.302 08 · 10^{-5}	1.250 99 · 10^{-4}	1.563 73 · 10^{-5}	1	0.016 666 7	2.083 33 · 10^{-3}	2.001 58 · 10^{-3}
1 UK fluid drachm	=	UK fl dr	1.007 87 · 10^{-4}	0.006 25	7.812 50 · 10^{-4}	7.505 93 · 10^{-3}	9.382 41 · 10^{-4}	60	1	0.125	0.120 095
1 UK fluid ounce	=	UK fl oz	8.062 93 · 10^{-4}	0.05	0.006 25	0.060 047 4	7.505 92 · 10^{-3}	480	8	1	0.960 759
1 US fluid ounce	=	US fl oz	8.392 26 · 10^{-4}	0.052 042 2	6.505 27 · 10^{-3}	0.062 5	7.812 50 · 10^{-3}	499.605	8.326 75	1.040 84	1

Example: 1 m³ = 10^6 cm³ = 1 000 000 cm³ ; 1 dm³ = 1.307 95 · 10^{-3} yd³ = 0.001 307 95 yd³ ;
1 UK pt = 568.261 · 10^{-3} dm³ = 0.568 261 dm³ ; 1 UK min = 1.302 08 · 10^{-5} UK gal = 0.000 013 020 8 UK gal.
Exact values are printed in bold type.
The United Kingdom units are denoted by the prefix UK.
The American units are denoted by the prefix US.
*Conforming to the decision of the General Conference of Weights and Measures, in 1964, the volume unit has been defined as follows: liter(1) = cubic decimeter = dm³ (exactly).

Table 106-2 The Metric Units of Volume and Capacity

Unit	Abbreviation		km^3	hm^3	dam^3	m^3	hl	dal	$dm^3 = l$	dl	cl	$cm^3 = ml$	$mm^3 = \mu l$	$\mu m^3 = nl$	nm^3	pm^3
1 cubic kilometer	km^3	=	1	10^3	10^6	10^9	10^{10}	10^{11}	10^{12}	10^{13}	10^{14}	10^{15}	10^{18}	10^{27}	10^{36}	10^{45}
1 cubic hectameter	hm^3	=	10^{-3}	1	10^3	10^6	10^7	10^8	10^9	10^{10}	10^{11}	10^{12}	10^{15}	10^{24}	10^{33}	10^{42}
1 cubic decameter	dam^3	=	10^{-6}	10^{-3}	1	10^3	10^4	10^5	10^6	10^7	10^8	10^9	10^{12}	10^{21}	10^{30}	10^{39}
1 cubic meter	m^3	=	10^{-9}	10^{-6}	10^{-3}	1	10^1	10^2	10^3	10^4	10^5	10^6	10^9	10^{18}	10^{27}	10^{36}
1 hectaliter	hl	=	10^{-10}	10^{-7}	10^{-4}	10^{-1}	1	10^1	10^2	10^3	10^4	10^5	10^8	10^{17}	10^{26}	10^{35}
1 decaliter	dal	=	10^{-11}	10^{-8}	10^{-5}	10^{-2}	10^{-1}	1	10^1	10^2	10^3	10^4	10^7	10^{16}	10^{25}	10^{34}
1 cubic decimeter = 1 liter	$dm^3 = l$	=	10^{-12}	10^{-9}	10^{-6}	10^{-3}	10^{-2}	10^{-1}	1	10^1	10^2	10^3	10^6	10^{15}	10^{24}	10^{33}
1 deciliter	dl	=	10^{-13}	10^{-10}	10^{-7}	10^{-4}	10^{-3}	10^{-2}	10^{-1}	1	10^1	10^2	10^5	10^{14}	10^{23}	10^{32}
1 centiliter	cl	=	10^{-14}	10^{-11}	10^{-8}	10^{-5}	10^{-4}	10^{-3}	10^{-2}	10^{-1}	1	10^1	10^4	10^{13}	10^{22}	10^{31}
1 cubic centimeter = 1 milliliter	$cm^3 = ml$	=	10^{-15}	10^{-12}	10^{-9}	10^{-6}	10^{-5}	10^{-4}	10^{-3}	10^{-2}	10^{-1}	1	10^3	10^{12}	10^{21}	10^{30}
1 cubic millimeter = 1 microliter	$mm^3 = \mu l$	=	10^{-18}	10^{-15}	10^{-12}	10^{-9}	10^{-8}	10^{-7}	10^{-6}	10^{-5}	10^{-4}	10^{-3}	1	10^9	10^{18}	10^{27}
1 cubic micrometer = 1 nanoliter	$\mu m^3 = nl$	=	10^{-27}	10^{-24}	10^{-21}	10^{-18}	10^{-17}	10^{-16}	10^{-15}	10^{-14}	10^{-13}	10^{-12}	10^{-9}	1	10^9	10^{18}
1 cubic nanometer	nm^3	=	10^{-36}	10^{-33}	10^{-30}	10^{-27}	10^{-26}	10^{-25}	10^{-24}	10^{-23}	10^{-22}	10^{-21}	10^{-18}	10^{-9}	1	10^9
1 cubic picometer	pm^3	=	10^{-45}	10^{-42}	10^{-39}	10^{-36}	10^{-35}	10^{-34}	10^{-33}	10^{-32}	10^{-31}	10^{-30}	10^{-27}	10^{-18}	10^{-9}	1

Example: $1\ m^3 = 10^{-3}\ dam^3 = 0.001\ dam^3$; $1\ dl = 10^{-4}\ m^3 = 0.001\ 1\ m^3$.

Table 106-3 Relationship Between UK (Imperial) and US Units of Volume (Capacity)

1 UK minim	=	0.960 759 US minim		1 US minim	=	1.040 84 UK minim
1 UK fluid drachm	=	0.960 759 US fluid dram		1 US fluid dram	=	1.040 84 UK fluid drachm
1 UK fluid ounce	=	0.960 759 US fluid ounce		1 US fluid ounce	=	1.040 84 UK fluid ounce
1 UK gill	=	1.200 95 US gill		1 US gill	=	0.832 675 UK gill
1 UK pint	=	1.200 95 US liquid pint		1 US liquid pint	=	0.832 675 UK pint
1 UK quart	=	1.200 95 US liquid quart		1 US liquid quart	=	0.832 675 UK quart
1 UK gallon	=	1.200 95 US gallon		1 US gallon	=	0.832 675 UK gallon
1 UK pint	=	1.032 06 US dry pint		1 US dry pint	=	0.968 940 UK pint
1 UK quart	=	1.032 06 US dry quart		1 US dry quart	=	0.968 940 UK quart
1 UK peck	=	1.032 06 US peck		1 US peck	=	0.968 940 UK peck
1 UK bushel	=	1.032 06 US bushel		1 US bushel	=	0.968 940 UK bushel

Table 107-1 Specific Volume Equivalents: Conversion Factors for Units of Specific Volume

Unit	Abbreviation		m^3/kg	$1/kg$	ft^3/lb	$in.^3/lb$	ft^3/ton	UK gal/lb
1 cubic meter per kilogram	m^3/kg	=	**1**	**1 000**	16.018 5	27 679.9	35 881.4	99.776 4
1 liter per kilogram	$1/kg$	=	**0.001**	1	0.016 018 5	27.679 9	35.881 4	0.099 776 4
1 cubic foot per pound	ft^3/lb	=	0.062 428 0	62.428 0	1	**1 728**	**2 240**	6.228 84
1 cubic inch per pound	$in.^3/lb$	=	$3.612\ 73 \cdot 10^{-5}$	0.036 127 3	$5.787\ 04 \cdot 10^{-4}$	1	1.296 30	$3.604\ 65 \cdot 10^{-3}$
1 cubic foot per UK ton	ft^3/ton	=	$2.786\ 96 \cdot 10^{-5}$	0.027 869 6	$4.464\ 29 \cdot 10^{-4}$	0.771 429	1	$2.780\ 73 \cdot 10^{-3}$
1 UK gallon per pound	UK gal/lb	=	0.010 022 4	10.022 4	0.160 544	277.419	359.618	1

Examples: $1\ m^3/kg = 16.0185\ ft^3/lb$; $1\ in.^3/lb = 0.036\ 127\ 3\ 1/kg$
Exact values are printed in bold type.
The United Kingdom units are denoted by the prefix UK.
The American units are denoted by the prefix US.

Velocity equivalents

The unit of velocity in the International System (SI) is the meter per second, abbreviation m/s; 1 m/s = 3.280 84 ft/s.
The Anglo-American unit of velocity is the foot per second, abbreviation ft/s; 1 ft/s = 0.304 8 m/s.

Table 108-1 Conversion Factors for Units of Velocity

Unit		Abbreviation	km/s	m/s	dm/s	cm/s	mm/s	km/h	m/h	in./s	ft/s
1 kilometer per second	=	km/s	1	10^3	10^4	10^5	10^6	3 600	$3\,600 \cdot 10^3$	$39.370\,1 \cdot 10^3$	$3.280\,84 \cdot 10^3$
1 meter per second	=	m/s	10^{-3}	1	10^1	10^2	10^3	$3\,600 \cdot 10^{-3}$	3 600	39.370 1	3.280 84
1 decimeter per second	=	dm/s	10^{-4}	10^{-1}	1	10^1	10^2	$3\,600 \cdot 10^{-4}$	$3\,600 \cdot 10^{-1}$	$39.370\,1 \cdot 10^{-1}$	$3.280\,84 \cdot 10^{-1}$
1 centimeter per second	=	cm/s	10^{-5}	10^{-2}	10^{-1}	1	10^1	$3\,600 \cdot 10^{-5}$	$3\,600 \cdot 10^{-2}$	$39.370\,1 \cdot 10^{-2}$	$3.280\,84 \cdot 10^{-2}$
1 millimeter per second	=	mm/s	10^{-6}	10^{-3}	10^{-2}	10^{-1}	1	$3\,600 \cdot 10^{-6}$	$3\,600 \cdot 10^{-3}$	$39.370\,1 \cdot 10^{-3}$	$3.280\,84 \cdot 10^{-3}$
1 kilometer per hour	=	km/h	$2.777\,78 \cdot 10^{-4}$	$2.777\,78 \cdot 10^{-1}$	2.777 78	$2.777\,78 \cdot 10^1$	$2.777\,78 \cdot 10^2$	1	10^3	$1.093\,61 \cdot 10^1$	$9.113\,44 \cdot 10^{-1}$
1 meter per hour	=	m/h	$2.777\,78 \cdot 10^{-7}$	$2.777\,78 \cdot 10^{-4}$	$2.777\,78 \cdot 10^{-3}$	$2.777\,78 \cdot 10^{-2}$	$2.777\,78 \cdot 10^{-1}$	10^{-3}	1	$1.093\,61 \cdot 10^{-2}$	$9.113\,44 \cdot 10^{-4}$
1 inch per second	=	in./s	$2.54 \cdot 10^{-5}$	$2.54 \cdot 10^{-2}$	$2.54 \cdot 10^{-1}$	2.54	$2.54 \cdot 10^1$	$9.144 \cdot 10^{-2}$	$9.144 \cdot 10^1$	1	0.083 333 3
1 foot per second	=	ft/s	$3.048 \cdot 10^{-4}$	$3.048 \cdot 10^{-1}$	3.048	$3.048 \cdot 10^1$	$3.048 \cdot 10^2$	1.097 28	$1.097\,28 \cdot 10^3$	12	1
1 yard per second	=	yd/s	$9.144 \cdot 10^{-4}$	$9.144 \cdot 10^{-1}$	9.144	$9.144 \cdot 10^1$	$9.144 \cdot 10^2$	3.291 84	$3.291\,84 \cdot 10^3$	36	3
1 mile per second	=	mile/s	1.609 344	$1.609\,344 \cdot 10^3$	$1.609\,344 \cdot 10^4$	$1.609\,344 \cdot 10^5$	$1.609\,344 \cdot 10^6$	$5.793\,64 \cdot 10^3$	$5.793\,64 \cdot 10^6$	63 360	5 280
1 inch per minute	=	in./min	$4.233\,33 \cdot 10^{-7}$	$4.233\,33 \cdot 10^{-3}$	$4.233\,33 \cdot 10^{-2}$	$4.233\,33 \cdot 10^{-1}$	$4.233\,33 \cdot 10^{-1}$	$1.524 \cdot 10^{-3}$	1.524	0.016 666 7	$1.388\,89 \cdot 10^{-3}$
1 foot per minute	=	ft/min	$5.08 \cdot 10^{-6}$	$5.08 \cdot 10^{-3}$	$5.08 \cdot 10^{-2}$	$5.08 \cdot 10^{-1}$	5.08	$1.828\,8 \cdot 10^{-2}$	$1.828\,8 \cdot 10^1$	0.2	0.016 666 7
1 yard per minute	=	yd/min	$1.524 \cdot 10^{-5}$	$1.524 \cdot 10^{-2}$	$1.524 \cdot 10^{-1}$	1.524	$1.524 \cdot 10^1$	$5.486\,4 \cdot 10^{-2}$	$5.486\,4 \cdot 10^1$	0.6	0.05
1 yard per hour	=	yd/h	$2.54 \cdot 10^{-7}$	$2.54 \cdot 10^{-4}$	$2.54 \cdot 10^{-3}$	$2.54 \cdot 10^{-2}$	$2.54 \cdot 10^{-1}$	$9.144 \cdot 10^{-4}$	$9.144 \cdot 10^{-1}$	0.01	$8.333\,33 \cdot 10^{-4}$
1 mile per hour	=	mile/h	$4.470\,4 \cdot 10^{-4}$	$4.470\,4 \cdot 10^{-1}$	4.470 4	$4470\,4 \cdot 10^1$	$4.470\,4 \cdot 10^2$	1.609 344	$1.609\,344 \cdot 10^3$	17.600	1.466 67
1 UK knot	=	UK kn	$5.147\,73 \cdot 10^{-4}$	$5.147\,73 \cdot 10^{-1}$	5.147 73	$5.147\,73 \cdot 10^1$	$5.147\,73 \cdot 10^2$	1.853 18	$1.853\,18 \cdot 10^3$	20.266 6	1.688 89
1 international knot	=	kn	$5.144\,44 \cdot 10^{-4}$	$5.144\,44 \cdot 10^{-1}$	5.144 44	$5.144\,44 \cdot 10^1$	$5.144\,44 \cdot 10^2$	1.852	$1.852 \cdot 10^3$	20.2537	1.687 81

Unit		Abbreviation	yd/s	mile/s	in./min	ft/min	yd/min	yd/h	mile/h	UK kn	kn
1 kilometer per second	=	km/s	$1.093\,61 \cdot 10^3$	$6.213\,71 \cdot 10^{-1}$	$2.362\,20 \cdot 10^6$	$1.968\,50 \cdot 10^5$	$6.561\,68 \cdot 10^4$	$3.937\,01 \cdot 10^6$	$2.236\,94 \cdot 10^3$	$1.942\,60 \cdot 10^3$	$1.943\,84 \cdot 10^3$
1 meter per second	=	m/s	1.093 61	$6.213\,71 \cdot 10^{-4}$	$2.362\,20 \cdot 10^3$	$1.968\,50 \cdot 10^2$	$6.561\,68 \cdot 10^1$	$3.937\,01 \cdot 10^3$	2.236 94	1.942 60	1.943 84
1 decimeter per second	=	dm/s	$1.093\,61 \cdot 10^{-1}$	$6.213\,71 \cdot 10^{-5}$	$2.362\,20 \cdot 10^2$	$1.968\,50 \cdot 10^1$	6.561 68	$3.937\,01 \cdot 10^2$	$2.236\,94 \cdot 10^{-1}$	$1.942\,60 \cdot 10^{-1}$	$1.943\,84 \cdot 10^{-1}$
1 centimeter per second	=	cm/s	$1.093\,61 \cdot 10^{-2}$	$6.213\,71 \cdot 10^{-6}$	$2.362\,20 \cdot 10^1$	1.968 50	$6.561\,68 \cdot 10^{-1}$	$3.937\,01 \cdot 10^1$	$2.236\,94 \cdot 10^{-2}$	$1.942\,60 \cdot 10^{-2}$	$1.943\,84 \cdot 10^{-2}$
1 millimeter per second	=	mm/s	$1.093\,61 \cdot 10^{-3}$	$6.213\,71 \cdot 10^{-7}$	2.362 20	$1.968\,50 \cdot 10^{-1}$	$6.561\,68 \cdot 10^{-2}$	3.937 01	$2.236\,94 \cdot 10^{-3}$	$1.942\,60 \cdot 10^{-3}$	$1.943\,84 \cdot 10^{-3}$
1 kilometer per hour	=	km/h	$3.037\,81 \cdot 10^{-1}$	$1.726\,03 \cdot 10^{-4}$	$6.561\,68 \cdot 10^2$	$5.468\,07 \cdot 10^1$	$1.822\,69 \cdot 10^1$	$1.093\,61 \cdot 10^3$	$6.213\,71 \cdot 10^{-1}$	$5.396\,12 \cdot 10^{-1}$	$5.399\,57 \cdot 10^{-1}$
1 meter per hour	=	m/h	$3.037\,81 \cdot 10^{-4}$	$1.726\,03 \cdot 10^{-7}$	$6.561\,68 \cdot 10^{-1}$	$5.468\,07 \cdot 10^{-2}$	$1.822\,69 \cdot 10^{-2}$	1.093 61	$6.213\,71 \cdot 10^{-4}$	$5.396\,12 \cdot 10^{-4}$	$5.399\,57 \cdot 10^{-4}$
1 inch per second	=	in./s	0.027 777 8	$1.578\,28 \cdot 10^{-5}$	60	5	1.666 67	100	0.056 818 2	0.049 342 2	0.049 373 7
1 foot per second	=	ft/s	0.333 333	$1.893\,94 \cdot 10^{-4}$	720	60	20	1 200	0.681 818	0.592 105	0.592 484
1 yard per second	=	yd/s	1	$5.681\,82 \cdot 10^{-4}$	2160	180	60	3 600	2.045 45	1.776 32	1.777 45
1 mile per second	=	mile/s	1 760	1	3 801 600	316 800	105 600	6 336 000	3 600	3 126.32	3 128.31
1 inch per minute	=	in./min	$4.629\,63 \cdot 10^{-4}$	$2.630\,47 \cdot 10^{-7}$	1	0.083 333 3	0.027 777 8	1.666 67	$9.469\,70 \cdot 10^{-4}$	$8.223\,70 \cdot 10^{-4}$	$8.228\,94 \cdot 10^{-4}$
1 foot per minute	=	ft/min	$5.555\,56 \cdot 10^{-3}$	$3.156\,57 \cdot 10^{-6}$	12	1	0.333 333	20	0.011 363 6	$9.868\,45 \cdot 10^{-3}$	$9.874\,73 \cdot 10^{-3}$
1 yard per minute	=	yd/min	0.016 666 7	$9.469\,70 \cdot 10^{-6}$	36	3	1	60	0.034 090 9	0.029 605 2	0.029 624 2
1 yard per hour	=	yd/h	$2.777\,78 \cdot 10^{-4}$	$1.578\,28 \cdot 10^{-7}$	0.6	0.05	0.016 666 7	1	$5.681\,82 \cdot 10^{-4}$	$4.934\,22 \cdot 10^{-4}$	$4.937\,36 \cdot 10^{-4}$
1 mile per hour	=	mile/h	0.488 889	$2.777\,78 \cdot 10^{-4}$	1056	88	29.333 3	1 760	1	0.868 421	0.868 976
1 UK knot	=	UK kn	0.562 963	$3.198\,65 \cdot 10^{-4}$	1 216.00	101.333	33.777 7	2 026.66	1.151 52	1	1.000 64
1 international knot	=	kn	0.562 603	$3.196\,61 \cdot 10^{-4}$	1 215.22	101.269	33.756 2	2 025.37	1.150 78	0.999 361	1

Examples: 1 m/s = $3.937\,01 \cdot 10^3$ yd/h = 3 937.01 yd/h; 1 mile/h = $4.470\,4 \cdot 10^{-1}$ m/s = 0.447 04 m/s.
Exact values are printed in bold type.
The United Kingdom units are denoted by the prefix UK.
The American units are denoted by the prefix US.

Angular velocity (angular speed) speed of rotation equivalents

The unit of angular velocity angular speed in the International System (SI)
is the radian per second, abbreviation rad/s.

Table 109-1 Conversion Factors for Units of Angular Velocity

Unit		Abbreviation	rad/s	rad/min	rev/s	rev/min	°/s
1 radian per second	=	rad/s	**1**	**60**	0.159 155	9.549 30	57.295 8
1 radian per minute	=	rad/min	0.016 666 7	**1**	0.002 652 58	0.159 155	0.954 930
1 revolution per second	=	rev/s	6.283 19	376.991	**1**	**60**	**360**
1 revolution per minute	=	rev/min	0.104 720	6.283 19	0.016 666 7	**1**	**6**
1 degree per second	=	°/s	0.017 453 3	1.047 20	0.002 777 78	0.166 667	**1**

Examples: 1 rad/s = 0.159 155 rev/s; 1 °/s = 0.017 453 3 rad/s.
Exact values are printed in bold type.

Equivalents of weight density

The unit of weight density in the International System (SI) is the newton per cubic meter, abbreviation N/m³; 1 N/m³ = 0.101 972 kp/m³ ≈ 102 kp/m³ .
In the metric gravitational system, the unit of weight density is the kilopond per cubic meter, abbreviation kp/m³ ; 1 kp/m³ = 9.806 65 N/m³ .

Table 110-1 Conversion Factors for Units of Weight Density

Unit		Abbreviation	N/m³	kN/m³	kp/m³	Mp/m³	kp/dm³	p/cm³
1 newton per cubic meter	=	N/m³	**1**	10^{-3}	$1.019\,72 \cdot 10^{-1}$	$1.019\,72 \cdot 10^{-4}$	$1.019\,72 \cdot 10^{-4}$	$1.019\,72 \cdot 10^{-4}$
1 kilonewton per cubic meter	=	kN/m³	10^3	**1**	$1.019\,72 \cdot 10^2$	$1.019\,72 \cdot 10^{-1}$	$1.019\,72 \cdot 10^{-1}$	$1.019\,72 \cdot 10^{-1}$
1 kilopond per cubic meter	=	kp/m³	**9.806 65**	$\mathbf{9.806\,65 \cdot 10^{-3}}$	**1**	10^{-3}	10^{-3}	10^{-3}
1 megapond per cubic meter	=	Mp/m³	$\mathbf{9.806\,65 \cdot 10^3}$	**9.806 65**	10^3	**1**	**1**	**1**
1 kilopond per cubic decimeter	=	kp/dm³	$\mathbf{9.806\,65 \cdot 10^3}$	**9.806 65**	10^3	**1**	**1**	**1**
1 pond per cubic centimeter	=	p/cm³	$\mathbf{9.806\,65 \cdot 10^3}$	**9.806 65**	10^3	**1**	**1**	**1**

Examples: 1 N/m³ = 1.019 72 · 10⁻¹ kp/m³ = 0.101 972 kp/m³ ; 1 Mp/m³ = 9.806 65 kN/m³ .
Exact values are printed in bold type.

Mass equivalents

The unit of mass in the International System (SI) is the kilogram, abbreviation kg; 1 kg = 2.204 62 lb.
The Anglo-American unit of mass is the pound, abbreviation lb; 1 lb = 0.453 592 37 kg.

Table 111-1 Conversion Factors for Units of Mass

Unit		Abbreviation	kg	dag (dkg)	g	dg	cg	mg
1 kilogram	=	kg	1	10^2	10^3	10^4	10^5	10^6
1 decagram	=	dag (kdg)	10^{-2}	1	10^1	10^2	10^3	10^4
1 gram	=	g	10^{-3}	10^{-1}	1	10^1	10^2	10^3
1 decigram	=	dg	10^{-4}	10^{-2}	10^{-1}	1	10^1	10^2
1 centigram	=	cg	10^{-5}	10^{-3}	10^{-2}	10^{-1}	1	10^1
1 milligram	=	mg	10^{-6}	10^{-4}	10^{-3}	10^{-2}	10^{-1}	1
1 microgram	=	µg	10^{-9}	10^{-7}	10^{-6}	10^{-5}	10^{-4}	10^{-3}
1 ton (= 1000 kg)	=	t	10^3	10^5	10^6	10^7	10^8	10^9
1 pound	=	lb	$4.535\ 923\ 7 \cdot 10^{-1}$	$4.535\ 923\ 7 \cdot 10^1$	$4.535\ 923\ 7 \cdot 10^2$	$4.535\ 923\ 7 \cdot 10^3$	$4.535\ 923\ 7 \cdot 10^4$	$4.535\ 923\ 7 \cdot 10^5$
1 hyl* (techma)	=	hyl*	$9.806\ 65$	$9.806\ 65 \cdot 10^2$	$9.806\ 65 \cdot 10^3$	$9.806\ 65 \cdot 10^4$	$9.806\ 65 \cdot 10^5$	$9.806\ 65 \cdot 10^6$
1 slug**	=	slug**	$14.593\ 9$	$14.593\ 9 \cdot 10^2$	$14.593\ 9 \cdot 10^3$	$14.593\ 9 \cdot 10^4$	$14.593\ 9 \cdot 10^5$	$14.593\ 9 \cdot 10^6$
1 grain	=	gr	$6.479\ 89 \cdot 10^{-5}$	$6.479\ 89 \cdot 10^{-3}$	$6.479\ 89 \cdot 10^{-2}$	$6.479\ 89 \cdot 10^{-1}$	$6.479\ 89$	$6.479\ 89 \cdot 10^1$
1 dram (avoir)	=	dr (av.)	$1.771\ 85 \cdot 10^{-3}$	$1.771\ 85 \cdot 10^{-1}$	$1.771\ 85$	$1.771\ 85 \cdot 10^1$	$1.771\ 85 \cdot 10^2$	$1.771\ 85 \cdot 10^3$
1 drachm (apoth)	=	drahm (apoth)	$3.887\ 93 \cdot 10^{-3}$	$3.887\ 93 \cdot 10^{-1}$	$3.887\ 93$	$3.887\ 93 \cdot 10^1$	$3.887\ 93 \cdot 10^2$	$3.887\ 93 \cdot 10^3$
1 ounce (avoir)	=	oz (av)	$28.349\ 5 \cdot 10^{-3}$	$28.349\ 5 \cdot 10^{-1}$	$28.349\ 5$	$28.349\ 5 \cdot 10^1$	$28.349\ 5 \cdot 10^2$	$28.349\ 5 \cdot 10^3$
1 ounce (troy; apoth)	=	oz (tr; ap)	$31.103\ 5 \cdot 10^{-3}$	$31.103\ 5 \cdot 10^{-1}$	$31.103\ 5$	$31.103\ 5 \cdot 10^1$	$31.103\ 5 \cdot 10^2$	$31.103\ 5 \cdot 10^3$
1 hundredweight	=	cwt	$5.080\ 23 \cdot 10^1$	$5.080\ 23 \cdot 10^3$	$5.080\ 23 \cdot 10^4$	$5.080\ 23 \cdot 10^5$	$5.080\ 23 \cdot 10^6$	$5.080\ 23 \cdot 10^7$
1 US short hundredweight	=	US sh cwt	$4.535\ 92 \cdot 10^1$	$4.535\ 92 \cdot 10^3$	$4.535\ 92 \cdot 10^4$	$4.535\ 92 \cdot 10^5$	$4.535\ 92 \cdot 10^6$	$4.535\ 92 \cdot 10^7$
1 UK ton	=	UK ton	$1.016\ 05 \cdot 10^3$	$1.016\ 05 \cdot 10^5$	$1.016\ 05 \cdot 10^6$	$1.016\ 05 \cdot 10^7$	$1.016\ 05 \cdot 10^8$	$1.016\ 05 \cdot 10^9$
US short ton	=	US sh ton	$9.071\ 85 \cdot 10^2$	$9.071\ 85 \cdot 10^4$	$9.071\ 85 \cdot 10^5$	$9.071\ 85 \cdot 10^6$	$9.071\ 85 \cdot 10^7$	$9.071\ 85 \cdot 10^8$

Unit		Abbreviation	µg	t	lb	hyl*	slug**	gr	dr (av)
1 kilogram	=	kg	10^9	10^{-3}	$2.204\ 62$	$1.019\ 72 \cdot 10^{-1}$	$6.852\ 18 \cdot 10^{-2}$	$15.432\ 4 \cdot 10^3$	$5.643\ 83 \cdot 10^2$
1 decagram	=	dag (dkg)	10^7	10^{-5}	$2.204\ 62 \cdot 10^{-2}$	$1.019\ 72 \cdot 10^{-3}$	$6.852\ 18 \cdot 10^{-4}$	$15.432\ 4 \cdot 10^1$	$5.643\ 83$
1 gram	=	g	10^6	10^{-6}	$2.204\ 62 \cdot 10^{-3}$	$1.019\ 72 \cdot 10^{-4}$	$6.852\ 18 \cdot 10^{-5}$	$15.432\ 4$	$5.643\ 83 \cdot 10^{-1}$
1 decigram	=	dg	10^5	10^{-7}	$2.204\ 62 \cdot 10^{-4}$	$1.019\ 72 \cdot 10^{-5}$	$6.852\ 18 \cdot 10^{-6}$	$15.432\ 4 \cdot 10^{-1}$	$5.643\ 83 \cdot 10^{-2}$
1 centigram	=	cg	10^4	10^{-8}	$2.204\ 62 \cdot 10^{-5}$	$1.019\ 72 \cdot 10^{-6}$	$6.852\ 18 \cdot 10^{-7}$	$15.432\ 4 \cdot 10^{-2}$	$5.643\ 83 \cdot 10^{-3}$
1 milligram	=	mg	10^3	10^{-9}	$2.204\ 62 \cdot 10^{-6}$	$1.019\ 72 \cdot 10^{-7}$	$6.852\ 18 \cdot 10^{-8}$	$15.432\ 4 \cdot 10^{-3}$	$5.643\ 83 \cdot 10^{-4}$
1 microgram	=	µg	1	10^{-12}	$2.204\ 62 \cdot 10^{-9}$	$1.019\ 72 \cdot 10^{-10}$	$6.852\ 18 \cdot 10^{-11}$	$15.432\ 4 \cdot 10^{-6}$	$5.643\ 83 \cdot 10^{-7}$
1 ton (= 1000 kg)	=	t	10^{12}	1	$2.204\ 62 \cdot 10^3$	$1.019\ 72 \cdot 10^2$	$6.852\ 18 \cdot 10^1$	$15.432\ 4 \cdot 10^6$	$5.643\ 83 \cdot 10^5$

Unit	Abbreviation	=	µg	t	lb	hyl*	slug**	gr	dr (av)
1 pound	lb	=	$4.535\,923\,7 \cdot 10^{8}$	$4.535\,923\,7 \cdot 10^{-4}$	**1**	0.046 253 5	0.031 081 0	**7 000**	**256**
1 hyl* (techma)	hyl*	=	$9.806\,65 \cdot 10^{9}$	$9.806\,65 \cdot 10^{-3}$	21.620 0	**1**	0.671 969	151 340	5 534.70
1 slug**	slug**	=	$14.593\,9 \cdot 10^{9}$	$14.593\,9 \cdot 10^{-3}$	32.174 0	1.488 16	**1**	225 218	8 236.53
1 grain	gr	=	$6.479\,89 \cdot 10^{4}$	$6.479\,89 \cdot 10^{-8}$	$1.428\,57 \cdot 10^{-4}$	$6.607\,65 \cdot 10^{-6}$	$4.440\,14 \cdot 10^{-6}$	**1**	0.036 5714
1 dram (avoir)	dr (av)	=	$1.771\,85 \cdot 10^{6}$	$1.771\,85 \cdot 10^{-6}$	$3.906\,25 \cdot 10^{-3}$	$1.806\,78 \cdot 10^{-4}$	$1.214\,10 \cdot 10^{-4}$	27.343 75	**1**
1 drachm (apoth)	drachm (ap)	=	$3.887\,93 \cdot 10^{6}$	$3.887\,93 \cdot 10^{-6}$	$8.571\,43 \cdot 10^{-3}$	$3.964\,59 \cdot 10^{-4}$	$2.664\,08 \cdot 10^{-4}$	60	2.194 29
1 ounce (avoir)	oz (av)	=	$28.349\,5 \cdot 10^{6}$	$28.349\,5 \cdot 10^{-6}$	**0.0625**	$2.890\,84 \cdot 10^{-3}$	$1.942\,56 \cdot 10^{-3}$	437.5	**16**
1 ounce (troy; apoth)	oz (tr; ap)	=	$31.103\,5 \cdot 10^{6}$	$31.103\,5 \cdot 10^{-6}$	$6.857\,14 \cdot 10^{-2}$	$3.171\,67 \cdot 10^{-3}$	$2.131\,27 \cdot 10^{-3}$	480	17.554 3
1 hundredweight	cwt	=	$5.080\,23 \cdot 10^{10}$	$5.080\,23 \cdot 10^{-2}$	**112**	5.180 39	3.481 07	784 000	28 672
1 US short hundredweight	US sh cwt	=	$4.535\,92 \cdot 10^{10}$	$4.535\,92 \cdot 10^{-2}$	**100**	4.625 35	3.108 10	700 000	25 600
1 UK ton	UK ton	=	$1.016\,05 \cdot 10^{12}$	1.016 05	**2 240**	103.608	69.621 4	15 680 000	573 440
US short ton	US sh ton	=	$9.071\,85 \cdot 10^{11}$	$9.071\,85 \cdot 10^{-1}$	**2 000**	92.506 9	62.162 0	14 000 000	512 000

Unit	Abbreviation	=	drachm (ap)	oz (av)	oz (tr; ap)	cwt	US sh cwt	UK ton	US sh ton
1 kilogram	kg	=	$2.572\,06 \cdot 10^{2}$	$3.527\,40 \cdot 10^{1}$	$3.215\,07 \cdot 10^{1}$	$19.684\,57 \cdot 10^{-3}$	$22.046\,2 \cdot 10^{-3}$	$9.842\,07 \cdot 10^{-4}$	$1.102\,31 \cdot 10^{-3}$
1 decagram	dag (dkg)	=	2.572 06	$3.527\,40 \cdot 10^{-1}$	$3.215\,07 \cdot 10^{-1}$	$19.684\,1 \cdot 10^{-5}$	$22.046\,2 \cdot 10^{-5}$	$9.842\,07 \cdot 10^{-6}$	$1.102\,31 \cdot 10^{-5}$
1 gram	gr	=	$2.572\,06 \cdot 10^{-1}$	$3.527\,40 \cdot 10^{-2}$	$3.215\,07 \cdot 10^{-2}$	$19.684\,1 \cdot 10^{-6}$	$22.046\,2 \cdot 10^{-6}$	$9.842\,07 \cdot 10^{-7}$	$1.102\,31 \cdot 10^{-6}$
1 decigram	dg	=	$2.572\,06 \cdot 10^{-2}$	$3.527\,40 \cdot 10^{-3}$	$3.215\,07 \cdot 10^{-3}$	$19.684\,1 \cdot 10^{-7}$	$22.046\,2 \cdot 10^{-7}$	$9.842\,07 \cdot 10^{-8}$	$1.102\,31 \cdot 10^{-7}$
1 centigram	cg	=	$2.572\,06 \cdot 10^{-3}$	$3.527\,40 \cdot 10^{-4}$	$3.215\,07 \cdot 10^{-4}$	$19.684\,1 \cdot 10^{-8}$	$22.046\,2 \cdot 10^{-8}$	$9.842\,07 \cdot 10^{-9}$	$1.102\,31 \cdot 10^{-8}$
1 milligram	mg	=	$2.572\,06 \cdot 10^{-4}$	$3.527\,40 \cdot 10^{-5}$	$3.215\,07 \cdot 10^{-5}$	$19.684\,1 \cdot 10^{-9}$	$22.046\,2 \cdot 10^{-9}$	$9.842\,07 \cdot 10^{-10}$	$1.102\,31 \cdot 10^{-9}$
1 microgram	µg	=	$2.572\,06 \cdot 10^{-7}$	$3.527\,40 \cdot 10^{-8}$	$3.215\,07 \cdot 10^{-8}$	$19.684\,1 \cdot 10^{-12}$	$22.046\,2 \cdot 10^{-12}$	$9.842\,07 \cdot 10^{-13}$	$1.102\,31 \cdot 10^{-12}$
1 ton (= 1000 kg)	t	=	$2.572\,06 \cdot 10^{5}$	$3.527\,40 \cdot 10^{4}$	$3.215\,07 \cdot 10^{4}$	19.684 1	22.046 2	$9.842\,07 \cdot 10^{-1}$	1.102 31
1 pound	lb	=	116.667	**16**	14.583 3	$8.928\,57 \cdot 10^{-3}$	**0.01**	$4.464\,29 \cdot 10^{-4}$	**0.000 5**
1 hyl* (techma)	hyl*	=	2 522.33	345.920	315.292	0.193 036	0.216 200	$9.651\,79 \cdot 10^{-3}$	0.010 810 0
1 slug**	slug**	=	3 753.64	514.785	469.204	0.287 268	0.321 740	0.014 363 4	0.016 087 0
1 grain	gr	=	0.016 666 7	0.002 285 71	0.002 083 33	$1.275\,51 \cdot 10^{-6}$	$1.428\,57 \cdot 10^{-6}$	$6.377\,55 \cdot 10^{-8}$	$7.142\,85 \cdot 10^{-8}$
1 dram (avoir)	dr (av)	=	0.455 729	**0.0625**	0.056 966 1	$3.487\,72 \cdot 10^{-5}$	$3.906\,25 \cdot 10^{-5}$	$1.743\,86 \cdot 10^{-6}$	$1.953\,13 \cdot 10^{-6}$
1 drachm (apoth)	drachm (ap)	=	**1**	0.137 143	**0.125**	$7.653\,06 \cdot 10^{-5}$	$8.571\,43 \cdot 10^{-5}$	$3.826\,53 \cdot 10^{-6}$	$4.285\,71 \cdot 10^{-6}$
1 ounce (avoir)	oz (av)	=	7.291 67	**1**	0.911 458	$5.580\,36 \cdot 10^{-4}$	**0.000 625**	$2.790\,17 \cdot 10^{-5}$	$3.125\,00 \cdot 10^{-5}$
1 ounce (troy; apoth)	oz (tr; ap)	=	**8**	1.097 14	**1**	$6.122\,45 \cdot 10^{-4}$	$6.857\,14 \cdot 10^{-4}$	$3.061\,22 \cdot 10^{-5}$	$3.428\,57 \cdot 10^{-5}$
1 hundredweight	cwt	=	13 066.7	**1 792**	1 633.33	**1**	**1.12**	**0.05**	0.056
1 US short hundredweight	US sh cwt	=	11 666.7	**1 600**	1 458.33	0.892 857	**1**	0.044 642 9	0.05
1 UK ton	UK ton	=	261 333	**35 840**	32 666.7	**20**	22.4	**1**	1.12
US short ton	US sh ton	=	233 333	**32 000**	29 166.7	17.857 1	**20**	0.892 857	**1**

Examples: $1\,\text{g} = 10^{-3}\,\text{kg} = 0.001\,\text{kg};\ 1\,\text{kg} = 6.852\,18 \cdot 10^{-2}\,\text{slug} = 0.068\,521\,8\,\text{sluga};$

$1\,\text{lb} = 4.535\,923\,7 \cdot 10^{2}\,\text{g} = 453.592\,37\,\text{g};\ 1\,\text{hyl} = 9.651\,79 \cdot 10^{-3}\,\text{UK ton} = 0.009\,651\,79\,\text{UK ton}.$

Exact values are printed in bold type.

The United Kingdom units are denoted by the prefix UK.

The American units are denoted by the prefix US.

*Technical unity of mass in the metric system (unity of mass in the gravitational metric system).

**Technical unity of mass in the Anglo-American system (unity of mass in the Anglo-American gravitational system).

Table 111-2 The Metric Units of Mass

Unit	Abbreviation	mt	kt	t	q	kg	hg	dag (dkg)	g	dg	mg	µg(γ)
1 megaton =	mt	1	10^3	10^6	10^7	10^9	10^{10}	10^{11}	10^{12}	10^{13}	10^{15}	10^{18}
1 kiloton =	kt	10^{-3}	1	10^3	10^4	10^6	10^7	10^8	10^9	10^{10}	10^{12}	10^{15}
1 ton =	t	10^{-6}	10^{-3}	1	10	10^3	10^4	10^5	10^6	10^7	10^9	10^{12}
1 deciton (quintal) =	q	10^{-7}	10^{-4}	10^{-1}	1	10^2	10^3	10^4	10^5	10^6	10^8	10^{11}
1 kilogram =	kg	10^{-9}	10^{-6}	10^{-3}	10^{-2}	1	10^1	10^2	10^3	10^4	10^6	10^9
1 hectogram =	hg	10^{-10}	10^{-7}	10^{-4}	10^{-3}	10^{-1}	1	10^1	10^2	10^3	10^5	10^8
1 decagram =	dag (dkg)	10^{-11}	10^{-8}	10^{-5}	10^{-4}	10^{-2}	10^{-1}	1	10^1	10^2	10^4	10^7
1 gram =	g	10^{-12}	10^{-9}	10^{-6}	10^{-5}	10^{-3}	10^{-2}	10^{-1}	1	10^1	10^3	10^6
1 decigram =	dg	10^{-13}	10^{-10}	10^{-7}	10^{-6}	10^{-4}	10^{-3}	10^{-2}	10^{-1}	1	10^2	10^5
1 milligram =	mg	10^{-15}	10^{-12}	10^{-9}	10^{-8}	10^{-6}	10^{-5}	10^{-4}	10^{-3}	10^{-2}	1	10^3
1 microgram (gama) =	µg (γ)	10^{-18}	10^{-15}	10^{-12}	10^{-11}	10^{-9}	10^{-8}	10^{-7}	10^{-6}	10^{-5}	10^{-3}	1

Examples: 1 kt $= 10^{-3}$ mt $= 0.001$ mt; 1 kg $= 10^3$ g $= 1\,000$ g

Table 111-3 Relationship Between Metric and Chemical Units of Mass

Unit	Abbreviation	kg	dag (dkg)	g	k	dg	cg	mg	µg / γ	ng / γγ	pg / γγγ
1 kilogram =	kg	1	10^2	10^3	$5 \cdot 10^3$	10^4	10^5	10^6	10^9	10^{12}	10^{15}
1 decagram =	dag (dkg)	10^{-2}	1	10^1	$5 \cdot 10^1$	10^2	10^3	10^4	10^7	10^{10}	10^{13}
1 gram =	g	10^{-3}	10^{-1}	1	5	10^1	10^2	10^3	10^6	10^9	10^{12}
1 carat =	k	$2 \cdot 10^{-4}$	$2 \cdot 10^{-2}$	$2 \cdot 10^{-1}$	1	2	$2 \cdot 10^1$	$2 \cdot 10^2$	$2 \cdot 10^5$	$2 \cdot 10^8$	$2 \cdot 10^{11}$
1 decigram =	dg	10^{-4}	10^{-2}	10^{-1}	$5 \cdot 10^{-1}$	1	10^1	10^2	10^5	10^8	10^{11}
1 centigram =	cg	10^{-5}	10^{-3}	10^{-2}	$5 \cdot 10^{-2}$	10^{-1}	1	10^1	10^4	10^7	10^{10}
1 milligram =	mg	10^{-6}	10^{-4}	10^{-3}	$5 \cdot 10^{-3}$	10^{-2}	10^{-1}	1	10^3	10^6	10^9
1 microgram = µg / 1 gama = γ		10^{-9}	10^{-7}	10^{-6}	$5 \cdot 10^{-6}$	10^{-5}	10^{-4}	10^{-3}	1	10^3	10^6
1 nanogram = ng / 1 milligama = γγ		10^{-12}	10^{-10}	10^{-9}	$5 \cdot 10^{-9}$	10^{-8}	10^{-7}	10^{-6}	10^{-3}	1	10^3
1 picogram = pg / 1 microgama = γγγ		10^{-15}	10^{-13}	10^{-12}	$5 \cdot 10^{-12}$	10^{-11}	10^{-10}	10^{-9}	10^{-6}	10^{-3}	1

Examples: 1 dag $= 10^4$ mg $= 10\,000$ mg; 1 ng (γγ) $= 10^{-3}$ µγ (γ) $= 10^{-3}$ µγ (γ) $= 0.001$ µγ (γ).

Density equivalents (mass/volume)

The unit of density in the International System (SI) is the kilogram per cubic meter, abbreviation kg/m³; $1\ \text{kg/m}^3 = 0.062\,428\ \text{lb/ft}^3$. The Anglo-American unit of density is the pound per cubic foot, abbreviation lb/ft³; $1\ \text{lb/ft}^3 = 16.018\,5\ \text{kg/m}^3$.

Table 112-1 Conversion Factors for Units of Density

Unit		Abbreviation	kg/m³	kg/dm³ (kg/l)	g/m³	g/dm³ (g/l)	g/cm³ (g/ml)	g/mm³ (g/µl)
1 kilogram per cubic meter	=	kg/m³	1	10^{-3}	10^3	1	10^{-3}	10^{-6}
1 kilogram per cubic decimeter (liter)	=	kg/dm³ (kg/l)	10^3	1	10^6	10^3	1	10^{-3}
1 gram per cubic meter	=	g/m³	10^{-3}	10^{-6}	1	10^{-3}	10^{-6}	10^{-9}
1 gram per cubic decimeter (liter)	=	g/dm³ (g/l)	1	10^{-3}	10^3	1	10^{-3}	10^{-6}
1 gram per cubic centimeter (milliliter)	=	g/cm³ (g/ml)	10^3	1	10^6	10^3	1	10^{-3}
1 gram per cubic millimeter (microliter)	=	g/mm³ (g/µl)	10^6	10^3	10^9	10^6	10^3	1
1 ton (1000 kg) per cubic meter	=	t/m³	10^3	1	10^6	10^3	1	10^{-3}
1 ton (1000 kg) per hectoliter	=	t/hl	10^4	10^1	10^7	10^4	10^1	10^{-2}
1 pound per cubic inch	=	lb/in.³	$2.767\,99 \cdot 10^4$	$2.767\,99 \cdot 10^1$	$2.767\,99 \cdot 10^7$	$2.767\,99 \cdot 10^4$	$2.767\,99 \cdot 10^1$	$2.767\,99 \cdot 10^{-2}$
1 pound per cubic foot	=	lb/ft³	16.018 5	$16.018\,5 \cdot 10^{-3}$	$16.018\,5 \cdot 10^3$	16.018 5	$16.018\,5 \cdot 10^{-3}$	$16.018\,5 \cdot 10^{-6}$
1 UK ton per cubic yard	=	UK ton/yd³	1 328.94	$1\,328.94 \cdot 10^{-3}$	$1\,328.94 \cdot 10^3$	1 328.94	$1\,328.94 \cdot 10^{-3}$	$1\,328.94 \cdot 10^{-6}$
1 pound per UK gallon	=	lb/UK gal	99.776 4	$99.776\,4 \cdot 10^{-3}$	$99.776\,4 \cdot 10^3$	99.776 4	$99.776\,4 \cdot 10^{-3}$	$99.776\,4 \cdot 10^{-6}$
1 pound per US gallon	=	lb/US gal	119.826	$119.826 \cdot 10^{-3}$	$119.826 \cdot 10^3$	119.826	$119.826 \cdot 10^{-3}$	$119.826 \cdot 10^{-6}$
1 grain per 100 cubic foot	=	gr/100 ft³	$22.883\,5 \cdot 10^{-6}$	$22.883\,5 \cdot 10^{-9}$	$22.883\,5 \cdot 10^{-3}$	$22.883\,5 \cdot 10^{-6}$	$22.883\,5 \cdot 10^{-9}$	$22.883\,5 \cdot 10^{-12}$
1 grain per UK gallon	=	gr/UK gal	$14.253\,8 \cdot 10^{-3}$	$14.253\,8 \cdot 10^{-6}$	14.253 8	$14.253\,8 \cdot 10^{-3}$	$14.253\,8 \cdot 10^{-6}$	$14.253\,8 \cdot 10^{-9}$
1 grain per US gallon	=	gr/US gal	$17.118\,1 \cdot 10^{-3}$	$17.118\,1 \cdot 10^{-6}$	17.118 1	$17.118\,1 \cdot 10^{-3}$	$17.118\,1 \cdot 10^{-6}$	$17.118\,1 \cdot 10^{-9}$
1 ounce per UK gallon	=	oz/UK gal	6.236 03	$5.236\,03 \cdot 10^{-3}$	$6.236\,03 \cdot 10^3$	6.236 03	$6.236\,03 \cdot 10^{-3}$	$6.236\,03 \cdot 10^{-6}$
1 ounce per US gallon	=	oz/US gal	7.489 15	$7.489\,15 \cdot 10^{-3}$	$7.489\,15 \cdot 10^3$	7.489 15	$7.489\,15 \cdot 10^{-3}$	$7.489\,15 \cdot 10^{-6}$

Unit		Abbreviation	t/m³	t/hl	lb/in.³	lb/ft³	UK ton/yd³	lb/UK gal
1 kilogram per cubic meter	=	kg/m³	10^{-3}	10^{-4}	$3.612\,73 \cdot 10^{-5}$	$6.242\,80 \cdot 10^{-2}$	$7.524\,80 \cdot 10^{-4}$	$1.002\,24 \cdot 10^{-2}$
1 kilogram per cubic decimeter (liter)	=	kg/dm³ (kg/l)	1	10^{-1}	$3.612\,73 \cdot 10^{-2}$	$6.242\,80 \cdot 10^1$	$7.524\,80 \cdot 10^{-1}$	$1.002\,24 \cdot 10^1$
1 gram per cubic meter	=	g/m³	10^{-6}	10^{-7}	$3.612\,73 \cdot 10^{-8}$	$6.242\,80 \cdot 10^{-5}$	$7.524\,80 \cdot 10^{-7}$	$1.002\,24 \cdot 10^{-5}$
1 gram per cubic decimeter (liter)	=	g/dm³ (g/l)	10^{-3}	10^{-4}	$3.612\,73 \cdot 10^{-5}$	$6.242\,80 \cdot 10^{-2}$	$7.524\,80 \cdot 10^{-4}$	$1.002\,24 \cdot 10^{-2}$
1 gram per cubic centimeter (milliliter)	=	g/cm³ (g/ml)	1	10^{-1}	$3.612\,73 \cdot 10^{-2}$	$6.242\,80 \cdot 10^1$	$7.524\,80 \cdot 10^{-1}$	$1.002\,24 \cdot 10^1$
1 gram per cubic millimeter (microliter)	=	g/mm³ (g/µl)	10^3	10^2	$3.612\,73 \cdot 10^1$	$6.242\,80 \cdot 10^4$	$7.524\,80 \cdot 10^2$	$1.002\,24 \cdot 10^4$
1 ton (1000 kg) per cubic meter	=	t/m³	1	10^{-1}	$3.612\,73 \cdot 10^{-2}$	$6.242\,80 \cdot 10^1$	$7.524\,80 \cdot 10^{-1}$	$1.002\,24 \cdot 10^1$
1 ton (1000 kg) per hectoliter	=	t/hl	10^1	1	$3.612\,73 \cdot 10^{-1}$	$6.242\,80 \cdot 10^2$	7.524 80	$1.002\,24 \cdot 10^2$
1 pound per cubic inch	=	lb/in.³	$2.767\,99 \cdot 10^1$	2.767 99	1	**1728**	20.828 6	277.419
1 pound per cubic foot	=	lb/ft³	$16.018\,5 \cdot 10^{-3}$	$16.018\,5 \cdot 10^{-4}$	$5.787\,04 \cdot 10^{-4}$	1	0.012 053 6	0.160 544
1 UK ton per cubic yard	=	UK ton/yd³	$1\,328.94 \cdot 10^{-3}$	$1\,328.94 \cdot 10^{-4}$	0.048 011 0	82.963 0	1	13.319 2
1 pound per UK gallon	=	lb/Uk gal	$99.776\,4 \cdot 10^{-3}$	$99.776\,4 \cdot 10^{-4}$	$3.604\,65 \cdot 10^{-3}$	6.228 84	0.075 079 8	1
1 pound per US gallon	=	lb/US gal	$119.826 \cdot 10^{-3}$	$119.826 \cdot 10^{-4}$	$4.329\,00 \cdot 10^{-3}$	7.480 52	0.090 167 0	1.200 95
1 grain per 100 cubic foot	=	gr/100 ft³	$22.883\,5 \cdot 10^{-9}$	$22.883\,5 \cdot 10^{-10}$	$8.267\,19 \cdot 10^{-10}$	$1.428\,57 \cdot 10^{-6}$	$1.721\,94 \cdot 10^{-8}$	$2.293\,48 \cdot 10^{-7}$
1 grain per UK gallon	=	gr/UK gal	$14.253\,8 \cdot 10^{-6}$	$14.253\,8 \cdot 10^{-7}$	$5.149\,51 \cdot 10^{-7}$	$8.898\,34 \cdot 10^{-4}$	$1.072\,57 \cdot 10^{-5}$	$1.428\,57 \cdot 10^{-4}$
1 grain per US gallon	=	gr/US gal	$17.118\,1 \cdot 10^{-6}$	$17.118\,1 \cdot 10^{-7}$	$6.184\,31 \cdot 10^{-7}$	$1.068\,65 \cdot 10^{-3}$	$1.288\,10 \cdot 10^{-5}$	$1.715\,65 \cdot 10^{-4}$
1 ounce per UK gallon	=	oz/UK gal	$6.236\,03 \cdot 10^{-3}$	$6.236\,03 \cdot 10^{-4}$	$2.252\,91 \cdot 10^{-4}$	0.389 302	$4.692\,48 \cdot 10^{-3}$	**0.062 5**
1 ounce per US gallon	=	oz/US gal	$7.489\,15 \cdot 10^{-3}$	$7.489\,15 \cdot 10^{-4}$	$2.705\,63 \cdot 10^{-4}$	0.467 531	$5.635\,43 \cdot 10^{-3}$	0.075 059 3

Unit		Abbreviation	lb/US gal	gr/100 ft³	gr/UK gal	gr/US gal	oz/UK gal	oz/US gal
1 kilogram per cubic meter	=	kg/m³	$8.345\,40 \cdot 10^{-3}$	$43.699\,6 \cdot 10^3$	$7.015\,68 \cdot 10^1$	$5.841\,78 \cdot 10^1$	$1.603\,59 \cdot 10^{-1}$	$1.335\,26 \cdot 10^{-1}$
1 kilogram per cubic decimeter (liter)	=	kg/dm³ (kg/l)	8.345 40	$43.699\,6 \cdot 10^6$	$7.015\,68 \cdot 10^4$	$5.841\,78 \cdot 10^4$	$1.603\,59 \cdot 10^2$	$1.335\,26 \cdot 10^2$
1 gram per cubic meter	=	g/m³	$8.345\,40 \cdot 10^{-6}$	43.699 6	$7.015\,68 \cdot 10^{-2}$	$5.841\,78 \cdot 10^{-2}$	$1.603\,59 \cdot 10^{-4}$	$1.335\,26 \cdot 10^{-4}$
1 gram per cubic decimeter (liter)	=	g/dm³ (g/l)	$8.345\,40 \cdot 10^{-3}$	$43.699\,6 \cdot 10^3$	$7.015\,68 \cdot 10^1$	$5.841\,78 \cdot 10^1$	$1.603\,59 \cdot 10^{-1}$	$1.335\,26 \cdot 10^{-1}$
1 gram per cubic centimeter (milliliter)	=	g/cm³ (g/ml)	8.345 40	$43.699\,6 \cdot 10^6$	$7.015\,68 \cdot 10^4$	$5.841\,58 \cdot 10^4$	$1.603\,59 \cdot 10^2$	$1.335\,26 \cdot 10^2$
1 gram per cubic millimeter (microliter)	=	g/mm³ (g/µl)	$8.345\,40 \cdot 10^3$	$43.699\,6 \cdot 10^9$	$7.015\,68 \cdot 10^7$	$5.841\,78 \cdot 10^7$	$1.603\,59 \cdot 10^5$	$1.335\,26 \cdot 10^5$
1 ton (1000 kg) per cubic meter	=	t/m³	8.345 40	$43.699\,6 \cdot 10^6$	$7.015\,68 \cdot 10^4$	$5.841\,78 \cdot 10^4$	$1.603\,59 \cdot 10^2$	$1.335\,26 \cdot 10^2$
1 ton (1000 kg) per hectoliter	=	t/hl	$8.345\,40 \cdot 10^1$	$43.699\,6 \cdot 10^7$	$7.015\,68 \cdot 10^5$	$5.841\,78 \cdot 10^5$	$1.603\,59 \cdot 10^3$	$1.335\,26 \cdot 10^3$
1 pound per cubic inch	=	lb/in.³	231	$120\,960 \cdot 10^4$	1 941 931	1 616.996	4 438.705	3 696
1 pound per cubic foot	=	lb/ft³	0.133 681	700 000	1 123.81	935.764	2.568 70	2.138 89
1 UK ton per cubic yard	=	UK ton/yd³	11.090 5	58 074 158	93 234.1	77 633.6	213.107	177.449
1 pound per UK gallon	=	lb/UK gal	0.832 675	$4.360\,19 \cdot 10^6$	7 000	5 828.708	16	13.322 79
1 pound per US gallon	=	lb/US gal	1	5 236 349	8 406.60	7 000	19.215 1	16
1 grain per 100 cubic foot	=	gr/100 ft³	$1.909\,73 \cdot 10^{-7}$	1	$1.605\,44 \cdot 10^{-3}$	$1.336\,81 \cdot 10^{-3}$	$3.669\,57 \cdot 10^{-6}$	$3.055\,56 \cdot 10^{-6}$
1 grain per UK gallon	=	gr/UK gal	$1.189\,54 \cdot 10^{-4}$	622.884	1	0.832 673	$2.285\,71 \cdot 10^{-3}$	$1.903\,26 \cdot 10^{-3}$
1 grain per US gallon	=	gr/US gal	$1.428\,57 \cdot 10^{-4}$	748.052	1.200 95	1	$2.745\,03 \cdot 10^{-3}$	$2.285\,71 \cdot 10^{-3}$
1 ounce per UK gallon	=	oz/UK gal	0.052 042 3	272 512	**437.5**	364.295	1	0.832 675
1 ounce per US gallon	=	oz/US gal	**0.062 5**	327 273	525.415	**437.5**	1.200 95	1

Examples: $1\ \text{kg/m}^3 = 10^{-3}\ \text{g/cm}^3 = 0.001\ \text{g/cm}^3$; $1\ \text{g/cm}^3 = 3.612\,73\ 10^{-2}\ \text{lb/in}^3 = 0.036\,127\,3\ \text{lb/in}^3$; $1\ \text{lb/ft}^3 = 16.018\,5 \cdot 10^{-3}\ \text{kg/dm}^3 = 0.016\,018\,5\ \text{kg/dm}^3$; $1\ \text{gr/US gal} = 2.745\,03\ 10^{-3}\ \text{oz/UK gal} = 0.002\,745\,03\ \text{oz/UK gal}$.

Exact values are printed in bold type.
The United Kingdom units are denoted by the prefix UK.
The American units are denoted by the prefix US.

Equivalents of masses per unit lengths (mass/length)

The unit of mass per unit length in the International System (SI) is the kilogram per meter, abbreviation kg/m; 1 kg/m = 2.015 91 lb/yd.
The Anglo-American unit of masses per unit lengths is the pound per yard, abbreviation lb/yd; 1 lb/yd = 0.496 055 kg/m.

Table 113-1 Conversion Factors for Units of Mass Per Unit Length

Unit	Abbreviation	kg/m	kg/dm	kg/cm	t/m	lb/in.	lb/ft	lb/yd	ton/1000 yd	ton/mile
1 kilogram per meter =	kg/m	1	10^{-1}	10^{-2}	10^{-3}	$5.599\,74 \cdot 10^{-2}$	$6.719\,69 \cdot 10^{-1}$	2.015 91	$8.999\,58 \cdot 10^{-1}$	1.583 93
1 kilogram per decimeter =	kg/dm	10^{1}	1	10^{-1}	10^{-2}	$5.599\,74 \cdot 10^{-1}$	6.719 69	$2.015\,91 \cdot 10^{1}$	8.999 58	$1.583\,93 \cdot 10^{1}$
1 kilogram per centimeter =	kg/cm	10^{2}	10^{1}	1	10^{-1}	5.599 74	$6.719\,69 \cdot 10^{1}$	$2.015\,91 \cdot 10^{2}$	$8.999\,58 \cdot 10^{1}$	$1.583\,93 \cdot 10^{2}$
1 ton (1000 kg) per meter =	t/m	10^{3}	10^{2}	10^{1}	1	$5.599\,74 \cdot 10^{1}$	$6.719\,69 \cdot 10^{2}$	$2.015\,91 \cdot 10^{3}$	$8.999\,58 \cdot 10^{2}$	$1.583\,93 \cdot 10^{3}$
1 pound per inch =	lb/in.	17.858 0	$17.858\,0 \cdot 10^{-1}$	$17.858\,0 \cdot 10^{-2}$	$17.858\,0 \cdot 10^{-3}$	1	12	36	16.071 4	28.285 7
1 pound per foot =	lb/ft	1.488 16	$1.488\,16 \cdot 10^{-1}$	$1.488\,16 \cdot 10^{-2}$	$1.488\,16 \cdot 10^{-3}$	0.083 333 3	1	3	1.339 29	2.357 14
1 pound per yard =	lb/yd	$4.960\,55 \cdot 10^{-1}$	$4.960\,55 \cdot 10^{-2}$	$4.960\,55 \cdot 10^{-3}$	$4.960\,55 \cdot 10^{-4}$	0.027 777 8	0.333 333	1	0.446 429	0.785 714
1 UK ton per 1000 yards =	ton/1000 yd	1.111 16	$1.111\,16 \cdot 10^{-1}$	$1.111\,16 \cdot 10^{-2}$	$1.111\,16 \cdot 10^{-3}$	0.062 222 2	0.746 667	2.24	1	1.76
1 UK ton per mile =	ton/mile	$6.313\,42 \cdot 10^{-1}$	$6.313\,42 \cdot 10^{-2}$	$6.313\,42 \cdot 10^{-3}$	$6.313\,42 \cdot 10^{-4}$	0.035 353 5	0.424 242	1.272 73	0.568 182	1

Examples: 1 kg/m = 6.719 69 · 10⁻¹ lb/ft = 0.671 969 lb/ft; 1 lb/yd = 4.960 55 · 10⁻¹ kg/m = 0.496 055 kg/m.
Exact values are printed in bold type.
The United Kingdom units are denoted by the prefix UK.

Equivalents of masses per unit areas (mass/area)

The unit of mass per unit area in the International System (SI) is the kilogram per square meter, abbreviation kg/m²; 1 kg/m² = 29.493 5 oz/yd².
The Anglo-American unit of masses per unit areas is the ounce per square yard, abbreviation oz/yd²; 1 oz/yd² = 0.033905 kg/m².

Table 114-1 Conversion Factors for Units of Mass Per Unit Area

Unit	Abbreviation	kg/m²	kg/ha	g/m²	mg/cm²	lb/1000 ft²	oz/yd²	oz/ft²	lb/acre	UK ton/sq mile
1 kilogram per square meter =	kg/m²	1	10^{4}	10^{3}	10^{2}	$2.048\,16 \cdot 10^{2}$	$2.949\,35 \cdot 10^{1}$	3.277 06	$8.921\,79 \cdot 10^{3}$	$2.549\,08 \cdot 10^{3}$
1 kilogram per hectare =	kg/ha	10^{-4}	1	10^{-1}	10^{-2}	$2.048\,16 \cdot 10^{-2}$	$2.949\,35 \cdot 10^{-3}$	$3.277\,06 \cdot 10^{-4}$	$8.921\,79 \cdot 10^{-1}$	$2.549\,08 \cdot 10^{-1}$
1 gram per square meter =	g/m²	10^{-3}	10^{1}	1	10^{-1}	$2.048 \cdot 10^{-1}$	$2.949\,35 \cdot 10^{-2}$	$3.277\,06 \cdot 10^{-3}$	8.921 79	2.549 08
1 milligram per square meter =	mg/cm²	10^{-2}	10^{2}	10^{1}	1	2.048 16	$2.949\,35 \cdot 10^{-1}$	$3.277\,06 \cdot 10^{-2}$	$8.921\,79 \cdot 10^{1}$	$2.549\,08 \cdot 10^{1}$
1 pound per thousand square feet =	lb/1000 ft²	$4.882\,43 \cdot 10^{-3}$	$4.882\,43 \cdot 10^{1}$	4.882 43	$4.882\,43 \cdot 10^{-1}$	1	**0.144**	0.016	43.56	12.445 7
1 ounce per square yard =	oz/yd²	$33.905\,7 \cdot 10^{-3}$	$33.905\,7 \cdot 10^{1}$	33.905 7	$33.905\,7 \cdot 10^{-1}$	6.944 44	1	0.111 111	302.5	86.428 6
1 ounce per square foot =	oz/ft²	$305.152 \cdot 10^{-3}$	$305.152 \cdot 10^{1}$	305.152	$305.152 \cdot 10^{-1}$	62.5	9	1	2 722.5	777.857
1 pound per acre =	lb/acre	$1.120\,85 \cdot 10^{-4}$	1.120 85	$1.120\,85 \cdot 10^{-1}$	$1.120\,85 \cdot 10^{-2}$	0.022 956 8	$3.305\,79 \cdot 10^{-3}$	$3.673\,09 \cdot 10^{-4}$	1	0.285 714
1 UK ton per square mile =	UK ton/sq mile	$3.922\,98 \cdot 10^{-4}$	3.922 98	$3.922\,98 \cdot 10^{-1}$	$3.922\,98 \cdot 10^{-2}$	0.080 348 9	0.011 570 2	$1.285\,58 \cdot 10^{-3}$	3.5	1

Examples: 1 g/m² = 2.949 35 · 10⁻² oz/yd² = 0.029 493 5 oz/yd²; 1 oz/ft² = 305.152 · 10⁻¹ mg/cm² = 30.515 2 mg/cm².
Exact values are printed in bold type.
The United Kingdom units are denoted by the prefix UK.

Mass rate of flow equivalents (mass/time)

The unit of mass rate of flow in the International System (SI) is the kilogram per second, abbreviation kg/s; 1 kg/s = 2.204 62 lb/s.
In the Anglo-American system the unit of mass rate of flow is the pound per second, abbreviation lb/s; 1 lb/s = 0.453 592 kg/s.

Table 115-1 Conversion Factors for Units of Mass Rate of Flow

Unit		Abbreviation	kg/s	g/s	t/s	kg/min	q/min	t/min
1 kilogram per second	=	kg/s	**1**	**10^3**	**10^{-3}**	**60**	**$60 \cdot 10^{-2}$**	**$60 \cdot 10^{-3}$**
1 gram per second	=	g/s	**10^{-3}**	**1**	**10^{-6}**	**$60 \cdot 10^{-3}$**	**$60 \cdot 10^{-5}$**	**$60 \cdot 10^{-6}$**
1 ton (1000 kg) per second	=	t/s	**10^3**	**10^6**	**1**	**$60 \cdot 10^3$**	**$60 \cdot 10^1$**	**60**
1 kilogram per minute	=	kg/min	$1.666\,67 \cdot 10^{-2}$	$1.666\,67 \cdot 10^1$	$1.666\,67 \cdot 10^{-5}$	**1**	**10^{-2}**	**10^{-3}**
1 deciton per minute	=	q/min	$1.666\,67$	$1.666\,67 \cdot 10^3$	$1.666\,67 \cdot 10^{-3}$	**10^2**	**1**	**10^{-1}**
1 ton (1000 kg) per minute	=	t/min	$1.666\,67 \cdot 10^1$	$1.666\,67 \cdot 10^4$	$1.666\,67 \cdot 10^{-2}$	**10^3**	**10^1**	**1**
1 kilogram per hour	=	kg/h	$2.777\,78 \cdot 10^{-4}$	$2.777\,78 \cdot 10^{-1}$	$2.777\,78 \cdot 10^{-7}$	$1.666\,67 \cdot 10^{-2}$	$1.666\,67 \cdot 10^{-4}$	$1.666\,67 \cdot 10^{-5}$
1 deciton per hour	=	q/h	$2.777\,78 \cdot 10^{-2}$	$2.777\,78 \cdot 10^1$	$2.777\,78 \cdot 10^{-5}$	$1.666\,67$	$1.666\,67 \cdot 10^{-2}$	$1.666\,67 \cdot 10^{-3}$
1 ton (1000 kg) per hour	=	t/h	$2.777\,78 \cdot 10^{-1}$	$2.777\,78 \cdot 10^2$	$2.777\,78 \cdot 10^{-4}$	$1.666\,67 \cdot 10^1$	$1.666\,67 \cdot 10^{-1}$	$1.666\,67 \cdot 10^{-2}$
1 pound per second	=	lb/s	$4.535\,92 \cdot 10^{-1}$	$4.535\,92 \cdot 10^2$	$4.535\,92 \cdot 10^{-4}$	$27.215\,5$	$27.215\,5 \cdot 10^{-2}$	$27.215\,5 \cdot 10^{-3}$
1 pound per hour	=	lb/h	$1.259\,98 \cdot 10^{-4}$	$1.259\,98 \cdot 10^{-1}$	$1.259\,98 \cdot 10^{-7}$	$7.559\,88 \cdot 10^{-3}$	$7.559\,88 \cdot 10^{-5}$	$7.559\,88 \cdot 10^{-6}$
1 UK ton per hour	=	UK ton/h	$2.822\,35 \cdot 10^{-1}$	$2.822\,35 \cdot 10^2$	$2.822\,35 \cdot 10^{-4}$	$16.934\,1$	$16.934\,1 \cdot 10^{-2}$	$16.934\,1 \cdot 10^{-3}$

Unit		Abbreviation	kg/h	q/h	t/h	lb/s	lb/h	UK ton/h
1 kilogram per second	=	kg/s	**$3\,600$**	**$3\,600 \cdot 10^{-2}$**	**$3\,600 \cdot 10^{-3}$**	$2.204\,62$	$7.936\,64 \cdot 10^3$	$3.543\,14$
1 gram per second	=	g/s	**$3\,600 \cdot 10^{-3}$**	**$3\,600 \cdot 10^{-5}$**	**$3\,600 \cdot 10^{-6}$**	$2.204\,62 \cdot 10^{-3}$	$7.936\,64$	$3.543\,14 \cdot 10^{-3}$
1 ton (1000 kg) per second	=	t/s	**$3\,600 \cdot 10^3$**	**$3\,600 \cdot 10^1$**	**$3\,600$**	$2.204\,62 \cdot 10^3$	$7.936\,64 \cdot 10^6$	$3.543\,14 \cdot 10^3$
1 kilogram per minute	=	kg/min	**60**	**$60 \cdot 10^{-2}$**	**$60 \cdot 10^{-3}$**	$3.674\,38 \cdot 10^{-2}$	132.277	$5.905\,25 \cdot 10^{-2}$
1 deciton per minute	=	q/min	**$60 \cdot 10^2$**	**60**	**$60 \cdot 10^{-1}$**	$3.674\,38$	$132.277 \cdot 10^2$	$5.905\,25$
1 ton (1000 kg) per minute	=	t/min	**$60 \cdot 10^3$**	**$60 \cdot 10^1$**	**60**	$3.674\,38 \cdot 10^1$	$132.277 \cdot 10^3$	$5.905\,25 \cdot 10^1$
1 kilogram per hour	=	kg/h	**1**	**10^{-2}**	**10^{-3}**	$6.123\,95 \cdot 10^{-4}$	$2.204\,62$	$9.842\,07 \cdot 10^{-4}$
1 deciton per hour	=	q/h	**10^2**	**1**	**10^{-1}**	$6.123\,95 \cdot 10^{-2}$	$2.204\,62 \cdot 10^2$	$9.842\,07 \cdot 10^{-2}$
1 ton (1000 kg) per hour	=	t/h	**10^3**	**10^1**	**1**	$6.123\,95 \cdot 10^{-1}$	$2.204\,62 \cdot 10^3$	$9.842\,07 \cdot 10^{-1}$
1 pound per second	=	lb/s	$1.632\,93 \cdot 10^3$	$1.632\,93 \cdot 10^1$	$1.632\,93$	**1**	**$3\,600$**	$1.607\,14$
1 pound per hour	=	lb/h	$4.535\,92 \cdot 10^{-1}$	$4.535\,92 \cdot 10^{-3}$	$4.535\,92 \cdot 10^{-4}$	$2.777\,78 \cdot 10^{-4}$	**1**	$4.464\,29 \cdot 10^{-4}$
1 UK ton per hour	=	UK ton/h	$1.016\,05 \cdot 10^3$	$1.016\,05 \cdot 10^1$	$1.016\,05$	$0.622\,222$	**$2\,240$**	**1**

Examples: 1 kg/s = 2.204 62 lb/s; 1 lb/h = 4.535 92 · 10^{-1} kg/h = 0.453 592 kg/h.
Exact values are printed in bold type.
The United Kingdom units are denoted by the prefix UK.
The American units are denoted by the prefix US.

Volume rate of flow equivalents (volume/time)

The unit volume rate of flow in the International System (SI) is the cubic meter per second, abbreviation m³/s; 1 m³/s = 35.3145 ft³/s.
The Anglo-American unit volume rate of flow is the cubic foot per second, abbreviation ft³/s; 1 ft³/s = 0.028 316 9 m³/s.

Table 116-1 Conversion Factors for Units of Volume Rate of Flow

Unit	Abbreviation	m³/s	hl/s	l/s	m³/min	hl/min	l/min
1 cubic meter per second =	m³/s	1	10^1	10^3	60	$60 \cdot 10^1$	$60 \cdot 10^3$
1 hectoliter per second =	hl/s	10^{-1}	1	10^2	$60 \cdot 10^{-1}$	60	$60 \cdot 10^2$
1 liter per second =	l/s	10^{-3}	10^{-2}	1	$60 \cdot 10^{-3}$	$60 \cdot 10^{-2}$	60
1 cubic meter per minute =	m³/min	$1.666\ 67 \cdot 10^{-2}$	$1.666\ 67 \cdot 10^{-1}$	$1.666\ 67 \cdot 10^1$	1	10^1	10^3
1 hectoliter per minute =	hl/min	$1.666\ 67 \cdot 10^{-3}$	$1.666\ 67 \cdot 10^{-2}$	$1.666\ 67$	10^{-1}	1	10^2
1 liter per minute =	l/min	$1.666\ 67 \cdot 10^{-5}$	$1.666\ 67 \cdot 10^{-4}$	$1.666\ 67 \cdot 10^{-2}$	10^{-3}	10^{-2}	1
1 cubic meter per hour =	m³/h	$2.777\ 78 \cdot 10^{-4}$	$2.777\ 78 \cdot 10^{-3}$	$2.777\ 78 \cdot 10^{-1}$	$1.666\ 67 \cdot 10^{-2}$	$1.666\ 67 \cdot 10^{-1}$	$1.666\ 67 \cdot 10^1$
1 hectoliter per hour =	hl/h	$2.777\ 78 \cdot 10^{-5}$	$2.777\ 78 \cdot 10^{-4}$	$2.777\ 78 \cdot 10^{-2}$	$1.666\ 67 \cdot 10^{-3}$	$1.666\ 67 \cdot 10^{-2}$	$1.666\ 67$
1 liter per hour =	l/h	$2.777\ 78 \cdot 10^{-7}$	$2.777\ 78 \cdot 10^{-6}$	$2.777\ 78 \cdot 10^{-4}$	$1.666\ 67 \cdot 10^{-5}$	$1.666\ 67 \cdot 10^{-4}$	$1.666\ 67 \cdot 10^{-2}$
1 UK gallon per hour =	UK gal/h	$1.262\ 80 \cdot 10^{-6}$	$1.262\ 80 \cdot 10^{-5}$	$1.262\ 80 \cdot 10^{-3}$	$7.576\ 82 \cdot 10^{-5}$	$7.576\ 82 \cdot 10^{-4}$	$7.576\ 82 \cdot 10^{-2}$
1 UK gallon per minute =	UK gal/min	$7.576\ 81 \cdot 10^{-5}$	$7.576\ 81 \cdot 10^{-4}$	$7.576\ 81 \cdot 10^{-2}$	$4.546\ 08 \cdot 10^{-3}$	$4.546\ 08 \cdot 10^{-2}$	$4.546\ 08$
1 cubic foot per second =	ft³/s	$2.831\ 69 \cdot 10^{-2}$	$2.831\ 69 \cdot 10^{-1}$	$2.831\ 69 \cdot 10^1$	$1.699\ 02$	$1.699\ 02 \cdot 10^1$	$1.699\ 02 \cdot 10^3$
1 UK gallon per second =	UK gal/s	$4.546\ 08 \cdot 10^{-3}$	$4.546\ 08 \cdot 10^{-2}$	$4.546\ 08$	$2.727\ 65 \cdot 10^{-1}$	$2.727\ 65$	$2.727\ 65 \cdot 10^2$

Unit	Abbreviation	m³/h	hl/h	l/h	UK gal/h	UK gal/min	ft³/s	UK gal/s
1 cubic meter per second =	m³/s	**3 600**	$\mathbf{3\ 600 \cdot 10^1}$	$\mathbf{3\ 600 \cdot 10^3}$	$791.889 \cdot 10^3$	$13.198\ 2 \cdot 10^3$	$3.531\ 45 \cdot 10^1$	$2.199\ 70 \cdot 10^2$
1 hectoliter per second =	hl/s	$\mathbf{3\ 600 \cdot 10^{-1}}$	**3 600**	$\mathbf{3\ 600 \cdot 10^2}$	$791.889 \cdot 10^2$	$13.198\ 2 \cdot 10^2$	$3.531\ 45$	$2.199\ 70 \cdot 10^1$
1 liter per second =	l/s	$\mathbf{3\ 600 \cdot 10^{-3}}$	$\mathbf{3\ 600 \cdot 10^{-2}}$	**3 600**	791.889	$13.198\ 2$	$3.531\ 45 \cdot 10^{-2}$	$2.199\ 70 \cdot 10^{-1}$
1 cubic meter per minute =	m³/min	**60**	$\mathbf{60 \cdot 10^1}$	$\mathbf{60 \cdot 10^3}$	$13.1982 \cdot 10^3$	$2.199\ 70 \cdot 10^2$	$5.885\ 76 \cdot 10^{-1}$	$3.666\ 16$
1 hectoliter per minute =	hl/min	$\mathbf{60 \cdot 10^{-1}}$	**60**	$\mathbf{60 \cdot 10^2}$	$13.1982 \cdot 10^2$	$2.199\ 70 \cdot 10^1$	$5.886\ 76 \cdot 10^{-2}$	$3.666\ 16 \cdot 10^{-1}$
1 liter per minute =	l/min	$\mathbf{60 \cdot 10^{-3}}$	$\mathbf{60 \cdot 10^{-2}}$	**60**	13.1982	$2.199\ 70 \cdot 10^{-1}$	$5.885\ 76 \cdot 10^{-4}$	$3.666\ 16 \cdot 10^{-3}$
1 cubic meter per hour =	m³/h	**1**	$\mathbf{10^1}$	$\mathbf{10^3}$	219.969	$3.666\ 16$	$9.809\ 63 \cdot 10^{-3}$	$6.110\ 26 \cdot 10^{-2}$
1 hectoliter per hour =	hl/h	$\mathbf{10^{-1}}$	**1**	$\mathbf{10^2}$	$219.969 \cdot 10^{-1}$	$3.666\ 16 \cdot 10^{-1}$	$9.809\ 63 \cdot 10^{-4}$	$6.110\ 26 \cdot 10^{-3}$
1 liter per hour =	l/h	$\mathbf{10^{-3}}$	$\mathbf{10^{-2}}$	**1**	$219.969 \cdot 10^{-3}$	$3.666\ 16 \cdot 10^{-3}$	$9.809\ 63 \cdot 10^{-6}$	$6.110\ 26 \cdot 10^{-5}$
1 UK gallon per hour =	UK gal/h	$\mathbf{4.546\ 09 \cdot 10^{-3}}$	$\mathbf{4.546\ 09 \cdot 10^{-2}}$	**4.546 09**	1	$0.016\ 666\ 7$	$4.459\ 54 \cdot 10^{-5}$	$2.777\ 78 \cdot 10^{-4}$
1 UK gallon per minute =	UK gal/min	$\mathbf{272.765 \cdot 10^{-3}}$	$\mathbf{272.765 \cdot 10^{-2}}$	**272.765**	60	1	$2.675\ 73 \cdot 10^{-3}$	$0.016\ 666\ 7$
1 cubic foot per second =	ft³/s	**101.941**	$\mathbf{191.941 \cdot 10^1}$	$\mathbf{101.941 \cdot 10^3}$	$2.242\ 38 \cdot 10^4$	373.730	1	$6.228\ 84$
1 UK gallon per second =	UK gal/s	**16.365 9**	$\mathbf{16.365\ 9 \cdot 10^1}$	$\mathbf{16.365\ 9 \cdot 10^3}$	$3\ 600$	60	$0.160\ 544$	1

Examples: 1 m³/h = 9.809 63 · 10⁻³ ft³/s = 0.009 809 63 ft³/s;
1 UK gal/h = 4.546 09 · 10⁻² hl/h = 0.045 460 9 hl/h.
Exact values are printed in bold type.
The United Kingdom units are denoted by the prefix UK.
The American units are denoted by the prefix US.

Force equivalents

The unit of force in the International System (SI) is the newton, abbreviation N; 1 N = 0.101 971 6 kp ≈ 0.102 kp = 0.224 09 lbf.
The unit of force in the metric gravitational system is the kilopond, abbreviation kp; 1 kp = 9.806 65 N = 2.204 62 lbf.
The Anglo-American unit of force is the pound-force, abbreviation lbf; 1 lbf = 4.448 22 N = 0.453 592 kp.

Table 117-1 Conversion Factors for Units of Force

Unit	Abbreviation	N	kN	MN	mN	dyne	sn	kp
1 newton =	N	**1**	**10^{-3}**	**10^{-6}**	**10^3**	**10^5**	**10^{-3}**	1.019 72 · 10^{-1}
1 kilonewton =	kN	**10^3**	**1**	**10^{-3}**	**10^6**	**10^8**	**1**	1.019 72 · 10^2
1 meganewton =	MN	**10^6**	**10^3**	**1**	**10^9**	**10^{11}**	**10^3**	1.019 72 · 10^5
1 millinewton =	mN	**10^{-3}**	**10^{-6}**	**10^{-9}**	**1**	**10^2**	**10^{-6}**	1.019 72 · 10^{-4}
1 dyne =	dyne	**10^{-5}**	**10^{-8}**	**10^{-11}**	**10^{-2}**	**1**	**10^{-8}**	1.019 2 · 10^{-6}
1 sthene =	sn	**10^3**	**1**	**10^{-3}**	**10^6**	**10^8**	**1**	1.019 2 · 10^2
1 kilopond =	kp	**9.806 65**	**9.806 65 · 10^{-3}**	**9.806 65 · 10^{-6}**	**9.806 65 · 10^3**	**9.806 65 · 10^5**	**9.806 65 · 10^{-3}**	**1**
1 megapond =	Mp	**9.806 65 · 10^3**	**9.806 65**	**9.806 65 · 10^{-3}**	**9.806 65 · 10^6**	**9.806 65 · 10^8**	**9.806 65**	**10^3**
1 pond =	p	**9.806 65 · 10^{-3}**	**9.806 65 · 10^{-6}**	**9.806 65 · 10^{-9}**	**9.806 65**	**9.806 65 · 10^2**	**9.806 65 · 10^{-6}**	**10^{-3}**
1 millipond =	mp	**9.806 65 · 10^{-6}**	**9.806 65 · 10^{-9}**	**9.806 65 · 10^{-12}**	**9.806 65 · 10^{-3}**	**9.806 65 · 10^{-1}**	**9.806 65 · 10^{-9}**	**10^{-6}**
1 poundal =	pdl	1.382 55 · 10^{-1}	1.382 55 · 10^{-4}	1.382 55 · 10^{-7}	1.382 55 · 10^2	1.382 55 · 10^4	1.382 55 · 10^{-4}	1.409 81 · 10^{-2}
1 pound-force =	lbf	4.448 22	4.448 22 · 10^{-3}	4.448 22 · 10^{-6}	4.448 22 · 10^3	4.448 22 · 10^5	4.448 22 · 10^{-3}	4.535 92 · 10^{-1}
1 UK ton-force =	UK tonf	9 964.02	9 964.02 · 10^{-3}	9 964.02 · 10^{-6}	9 964.02 · 10^3	9 964.02 · 10^5	9 964.02 · 10^{-3}	1 016.05
1 ounce-force =	ozf	2.780 14 · 10^{-1}	2.780 14 · 10^{-4}	2.780 14 · 10^{-7}	2.780 14 · 10^2	2.780 14 · 10^4	2.780 14 · 10^{-4}	2.834 95 · 10^{-2}

Unit	Abbreviation	Mp	p	mp	pdl	lbf	UK tonf	ozf
1 newton =	N	1.019 72 · 10^{-4}	1.019 72 · 10^2	1.019 72 · 10^5	7.233 01	2.248 09 · 10^{-1}	1.003 61 · 10^{-4}	3.596 94
1 kilonewton =	kN	1.019 72 · 10^{-1}	1.019 72 · 10^5	1.019 72 · 10^8	7.233 01 · 10^3	2.248 09 · 10^2	1.003 61 · 10^{-1}	3.596 94 · 10^3
1 meganewton =	MN	1.019 72 · 10^2	1.019 72 · 10^8	1.019 72 · 10^{11}	7.233 01 · 10^6	2.248 09 · 10^5	1.003 61 · 10^2	3.596 94 · 10^6
1 millinewton =	mN	1.019 72 · 10^{-7}	1.019 72 · 10^{-1}	1.019 72 · 10^2	7.233 01 · 10^{-3}	2.248 09 · 10^{-4}	1.003 61 · 10^{-7}	3.596 94 · 10^{-3}
1 dyne =	dyne	1.019 72 · 10^{-9}	1.019 72 · 10^{-3}	1.019 72	7.233 01 · 10^{-5}	2.248 09 · 10^{-6}	1.003 61 · 10^{-9}	3.596 94 · 10^{-5}
1 sthene =	sn	1.019 72 · 10^{-1}	1.019 72 · 10^5	1.019 72 · 10^8	7.233 01 · 10^3	2.248 09 · 10^2	1.003 61 · 10^{-1}	3.596 94 · 10^3
1 kilopond =	kp	**10^{-3}**	**10^3**	**10^6**	70.931 6	2.204 62	9.842 07 · 10^{-4}	35.274 0
1 megapond =	Mp	**1**	**10^6**	**10^9**	70.931 6 · 10^3	2.204 62 · 10^3	9.842 07 · 10^{-1}	35.274 0 · 10^3
1 pond =	p	**10^{-6}**	**1**	**10^3**	70.931 6 · 10^{-3}	2.204 62 · 10^{-3}	9.842 07 · 10^{-7}	35.274 0 · 10^{-3}
1 millipond =	mp	**10^{-9}**	**10^{-3}**	**1**	70.931 6 · 10^{-6}	2.204 62 · 10^{-6}	9.842 07 · 10^{-10}	35.274 0 · 10^{-6}
1 poundal =	pdl	1.409 81 · 10^{-5}	1.409 81 · 10^1	1.409 81 · 10^4	**1**	0.031 081 0	1.387 54 · 10^{-5}	0.497 295
1 pound-force =	lbf	4.535 92 · 10^{-4}	4.535 92 · 10^2	4.535 92 · 10^5	32.174 0	**1**	4.464 29 · 10^{-4}	**16**
1 UK ton-force =	UK tonf	1 016.05 · 10^{-3}	1 016.05 · 10^3	1 016.05 · 10^6	72 069.9	**2 240**	**1**	**35 840**
1 ounce-force =	ozf	2.834 95 · 10^{-5}	2.834 95 · 10^1	2.834 95 · 10^4	2.010 88	**0.062 5**	2.790 18 · 10^{-5}	**1**

Examples: 1 N = 0.101 972 kp; 1 kN = 1.003 61 · 10^{-1} tonf = 0.100 361 tonf;
1 kp = 9.806 65 · 10^{-3} kN = 0.009 806 65 kN; 1 lbf = 4.464 29 · 10^{-4} tonf = 0.000 446 429 tonf.

Exact values are printed in bold type.
The United Kingdom units are denoted by the prefix UK.
The American units are denoted by the prefix US.

Table 117-2 Relations Between the Units of Force: Newton, Sthene and Dyne

Unit	Abbreviation	mN	N	kN	MN	msn	csn	dsn	sn	dasn	hsn	ksn	dyne
1 millinewton =	mN	1	10^{-3}	10^{-6}	10^{-9}	10^{-3}	10^{-4}	10^{-5}	10^{-6}	10^{-7}	10^{-8}	10^{-9}	10^2
1 newton =	N	10^3	1	10^{-3}	10^{-6}	1	10^{-1}	10^{-2}	10^{-3}	10^{-4}	10^{-5}	10^{-6}	10^5
1 kilonewton =	kN	10^6	10^3	1	10^{-3}	10^3	10^2	10^1	1	10^{-1}	10^{-2}	10^{-3}	10^8
1 meganewton =	MN	10^9	10^6	10^3	1	10^6	10^5	10^4	10^3	10^2	10^1	1	10^{11}
1 millisthene =	msn	10^3	1	10^{-3}	10^{-6}	1	10^{-1}	10^{-2}	10^{-3}	10^{-4}	10^{-5}	10^{-6}	10^5
1 centisthene =	csn	10^4	10^1	10^{-2}	10^{-5}	10^1	1	10^{-1}	10^{-2}	10^{-3}	10^{-4}	10^{-5}	10^6
1 decisthene =	dsn	10^5	10^2	10^{-1}	10^{-4}	10^2	10^1	1	10^{-1}	10^{-2}	10^{-3}	10^{-4}	10^7
1 sthene =	sn	10^6	10^3	1	10^{-3}	10^3	10^2	10^1	1	10^{-1}	10^{-2}	10^{-3}	10^8
1 decasthene =	dasn	10^7	10^4	10^1	10^{-2}	10^4	10^3	10^2	10^1	1	10^{-1}	10^{-2}	10^9
1 hectosthene =	hsn	10^8	10^5	10^2	10^{-1}	10^5	10^4	10^3	10^2	10^1	1	10^{-1}	10^{10}
1 kilosthene =	ksn	10^9	10^6	10^3	1	10^6	10^5	10^4	10^3	10^2	10^1	1	10^{11}
1 dyne =	dyne	10^{-2}	10^{-5}	10^{-8}	10^{-11}	10^{-5}	10^{-6}	10^{-7}	10^{-8}	10^{-9}	10^{-10}	10^{-11}	1

Examples: 1 N = 10^{-1} csn = 0.1 csn; 1 hsn = 10^2 kN = 100 kN.

Moment of force (torque) equivalents

The unit of torque (moment of force) in the International System (SI) is the newton meter, abbreviation N m;
1 N m = 0.101 971 6 kp m ≈ 0.102 kp m = 0.737 562 lbf ft.
The unit of torque (moment of force) in the metric gravitational system is the kilopond meter, abbreviation kp m;
1 kp m = 9.806 65 N m = 7.233 01 lbf ft.
The Anglo-American unit of torque (moment of force) is the pound-force foot, abbreviation lbf ft;
1 lbf ft = 1.355 82 N m = 0.138 255 kp m.

Table 118-1 Conversion Factors for Units of Torque (Moment Force)

Unit	Abbreviation	N m	dyne cm	kp m	pdl ft
1 newton meter =	N m	**1**	**10^7**	$1.019\ 72 \cdot 10^{-1}$	23.730 4
1 dyne centimeter =	dyne cm	**10^{-7}**	**1**	$1.019\ 72 \cdot 10^{-8}$	$23.730\ 4 \cdot 10^{-7}$
1 kilopond meter =	kp m	**9.806 65**	**$9.806\ 65 \cdot 10^7$**	**1**	232.715
1 poundal foot =	pdl ft	$4.214\ 01 \cdot 10^{-2}$	$4.214\ 01 \cdot 10^5$	$4.297\ 10 \cdot 10^{-3}$	**1**
1 pound-force foot =	lbf ft	**1.355 82**	**$1.355\ 82 \cdot 10^7$**	0.138 255	32.174 0
1 pound-force inch =	lbf in.	$1.129\ 85 \cdot 10^{-1}$	$1.129\ 85 \cdot 10^6$	0.011 521 2	2.681 17
1 UK ton-force foot =	UK tonf ft	3 037.03	$3\ 037.03 \cdot 10^7$	309.691	72 069.9
1 ounce-force inch =	ozf in.	$7.061\ 55 \cdot 10^{-3}$	$7.061\ 55 \cdot 10^4$	$7.200\ 78 \cdot 10^{-4}$	0.167 573
Unit	Abbreviation	lbf ft	lbf in.	UK tonf ft	ozf in.
1 newton meter =	N m	$7.375\ 62 \cdot 10^{-1}$	8.850 75	$3.292\ 69 \cdot 10^{-4}$	141.612
1 dyne centimeter =	dyne cm	$7.375\ 62 \cdot 10^{-8}$	$8.850\ 75 \cdot 10^{-7}$	$3.292\ 69 \cdot 10^{-11}$	$141.612 \cdot 10^{-7}$
1 kilopond meter =	kp m	7.233 01	86.796 2	$3.229\ 02 \cdot 10^{-3}$	1 388.74
1 poundal foot =	pdl ft	0.031 081 0	0.372 971	$1.387\ 54 \cdot 10^{-5}$	5.967 54
1 pound-force foot =	lbf ft	**1**	**12**	$4.464\ 29 \cdot 10^{-4}$	**192**
1 pound-force inch =	lbf in.	0.083 333 3	**1**	$3.270\ 24 \cdot 10^{-5}$	**16**
1 UK ton-force foot =	UK tonf ft	**2 240**	**26 880**	**1**	**430 080**
1 ounce-force inch =	ozf in.	$5.208\ 33 \cdot 10^{-3}$	**0.062 5**	$2.325\ 15 \cdot 10^{-6}$	**1**

Examples: 1 N m = $7.375\ 62 \cdot 10^{-1}$ lbf ft = 0.737 562 lbf ft; 1 pdl ft = $4.214\ 01 \cdot 10^{-2}$ N m = 0.042 140 1 N m.
Exact values are printed in bold type.
The United Kingdom units are denoted by the prefix UK.

Pressure and stress equivalents

The unit of pressure and stress in the International System (SI) is the pascal, abbreviation Pa; 1 Pa = 1 N/m² = 0.101 971 6 kp/m² ≈ 0.102 kp/m² = 0.020 885 54 lbf/ft².
The unit of pressure and stress in the metric gravitational system is the kilopond per square meter, abbreviation kp/m²; 1 kp/m² = 9.806 65 Pa = 9.806 65 N/m² = 0.204 816 lbf/ft².
The Anglo-American unit of pressure and stress is the pound-force per square foot, abbreviation lbf/ft²; 1 lbf/ft² = 47.880 3 Pa = 47.880 3 N/m² = 4.882 43 kp/m².

Table 119-1 Conversion Factors for Units of Pressure and Stress

Unit	Abbreviation	Pa N/m²	N/cm²	N/mm²	bar	mbar	μbar	kp/m²	kp/cm² at	kp/mm²	mmH₂O*
1 pascal / 1 newton per square meter	= Pa = N/m²	1	10^{-4}	10^{-6}	10^{-5}	10^{-2}	10^{1}	$1.019\,72 \cdot 10^{-1}$	$1.019\,72 \cdot 10^{-5}$	$1.019\,72 \cdot 10^{-7}$	$1.019\,72 \cdot 10^{-1}$
1 newton per square centimeter	= N/cm²	10^{4}	1	10^{-2}	10^{-1}	10^{2}	10^{5}	$1.019\,72 \cdot 10^{3}$	$1.019\,72 \cdot 10^{-1}$	$1.019\,72 \cdot 10^{-3}$	$1.019\,72 \cdot 10^{3}$
1 newton per square millimeter	= N/mm²	10^{6}	10^{2}	1	10^{1}	10^{4}	10^{7}	$1.019\,72 \cdot 10^{5}$	$1.019\,72 \cdot 10^{1}$	$1.019\,72 \cdot 10^{-1}$	$1.019\,72 \cdot 10^{5}$
1 bar	= bar	10^{5}	10^{1}	10^{-1}	1	10^{3}	10^{6}	$1.019\,72 \cdot 10^{4}$	$1.019\,72$	$1.019\,72 \cdot 10^{-2}$	$1.019\,72 \cdot 10^{4}$
1 millibar	= mbar	10^{2}	10^{-2}	10^{-4}	10^{-3}	1	10^{3}	$1.019\,72 \cdot 10^{1}$	$1.019\,72 \cdot 10^{-3}$	$1.019\,72 \cdot 10^{-5}$	$1.019\,72 \cdot 10^{1}$
1 microbar	= μbar	10^{-1}	10^{-5}	10^{-7}	10^{-6}	10^{-3}	1	$1.019\,72 \cdot 10^{-2}$	$1.019\,72 \cdot 10^{-6}$	$1.019\,72 \cdot 10^{-8}$	$1.019\,72 \cdot 10^{-2}$
1 kilopond per square meter	= kp/m²	$9.806\,65$	$9.806\,65 \cdot 10^{-4}$	$9.806\,65 \cdot 10^{-6}$	$9.806\,65 \cdot 10^{-5}$	$9.806\,65 \cdot 10^{-2}$	$9.806\,65 \cdot 10^{1}$	1	10^{-4}	10^{-6}	1
1 kilopond per square centimeter / 1 technical atmosphere	= kp/cm² = at	$9.806\,65 \cdot 10^{4}$	$9.806\,65$	$9.806\,65 \cdot 10^{-2}$	$9.806\,65 \cdot 10^{-1}$	$9.806\,65 \cdot 10^{2}$	$9.806\,65 \cdot 10^{5}$	10^{4}	1	10^{-2}	10^{4}
1 kilopond per square millimeter	= kp/mm²	$9.806\,65 \cdot 10^{6}$	$9.806\,65 \cdot 10^{2}$	$9.806\,65$	$9.806\,65 \cdot 10^{1}$	$9.806\,65 \cdot 10^{4}$	$9.806\,65 \cdot 10^{7}$	10^{6}	10^{2}	1	10^{6}
1 millimeter of water	= mm H₂O*	$9.806\,65$	$9.806\,65 \cdot 10^{-4}$	$9.806\,65 \cdot 10^{-6}$	$9.806\,65 \cdot 10^{-5}$	$9.806\,65 \cdot 10^{-2}$	$9.806\,65 \cdot 10^{1}$	1	10^{-4}	10^{-6}	1
1 millimeter of mercury / 1 torr**	= mm Hg = 1 torr**	133.322	$133.322 \cdot 10^{-4}$	$133.322 \cdot 10^{-6}$	$133.322 \cdot 10^{-5}$	$133.322 \cdot 10^{-2}$	$133.322 \cdot 10^{1}$	$13.595\,1$	$13.595\,1 \cdot 10^{-4}$	$13.595\,1 \cdot 10^{-6}$	$13.595\,1$
1 standard atmosphere	= atm	$1.013\,25 \cdot 10^{5}$	$1.013\,25 \cdot 10^{1}$	$1.013\,25 \cdot 10^{-1}$	$1.013\,25$	$1.013\,25 \cdot 10^{3}$	$1.013\,25 \cdot 10^{6}$	$1.033\,23 \cdot 10^{4}$	$1.033\,23$	$1.033\,23 \cdot 10^{-2}$	$1.033\,23 \cdot 10^{4}$
1 poundal per square foot	= pdl/ft²	$1.488\,16$	$1.488\,16 \cdot 10^{-4}$	$1.488\,16 \cdot 10^{-6}$	$1.488\,16 \cdot 10^{-5}$	$1.488\,16 \cdot 10^{-2}$	$1.488\,16 \cdot 10^{1}$	$1.517\,50 \cdot 10^{-1}$	$1.517\,50 \cdot 10^{-5}$	$1.517\,50 \cdot 10^{-7}$	$1.517\,50 \cdot 10^{-1}$
1 pound-force per square inch	= lbf/in.²	$6.894\,76 \cdot 10^{3}$	$6.894\,76 \cdot 10^{-1}$	$6.894\,76 \cdot 10^{-3}$	$6.894\,76 \cdot 10^{-2}$	$6.894\,76 \cdot 10^{1}$	$6.894\,76 \cdot 10^{4}$	$7.030\,70 \cdot 10^{2}$	$7.030\,70 \cdot 10^{-2}$	$7.030\,70 \cdot 10^{-4}$	$7.030\,70 \cdot 10^{2}$
1 pound-force per square foot	= lbf/ft²	$47.880\,3$	$47.880\,3 \cdot 10^{-4}$	$47.880\,3 \cdot 10^{-6}$	$47.880\,3 \cdot 10^{-5}$	$47.880\,3 \cdot 10^{-2}$	$47.880\,3 \cdot 10^{1}$	$4.882\,43$	$4.882\,43 \cdot 10^{-4}$	$4.882\,43 \cdot 10^{-6}$	$4.882\,43$
1 UK ton-force per square inch	= UK tonf/in.²	$1.544\,43 \cdot 10^{7}$	$1.544\,43 \cdot 10^{3}$	$1.544\,43 \cdot 10^{1}$	$1.544\,43 \cdot 10^{2}$	$1.544\,43 \cdot 10^{5}$	$1.544\,43 \cdot 10^{8}$	$157.488 \cdot 10^{4}$	157.488	$157.488 \cdot 10^{-2}$	$157.488 \cdot 10^{4}$
1 UK ton-force per square foot	= UK tonf/ft²	$1.072\,52 \cdot 10^{5}$	$1.072\,52 \cdot 10^{1}$	$1.072\,52 \cdot 10^{-1}$	$1.072\,52$	$1.072\,52 \cdot 10^{3}$	$1.072\,52 \cdot 10^{6}$	$1.093\,66 \cdot 10^{4}$	$1.093\,66$	$1.093\,66 \cdot 10^{-2}$	$1.093\,66 \cdot 10^{4}$
1 inch of water	= in. H₂O	$2.490\,89 \cdot 10^{2}$	$2.490\,89 \cdot 10^{-2}$	$2.490\,89 \cdot 10^{-4}$	$2.490\,89 \cdot 10^{-3}$	$2.490\,89$	$2.490\,89 \cdot 10^{3}$	25.4	$25.4 \cdot 10^{-4}$	$25.4 \cdot 10^{-6}$	25.4
1 foot of water	= ft H₂O	$29.890\,7 \cdot 10^{2}$	$29.890\,7 \cdot 10^{-2}$	$29.890\,7 \cdot 10^{-4}$	$29.890\,7 \cdot 10^{-3}$	$29.890\,7$	$29.890\,7 \cdot 10^{3}$	304.8	$304.8 \cdot 10^{-4}$	$304.8 \cdot 10^{-6}$	304.8
1 barometric inch of mercury	= in. Hg	$33.863\,9 \cdot 10^{2}$	$33.863\,9 \cdot 10^{-2}$	$33.863\,9 \cdot 10^{-4}$	$33.863\,9 \cdot 10^{-3}$	$33.863\,9$	$33.863\,9 \cdot 10^{3}$	345.316	$345.316 \cdot 10^{-4}$	$345.316 \cdot 10^{-6}$	345.316

Table 119-1 Conversion Factors for Units of Pressure and Stress (Continued)

Unit	Abbreviation	mm Hg torr**	atm	pdl/ft²	lbf/in.²	lbf/ft²	UK tonf/in.²	UK tonf/ft²	in. H₂O	ft H₂O	in. Hg
1 pascal / 1 newton per square meter	= Pa / = N/m²	$7.500\ 62 \cdot 10^{-3}$	$9.869\ 23 \cdot 10^{-6}$	$6.719\ 69 \cdot 10^{-1}$	$1.450\ 38 \cdot 10^{-4}$	$2.088\ 54 \cdot 10^{-2}$	$6.474\ 90 \cdot 10^{-8}$	$9.323\ 85 \cdot 10^{-6}$	$4.014\ 63 \cdot 10^{-3}$	$3.345\ 53 \cdot 10^{-4}$	$2.953\ 00 \cdot 10^{-4}$
1 newton per square centimeter	= N/cm²	$7.500\ 62 \cdot 10^{1}$	$9.869\ 23 \cdot 10^{-1}$	$6.719\ 69 \cdot 10^{3}$	$1.450\ 38$	$2.088\ 54 \cdot 10^{2}$	$6.474\ 90 \cdot 10^{-4}$	$9.323\ 85 \cdot 10^{-2}$	$4.014\ 63 \cdot 10^{1}$	$3.345\ 53$	$2.953\ 00$
1 newton per square millimeter	= N/mm²	$7.500\ 62 \cdot 10^{3}$	$9.869\ 23$	$6.719\ 69 \cdot 10^{5}$	$1.450\ 38 \cdot 10^{2}$	$2.088\ 54 \cdot 10^{4}$	$6.474\ 90 \cdot 10^{-2}$	$9.323\ 85$	$4.014\ 63 \cdot 10^{3}$	$3.345\ 53 \cdot 10^{2}$	$2.953\ 00 \cdot 10^{2}$
1 bar	= bar	$7.500\ 62 \cdot 10^{2}$	$9.869\ 23 \cdot 10^{-1}$	$6.719\ 69 \cdot 10^{4}$	$1.450\ 38 \cdot 10^{1}$	$2.088\ 54 \cdot 10^{3}$	$6.474\ 90 \cdot 10^{-3}$	$9.323\ 85 \cdot 10^{-1}$	$4.014\ 63 \cdot 10^{2}$	$3.345\ 53 \cdot 10^{1}$	$2.953\ 00 \cdot 10^{1}$
1 millibar	= mbar	$7.500\ 62 \cdot 10^{-1}$	$9.869\ 23 \cdot 10^{-4}$	$6.719\ 69 \cdot 10^{1}$	$1.450\ 38 \cdot 10^{-2}$	$2.088\ 54$	$6.474\ 90 \cdot 10^{-6}$	$9.323\ 85 \cdot 10^{-4}$	$4.014\ 63 \cdot 10^{-1}$	$3.345\ 53 \cdot 10^{-2}$	$2.953\ 00 \cdot 10^{-2}$
1 microbar	= μbar	$7.500\ 62 \cdot 10^{-4}$	$9.869\ 23 \cdot 10^{-7}$	$6.719\ 69 \cdot 10^{-2}$	$1.450\ 38 \cdot 10^{-5}$	$2.088\ 54 \cdot 10^{-3}$	$6.474\ 90 \cdot 10^{-9}$	$9.323\ 85 \cdot 10^{-7}$	$4.014\ 63 \cdot 10^{-4}$	$3.345\ 53 \cdot 10^{-5}$	$2.953\ 00 \cdot 10^{-5}$
1 kilopond per square meter	= kp/m²	$735.559 \cdot 10^{-4}$	$9.678\ 41 \cdot 10^{-5}$	$65\ 897.6 \cdot 10^{-4}$	$14.223\ 3 \cdot 10^{-4}$	$2.048\ 16 \cdot 10^{-1}$	$6.349\ 71 \cdot 10^{-7}$	$9.143\ 58 \cdot 10^{-5}$	$3.937\ 01 \cdot 10^{-2}$	$3.280\ 84 \cdot 10^{-3}$	$28.959\ 0 \cdot 10^{-4}$
1 kilopond per square centimeter / 1 technical atmosphere	= kp/cm² / = at	735.559	$9.678\ 41 \cdot 10^{-1}$	$65\ 897.6$	$14.223\ 3$	$2.048\ 16 \cdot 10^{3}$	$6.349\ 71 \cdot 10^{-3}$	$9.143\ 58 \cdot 10^{-1}$	$3.937\ 01 \cdot 10^{2}$	$3.280\ 84 \cdot 10^{1}$	$28.959\ 0$
1 kilopond per square millimeter	= kp/mm²	$735.559 \cdot 10^{2}$	$9.678\ 41 \cdot 10^{1}$	$65\ 897.6 \cdot 10^{2}$	$14.223\ 3 \cdot 10^{2}$	$2.048\ 16 \cdot 10^{5}$	$6.349\ 71 \cdot 10^{-1}$	$9.143\ 58 \cdot 10^{1}$	$3.937\ 01 \cdot 10^{4}$	$3.280\ 84 \cdot 10^{3}$	$28.959\ 0 \cdot 10^{2}$
1 millimeter of water	= mm H₂O*	$735.559 \cdot 10^{-4}$	$9.678\ 41 \cdot 10^{-5}$	$65\ 897.6 \cdot 10^{-4}$	$14.233\ 3 \cdot 10^{-4}$	$2.048\ 16 \cdot 10^{-1}$	$6.349\ 71 \cdot 10^{-7}$	$9.145\ 38 \cdot 10^{-5}$	$3.937\ 01 \cdot 10^{-2}$	$3.280\ 84 \cdot 10^{-3}$	$28.959\ 0 \cdot 10^{-4}$
1 millimeter of mercury / 1 torr**	= mm Hg / 1 torr**	1	$1.315\ 79 \cdot 10^{-3}$	$89.588\ 5$	$0.019\ 336\ 8$	$2.784\ 50$	$8.632\ 47 \cdot 10^{-6}$	$1.243\ 08 \cdot 10^{-3}$	$0.535\ 240$	$0.044\ 603\ 3$	$0.039\ 370\ 1$
1 standard atmosphere	= atm	760	1	$68\ 087.4$	$14.695\ 9$	$2\ 116.22$	$6.560\ 67 \cdot 10^{-3}$	$0.944\ 738$	406.782	$33.898\ 5$	$29.921\ 3$
1 poundal per square foot	= pdl/ft²	$0.011\ 162\ 1$	$1.468\ 70 \cdot 10^{-5}$	1	$2.158\ 40 \cdot 10^{-4}$	$0.031\ 081\ 0$	$9.635\ 71 \cdot 10^{-8}$	$1.387\ 54 \cdot 10^{-5}$	$5.974\ 41 \cdot 10^{-3}$	$4.978\ 67 \cdot 10^{-4}$	$4.394\ 53 \cdot 10^{-4}$
1 pound-force per square inch	= lbf/in.²	$51.714\ 9$	$0.068\ 046\ 0$	$4\ 633.06$	1	144	$4.464\ 29 \cdot 10^{-4}$	$0.064\ 285\ 7$	$27.679\ 9$	$2.306\ 66$	$2.036\ 02$
1 pound-force per square foot	= lbf/ft²	$0.359\ 131$	$4.725\ 42 \cdot 10^{-4}$	$32.174\ 0$	$6.944\ 44 \cdot 10^{-3}$	1	$3.100\ 20 \cdot 10^{-6}$	$4.464\ 29 \cdot 10^{-4}$	$0.192\ 222$	$0.016\ 018\ 5$	$0.014\ 139\ 0$
1 UK ton-force per square inch	= UK tonf/in.²	115.842	152.423	$1.037\ 81 \cdot 10^{7}$	$2\ 240$	$322\ 560$	1	144	$62\ 003.1$	$5\ 166.93$	$4\ 560.69$
1 UK ton-force per square foot	= UK tonf/ft²	804.452	$1.058\ 49$	$72\ 069.9$	$15.555\ 6$	$2\ 240$	$6.944\ 44 \cdot 10^{-3}$	1	430.575	$35.881\ 2$	$31.671\ 3$
1 inch of water	= in. H₂O	$1.868\ 32$	$2.458\ 32 \cdot 10^{-3}$	167.381	$0.036\ 127\ 3$	$5.202\ 33$	$1.612\ 82 \cdot 10^{-5}$	$2.322\ 48 \cdot 10^{-3}$	1	$0.083\ 333\ 3$	$0.073\ 555\ 9$
1 foot of water	= ft H₂O	$22.419\ 8$	$0.029\ 499\ 8$	$2\ 008.57$	$0.433\ 527$	$62.428\ 0$	$1.935\ 39 \cdot 10^{-4}$	$0.027\ 869\ 7$	12	1	$0.882\ 671$
1 barometric inch of mercury	= in. Hg	**25.4**	$0.033\ 421\ 1$	$2\ 275.56$	$0.491\ 154$	$70.726\ 2$	$2.192\ 65 \cdot 10^{-4}$	$0.031\ 574\ 3$	$13.595\ 1$	$1.132\ 92$	1

Examples: $1\ \text{N/m}^2 = 10^{-5}\ \text{bar} = 0.000\ 01\ \text{bar}$; $1\ \text{kp/cm}^2 = 1\ \text{at} = 6.349\ 71 \cdot 10^{-3}\ \text{tonf/in.}^2 = 0.006\ 349\ 71\ \text{tonf/in.}^2$;

$1\ \text{pdl/ft}^2 = 1.488\ 16 \cdot 10^{1}\ \mu\text{bar} = 14.881\ 6\ \mu\text{bar}$; $1\ \text{lbf/ft}^2 = 0.014\ 139\ 0\ \text{in. Hg}$.

Exact values are printed in bold type.

The United Kingdom units are denoted by the prefix UK.
The American units are denoted by the prefix US.
*Exactly: $1.000\ 028\ \text{mm H}_2\text{O} = 1\ \text{kp/m}^2$
**The torr is equal to the millimeter of mercury with one part in 7 million.

Relations between the pressure units bar, dyne/cm², pascal, and pieze and their derivatives

bar = 10^5 N/m² = hectopieze = 10^6 dyne/cm²
barye = 0.1 N/m² = decipascal = dyne/cm²
centibar (cbar) = 1000 N/m² = pieze
centipieze (cpz) = 10 N/m² = decapascal
decabar (dabar) = 10^6 N/m² = 10 bar = 10^7 dyne/cm²
decapascal (daPa) = 10 N/m² = centipieze
decibar (dbar) = 10^4 N/m² = 0.1 bar = 10^5 dyne/cm²
decipascal (dPa) = 0.1 N/m² = barye = dyne/cm²
dyne/cm² = 0.1 N/m² = barye = decipascal

hectobar (hbar) = 10^7 N/m² = 10^6 dyne/cm²
hectopascal (hPa) = 100 N/m² = millibar = 1000 dyne/cm²
hectopieze (hpz) = 10^5 N/m² = bar = 10^6 dyne/cm²
megadyne/cm² = 10^5 N/m² = bar = hectopieze
microbar (μbar) = 0.1 N/m² = barye = dyne/cm² = decipascal
millibar (mbar) = 100 N/m² = hectopascal = 1000 dyne/cm²
millipieze (mpz) = N/m² = 10 barye = pascal = 10 dyne/cm²
myriapieze (mpaz) = 10^7 N/m² = 100 bar = 10^8 dyne/cm²
pascal (Pa) = N/m² = 10^{-5} bar = millipieze
pieze (pz) = sthene/m² = 1000 N/m² = centibar

Mass moment of inertia equivalents

The unit of mass moment of inertia in the International System (SI) is the kilogram meter squared, abbreviation kg m²; 1 kg m² = 23.730 4 lb ft². The Anglo-American unit of moment of inertia is the pound foot squared, abbreviation lb ft²; 1 lb ft² = 0.042 140 1 kg m².

Table 120-1 Conversion Factors for Units of Mass Moment of Inertia

Unit	Abbreviation	kg m²	kg cm²	g cm²	lb ft²	lb in.²	oz in.²
1 kilogram meter squared =	kg m²	**1**	**10^4**	**10^7**	23.730 4	3.417 17 · 10^3	54 674.8
1 kilogram centimeter squared =	kg cm²	**10^{-4}**	**1**	**10^3**	23.730 4 · 10^{-4}	3.417 17 · 10^{-1}	54 674.8 · 10^{-4}
1 gram centimeter squared =	g cm²	**10^{-7}**	**10^{-3}**	**1**	23.730 4 · 10^{-7}	3.417 17 · 10^{-4}	54 674.8 · 10^{-7}
1 pound foot squared =	lb ft²	4.214 01 · 10^{-2}	4.214 01 · 10^2	4.214 01 · 10^5	**1**	**144**	**2 304**
1 pound inch squared =	lb in.²	2.926 40 · 10^{-4}	2.926 40	2.926 40 · 10^3	6.944 44 · 10^{-3}	**1**	**16**
1 ounce inch squared =	oz in.²	1.829 00 · 10^{-5}	1.829 00 · 10^{-1}	1.829 00 · 10^2	4.340 28 · 10^{-4}	**0.062 5**	**1**

Examples: 1 kg m² = 23.730 4 lb ft²
Exact values are printed in bold type.

Dynamic viscosity equivalents

The unit of dynamic viscosity in the International System (SI) is the pascal second, abbreviation Pa s;

$$1 \text{ Pa s} = 1 \text{ N s/m}^2 = 0.101\,972 \text{ kp s/m}^2 = 0.671\,969 \text{ pdl s/ft}^2 .$$

The unit of dynamic viscosity in the metric gravitational system is the kilopond second per square meter, abbreviation kp s/m² ;

$$1 \text{ kp s/m}^2 = 9.806\,65 \text{ Pa s} = 9.806\,65 \text{ N s/m}^2 = 6.589\,76 \text{ pdl s/ft}^2 .$$

The Anglo-American unit of dynamic viscosity is the poundal second per square foot, abbreviation pdl s/ft² ;

$$1 \text{ pdl s/ft}^2 = 1.488\,16 \text{ Pa s} = 1.488\,16 \text{ N s/m}^2 = 0.151\,750 \text{ kp s/m}^2 .$$

Table 121-1 Conversion Factors for Units of Dynamic Viscosity

Unit	Abbreviation	Pa s / N s/m²	mPa s / mN s/m²	μPa s / μN s/m²	P	cP	mP
1 pascal second = / 1 newton second per square meter =	Pa s / N s/m²	1	10^3	10^6	10^1	10^3	10^4
1 millipascal second = / 1 millinewton second per square meter =	mPa s / mN s/m²	10^{-3}	1	10^3	10^{-2}	1	10^1
1 micropascal second = / 1 micronewton second per square m. =	μPa s / μN s/m²	10^{-6}	10^{-3}	1	10^{-5}	10^{-3}	10^{-2}
1 poise =	P	10^{-1}	10^2	10^5	1	10^2	10^3
1 centipoise =	cP	10^{-3}	1	10^3	10^{-2}	1	10^1
1 millipoise =	mP	10^{-4}	10^{-1}	10^2	10^{-3}	10^{-1}	1
1 kilopond second per square meter =	kp s/m²	9.806 65	9.806 65 · 10^3	9.806 65 · 10^6	9.806 65 · 10^1	9.806 65 · 10^3	9.806 65 · 10^4
1 kilopond hour per square meter =	kp h/m²	35 303.94	35 303.94 · 10^3	35 303.94 · 10^6	35 303.94 · 10^1	35 303.94 · 10^3	35 303.94 · 10^4
1 poundal second per square foot =	pdl s/ft²	1.488 16	1.488 16 · 10^3	1.488 16 · 10^6	1.488 16 · 10^1	1.488 16 · 10^3	1.488 16 · 10^4
1 pound-force second per square foot =	lbf s/ft²	47.880 3	47.880 3 · 10^3	47.880 3 · 10^6	47.880 3 · 10^1	47.880 3 · 10^3	47.880 3 · 10^4
1 pound-force hour per square foot =	lbf h/ft²	1.723 69 · 10^5	1.723 69 · 10^8	1.723 69 · 10^{11}	1.723 69 · 10^6	1.723 69 · 10^8	1.723 69 · 10^9

Unit	Abbreviation	kp s/m²	kp h/m²	pdl s/ft²	lbf s/ft²	lbf h/ft²
1 pascal second / 1 newton second per square meter	Pa s / N s/m²	1.019 72 · 10^{-1}	2.832 54 · 10^{-5}	6.719 69 · 10^{-1}	2.088 54 · 10^{-2}	5.801 51 · 10^{-6}
1 millipascal second / 1 millinewton second per square meter	mPa s / mN s/m²	1.019 72 · 10^{-4}	2.832 54 · 10^{-8}	6.719 69 · 10^{-4}	2.088 54 · 10^{-5}	5.801 51 · 10^{-9}
1 micropascal second / 1 micronewton second per square m.	μPa s / μN s/m²	1.019 72 · 10^{-7}	2.832 54 · 10^{-11}	6.719 69 · 10^{-7}	2.088 54 · 10^{-8}	5.801 51 · 10^{-12}
1 poise	P	1.019 72 · 10^{-2}	2.832 54 · 10^{-6}	6.719 69 · 10^{-2}	2.088 54 · 10^{-3}	5.801 51 · 10^{-7}
1 centipoise	cP	1.019 72 · 10^{-4}	2.832 54 · 10^{-8}	6.719 69 · 10^{-4}	2.088 54 · 10^{-5}	5.801 51 · 10^{-9}
1 millipoise	mP	1.019 72 · 10^{-5}	2.832 54 · 10^{-9}	6.719 69 · 10^{-5}	2.088 54 · 10^{-6}	5.801 51 · 10^{-10}
1 kilopond second per square meter	kp s/m²	1	2.777 78 · 10^{-4}	6.589 76	0.204 816	5.689 34 · 10^{-5}
1 kilopond hour per square meter	kp h/m²	3 600	1	23 723.2	737.338	0.204 816
1 poundal second per square foot	pdl s/ft²	0.151 750	4.215 28 · 10^{-5}	1	0.031 081 0	8.633 60 · 10^{-6}
1 pound-force second per square foot	lbf s/ft²	4.882 43	1.356 23 · 10^{-3}	32.174 0	1	2.777 78 · 10^{-4}
1 pound-force hour per square foot	lbf h/ft²	17 576.7	4.882 43	115 827	3 600	1

Examples: 1 N s/m² = 2.088 54 · 10^{-2} lbf s/ft² = 0.020 8854 lbf s/ft² ; 1 cP = 10^{-2} P = 0.01 P;
1 pdl s/ft² = 1.488 16 N s/m² ; 1 lbf h/ft² = 4.882 42 kp h/m² .

Exact values are printed in bold type.

Kinematic viscosity equivalents

The unit of kinematic viscosity in the International System (SI) is the meter squared per second, abbreviation m^2/s; $1\ m^2/s = 10.763\ 9\ ft^2/s$.
The Anglo-American unit of kinematic viscosity is the foot squared per second, abbreviation ft^2/s; $1\ ft^2/s = 0.092\ 903\ m^2/s$.

Table 122-1 Conversion Factors for Units of Kinematic Viscosity

Unit	Abbreviation	m^2/s	St cm^2/s	cSt mm^2/s	mSt	m^2/h
1 meter squared per second =	m^2/s	**1**	**10^4**	**10^6**	**10^7**	3 600
1 stokes = 1 centimeter squared per second =	St cm^2/s	**10^{-4}**	**1**	**10^2**	**10^3**	$3\ 600 \cdot 10^{-4}$
1 centistokes = 1 millimeter squared per second =	cSt mm^2/s	**10^{-6}**	**10^{-2}**	**1**	**10^1**	$3\ 600 \cdot 10^{-6}$
1 millistokes =	mSt	**10^{-7}**	**10^{-3}**	**10^{-1}**	**1**	$3\ 600 \cdot 10^{-7}$
1 meter squared per hour =	m^2/h	$2.777\ 78 \cdot 10^{-4}$	$2.777\ 78$	$2.777\ 78 \cdot 10^2$	$2.777\ 78 \cdot 10^3$	1
1 inch squared per second =	$in.^2/s$	**$6.451\ 6 \cdot 10^{-4}$**	**$6.451\ 6$**	**$6.451\ 6 \cdot 10^2$**	**$6.451\ 6 \cdot 10^3$**	2.322 58
1 foot squared per second =	ft^2/s	**$9.290\ 30 \cdot 10^{-2}$**	**$9.290\ 30 \cdot 10^2$**	**$9.290\ 30 \cdot 10^4$**	**$9.290\ 30 \cdot 10^5$**	334.451
1 inch squared per hour =	$in.^2/h$	$1.792\ 11 \cdot 10^{-7}$	$1.792\ 11 \cdot 10^{-3}$	$1.792\ 11 \cdot 10^{-1}$	$1.792\ 11$	**$6.451\ 6 \cdot 10^{-4}$**
1 foot squared per hour =	ft^2/h	**$2.580\ 64 \cdot 10^{-5}$**	**$2.580\ 64 \cdot 10^{-1}$**	**$2.580\ 64 \cdot 10^1$**	**$2.580\ 64 \cdot 10^2$**	0.092 903 0

Unit	Abbreviation	$in.^2/s$	ft^2/s	$in.^2/h$	ft^2/h	
1 meter squared per second =	m^2/s	$1.550\ 00 \cdot 10^3$	$10.763\ 9$	$5.580\ 01 \cdot 10^6$	$3.875\ 01 \cdot 10^4$	
1 stokes = 1 centimeter squared per second =	St cm^2/s	$1.550\ 00 \cdot 10^{-1}$	$10.763\ 9 \cdot 10^{-4}$	$5.580\ 01 \cdot 10^2$	$3.875\ 01$	
1 centistokes = 1 millimeter squared per second =	cSt mm^2/s	$1.550\ 00 \cdot 10^{-3}$	$10.763\ 9 \cdot 10^{-6}$	$5.580\ 01$	$3.875\ 01 \cdot 10^{-2}$	
1 millistokes =	mSt	$1.550\ 00 \cdot 10^{-4}$	$10.763\ 9 \cdot 10^{-7}$	$5.580\ 01 \cdot 10^{-1}$	$3.875\ 01 \cdot 10^{-3}$	
1 meter squared per hour =	m^2/h	0.430 556	$2.989\ 98 \cdot 10^{-3}$	1550.00	10.763 9	
1 inch squared per second =	$in.^2/s$	**1**	$6.944\ 44 \cdot 10^{-3}$	**3 600**	**25**	
1 foot squared per second =	ft^2/s	**144**	**1**	**518 400**	**3 600**	
1 inch squared per hour =	$in.^2/h$	$2.777\ 78 \cdot 10^{-4}$	$1.929\ 01 \cdot 10^{-6}$	**1**	$6.944\ 44 \cdot 10^{-3}$	
1 foot squared per hour =	ft^2/h	**0.04**	$2.777\ 78 \cdot 10^{-4}$	**144**	**1**	

Examples: $1\ m^2/s = 10^4\ cm^2/s = 10^4\ St = 10\ 000\ cm^2/s = 10\ 000\ St$;
$1\ mm^2/s = 1\ cSt = 3.875\ 01 \cdot 10^{-2}\ ft^2/h = 0.038\ 750\ 1\ ft^2/h$.

Exact values are printed in bold type.

The unit of energy in the International System (SI) is the joule, abbreviation J; 1 J = 0.101 971 6 kp m ≈ 0.102 kp m = 0.737 562 ft lbf.
The unit of energy in the metric gravitational system is the kilopond meter, abbreviation kp m; 1 kp m = 9.806 65 J = 7.233 01 ft lbf.
The Anglo-American unit of energy is the foot pound-force, abbreviation ft lbf; 1 ft lbf = 1.355 82 J = 0.138 255 kp m.

Table 123-1 Conversion Factors for Units of Energy

Unit		Abbreviation	J*	kJ	MJ	erg	Wh
1 joule*	=	J*	1	10^{-3}	10^{-6}	10^{7}	$2.777\,78 \cdot 10^{-4}$
1 kilojoule	=	kJ	10^{3}	1	10^{-3}	10^{10}	$2.777\,78 \cdot 10^{-1}$
1 megajoule	=	MJ	10^{6}	10^{3}	1	10^{13}	$2.777\,78 \cdot 10^{2}$
1 erg	=	erg	10^{-7}	10^{-10}	10^{-13}	1	$2.777\,78 \cdot 10^{-11}$
1 watt hour	=	Wh	3 600	$3\,600 \cdot 10^{-3}$	$3\,600 \cdot 10^{-6}$	$3\,600 \cdot 10^{7}$	1
1 kilowatt hour	=	kWh	$3\,600 \cdot 10^{3}$	3 600	$3\,600 \cdot 10^{-3}$	$3\,600 \cdot 10^{10}$	10^{3}
1 megawatt hour	=	MWh	$3\,600 \cdot 10^{6}$	$3\,600 \cdot 10^{3}$	3 600	$3\,600 \cdot 10^{13}$	10^{6}
1 kilopond meter	=	kp m	9.806 65	$9.806\,65 \cdot 10^{-3}$	$9.806\,65 \cdot 10^{-6}$	$9.806\,65 \cdot 10^{7}$	$2.724\,069 \cdot 10^{-3}$
1 pond centimeter	=	p cm	$9.806\,65 \cdot 10^{-5}$	$9.806\,65 \cdot 10^{-8}$	$9.806\,65 \cdot 10^{-11}$	$9.806\,65 \cdot 10^{2}$	$2.724\,069 \cdot 10^{-8}$
1 calorie	=	cal	4.186 8	$4.186\,8 \cdot 10^{-3}$	$4.186\,8 \cdot 10^{-6}$	$4.186\,8 \cdot 10^{7}$	$1.163 \cdot 10^{-3}$
1 kilocalorie	=	kcal	$4.186\,8 \cdot 10^{3}$	4.186 8	$4.186\,8 \cdot 10^{-3}$	$4.186\,8 \cdot 10^{10}$	1.163
1 metric horsepower second =		KS s	735.499	$735.499 \cdot 10^{-3}$	$735.499 \cdot 10^{-6}$	$735.499 \cdot 10^{7}$	$2.043\,05 \cdot 10^{-1}$
1 metric horsepower hour	=	KS h	$2.647\,796 \cdot 10^{6}$	$2.647\,796 \cdot 10^{3}$	2.647 796	$2.647\,796 \cdot 10^{13}$	735.499
1 electronvolt	=	eV	$1.602 \cdot 10^{-19}$	$1.602 \cdot 10^{-22}$	$1.602 \cdot 10^{-25}$	$1.602 \cdot 10^{-12}$	$4.450 \cdot 10^{-23}$
1 foot poundal	=	ft pdl	$4.214\,01 \cdot 10^{-2}$	$4.214\,01 \cdot 10^{-5}$	$4.214\,01 \cdot 10^{-8}$	$4.214\,01 \cdot 10^{5}$	$1.170\,56 \cdot 10^{-5}$
1 foot pound-force	=	ft lbf	1.355 82	$1.355\,82 \cdot 10^{-3}$	$1.355\,82 \cdot 10^{-6}$	$1.355\,82 \cdot 10^{7}$	$3.766\,16 \cdot 10^{-4}$
1 horsepower hour	=	hp h	$2.684\,52 \cdot 10^{6}$	$2.684\,52 \cdot 10^{3}$	2.684 52	$2.684\,52 \cdot 10^{13}$	745.700
1 British thermal unit	=	Btu	$1.055\,06 \cdot 10^{3}$	1.055 06	$1.055\,06 \cdot 10^{-3}$	$1.055\,06 \cdot 10^{10}$	$2.930\,71 \cdot 10^{-1}$
1 thermie	=	th	$4.185\,5 \cdot 10^{6}$	$4.185\,5 \cdot 10^{3}$	4.185 5	$4.185\,5 \cdot 10^{13}$	$1.162\,64 \cdot 10^{3}$
Unit		Abbreviation	kWh	MWh	kp m	p cm	cal
1 joule*	=	J*	$2.777\,78 \cdot 10^{-7}$	$2.777\,78 \cdot 10^{-10}$	$1.019\,72 \cdot 10^{-1}$	$1.019\,72 \cdot 10^{4}$	$2.388\,46 \cdot 10^{-1}$
1 kilojoule	=	kJ	$2.777\,78 \cdot 10^{-4}$	$2.777\,78 \cdot 10^{-7}$	$1.019\,72 \cdot 10^{2}$	$1.019\,72 \cdot 10^{7}$	$2.388\,46 \cdot 10^{2}$
1 megajoule	=	MJ	$2.777\,78 \cdot 10^{-1}$	$2.777\,78 \cdot 10^{-4}$	$1.019\,72 \cdot 10^{5}$	$1.019\,72 \cdot 10^{10}$	$2.388\,46 \cdot 10^{5}$
1 erg	=	erg	$2.777\,78 \cdot 10^{-14}$	$2.777\,78 \cdot 10^{-17}$	$1.019\,72 \cdot 10^{-8}$	$1.019\,72 \cdot 10^{-3}$	$2.388\,46 \cdot 10^{-8}$
1 watt hour	=	Wh	10^{-3}	10^{-6}	$3.670\,98 \cdot 10^{2}$	$3.670\,98 \cdot 10^{7}$	859.845
1 kilowatt hour	=	kWh	1	10^{-3}	$3.670\,98 \cdot 10^{5}$	$3.670\,98 \cdot 10^{10}$	$859.845 \cdot 10^{3}$
1 megawatt hour	=	MWh	10^{3}	1	$3.670\,98 \cdot 10^{8}$	$3.670\,98 \cdot 10^{13}$	$859.845 \cdot 10^{6}$
1 kilopond meter	=	kp m	$2.724\,069 \cdot 10^{-6}$	$2.724\,069 \cdot 10^{-9}$	1	10^{5}	2.342 28
1 pond centimeter	=	p cm	$2.724\,069 \cdot 10^{-11}$	$2.724\,069 \cdot 10^{-14}$	10^{-5}	1	$2.342\,28 \cdot 10^{-5}$
1 calorie	=	cal	$1.163 \cdot 10^{-6}$	$1.163 \cdot 10^{-9}$	$426.935 \cdot 10^{-3}$	$426.935 \cdot 10^{2}$	1
1 kilocalorie	=	kcal	$1.163 \cdot 10^{-3}$	$1.163 \cdot 10^{-6}$	426.935	$426.935 \cdot 10^{5}$	10^{3}
1 metric horsepower second =		KS s	$2.043\,05 \cdot 10^{-4}$	$2.043\,05 \cdot 10^{-7}$	75	$75 \cdot 10^{5}$	$1.7567 \cdot 10^{2}$
1 metric horsepower hour	=	KS h	$735.499 \cdot 10^{-3}$	$735.499 \cdot 10^{-6}$	270 000	$270\,000 \cdot 10^{5}$	$632.415 \cdot 10^{3}$
1 electronvolt	=	eV	$4.450 \cdot 10^{-26}$	$4.450 \cdot 10^{-29}$	$1.634 \cdot 10^{-20}$	$1.634 \cdot 10^{-15}$	$3.826 \cdot 10^{-20}$
1 foot poundal	=	ft pdl	$1.170\,56 \cdot 10^{-8}$	$1.170\,56 \cdot 10^{-11}$	$4.297\,10 \cdot 10^{-3}$	$4.297\,10 \cdot 10^{2}$	$1.006\,50 \cdot 10^{-2}$
1 foot pound-force	=	ft lbf	$3.766\,16 \cdot 10^{-7}$	$3.766\,16 \cdot 10^{-10}$	$1.382\,55 \cdot 10^{-1}$	$1.382\,55 \cdot 10^{4}$	$3.238\,32 \cdot 10^{-1}$
1 horsepower hour	=	hp h	$745.700 \cdot 10^{-3}$	$745.700 \cdot 10^{-6}$	$2.737\,45 \cdot 10^{5}$	$2.737\,45 \cdot 10^{10}$	$641.186 \cdot 10^{3}$
1 British thermal unit	=	Btu	$2.930\,71 \cdot 10^{-4}$	$2.930\,71 \cdot 10^{-7}$	$1.075\,862 \cdot 10^{2}$	$1.075\,862 \cdot 10^{7}$	$2.519\,96 \cdot 10^{2}$
1 thermie	=	th	1.162 64	$1.162\,64 \cdot 10^{-3}$	$4.268\,02 \cdot 10^{5}$	$4.268\,02 \cdot 10^{10}$	$999.690 \cdot 10^{3}$

*1 J = 1 W s = 1 Nm

Table 123-2 Conversion Factors for Units of Energy (*Continued*)

Unit		Abbreviation	kcal	KS s	KS h	eV	ft pdl	ft lbf	hp h	Btu	th
1 joule*	=	J*	$2.388\,46 \cdot 10^{-4}$	$1.359\,62 \cdot 10^{-3}$	$3.776\,726 \cdot 10^{-7}$	$6.242 \cdot 10^{18}$	$23.730\,4$	$7.375\,62 \cdot 10^{-1}$	$3.725\,06 \cdot 10^{-7}$	$9.478\,17 \cdot 10^{-4}$	$2.389\,20 \cdot 10^{-7}$
1 kilojoule	=	kJ	$2.388\,46 \cdot 10^{-1}$	$1.359\,62$	$3.776\,726 \cdot 10^{-4}$	$6.242 \cdot 10^{21}$	$23.730\,4 \cdot 10^{3}$	$7.375\,62 \cdot 10^{2}$	$3.725\,06 \cdot 10^{-4}$	$9.478\,17 \cdot 10^{-1}$	$2.389\,20 \cdot 10^{-4}$
1 megajoule	=	MJ	$2.388\,46 \cdot 10^{2}$	$1.359\,62 \cdot 10^{3}$	$3.776\,726 \cdot 10^{-1}$	$6.242 \cdot 10^{24}$	$23.730\,4 \cdot 10^{6}$	$7.375\,62 \cdot 10^{5}$	$3.725\,06 \cdot 10^{-1}$	$9.478\,17 \cdot 10^{2}$	$2.389\,20 \cdot 10^{-1}$
1 erg	=	erg	$2.388\,46 \cdot 10^{-11}$	$1.359\,62 \cdot 10^{-10}$	$3.776\,726 \cdot 10^{-14}$	$6.242 \cdot 10^{11}$	$23.730\,4 \cdot 10^{-7}$	$7.375\,62 \cdot 10^{-8}$	$3.725\,06 \cdot 10^{-14}$	$9.478\,17 \cdot 10^{-11}$	$2.389\,20 \cdot 10^{-14}$
1 watt hour	=	Wh	$859.845 \cdot 10^{-3}$	$4.894\,637$	$1.359\,621 \cdot 10^{-3}$	$2.247 \cdot 10^{22}$	$8.542\,93 \cdot 10^{4}$	$2.655\,22 \cdot 10^{3}$	$1.341\,02 \cdot 10^{-3}$	$3\,412.14 \cdot 10^{-3}$	$8.601\,12 \cdot 10^{-4}$
1 kilowatt hour	=	kWh	859.845	$4.894\,637 \cdot 10^{3}$	$1.359\,621$	$2.247 \cdot 10^{25}$	$8.542\,93 \cdot 10^{7}$	$2.655\,22 \cdot 10^{6}$	$1.341\,02$	$3\,412.14$	$8.601\,12 \cdot 10^{-1}$
1 megawatt hour	=	MWh	$859.845 \cdot 10^{3}$	$4.894\,637 \cdot 10^{6}$	$1.359\,621 \cdot 10^{3}$	$2.247 \cdot 10^{28}$	$8.542\,93 \cdot 10^{10}$	$2.655\,22 \cdot 10^{9}$	$1.341\,02 \cdot 10^{3}$	$3\,412.14 \cdot 10^{3}$	$8.601\,12 \cdot 10^{2}$
1 kilopond meter	=	kp m	$2.342\,28 \cdot 10^{-3}$	$1.3333\,3 \cdot 10^{-2}$	$3.703\,704 \cdot 10^{-6}$	$6.122 \cdot 10^{19}$	232.715	$7.233\,01$	$3.653\,04 \cdot 10^{-6}$	$9.294\,87 \cdot 10^{-3}$	$2.343\,00 \cdot 10^{-6}$
1 pond centimeter	=	p cm	$2.342\,28 \cdot 10^{-8}$	$1.3333\,3 \cdot 10^{-7}$	$3.703\,704 \cdot 10^{-11}$	$6.122 \cdot 10^{14}$	$232.715 \cdot 10^{-5}$	$7.233\,01 \cdot 10^{-5}$	$3.653\,04 \cdot 10^{-11}$	$9.294\,87 \cdot 10^{-8}$	$2.343\,00 \cdot 10^{-11}$
1 calorie	=	cal	**10^{-3}**	$5.692\,46 \cdot 10^{-3}$	$1.581\,24 \cdot 10^{-6}$	$2.614 \cdot 10^{19}$	$99.354\,3$	$3\,088.03 \cdot 10^{-3}$	$1.559\,61 \cdot 10^{-6}$	$3.968\,32 \cdot 10^{-3}$	$1.000\,31 \cdot 10^{-6}$
1 kilocalorie	=	kcal	**1**	$5.692\,46$	$1.581\,24 \cdot 10^{-3}$	$2.614 \cdot 10^{22}$	$99.354\,3 \cdot 10^{3}$	$3\,088.03$	$1.559\,61 \cdot 10^{-3}$	$3.968\,32$	$1.000\,31 \cdot 10^{-3}$
1 metric horsepower second	=	KS s	$1.7567 \cdot 10^{-1}$	**1**	$2.777\,78 \cdot 10^{-4}$	$4.591\,13 \cdot 10^{21}$	$17\,453.6$	542.476	$2.739\,78 \cdot 10^{-4}$	$0.697\,116$	$1.757\,25 \cdot 10^{-4}$
1 metric horsepower hour	=	KS h	632.415	**$3\,600$**	1	$1.653 \cdot 10^{25}$	$6.283\,31 \cdot 10^{7}$	$1\,952\,913$	$0.986\,319$	$2\,509.63$	$0.632\,611$
1 electronvolt	=	eV	$3.826 \cdot 10^{-23}$	$2.178 \cdot 10^{-22}$	$6.050 \cdot 10^{-26}$	1	$3.801\,60 \cdot 10^{-18}$	$1.181\,57 \cdot 10^{-19}$	$5.967\,53 \cdot 10^{-26}$	$1.518\,40 \cdot 10^{-22}$	$3.827\,50 \cdot 10^{-26}$
1 foot poundal	=	ft pdl	$1.006\,50 \cdot 10^{-5}$	$5.729\,46 \cdot 10^{-5}$	$1.591\,52 \cdot 10^{-8}$	$2.630\,5 \cdot 10^{17}$	1	$0.031\,081\,0$	$1.569\,74 \cdot 10^{-8}$	$3.994\,10 \cdot 10^{-5}$	$1.006\,81 \cdot 10^{-8}$
1 foot pound-force	=	ft lbf	$3.238\,32 \cdot 10^{-4}$	$1.843\,40 \cdot 10^{-3}$	$5.120\,56 \cdot 10^{-7}$	$8.463\,30 \cdot 10^{18}$	$32.174\,0$	1	$5.050\,51 \cdot 10^{-7}$	$1.285\,07 \cdot 10^{-3}$	$3.239\,32 \cdot 10^{-7}$
1 horsepower hour	=	hp h	641.186	$3\,649.93$	$1.013\,87$	$1.675\,73 \cdot 10^{25}$	$6.370\,46 \cdot 10^{7}$	**$1.98 \cdot 10^{6}$**	1	$2\,544.43$	$0.641\,386$
1 British thermal unit	=	Btu	$2.519\,96 \cdot 10^{-1}$	$1.434\,48$	$3.984\,66 \cdot 10^{-4}$	$6.585\,89 \cdot 10^{21}$	$2.503\,70 \cdot 10^{4}$	778.169	$3.930\,15 \cdot 10^{-4}$	1	$2.520\,74 \cdot 10^{-4}$
1 thermie	=	th	999.690	$5\,690.69$	$1.580\,75$	$2.612\,67 \cdot 10^{25}$	$9.932\,34 \cdot 10^{7}$	$3.087\,07 \cdot 10^{6}$	$1.559\,12$	$3\,967.09$	1

Examples: $1\ kJ = 1.019\,72 \cdot 10^{2}\ kp\,m = 101.972\ kp\,m \approx 102\ kp\,m$;
$1\ ft\,lbf = 3.238\,32 \cdot 10^{-4}\ kcal = 0.000\,323\,832\ kcal$.

Exact values are printed in bold type.

Liter standard atmosphere $= 1\ atm = 101.327\,837\ J = 101.327\,837 \cdot 10^{7}\ erga = 24.201\,5\ cal = 6.325 \cdot 10^{20}\ eV$.

Inch ounce-force $=$ in. ozf $= ft\,lbf/192 = 1.355\,82 J/192 = 7.061\,56 \cdot 10^{-3}\ J$.

*$1\ J = 1\ W\,s = 1\ Nm$.

Table 123-3 Equivalents of Energy Units: Joule, Watt Hour, and Their Multiples

Unit		Abbreviation	J	kJ	MJ	GJ	TJ
1 joule	=	J	1	10^{-3}	10^{-6}	10^{-9}	10^{-12}
1 kilojoule	=	kJ	10^3	1	10^{-3}	10^{-6}	10^{-9}
1 megajoule	=	MJ	10^6	10^3	1	10^{-3}	10^{-6}
1 gigajoule	=	GJ	10^9	10^6	10^3	1	10^{-3}
1 terajoule	=	TJ	10^{12}	10^9	10^6	10^3	1
1 watt hour	=	Wh	$3.6 \cdot 10^3$	3.6	$3.6 \cdot 10^{-3}$	$3.6 \cdot 10^{-6}$	$3.6 \cdot 10^{-9}$
1 kilowatt hour	=	kWh	$3.6 \cdot 10^6$	$3.6 \cdot 10^3$	3.6	$3.6 \cdot 10^{-3}$	$3.6 \cdot 10^{-6}$
1 megawatt hour	=	MWh	$3.6 \cdot 10^9$	$3.6 \cdot 10^6$	$3.6 \cdot 10^3$	3.6	$3.6 \cdot 10^{-3}$
1 gigawatt hour	=	GWh	$3.6 \cdot 10^{12}$	$3.6 \cdot 10^9$	$3.6 \cdot 10^6$	$3.6 \cdot 10^3$	3.6
1 terawatt hour	=	TWh	$3.6 \cdot 10^{15}$	$3.6 \cdot 10^{12}$	$3.6 \cdot 10^9$	$3.6 \cdot 10^6$	$3.6 \cdot 10^3$

Unit		Abbreviation	Wh	kWh	MWh	GWh	TWh
1 joule	=	J	$2.777\ 78 \cdot 10^{-4}$	$2.777\ 78 \cdot 10^{-7}$	$2.777\ 78 \cdot 10^{-10}$	$2.777\ 78 \cdot 10^{-13}$	$2.777\ 78 \cdot 10^{-16}$
1 kilojoule	=	kJ	$2.777\ 78 \cdot 10^{-1}$	$2.777\ 78 \cdot 10^{-4}$	$2.777\ 78 \cdot 10^{-7}$	$2.777\ 78 \cdot 10^{-10}$	$2.777\ 78 \cdot 10^{-13}$
1 megajoule	=	MJ	$2.777\ 78 \cdot 10^{2}$	$2.777\ 78 \cdot 10^{-1}$	$2.777\ 78 \cdot 10^{-4}$	$2.777\ 78 \cdot 10^{-7}$	$2.777\ 78 \cdot 10^{-10}$
1 gigajoule	=	GJ	$2.777\ 78 \cdot 10^{5}$	$2.777\ 78 \cdot 10^{2}$	$2.777\ 78 \cdot 10^{-1}$	$2.777\ 78 \cdot 10^{-4}$	$2.777\ 78 \cdot 10^{-7}$
1 terajoule	=	TJ	$2.777\ 78 \cdot 10^{8}$	$2.777\ 78 \cdot 10^{5}$	$2.777\ 78 \cdot 10^{2}$	$2.777\ 78 \cdot 10^{-1}$	$2.777\ 78 \cdot 10^{-4}$
1 watt hour	=	Wh	1	10^{-3}	10^{-6}	10^{-9}	10^{-12}
1 kilowatt hour	=	kWh	10^3	1	10^{-3}	10^{-6}	10^{-9}
1 megawatt hour	=	MWh	10^6	10^3	1	10^{-3}	10^{-6}
1 gigawatt hour	=	GWh	10^9	10^6	10^3	1	10^{-3}
1 terawatt hour	=	TWh	10^{12}	10^9	10^6	10^3	1

Specific energy (mass basis) equivalents

The unit of specific energy (mass basis) in the International System (SI) is the joule per kilogram, abbreviation J/kg;
1 J/kg = 0.101 972 kp m/kg = 0.334 553 ft lbf/lb.
The unit of specific energy (mass basis) in the metric gravitational system is the kilopond meter per kilogram, abbreviation kp m/kg;
1 kp m/kg = 9.806 65 J/kg = 3.280 84 ft lbf/lb.
The Anglo-American unit of specific energy (mass basis) is the foot pound-force per pound, abbreviation ft lbf/lb;
1 ft lbf/lb = 2.989 07 J/kg = 0.304 8 kp m/kg.

Table 124-1 Conversion Factors for Units of Specific Energy (Mass Basis)

Unit		Abbreviation	J/g	kJ/kg	kcal/kg	kp m/kg	Btu/lb	ft lbf/lb
1 joule per gram	=	J/g	**1**	**1**	0.238 846	101.972	0.429 923	334.553
1 kilojoule per kilogram	=	kJ/kg	**1**	**1**	0.238 846	101.972	0.429 923	334.553
1 kilocalorie per kilogram	=	kcal/kg	**4.186 8**	**4.186 8**	1	426.935	**1.8**	1 400.70
1 kilopond meter per kilogram	=	kp m/kg	$9.806\ 65 \cdot 10^{-3}$	$9.806\ 65 \cdot 10^{-3}$	$2.342\ 28 \cdot 10^{-3}$	**1**	$4.216\ 10 \cdot 10^{-3}$	3.280 84
1 British thermal unit per pound	=	Btu/lb	**2.326**	**2.326**	0.555 556	237.186	1	778.169
1 foot pound-force per pound	=	ft lbf/lb	$2.989\ 07 \cdot 10^{-3}$	$2.989\ 07 \cdot 10^{-3}$	$7.139\ 26 \cdot 10^{-4}$	**0.304 8**	$1.285\ 07 \cdot 10^{-3}$	1

Examples: 1 kJ/kg = 101.972 kp m/kg $\approx$ 102 kp m/kg;
1 kcal/kg = 1.8 Btu/lb; 1 ft lbf/lb = 0.304 8 kp m/kg.
Exact values are printed in bold type.

The unit of specific heat (capacity) in the International System (SI) is the joule per kilogram kelvin, abbreviation J/kg K;
$$1 \text{ J/kg K} = 2.388\ 46 \cdot 10^{-4} \text{ kcal/kg K.}$$

The unit of specific heat (capacity) in the metric gravitational system is the kilocalorie per kilogram kelvin,
abbreviation kcal/kg K; 1 kcal/kg K = 4186.8 J/kg K.

Table 125-1 Conversion Factors for Units of Specific Heat

Unit		Abbreviation	J/kg K	kJ/kg K	cal/kg K	kcal/kg K	cal/g K
1 joule per kilogram kelvin	=	J/kg K	**1**	10^{-3}	$2.388\ 46 \cdot 10^{-1}$	$2.388\ 46 \cdot 10^{-4}$	$2.388\ 46 \cdot 10^{-4}$
1 kilojoule per kilogram kelvin	=	kJ/kg K	10^3	**1**	$2.388\ 46 \cdot 10^2$	$2.388\ 46 \cdot 10^{-1}$	$2.388\ 46 \cdot 10^{-1}$
1 calorie per kilogram kelvin	=	cal/kg K	**4.186 8**	$4.186\ 8 \cdot 10^{-3}$	**1**	10^{-3}	10^{-3}
1 kilocalorie per kilogram kelvin	=	kcal/kg K	$4.186\ 8 \cdot 10^3$	**4.186 8**	10^3	**1**	**1**
1 calorie per gram kelvin	=	cal/g K	$4.186\ 8 \cdot 10^3$	**4.186 8**	10^3	**1**	**1**
1 watt hour per kilogram kelvin	=	Wh/kg K	**3 600**	$3\ 600 \cdot 10^{-3}$	859.845	$859.845 \cdot 10^{-3}$	$859.845 \cdot 10^{-3}$
1 kilowatt hour per kilogram kelvin	=	kWh/kg K	$3\ 600 \cdot 10^3$	**3 600**	$859.845 \cdot 10^3$	859.845	859.845
1 joule per kilomole kelvin	=	J/kmol K	$1/M$	$10^{-3}/M$	$2.388\ 46 \cdot 10^{-1}/M$	$2.388\ 46 \cdot 10^{-4}/M$	$2.388\ 46 \cdot 10^{-4}/M$
1 kilojoule per kilomole kelvin	=	kJ/kmol K	$10^3/M$	$1/M$	$2.388\ 46 \cdot 10^2/M$	$2.388\ 46 \cdot 10^{-1}/M$	$2.388\ 46 \cdot 10^{-1}/M$
1 calorie per kilomole kelvin	=	cal/kmol K	$4.186\ 8/M$	$4.186\ 8 \cdot 10^{-3}/M$	$1/M$	$10^{-3}/M$	$10^{-3}/M$
1 kilocalorie per kilomole kelvin	=	kcal/kmol K	$4.186\ 8 \cdot 10^3/M$	$4.186\ 8/M$	$10^3/M$	$1/M$	$1/M$

Unit		Abbreviation	Wh/kg K	kWh/kg K	J/kmol K	kJ/kmol K	cal/kmol K	kcal/kmol K
1 joule per kilogram kelvin	=	J/kg K	$2.777\ 78 \cdot 10^{-4}$	$2.777\ 78\ 10^{-7}$	M	$M \cdot 10^{-3}$	$2.388\ 46 \cdot 10^{-1} \cdot M$	$2.388\ 46\ 10^{-4}\ M$
1 kilojoule per kilogram kelvin	=	kJ/kg K	$2.777\ 78 \cdot 10^{-1}$	$2.777\ 78\ 10^{-4}$	$M \cdot 10^3$	M	$2.388\ 46 \cdot 10^2 \cdot M$	$2.388\ 46\ 10^{-1}\ M$
1 calorie per kilogram kelvin	=	cal/kg K	**1.163** $\cdot 10^{-3}$	**1.163** 10^{-6}	$4.186\ 8 \cdot M$	$4.186\ 8 \cdot 10^{-3} \cdot M$	M	$10^{-3}\ M$
1 kilocalorie per kilogram kelvin	=	kcal/kg K	**1.163**	**1.163** 10^{-3}	$4.186\ 8 \cdot 10^3 \cdot M$	$4.186\ 8 \cdot M$	$10^3 \cdot M$	M
1 calorie per gram kelvin	=	cal/g K	**1.163**	**1.163** 10^{-3}	$4.186\ 8 \cdot 10^3 \cdot M$	$4.186\ 8 \cdot M$	$10^3 \cdot M$	M
1 watt hour per kilogram kelvin	=	Wh/kg K	**1**	10^{-3}	$3\ 600 \cdot M$	$3\ 600 \cdot 10^{-3} \cdot M$	$859.845 \cdot M$	$859.845\ 10^{-3}\ M$
1 kilowatt hour per kilogram kelvin	=	kWh/kg K	10^3	**1**	$3\ 600 \cdot 10^3 \cdot M$	$3\ 600 \cdot M$	$859.845 \cdot 10^3 \cdot M$	$859.845\ M$
1 joule per kilomole kelvin	=	J/kmol K	$2.777\ 78 \cdot 10^{-4}/M$	$2.777\ 78\ 10^{-7}/M$	**1**	10^{-3}	$2.388\ 46 \cdot 10^{-1}$	$2.388\ 46\ 10^{-4}$
1 kilojoule per kilomole kelvin	=	kJ/kmol K	$2.777\ 78 \cdot 10^{-1}/M$	$2.777\ 78\ 10^{-4}/M$	10^3	**1**	$2.388\ 46 \cdot 10^2$	$2.388\ 46\ 10^{-1}$
1 calorie per kilomole kelvin	=	cal/kmol K	$1.163 \cdot 10^{-3}/M$	$1.163 \cdot 10^{-6}/M$	**4.186 8**	$4.186\ 8 \cdot 10^{-3}$	**1**	10^{-3}
1 kilocalorie per kilomole kelvin	=	kcal/kmol K	$1.163/M$	$1.163 \cdot 10^{-3}/M$	$4.186\ 8 \cdot 10^3$	**4.186 8**	10^3	**1**

Examples: 1 kJ/kg K = 2.777 78 10⁻⁴ kWh/kg K = 0.000 277 778 kWh/kg K;
1 kcal/kmol K = 4.186 8 kJ/mol K.

Exact values are printed in bold type.

M — denotes the relative molecular mass ("molecular weight"), in kg/kmol.

Table 125-2 Relations between the Units Kilomole, Standard Cubic Meter, and Kilogram

Unit		Abbreviation	kmol	m_n^3	kg
1 kilomole	=	kmol	1	22.414	M
1 standard cubic meter	=	m_n^3	1/22.414	1	M/22.414
1 kilogram	=	kg	1/M	22.414/M	1

M – denotes the relative molecular mass (molecular weight) in kg/kmol.

Specific energy (volume basis) equivalents

The unit of specific energy (volume basis) in the International System (SI) is the joule per cubic meter, abbreviation J/m^3 ; $1\ J/m^3 = 0.101\ 972\ kpm/m^3$.
The unit of specific energy (volume basis) in the metric gravitational system is the kilopond meter per cubic meter; abbreviation kpm/m^3 ; $1\ kpm/m^3 = 9.806\ 65\ J/m^3$.

Table 126-1 Conversion Factors for Units of Specific Energy (Volume Basis)

Unit	Abbreviation	J/m^3	J/dm^3	J/cm^3	kJ/m^3	$kp\ m/m^3$	$kcal/m^3$	Btu/ft^3	thr/UK gal	th/l
1 joule per cubic meter	=	1	10^{-3}	10^{-6}	10^{-3}	$1.019\ 72 \cdot 10^{-1}$	$2.388\ 46 \cdot 10^{-4}$	$2.683\ 92 \cdot 10^{-5}$	$4.308\ 86 \cdot 10^{-11}$	$2.389\ 27 \cdot 10^{-10}$
1 joule per cubic decimeter	=	10^3	1	10^{-3}	1	$1.019\ 72 \cdot 10^{2}$	$2.388\ 46 \cdot 10^{-1}$	$2.683\ 92 \cdot 10^{-2}$	$4.308\ 86 \cdot 10^{-8}$	$2.389\ 27 \cdot 10^{-7}$
1 joule per cubic centimeter	=	10^6	10^3	1	10^3	$1.019\ 72 \cdot 10^{5}$	$2.388\ 46 \cdot 10^{2}$	$2.683\ 92 \cdot 10^{1}$	$4.308\ 86 \cdot 10^{-5}$	$2.389\ 27 \cdot 10^{-4}$
1 kilojoule per cubic meter	=	10^3	1	10^{-3}	1	$1.019\ 72 \cdot 10^{2}$	$2.388\ 46 \cdot 10^{-1}$	$2.683\ 92 \cdot 10^{-2}$	$4.308\ 86 \cdot 10^{-8}$	$2.389\ 27 \cdot 10^{-7}$
1 kilopond per cubic meter	=	**9.806 65**	**9.806 65** $\cdot 10^{-3}$	**9.806 65** $\cdot 10^{-6}$	**9.806 65** $\cdot 10^{-3}$	1	$2.342\ 28 \cdot 10^{-3}$	$2.632\ 03 \cdot 10^{-4}$	$4.225\ 55 \cdot 10^{-10}$	$2.343\ 07 \cdot 10^{-9}$
1 kilocalorie per cubic meter	=	**4.186 8** $\cdot 10^{3}$	**4.186 8**	**4.186 8** $\cdot 10^{-3}$	**4.186 8**	426.935	1	0.112 370	$1.804\ 03 \cdot 10^{-7}$	$1.000\ 34 \cdot 10^{-6}$
1 British thermal unit per cubic foot	=	$3.725\ 89 \cdot 10^{4}$	$3.725\ 89 \cdot 10^{1}$	$3.725\ 89 \cdot 10^{-2}$	$3.725\ 89 \cdot 10^{1}$	3 799.35	8.899 15	1	$1.605\ 44 \cdot 10^{-6}$	$8.902\ 16 \cdot 10^{-6}$
1 therm per UK gallon	=	$2.320\ 80 \cdot 10^{10}$	$2.320\ 80 \cdot 10^{7}$	$2.320\ 80 \cdot 10^{4}$	$2.320\ 80 \cdot 10^{7}$	$2.366\ 56 \cdot 10^{9}$	$5.543\ 14 \cdot 10^{6}$	$6.228\ 84 \cdot 10^{5}$	1	5.545 01
1 thermie per litre	=	$4.185\ 38 \cdot 10^{9}$	$4.185\ 38 \cdot 10^{6}$	$4.185\ 38 \cdot 10^{3}$	$4.185\ 38 \cdot 10^{6}$	$4.267\ 90 \cdot 10^{8}$	$9.996\ 62 \cdot 10^{5}$	$1.123\ 32 \cdot 10^{5}$	0.180 342	1

Examples: $1\ J/m^3 = 2.683\ 92 \cdot 10^{-5}\ Btu/ft^3 = 0.000\ 026\ 839\ 2\ Btu/ft^3$; $1\ th/l = 4.185\ 38 \cdot 10^3\ J/cm^3 = 4185.38\ J/cm^3$.
Exact values are printed in bold type.
The United Kingdom units are denoted by the prefix UK.

The unit of power in the International System (SI) is the watt, abbreviation W; 1 W = 0.101 971 6 kp m/s = 0.737 562 ft lbf/s.
The unit of power in the metric gravitational system is the kilopond meter per second, abbreviation kp m/s; 1 kp m/s = 9.806 65 W = 7.233 01 ft lbf/s.
The Anglo-American unit of power is the foot pound-force per second, abbreviation ft lbf/s; 1 ft lbf/s = 1.355 82 W = 0.138 255 kp m/s.

Table 127-1 Conversion Factors for Units of Power

Unit		Abbreviation	W*	kW	MW	erg/s	kp m/s	prony	poncelet
1 watt*	=	W*	1	10^{-3}	10^{-6}	10^{7}	$1.019\,72 \cdot 10^{-1}$	$1.019\,72 \cdot 10^{-2}$	$1.019\,72 \cdot 10^{-3}$
1 kilowatt	=	kW	10^{3}	1	10^{-3}	10^{10}	$1.019\,72 \cdot 10^{2}$	$1.019\,72 \cdot 10^{1}$	$1.019\,72$
1 megawatt	=	MW	10^{6}	10^{3}	1	10^{13}	$1.019\,72 \cdot 10^{5}$	$1.019\,72 \cdot 10^{4}$	$1.019\,72 \cdot 10^{3}$
1 erg per second	=	erg/s	10^{-7}	10^{-10}	10^{-13}	1	$1.019\,72 \cdot 10^{-8}$	$1.019\,72 \cdot 10^{-9}$	$1.019\,72 \cdot 10^{-10}$
1 kilopond meter per second	=	kp m/s	$9.806\,65$	$9.806\,65 \cdot 10^{-3}$	$9.806\,65 \cdot 10^{-6}$	$9.806\,65 \cdot 10^{7}$	1	10^{-1}	10^{-2}
1 prony	=	prony	$9.806\,65 \cdot 10^{1}$	$9.806\,65 \cdot 10^{-2}$	$9.806\,65 \cdot 10^{-5}$	$9.806\,65 \cdot 10^{8}$	10^{1}	1	10^{-1}
1 poncelet	=	poncelet	$9.806\,65 \cdot 10^{2}$	$9.806\,65 \cdot 10^{-1}$	$9.806\,65 \cdot 10^{-4}$	$9.806\,65 \cdot 10^{9}$	10^{2}	10^{1}	1
1 metric horsepower	=	KS	735.499	$735.499 \cdot 10^{-3}$	$735.499 \cdot 10^{-6}$	$735.499 \cdot 10^{7}$	75	$75 \cdot 10^{-1}$	$75 \cdot 10^{-2}$
1 calorie per second	=	cal/s	$4.186\,8$	$4.186\,8 \cdot 10^{-3}$	$4.186\,8 \cdot 10^{-6}$	$4.186\,8 \cdot 10^{7}$	$4.269\,35 \cdot 10^{-1}$	$4.269\,35 \cdot 10^{-2}$	$4.269\,35 \cdot 10^{-3}$
1 kilocalorie per hour	=	kcal/h	1.163	$1.163 \cdot 10^{-3}$	$1.163 \cdot 10^{-6}$	$1.163 \cdot 10^{7}$	$1.185\,93 \cdot 10^{-1}$	$1.185\,93 \cdot 10^{-2}$	$1.185\,93 \cdot 10^{-3}$
1 foot poundal per second	=	ftpdl/s	$4.214\,01 \cdot 10^{-2}$	$4.214\,01 \cdot 10^{-5}$	$4.214\,01 \cdot 10^{-8}$	$4.214\,01 \cdot 10^{5}$	$4.297\,10 \cdot 10^{-3}$	$4.297\,10 \cdot 10^{-4}$	$4.297\,10 \cdot 10^{-5}$
1 foot pound-force per minute	=	ftfb/min	$2.259\,70 \cdot 10^{-2}$	$2.259\,70 \cdot 10^{-5}$	$2.259\,70 \cdot 10^{-8}$	$2.259\,70 \cdot 10^{5}$	$2.304\,25 \cdot 10^{-3}$	$2.304\,25 \cdot 10^{-4}$	$2.304\,25 \cdot 10^{-5}$
1 foot pound-force per second	=	ftlbf/s	$1.355\,82$	$1.355\,82 \cdot 10^{-3}$	$1.355\,82 \cdot 10^{-6}$	$1.355\,82 \cdot 10^{7}$	$1.382\,55 \cdot 10^{-1}$	$1.382\,55 \cdot 10^{-2}$	$1.382\,55 \cdot 10^{-3}$
1 horsepower	=	hp	745.700	$745.700 \cdot 10^{-3}$	$745.700 \cdot 10^{-6}$	$745.700 \cdot 10^{7}$	$76.040\,2$	$76.040\,2 \cdot 10^{-1}$	$76.040\,2 \cdot 10^{-2}$
1 British thermal unit per hour	=	Btu/h	$2.930\,71 \cdot 10^{-1}$	$2.930\,71 \cdot 10^{-4}$	$2.930\,71 \cdot 10^{-7}$	$2.930\,71 \cdot 10^{6}$	$2.988\,49 \cdot 10^{-2}$	$2.988\,49 \cdot 10^{-3}$	$2.988\,49 \cdot 10^{-4}$

Unit		Abbreviation	KS	cal/s	kcal/h	ft pdl/s	ft lbf/min	ft lbf/s	hp	Btu/h
1 watt*	=	W*	$1.359\,62 \cdot 10^{-3}$	$2.388\,46 \cdot 10^{-1}$	$859.845 \cdot 10^{-3}$	$2.373\,04 \cdot 10^{1}$	$4.425\,37 \cdot 10^{1}$	$7.375\,62 \cdot 10^{-1}$	$1.341\,02 \cdot 10^{-3}$	$3.412\,14$
1 kilowatt	=	kW	$1.359\,62$	$2.388\,46 \cdot 10^{2}$	859.845	$2.373\,04 \cdot 10^{4}$	$4.425\,37 \cdot 10^{4}$	$7.375\,62 \cdot 10^{2}$	$1.341\,02$	$3.412\,14 \cdot 10^{3}$
1 megawatt	=	MW	$1.359\,62 \cdot 10^{3}$	$2.388\,46 \cdot 10^{5}$	$859.845 \cdot 10^{3}$	$2.373\,04 \cdot 10^{7}$	$4.425\,37 \cdot 10^{7}$	$7.375\,62 \cdot 10^{5}$	$1.341\,02 \cdot 10^{3}$	$3.412\,14 \cdot 10^{6}$
1 erg per second	=	erg/s	$1.359\,62 \cdot 10^{-10}$	$2.388\,46 \cdot 10^{-8}$	$859.845 \cdot 10^{-10}$	$2.373\,04 \cdot 10^{-6}$	$4.425\,37 \cdot 10^{-6}$	$7.375\,62 \cdot 10^{-8}$	$1.341\,02 \cdot 10^{-10}$	$3.412\,14 \cdot 10^{-7}$
1 kilopond meter per second	=	kp m/s	$1.333\,33 \cdot 10^{-2}$	$2.342\,28$	$8.432\,20$	232.715	433.980	$7.233\,01$	$1.315\,09 \cdot 10^{-2}$	$33.461\,7$
1 prony	=	prony	$1.333\,33 \cdot 10^{-1}$	$2.342\,28 \cdot 10^{1}$	$8.432\,20 \cdot 10^{1}$	$232.715 \cdot 10^{1}$	$433.980 \cdot 10^{1}$	$7.233\,01 \cdot 10^{1}$	$1.315\,09 \cdot 10^{-1}$	$33.461\,7 \cdot 10^{1}$
1 poncelet	=	poncelet	$1.333\,33$	$2.342\,28 \cdot 10^{2}$	$8.432\,20 \cdot 10^{2}$	$232.715 \cdot 10^{2}$	$433.980 \cdot 10^{2}$	$7.233\,01 \cdot 10^{2}$	$1.315\,09$	$33.461\,7 \cdot 10^{2}$
1 metric horsepower	=	KS	1	175.671	632.415	$17\,453.6$	$32\,548.6$	542.476	$0.986\,320$	$2\,509.63$
1 calorie per second	=	cal/s	$5.692\,46 \cdot 10^{-3}$	1	3.6	$99.354\,3$	185.281	$3.088\,03$	$5.614\,60 \cdot 10^{-3}$	$14.286\,0$
1 kilocalorie per hour	=	kcal/h	$1.581\,24 \cdot 10^{-3}$	$0.277\,778$	1	$27.598\,4$	$51.467\,0$	$0.857\,783$	$1.559\,61 \cdot 10^{-3}$	$3.968\,32$
1 foot poundal per second	=	ftpdl/s	$5.729\,46 \cdot 10^{-5}$	$0.010\,0650$	$0.036\,2340$	1	$1.864\,85$	$0.031\,081$	$5.651\,08 \cdot 10^{-5}$	$0.143\,788$
1 foot pound-force per minute	=	ftlbf/min	$3.072\,33 \cdot 10^{-5}$	$5.397\,20 \cdot 10^{-3}$	$0.019\,4299$	$0.536\,235$	1	$1.666\,67 \cdot 10^{-2}$	$3.030\,3 \cdot 10^{-5}$	$7.710\,42 \cdot 10^{-2}$
1 foot pound-force per second	=	ftlbf/s	$1.843\,40 \cdot 10^{-3}$	$0.323\,832$	$1.165\,80$	$32.174\,0$	60	1	$1.818\,18 \cdot 10^{-3}$	$4.626\,25$
1 horsepower	=	hp	$1.013\,87$	178.107	641.186	$17\,695.73$	$33\,000$	550	1	$2\,544.43$
1 British thermal unit per hour	=	Btu/h	$3.984\,66 \cdot 10^{-4}$	$0.069\,998\,8$	$0.251\,996$	$6.954\,68$	$12.969\,5$	$0.216\,158$	$3.930\,15 \cdot 10^{-4}$	1

Examples: 1 W = $1.019\,72\ 10^{-1}$ kp m/s = 0.101 972 kp m/s ≈ 0.102 kp m/s;
1 ft lbf/s = $1.355\,82\ 10^{-3}$ kW = 0.001 355 82 kW.
Exact values are printed in bold type.
*1 W = 1 J/s = 1 Nm/s.
Inch ounce-force per second = in ozf/s = 0.007 061 56 W.
British commercial ton of refrigeration = CTR = 1.1154 CTR (US)
American commercial ton of refrigeration = CTR (US) = 12 000 Btu/h

Table 127-2 Heat Flow Rate Equivalents

Unit		Abbreviation	W	kW	kcal/min	kcal/h	Btu/min	Btu/h	CTR(USA)
1 watt	=	W	1	10^{-3}	$14.330\ 7 \cdot 10^{-3}$	$859.845 \cdot 10^{-3}$	$56.869 \cdot 10^{-3}$	3.412 14	$2.843\ 45 \cdot 10^{-4}$
1 kilowatt	=	kW	10^{3}	1	14.330 7	859.845	56.869	$3.412\ 14 \cdot 10^{3}$	$2.843\ 45 \cdot 10^{-1}$
1 kilocalorie per minute	=	kcal/min	69.78	$69.78 \cdot 10^{-3}$	1	60	3.968 32	238.099	$1.984 \cdot 10^{-2}$
1 kilocalorie per hour	=	kcal/h	1.163	$1.163 \cdot 10^{-3}$	0.016 667	1	0.066 139	3.968 32	$3.3069 \cdot 10^{-4}$
1 British thermal unit per minute	=	Btu/min	17.584	$17.584 \cdot 10^{-3}$	0.251 996	15.119 76	1	60	0.005
1 British thermal unit per hour	=	Btu/h	$2.930\ 71 \cdot 10^{-1}$	$2.930\ 71 \cdot 10^{-4}$	0.004 1999	0.251 996	0.016 666 7	1	$8.333\ 33 \cdot 10^{-5}$
1 Commercial Ton of Refrigeration	=	CTR(USA)	3 516.85	$3\ 516.85 \cdot 10^{-3}$	50.399	3023.95	200	12 000	1

Examples: $1\ W = 56.869 \cdot 10^{-3}$ Btu/min $= 0.056\ 869$ Btu/min; 1 Btu/h $= 2.930\ 71 \cdot 10^{-1}$ W $= 0.293\ 071$ W.

Remark: kcal $=$ kcal$_{IT} = 4\ 186.8$ J; Btu $=$ Btu$_{IT} = 1\ 055.06$ J; CTR (USA) $= 12\ 000$ Btu/h.

Exact values are printed in bold type.

The unit of intensity of heat flow (rate) in the International System (SI) is the watt per square meter, abbreviation W/m²; 1 W/m² = 0.859 845 kcal/m² h ≈ 0.86 kcal/hm².
The unit of intensity of heat flow (rate) in the metric gravitational system is the kilocalorie per square hour meter, abbreviation kcal/hm²;
1 kcal/hm² = 1.163 W/m².

Table 128-1 Conversion Factors for Units of Heat Flow (Rate)

Unit	Abbreviation	W/m²	kW/m²	W/cm²	kW/cm²	kcal/m² h	cal/cm² s	Btu/in² s	Btu/ft² s	Btu/ft² h
1 watt per square meter	= W/m²	**1**	10^{-3}	10^{-4}	10^{-7}	$8.598\ 45 \cdot 10^{-1}$	$238.846 \cdot 10^{-7}$	$6.114\ 93 \cdot 10^{-7}$	$880.55 \cdot 10^{-7}$	$3.169\ 98 \cdot 10^{-1}$
1 kilowatt per square meter	= kW/m²	10^{3}	**1**	10^{-1}	10^{-4}	$8.598\ 45 \cdot 10^{2}$	$238.846 \cdot 10^{-4}$	$6.114\ 93 \cdot 10^{-4}$	$880.55 \cdot 10^{-4}$	$3.169\ 98 \cdot 10^{2}$
1 watt per square centimeter	= W/cm²	10^{4}	10^{1}	**1**	10^{-3}	$8.598\ 45 \cdot 10^{3}$	$238.846 \cdot 10^{-3}$	$6.114\ 93 \cdot 10^{-3}$	$880.55 \cdot 10^{-3}$	$3.169\ 98 \cdot 10^{3}$
1 kilowatt per square centimeter	= kW/cm²	10^{7}	10^{4}	10^{3}	**1**	$8.598\ 45 \cdot 10^{6}$	238.846	$6.114\ 93$	880.55	$3.169\ 98 \cdot 10^{6}$
1 kilocalorie per square meter hour	= kcal/m² h	**1.163**	$\mathbf{1.163 \cdot 10^{-3}}$	$\mathbf{1.163 \cdot 10^{-4}}$	$\mathbf{1.163 \cdot 10^{-7}}$	**1**	$2.777\ 78 \cdot 10^{-5}$	$7.111\ 67 \cdot 10^{-7}$	$1.024\ 08 \cdot 10^{-4}$	$0.368\ 669$
1 kilocalorie per square centimeter second	= cal/cm² s	$\mathbf{4.186\ 8 \cdot 10^{4}}$	$\mathbf{4.186\ 8 \cdot 10^{1}}$	**4.186 8**	$\mathbf{4.186\ 8 \cdot 10^{-3}}$	**36 000**	**1**	$2.560\ 20 \cdot 10^{-2}$	$3.686\ 69$	$13\ 272.1$
1 British thermal units per square inch and second	= Btu/in² s	$16.353\ 4 \cdot 10^{5}$	$16.353\ 4 \cdot 10^{2}$	$16.353\ 4 \cdot 10^{1}$	$16.353\ 4 \cdot 10^{-2}$	$1.406\ 14 \cdot 10^{6}$	$39.059\ 4$	**1**	**144**	**518 400**
1 British thermal units per square foot and second	= Btu/ft² s	$1.135\ 65 \cdot 10^{4}$	$1.135\ 65 \cdot 10^{1}$	$1.135\ 65$	$1.135\ 65 \cdot 10^{-3}$	$9.764\ 86 \cdot 10^{3}$	$0.271\ 246$	$6.944\ 44 \cdot 10^{-3}$	**1**	**3 600**
1 British thermal units per square foot and hour	= Btu/ft² h	$3.154\ 59$	$3.154\ 59 \cdot 10^{-3}$	$3.154\ 59 \cdot 10^{-4}$	$3.154\ 59^{-7}$	$2.712\ 46$	$7.534\ 61 \cdot 10^{-5}$	$1.929\ 01 \cdot 10^{-6}$	$2.777\ 78 \cdot 10^{-4}$	**1**

Examples: 1 kW/m² = 8.598 45 10² kcal/m² h ≙ 859.845 kcal/m²h;
1 Btu/ft² s ≙ 1.135 65 10⁴ W/m² = 11 356.5 W/m².
Exact values are printed in bold type.

The unit of thermal conductivity in the International System (SI) is the watt per meter and kelvin, abbreviation W/m K; 1 W/m K = 0.859 845 kcal/h m K ≈ 0.86 kcal/h m K.
The unity of conductivity in the metric gravitational system is the kilocalorie per hour meter and kelvin, abbreviation kcal/h m K; 1 kcal/h m K = 1.163 W/m K.

Table 129-1 Conversion Factors for Units of Thermal Conductivity

Unit	Abbreviation	W/m K	W/cm K	kcal/h m K	cal/s cm K	Btu in/ft² h deg F	Btu/ft h deg F	Btu/in h deg F
1 watt per meter and kelvin	= W/m K	**1**	10^{-2}	$85.984\ 5 \cdot 10^{-2}$	$2.388\ 46 \cdot 10^{-3}$	$693.347 \cdot 10^{-2}$	$57.778\ 9 \cdot 10^{-2}$	$4.814\ 91 \cdot 10^{-2}$
1 watt per centimeter and kelvin	= W/cm K	10^{2}	**1**	$85.984\ 5$	$2.388\ 45 \cdot 10^{-1}$	693.347	$57.778\ 9$	$4.814\ 91$
1 kilocalorie per meter, hour and kelvin	= kcal/h m K	**1.163**	$\mathbf{1.163 \cdot 10^{-2}}$	**1**	$2.777\ 78 \cdot 10^{-3}$	$8.063\ 63$	$0.671\ 969$	$0.055\ 997\ 4$
1 calorie per centimeter, second and kelvin	= cal/s cm K	$\mathbf{4.186\ 8 \cdot 10^{2}}$	**4.186 8**	**360**	**1**	$2\ 902.91$	241.909	$20.159\ 1$
1 British thermal units inch per square foot, hour and degree F	= Btu in/ft² h deg F	$1.442\ 28 \cdot 10^{-1}$	$1.442\ 28 \cdot 10^{-3}$	$0.124\ 014$	$3.444\ 82 \cdot 10^{-4}$	**1**	$0.083\ 333$	$6.944\ 44 \cdot 10^{-3}$
1 British thermal units per foot, hour, and degree F	= Btu/ft h deg F	$1.730\ 73$	$1.730\ 73 \cdot 10^{-2}$	$1.488\ 16$	$4.133\ 79 \cdot 10^{-3}$	**12**	**1**	$0.083\ 333$
1 British thermal units per inch, hour, and degree F	= Btu/in h deg F	$2.076\ 88 \cdot 10^{1}$	$2.076\ 88 \cdot 10^{-1}$	$17.858\ 0$	$4.960\ 54 \cdot 10^{-2}$	**144**	**12**	**1**

Examples: 1 W/m K = 693.347 · 10⁻² Btu in/ft² h deg F = 6.933 47 Btu in/ft² h deg F;
1 Btu/ft h deg F = 1.730 73 · 10⁻² W/cm K = 0.017 3073 W/cm K.
Exact values are printed in bold type.

Thermal conductance and transmittance (heat transfer coefficient) equivalents κ

The unit of thermal conductance and transmittance in the International System (SI) is the watt per square meter and kelvin, abbreviation W/m² K;

$$1 \text{ W/m}^2\text{ K} = 0.859\ 845 \text{ kcal/m}^2\text{ h K} \approx 0.86 \text{ kcal/m}^2\text{ h K}.$$

The unit of thermal conductance and transmittance in the metric gravitational system is the kilocalorie per square meter, hour and kelvin, abbreviation kcal/m² h K; 1 kcal/m² h K = 1.163 W/m² K.

Table 130-1 Conversion Factors for Units of Heat Conductance and Transmittance κ

Unit	Abbreviation	W/m² K	kW/m² K	W/cm² K	kcal/m² h K	cal/cm² s K	Btu/ft² h deg F
1 watt per square meter and kelvin =	W/m² K	**1**	10^{-3}	10^{-4}	$8.598\ 45 \cdot 10^{-1}$	$2.388\ 46 \cdot 10^{-5}$	$1.761\ 10 \cdot 10^{-1}$
1 kilowatt per square meter and kelvin =	kW/m² K	10^3	**1**	10^{-1}	$8.598\ 45 \cdot 10^2$	$2.388\ 46 \cdot 10^{-2}$	$1.761\ 10 \cdot 10^2$
1 watt per square centimeter and kelvin =	W/cm² K	10^4	10^1	**1**	$8.598\ 45 \cdot 10^3$	$2.388\ 46 \cdot 10^{-1}$	$1.761\ 10 \cdot 10^3$
1 kilocalorie per square meter, hour and kelvin =	kcal/m² h K	**1.163**	$\mathbf{1.163 \cdot 10^{-3}}$	$\mathbf{1.163 \cdot 10^{-4}}$	**1**	$2.777\ 78 \cdot 10^{-5}$	0.204 816
1 calorie per square centimeter, second and kelvin =	cal/cm² s K	$\mathbf{4.186\ 8 \cdot 10^4}$	$\mathbf{4.186\ 8 \cdot 10^1}$	**4.186 8**	36 000	**1**	7 373.38
1 British thermal unit per square foot, hour, and degree F =	Btu/ft² h deg F	5.678 26	$5.678\ 26 \cdot 10^{-3}$	$5.678\ 26 \cdot 10^{-4}$	4.882 43	$1.356\ 23 \cdot 10^{-4}$	**1**

Examples: 1 W/cm² K = 2.388 46 · 10⁻¹ cal/cm² s K = 0.238 846 cal/cm² s K;
1 Btu/ft² h deg F = 5.678 26 · 10⁻³ kW/m² K = 0.005 678 26 kW/m² K.
Exact values are printed in bold type.

Radiation constant equivalents C

In the International System (SI) the radiation constant is expressed in watts per square meter and kelvin to the fourth, abbreviation W/m² K⁴;

$$1 \text{ W/m}^2\text{ K}^4 = 0.859\ 845 \text{ kcal/m}^2\text{ h K}^4 \approx 0.86 \text{ kcal/m}^2\text{ h K}^4.$$

In the metric gravitational system the radiation constant is expressed in kilocalories per hour, square meter, and kelvin to the fourth, abbreviation kcal/h m² K⁴; 1 kcal/h m² K⁴ = 1.163 W/m² K⁴.

Table 131-1 Conversion Factors for Units of the Radiation Constant

Unit	Abbreviation	W/m² K⁴	kW/m² K⁴	W/cm² K⁴	kcal/m² h K⁴	cal/cm² s K⁴	Btu/ft² h deg F⁴
1 watt per square meter and kelvin to the fourth =	W/m² K⁴	**1**	10^{-3}	10^{-4}	$8.598\ 45 \cdot 10^{-1}$	$2.388\ 46 \cdot 10^{-5}$	$3.020 \cdot 10^{-2}$
1 kilowatt per square meter and kelvin to the fourth =	kW/m² K⁴	10^3	**1**	10^{-1}	$8.598\ 45 \cdot 10^2$	$2.388\ 46 \cdot 10^{-2}$	$3.020 \cdot 10^1$
1 watt per square centimeter and kelvin to the fourth =	W/cm² K⁴	10^4	10^1	**1**	$8.598\ 45 \cdot 10^3$	$2.388\ 46 \cdot 10^{-1}$	$3.020 \cdot 10^2$
1 kilocalorie per square meter, hour. and kelvin to the fourth =	kcal/m² h K⁴	**1.163**	$\mathbf{1.163 \cdot 10^{-3}}$	$\mathbf{1.163 \cdot 10^{-4}}$	**1**	$2.777\ 78 \cdot 10^{-5}$	$3.512 \cdot 10^{-2}$
1 calorie per square centimeter, second, and kelvin to the fourth =	cal/cm² s K⁴	$\mathbf{4.186\ 8 \cdot 10^4}$	$\mathbf{4.186\ 8 \cdot 10^1}$	**4.186 8**	3 600	**1**	$1.264 \cdot 10^3$
1 British thermal unit per square foot, hour and degree F⁴ =	Btu/ft² h deg F⁴	$3.311 \cdot 10^1$	$3.311 \cdot 10^{-3}$	$3.311 \cdot 10^{-3}$	28.49	$7.908 \cdot 10^{-4}$	**1**

Examples: 1 W/m² K⁴ = 8.598 45 · 10⁻¹ kcal/m² h K⁴ = 0.859 845 kcal/m² h K⁴;
1 cal/cm² s K⁴ = 4.186 8 · 10¹ kW/m² K⁴ = 41.868 kW/m² K⁴.
Exact values are printed in bold type.

TEMPERATURE UNITS

The temperature unit of the International System of Units (SI) is the Kelvin, symbol K. The Kelvin temperature T[K] is defined by the triple point of pure water to which the quantity 273.16 (exactly) is assigned;

the triple point of pure water = 273.16 K (exactly).

Another temperature unit is the degree Celsius, symbol °C. The Celsius temperature t_C [°C] is defined by designating 0 °C as the freezing point and 100 °C as the boiling point of water, both at the pressure of 1 atm;

the freezing point of pure water 0 °C $\triangleq$ 273.15 K; the boiling point of pure water 100 °C $\triangleq$ 373.15 K.

The Réaumur temperature $t_{Ré}$ [Ré] is defined by designating 0 [°Ré] as the freezing point and 80 [°Ré] as the boiling point of water;

the freezing point of pure water = 0 °Ré; the boiling point of pure water = 80 °Ré.

One Anglo-American temperature unit is the degree Fahrenheit, symbol °F or deg F. On the Fahrenheit temperature scale t_F [°F] the freezing point of pure water is 32 °F and the boiling point is 212 °F;

the freezing point of pure water = 32 °F; the boiling point of pure water = 212 °F.

Another Anglo-American temperature unit is the degree Rankine, symbol °R, or deg R. This unit is used to measure thermodynamic temperature t_R [°R] ;

1 °R = 5/9 K, 1 K = 9/5 °R; as a unit of temperature difference 1 °R = 1 °F.

The Fahrenheit temperature 0°F corresponds to the Rankine temperature 459.67°R, so $t_R = t_F + 459.67$.

Table 132-1 Equivalents of Temperature Units

Unit		Symbol	K	°C	°Rè	°F	°R
1 kelvin	=	K	1	1	4/5	9/5	9/5
1 degree Celsius	=	°C	1	1	4/5	9/5	9/5
1 degree Reaumur	=	°Rè	5/4	5/4	1	9/4	9/4
1 degree Fahrenheit	=	°F	5/9	5/9	4/9	1	1
1 degree Rankine	=	°R	5/9	5/9	4/9	1	1

Table 133-1 Some Characteristic Temperatures

		K	°C	°Rè	°F	°R
Absolute zero	=	0	−273.15	−218.52	−459.67	0
Freezing point of pure water	=	273.15	0	0	+32	491.67
Triple point of pure water	=	273.15	+0.01	+0.008	+32.0183	491.688
Boiling point of pure water	=	373.15	+100	+80	+212	671.67

Table 134-1 Temperature Conversion Table

		$T[K]$	$t_C[°C]$	$t_{Rè}[°Rè]$	$t_F[°F]$	$t_R[°R]$
Kelvin temperature $T[K]$	$=$	T	$T - 273.15$	$\frac{4}{5}(T - 273.15)$	$\frac{9}{5}(T - 273.15) + 32$	$\frac{9}{5}T$
Celsius temperature $t_C[°C]$	$=$	$t_C + 273.15$	t_C	$\frac{4}{5}t_C$	$\frac{9}{5}t_C + 32$	$\frac{9}{5}t_C + 491.67$
Reaumur temperature $t_{Rè}[°Rè]$	$=$	$\frac{5}{4}t_{Rè} + 273.15$	$\frac{5}{4}t_{Rè}$	$t_{Rè}$	$\frac{9}{5}t_{Rè} + 32$	$\frac{9}{4}t_{Rè} + 491.67$
Fahrenheit temperature $t_F[°F]$	$=$	$\frac{5}{9}(t_F - 32) + 273.15$	$\frac{5}{9}(t_F - 32)$	$\frac{4}{9}(t_F - 32)$	t_F	$t_F + 459.67$
Rankine temperature $t_R[°R]$	$=$	$\frac{5}{9}t_R$	$\frac{5}{9}(t_R - 491.67)$	$\frac{4}{9}t_R - 491.67$	$t_R - 459.67$	t_R

Table 135-1 Temperature Conversion Table Celsius (°C) and Fahrenheit (°F)

C	F	C	F	C	F	C	F	C	F	C	F
−20	−4.0	+30	+86.0	+80	+176.0	+130	+266.0	+180	+356.0	+500	+913
−19	−2.2	31	87.8	81	177.8	131	267.8	181	357.8	550	1022
−18	−0.4	32	89.6	82	179.6	132	269.6	182	359.6	600	1122
−17	+1.4	33	91.4	83	181.4	133	271.4	183	361.4	650	1202
−16	3.2	34	93.2	84	183.2	134	273.2	184	363.2	700	1292
−15	5.0	35	95.0	85	185.0	135	275.0	185	365.0	750	1382
−14	6.8	36	96.8	86	186.8	136	276.8	186	366.8	800	1472
−13	8.6	37	98.6	87	188.6	137	278.6	187	368.6	850	1562
−12	10.4	38	100.4	88	190.4	138	280.4	188	370.4	900	1652
−11	12.2	39	102.2	89	192.2	139	282.2	189	372.2	950	1742
−10	14.0	40	104.0	90	194.0	140	284.0	190	374.0	1000	1832
− 9	15.8	41	105.8	91	195.8	141	285.8	191	375.8	1050	1922
− 8	17.6	42	107.6	92	197.6	142	287.6	192	377.6	1100	2012
− 7	19.4	43	109.4	93	199.4	143	289.4	193	379.4	1150	2102
− 6	21.2	44	111.2	94	201.2	144	291.2	194	381.2	1200	2192
− 5	23.0	45	113.0	95	203.0	145	293.0	195	383.0	1250	2282
− 4	24.8	46	114.8	96	204.8	146	294.8	196	384.8	1300	2372
− 3	26.6	47	116.6	97	206.6	147	296.6	197	386.6	1350	2462
− 2	28.4	48	118.4	98	208.4	148	298.4	198	388.4	1400	2552
− 1	30.2	49	120.2	99	210.2	149	300.2	199	390.2	1450	2642
0	32.0	50	122.0	100	212.0	150	302.0	200	392	1500	2732
+ 1	33.8	51	123.8	101	213.8	151	303.8	210	410	1550	2822
2	35.6	52	125.6	102	215.6	152	305.6	220	428	1600	2912
3	37.4	53	127.4	103	217.4	153	307.4	230	446	1650	3002
4	39.2	54	129.2	104	219.2	154	309.2	240	464	1700	3092
5	41.0	55	131.0	105	221.0	155	311.0	250	482	1750	3182
6	42.8	56	132.8	106	222.8	156	312.8	260	500	1800	3272
7	44.6	57	134.6	107	224.6	157	314.6	270	518	1850	3362
8	46.4	58	136.4	108	226.4	158	316.4	280	536	1900	3452
9	48.2	59	138.2	109	228.2	159	318.2	290	554	1950	3542
10	50.0	60	140.0	110	230.0	160	320.0	300	572	2000	3632
11	51.8	61	141.8	111	231.8	161	321.8	310	590	2050	3722
12	53.6	62	143.6	112	233.6	162	323.6	320	608	2100	3812
13	55.4	63	145.4	113	235.4	163	325.4	330	626	2150	3902
14	57.2	64	147.2	114	237.2	164	327.2	340	644	2200	3992
15	59.0	65	149.0	115	239.0	165	329.0	350	662	2250	4082
16	60.8	66	150.8	116	240.8	166	330.8	360	680	2300	4172
17	62.6	67	152.6	117	242.6	167	332.6	370	698	2350	4262
18	64.4	68	154.4	118	244.4	168	334.4	380	716	2400	4352
19	66.2	69	156.2	119	246.2	169	336.2	390	734	2450	4442
20	68.0	70	158.0	120	248.0	170	338.0	400	752	2500	4532
21	69.8	71	159.8	121	249.8	171	339.8	410	770	2550	4622
22	71.6	72	161.6	122	251.6	172	341.6	420	788	2600	4712
23	73.4	73	163.4	123	253.4	173	343.4	430	806	2650	4802
24	75.2	74	165.2	124	255.2	174	345.2	440	824	2700	4892
25	77.0	75	167.0	125	257.0	175	347.0	450	842	2750	4982
26	78.8	76	168.8	126	258.8	176	348.8	460	860	2800	5072
27	80.6	77	170.6	127	260.6	177	350.6	470	878	2850	5162
28	82.4	78	172.4	128	262.4	178	352.4	480	896	2900	5252
29	84.2	79	174.2	129	264.2	179	354.2	490	914	2950	5342

Table 136-1   Fourth Power of Absolute Temperature $\left(\dfrac{T}{100}\right)^4$

Temperature		Potential	Temperature		Potential	Temperature		Potential	Temperature		Potential	Temperature		Potential	Temperature		Potential
t	T	$\left(\dfrac{T}{100}\right)^4$	t	T	$\left(\dfrac{T}{100}\right)^4$	t	T	$\left(\dfrac{T}{100}\right)^4$	t	T	$\left(\dfrac{T}{100}\right)^4$	t	T	$\left(\dfrac{T}{100}\right)^4$	t	T	$\left(\dfrac{T}{100}\right)^4$
°C	°K		°C	°K		°C	°K		°C	°K		°C	°K		°C	°K	
0.0	273.15	55.627	3.1	276.25	58.196	6.1	279.25	60.766	9.1	282.25	63.240	12.1	285.25	66.160			
0.1	273.25	55.709	3.2	276.35	58.280	6.2	279.35	60.853	9.2	282.35	63.510	12.2	285.35	66.253			
0.2	273.35	55.790	3.3	276.45	58.365	6.3	279.45	60.940	9.3	282.45	63.600	12.3	285.45	66.346			
0.3	273.45	55.872	3.4	276.55	58.449	6.4	279.55	61.028	9.4	282.55	63.691	12.4	285.55	66.439			
0.4	273.55	55.945	3.5	276.65	58.534	6.5	279.65	61.115	9.5	282.65	63.781	12.5	285.65	66.532			
0.5	273.65	56.036	3.6	276.75	58.619	6.6	279.75	61.202	9.6	282.75	63.871	12.6	285.75	66.625			
0.6	273.75	56.118	3.7	276.85	58.704	6.7	279.85	61.290	9.7	282.85	63.961	12.7	285.85	66.719			
0.7	273.85	56.200	3.8	276.95	58.789	6.8	279.95	61.378	9.8	282.95	64.052	12.8	285.95	66.812			
0.8	273.95	56.282	3.9	277.05	58.874	6.9	280.05	61.466	9.9	283.05	64.142	12.9	286.05	66.906			
0.9	274.05	56.364	4.0	277.15	58.959	7.0	280.15	61.553	10.0	283.15	64.233	13.0	286.15	67.000			
1.0	274.15	56.446	4.1	277.25	59.044	7.1	280.25	61.641	10.1	283.25	64.324	13.1	286.25	67.094			
1.1	274.25	56.529	4.2	277.35	59.129	7.2	280.35	61.729	10.2	283.35	64.415	13.2	286.35	67.187			
1.2	274.35	56.611	4.3	277.45	59.214	7.3	280.45	61.818	10.3	283.45	64.506	13.3	286.45	67.281			
1.3	274.45	56.694	4.4	277.55	59.300	7.4	280.55	61.906	10.4	283.55	64.597	13.4	286.55	67.375			
1.4	274.55	56.777	4.5	277.65	59.385	7.5	280.65	61.994	10.5	283.65	64.688	13.5	286.65	67.469			
1.5	274.65	56.860	4.6	277.75	59.471	7.6	280.75	62.082	10.6	283.75	64.779	13.6	286.75	67.563			
1.6	274.75	56.942	4.7	277.85	59.556	7.7	280.85	62.171	10.7	283.85	64.871	13.7	286.85	67.658			
1.7	274.85	57.025	4.8	277.95	59.642	7.8	280.95	62.260	10.8	283.95	64.962	13.8	286.95	67.753			
1.8	274.95	57.108	4.9	278.05	59.728	7.9	281.05	62.348	10.9	284.05	65.054	13.9	287.05	67.847			
1.9	275.05	57.192	5.0	278.15	59.814	8.0	281.15	62.437	11.0	284.15	65.145	14.0	287.15	67.941			
2.0	275.15	57.275	5.1	278.25	59.900	8.1	281.25	62.526	11.1	284.25	65.237	14.1	287.25	68.036			
2.1	275.25	57.358	5.2	278.35	59.986	8.2	281.35	62.615	11.2	284.35	65.329	14.2	287.35	68.131			
2.2	275.35	57.441	5.3	278.45	60.072	8.3	281.45	62.704	11.3	284.45	65.421	14.3	287.45	68.226			
2.3	275.45	57.525	5.4	278.55	60.159	8.4	281.55	62.793	11.4	284.55	65.513	14.4	287.55	68.321			
2.4	275.55	57.609	5.5	278.65	60.245	8.5	281.65	62.883	11.5	284.65	65.606	14.5	287.65	68.416			
2.5	275.65	57.692	5.6	278.75	60.332	8.6	281.75	62.972	11.6	284.75	65.689	14.6	287.75	68.511			
2.6	275.75	57.776	5.7	278.85	60.419	8.7	281.85	63.061	11.7	284.85	65.790	14.7	287.85	68.607			
2.7	275.85	57.860	5.8	278.95	60.505	8.8	281.95	63.151	11.8	284.95	65.882	14.8	287.95	68.702			
2.8	275.95	57.944	5.9	279.05	60.592	8.9	282.05	63.241	11.9	285.05	65.975	14.9	288.05	68.798			
2.9	276.05	58.028	6.0	279.15	60.679	9.0	282.15	63.330	12.0	285.15	66.067	15.0	288.15	68.893			
3.0	276.15	58.112															

Table 136-2 Fourth Power of Absolute Temperature $\left(\dfrac{T}{100}\right)^4$ *(Continued)*

Temperature		Potential	Temperature		Potential	Temperature		Potential	Temperature		Potential	Temperature		Potential	Temperature		Potential
t	T	$\left(\dfrac{T}{100}\right)^4$	t	T	$\left(\dfrac{T}{100}\right)^4$	t	T	$\left(\dfrac{T}{100}\right)^4$	t	T	$\left(\dfrac{T}{100}\right)^4$	t	T	$\left(\dfrac{T}{100}\right)^4$	t	T	$\left(\dfrac{T}{100}\right)^4$
°C	°K		°C	°K		°C	°K		°C	°K		°C	°K		°C	°K	
15.1	288.25	68.989	19.1	292.25	72.899	23.1	296.25	76.974	27.1	300.25	81.216	32.0	305.15	86.65			
15.2	288.35	69.084	19.2	292.35	72.999	23.2	296.35	77.078	27.2	300.35	81.324	33.0	306.15	87.79			
15.3	288.45	69.180	19.3	292.45	73.099	23.3	296.45	77.182	27.3	300.45	81.433	34.0	307.15	88.94			
15.4	288.55	69.276	19.4	292.55	73.199	23.4	296.55	77.286	27.4	300.55	81.541	35.0	308.15	90.11			
15.5	288.65	69.373	19.5	292.65	73.299	23.5	296.65	77.391	27.5	300.65	81.650	36.0	309.15	91.28			
15.6	288.75	69.469	19.6	292.75	73.399	23.6	296.75	77.495	27.6	300.75	81.758	37.0	310.15	92.47			
15.7	288.85	69.565	19.7	292.85	73.500	23.7	296.85	77.599	27.7	300.85	81.867	38.0	311.15	93.67			
15.8	288.95	69.661	19.8	292.95	73.600	23.8	296.95	77.703	27.8	300.95	81.976	39.0	312.15	94.88			
15.9	289.05	69.758	19.9	293.05	73.701	23.9	297.05	77.808	27.9	301.05	82.085	40.0	313.15	96.10			
16.0	289.15	69.854	20.0	293.15	73.801	24.0	297.15	77.913	28.0	301.15	82.195	41.0	314.15	97.33			
16.1	289.25	69.950	20.1	293.25	73.902	24.1	297.25	78.018	28.1	301.25	82.304	42.0	315.15	98.57			
16.2	289.35	70.047	20.2	293.35	74.003	24.2	297.35	78.123	28.2	301.35	82.413	43.0	316.15	99.82			
16.3	289.45	70.145	20.3	293.45	74.104	24.3	297.45	78.229	28.3	301.45	82.522	44.0	317.15	101.10			
16.4	289.55	70.242	20.4	293.55	74.205	24.4	297.55	78.334	28.4	301.55	82.632	45.0	318.15	102.39			
16.5	289.65	70.339	20.5	293.65	74.306	24.5	297.65	78.440	28.5	301.65	82.742	46.0	319.15	103.68			
16.6	289.75	70.436	20.6	293.75	74.407	24.6	297.75	78.545	28.6	301.75	82.851	47.0	320.15	104.99			
16.7	289.85	70.533	20.7	293.85	74.509	24.7	297.85	78.650	28.7	301.85	82.961	48.0	321.15	106.31			
16.8	289.95	70.630	20.8	293.95	74.611	24.8	297.95	78.755	28.8	301.95	83.072	49.0	322.15	107.64			
16.9	290.05	70.728	20.9	294.05	74.713	24.9	298.05	78.861	28.9	302.05	83.182	50.0	323.15	108.98			
17.0	290.15	70.826	21.0	294.15	74.814	25.0	298.15	78.967	29.0	302.15	83.292						
17.1	290.25	70.924	21.1	294.25	74.916	25.1	298.25	79.073	29.1	302.25	83.402						
17.2	290.35	71.021	21.2	294.35	75.017	25.2	298.35	79.179	29.2	302.35	83.513						
17.3	290.45	71.119	21.3	294.45	75.119	25.3	298.45	79.268	29.3	302.45	83.623						
17.4	290.55	71.217	21.4	294.55	75.221	25.4	298.55	79.392	29.4	302.55	83.734						
17.5	290.65	71.315	21.5	294.65	75.324	25.5	298.65	79.499	29.5	302.65	83.845						
17.6	290.75	71.413	21.6	294.75	75.426	25.6	298.75	79.605	29.6	302.75	83.956						
17.7	290.85	71.512	21.7	294.85	75.529	25.7	298.85	79.712	29.7	302.85	84.067						
17.8	290.95	71.610	21.8	294.95	75.631	25.8	298.95	79.819	29.8	302.95	84.178						
17.9	291.05	71.709	21.9	295.05	75.734	25.9	299.05	79.926	29.9	303.05	84.289						
18.0	291.15	71.807	22.0	295.15	75,836	26.0	299.15	80.033	30.0	303.15	84.401						
18.1	291.25	71.906	22.1	295.25	75.939	26.1	299.25	80.140	30.1	303.25	84.512						
18.2	291.35	72.005	22.2	295.35	76.042	26.2	299.35	80.247	30.2	303.35	84.623						
18.3	291.45	72.104	22.3	295.45	76.146	26.3	299.45	80.354	30.3	303.45	84.735						
18.4	291.55	72.203	22.4	295.55	76.249	26.4	299.55	80.462	30.4	303.55	84.847						
18.5	291.65	72.302	22.5	295.65	76.352	26.5	299.65	80.569	30.5	303.65	84.959						
18.6	291.75	72.401	22.6	295.75	76.455	26.6	299.75	80.677	30.6	303.75	85.071						
18.7	291.85	72.501	22.7	295.85	76.559	26.7	299.85	80.784	30.7	303.85	85.183						
18.8	291.95	72.600	22.8	295.95	76.662	26.8	299.95	80.892	30.8	303.95	85.295						
18.9	292.05	72.700	22.9	296.05	76.766	26.9	300.05	81.000	30.9	304.05	85.407						
19.0	292.15	72.799	23.0	296.15	76.870	27.0	300.15	81.108	31.0	304.15	85.520						

Table 137-1 Some Dimensionless Numbers

| Number | | Formula | | |
Name	Symbol	Mass basis	Weight basis	Application and Nomenclature
Clausius	Cl	$Cl = \dfrac{v^3 \, l \, \rho}{\lambda \Delta T}$	$Cl = \dfrac{v^3 \, l \, \gamma}{g \lambda \Delta T}$	Heat conduction; v = velocity, l = length, ρ = density, λ = conductivity, ΔT = temperature difference, γ = weight density, g = acceleration due to gravity.
Dulong	Du	$Du = \dfrac{v^2}{c \Delta T}$	$Du = \dfrac{v^2}{g c' \Delta T}$	Accumulation of energy in flowing fluids; v = velocity, c = specific heat at constant pressure, mass basis, ΔT = temperature difference, g = acceleration due to gravity, c' = specific heat at constant pressure, weight basis.
Euler	Eu	$Eu = \dfrac{\Delta p}{\frac{1}{2}\rho v^2}$	$Eu = \dfrac{\Delta p}{\gamma v^2 / 2g}$	Fluid friction in conduits; Δp = pressure difference, ρ = density, v = velocity, γ = weight density, g = acceleration due to gravity.
Fourier	Fo	$Fo = \dfrac{t\lambda}{l^2 \rho c}$	$Fo = \dfrac{t\lambda}{l^2 \gamma c'}$	Unsteady state heat conduction; t = time, λ = conductivity, l = length, ρ = density, c = specific heat at constant pressure, mass basis, γ = weight density, c' = specific heat at constant pressure, weight basis.
Froude	Fr	$Fr = \dfrac{v^2}{gl}$	$Fr = \dfrac{v^2}{gl}$	Wave and surface behavior; v = velocity, g = acceleration due to gravity, l = length.
Gay-Lussac	Ga	$Ga = \dfrac{1}{\beta \Delta T}$	$Ga = \dfrac{1}{\beta \Delta T}$	Thermal expansion; β = coefficient of cubical expansion, ΔT = temperature difference.
Grashof	Gr	$Gr = \dfrac{g \beta l^3 \Delta T}{(\eta/\rho)^2}$ $= \dfrac{g l^3 \Delta T}{\nu^2}$	$Gr = \dfrac{\beta l^3 \Delta T}{g(\eta/\gamma)^2}$	Heat transfer by free convection; g = acceleration due to gravity, β = coefficient of cubical expansion, l = length, ΔT = temperature difference, η = dynamic viscosity, ρ = density, γ = weight density, ν = kinematic viscosity ($\nu = \eta/\rho$). The relation $Gr \cdot Ga \cdot Fr = Re^2$ is valid.
Graetz	Gz	$Gz = \dfrac{m'c}{\lambda l}$	$Gz = \dfrac{G'c'}{\lambda l}$	Heat conduction in streamline flow; m' = mass rate of flow, c = specific heat at constant pressure, mass basis, λ = conductivity, l = length, G' = weight rate of flow, c' = specific heat at constant pressure, weight basis.
Hooke	Ho	$Ho = \dfrac{\rho v^2}{E}$	$Ho = \dfrac{\gamma v^2}{gE}$	Compressible flow; ρ = density, v = velocity, E = model of elasticity, γ = weight density, g = acceleration due to gravity. The Hooke number is also called Cauchy number, $Ho = Ca$.
Lewis	Le	$Le \dfrac{\alpha}{\sigma c}$	$Le \dfrac{\alpha}{\sigma' c'}$	Heat and mass transfer; α = conductance (heat transfer coefficient), σ = mass transfer coefficient (kg/m^2 h), c = specific heat at constant pressure, mass basis, σ' = weight transfer coefficient (kp/m^2 h), c' = specific heat at constant pressure, weight basis.
Mach	Ma	$Ma = \dfrac{v}{u}$	$Ma = \dfrac{v}{u}$	Compressible flow; v = fluid velocity, u = sound velocity.
Newton	Ne	$Ne = \dfrac{R}{\rho v^2 l^2}$	$Ne = \dfrac{gR}{\gamma v^2 l^2}$	Agitation; R = force of resistance (imposed force), ρ = density, v = velocity, l = length, g = acceleration due to gravity, γ = weight density.

Table 137-2 Some Dimensional Numbers (*Continued*)

Number		Formula		
Name	Symbol	Mass basis	Weight basis	Application and Nomenclature
Nusselt	Nu	$Nu = \dfrac{\alpha l}{\lambda}$	$Nu = \dfrac{\alpha l}{\lambda}$	Heat transfer by forced convection; α = conductance (heat transfer coefficient), l = length, λ = conductivity. The Nusselt number (Nu) is also called Biot (Bi) number, Nu = Bi.
Péclet	Pe	$Pe = \dfrac{\rho c v l}{\lambda}$	$Pe = \dfrac{\gamma c' v l}{\lambda}$	Diffusion in packed beds; ρ = density, c = specific heat at constant pressure, mass basis, v = velocity, l = length, λ = conductivity, γ = weight density, c' = specific heat at constant pressure, weight basis. The relation Pe/Re = Pr is valid.
Prandtl	Pr	$Pr = \dfrac{c\eta}{\lambda} = \dfrac{\rho c v}{\lambda}$	$Pr = \dfrac{g c' \eta}{\lambda}$	Heat conduction in streamline flow; c = specific heat at constant pressure, mass basis, η = dynamic viscosity, λ = conductivity, ρ = density, v = kinematic viscosity, c' = specific heat at constant pressure, weight basis, g = acceleration due to gravity. The quantity $\lambda/c\rho = a$ is called thermal diffusivity.
Rayleigh	Ra	$Ra = \dfrac{g\beta l^3 \Delta T \rho^2 c}{\eta\lambda}$	$Ra = \dfrac{\beta l^3 \Delta T \gamma^2 c'}{\eta\lambda}$	Heat transfer by free convection; g = acceleration due to gravity, β = coefficient of cubical expansion, l = length, ΔT = temperature difference, ρ = density, c = specific heat at constant pressure, mass basis, η = dynamic viscosity, λ = conductivity, γ = weight density, c' = specific heat at constant pressure, weight basis. The relation Ra = Gr · Pr is valid.
Reynolds	Re	$Re = \dfrac{v l \rho}{\eta} = \dfrac{v l}{\nu}$	$Re = \dfrac{v l \gamma}{g\eta}$	Dynamic similarity; v = velocity, l = length, ρ = density, η = dynamic viscosity, ν = kinematic viscosity ($\nu = \eta/\rho$), γ = weight density, g = acceleration due to gravity. The following relations are valid: Re = Pe/Pr, Re^2 = Gr · Ga · Fr.
Schmidt	Sc	$Sc = \dfrac{\eta}{\rho k} = \dfrac{\nu}{k}$	$Sc = \dfrac{\eta g}{\gamma k}$	Mass transfer; η = dynamic viscosity, ρ = density, k = diffusion coefficient, ν = kinematic viscosity ($\nu = \eta/\rho$) g = acceleration due to gravity, γ = weight density.
Stanton	St	$St = \dfrac{\alpha}{\rho c v}$	$St = \dfrac{\alpha}{\gamma c' v}$	Forced convection; α = conductance (heat transfer coefficient), ρ = density, c = specific heat at constant pressure, mass basis, v = velocity, γ = weight density, c' = specific heat at constant pressure, weight basis. The relation St = Nu/(Re · Pr) is valid.
Strouhal	Sh	$Sh = \dfrac{v}{f l}$	$Sh = \dfrac{v}{f l}$	Vortex streets; f = frequency of vortex, l = length, v = velocity.
Weber	We	$We = \dfrac{v^2 l \rho}{\sigma}$	$We = \dfrac{v^2 l \gamma}{g\sigma}$	Bubble formation, breaking of liquid jets; v = velocity, l = length, δ = density, σ = surface tension, γ = weight density, g-acceleration.

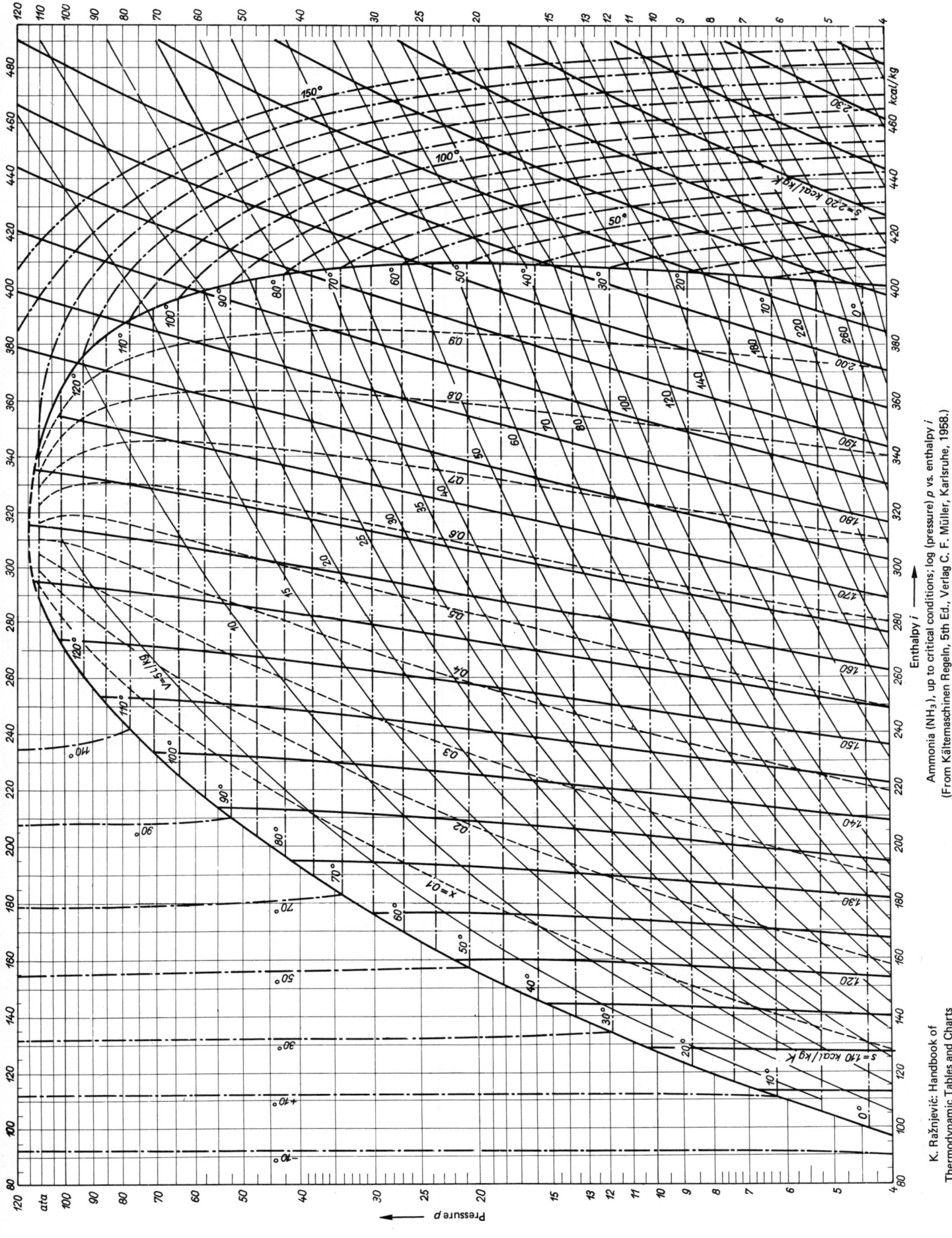

Ammonia (NH_3), up to critical conditions; log (pressure) p vs. enthalpy i
(From Käitemaschinen Regeln, 5th Ed., Verlag C. F. Müller, Karlsruhe, 1958.)

K. Ražnjević: Handbook of
Thermodynamic Tables and Charts

Ammonia (NH_3), at $0°C$ ($p = 4.29$ bar);
$i_0' = 0.5$ MJ/kg, $s_0' = 2.0$ kJ $°K$;
log (pressure) p vs. enthalpy i

Pressure p →

Enthalpy i MJ/kg →

(From Kältemachinen Regeln, 5th Ed., Verlag C. F. Müller, Karlsruhe, 1958.)

Chart 2-b

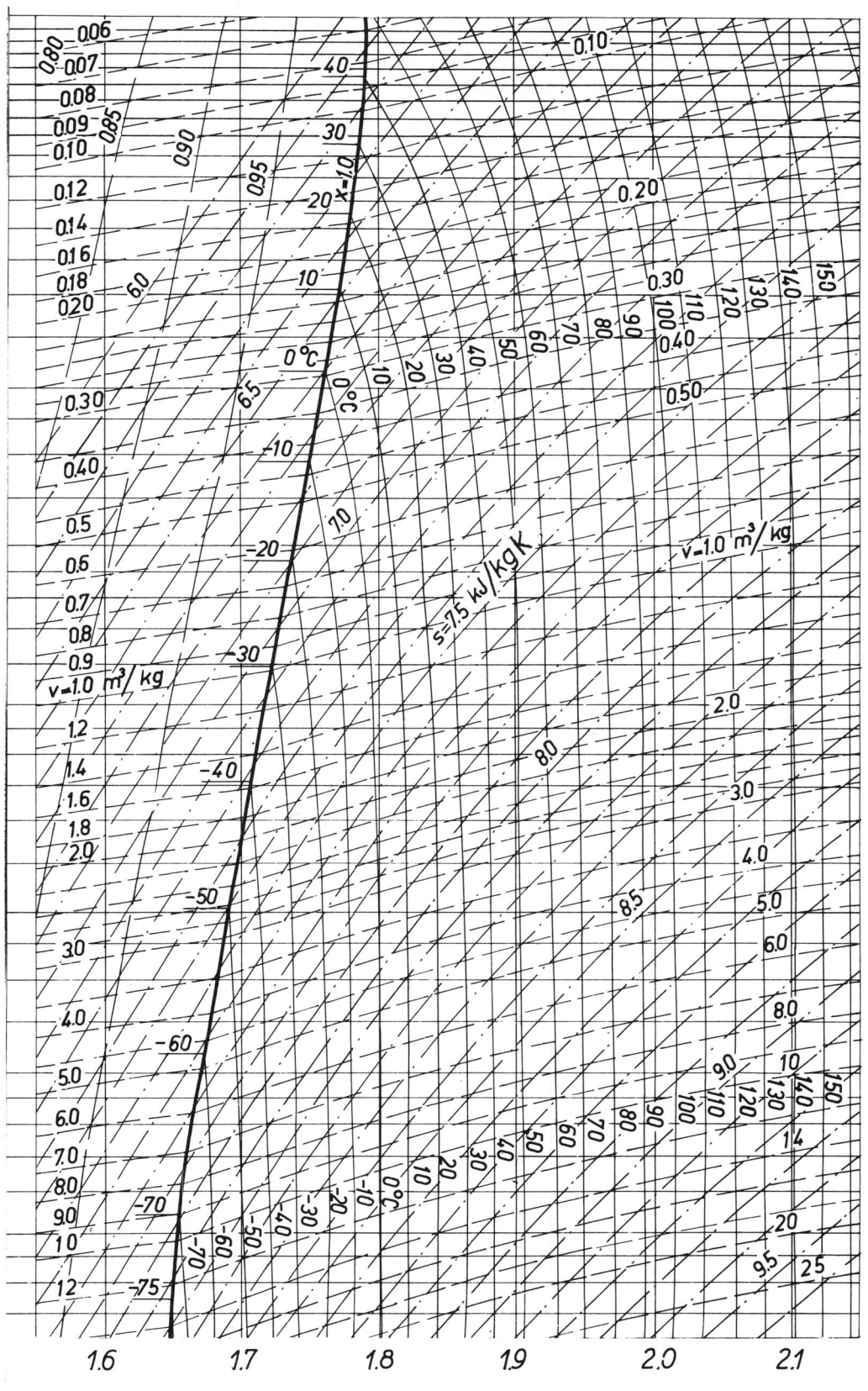

K. Ražnjević : Handbook of
Thermodynamic Tables and Charts

Chart 3-a (*Continued*)

Enthalpy *i* ⟶

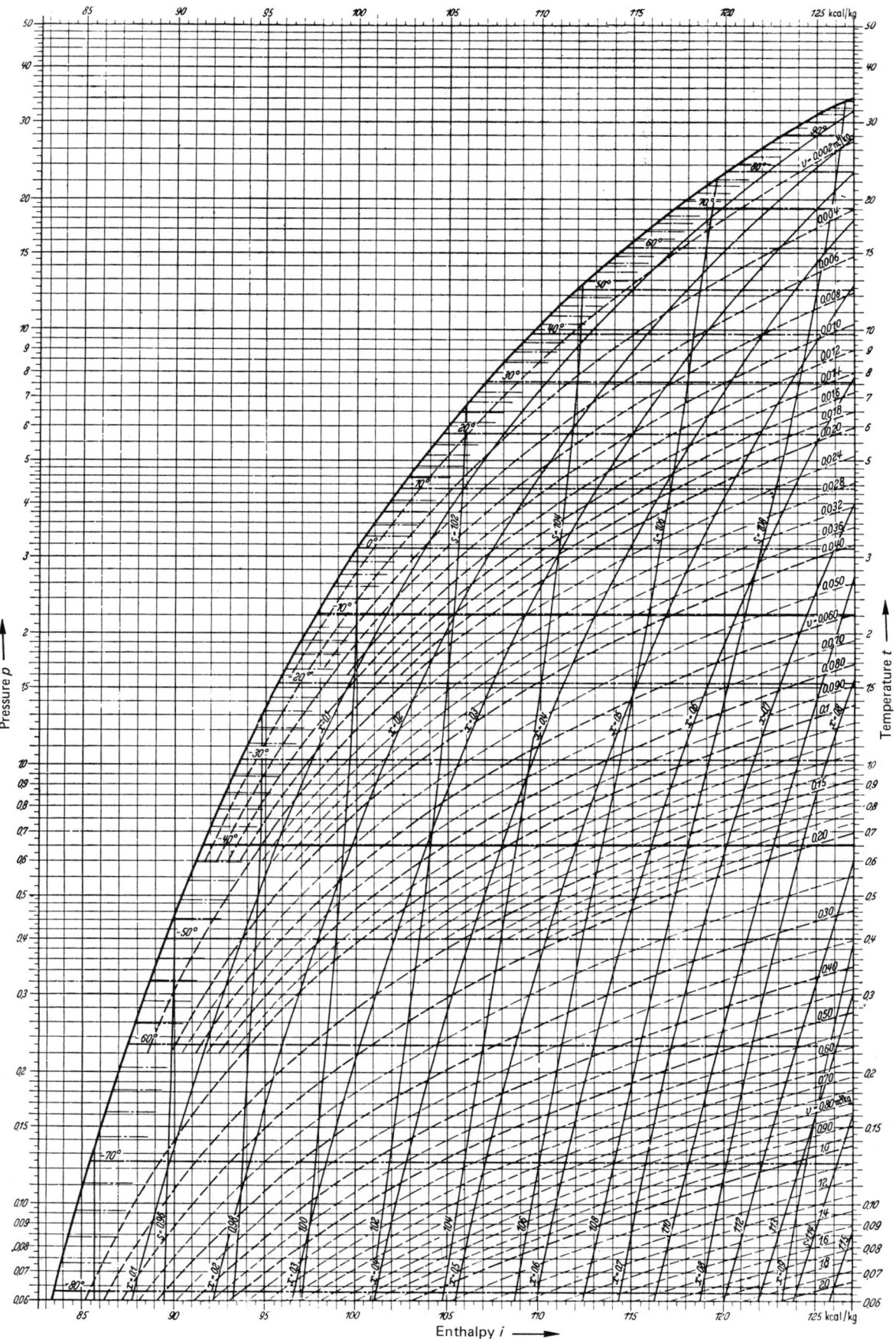

Enthalpy *i* ⟶

Pressure *p* ⟶

Temperature *t* ⟶

K. Ražnjević: Handbook of
Thermodynamic Tables and Charts

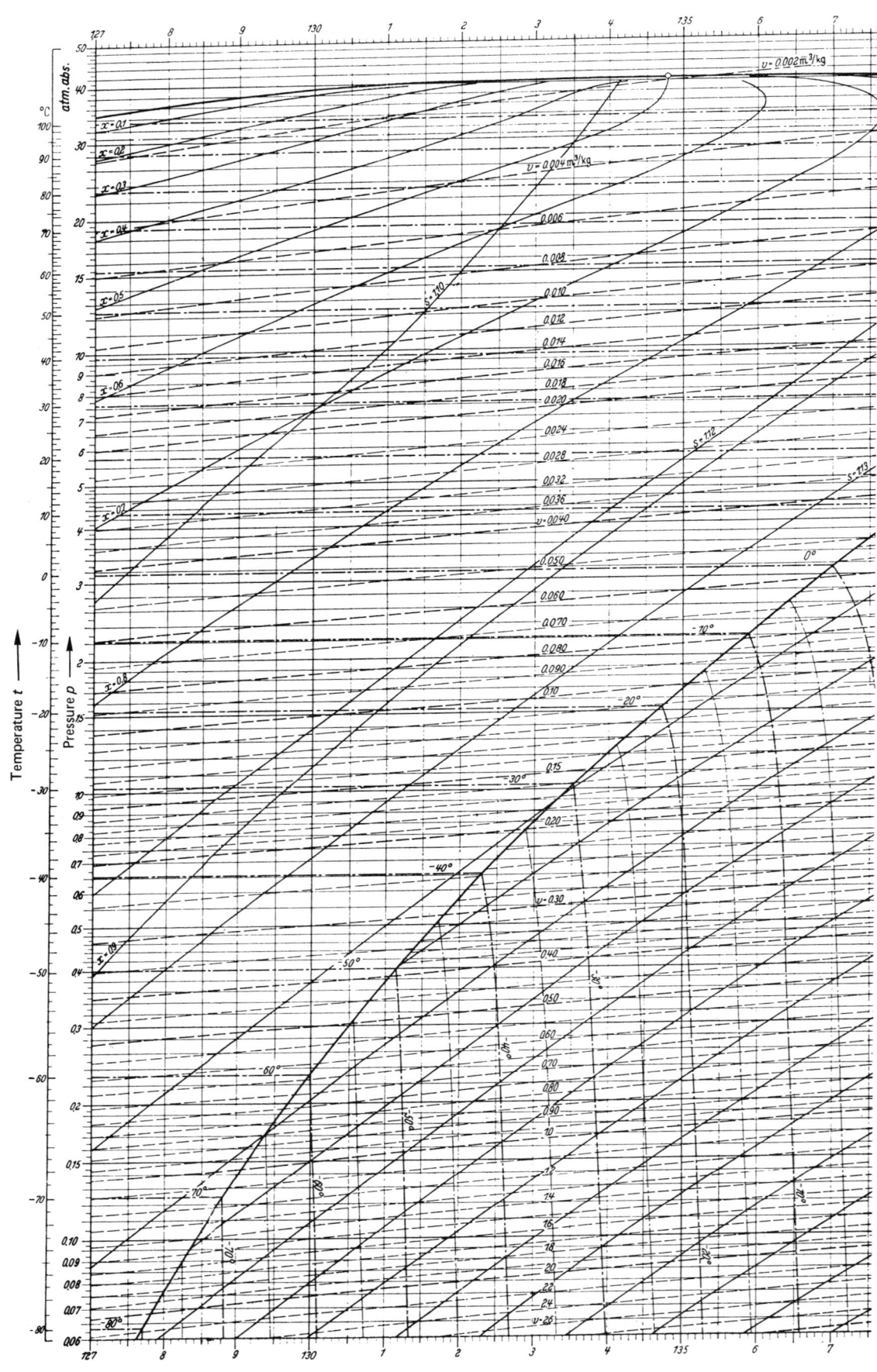

Dichlorodifluoromethane, "Freon 12"; log (pressure) p vs. enthalpy i
(From Kältemachinen Regeln, 5th Ed., Verlag C. F. Müller, Karlsruhe, 1958.)

Chart 3-a

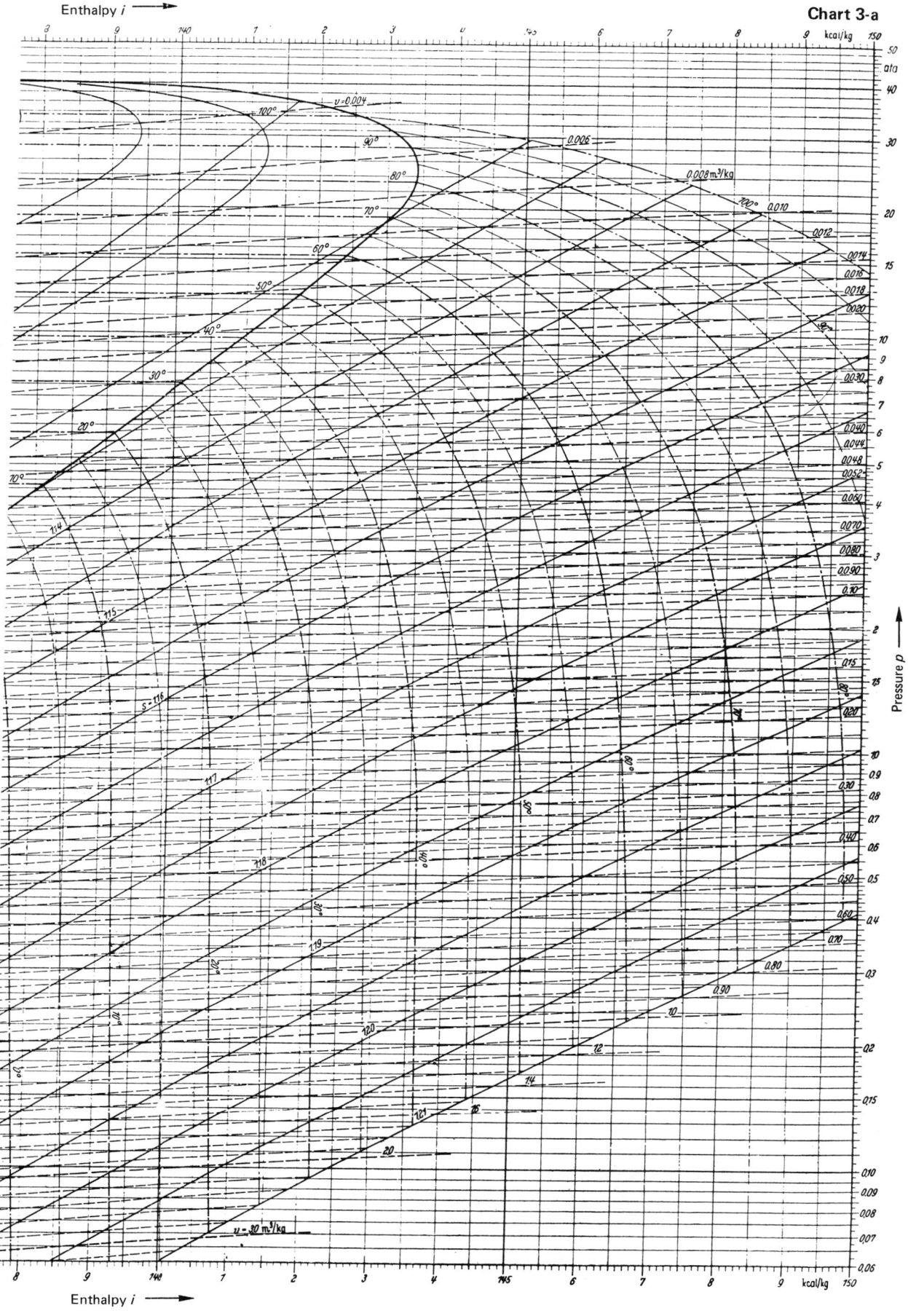

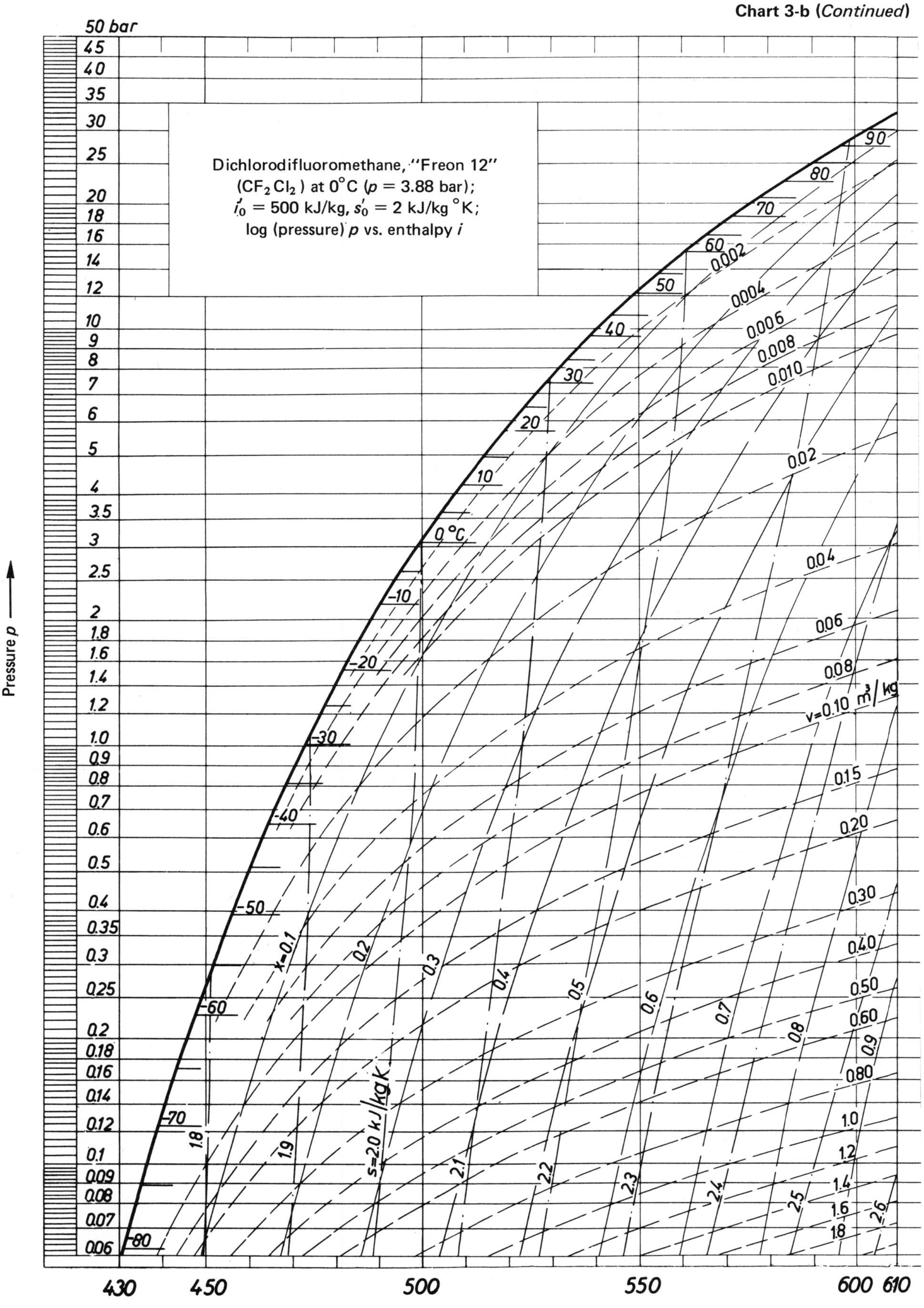

Chart 3-b (*Continued*)

50 bar

Dichlorodifluoromethane, "Freon 12"
(CF_2Cl_2) at $0°C$ ($p = 3.88$ bar);
$i'_0 = 500$ kJ/kg, $s'_0 = 2$ kJ/kg $°K$;
log (pressure) p vs. enthalpy i

Pressure p ⟶

$v = 0.10$ m^3/kg

$s = 2.0$ kJ/kg K

$x = 0.1$

K. Ražnjević: Handbook of
Thermodynamic Tables and Charts

367

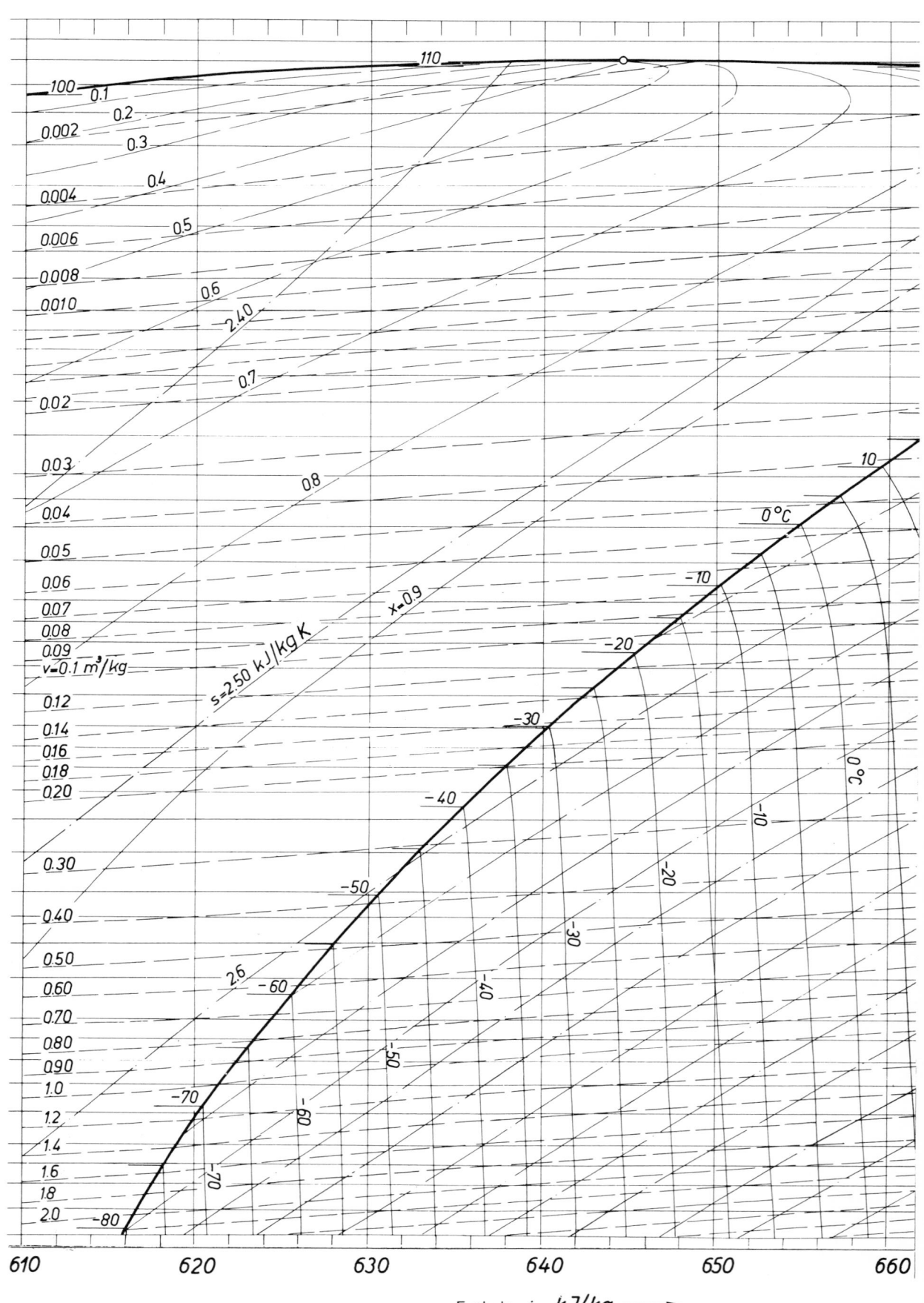

Enthalpy *i* **kJ/kg** ⟶

(From Kältemachinen Regeln, 5th Ed., Verlag C. F. Müller, Karlsruhe, 1958.)

Chart 3-b

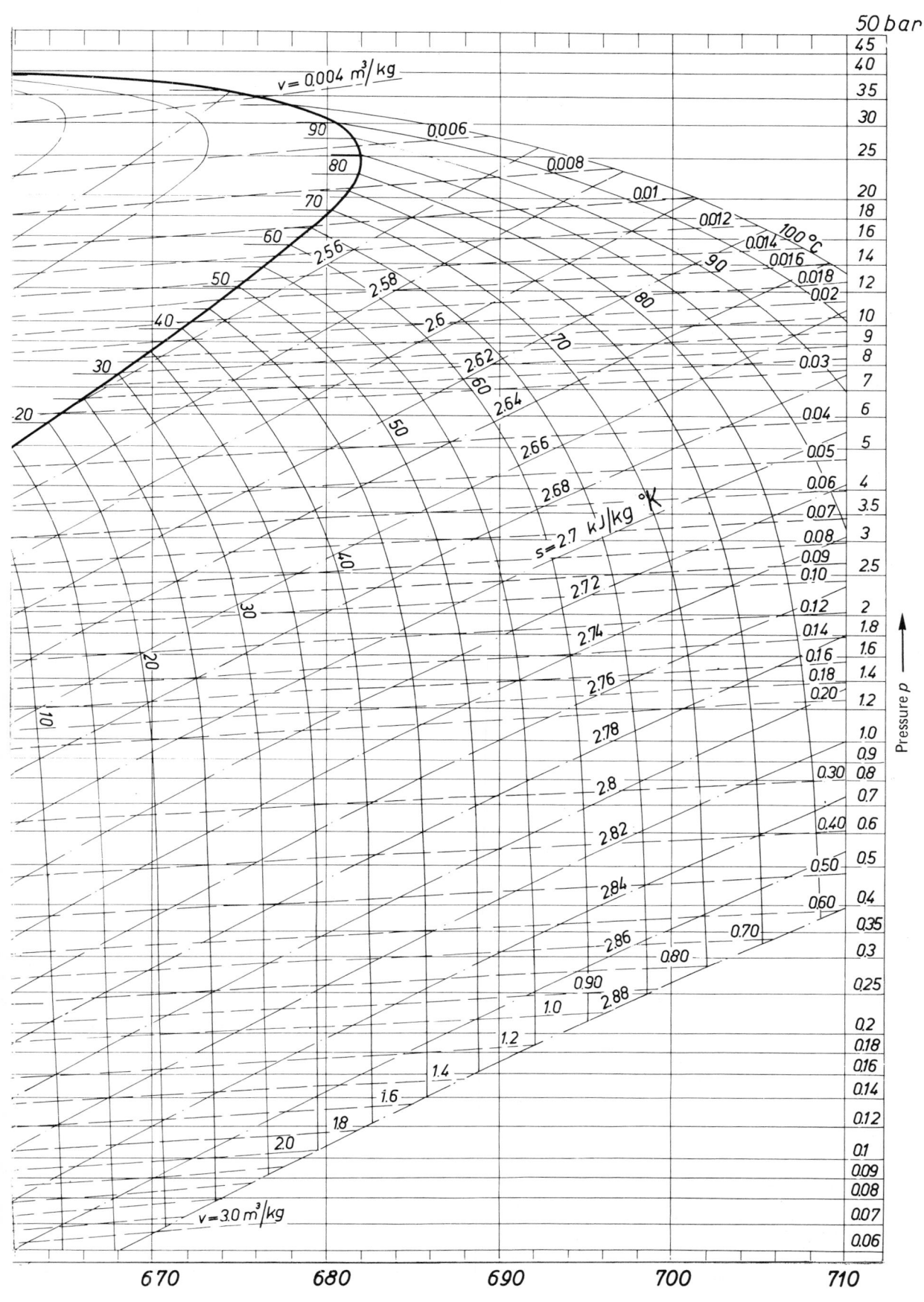

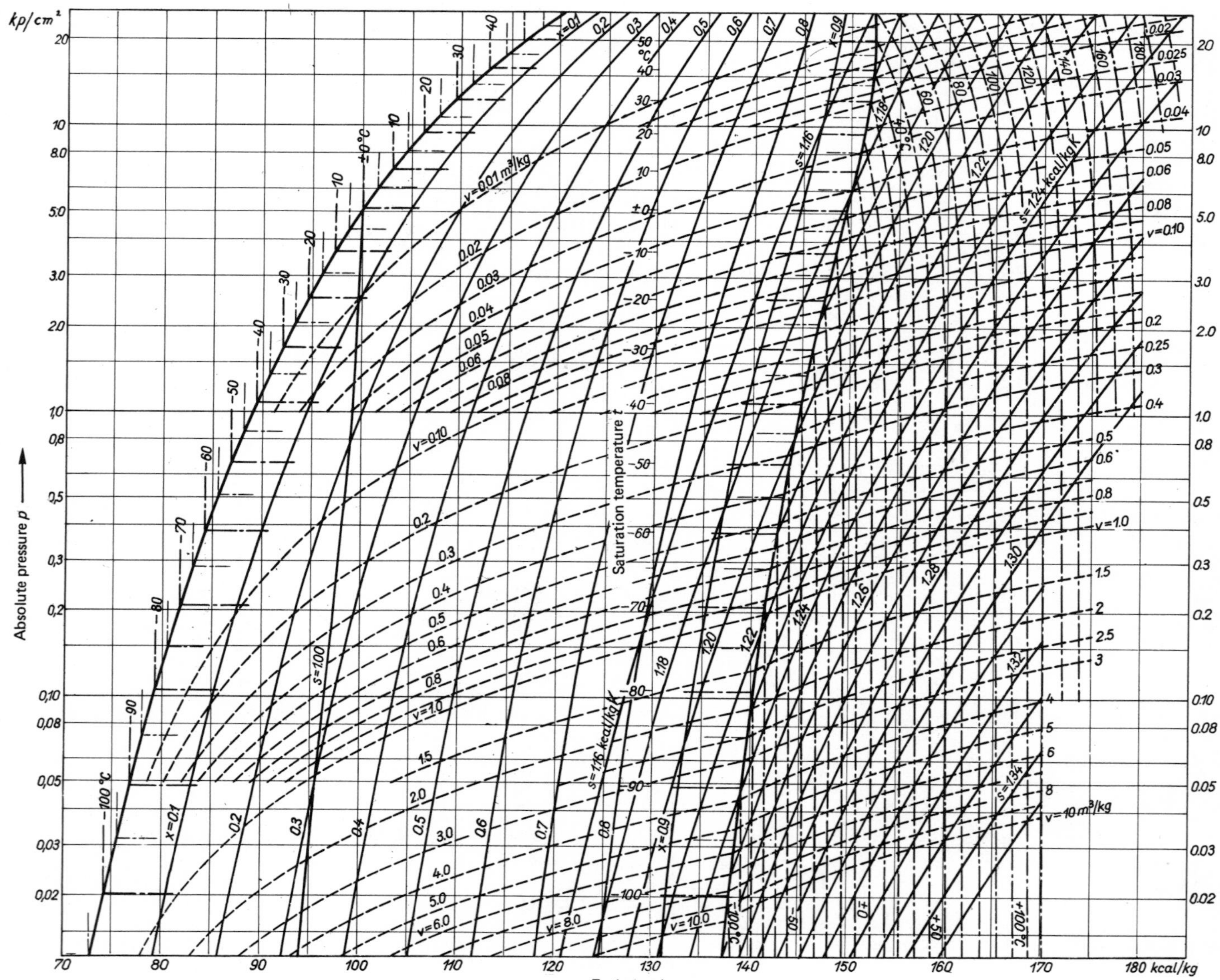

Chart 4

Monochlorodifluoromethane, "Freon 22" (CHF$_2$Cl) at 0°C, $i'_0 = 100$ kcal/kg, $s'_0 = 1$
kcal/kg °K; log (absolute pressure) p vs. enthalpy i
(From Kältemaschinen Regeln, 5th Ed., Verlag C. F. Müller, Karlsruhe, 1958.)

K. Ražnjević: Handbook of
Thermodynamic Tables and Charts

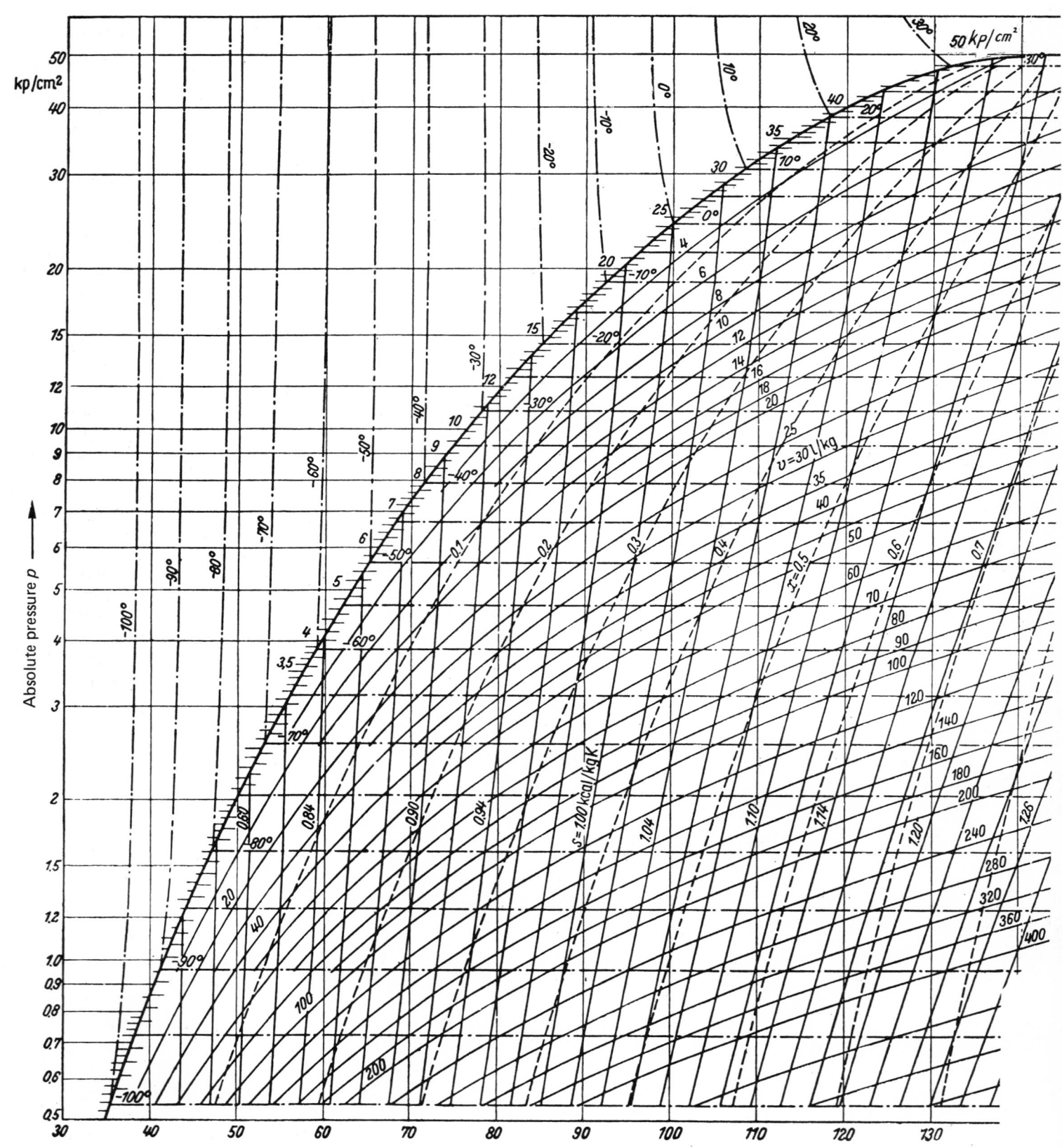

Chart 5

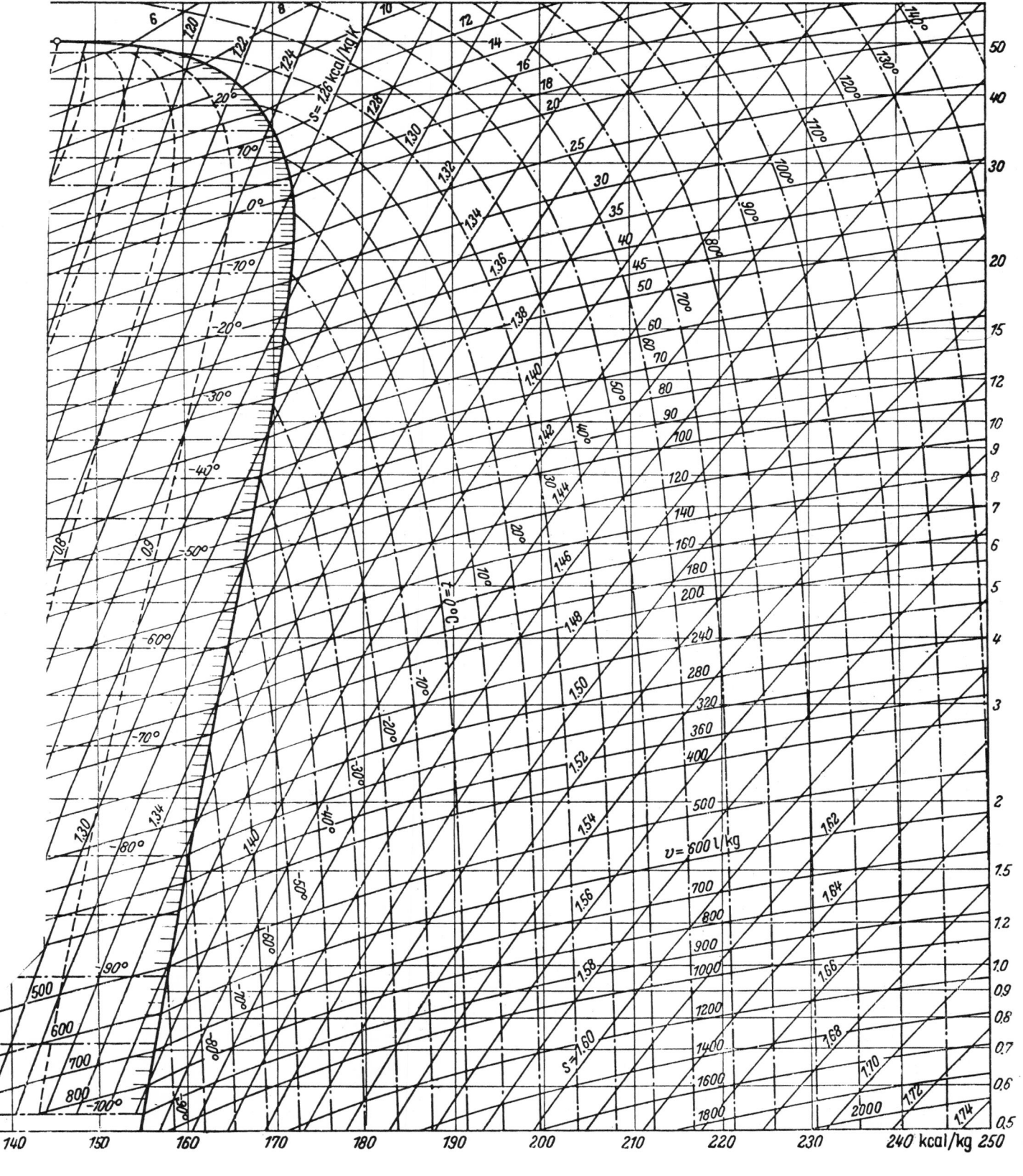

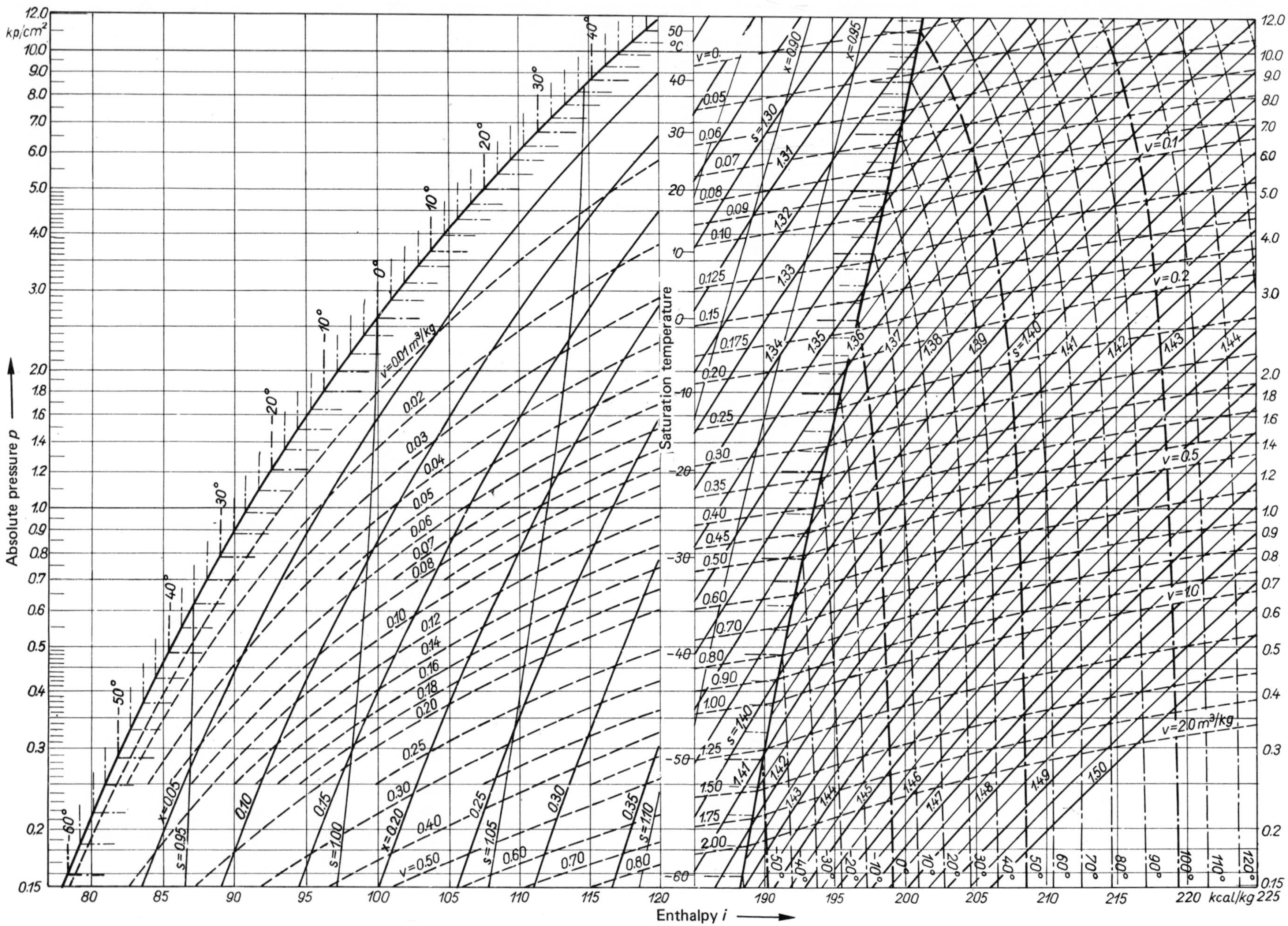

Chart 6

Methyl chloride (CH_3Cl) at 0°C, $i'_0 = 100$ kcal/kg, $s'_0 = 1$ kcal/kg °K; log (absolute pressure) p vs. enthalpy i

(From Kältemaschinen Regeln, 5th Ed., Verlag C. F. Müller, Karlsruhe, 1958.)

K. Ražnjević: Handbook of
Thermodynamic Tables and Charts

Chart 7

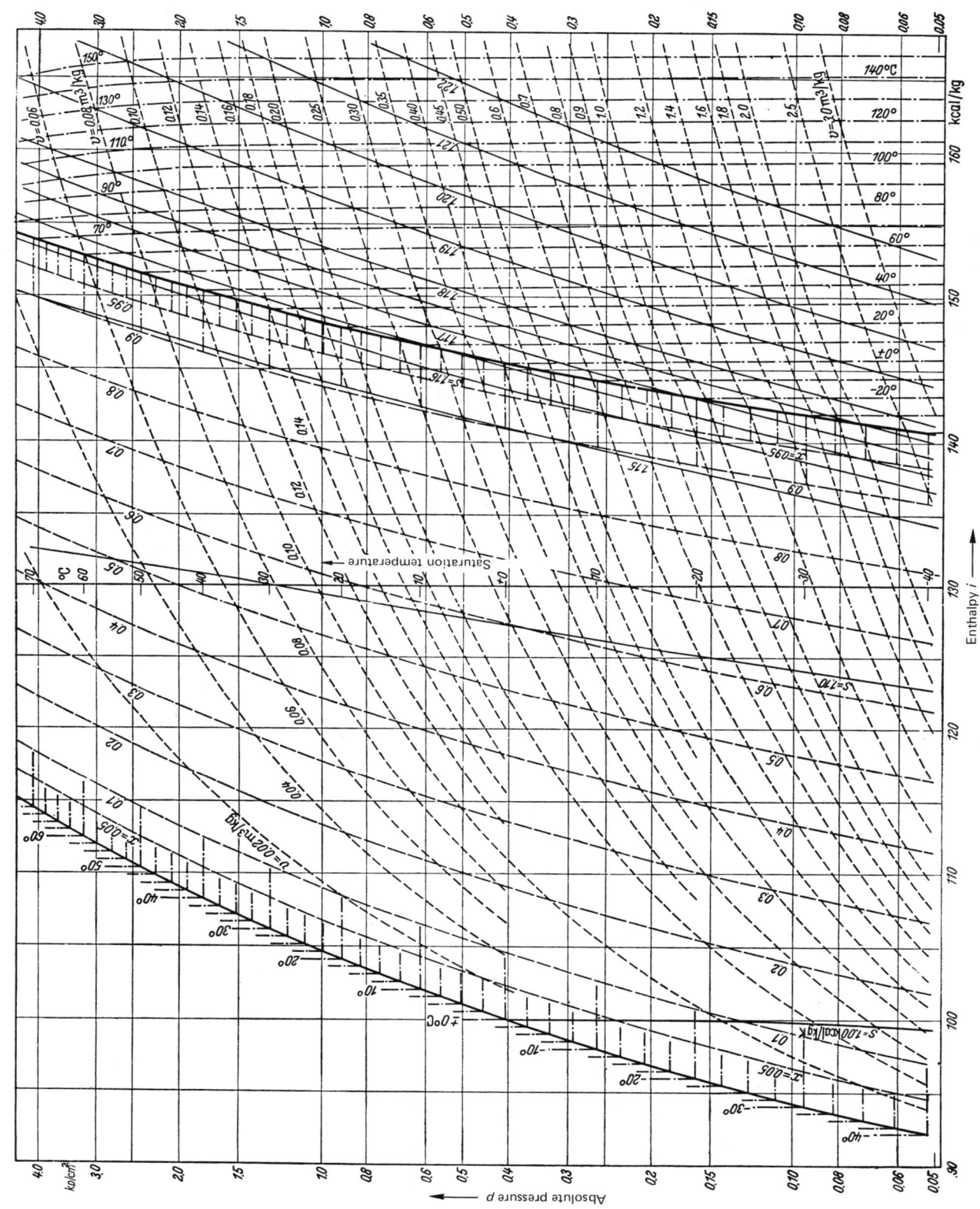

Trichloromonofluoromethane, "Freon 11'' (CFCl$_3$) at 0°C, i_0' = 100 kcal/kg,
s_0' = 1 kcal/kg °K; log (absolute pressure) p vs. enthalpy i
(From Kältemaschinen Regeln, 5th Ed., Verlag C. F. Müller, Karlsruhe, 1958.)

K. Ražnjević: Handbook of
Thermodynamic Tables and Charts

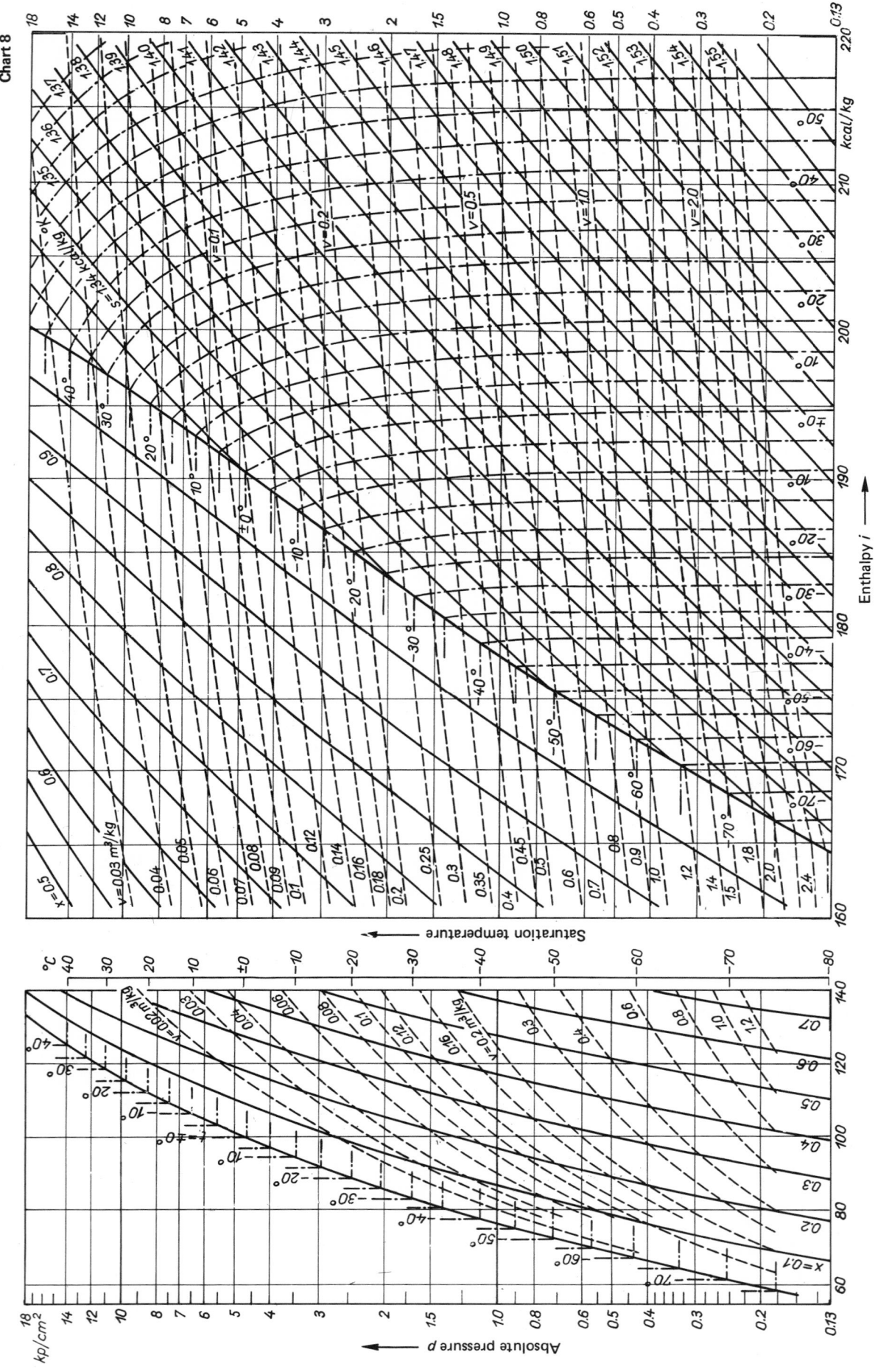

Chart 8

Enthalpy i ⟶

Propane (C_3H_8) at 0°C, $i'_0 = 100$ kcal/kg, $s'_0 = 1$ kcal/kg °K;
log (absolute pressure) p vs. enthalpy i
(From Kältemaschinen Regeln, 5th Ed., Verlag C. F. Müller, Karlsruhe, 1958.)

K. Ražnjević: Handbook of
Thermodynamic Tables and Charts

376

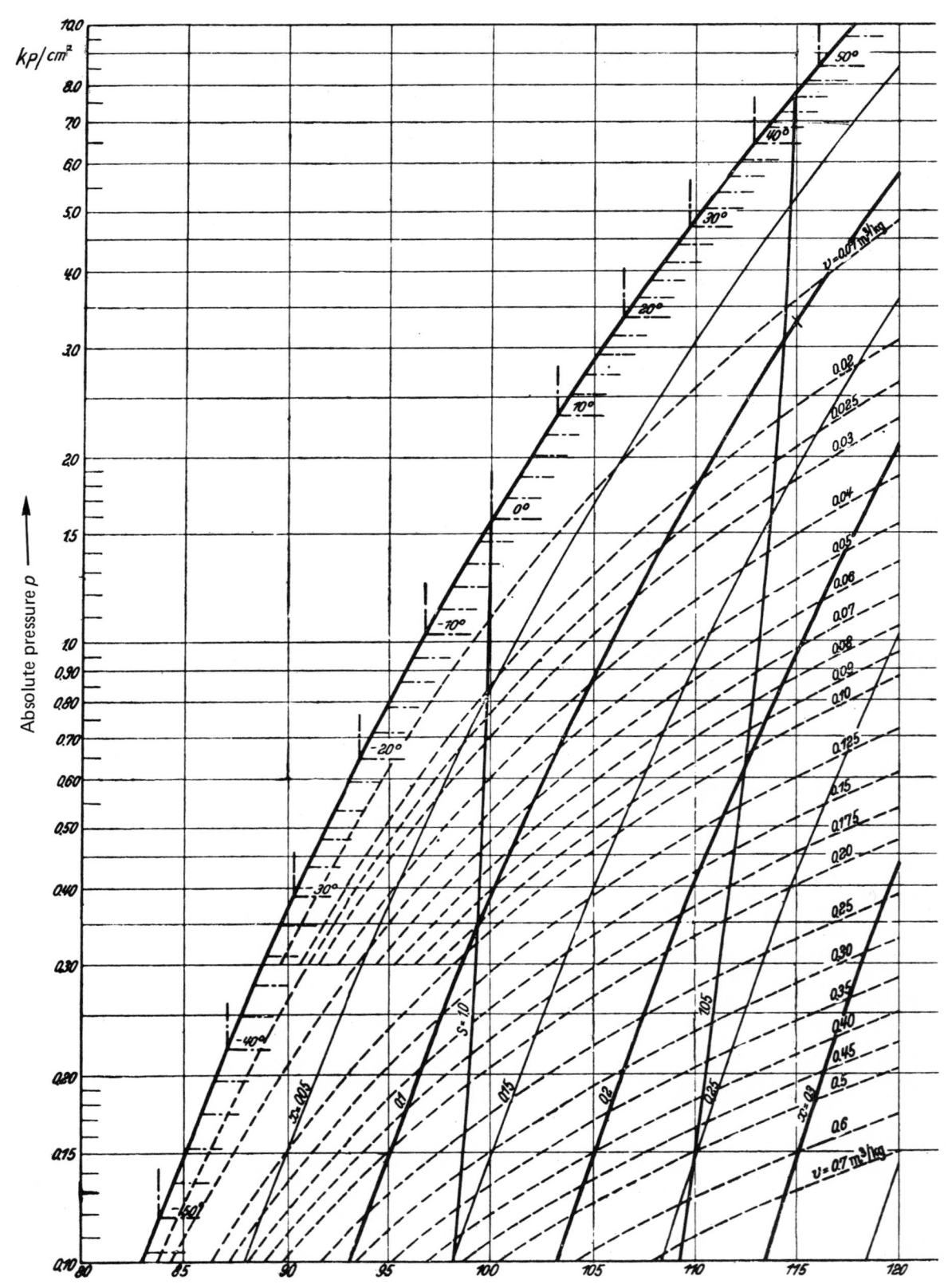

K. Ražnjević: Handbook of
Thermodynamic Tables and Charts

Chart 9

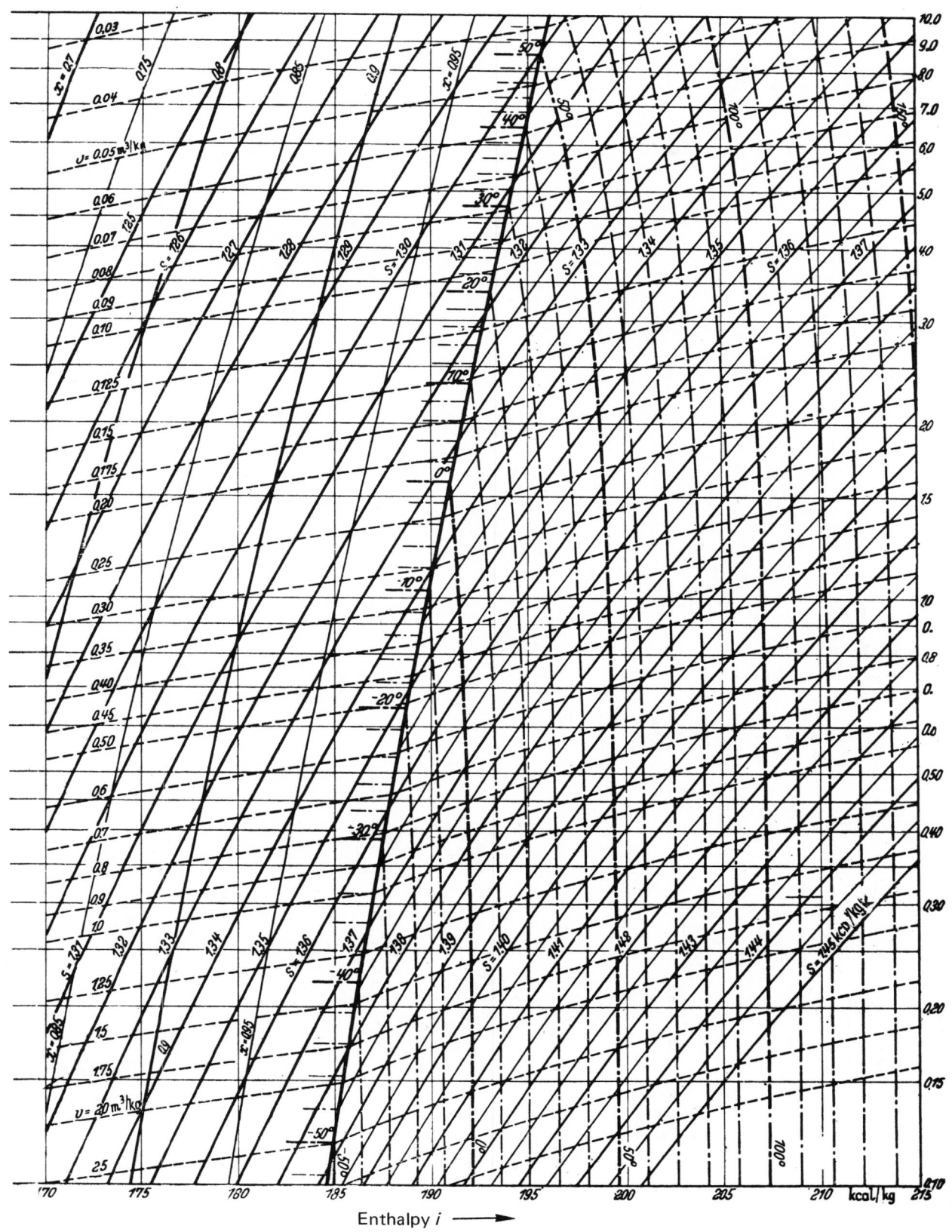

Enthalpy *i* ⟶

Sulfur dioxide (SO$_2$) at 0°C, $i'_0 = 100$ kcal/kg, $s'_0 = 1$ kcal/kg °K;
log (absolute pressure) *p* vs. enthalpy *i*
(From Kältemaschinen Regeln, 5th Ed., Verlag C. F. Müller, Karlsruhe, 1958.)

Chart 10

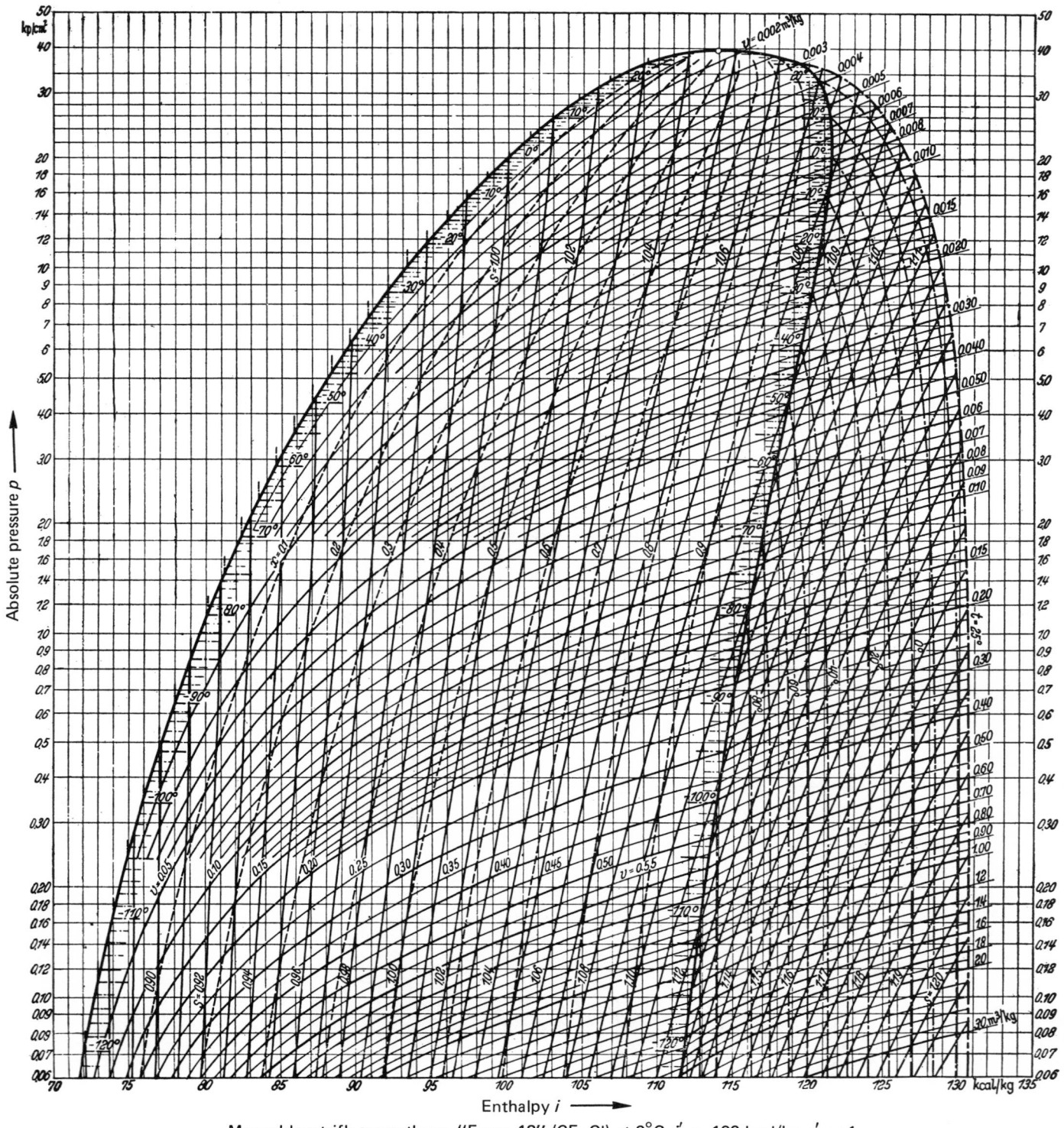

Monochlorotrifluoromethane, "Freon 13" (CF$_3$Cl) at 0°C, $i'_0 = 100$ kcal/kg, $s'_0 = 1$
kcal/kg °K; log (absolute pressure) p vs. enthalpy i
(From Kältemaschinen Regeln, 5th Ed., Verlag C. F. Müller, Karlsruhe, 1958.)

K. Ražnjević: Handbook of
Thermodynamic Tables and Charts

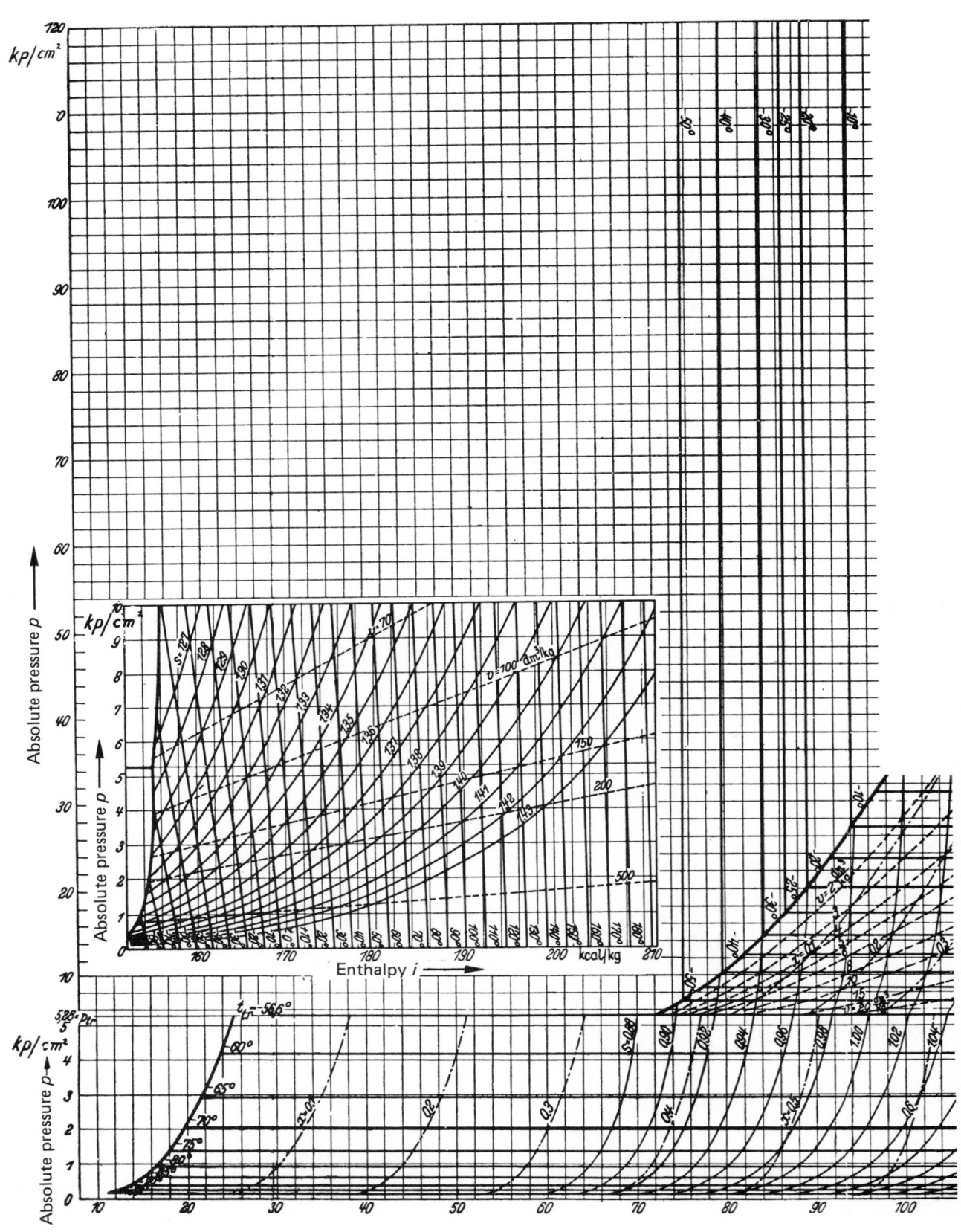

K. Ražnjević: Handbook of
Thermodynamic Tables and Charts

Chart 11

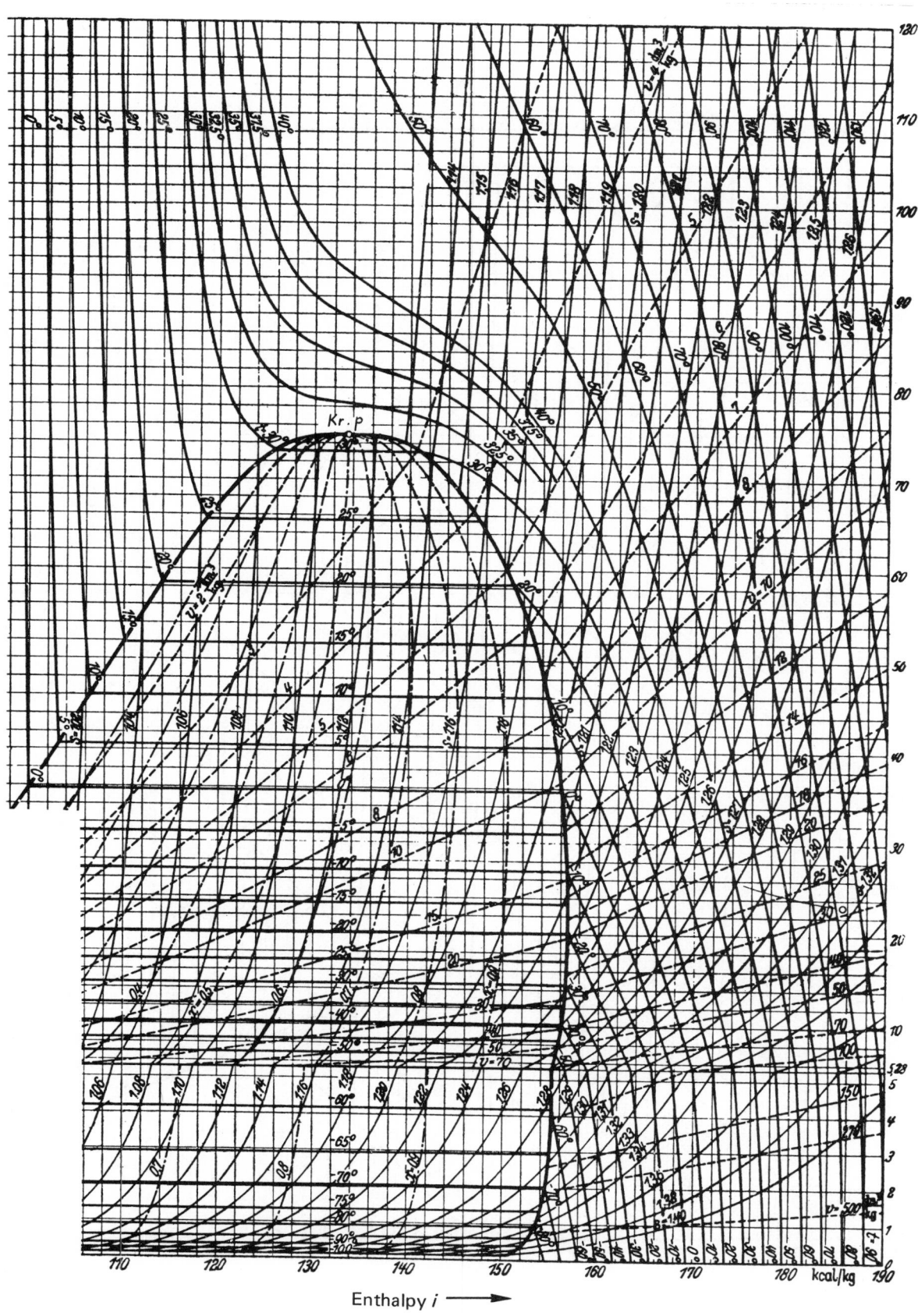

Enthalpy i ⟶

Carbon dioxide (CO_2) at $0°C$, $i'_0 = 100$ kcal/kg, $s'_0 = 1$ kcal/kg $°K$;
absolute pressure p vs. enthalpy i
(From Kältemaschinen Regeln, 5th Ed., Verlag C. F. Müller, Karlsruhe, 1958.)

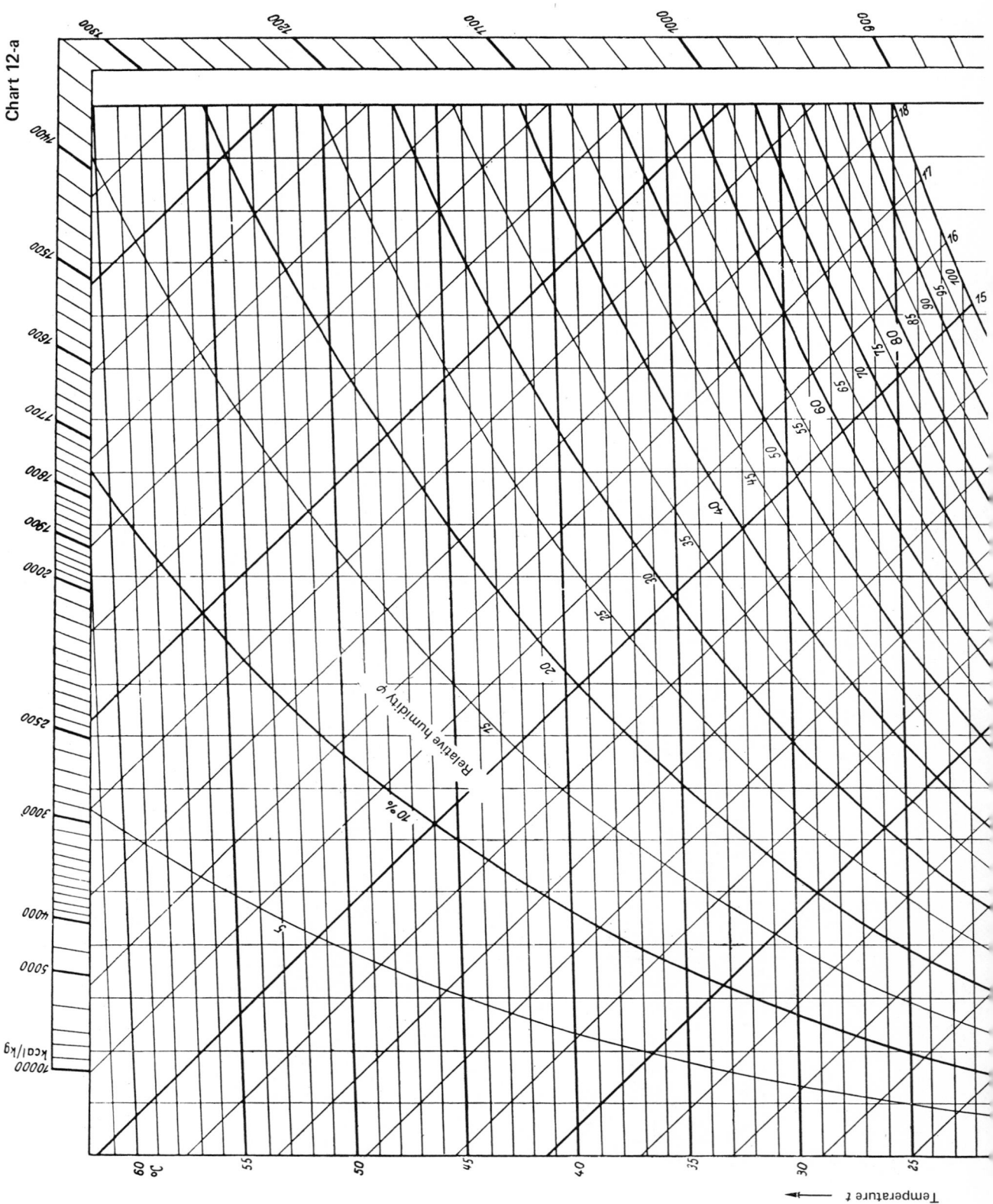

Chart 12-a

Relative humidity φ

Temperature t

384

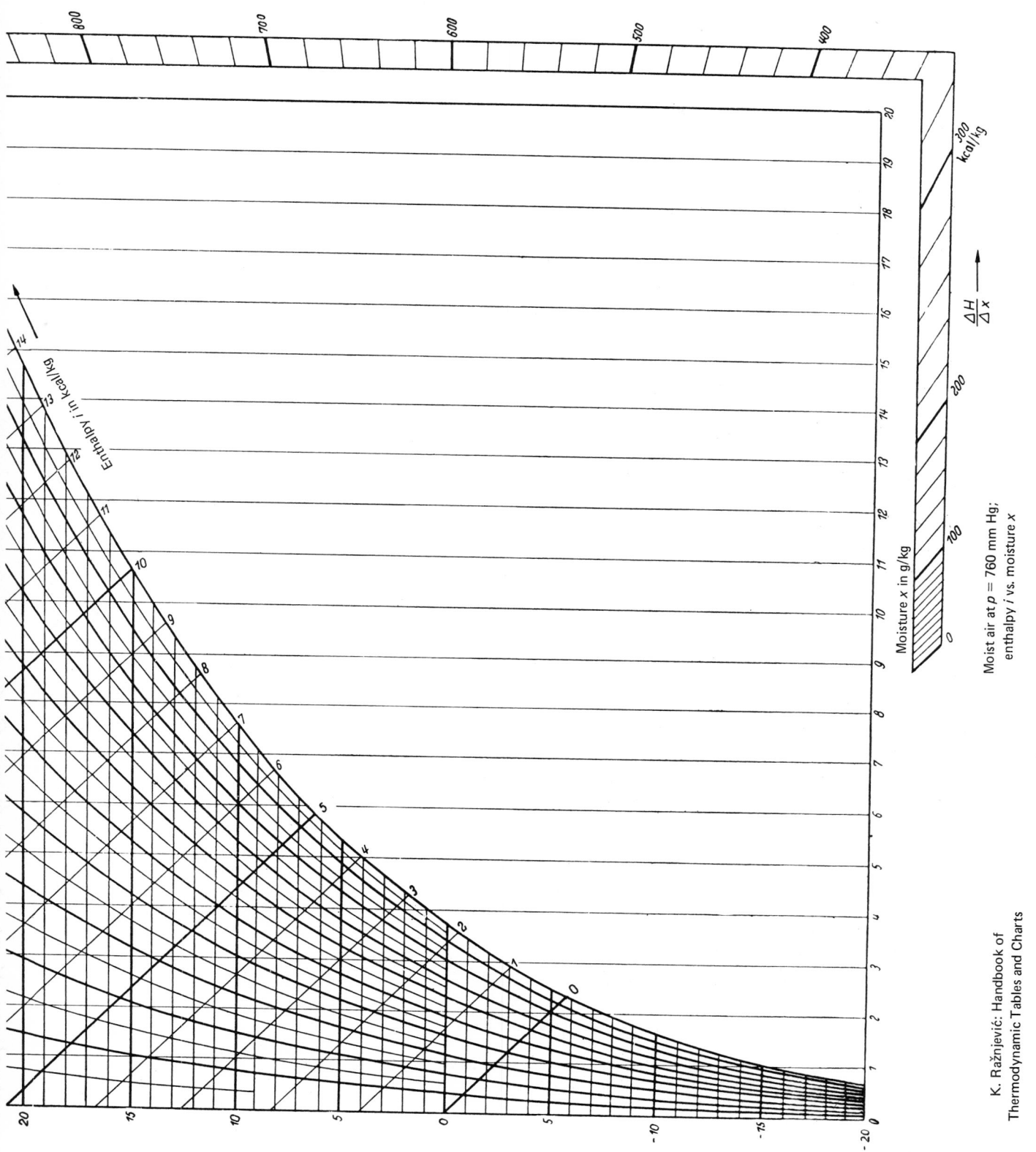

Moist air at p = 760 mm Hg;
enthalpy i vs. moisture x

K. Ražnjević: Handbook of
Thermodynamic Tables and Charts

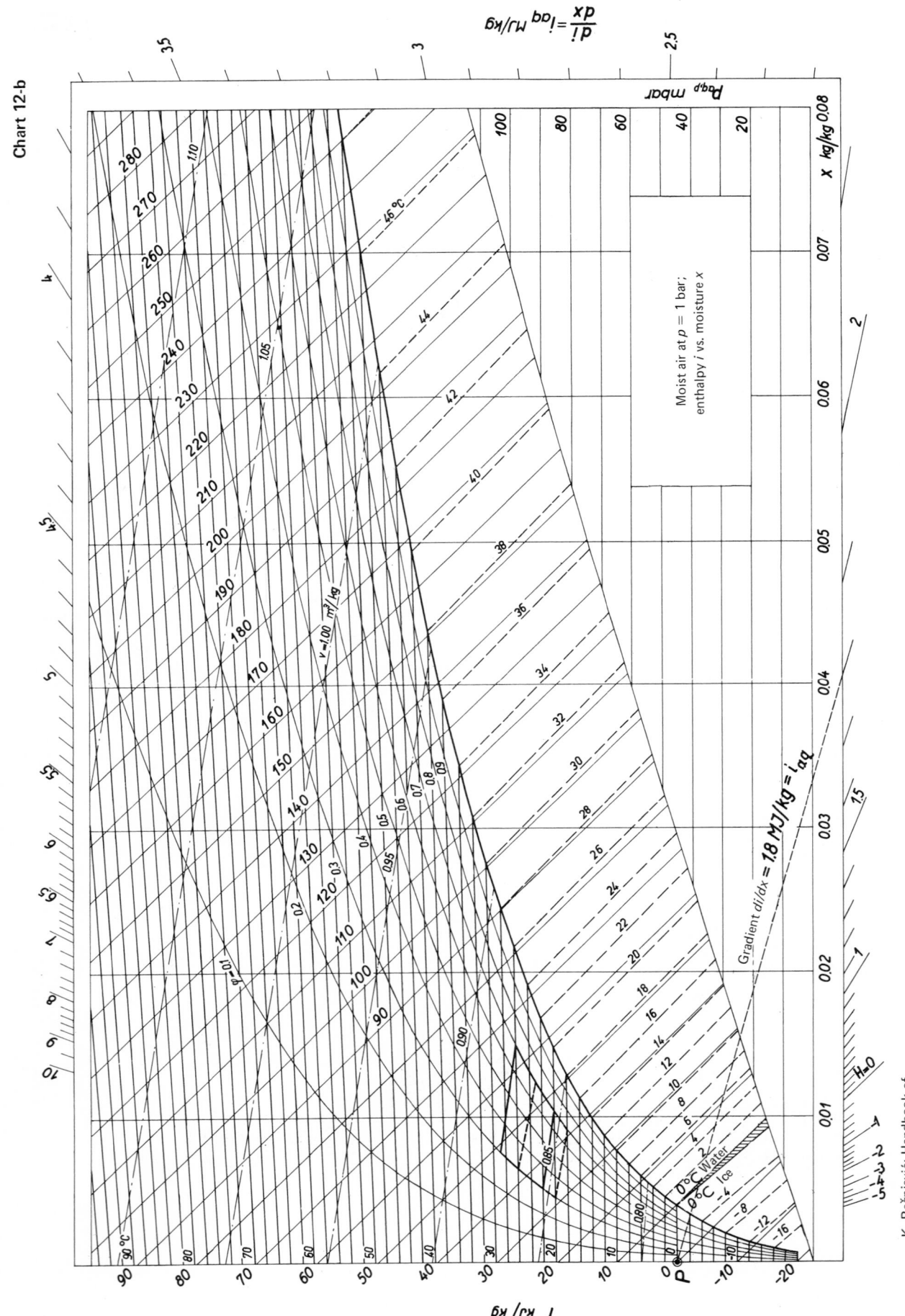

Chart 12-b

Moist air at $p = 1$ bar;
enthalpy i vs. moisture x

$\frac{di}{dx} = i_{aq}$ MJ/kg

p_{aq}, mbar

x kg/kg

i kJ/kg

Gradient $di/dx = 1.8$ MJ/kg $= i_{aq}$

$v = 1.00$ m³/kg

K. Ražnjević: Handbook of
Thermodynamic Tables and Charts

INDEX

SOLIDS

IMPORTANT NOTE: Users of Liquids, Vapors, and Gases tables are alerted to the fact that the symbol for pressure, *at*, refers to the **technical atmosphere** and not to the **standard atmosphere**, which is designated by *atm*.

The relationship of conversion factors between the two atmospheres is clearly explained in Table 119 on page 337.

LIQUIDS

VAPORS

GASES

UNITS AND MEASURES

IMPORTANT NOTE: Users of Liquids, Vapors, and Gases tables are alerted to the fact that the symbol for pressure, *at*, refers to the **technical atmosphere** and not to the **standard atmosphere**, which is designated by *atm*.
The relationship of conversion factors between the two atmospheres is clearly explained in Table 119 on page 337.

DATE DUE